# Culinary Arts

## PRINCIPLES AND APPLICATIONS

AMERICAN TECHNICAL PUBLISHERS, INC.
HOMEWOOD, ILLINOIS 60430-4600

Chef Michael J. McGreal

*Culinary Arts Principles and Applications* contains procedures commonly practiced in the industry. Specific procedures must be performed by a qualified person. For maximum safety, always refer to manufacturer recommendations, specific job site procedures, applicable federal, state, and local regulations, and any authority having jurisdiction. The material contained is intended to be an educational resource for the user. American Technical Publishers, Inc. assumes no responsibility or liability in connection with this material or its use by any individual or organization.

American Technical Publishers, Inc. Editorial Staff

Editor in Chief:
    Jonathan F. Gosse
Vice President—Production:
    Peter A. Zurlis
Art Manager:
    James M. Clarke
Technical Editors:
    Cathy A. Scruggs
    Aimée M. Gurski
Copy Editor:
    Valerie A. Deisinger
Cover Design:
    Carl R. Hansen

Illustration/Layout:
    Mark S. Maxwell
    Peter J. Jurek
    Samuel T. Tucker
    Jennifer M. Hines
    Nicole S. Polak
CD-ROM Development:
    Carl R. Hansen
    Peter J. Jurek
    Chris J. Bell

1 2 3 4 5 6 7 8 9 – 08 – 9 8 7 6 5 4 3 2 1

Printed in the United States of America

ISBN 978-0-8269-4200-5

 This book is printed on 10% recycled paper.

# Acknowledgments

The author and publisher are grateful for the technical information and assistance provided by the following companies and organizations.

Advance Tabco
Alaska Seafood Marketing Institute
All-Clad Metalcrafters
Alpha Baking Co., Inc.
Amana Commercial Products
American Egg Board
American METALCRAFT, Inc
Atlantic Marine Aquaculture Center, University of New Hampshire
Anchor Food Products, Inc.
Ansul Fire Protection
Badger Fire Protection
Barker Company
The Beef Checkoff
Blodgett Oven Company
Browne-Halco (NJ)
Bunn Corporation
California Fresh Apricot Council
Carlisle FoodService Products
Charlie Trotter's
Chef's Choice® by EdgeCraft Corporation
Collins Caviar Company
Cooper-Atkins Corporation
Cres Cor
CSI Hospitality Systems, Inc.
Cuisinart
Dakota Pasta Growers Co.
Daniel NYC
Detecto, A Division of Cardinal Scale Manufacturing Co.
Dexter-Russell, Inc.
Earthstone Ovens, Inc.
Edlund Co.
Emu Today & Tomorrow
Entourage
Florida Department of Agriculture and Consumer Services, Bureau of Seafood and Aquaculture Marketing
Florida Department of Citrus
Florida Tomato Committee
Fluke Corporation
Frieda's, Inc.
Frymaster
Harbor Seafood
Henny Penny Corporation
HerbThyme Farms
Hobart
Idaho Potato Commission
Indian Harvest Specialty Foods, Inc.
In-Sink-Erator
InterMetro Industries Corporation
Joliet Junior College
Kidde

Kolpak
Kyocera Tycom Corporation
Lincoln Foodservice Products, Inc.
Lodge Manufacturing, Tennessee, U.S.A.
MacArthur Place Hotel, Sonoma
ManYee DeSandies/Pics 4 Learning, Inc.
Matfer Bourgeat USA
McFarlane Pheasants, Inc.
Messermeister
NASA
National Broiler Council
National Cancer Institute
National Cattlemen's Beef Association
National Cherry Growers and Industries Foundation
National Chicken Council
National Fisheries Institute
National Oceanic and Atmospheric Administration
National Onion Association
National Pasta Association
National Pork Producers Council
National Potato Promotion Board
National Restaurant Association
National Turkey Federation
New Zealand Mussel Industry Council
NSF International
The Pasta Factory
Perdue Foodservice, Perdue Farms Incorporated
Planet Hollywood International, Inc.
The Plitt Company
Pride of San Juan
Russell Harrington Cutlery, Inc.
R. Whittingham & Sons Meat Co.
San Jamar
Service Ideas, Inc.
Specials Inc. Produce and More
The Spice House.com
Trail End Chestnuts
Tru, Chicago
True FoodService Equipment, Inc.
USA Rice Federation
U.S. Department of Agriculture
U.S. Fish and Wildlife Service
U.S. Geological Survey
U.S. National Oceanic and Atmospheric Administration
U.S. Wellness Meats
The Vollrath Company, LLC
Vulcan-Hart, a division of the ITW Food Equipment Group LLC
Wisconsin Milk Marketing Board
WholeSpice.com
World Cuisine, Inc.

# Contents

# Contents

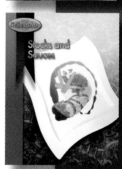

# Contents

## CD-ROM Contents

- **Quick Quizzes®**
- **Illustrated Glossary**
- **Media Clips**
- **Culinary Flash Cards**
- **Culinary Math**
- **ATPeResources**

# About the Author

**Chef Michael J. McGreal** has worked in the food service industry for over 25 years, holding chef positions at some of Chicago's premier restaurants and hotels. He earned his chef training degree from Washburne Culinary Institute in Chicago and his bachelor's degree in hospitality organizational management from the University of St. Francis. Chef McGreal joined the prestigious Culinary Arts and Hospitality Management degree programs at Joliet Junior College as an instructor in 1996, and currently serves as department chair.

Throughout his career Chef McGreal has earned many industry certifications, including Certified Culinary Educator, Certified Hospitality Educator, Certified Food Service Management Professional, Certified Hotel Administrator, Certified Executive Chef, and Certified Master Foodservice Executive. He is an active member of the American Culinary Federation, the International Association of Culinary Professionals, the Council on Hospitality, Restaurant and Institutional Education, the International Association of Food Service Executives, the National Tech Prep Network, and both the Illinois and the American Association of Family and Consumer Sciences. He also serves as a trustee for the National Restaurant Association Educational Foundation.

Chef McGreal has received many awards over the years, including being recognized as an American Culinary Federation local chapter *Culinary Educator of the Year* and *Chef of the Year*, as an *Illinois Pork Producer's Taste of Elegance* competition winner, and as a *ProStart Mentor of the Year* by the National Restaurant Association. Chef McGreal has also received the Illinois Federation of Teachers *Everyday Hero* award, and the *Professional Achievement Award* from the University of St. Francis.

Chef McGreal serves on numerous college and high school advisory committees and frequently gives presentations and demonstrations on healthy cooking, nutrition, and using food to fight illness to members of professional organizations and at conferences, as well as to cancer survivor groups. He has also been featured as a guest chef on cruise lines, offering cooking and wine-tasting demonstrations for passengers at sea.

This textbook is the culmination of many years of effort, and many people have provided support and assistance in its development, including the following individuals:

**Linda J. Trakselis**
Assistant Professor, Culinary Arts
Illinois Institute of Art – Chicago
Chicago, IL

**Michelle Hassan**
Culinary Arts Program Manager
Chicago Public Schools
Chicago, IL

**Brigitta L. McGreal, M.Ed.**
Applied Academics Department Chair
Bolingbrook High School
Bolingbrook, IL

**John Johnson, CEC, CCE**
Culinary Arts Instructor
Madison Area Technical College
Madison, WI

**Joliet Junior College**
Culinary Arts Department
Joliet, IL

- Chef Keith Vonhoff, CEPC, CCE, CHE, FMP, CHA
- Chef Tim Bucci, CEC, CCE, CHA, CCJ
- Chef Kyle Richardson, CEC, CCE, CHA
- Chef Mark Muszynski, CEPC

Chef McGreal's true passion for cooking and teaching is reflected in this foundational reference for the culinary field. It is our hope that this textbook will allow a sharing of his knowledge with current as well as new and aspiring professionals in the industry.

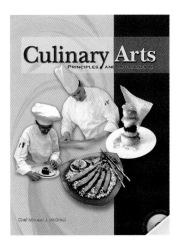

*Culinary Arts Principles and Applications* by Chef Michael J. McGreal focuses on providing learners with a solid foundation of proven culinary theory on which they can build a repertoire of professional skills. Emphasis is placed on classical cooking methods and techniques. More than 300 recipes are included, highlighting the use of different cooking methods and techniques while providing experience preparing a comprehensive variety of menu items.

The textbook includes current industry information on safety and sanitation, food service equipment, cooking methods and techniques, nutrition and menu planning, and standardized recipes and cost control. Detailed illustrations throughout the textbook provide visual references for procedures and techniques. Also included are special features that enhance the learner's experience in context with content covered.

- *Culinary Procedures* present key preparation methods and techniques in a clear, step-by-step format.

- *Helpful Tips* provide suggestions to ensure successful execution of culinary procedures.

- *Action Labs* offer hands-on activities to reinforce key concepts.

- *Chef's Tips* highlight additional information to enhance understanding of concepts and techniques.

- *Nutrition Notes* suggest healthier ways of preparing food and provide tips for planning healthy menus.

- *Historical Notes* provide background and historical insight on culinary topics.

Learning theory and instructional design have been incorporated in each chapter. Objectives at the beginning of each chapter identify the main concepts covered in the chapter. Vocabulary terms within each chapter are italicized, and the number of vocabulary terms far exceeds the number of key terms listed at the beginning of each chapter. The summary at the end of each chapter serves as a reinforcement tool, and the review questions at the end of each chapter correlate to chapter objectives and serve as a knowledge check.

The CD-ROM included at the back of this book complements content covered throughout the chapters. Detailed information about using the *Culinary Arts Principles and Applications* CD-ROM is included on the last page of this book. To obtain information about related products from American Technical Publishers, visit www.go2atp.com.

The Publisher

# Features

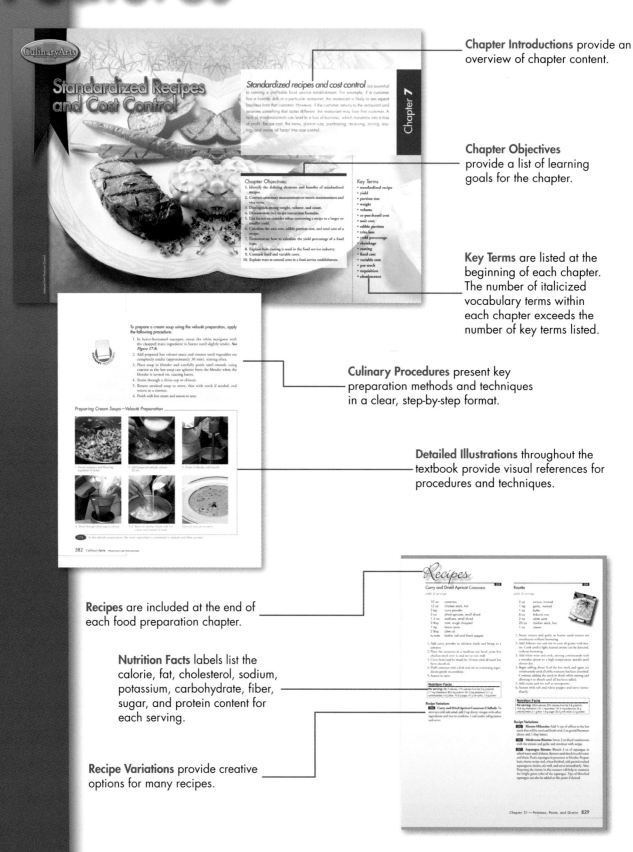

**Chapter Introductions** provide an overview of chapter content.

**Chapter Objectives** provide a list of learning goals for the chapter.

**Key Terms** are listed at the beginning of each chapter. The number of italicized vocabulary terms within each chapter exceeds the number of key terms listed.

**Culinary Procedures** present key preparation methods and techniques in a clear, step-by-step format.

**Detailed Illustrations** throughout the textbook provide visual references for procedures and techniques.

**Recipes** are included at the end of each food preparation chapter.

**Nutrition Facts** labels list the calorie, fat, cholesterol, sodium, potassium, carbohydrate, fiber, sugar, and protein content for each serving.

**Recipe Variations** provide creative options for many recipes.

**Nutrition Notes** suggest healthier ways of preparing food and provide tips for planning healthy menus.

**Chef's Tips** highlight additional information to enhance understanding of concepts and techniques.

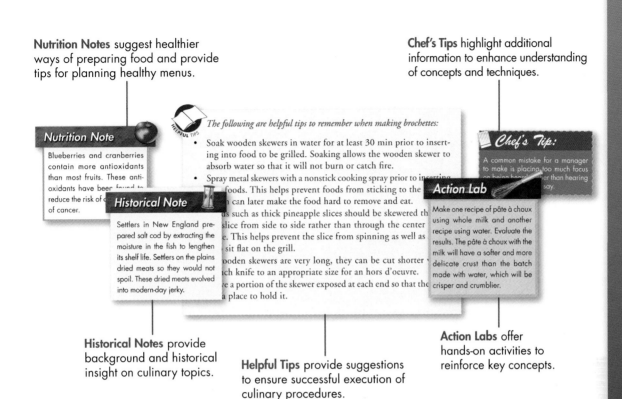

**Nutrition Note**

Blueberries and cranberries contain more antioxidants than most fruits. These antioxidants have been found to reduce the risk of of cancer.

**Historical Note**

Settlers in New England prepared salt cod by extracting the moisture in the fish to lengthen its shelf life. Settlers on the plains dried meats so they would not spoil. These dried meats evolved into modern-day jerky.

**HELPFUL TIPS** *The following are helpful tips to remember when making brochettes:*

- Soak wooden skewers in water for at least 30 min prior to inserting into food to be grilled. Soaking allows the wooden skewer to absorb water so that it will not burn or catch fire.
- Spray metal skewers with a nonstick cooking spray prior to inserting foods. This helps prevent foods from sticking to the can later make the food hard to remove and eat.
- such as thick pineapple slices should be skewered th slice from side to side rather than through the center . This helps prevent the slice from spinning as well as sit flat on the grill.
- ooden skewers are very long, they can be cut shorter ch knife to an appropriate size for an hors d'oeuvre. e a portion of the skewer exposed at each end so that the a place to hold it.

**Chef's Tip:**

A common mistake for a manager to make is placing too much focus on being heard rather than hearing say.

**Action Lab**

Make one recipe of pâte à choux using whole milk and another recipe using water. Evaluate the results. The pâte à choux with the milk will have a softer and more delicate crust than the batch made with water, which will be crispier and crumblier.

**Historical Notes** provide background and historical insight on culinary topics.

**Helpful Tips** provide suggestions to ensure successful execution of culinary procedures.

**Action Labs** offer hands-on activities to reinforce key concepts.

**Using This CD-ROM** provides information about components included on the CD-ROM.

**Quick Quizzes®** reinforce fundamental concepts, with 10 questions per chapter.

**Illustrated Glossary** links common culinary terms to select illustrations and media clips.

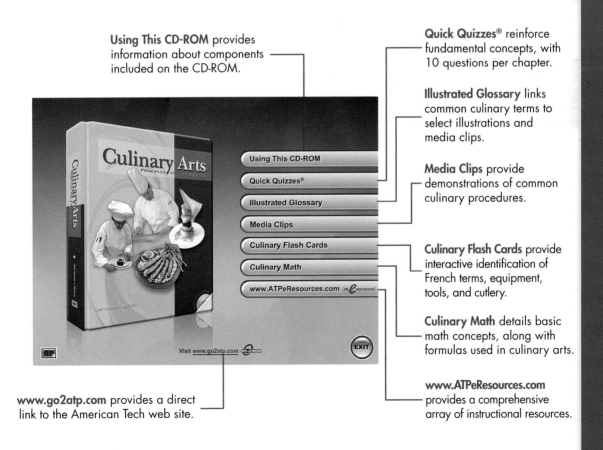

**Media Clips** provide demonstrations of common culinary procedures.

**Culinary Flash Cards** provide interactive identification of French terms, equipment, tools, and cutlery.

**Culinary Math** details basic math concepts, along with formulas used in culinary arts.

**www.ATPeResources.com** provides a comprehensive array of instructional resources.

**www.go2atp.com** provides a direct link to the American Tech web site.

# The Food Service Industry

## Chapter 1

*The food service industry* employs approximately 13 million people in the United States alone. With the average household spending more than 40% of its food dollar on food prepared away from home, it is hard to imagine life without restaurants. The first restaurants opened in France in the 1800s, and the industry soon spread to other countries. Today, the cuisine served in many restaurants reflects a blending of cultures from around the world. Choosing a career in the food service industry can be challenging and rewarding.

## Chapter Objectives:

1. Explain why each culture is associated with particular foods.
2. Chronicle the origin and evolution of the food service industry.
3. Describe nouvelle, New American, and fusion cuisines.
4. Describe reasons for the dramatic growth of the food service industry in the United States over the past 50 years.
5. Identify differences among types of dining environments.
6. Describe food service career opportunities in customer service, food production, and management.
7. Describe various education and training options in culinary arts.
8. Explain the role of communication, math, and technology skills in the food service industry.
9. Identify the qualities of an effective leader.
10. Describe tips for developing an effective portfolio and résumé.
11. List job search tools to use when looking for work in the food service industry.
12. Describe strategies to use when filling out a job application and interviewing for a job.
13. Identify the protocol to follow after a job offer has been made.

## Key Terms

- guild
- grande cuisine
- classical cuisine
- nouvelle cuisine
- New American cuisine
- fusion cuisine
- kitchen brigade system
- restaurateur
- chef
- apprentice
- active listener
- toque
- portfolio
- résumé
- networking

Survival required foraging for food and the most basic methods of preparation.

Explorers often brought foods and spices back to their countries of origin.

## A HISTORICAL LOOK AT FOOD PREPARATION

Foods and the food service industry have changed dramatically over time. In prehistoric times, foods were consumed more as a means of survival than as entertainment. Food was sought by hunting animals in the wild and searching for edible plants. *See Figure 1-1.* As the means to obtaining food progressed, so did the manner in which food was prepared. A closer examination reveals some of the changes that have occurred throughout history in this process.

### Evolution of Food and Culture

As sea travel improved, explorers ventured great distances across the water in search of undiscovered lands. As these explorers encountered different lands and peoples, they also shared in local foods. These ingredients, their uses, and the methods of preparing the ingredients from the newly discovered cultures were absorbed into a new cuisine that reflected influences from both lands. *See Figure 1-2.*

People come from diverse backgrounds and experiences. The environment someone is raised in, culture, and ethnic differences are just some of the things that make an individual unique. At the same time, members of a group who share much of the same upbringing and culture often have a great deal in common. This is especially true when it comes to food. People from a particular culture or ethnic background might enjoy some of the same types of food, flavors, or ingredients.

Each culture has a unique history of foods and ingredients that are traditional to that heritage. Some common examples of this are the Irish, known for their stews, potatoes, and poached meats such as corned beef; and the Italians, who are associated with pastas, tomatoes, fresh herbs, wine, and cheeses. Each culture developed dishes with ingredients that were native, or local, to its specific part of the world and developed its own methods of cooking them.

### Restaurant Origins

The restaurant industry and the preparing of foods as a profession initiated in France. Royalty throughout Europe had many servants that worked in their enormous kitchens. Royal celebrations were all-day events, with meals often lasting half the day. Guilds were commissioned by the French royalty to prepare foods for these celebrations. A *guild* is an organization of craftsmen having exclusive control of the production of a particular craft and the distribution of its products. There were specific guilds for everything from meat roasters to bakers and caterers.

In 1765, a French pub owner by the name of Monsieur Boulanger decided to take advantage of the business opportunities available in food preparation. He was a good cook and thought that people might want to buy a meal when they came into his pub. He put a sign on the front of his pub advertising a meal of sheep's feet in white wine sauce, calling the hearty dish a *restaurant*. The word "restaurant" comes from the French word *restaurer*, which means, "to restore." At that time, many people considered hearty foods such as soups and stews to be "restoratives," believing that such foods restored energy, health, strength, and vitality to those who ate them.

Members of a competing guild were outraged and filed a lawsuit, forcing the pub to close. Monsieur Boulanger eventually went to court and argued that if customers in his pub wanted to buy a meal while they were there, he should be able to prepare and sell it. He won his case and shortly thereafter reopened his pub, offering different meals each day. It is believed that Monsieur Boulanger operated the first modern restaurant in the world.

After other pub, inn, and tavern owners witnessed the success of Monsieur Boulanger's new restaurant concept, some of them decided to prepare meals for sale also. ***See Figure 1-3.*** The quality of the food inside many establishments was not very good, as many of the restaurant owners did not have any experience cooking.

During the French Revolution in the late 1700s, the upper class and nobility of France began to fall apart. The revolution put an end to guilds in France, and when many of the private cooks lost their jobs, they decided to open their own restaurants. As these businesses began, French commoners were able to obtain work under elite chefs. Thus, the restaurant industry and training in the preparation of French cuisine were born.

## Influential Chefs and Trends in Cuisine

Throughout the rest of the 18th century, restaurants opened throughout Europe. Many of the finest hotels sought the most influential and meticulous chefs to help them stand above the rest. Chefs began to experiment with different foods and methods of preparation, but there were no industry standards for food preparation or cooking methods. As demand and interest in the restaurant industry grew, so did the complexity and detail of the presentation of food. Chefs were continually trying to outdo each other with artistic presentations. Many foods were intricately detailed that they were appreciated as works of art.

**Marie-Antoine (Antonin) Carême.** One of the most influential chefs at the start of the 19th century was Marie-Antoine (Antonin) Carême. Carême spent a great deal of his career working in the households of the nobility. He was responsible for creating many of the cooking methods and standards that are still used today. He did not specialize in any one area, but rather mastered many. He is mainly known for his influence on the French culinary movement known as grande cuisine. *Grande cuisine* is a style of food preparation involving intricate, elaborate cuisine

**1-3** *Pubs, inns, and taverns were the first public businesses to offer prepared foods to hungry customers.*

Carême

## Grande Cuisine

**1-4** Grande cuisine is a method of food preparation involving elaborate, artful displays of food.

and strict adherence to elaborate preparation methods and culinary principles. ***See Figure 1-4.*** Food was not simply prepared and served to the guest. Every item was decorated, sculpted, carved, sauced, and arranged so beautifully that guests were unsure whether to admire it as artwork or actually eat it.

This style of cuisine is still popular today at elite hotels around the world and in many professional culinary competitions. Carême also wrote some of the most influential books on cooking, which are still used by professional chefs today. He was one of the first chefs to standardize recipes, cooking methods, and techniques used in professional kitchens all around the world.

**Auguste Escoffier.** In the latter half of the 19th century, another influential French chef took the stage. Escoffier simplified many of Carême's processes and grouped professional methods into fewer categories. For example, Carême was a master of sauce making and wrote volumes about the many varieties of sauces that could be prepared. Escoffier simplified Carême's work on sauces by classifying all sauces into five categories he termed "the mother sauces," as all sauces stem from one of these mother sauces.

Escoffier is credited with the next culinary movement in France, called classical cuisine. *Classical cuisine* is a cooking style where the quality of ingredients and the use of refined preparation techniques are emphasized. In classical cuisine, dishes are not as elaborate as they are exquisitely prepared. Rather than focusing on artful presentation, Escoffier stressed the use of the finest and freshest of ingredients and the most appropriate preparation method to produce the best results.

Auguste Escoffier also developed the *kitchen brigade system*, a structured chain of command with specific job titles and duties that streamline food production in the professional kitchen. ***See Figure 1-5.*** Before Escoffier's development of the kitchen brigade system, food production was poorly organized, and guests would sometimes be seated for up to 14 hours to dine on cold or improperly cooked food. The various chefs and production personnel each acted independently, without regard for the order or timing of dishes. As a result, there was no telling whether the next course was going to be a dessert, salad, or entrée. Needless to say, these meals often turned out to be disasters. Escoffier saw the brigade structure as an organizational tool for kitchens of the aristocracy. Over time, Escoffier's system was adopted by professional kitchens throughout Europe. A simplified version of the brigade system is still used today in professional kitchens around the world.

Escoffier is often referred to as the father of culinary education and training, as he emphasized that all who entered the culinary profession should know and understand professional cooking methods and techniques. ***See Figure 1-6.***

| Kitchen Brigade Positions | |
|---|---|
| French Terms | English Meaning |
| **Boucher** (boo-SHAY) | Butcher of meats and poultry |
| **Boulanger** (boo-lawn-ZHAY) | Bread baker |
| **Chef de partis** (chef-duh-par-TEE) | Station chef |
| **Commis** (co-MEE) | Apprentice cook |
| **Confiseur** (cone-fiss-UHR) | Petits fours and specialty candy maker |
| **Decorateur** (duh-kur-AHTUR) | Showpiece and specialty cake maker |
| **Entremetier** (ehn-tra-meh-tee-YAY) | Vegetable and soup station chef combined into one station |
| **Friturier** (free-too-ree-YAY) | Fry station chef (all fried foods) |
| **Garde manger** (gahrd-mahn-ZHAY) | Pantry chef (responsible for cold food preparations) |
| **Glacier** (GLAH-see-yay) | Chilled and frozen dessert chef |
| **Grillardin** (gree-yar-DAHN) | Grill station chef |
| **Legumier** (lay-GOO-meeyay) | Vegetable station chef |
| **Pâtissier** (pah-tees-ee-YAY) | Pastry chef |
| **Poissonier** (pwah-sawng-ee-YAY) | Fish station chef |
| **Potager** (poh-tah-ZHAY) | Soup station chef |
| **Rotisseur** (roh-tees-UHR) | Roast station chef also responsible for related sauces |
| **Saucier** (SAW-see-yay) | Responsible for all sautéed items and most fine sauces (one of the most demanding positions in the kitchen) |
| **Sous chef** (soo-chef) | Second chef, person second in command after the chef |
| **Tournant** (toor-NAHN) | Rounds-man or swing cook; works wherever needed in the kitchen |

**1-5**   *French terms are used to describe the classic kitchen brigade positions.*

Escoffier

Refer to the CD-ROM for the **French Terms** Flash Cards.

**Nouvelle Cuisine.** As the world entered the 20th century, nouvelle cuisine adapted the methods of Carême and Escoffier to create even lighter dishes. *Nouvelle cuisine,* or new cuisine, is a cooking style where foods are cooked quickly, seasoned lightly, and presented simply. French chefs Paul Bocuse and Michel Guérard were leaders in this movement, which gained popularity in the United States in the early 1970s. Rejecting a reliance on rich cream sauces and butter, this culinary movement stemmed from the trend of eating lower-fat foods and an interest in letting the true flavors of food stand out.

## Culinary Education and Training

**1-6** Culinary education and training programs continue to offer students opportunities to become professional chefs.

**New American Cuisine.** The 1970s and 1980s saw increased interest in culinary education in the United States. As large numbers of American chefs were trained and entered the industry, the next culinary movement, New American cuisine, emerged. *New American cuisine* is a food preparation style that emphasizes the use of locally grown foods native to America.

**Fusion Cuisine.** Chefs soon discovered that foods, ingredients, and cooking styles from around the world could be combined with exciting results, and in the 1980s, a fusion style of cooking developed. *Fusion cuisine* is a cooking style that blends characteristics of two or more ethnic cuisines. Fusion cuisine is one of the emerging trends in restaurants in America and around the world. By combining ingredients and cooking methods from different cultures, creative and skilled chefs are able to create unique dishes that represent aspects of diverse culinary traditions. ***See Figure 1-7.***

## THE FOOD SERVICE INDUSTRY TODAY

The food service industry is directly affected by changes in lifestyles. For instance, lifestyle changes in the American family over the past 50 years have led to an increased demand for food service. In the past, one parent, often the mother, stayed at home to raise the family and take care of the household. Partially due to the increasing cost of living, in many families today both parents work outside the home. Additionally, on average people are working longer hours than past generations. As a result, people are dining out or picking up food much more often today than in the past. This trend toward dining out and eating carryout has led to a boom in the food service industry. ***See Figure 1-8.***

## Fusion Cuisine

National Chicken Council

**1-7**  Fusion cuisine blends characteristics of two cultures' foods into one dish.

## Restaurant Industry Sales Growth

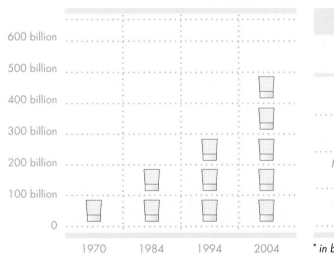

| Food and Drink Sales | | |
|---|---|---|
| Establishment Type | Sales* | Compound Annual Growth |
| Commercial restaurant | $ 403 | 3.9% |
| Non-commercial restaurant | $ 35 | 2.0% |
| Military restaurant services | $ 1.7 | 5.2% |
| Grand total | $ 440 | 3.8% |

\* in billions of dollars

Information reported by the National Restaurant Association

**1-8**  Annual sales figures for the U.S. restaurants exceed $440 billion, which averages out to over $1.2 billion per day.

Cooking shows have had quite an impact on the food service industry over the past 30 years. Talented chefs such as Julia Child, Jacques Pépin, Wolfgang Puck, and Emeril Lagasse have brought a great deal of exposure to the culinary profession by preparing exciting dishes while people watch from home. Celebrity chefs cooking on the Food Network are watched daily by people around the world. These chefs have inspired many young people to consider the culinary arts as a profession.

## Modern Dining Environments

Different types of food service operations may have different titles for workers in each position. A fine dining establishment may follow the traditional hierarchy of executive chef, sous chef, and prep chefs, while a casual restaurant may simply call all of the workers on the cooking line "cooks."

Interior décor, menu items, prices, and service styles may also be very different from one dining establishment to the next. An examination of popular dining establishments reveals the differences between common types of modern dining environments.

**Themed Restaurants.** Themed restaurants are establishments that make the patron feel like they are in a different time or place. ***See Figure 1-9.*** Some may have a fun, sports-themed atmosphere where guests can enjoy sporting events on television while surrounded by sports memorabilia. Other themed restaurants may have ethnic décor such as frescoed walls and overhead grapevine trellises that create the feeling of being in another country. Whatever the theme, this sort of establishment offers diners fun and entertainment. Prices for menu items in themed restaurants vary depending on the foods offered.

**Casual Dining.** Rather than emphasizing a theme, casual dining restaurants concentrate on good food at reasonable prices. Many casual restaurants boast a family-friendly atmosphere. Casual dining restaurant menus usually focus on appetizers and mainstream foods that appeal to the general public, such as club sandwiches, pasta dishes, and steak.

## Themed Restaurants

*Planet Hollywood International, Inc.*

**1-9** *Themed restaurants offer a fun and entertaining atmosphere for dining.*

**Quick Service.** Quick service restaurants offer prepared and wrapped foods across counters or through a drive-up window. Due to their convenience, the food offerings are often called "fast food." McDonald's, the largest quick-service chain, can be found in nearly every country around the world. The popularity of quick service reflects a fast pace of life and acceptance of mass standardization, food production, and marketing as mainstream. Although hamburgers and french fries remain the staples of quick service, a wide variety of food products are offered. Specialty foods such as sushi, ribs, and "home-cooked meals" have joined fish and chips, chicken, pizza, hot dogs, and pasta as common quick service offerings. *See Figure 1-10.*

**Fine Dining Restaurants.** Fine dining restaurants are very expensive but offer food beautifully presented and cooked to perfection. The ingredients used in each dish are usually of a better quality than those used in a casual restaurant, and this quality is reflected in higher menu prices. These restaurants offer elegant décor, often including linen tablecloths, elegant stemware, and beautiful china. *See Figure 1-11.* Fine dining restaurants are often found in upscale hotels or in the central area of most major cities.

Quick Service

1-10 *Drive-thru food pickup windows help the customer to purchase food quickly, without ever having to leave the vehicle.*

Fine Dining

*MacArthur Place Hotel, Sonoma*

1-11 *Fine dining restaurants often have linen tablecloths and beautiful china.*

**Institutional Food Service.** *Institutional food service* is a type of food service that features a cafeteria-style operation or a limited menu and is usually found in a university, hospital, or school. This food service operation generally has a very limited menu of foods that have been prepared in advance or convenience foods that require minimal on-site preparation. Examples of convenience foods are chicken portions that are prebreaded and ready for deep-frying, or lasagna that is ready to be baked and served.

**Catering Services.** Catering services are a rapidly growing segment of the food service industry. A *catering service* is a food service operation that brings prepared meals and service staff to a client's location. Meals are catered to many different settings. Airlines, reception halls, and hotel restaurants may opt to hire a catering service to prepare meals and serve the dining needs of guests. In catered events, food brought in from another location could vary from a Danish-and-coffee service to a formal six-course dinner.

## Careers in the Food Service Industry

Food service establishments vary in size, in products, and in services offered. Careers in the food service industry are as varied as the types of food that are served. Opportunities exist in management, production, customer service, sales, and education—from entry-level positions to ownership. An individual's level of success in the food service industry depends upon desire to improve and move ahead and the level of dedication and commitment demonstrated on the job.

There are many food service occupations from which to choose. The best approach when seeking a career in food service is to acquire basic training in many food service areas before specializing in a particular area. This enables students to acquire a variety of knowledge and skills, making them more valuable to a prospective employer. A person with a broad base of knowledge and skills may advance in position within a food service establishment from employee to manager and eventually to employer while demonstrating good leadership qualities, dedication, and professionalism. With the proper skills, work ethic, and commitment, a person may eventually open and own a food service establishment. ***See Figure 1-12.***

The food service operation is divided into a management team and two main groups of employees commonly referred to as customer service staff and food production staff. Customer service (also known as front-of-the-house) employees work in direct contact with customers.

Customer service positions include maître d', host or hostess, server, cashier, bus person, and bartender. Food production (also known as back-of-the-house) employees perform food production work behind the scenes.

Food production positions include sous chef, chef saucier, banquet chef, pastry chef, tournant, fry cook, broiler cook, garde manger, baker, chef de partie (line cook), institutional cook, prep cook, commis, expediter, and dishwasher. Customer service employees and food production employees

*Restaurant Owner*

Charlie Trotter's

**1-12** *Owning a restaurant takes hard work but can be very rewarding.*

are equally important. If the customer is not greeted by a friendly face and a personable server, the meal could turn out to be a not-so-enjoyable experience. If the kitchen staff puts out food that is cold, improperly cooked, or sloppy, the meal cannot be enjoyed. Every position in the operation can affect the customer's overall experience. Successful food service employees look at their positions not just as jobs, but as careers.

In addition to customer service and food production employees, there are the members of the management team. These individuals are responsible for day-to-day operations of each food service facility, supervision of the staff, and the overall guest experience. Selecting the right manager for the job is crucial, as this person must be the driving force behind the staff. A successful manager can be the key to an operation running smoothly.

Management positions include owner, general manager, food service director, assistant manager, catering director, banquet manager, supervisor, kitchen manager, and chef. Additional opportunities related to the food service industry include areas such as food service sales, food service buyer, and instruction. An examination of these areas and the corresponding positions provides a better understanding of food service operation.

**Customer Service Positions.** The customer service area includes personnel required to provide service to the customer. The quality of service may determine how the overall meal quality is perceived. Good food with poor service still results in a bad impression.

The *maître d'* is the head of the dining room service. Responsibilities of the maître d' include overseeing the dining room, assigning stations, and directing the host or hostess and the service staff.

The *host* or *hostess* is the person responsible for welcoming and seating customers in the food service establishment. A host or hostess must have a welcoming smile and a warm personality to greet guests as they enter the establishment. ***See Figure 1-13.*** Particular attention must be paid to seating patrons in their choice of a smoking or nonsmoking area (although many restaurants have now banned smoking in the restaurant). The host or hostess assigns a server to each table and carefully watches to make sure servers attend to guests' needs.

The *server* is the person responsible for serving all food and beverages to the guests. No other employee is in a position to have more influence on the guests than the server. A server must always act in a professional manner and have a welcoming and pleasant attitude. ***See Figure 1-14.*** After greeting the guests, the server takes the beverage order followed by the food order. It is important for the server to repeat the order to each guest to ensure that the information is correct.

Most restaurants use computerized ordering systems that allow the server to enter all food orders into a computer that sends the food orders to the kitchen and totals the bill when the guests are finished. Computerized ordering systems help to minimize communication er-

*Host* _____

Badger Fire Protection

1-13    A host greets customers as they enter the establishment.

*Server* _____

1-14    A pleasant server makes customers feel welcomed.

## Banquet Chef

**1-15** A banquet chef is in charge of large functions, such as banquets and special events.

## Pastry Chef

**1-16** A pastry chef can create decorations from pulled sugar to fit the theme of a special event.

rors between the server and the kitchen as well as billing errors when calculating the guest's bill.

The *cashier* is the person responsible for handling payment in situations where the server does not collect on the bill. There is usually a cashier in casual restaurants such as breakfast establishments. The cashier receives payment of the bill, makes change, and has final contact with the customer.

The *bus person* is the person responsible for removing dirty dishes from tables, setting tables, and taking dirty dishes to the dishwashing area. Bus persons assist the server by carrying trays of food to the dining room, filling glasses of water, and assuming other related duties. An efficient bus person can make the dining room run smoothly.

**Food Production Positions.** The food production area includes all personnel required to prepare or handle food prior to customer service. Today, kitchens are organized using a system similar to Escoffier's kitchen brigade system. In a brigade system, each member is assigned a specific task or job to do. Each person is then assigned a title designating them as the person responsible for that specific task.

A sous chef is the chef's first assistant. A *sous chef* is a person responsible for carrying out the chef's orders each day, instructing personnel in the preparation of some foods, and assisting the chef in directing all kitchen production and service. A sous chef usually has a more active cooking role than the chef does.

A chef saucier, or sauté and sauce station chef, is an all-around experienced worker and the lead person in the production crew. A *chef saucier* is a person responsible for preparing boiled, stewed, braised, and sautéed dishes as well as sauce making. A chef saucier follows the sous chef in order of authority.

A *banquet chef* is a person in charge of all parties and banquet functions. A banquet chef supervises the preparation of all foods for functions, directs the food plate-up, and is under direct supervision of the chef. A banquet chef must understand timing when plating up large functions. ***See Figure 1-15.***

A *pastry chef* is a person who supervises the pastry department, prepares dessert menus and baked goods, schedules work performed in the pastry department, and often decorates cakes and special pastries. ***See Figure 1-16.*** A pastry chef is under the direct supervision of the executive chef.

A *garde manger,* or pantry and cold kitchen chef, is a person responsible for the cold food department. A garde manger chef oversees preparation of sandwiches, salads, salad dressings and other cold sauces, cold appetizers and canapés, and other preparations featuring cold meats, vegetables, or other ingredients. A garde manger chef must also be experienced in decorating foods for buffets, so it is not uncommon to see garde manger chefs who carve elaborate ice carvings.

A *baker* is a person responsible for the operation of the bakery, usually under the supervision of a pastry chef. The head baker prepares all yeast breads and quick breads.

**14** CulinaryArts • PRINCIPLES AND APPLICATIONS

A *chef de partie* is any station chef that is responsible for a particular area. In a casual restaurant, they may simply be called line cooks.

A *tournant,* or swing cook, is a person who relieves cooks of their stations on their day off. A swing cook must cover a different job each day of the week. A swing cook possesses a variety of skills and is able to adapt to irregular working hours and different tasks.

A *broiler cook* is a person responsible for preparing all broiled foods, such as steaks, fish, and chicken. This person must learn the proper timing and heat control on a commercial broiler and must be experienced in judging the doneness of items that are broiled. It is this person's responsibility to ensure that all broiled items are cooked to the appropriate degree of doneness. *See Figure 1-17.*

A *fry cook,* or short order cook, is a person responsible for work performed around the range and deep fat fryer. Responsibilities include the preparation of eggs, omelets, crepes (pancakes), potatoes, and other fried items that appear on the menu. A fry cook is also in charge of casual foods and can often be found preparing foods in a room service kitchen or fast food operation.

An *institutional cook* is a person who prepares large quantities of prepackaged and prepared foods in an institutional setting such as a school cafeteria, hospital, or the military. *Institutional cooking* is cooking at an operation where all food is prepared in large quantities. *See Figure 1-18.* An institutional cook must be well versed in converting recipes and figuring out appropriate portion controls.

## Broiler Cook

1-17   *A broiler cook must know how to determine the doneness of broiled items so as not to over- or undercook them.*

## Institutional Cooking

1-18   *Institutional cooking produces large quantities of foods at one time.*

## Prep Cook

*Charlie Trotter's*

**1-19** A prep cook "preps" a majority of the foods used in the professional kitchen.

A prep cook is sometimes referred to as the backbone of the line cooks. A *prep cook* is a person who prepares or "preps" a majority of the foods used on the line. ***See Figure 1-19.*** A prep cook might perform tasks such as cleaning, trimming, and cutting up vegetables for the evening's service and trimming meats. A prep cook assists with any work that needs to be done.

A *commis,* or apprentice cook, is a person who is currently enrolled in a culinary school and who desires to excel in the food service and hospitality industry. This person works under close supervision of a station chef to learn from that person. At the end of a predetermined period, the commis is reviewed and, if deserving, may be offered a full-time position.

An *expediter,* or food checker, is a person responsible for all food that leaves the kitchen and the look of each plate. An expediter must organize the food orders received, check all trays leaving the kitchen, and ensure that only the foods that are ordered are on the tray. An expediter also keeps track of the number of each item served so an inventory of items that are selling is always available. One of the key duties of an expediter is to announce food orders as they come into the kitchen in order to reduce the time it takes for orders to be served.

A *dishwasher* is a person who operates the dishwashing machine, which washes all china, glassware, and silverware (called flatware), while keeping breakage to a minimum. This person also usually washes all pots and pans used by the kitchen staff. A dishwasher is one of the most important positions in the kitchen, as this person is responsible for providing clean and sanitary eating utensils and plates for the guests, as well as returning clean pots and pans to the kitchen.

**Management Positions.** Many food service employees aspire to be in food service management. However, the road to management takes years of hard work and continuing education. Successful food service managers put in countless hours and have a lot of work experience. There are several levels of management within food service.

A *restaurateur* is the term used to describe the manager or owner of a restaurant.

A *general manager* is a person who conducts and directs all affairs of an operation and oversees food production, beverage sales, and customer service. A general manager is responsible for inventory, scheduling, ordering, personnel, determining food and labor costs, and other related duties. In some cases, a general manager also orders equipment and determines the budget.

An *owner,* or proprietor, is a person who has the legal title to or sole possession of a food service establishment. The owner may have an active role in running the operation by also assuming the duties of the manager or may give many of the responsibilities of the operation to a general manager.

A *food service director* is a person responsible for all budgets and top management decisions, as well as supervision of other managers and supervisors employed in the establishment. The food service director position is similar to the general manager position but is usually held in institutional cooking operations or cafeteria-style operations.

An *assistant manager* is a person who helps the general manager carry out all the affairs of the operation and assists in many of the functions of the manager. The main difference between a general manager and an assistant manager is that an assistant manager enforces the rules and policies but is not as responsible for making them.

A *catering director* is a person responsible for all functions that may be taking place in the operation. This person handles all menu planning, pricing, sales, coordination, and customer relations for all events. A catering director usually meets with people planning an event to make sure that all of the guests' needs have been addressed as well as to initiate the signing of any booking agreements or contracts. ***See Figure 1-20.***

A *banquet manager* is a person responsible for the management of all banquets and food-related functions in the operation. This person works closely with the banquet chef and kitchen staff to ensure smooth plating and serving of each course. A banquet manager assigns duties to all staff for each event and is the main person in charge of all timing between courses. This person is in direct contact with guests during the event.

A *supervisor* is a person who oversees the overall operation of a food service establishment and confers with and directs managers of each operation. The title "supervisor" is often used in franchised (chain) establishments where a supervisor is a person in charge of two or more franchised establishments.

*Chef* is a general term that may refer to many different specialty and culinary management positions in a food service operation. In the past, the chef was the person of authority in the kitchen and had complete charge of all food preparation and serving of food. Today's chef must also be trained in cost control, accounting, management, nutrition, sanitation, and customer relations. In addition to these skills, a chef must continually be aware of current health concerns, cooking trends, and customer needs. Some of the certifications that professional chefs can earn are certified sous chef, certified working chef, certified executive chef, certified master chef, certified baker, certified pastry chef, certified executive pastry chef, certified master pastry chef, and certified culinary educator.

An *executive chef,* or head chef, is the person in charge of the kitchen in a large operation. The executive chef is responsible for menu planning, recipe development, kitchen staff supervision, budgets, ordering, and purchasing. This person is also responsible for establishing and maintaining all quality standards. Executive chefs are the ultimate food production authority and must have a good working relationship with all staff who report to them.

In small hotels, restaurants, and cafeterias, food production is on a smaller scale and requires a smaller crew. Under these conditions, a *working chef* is a person who assists in production by working where

## Catering Director

1-20 *A catering director coordinates all aspects of a special event.*

## Working Chef

*Advance Tabco*

**1-21** *In a smaller operation, a working chef cooks and assists in production.*

## Food Service Distributor

*Kolpak*

**1-22** *A food service distributor works directly with a chef or manager to deliver food, supplies, or equipment.*

needed in addition to having the regular responsibilities of a chef. ***See Figure 1-21.*** During hours of service, a working chef usually cooks as well as supervises kitchen production.

**Other Food Service Job Opportunities.** There are many opportunities available to someone interested in a food service career. Customer service positions or food production positions are both rewarding, and entry-level staff employment opportunities can be found in both areas. In addition to these restaurant staff positions, other areas related to food service can be considered as possible career choices.

*Food service sales* is a career area that allows someone interested in the food service industry to work for a distributor. A *food service distributor* is an organization that sells food, supplies, or equipment to food service clients. A salesperson works directly with a chef or manager who is interested in purchasing these products for the establishment. ***See Figure 1-22.***

A *food service buyer* is a person at a food service establishment who handles the selection and acquisition of food service items. The buyer sets up specifications, or requirements, for each item to be purchased in order to ensure that products meet standards.

A *chef instructor* is a person who teaches culinary arts after having worked in the industry as a chef for many years. A chef instructor usually teaches in a college culinary arts program and teaches more advanced methods and techniques than are taught in a high school setting. A chef instructor can relate lessons to real life from personal experience working in the industry. Being an instructor is a rewarding and challenging career and offers opportunities to help others pursue their goals of working in the food service industry. ***See Figure 1-23.***

## Chef Instructor

**1-23** *A chef instructor teaches many culinary skills and proper methods to interested culinary students.*

## CHOOSING A FOOD SERVICE CAREER PATH

A career in the food service industry can be challenging and rewarding. Hard work and a desire to learn can result in advancement with better pay, more responsibility, and increased benefits. A great deal of pride can come from knowing others appreciate the effort it takes to produce a quality meal. In food service a person has an opportunity to utilize artistic or creative ability. Many people consider an outstanding chef to be an artist, creating beauty in the preparation and presentation of foods.

The food service industry continues to change, and the need for skilled professionals continues to grow. This industry offers exciting career opportunities to individuals who work hard, practice proper techniques, and have a professional attitude. If a person demonstrates these qualities, great opportunities and personal satisfaction will be experienced along the journey.

*ManYee DeSandies/
Pics 4 Learning, Inc.*

### Service Counts

The first thing that someone entering the food service industry needs to understand is that success is all about service. This doesn't just refer to waiting on customers and taking their orders; it is much broader than that. Whether someone is cooking meals for guests or greeting them at the door of the restaurant, good customer service is imperative. It is easy for a food production employee to think, "I work in the kitchen. The customer does not see me." But the customer sees the employee through the food. If the food is cold when it should be hot, if it is not cooked properly, if it has sloppy presentation, the employee did not demonstrate good customer service and the customer leaves unhappy. Creating an atmosphere where customer service is the prime objective is the key to customer satisfaction and success in the food service industry.

## TRAINING TO BE THE BEST

Enrolling in a culinary education or training program is the first step to achieving a great food service career. Developing the skills, qualities, and work ethic that are most desired in this industry can help employees be recognized as future leaders and successful professionals.

Training for a career in food service was once only accomplished through on-the-job training. Today, advances in technology, food products, food preparation techniques, equipment, and management methods make it almost a necessity to attend a formal education or training program in order to move up through the ranks. There are many training and educational options available in the culinary arts. ***See Figure 1-24.*** Attending a culinary arts or hospitality program is only the beginning of a food service professional's training. A career in food service offers exciting challenges but requires constant updating to stay current in the field.

## Culinary Students

**1-24** *Students in a culinary arts program learn basic culinary principles.*

### On-the-Job Training

Prior to the development of culinary arts training in high schools and colleges, people who entered the food service industry learned job skills primarily through on-the-job training. An employee would usually train the new individual in the methods, recipes, and workings of that particular food service establishment. The problem with on-the-job training was that often the individual responsible for training new employees had no formal training or culinary education either. Although on-the-job training can be completely acceptable training for career advancement in many establishments, there is a greater risk of the trainee learning shortcuts and undesirable practices from someone with no formal education or training.

### Apprenticeship Programs

An apprenticeship program is a structured training program for someone desiring to enter a specialized career. Students in an apprenticeship program are called apprentices. An *apprentice* is someone in a formal training program who shadows a professional to learn directly under the professional's supervision and guidance. By job-shadowing an experienced professional such as a chef, an individual gains valuable experience and guidance to succeed. Apprenticeship programs consist of hands-on work and classroom instruction. Many restaurant chefs develop an apprenticeship program with a local culinary arts program as a means of finding dedicated students with potential. This type of program may also be called an internship.

## High School Culinary Programs

Basic food service and culinary arts classes are offered in many high schools. These programs are designed to prepare students for entry-level positions in the food service industry and to encourage the pursuit of further education and training. Students who excel at culinary arts in high school often pursue further education and training at the college level.

## Certificate Programs

Some college programs offer certificate programs instead of a degree. These certificate programs offer individuals an opportunity to receive specialized training in a particular area or focus of the industry. Certificate programs introduce students to specialized areas of the industry and provide students with the necessary skills to earn an entry-level position. Some individuals who have already earned a degree may sign up for a certificate program simply to acquire a new skill or sharpen an area where they may be weak.

*The Vollrath Co., LLC*

## Culinary Arts Degree Programs

Having a college degree is a valuable achievement, as many corporations now require a college degree for kitchen management positions. Earning a college degree is important not just for achieving a certain level of knowledge in culinary arts, but also for providing a well-rounded education.

There are primarily two types of students who attend postsecondary culinary arts programs. The first is the student who completes high school and enters to a college program. Many colleges even offer some college credit to high school students who earned good grades in high school culinary classes. The other typical student seen in a postsecondary culinary arts program is a working adult who has realized a strong desire to pursue a culinary career. Many of these students state that they were never really content in their former careers and that they have always dreamed of being a chef or restaurateur. The fast-paced lab classes and hands-on activities in a college program make pursuing a culinary arts degree an exciting step in planning for the future and achieving those dreams. ***See Figure 1-25.***

Postsecondary culinary arts programs range in length from two years to four years and offer training in the areas of management, production, cooking, baking, sales, service, nutrition, cost control, and sanitation. A two year-degree is called an associate's degree and a four-year degree is called a bachelor's degree. It is important to check out a school's reputation, accreditation, and cost and the background of its faculty before signing up for classes. In addition, the American Culinary Federation (ACF) initiated an apprenticeship program where students can earn a certification of Certified Culinarian (CC) upon successful completion of the degree from an ACF-accredited degree programs. Earning a certification tells an employer that the graduate is a dedicated professional who has worked hard to excel in the food service industry.

## Culinary Degree Programs

**1-25**  *Students of all ages pursue a degree in the culinary arts.*

## SHARPENING MORE THAN JUST KNIVES

A chef recognizes that the main tool of the trade is a sharp knife and that it is important to sharpen and hone knives after each use. A person's basic skills should be thought of in the same way. Having abilities in areas such as mathematics, computers, and technology, and communication skills such as listening, speaking, and writing, can make the difference between obtaining and advancing in a position. Effective employees are strong in these areas and capable of handling situations that may arise as well as working with a diverse workforce.

### Honing Basic Skills

Honing a knife is not actually putting an edge on a dull knife. Honing is done at two different times. First, it is done immediately after a knife has been sharpened to remove any small burrs from the edge of the knife. Secondly, a sharp knife is honed after each use to maintain the edge. Honing of basic skills can be looked at in the same way. In order to be successful, professionals need to continually refine their understanding of the industry and the basic skills needed to work successfully in that industry. Maintaining strong communication, math, and technology skills helps professionals work more efficiently and effectively.

**Communication Skills.** Effective communication skills are vital in the food service industry. In order for communication to be successful, the message must move clearly between a sender (speaker) and the receiver (listener). Finally, feedback must occur to indicate that the listener has understood the message that the speaker intended to send. There are four forms that can be used to create effective communication. They are listening, speaking, writing, and reading.

Charlie Trotter's

Listening can be a challenging communication skill. Have you ever been in a conversation with someone and been asked, "Are you even listening to me?" This may have been because the intended listener was there in body, "listening," but not really paying attention to what the speaker was saying. To be an effective listener, a person must be an active listener. An *active listener* is a person who provides feedback to the speaker via slight gestures such as direct eye contact, nodding the head to show understanding, and not interrupting until the person speaking is finished.

Effective speaking skills are imperative for giving instruction, sharing information, and providing feedback. People can improve their speaking skills by speaking clearly, not using slang or foul language, and speaking with the appropriate tone and at the correct level. If a person shouts orders in a harsh tone, that person is not demonstrating good communication skills.

Writing skills are a form of communication that is used to indicate something to someone in writing. Company memos, résumés, business letters, and job applications are all examples of written communication. In written communication, readers do not have the benefit of seeing facial expressions and body language, or hearing the tone of the speaker, so it is important to write clearly to avoid misinterpretation. To do this, decide what the purpose of the information that you are writing is, and then set the tone with the appropriate words.

Reading is a communication skill in which the reader obtains information from a writer through print. In order to remain competitive, professionals must continually read information that pertains to the profession. From cookbooks to magazines, there are many forms of written communication from which to learn and receive valuable information.

**Math Skills.** Math skills are essential for individuals entering the restaurant industry. For example, customer service staff collect payments for service, make change, and balance cash registers at the end of shifts. In food production areas, recipes are adjusted for the number of portions needed. Ingredients used in the recipes are weighed or measured, and if the yield of a recipe changes, the weights and measurements are adjusted. Food and supplies are ordered, received, and entered into inventory.

Management has to pay bills, figure payroll, calculate food costs, and generate sales reports. All of these tasks require strong math skills. Employees need to understand the principles of multiplication, division, subtraction, addition, fractions, and decimals if they are to be successful on the job.

**Technology Skills.** Food service employees use technology and computer software to place customer food orders, take inventory, order supplies, and handle the accounting. In the past, many errors occurred as the kitchen staff attempted to read servers' handwritten orders. The use of computers has greatly reduced the occurrence of human error, leading to clearer communication between customer service and food production staff. Handheld PDAs equipped with specialized ordering software can eliminate errors resulting from kitchen staff misreading a server's handwriting. *See Figure 1-26.*

Managers must continually update their technology skills if they are to be efficient on the job. Local community colleges offer an array of short courses allowing management to continually train on new technology applications. Trade shows such as those sponsored by the National Restaurant Association expose food service operators to new equipment, new foods, and other innovations in the industry. Through the Internet, managers can research ingredients, recipes, nutritional information, and other important facts related to daily operations.

## Developing Leadership Qualities

In addition to communication, math, and technology skills, successful employees need to develop leadership skills. Persons demonstrating leadership qualities are destined to be the chefs, managers, and restaurant owners of tomorrow. Although a good culinary arts program can provide the tools to develop basic skills, leadership qualities are most often learned by witnessing the following qualities in persons around us.

**Positive Attitude.** Workers must be willing to follow instructions and accept constructive criticism. They must also be able to leave personal problems at home and not allow them to interfere with their job. Remember that this industry is a service industry, and that people go to restaurants to relax and be waited on. Service should be performed with a pleasant and positive attitude.

*Food Service Technology*

Handheld computerized ordering system

CSI Hospitality Systems, Inc.

1-26 *Technology such as ordering software and the handheld PDA has reduced human error and increased staff efficiency.*

**Dependability.** Workers must report to the job on time, every day, without fail. Supervisors and managers count on employees to be on time and on task in order to run a successful food service operation. Dependability is often a major consideration in promotion decisions.

**Teamwork.** Teamwork is an important part of food preparation and service, and individuals that do not display teamwork do not belong in this industry. When you work in food service, no matter what your job is, you are part of a team. *See Figure 1-27.* If one member of the team does not perform, the rest of the team will suffer.

**Willingness to Work.** Hard work is a way of showing desire and is a sure way to move ahead. When directed to do a task, an employee must do it quickly and do it well. A chef arrives at that position by working hard and facing every challenge along the way.

1-27 *Culinary students often work together as a team to compete against other culinary schools.*

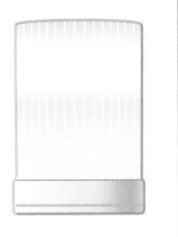

 The number of pleats in a chef's toque denotes accomplishment. The highest ranking chef wears a toque with 100 pleats.

### Historical Note

As a person's rank in the professional kitchen increases, so does the height of the hat. The sous chef and saucier often wear a chef's toque 6″ high while the chef or executive chef wears a toque as high as 12″. These tall hats were designed to resemble the old stove pipes that vented smoke from a wood-burning oven up and out through the wall. The pleats in the toque originally represented the 100 ways to properly cook an egg.

**Initiative.** Workers must be willing to do tasks without being told. Good employees see something that needs to be done and take the initiative to get the job done. For example, helping a coworker at another work station shows initiative. Supervisors look for workers with initiative when promoting someone from within the organization.

**Personal Hygiene.** Employees in any position that works closely with the public must take care of their personal hygiene. Workers should keep fingernails clean and trimmed, and hair should be clean and neat. Professionals represent the food service operation where they work. If they are not clean and tidy, customers will think poorly of the operation. Customers also place their trust in the kitchen staff that the food was prepared in a clean and organized kitchen and is safe to eat. The job of a professional chef is surrounded with proud traditions such as wearing a clean and pressed white coat and white toque. A *toque* is the traditional tall, pleated hat worn by chefs. ***See Figure 1-28.***

**Health.** Good health is necessary for optimum performance on any job. A career in food service can be very demanding, so employees should try to maintain a healthy lifestyle. It is important for people to exercise daily and eat a well-balanced diet. Because food service employees are around food for most of the day, they often fall prey to temptation and develop poor eating habits. Being knowledgeable about healthy lifestyles, eating habits, and exercise can help people stay healthy and productive on the job.

**Ethics.** A person who is ethical demonstrates honesty, trustworthiness, loyalty, character, and responsibility. These qualities are all part of ethical behavior. Employers can often identify persons with poor ethics, and these people do not keep their jobs for long.

## PREPARING FOR SUCCESSFUL EMPLOYMENT

Applying and interviewing for a job in the food service industry can be overwhelming. Arriving at the interview well prepared, organized, and confident can help to make the experience less frightening. Applicants should have a neat and professional résumé, demonstrate proper interviewing skills, and handle acceptance or rejection of employment in a respectful manner. The interview and application process should come only after the applicant has developed a portfolio and résumé.

### Developing a Portfolio and Résumé

A *portfolio* is a collection of items that depict skills, accomplishments, and overall abilities. It is a review of an applicant's best work and accomplishments. It should include letters of recommendation, any certificates earned, and photos of the applicant's best work relating to the position being applied for. For example, if a candidate is looking for a job in a pastry department, the candidate might want to include photos of cakes he or she had decorated or pastries he or she had made. Portfolio items should give the potential employer an overview of the skills and abilities that the applicant will bring to the job.

A portfolio should start off with an introductory cover sheet listing all contact information for the candidate and a short outline of the portfolio contents. It is imperative that all material be well organized, neatly presented, and complete. Professionals should constantly update their portfolios.

A *résumé* is a document listing an applicant's education, professional experience, and interests. It is probably the most important document in the portfolio. The purpose of a résumé is to convince the employer to hire the applicant for the job. A résumé must be short, factual, and accurate. It should outline all work experience, accomplishments, and special skills that relate to the position being sought. This outline format is designed to inform a potential employer of why a person is the best candidate for the position. If someone has not had very much work experience, the résumé should be limited to a single page. If the applicant has been in the industry for a while, it is fine to make the résumé a bit longer, but never more than two pages.

Daniel NYC

A résumé should include a heading with the applicant's name, address, phone number, and e-mail address. The applicant's name should be in a slightly larger font than the rest of the information, and in boldface type. Creative or artistic fonts are not recommended as the résumé is a business document and should look professional. A résumé should also include the following headings:

- **Career Objective.** A *career objective* is a single sentence stating what position a candidate is interested in. This statement should be basic and direct. For example, someone who has no experience in a restaurant but is looking for a first job might write the following: *Career Objective:* To acquire a position in a quality food service establishment where I can gain real-life work experience.

- **Education.** *Education* is a résumé category that lists the education a job candidate has obtained. The highest level of education should be listed first, with date of graduation, school or college name, and diploma or certificate title.

- **Work Experience.** *Work experience* is a résumé category that describes professional experience related to the job being applied for. Starting with the most recent employment, the résumé should list each position, start and end date of employment, name of work place, and two to four primary duties of the position. Remember, the purpose of this document is to sell the applicant to a potential employer, so how the information is written is as important as the information itself.

- **Related Coursework and Accomplishments.** The related coursework and accomplishments category lists any specialized courses taken that relate to the position being applied for as well as any

related accomplishments such as winning a competition or receiving an award. The résumé should list dates, courses, and titles as appropriate.

Before an employer even meets a job candidate, a résumé creates a first impression of that candidate. Misspellings, improper grammar, or a poorly written résumé send a negative message to the employer. A neat and professional-looking résumé will show the employer that the applicant is prepared for the job interview and is probably a neat and organized worker.

The information included on a résumé must be truthful and accurate. Providing false information, inaccurate dates, or inaccurate places of employment could lead to not getting the job or being terminated from the job after hiring. ***See Figure 1-29.***

## Résumés

**Alison Altmann**
10400 Parkside Drive
Little City, IL 02345
987-654-3210
a.altmann@inet.com

**Career Objective**
To acquire an entry-level position in the food service industry

**Education**
- Little City High School, 2004 Honors Graduate

**Work Experience**

*March 2003-present, Prep Cook, The Pancake Shop, Little City, IL*
- Prepares all pancake batters
- Cleans and prepares all produce for service
- Assists on hot line as needed

*March 2002-2003, Order Taker, The Pizza Palace, Little City, IL*
- Answered phones and took pizza orders from customers
- Ran the cash register for pick-up/take-out orders
- Organized pizza delivery orders for delivery drivers
- Re-stocked soda cooler inventory at close of business

**Related Coursework and Accomplishments**
- Holds a valid state sanitation license
- Member of the Little City Culinary Team
- Won 2nd place in the state chili cook-off
- Motivated and self-directed employee

 *A well-written résumé is used as the primary tool to market a person to a potential employer.*

## Job Search

The restaurant industry is booming and there are a vast number of entry-level positions available. However, searching for a job can be a very time-consuming task. How does a candidate find a job in an industry that has millions of employees already? Job search tools such as networking, classified ads, trade publications, and the Internet can all lead to potential careers in food service.

**Networking.** Networking is a common means of finding employment in the restaurant industry. *Networking* is a means of using personal connections through friends, teachers, or people in the industry to locate possible employment opportunies. Culinary students who enter competitions often network without even being aware of it. Professional chefs at the competition observe how well these students perform in the competition. It is not uncommon for a chef to give a competing student a business card and say, "Give me a call when you are looking for a job." This is an example of how networking through professional contacts can lead to rewarding careers.

**Classified Ads.** The classified ads are another good source of employment opportunities. These ads in the newspaper are where employers let the public know they are seeking new employees. ***See Figure 1-30.***

**Trade Publications.** Trade magazines and newspapers are directed at a specific industry or profession. These publications often contain job postings. Food service publications are written for members of the food service industry and provide valuable information about many different areas of the industry.

**Internet.** The Internet is an accessible and quick way to search for job opportunities. Anyone can search by city, state, or position and can even post a résumé for potential employers to review. A job seeker can look at newspaper classifieds online, access job-opportunity web sites, network through e-mail to friends and businesses, and search through trade publications on their respective web sites.

## The Application

Once a job of interest has been identified, it is time to fill out a job application. Many employers say that this step weeds out applicants. This is because some applicants do not fill out the application completely, accurately, or neatly. The way an application is completed tells the employer a great deal about the job applicant. If the application is filled out completely and neatly, it suggests that the applicant is a thorough person who strives to do each task well. If the application is sloppy, contains errors or unanswered questions, it shows that the applicant lacks focus.

In order to complete the application, the applicant needs to have some information either memorized or readily available. Information such as social security number, address, and previous employers' addresses and telephone numbers will be required on a job application. If there is a section that does not apply, it should not be left blank. Instead, "N/A" should be written in the space to signify "not applicable." It is always best to be truthful when filling out a job application. Falsifying information is illegal and can be grounds for termination.

## Job Openings

**RESTAURANT / CATERING MANAGER**
Northwest suburban catering company seeks Restaurant/Catering Mgr. Duties include direct and supervise employee activities, control and monitor quality of environment, facilitate coprorate events. Selling banquet functions, budgeting. Requirements minimum 3 years experience in fast paced kitchen operation. Salary commensurate with experience plus bonus, 401K, medical.
Mail resume with salary history to:
Joe's Catering
1000 Broadway
Chefsville, IL 60605

**RESTAURANT HIRING**
South Beach area country club now hiring Asst Bar Manager. Position requires min 3 years experience bar tending with short order cook experience, strong front house skills, procure music entertainment, ability to work weekend hours. Smoke-free environment, competitive salary.
REF # 78455783
Employer paid ad

**RESTAURANT OPENINGS**

**Angelo's Family Dining**
OPENING SOON AT:
Small Town Square

Now Hiring: •Servers •Bartenders •Host Staff •Espresso Bar •Bakery/Cashiers •Line & Prep Cooks •Asst Kitchen Manager
*Full & Part Time, Top $*
APPLY IN PERSON
Mon-Sat 9am to 6pm at our temporary interviewing site:
PARAGON INN & SUITES
4601 W. Main Street
Small Town, IL 60453

**RESTAURANT COOK**
Specialty cook trainer, Italian cuisine. 40-hours per week, M-F, 6a-2p. HS grad, 2 yr experience in position or as Chef of Italian Cuisine; overseeing & implementing training program in preparing & cooking in traditional Italian cuisine according to original recipes. Educating employees in fundamentals of preparing and cooking in fine cuisine, testing ingredients and final products. Applicants must show proof of legal auth to work in U.S.
REF # 22568413
Equal Opportunity Employer

**1-30** *The classified ads of a newspaper feature job openings listed in alphabetical order.*

## Interviewing Skills

One of the simplest ways to make a good impression at a job interview is to look neat and professional. Appearance is the first impression an applicant makes on the potential employer, and it will be remembered. ***See Figure 1-31.*** A candidate should try to arrive for the interview 15 minutes early. If the appointment is somewhere unfamiliar, it is a good idea to find the location the day before, to identify the best way to get there and how long it will take to travel. It is important to go into an interview confident and relaxed. The following tips can help job seekers interview with confidence:

• Maintain appropriate eye contact throughout the interview. It is important to pay attention to the interviewer and learn about the potential position.

• Think before speaking. Do not answer questions too quickly or inaccurately.

• Ask questions about the operation and working conditions. Employers expect qualified applicants to ask some questions about the position or the overall operation.

• Thank the interviewer when the interview comes to an end.

• After the interview, send a thank-you letter to the interviewer.

*Professionalism*

1-31 *Looking and acting professional in a job interview helps make a good first impression.*

## Making a Career Decision

Whether a job is offered or not, it is important to react professionally. In the event that the job is offered, an applicant can either accept the position, decline the position, or ask for some time to consider the offer. If the position is declined, it must be done tactfully. This means that the applicant must actually tell the employer that he or she will not be accepting the position at this time. Just as job seekers are anxious to get an offer, employers deserve a prompt response to an offer. A professional will inform the employer as soon as he or she has made a decision about the position offered.

If time is needed to consider a job offer, the applicant must still decide in a timely manner. If the offer is accepted, the applicant may still have an obligation to a current employer who will need to be given notice. The standard notification for leaving a job is two weeks for an hourly employee. This means that the employee will provide the current employer with a written notice of intention to end employment in two weeks. A person in a managerial position usually would provide a minimum of three weeks' notice.

## SUMMARY

Food preparation has a history rich in exploration and innovation. As explorers traveled the world searching for new lands, they encountered unfamiliar foods and ingredients. Many of these exotic ingredients were brought back to the explorers' homelands, resulting in a mingling of foods and flavors from all over the world.

Following the French Revolution, chefs who had lost their positions working for nobility opened fine restaurants serving high-quality food. Many culinary trends began in France, where influential chefs had a dramatic impact on the way food was prepared and served. Marie-Antoine (Antonin) Carême standardized many recipes and techniques and developed a culinary style, known as grande cuisine, that emphasized extravagant presentations. Auguste Escoffier simplified Carême's methods into practical, professional techniques used by professional chefs today.

The restaurant industry is a fast-paced and ever-changing industry. Customer service, food production, and management positions all have an effect a customer's dining experience. Additional career opportunities are also available in sales and culinary education. Whichever path is chosen, a career in the food service industry is likely to prove both challenging and rewarding.

Training for a successful career in the food service industry has many options. Whether a person chooses to learn on the job or enroll in a quality culinary arts program, proper education and training are the keys to success. In addition, a food service employee needs sharp skills in math, communication, and technology. Demonstrating qualities such as ethics, dependability, flexibility, and dedication are what separates the best employees from the merely good. Only professionals who believe in continual education and training will advance to the top.

As a person prepares for employment, a few steps need to be taken prior to looking for a job. A résumé and portfolio need to be developed to highlight accomplishments, strengths, and experience to a potential employer. These tools, when presented by a neat and organized candidate for employment, will make a great first impression on a potential employer.

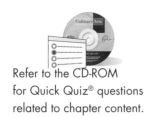

Refer to the CD-ROM for Quick Quiz® questions related to chapter content.

## Review Questions

1. How did early European exploration influence the cuisines of different cultures?

2. Why was Monsieur Boulanger sued by the guilds in 1765, and what was the outcome of this conflict?

3. How did the French Revolution influence the development of the restaurant industry?

4. Describe Marie-Antoine Carême's influence and culinary legacy.

5. What were Auguste Escoffier's main contributions to the restaurant industry?

6. Differentiate between nouvelle cuisine, New American cuisine, and fusion cuisine.

7. What specific factors in American society led to a boom in the food service industry?

8. Identify defining characteristics of the six common types of modern dining environments.

9. Identify the five jobs classified as customer service positions and the main responsibilities of each.

10. What are the main positions within the kitchen brigade system?

11. What is the difference between a restaurateur and a food service director?

12. Contrast the main responsibilities of an executive chef and a working chef.

13. What are some of the food service career opportunities other than customer service, food production, and management positions?

14. Describe how apprenticeship programs work.

15. What is the difference between a certificate program and a culinary arts degree program?

16. Explain what constitutes strong communication skills.

17. Provide examples of how math skills are used in the food service industry.

18. Identify common ways that technology is used in the food service industry.

19. List major leadership qualities that workers should possess in the food service industry.

20. Describe the main contents of a portfolio that a job candidate may submit to a potential employer.

21. List common job search tools that a job candidate can use when looking for employment.

22. Identify the steps that should be taken in preparing for a job interview.

23. Describe the proper actions a job candidate should take if he or she accepts a job offer.

# Management and Customer Service

Daniel NYC

*Management and customer service* is more challenging than ever before. A successful manager needs to have exceptional skills in communication, leadership, and customer service, as well as the business and technical skills needed to run and maintain a food service establishment. For an entry-level food service employee, earning a managerial position may be a few years away. However, it is important to be aware of and practice the skills needed to obtain a higher position. Each day that employees make the right decisions, treat customers with respect, and provide great customer service, they are a step closer to becoming successful food service managers.

## Chapter Objectives:

1. Compare and contrast the responsibilities of first-line, middle, and top managers.
2. List the types of skills required for effective management.
3. Describe a food service manager's obligations to employees, customers, and owners.
4. Explain what issues are involved in hiring, training, supervising, and evaluating employees.
5. Describe the main tasks involved in managing a food service operation.
6. Identify the key elements of exceptional customer service.
7. Compare and contrast common styles of meal service.
8. Describe the guidelines commonly used to handle customer complaints.

## Key Terms

- **first-line manager**
- **middle manager**
- **top manager**
- **technical skill**
- **interpersonal skill**
- **leadership skill**
- **job description**
- **mentor**
- **harassment**
- **discrimination**
- **booth service**
- **banquette service**
- **modern American plated service**
- **butler service**
- **buffet service**
- **family-style service**
- **Russian service**
- **classical French service**

## LEVELS OF MANAGEMENT

The food service and hospitality industry is a fast-paced and challenging industry that employs primarily hourly workers. In food service, both customer service and food production employees are paid by the hour. This industry depends heavily on hourly employees to clean and maintain facilities, prepare and cook food, and provide customer service. Typical levels of management structure in a food service operation include the employees, first-line manager, middle managers, and top manager. ***See Figure 2-1.***

Employees make up the primary workforce in a food service operation. They are typically hourly workers such as cooks, servers, bussers, dishwashers, cleaning staff, and bartenders. Employees are responsible for making products or performing services. They are an important group in the food service operation, as their performance directly affects the experience of the customer. A skillful employee with a pleasant attitude can create a positive dining experience for customers. On the other hand, an irritable employee with poor skills can negatively impact a customer's dining experience.

*Food Service Management Levels*

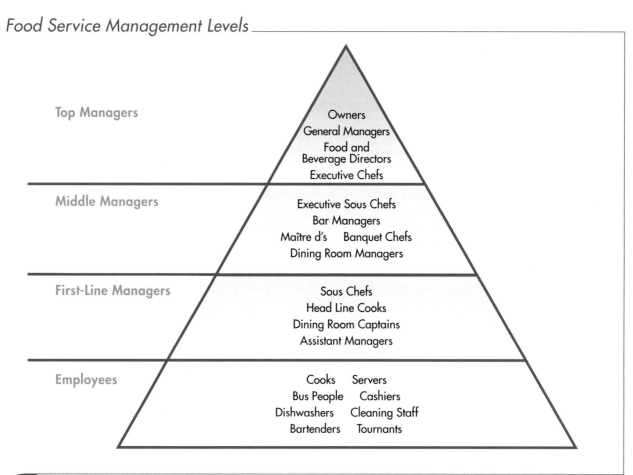

**2-1** *Typical levels of food service management include employees, first-line managers, middle managers, and top managers.*

A *first-line manager* is an employee responsible for the day-to-day supervision of hourly employees. In most operations first-line managers work side by side with the employees they supervise. Because of this, first-line managers are also commonly called working managers. Common examples of first-line managers include head line cooks, sous chefs, dining room captains, and assistant managers. ***See Figure 2-2.***

A *middle manager* is a person who directly supervises first-line managers. Middle managers usually ensure policies are being enforced, standards are being met, and that first-line managers are adequately motivating and managing employees. One of the primary duties of a middle manager is to be a liaison, or means for communication, between top managers and first-line managers. In this capacity, middle managers ensure that new policies and procedures dictated from top management are thoroughly explained and properly executed by employees. Examples of middle managers include executive sous chefs, bar managers, maître d's, and dining room managers. ***See Figure 2-3.***

A *top manager* is an administrator or owner of a food service establishment. Top managers have the ultimate control of policies, procedures, and standards that employees must adhere to, and establish and enforce the mission of an operation. Examples of top managers include owners, head chef or executive chef, general manager, and food and beverage director. ***See Figure 2-4.***

An effective food service manager must be able to both operate and maintain the food service operation while training and motivating staff to perform at their best. Many people understand the role of the food service manager to be nothing more than watching over the employees, but this is far from accurate. In the food service industry, many managers work side by side with the employees that they supervise. This is quite different from other industries where managers may work primarily in private offices away from the staff they supervise.

## MANAGERIAL SKILLS

All managers need to possess and master the basic skills necessary for interacting with employees, customers, and other managers. In order to master these skills, a manager must practice and successfully demonstrate them daily. Three major areas of managerial skills are technical skills, interpersonal skills, and leadership skills.

A *technical skill* is an ability to successfully perform a task of the trade. In food service, these tasks include the primary tasks that employees perform on a daily basis, such as culinary skills, customer service skills, computer skills, food costing skills, and budgeting and purchasing skills. Having the technical skills that employees are required to perform earns a manager more credibility and respect from staff. It is difficult for managers to properly evaluate performance of a task that they themselves

*First-Line Managers*

Line cook

Head cook

Hobart

**2-2** A first-line manager supervises hourly employees.

*Middle Managers*

Advance Tabco

**2-3** Middle managers, such as sous chefs or dining room managers, supervise first-line managers.

cannot perform, and employees have more acceptance of criticism from a manager who can effectively demonstrate a task in question. Having technical skills also allows a manager to more effectively train employees through demonstration. ***See Figure 2-5.***

## Top Managers

Cres Cor

**2-4** *Top managers have control of the policies, procedures, and standards that establish and maintain food service operations.*

An *interpersonal skill* is an ability to work with and treat people with respect and as individuals. Interpersonal skills include relating to others, conflict resolution, relationship building, and active listening. In the workplace, pretty much everyone is supervised and evaluated. Even top managers are evaluated on how well an establishment performs in terms of profitability and growth. To be effective and successful, managers need to learn about and appreciate the differences in their employees, respect cultures different than their own, and promote diversity in the workplace. No employee wants to be bullied, mistreated, or looked at as an inferior employee by a manager. Such treatment often results in poor performance, or employees leaving the operation for better treatment at another job. Managers that have patience, understanding, and compassion quickly earn the respect and loyalty of the people they supervise.

## Demonstrating Technical Skills

NSF International

**2-5**  Having technical skills allows a manager to effectively train employees through demonstration.

A *leadership skill* is an ability to influence others in a way that results in changes reflecting the values, goals, and vision of an individual or organization. Leadership skills include abilities such as teamwork, time management, communication skills, and the ability to motivate others. These skills can be achieved through practice.

Managers are leaders of teams. By setting a good example and treating each employee equally, managers demonstrate teamwork. Effective teams accomplish more than any one individual could, yet effective teams respect each team member.

Time management enables a person to use time effectively in order to accomplish required tasks within a given amount of time. In addition to getting tasks done on time, practicing effective time management enables a manager to have adequate time to supervise, motivate, and train employees. Time management involves effective use of one's own time as well as planning for efficient use of others' time. For example, time management skills are used to schedule the correct number of employees so there are neither too many nor too few workers assigned to a shift. ***See Figure 2-6.***

Communication skills allow messages to be conveyed clearly and ensure that the messages are understood. A manager needs to have good communication skills in order to effectively work with staff, owners, and customers. Communication involves both a sender and a receiver of a message. A sender uses verbal and nonverbal communication skills to convey a message to a receiver. Communication skills include listening skills, speaking skills, observation skills, and writing skills.

| | | Sunday 10A - 10P | Monday 4P - 9P | Tuesday 4P - 9P | Wednesday 4P - 9P | Thursday 4P - 11P | Friday 4P 11P | Saturday 10A - 11P | |
|---|---|---|---|---|---|---|---|---|---|
| **Employee Schedule Week of _____** | | | | | | | | | |
| **Name** | **Position** | Sunday 10A - 10P | Monday 4P - 9P | Tuesday 4P - 9P | Wednesday 4P - 9P | Thursday 4P - 11P | Friday 4P 11P | Saturday 10A - 11P | Total Hours |
| **Managers** | | | | | | | | | |
| | General Manager | | 2 P - 9 P | 2 P - 9 P | 8 A - 3 P | 8A - 4 P | 8 A - 5 P | | 40 |
| | Assistant Manager | 12 A - 5 P | | | 2 P - 9 P | 2 P - 11 P | 2 P - 11 P | 12 A - 8 P | 40 |
| | Executive Chef | 12 P - 8 P | | | 12 P - 8 P | 12 P - 8 P | 12 P - 8 P | 12 P - 8 P | 40 |
| | Banquet Manager | 8 A - 5 P | | 9 A - 5 P | | 4 P - 11 P | 4 P - 11 P | 8 A - 5 P | 40 |
| | Catering/Events Manager | 12 P - 8 P | 9 A - 5 P | | 9 A - 5 P | | 2 P - 11 P | 2 P - 11 P | 40 |
| **Service Staff** | | | | | | | | | |
| | Maître d' | 2 P - 10 P | 4 P - 9 P | 4 P - 9 P | | | 4 P - 11 P | | 25 |
| | Maître d' | | | | 4 P - 9 P | 4 P - 11 P | | 2 P - 10 P | 20 |
| | Host | 10 A - 2 P | 4 P - 9 P | | | | 4 P - 11 P | VACATION | 16 |
| | Hostess | | | | 4 P - 9 P | 4 P - 11 P | | 3 P - 11 P | 20 |
| | Hostess | 2 P - 10 P | | 4 P - 9 P | | VACATION | | 10 A - 3 P | 18 |
| | Server | 2 P - 10 P | 4 P - 9 P | 4 P - 9 P | | | 4 P - 11 P | | 25 |
| | Server | 10 A - 6 P | 4 P - 9 P | 4 P - 9 P | | | 4 P - 11 P | | 25 |
| | Server | | | | 4 P - 9 P | 4 P - 11 P | | 3 P - 11 P | 20 |
| | Server | | | | 4 P - 9 P | 4 P - 11 P | | 10 A - 6 P | 20 |
| | Server | 12 P - 8 P | | | | 4 P - 11 P | 4 P - 11 P | 12 P - 8 P | 30 |
| | Server | 12 P - 8 P | | | | 4 P - 11 P | 4 P - 11 P | 12 P - 8 P | 30 |
| | Bus person | 2 P - 10 P | 4 P - 9 P | 4 P - 9 P | | | 4 P - 11 P | | 25 |
| | Bus person | 10 A - 6 P | 4 P - 9 P | 4 P - 9 P | | | 4 P - 11 P | | 25 |
| | Bus person | | | | 4 P - 9 P | 4 P - 11 P | | 3 P - 11 P | 20 |
| | Bus person | | | | 4 P - 9 P | 4 P - 11 P | | 10 A - 6 P | 20 |
| | Bus person | 12 P - 8 P | | | | 4 P - 11 P | 4 P - 11 P | 12 P - 8 P | 30 |
| **Kitchen Staff** | | | | | | | | | |
| | Sous Chef | 8 A - 5 P | 2 P - 9 P | | | 2 P - 11 P | | 8 A - 5 P | 34 |
| | Sous Chef | 2 P - 10 P | | | 2 P - 9 P | 6 P - 11 P | 2 P - 11 P | | 29 |
| | Sous Chef | | | 2 P - 9 P | VACATION | | 6 P - 11 P | 6 P - 11 P | 17 |
| | Garde Manger | 12 P - 8 P | | | 12 P - 8 P | 12 P - 8 P | 12 P - 8 P | 12 P - 8 P | 40 |
| | Line Cook/Expediter | 8 A - 5 P | 2 P - 9 P | | | 2 P - 11 P | | 8 A - 5 P | 34 |
| | Line Cook/Expediter | | 2 P - 9 P | 2 P - 9 P | 2 P - 9 P | | 2 P - 8 P | 2 P - 9 P | 28 |
| | Line Cook/Expediter | 2 P - 10 P | | 2 P - 9 P | 2 P - 9 P | 2 P - 11 P | 2 P - 11 P | | 33 |
| | Tournant/Expediter | 3 P - 10 P | | 2 P - 9 P | | | 6 P - 11 P | 6 P - 11 P | 22 |
| | Prep Cook | 8 A - 5 P | 2 P - 9 P | | | 2 P - 11 P | | 8 A - 5 P | 34 |
| | Prep Cook | 2 P - 10 P | | | 2 P - 9 P | 2 P - 11 P | 2 P - 11 P | | 33 |
| | Prep Cook | VACATION | | 2 P - 9 P | | | 6 P - 11 P | 2 P - 11 P | 19 |
| | Pastry Chef | 12 P - 8 P | | | 12 P - 8 P | 12 P - 8 P | 12 P - 8 P | 12 P - 8 P | 40 |
| | Baker | | 12 P - 8 P | 12 P - 8 P | | 12 P - 8 P | 12 P - 8 P | 8 A - 5 P | 37 |
| | Baker | 8 A - 5 P | 12 P - 8 P | 12 P - 8 P | 12 P - 8 P | | 12 P - 8 P | | 37 |
| | Dishwasher | 10 A - 6 P | 4 P - 9 P | | | | | 10 A - 6 P | 21 |
| | Dishwasher | | | 4 P - 9 P | | 4 P - 11 P | | 3 P - 11 P | 20 |
| | Dishwasher | 2 P - 10 P | | | 4 P - 9 P | | 4 P - 11 P | | 20 |

**2-6** *Time management includes the ability to effectively plan for the efficient use of every employee's time.*

Verbal communication uses words to convey a message. Successful communication requires both speaking skills and listening skills. Speaking skills include not only pronouncing words clearly and at an adequate volume, but also choosing words that are specific so that the chance of misinterpretation is minimized. For example, the direction "Wipe off that plate," is less clear than "Wipe off the extra sauce from the platter of ribs." Time is a consideration in a professional kitchen during a meal service, so the ability to express specific directions clearly in a brief amount of time is a valuable skill. By listening closely to others, managers can gather information that can be used to help deliver messages more clearly. ***See Figure 2-7.***

Communication can also be nonverbal, such as a facial expression or a gesture. A nonverbal message that is consistent with a verbal message reinforces meaning. Observing others' nonverbal communication can also provide useful information. For example, if a manager observes expressions or gestures of frustration from an employee, he or she can approach the employee more effectively with encouragement rather than with criticism.

Written communication involves writing clearly, using proper spelling and grammar, and understanding the intended audience. For example, a piece of business correspondence to a vendor should be presented in a professional letter format, while a menu should be written to clearly describe the items in an appealing manner.

Managers often need to motivate employees to work toward common goals. Employees in entry-level positions may need motivation to work hard and strive for results. Higher-level employees may need to be motivated as they can become bored or lose interest in their work. Employees who are motivated to work effectively as part of a successful team strive for excellence and perform better. One of the easiest ways to motivate employees is to treat them with respect and let them know they are valued by management.

Chef's Tip:

A common mistake for a manager to make is placing too much focus on being heard, rather than hearing what others have to say.

## Communication Skills

- *Listening skills ensure a message is clearly heard and understood.*

- *Speaking skills convey a message clearly.*

- *Observation skills are used to gather visual information related to a message.*

Verbal and nonverbal communication of message

Sender → Receiver

Feedback confirms to sender that message was recieved and understood

**2-7** *A manager needs to have good communication skills in order to effectively work with staff, owners, and customers.*

The telephone operator game demonstrates the effectiveness of oral communication. With a group of five or more people, try this experiment. Designate one person as the manager and a second person as the employee. Everyone else has the role of a communicator. Have the manager give a set of directions to one communicator. The communicators pass the message orally until all communicators have heard the message. The last communicator to receive the message passes it to the employee. The employee executes the directions and the manager reviews the task. Was the result what the manager intended?

## RESPONSIBILITIES OF THE MANAGER

A food service manager has to fulfill many responsibilities to employees and customers on a daily basis. A manager has responsibilities to train, lead, and motivate employees to do their best and meet the expectations of customers. Because a manager is responsible for the performance of employees, a manager is also ultimately responsible for the quality of products and services employees provide to customers. Customer satisfaction has a direct impact on sales and profitability of a business, so a manager is also responsible for making the business profitable to the owners through repeat customers, operating efficiency, and increased sales. ***See Figure 2-8.***

If the quality of the food, beverage, or service coming from employees is poor, customers may not return. When a restaurant is losing sales and revenue, owners look to the manager for answers. On the other hand, if employees are producing high-quality food, beverage, and service, customers may develop a loyalty to the establishment and promote it through recommendations to others. When a restaurant gains sales and revenue from repeat customers and new customers, owners are likely to attribute a good portion of that success to the effectiveness of the manager.

## Responsibilities of the Manager

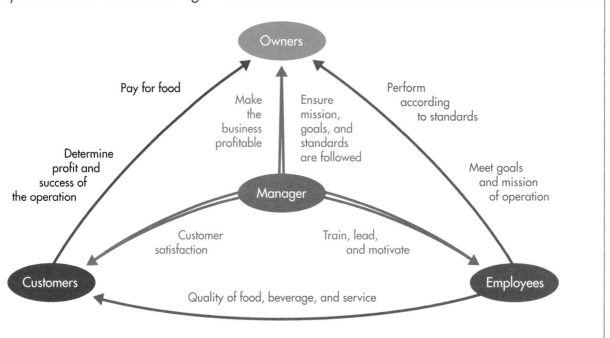

2-8 *A food service manager has responsibilities to employees, customers, and owners.*

## Obligations to Owners

A manager has an obligation to owners to make the business profitable. From an owner's standpoint, if the business is making money, things are running smoothly. An owner invests time and money into a food service establishment and relies on a manager to produce a good return on that investment. A manager can help ensure profitability through effective supervision, training, and motivation of employees, who in turn provide quality food and service to customers. Satisfied customers spend more money through repeat business, and owners see increased sales and profit. *See Figure 2-9.*

**Chef's Tip:**
When employees fail to provide quality food or service, it reflects on the manager's ability to lead and motivate. Managers must ask themselves what they need to do differently in order to get the desired results from employees.

Tru, Chicago

**2-9**  A manager's main obligation to owners is to make the business profitable.

Besides making a profit, another obligation of a manager to an owner is to ensure the operation's mission and goals are met and that standards are followed. In other words, a manager should run an operation the way its owner wants it to be run. A manager should never change the way that something is prepared or a policy that an owner has established without approval, even if it seems there is a better way to do it. Any changes or ideas need to be approved by the top manager or owner. Many operations have failed because a cook changed recipes that were standard to the operation. For example, a cook may think that there should be more cheese on a pizza, but changing the amount of an ingredient used in a dish has an impact on both the customer and the operating budget of the establishment. Using more cheese increases the cost to prepare the food being served, and can add up to a substantial increase in cost to the establishment. Although some customers may like the change, other customers may not and may choose to go to a different restaurant in the future as a result.

A manager is also responsible for accounting and management of operational costs, including food costs, payroll costs, supply costs, balancing cash register drawers, and counting the servers' money at the end of a shift. A manager needs to have good math skills to perform the various bookkeeping and administrative duties required by the job. **See Figure 2-10.**

## Accounting and Operational Costs

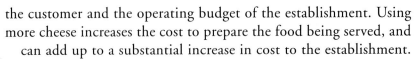

**Recipe Costing Worksheet**

| Item | grilled asparagus | | | | |
|------|------|------|------|------|------|
| Batch size | 8 servings | | | | |
| **Ingredient** | **Number of Units** | **Type of Units** | **Unit Cost** | **Total Cost** | **Cost %** |
| asparagus | 1 | lb | $1.90 | $1.90 | 29.78% |
| olive oil | 3 | tbsp | $0.68 | $2.05 | 32.24% |
| garlic, minced | 2 | cloves | $0.36 | $0.72 | 11.25% |
| lemon zest | 1 | tsp | $0.15 | $0.15 | 2.40% |
| rosemary | 2 | tsp | $0.76 | $1.52 | 23.82% |
| salt | 0.5 | tsp | $0.01 | $0.00 | 0.06% |
| black pepper | 0.5 | tsp | $0.06 | $0.03 | 0.44% |
| | | | **Potential Cost** | $6.36 | 100.00% |
| | | | **Unit Cost** | $0.80 | |
| | | | **Menu Price** | $2.50 | |
| | | | **Potential Percentage** | 100.00% | |
| | | | **Target Food Cost (%)** | 35.00% | |
| | | | **Target Menu Price ($)** | $2.27 | |

*Cost to prepare 8 servings*

*Cost per serving*

*Actual menu price*

**2-10** *A manager is responsible for accounting and must manage operational costs, including food costs.*

## Obligations to Employees

A manager has many obligations to employees, and employees are vital to achieving customer satisfaction. Since employees are responsible for preparing all products and providing all services in a food service establishment, a smart manager knows that employees are the key to his or her success and are the greatest reflection of a manager's abilities. It is the responsibility of a manager to provide employees with a working environment that encourages productivity and a commitment to quality.

There is a saying in the food service industry: "If you take care of your employees, the employees will take care of the customers, and the profits will take care of themselves." This means that if a manager has respect for employees and fair expectations of their performance, they likely will return that respect and be committed to performing well in the workplace. Additionally, the way a manager treats employees often is reflected in the way that the employees treat customers in turn. A manager who is respectful, professional, and courteous to employees increases the chance that employees will treat customers likewise. *See Figure 2-11.*

People have a need to feel challenged and to feel that they are learning and improving. A manager needs to be conscious of employees' needs to improve and learn. An employee who is bored in a job performs less effectively and may begin to look elsewhere for employment. When employees learn new skills and overcome new challenges at work, they are more interested in their job, feel more satisfaction, and become more invested in the company. Providing employees with new challenges and opportunities for accomplishment keeps employees interested and motivated.

## Obligations to Customers

Customers are the sole source of profit for a food service establishment. They are the reason a restaurant stays in business. If customers are unhappy, they may not come back, and may instead spend their money elsewhere. While a food service manager may not have direct contact with customers, managers are still ultimately responsible for customer satisfaction. Because almost all customer contact is handled by hourly employees, it is extremely important to have well-trained, committed, and personable employees. Typically, customers interact with managers only when a problem has occurred. In such cases, damage has already been done. A manager faced with a customer complaint can only reconcile the situation and work to ensure the problem is not repeated with another customer.

Customers come to a food service operation for many reasons. Whether it is to not have to cook for themselves, to rest and have someone else take care of them, or because they like the atmosphere, service, and food, customers have expectations that need to be met. A manager has three basic responsibilities to customers: to serve safe and wholesome food, to serve the highest-quality food, and to provide the highest level of customer service.

Customers expect and trust that the staff of a food service establishment is serving food that is safe to consume. Customers do not want to worry about whether the restaurant's refrigerator is working, or if the fish was stored properly. For safety as well as customer satisfaction, hot foods must be served hot, and cold foods must be served cold. Customers should be able to assume that the highest standards of cleanliness and wholesomeness are being followed.

*Treating Employees with Respect*

2-11 *Treating employees with respect encourages employees to perform well and treat customers with similar respect.*

## Quality Products

2-12 *All food items served should be of high quality according to USDA standards.*

A manager also has a responsibility to ensure that customers receive fresh, high-quality food. Customers do not want to go to a restaurant to get food that is of a lower quality than what they would buy at the supermarket and prepare for themselves. They expect that an operation is as conscious of quality and freshness as they are at home. Meat, poultry, seafood, produce, dairy products, and other food items ordered by a food service establishment should all be of high quality according to USDA standards. ***See Figure 2-12.***

Customer service is often the determining factor in the success or failure of a food service operation. When people dine out, they want to feel welcomed by the staff, be approached politely by the server, and be treated professionally throughout the meal. If the service is poor, the entire experience can be ruined, even if the food is excellent. Customers want to be pampered when they dine out. They want to be served by staff members with pleasant attitudes who are knowledgeable about the food and menu items, and who are attentive to their needs. ***See Figure 2-13.***

## MANAGING EMPLOYEES

Managing employees is one of the primary functions of a food service manager. Key factors that contribute to effective management of employees are hiring the right people for the job, thorough training, daily supervision, and periodic evaluations. When all of these factors have been applied properly, the manager and the employees have a clear understanding of what is expected and are more successful.

### Hiring Employees

One of the main concerns for any food service manager is finding the right people to do the required jobs. Because the hospitality industry has many entry-level positions, many of the applicants that apply for positions have little or no work experience. Some of the skills or qualities that managers look for in an applicant are not easily measured. Qualities such as working well with others, communication skills, punctuality, and honesty are difficult to assess until after someone has been hired. Having previous work experience or specific industry-related education such as culinary or hospitality coursework generally makes a person a stronger candidate for a particular job opening.

When a position at an establishment needs to be filled, the food service manager begins the hiring process. The manager must identify the skills that are necessary to perform the job and the qualities that are most desirable in a potential employee, and then must evaluate potential employees against these criteria. Three tools a manager can use to help find the ideal worker are a job description, an employment application, and an interview.

**2-13** *Customers want to be served by knowledgeable staff members with pleasant attitudes.*

A *job description* is a summary of an open position that accurately de-scribes skills, qualities, education, and previous work experience desired in an applicant, as well as a summary of the duties and responsibilities of the job. A job description helps job seekers determine if they would be interested in the job, and can serve as a way to encourage people who have the desired qualities to apply for the job. A manager can use the job description as a reference for comparing the attributes of potential candidates to the ideal attributes listed in the job description. In addition, a job description can be used at a performance evaluation as a list of duties and responsibilities that a person was hired to perform. ***See Figure 2-14.***

An *employment application* is a standard form that requests basic informa-tion from a job applicant, such as name, address, relevant work experience, educational background, position desired, and availability. Many entry-level employees do not have a résumé, so the employment application is used by the manager as the first look at the qualifications of a potential candidate. It can also serve as the first indication of a person's neatness and attention to detail. If the application is sloppy or missing important information, the candidate may not be a good choice for the job.

In addition to requesting information about background and experi-ence, the application also asks why a person left his or her last job. This question is significant as it can indicate something about a candidate's character and past job performance. For example, reasons for leaving a job such as "disagreement with boss" or "issues with co-workers" may be signs that a person has problems with authority or does not work well with

others. The amount of time that a person has worked at each previous job or the amount of time between jobs may also be indications of how committed a person is to staying with an organization or how serious they are about working. Of course, there may be good explanations for leaving a job after a short period or having a gap between jobs, such as a job being seasonal employment, an internship, or needing time between jobs to take care of family. ***See Figure 2-15.***

*Job Description*

## Job Description—Cook

Independence Senior Living is a leading owner and operator of senior living facilities throughout the United States. The company is committed to providing an exceptional living experience through properties that are designed and operated to provide the highest-quality service, care, and living accommodations for residents. We are currently seeking an experienced cook for our Fort Wayne, IN, facility.

This position is responsible for the preparation of meals for residents, cafeteria patrons, and guests at special functions. The candidate will follow all regulatory agencies' sanitation guidelines and maintain standards of kitchen cleanliness. All food will be prepared according to the menu in a safe, sanitary manner under the direction of the dining services coordinator/manager.

Candidates must possess the following:
• good attitude
• professional demeanor
• ability to perform in stressful situations
• mental strength and ability to take on challenges

**Responsibilities**
• Preparing quality food products according to standardized recipes and menus in a timely manner at specified meal times
• Following company procedures including sanitation, safety, and cleaning schedules
• Ensuring all food in refrigerator, freezer, and dry storage are labeled and dated
• Cleaning all food service equipment, kitchen, and food storage areas
• Coordinating work of dining services aides
• Supervision of dining services aides if the dining services coordinator/manager is not available.

**Physical Requirements**
Bending, stooping, lifting up to 50 lb, standing, twisting, reaching, walking; continuous exposure to chemicals and cleansers

**Qualifications**
• High school diploma (or GED)
• One to three years related professional experience and/or training
• Knowledge of special diets
• Board of Health Sanitation Certificate

2-14   *A job description accurately summarizes job requirements and the duties and responsibilities of the job.*

## Employment Application

| Name _____ | Date _____ |
| Address _____ | Telephone No. ( ) _____ |
| _____ | Social Security No. _____ |

Are you legally eligible for employment in the U.S.A.?  Yes____  No____
Are you of legal age to work?  Yes____  No____
Position(s) applying for_____
Were you previously employed by us?  Yes____  No____  If yes, when?_____
On what date would you be available to begin work?_____

### PERSONAL REFERENCES (not former employers or relatives)

| Name and Occupation | Address | Telephone No. |
|---|---|---|
| | | ( ) |
| | | ( ) |

### EDUCATION

| School | Name and Address | Course of Study | Last Year Completed | Did you graduate? | Diploma or Degree |
|---|---|---|---|---|---|
| High | | | | | |
| College | | | | | |
| Other | | | | | |

### EMPLOYMENT

| Name and Address of Company and Type of Business | From Month/Year | To Month/Year | Starting Salary | Last Salary |
|---|---|---|---|---|
| | | | | |

| Telephone No. ( ) | Duties |
|---|---|
| Supervisor | |
| | Reason for Leaving |

List any office and computer skills you have that are relevant to the position(s) for which you are applying.
_____
_____

List any experiences or qualifications that would be of benefit in the position(s) for which you are applying.
_____
_____

I hereby give my permission to contact the employers listed above concerning my work experience.
The information submitted above is true and accurate to the best of my knowledge.

Signed _____

Date _____

**2-15** *An employment application is a standard form that requests basic information from a job applicant.*

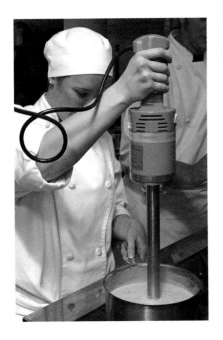

A *job interview* is a face-to-face meeting between a manager and a job candidate that allows the manager to find out more about the candidate, provide information about the job, and answer any questions that the candidate may have. This is probably the most important step in the hiring process, as it gives a manager the opportunity to actually speak with and ask questions of potential candidates. The job interview must be planned and well organized to ensure that all important questions are asked. In addition to providing information relevant to how the abilities of the candidate match the position, the interview also reveals a great deal about the candidate's communication skills, confidence, honesty, and ability to be a team player. The way an individual carries himself or herself is also important if the job puts the employee in direct contact with customers. If the candidate does not impress the manager in an interview, that person probably will not impress customers either.

It is important that interview questions be open-ended questions. Open-ended questions require an explanation as an answer, rather than a simple yes, no, or other one-word response. For example, a question such as "Did you leave your last job on good terms?" typically results in a one-word answer of yes or no. Asking the question "Why did you leave your last position?" requires a detailed explanation that allows a manager to evaluate a candidate on factors such as character, work ethic, and ability to get along with others.

What many managers forget about the interview process is that it is the candidate's one opportunity to sell himself or herself, not the manager's chance to sell the position. A manager may be desperate to fill the position as quickly as possible. However, managers must remember that they are trying to find the best candidate, not just fill the position. Many poorly conducted interviews focus primarily on how soon a person can start work and the salary offered, with the manager forgetting to inquire about reasons for that person wanting the position or what that person can bring to the organization. Gathering information about the desires and long-term goals of a candidate can be useful later in encouraging the employee to learn and grow with the establishment.

## Training Employees

During training, a new employee (or a recently promoted employee) learns how to perform required tasks and any expected standards for performance. A *training program* is a comprehensive plan of study containing all materials and information necessary for an employee to learn the requirements of a job. Having a documented training program ensures that staff are properly trained to be successful in their positions, and that customers receive food, beverages, and service according to the standards of the establishment.

A training program should be conducted by either a manager or a mentor. A *mentor* is a trusted and experienced person who can coach and guide a new employee. A mentor should be a person on the team who is best at performing a particular job and has a thorough understanding of how duties need to be performed. For a training program to be successful, it is essential that the tasks are demonstrated properly and that accurate information is provided by the manager or mentor. Proper instruction reduces the chance that an employee will need to be retrained later. ***See Figure 2-16.***

Many training programs begin with an orientation. An *orientation* is a period during which new employees are introduced to the policies and procedures of an establishment, the roles and responsibilities of a job, the people they will be working with, and any other information needed to be successful in the new job. An orientation period can last a few hours to a few days, depending on the position and the person being trained. After orientation is complete, a new employee should be given a written or practical test to determine if policies, procedures, roles, and responsibilities are understood. At the end of a training program, another test can be given to ensure the employee is ready to perform duties of the job in a manner that coincides with the goals of the organization.

Chef's Tip:

An orientation checklist can serve as a tool for managers, trainers, and new employees. Following a checklist ensures that all essential topics are addressed. It also indicates to a new employee the kind of information that management feels is most important to the job.

## Training Programs

2-16  *It is essential that tasks be demonstrated properly and that accurate information be provided by the manager or mentor.*

### Supervising and Evaluating Employees

Once employees have been hired and trained, day-to-day supervision begins. A manager does not supervise to act as a company police officer, but rather to help ensure employees are performing as required and following company policies. Most people perform best with positive reinforcement while they are learning new tasks. *Positive reinforcement* is a technique of giving positive feedback to an employee when he or she accomplishes a task to expected standards. Employees who receive positive reinforcement quickly realize that performing to expectations yields praise and encouraging feedback from the manager or mentor. This positive reinforcement encourages employees to continue to perform well.

Certain policies need to be explained in writing to all employees to ensure safety, well being, and fair treatment to employees as well as others they may be in contact with as part of their job. Policies regarding drugs and alcohol, harassment, discrimination, work schedule, and conduct should be documented in writing. An *employee handbook* is a written document containing all of the official policies and procedures of an establishment. In addition to policies and procedures, the handbook may contain information about employee benefits and the goals and mission of the establishment. These written policies should be read by all employees, and a sign-off sheet should be used to document that each employee has read and understood the policies. *See Figure 2-17.*

It is important that managers enforce written policies for the protection of other employees and customers. Drugs, alcohol, harassment, and discrimination have no place in a work environment and should not be tolerated.

Zero-tolerance drug and alcohol policies clearly prohibit the use of such substances in the workplace. Working with kitchen equipment, open flames, cutlery, and other items while under the influence of such substances is dangerous.

Harassment policies should be clearly defined. *Harassment* is behavior that is found to be threatening or disturbing and that is not considered acceptable by the general public. Harassment can range from words or actions that make a person feel demeaned or uncomfortable to situations that make a person feel threatened or endangered. *Sexual harassment* is unwelcome verbal or physical conduct of a sexual nature. Harassment of any kind creates an unfriendly work environment and should never be tolerated.

Discrimination policies should also be clearly defined. *Discrimination* is unfair treatment of people based on characteristics such as gender, race, ethnicity, age, religion, appearance, or disability. Examples of discrimination include treating someone of a different race differently than others or denying promotions to members of one gender while promoting people of the other gender. It must be made clear to employees that discrimination against customers or other employees is prohibited. Discrimination can be avoided by treating all people equally and rewarding employees based on performance and merit.

An employee work schedule and policies associated with employment should be made available at the time of hire. Policies regarding tardiness, requests for absence, emergency call-offs, overtime, and any paid time off need to be available to employees and should be provided in a written handbook. Only written work schedule policies can be used in employee evaluations or as grounds for termination.

The policies and procedures in this EMPLOYEE POLICY HANDBOOK are not intended to constitute contractual terms or conditions of employment or to create any legal rights. They may be modified to suit the circumstances, changed from time to time, or terminated.

**I have read and understand all of the written policies regarding drugs and alcohol, harassment, discrimination, work schedule, and conduct in the Employee Policy Handbook.**

Name

Date

*Sign-off sheet to document employee awareness of policies and procedures*

2-17 *A sign-off sheet should document that all employees have read and understood the policies in an employee handbook.*

Evaluations of employees can be either informal or formal, but thorough evaluation involves both informal and formal evaluations. An *informal evaluation* is an evaluation done on a daily basis by a manager, providing constructive feedback on daily skills, duties, and overall performance. A *formal evaluation* is a scheduled evaluation conducted at a predetermined time, such as every six months, where an employee meets with the manager for an evaluation that is documented and becomes part of an employee's permanent record. When it is time to conduct a formal evaluation, a well-written employee handbook can provide the basis for an evaluation of predetermined expectations, goals, and standards. ***See Figure 2-18.*** Both informal and formal evaluations should be used as a means of providing constructive feedback on the strengths and weaknesses of an employee. Evaluations can help employees improve skills and increase their value as part of the team.

Once employees are evaluated, it is determined whether or not they are accomplishing the expected goals and duties. If an employee is not meeting expectations, the manager needs to determine whether the employee needs to be retrained or if corrective action needs to be taken. If it is believed that an underperforming employee has the potential to improve, that person should be given an opportunity to be retrained and reevaluated.

## MANAGING THE OPERATION

A manager has many responsibilities in a food service establishment. Responsibilities can range from managing daily tasks to overseeing the long-term operation of all aspects of the business. Some of these responsibilities include managing safety and sanitation issues, as well as operations such as purchasing, receiving, stocking and storing inventory, and scheduling employees.

A food service establishment requires daily maintenance and upkeep in order for the operation to work smoothly. The following are examples of tasks that must be completed daily in a food service establishment:

- Restrooms must be cleaned thoroughly.

- Dinnerware, glassware, and flatware must be washed and sanitized.

- Inventory of food and supplies needs to be checked daily and orders must be placed as needed to ensure kitchen staff has all ingredients necessary to prepare menu items.

- Employee attendance needs to be verified to ensure the necessary staff are present for each shift.

It is a manager's responsibility to make sure all required daily tasks are completed. Managing involves coordinating all the activities and daily responsibilities of the food service establishment. Hourly employees may be concerned only with their own jobs and duties, so a manager must ensure each employee accomplishes his or her individual tasks as required.

*Charlie Trotter's*

**EMPLOYEE NAME** _____ **DATE OF HIRE** _____

**DEPARTMENT** _____ **POSITION** _____

| GENERAL FACTORS | RATINGS<br>Needs Improvement–Excellent | SUPPORTING DETAILS or COMMENTS |
|---|---|---|
| **1. Quality** — The extent to which an employee's work is accurate, thorough, and neat | 1  2  3  4  5 | |
| **2. Productivity** — The extent to which an employee produces a significant volume of work efficiently in a specified period of time | 1  2  3  4  5 | |
| **3. Job Knowledge** — The extent to which an employee possesses the practical/technical knowledge required on the job | 1  2  3  4  5 | |
| **4. Reliability** — The extent to which an employee can be relied upon regarding task completion and follow-up | 1  2  3  4  5 | |
| **5. Attendance** — The extent to which an employee is punctual, observes prescribed work break/meal periods, and has an acceptable overall attendance record | 1  2  3  4  5 | |
| **6. Initiative** — The extent to which an employee seeks out new assignments and assumes additional duties when necessary | 1  2  3  4  5 | |
| **7. Interpersonal Relationships** — The extent to which an employee is willing and demonstrates the ability to cooperate, work, and communicate with co-workers, supervisors, subordinates, and/or outside contacts | 1  2  3  4  5 | |
| **8. Judgment** — The extent to which an employee demonstrates proper judgment and decision-making skills when necessary | 1  2  3  4  5 | |

**EMPLOYEE DEVELOPMENT SECTION**

**In what areas has the employee improved since the last evaluation?**

_____

_____

**What should be done over the next six months to improve the employee's performance?**

_____

_____

**Manager and/or supervisor's comments:**

_____

**Employee's comments:** _____

_____

**Manager's Signature:** _____ **Date:** _____

**Supervisor's Signature:** _____ **Date:** _____

**Employee's Signature:** _____ **Date:** _____

**2-18** *During a formal evaluation, an employee meets with the manager for an evaluation that becomes part of an employee's permanent record.*

## Safety Training

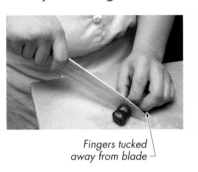

Fingers tucked away from blade

**2-19** *Safety training by a manager can help reduce the occurrence of accidents.*

## Managing Safety

Managing safety is important not only to provide a safe workplace for employees, but also to ensure a safe environment for customers. The most common safety concerns in a food service operation include cuts, burns, slips, and falls. A manager can help reduce the occurrence of these accidents by training staff in the following areas:

- mopping up spills or wet areas immediately and putting caution signs on a wet area to warn others that the floor may be slippery

- training staff on safe knife handling, washing of cutlery, storing of knives, and walking with knives; ***See Figure 2-19.***

- proper use of equipment such as steamers, ovens, deep fryers, or other potentially dangerous cooking equipment

- training staff on how to operate all kitchen machinery, such as mixers, food processors, slicers, and dough sheeters

## Managing Purchases

Managing purchases is an important responsibility of a manager. In a food service operation, purchasing involves much more than just buying goods. *Purchasing* is the selection and procurement of goods, including all steps from selecting goods, to verifying the correct goods are received, and finally paying for the goods. It is important for the manager to understand the purpose for which an item is being purchased so that the best-possible product is selected at an appropriate price. A manager must also know how much of an item to order.

*Par stock* is the amount of inventory that is used by an operation from the time of delivery to the time of the next scheduled delivery. Once an item has been selected, a manager needs to determine appropriate means of payment, such as with credit or cash on delivery (COD). A manager must also determine when an item is to be delivered. All of this information is then recorded on a purchase order. Purchasing is a time-consuming task because of the many factors involved. ***See Figure 2-20.***

## Managing Receiving

Managing receiving is important to ensure that all the correct items have been delivered at the agreed upon price. When ordered products arrive at an operation, a manager is responsible for checking the delivery. The items delivered should match those requested on the original purchase order in type, quantity, and size. A manager must also check delivered items for freshness, spoilage, proper temperature, and handling methods to ensure money is not wasted on goods that are out of date, spoiled, or mishandled.

## Purchase Orders

| DATE | PURCHASE ORDER NO. | PAGE NO. | DELIVERY DATE |
|---|---|---|---|
| **SHIP TO:** | | VENDOR | |
| **CHARGE & INVOICE TO:** | | | |

| QUANTITY | UNIT | ITEM ID — DESCRIPTION | UNIT PRICE | ITEM TOTAL |
|---|---|---|---|---|
| | | | | |

I certify that sufficient funds are available for this purchase.

| | Tax (if applicable) | |
|---|---|---|
| | Shipping | |
| | Total | |

_____ Signature

_____ Date

_____ Title

**Type of Payment**
☐ Payment Enclosed
☐ COD
☐ Credit

**2-20** *A purchase order specifies the number and type of items requested, the type of payment, and the delivery date required.*

Refer to the CD-ROM for the **Stock Rotation** Media Clip.

## Managing Inventory

Managing inventory is imperative to ensure an operation uses the oldest items before using newer stock. *First in, first out (FIFO)* is the process of dating items as they are received, and rotating older items to the front while placing the new items behind the old. **See Figure 2-21.** It is important that a manager is present at the time of deliveries. If a manager is not present at the time of a delivery, new items may be placed in front of current inventory, resulting in older products staying on the shelf longer and potential spoilage and loss of product. A manager must also ensure all refrigerated, frozen, and dry-goods storage areas are maintained at the appropriate temperatures.

## Managing the Schedule

Managing the schedule helps ensure the right numbers of staff are scheduled for each shift. Scheduling too few employees for a shift could result in a struggle to meet customer demand and poor service to customers. Scheduling too many employees for a shift could result in employees being bored without enough work to do to stay busy. Having too many employees scheduled also greatly increases the labor expenses for the shift, which cuts into the profit of the operation.

## First In, First Out (FIFO)

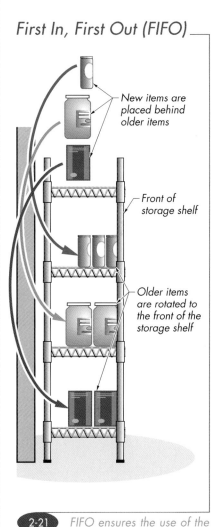

New items are placed behind older items

Front of storage shelf

Older items are rotated to the front of the storage shelf

**2-21** *FIFO ensures the use of the oldest items first.*

## Hosts

Charlie Trotter's

**2-22** *A host is the first person that customers encounter at a food service establishment.*

## CUSTOMER SERVICE

Exceptional customer service is essential to the success of a food service establishment. Even if a restaurant serves amazing food, if the quality of service is poor, customers may not enjoy themselves and may not return as a result. Many customers choose to go to a restaurant so they can relax and enjoy themselves. These customers want to be greeted with a pleasant smile and waited on by efficient and polite servers. Each member of a service staff has an impact on the overall experience of a guest. Every member of the service staff who has contact with customers must understand the importance of creating a good first impression. In addition, the host, server, and bus person each must know how the performance of their respective duties and responsibilities affects the service a customer receives.

A host is the first person that customers encounter when they enter a food service establishment. To establish a pleasant and hospitable atmosphere, a host should always speak to guests in a pleasant and welcoming tone of voice. If reservations are taken, the host will ask if the customers made a reservation and, if so, under what name. If no reservation was made, a host must determine if a table is available immediately or if the customers need to be placed on a waiting list for a table. When the table is ready, a host should guide the customers to their table, provide menus, and inform them of who their server will be. Later, as the customers are leaving the operation, a host is the last employee to thank them and say farewell. ***See Figure 2-22.***

A server is the second person that customers are in contact with at a restaurant. Servers have the greatest amount of contact with the customers, so servers must have very good interpersonal skills. In a food service establishment, servers can be compared to sales staff, in that they are responsible for suggesting and selling food and beverages to the customers.

Exceptional, well-trained servers can sell anything on a menu, because they know the ingredients in each dish and the preparation methods used. The more information about the menu and the greater the communication skills that a server has, the better that server will be at selling to customers. It is also the responsibility of a server to anticipate a customer's needs, check to make sure the customers are satisfied and have everything needed to enjoy their meal, and be available in the event that a customer should suddenly need something during the meal. A server must also be coordinated and able to present food and beverages to the customer neatly and gracefully without spilling. Besides representing an operation and selling food and beverages, the server is responsible for collecting payment for the food and beverages ordered. Since servers work for gratuity (tips), the better they are at their job, the more money they can earn. ***See Figure 2-23.***

A bus person is responsible for clearing tables after each course (bussing), removing dirty dishes, and clearing and re-setting the tables after customers have left. All of the bus person's tasks should be executed quietly and discretely to not interrupt the customers' meal.

## Personal Qualities

Exceptional front of the house service staff are a main reason that guests return to a restaurant. However, a pleasant personality and welcoming smile are only two characteristics of a great service staff member. Some additional personal qualities need to be developed in all staff for them to be good ambassadors of an operation. Qualities such as having a positive attitude, being a team player, dressing appropriately, maintaining good personal hygiene and health, and practicing excellent communication skills are some of the qualities that separate an average staff from a superior one.

**Positive Attitude.** Having a positive attitude is one of the most important elements for providing customer service. A service staff member's attitude can have a huge impact on a customer's overall dining experience. If a server has an upset or angry customer at one table, the server should not allow the problem to spread to other tables. This includes not approaching other tables in a manner that would suggest there is a problem within the dining area. It is important for service staff to be positive and pleasant when greeting and serving customers. Customers come to a restaurant to escape from stress and do not want to be affected by the stress of restaurant employees or other customers.

**Teamwork.** Teamwork is as important in the food service industry as it is in any sport. A restaurant depends on the teamwork of the staff to run efficiently. Teamwork means that all staff work together to ensure customers have a positive experience. For example, if server A is refilling coffee at one of his tables and notices that one of server B's customers needs a refill at a nearby table, server A should refill the cup of server B's customer. Working together to provide excellent service helps to ensure that customers have a pleasant experience.

**Appropriate Dress.** Dressing appropriately is a key not only to looking professional at work, but also to representing the professionalism of an operation. Uniforms or clothes worn to work, referred to as work attire, should be clean, free of stains, and neatly pressed. Clothes should fit properly. Work shoes should be clean, and polished if appropriate. The level of care and attention a person pays to his or her appearance is often seen as a reflection of the level of care and attention that person pays to his or her work. ***See Figure 2-24.***

## Servers

2-23 *A server must have very good communication skills.*

## Professional Attire

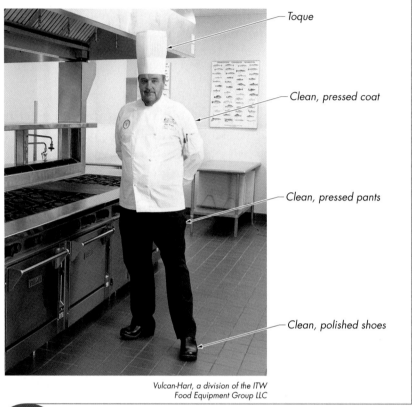

Toque

Clean, pressed coat

Clean, pressed pants

Clean, polished shoes

*Vulcan-Hart, a division of the ITW
Food Equipment Group LLC*

**2-24**  *Work attire should be clean, free of stains, and neatly pressed.*

**Personal Hygiene.** Having good personal hygiene goes hand in hand with wearing appropriate attire. Practicing proper personal hygiene is not only important for employees who interact with customers, but should be a part of the daily personal hygiene habits of all employees. Customers expect the service staff to be clean and well groomed. Employees must shower and use deodorant every day. Hair should always be clean and neat. Long hair should be tied back in compliance with safety and sanitation guidelines. Hands should be washed frequently, as food service employees touch flatware, glasses, plates, and food continuously throughout the day. Employees should always wash their hands thoroughly after sneezing, coughing, or touching the mouth, nose, hair, or face while working with food. Fingernails should be kept trimmed and very clean. Teeth should be clean and breath should not be offensive.

**Personal Health.** Maintaining personal health is essential in a food service establishment to avoid spreading illnesses to customers or other employees. Working in the hospitality industry is physically and mentally demanding. Therefore, it is important that employees take the necessary steps to maintain good heath, such as eating well, drinking plenty of water, exercising, and obtaining adequate rest.

Maintaining good health requires that the body receive nutrients daily. A poor diet can lead to poor health because the body does not receive nutrients essential for maintaining systems and defenses such as the immune system. Food service employees are surrounded by food all day in the operation, which can make them very hungry, and they are usually very busy. As a result, food service employees often skip meals or consume quick, empty-calorie foods. Empty-calorie foods are high in fat, salt, and simple carbohydrates, rather than the proteins, complex carbohydrates, vitamins, and minerals that the body needs to function efficiently. Quick but nutritious foods such as fresh fruit, vegetables, or sandwiches are good food choices when time is limited.

Employees may also drink caffeinated beverages such as coffee or soft drinks in place of pure drinking water. Caffeine is a stimulant that affects the nervous system and helps a person to feel more alert and awake. However, caffeine has several negative effects on the body, including dilating the blood vessels and increasing urine volume, which can lead to mild dehydration. People can also become dependent on caffeine, and can suffer from withdrawal symptoms such as headaches if they do not receive their typical daily intake.

A majority of positions within the food service industry require heavy lifting and long hours of standing. Exercising can help to protect the body from injury and can condition the body to be able to handle the physical demands of the job. Additionally, exercise can help to alleviate the stress that often results from working in a fast-paced, demanding job.

Just as exercise is important to maintaining overall body health, so is having adequate rest. When the body is fatigued, the immune system is weakened and people are at greater risk for illness. Employees in food service must make sure they get enough rest and sleep in order to recharge and maintain overall good health.

**Communication Skills.** Communication skills are essential not only for managers in leadership positions, but for employees at any level. Employees should always speak in a pleasant tone, clearly and loudly enough to be heard, while not being so loud as to disturb others. To develop speaking skills, a person should practice speaking slowly and clearly so that they are easily understood. The tone of a person's voice also conveys a great deal of information about that person's mood. A customer should never feel that an employee is anything but happy and professional, so it is important for employees to monitor their speaking tone as well.

Nonverbal communication should indicate that an employee is attentive to and focused on the customer's needs. For example, a person who makes appropriate eye contact when listening is indicating that they are paying attention. Looking around, watching others, or interrupting a speaker indicates that the listener is not focused on what the other person has to say. Written messages, such as writing "Thank you for dining with us" on a guest's bill, are another key form of communication. When communicating in writing, it is important to write clearly so that the message is not misunderstood.

## Meal Service Styles

The type of dining operation often dictates the type and level of service that needs to be provided by the staff. Different service styles found in the food service industry include booth, banquette, modern American plated, butler, buffet, family-style, Russian, and classical French. Although some of these service styles have names affiliated with countries, these styles are commonly used around the world.

*Booth service* is a term that refers to serving customers who are seated in a booth. A booth consists of two benchlike seats with a table between them where one end of each bench is usually affixed to a wall. Because the server cannot work on the side that is attached to the wall, they must provide service from one end of the booth. A server should excuse or pardon him or herself before reaching toward anyone seated at the far end of the booth. The customers on the right side of the booth should be served from the left with the server's left hand, while the customers on the left should be served from their right with the server's right hand. ***See Figure 2-25.***

## Booth Service

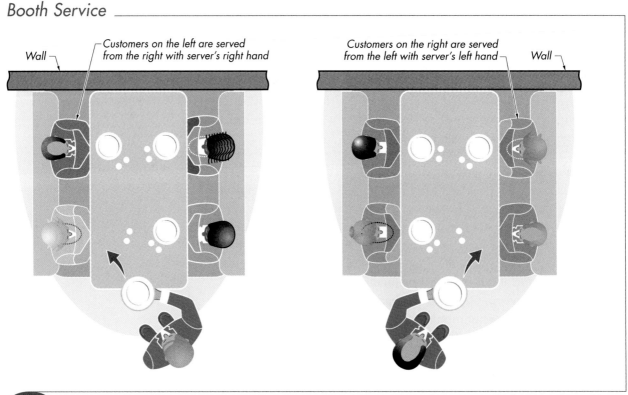

2-25   Booth service involves serving guests who are seated at tables with benchlike seats on both sides of the table.

*Banquette service* is a term that refers to serving customers seated in a banquette. A banquette is a long, benchlike seat that is affixed to a wall with a table or tables in front of it opposite individual chairs. In banquette service, the server serves all food and beverages with the right hand and clears the table with the left hand. *See Figure 2-26.*

## Banquette Service

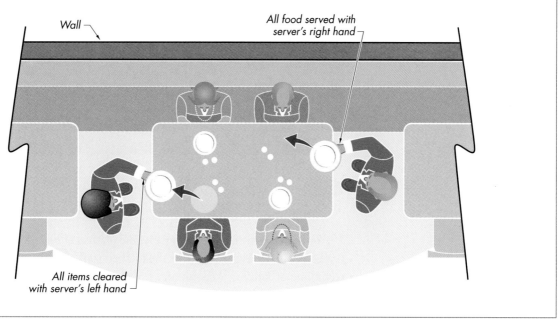

Wall

All food served with server's right hand

All items cleared with server's left hand

2-26   Banquette service involves serving guests who are seated at tables with benchlike seats on one side of the table.

*Modern American plated service* is a style of service where all food is prepared, portioned, plated, and garnished by the kitchen staff. This is the style of service most commonly found in casual restaurants because it is easy to execute and allows the greatest control over the presentation and amount of food served to customers. The chef has complete control over the final presentation of each plate.

*Family-style service* is a style of service where food is served on platters and placed directly on the table. Each dish is served with its own serving utensil. For example, a typical family-style meal may include a platter of sliced roasted beef, a platter of fried chicken, and several platters of vegetables. The platters are passed around the table by the guests as they serve themselves.

*Buffet service* is a style of service where all the food is arranged on a table and customers walk up to the table to serve themselves. It is important to provide an ample supply of clean plates near the buffet and to ensure that customers use a clean plate each time they take food. The reuse of a dirty plate at a buffet could cause cross-contamination. Although buffets are commonly self-service, some buffets have action stations where chefs prepare, carve, or serve individual items to guests. Action stations give a buffet an upscale appearance and create the potential for greater customer service through increased interaction between the customers and staff. *See Figure 2-27.*

*Chef's Tip:*
Always use the same hand as the side you are serving from, using the left hand when serving from the left and the right hand when serving from the right, so that you do not cross in front of the customer.

## Buffet Service

Food served on a buffet is arranged on a table from which the customers serve themselves.

*Russian service* is a style of service where platters of food are prepared, portioned, plated, and garnished by the kitchen staff and served using a special technique of holding a spoon and fork in one hand to serve food directly to the customers' empty plates. The server has a service napkin over the left forearm to rest platters on while serving. Food is served from the left with the serving utensils held in the server's right hand. The table is cleared from the customer's right side with the server's right hand. ***See Figure 2-28.***

## Russian Service

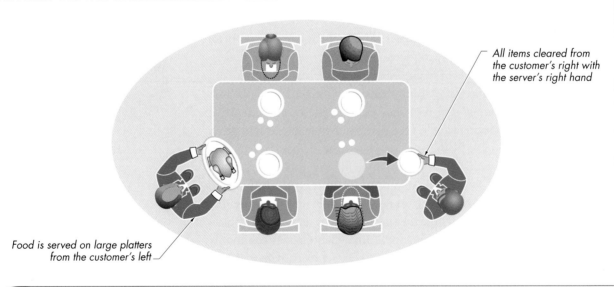

All items cleared from the customer's right with the server's right hand

Food is served on large platters from the customer's left

 2-28 Russian service involves a technique of holding a spoon and fork in one hand to serve food directly onto customers' empty plates.

*Classical French service* is a style of service where some of the food items are either finished or fully prepared tableside in front of the guest. This is commonly referred to as tableside preparation. Because the staff is preparing the dish directly in front of the customer, they must have a thorough understanding of the food item being prepared and the cooking methods being used, as well as good communication skills. Tableside preparation can range from simply finishing a nearly completed dish to complete tableside cooking. Common types of tableside preparation include the following:

- Assembling—Items such as salad and dressing are assembled in front of the customer. No cooking is required.

- Finishing or saucing—Occasionally, items are cooked in the kitchen and need to be finished, heated, or sauced in front of the guest. Items are typically placed over heat, basted with a sauce, and then served.

- Sautéing and flambéing—*Flambé* (pronounced flahm-bay) is a French term meaning "to flame" and refers to the procedure of heating alcohol and igniting it to deglaze a pan. Items for flambéing may be fully cooked ahead of time in the kitchen or they may be sautéed directly in front of the customer. When an item to be flambéed is nearly finished, a small amount of alcohol can be added to the pan and ignited. Items commonly sautéed and flambéed include cherries jubilee, bananas Foster, and steak Dianne.

- Deboning and carving—Many items are finished by deboning or carving them in the dining room at a table. Items such as cooked fish are often filleted in front of the guests, while meats, poultry, and game are often carved and served tableside. Carving and deboning tableside not only show the guests the freshness of an item, but also provide the guests with entertainment, allowing them to observe the chef or server's skill. Some restaurants may slice cheeses tableside after offering customers a selection of cheeses presented on a platter.

Most tableside preparations are usually done on a gueridon. A *gueridon* is a cart that can be wheeled to a table and is typically equipped with a carving station or a sautéing and flambéing unit called a réchaud (pronounced ray-show).

*Butler service* is a style of service where servers carry beautifully arranged hors d'œuvres on a silver tray or small elegant platter. This style of service is typically used for hors d'oeuvre receptions. Servers approach guests and offer an hors d'oeuvre from the tray. The server usually offers napkins to guests with the hand that is not holding the tray.

## Handling Customer Complaints

No matter how well the employees are trained and how high the quality of the food is, there will be occasions when things go wrong. It is important for both service staff and management to be attentive to customers and be aware of their needs at all times. Most customers who have a problem or make a complaint are not troublemakers. In most cases when a complaint is made, the customer has a legitimate reason to be dissatisfied. If the service staff avoids a customer because they do not want to deal with a problem situation, that customer is often lost forever. Failure to remedy a problem may result in other current or potential customers hearing negative feedback about the restaurant, and may lead to lost business.

Customers are the sole reason a food service establishment stays in business, as they provide its sole source of income. The popular saying "the customer is always right" does not mean that a customer never makes a mistake. Rather, it means that it does not matter whether the customer is right or wrong, because it is an employee's job to satisfy the customer. It is much easier to prevent problems from occurring in the first place, by having a well-trained staff and well-run operation, than it is to fix problems after they have occurred or to make an upset customer happy again.

It is obvious that a customer has a complaint when he is upset and asks to see a manager. Unfortunately, most customers do not say anything when they are upset and instead leave unhappy. If the staff is unaware of a problem, they may continue to make the same mistake and possibly upset other customers. Service staff should be trained in how to determine if a problem has occurred or if a customer is unhappy. Asking questions and observing nonverbal communication are two basic ways of identifying an unhappy customer.

One of the best ways to determine whether customers are happy is by asking them. After each food course, servers should ask open-ended questions about the items served. For example, a server could ask an open-ended question such as "How did you like the seafood special?" Open-ended questions encourage the person responding to provide a detailed answer rather than a one-word response. A detailed answer may help to expose any problems or underlying issues.

Another way of determining if a customer is unhappy is by observing the customer's nonverbal communication. If a guest looks irritated or disappointed or is simply picking at a dish rather than eating it, the guest may be signaling dissatisfaction.

If a customer is dissatisfied, the problem should be identified and resolved. When a problem occurs or a customer is unhappy about something, the following guidelines can be used to deal with the situation:

- Listen to the customer's concern. Listening is the first and most important step in resolving a negative situation. A customer should be given undivided attention when voicing concerns, and service staff should listen closely to fully understand any issues or problems a customer may have. Most customers do not want to draw complaints to the attention of other guests, but they do want their concerns to be heard by the appropriate members of the staff. Listening closely to customer concerns usually allows the staff to find a solution that is acceptable to both the customer and the staff. *See Figure 2-29.*

- Never ignore a dissatisfied guest or avoid dealing with a problem. A problem cannot solve itself, and if guests leave unhappy, they may never return. Any problem needs to be addressed immediately.

- It is important for service staff to remain calm. Remaining calm increases the chance that customers will also remain calm.

## Handling Customer Complaints

2-29 *If a customer is dissatisfied, the problem should be identified and resolved quickly and calmly.*

- Do not become defensive. Service staff must not take negative comments personally. Many customer complaints concern the service staff or the food. Any critique, positive or negative, should be viewed as a means of improving the overall operation of the establishment.

- Never blame a customer or another employee for a problem. Regardless of who is believed to be at fault, the important thing is to make the situation right. Whether the customer misspoke when ordering, the kitchen staff cooked an item improperly, or the server gave the customer the wrong order, what matters is finding an acceptable solution and correcting the problem. When service staff corrects problems quickly and calmly, customers leave satisfied and everyone wins.

- Finish by following up and apologizing. Once a solution has been implemented, service staff should follow up with the customer to ensure that he or she is satisfied with the outcome. When the problem has been reconciled to the guest's satisfaction, it is important to apologize again for the situation that occurred (again, no matter whose fault it was) and thank the customer for being understanding and patient.

## SUMMARY

Becoming a successful food service manager is not an easy task. It may take years to master the many tasks, obligations, and skills that a food service manager is responsible for on a daily basis. In addition to the basic technical and business skills needed to successfully run an operation, managers need to have exceptional interpersonal skills.

After the right candidates are found, managers need to make sure employees receive adequate training for the positions they are hired to fill. A manager needs to be able to inspire staff to do their best and provide the highest level of customer service.

All employees must understand that customer satisfaction is truly the secret to success. Even if the food is wonderful, if the service is poor customers will likely choose to go elsewhere. Providing good customer service involves meeting customer expectations, anticipating customer needs, and resolving customer concerns calmly and quickly.

The skills needed to succeed as a manager are developed through years of practice. This is why it is important to begin training and practicing managerial skills whenever possible early in one's career. With practice and a commitment to integrity and compassion, an employee can hone the skills necessary to be a successful food service manager.

Refer to the CD-ROM for Quick Quiz® questions related to chapter content.

1. Describe the supervisory responsibilities of first-line, middle, and top managers.

2. Summarize the three major areas of managerial skills that a food service manager should possess.

3. Provide three reasons why managers often request that job candidates complete a job application during the hiring process.

4. Explain why interview questions should be open-ended questions.

5. List four tasks that must be performed daily in a food service establishment that require supervision or completion by a manager.

6. Explain the process of first in, first out (FIFO) inventory control.

7. List the six personal qualities that customer service staff must possess in order to provide excellent customer service.

8. Describe the distinguishing characteristics of the following meal service styles: booth, banquette, modern American plated, butler, buffet, family-style, Russian, and classical French.

9. Describe how a server should handle a customer who complains that he was served the wrong entrée.

# Food Safety and Sanitation

## Chapter 3

*Safety and sanitation* are paramount to successful food service. Restaurants provide a gathering place for many important occasions, such as birthday celebrations, anniversary dinners, and business meetings. No matter what the occasion, customers expect to be served safe, tasty, and nutritious food, presented by a pleasant staff in a safe and enjoyable environment. It is the duty of every food service establishment to enforce food safety and sanitation standards and provide good customer service in a clean environment.

## Chapter Objectives:

1. List the main agencies that are involved in the establishment and regulation of food safety.
2. Describe three types of hazards that can cause direct contamination of food.
3. Identify common pathogens that cause foodborne illnesses.
4. Describe the six environmental conditions that contribute to bacterial growth, possibly resulting in foodborne illness.
5. Define potentially hazardous food.
6. Describe common food service precautions to take for highly susceptible populations.
7. List the general guidelines for preventing food contamination.
8. Identify the steps for developing a Hazard Analysis and Critical Control Point (HACCP) plan.
9. Describe the use of sanitizers for warewashing.
10. Demonstrate proper handwashing procedure.
11. Explain the importance of heating and cooling food.
12. Address common injuries resulting from accidents in the professional kitchen.

## Key Terms

- **food safety**
- **Food Code**
- **contamination**
- **microorganism**
- **direct contamination**
- **biological hazard**
- **chemical hazard**
- **physical hazard**
- **cross-contamination**
- **foodborne illness**
- **pathogens**
- **bacteria**
- **FAT TOM**
- **temperature danger zone**
- **potentially hazardous food**
- **HACCP**
- **sanitizing**
- **sanitizers**

## FOOD SAFETY

The reputation of any food service establishment is earned by the quality of the food and service provided to guests. *Food safety* is the practice of handling food in ways that prevent contamination or spoilage. Food must be carefully inspected, properly stored, and safely handled during the entire journey from purveyor to plate. A *purveyor* is a company or person who makes food, ingredients, or supplies available to a business for purchase. Food safety standards are established by local and national organizations and government agencies and enforced by local health departments and other authorities having jurisdiction. *See Figure 3-1.* The following agencies are involved in the establishment and regulation of food safety standards:

- The U.S. Food and Drug Administration (FDA) ensures the safety of all food except meat, poultry, and egg products. The FDA is also responsible for publishing and revising the *Food Code*. The *Food Code* establishes standards that assist food control jurisdictions at the national, state, and local levels in regulating the food service industry. It also provides the food service industry with guidelines for prevention of foodborne illness.

- The Food Safety and Inspection Service (FSIS) ensures the safety of meat, poultry, and egg products.

- The Animal and Plant Health Inspection Service (APHIS) ensures protection of animals and plants from diseases or pests.

- The Environmental Protection Agency (EPA) establishes levels of pesticide residue that can be tolerated by humans.

- The Centers for Disease Control (CDC) investigates foodborne illnesses.

The reputation of any food service establishment can be quickly ruined by inadequate or improper personal hygiene and sanitation practices. Failure to follow food safety guidelines can result in serious consequences. Poor sanitation procedures and improperly handled food can result in sickness or death. For this reason, organizations that enforce food safety codes have the authority to shut down food service establishments that fail to meet local safety codes. If there is any doubt regarding the quality or safety of a food item, the item must be thrown out immediately and the disposal recorded. Improper food handling can lead to food being unsafe for consumption.

**3-1** Food service establishments must follow food safety codes and guidelines enforced by local health departments and other authorities having jurisdiction.

## CONTAMINATION

In the food service industry, sanitation refers to creating an environment free of foodborne illnesses, harmful microorganisms, and harmful substances that can put food at risk of contamination. *Contamination* is the state of food or equipment being hazardous as a result of unsafe organisms or other items coming in contact with food or preparation equipment. Anyone who eats contaminated food is at risk for illness. A *contaminant* is any microorganism or substance that can contaminate food or preparation equipment. Contaminants can be present in foods in three different forms: biological, chemical, and physical.

Biological contaminants are called microorganisms. A *microorganism* is a single-celled plant or animal that can only be seen through a microscope. Because microorganisms cannot be seen by the human eye, they are an invisible hazard in the professional kitchen. Microorganisms present the greatest peril to the safety of food. Chemical contaminants include any hazardous substance in chemical form. A physical contaminant is any hazardous substance that is not biological or chemical.

Food items can become contaminated through either direct contamination or cross-contamination. Understanding the causes of contamination can lead to better prevention. A food service establishment should follow all local codes to prevent contamination of food and food preparation equipment in the professional kitchen and dining areas.

*Direct contamination* is contamination that occurs when uncooked foods, or the plants or animals that the foods are made from are contaminated in their natural environment. For example, crops can become contaminated by pesticides during the growing process or by naturally occurring bacteria in the fertilized soil. Direct contamination includes biological, chemical, and physical hazards.

**Biological Hazards.** The term *biological hazard* refers to various forms of microorganisms such as bacteria, parasites, viruses, and fungi that can cause a foodborne illness. All foods contain bacteria and bacteria can lead to outbreaks. Bacteria do not need a host to grow or survive. However, bacteria cannot move about on their own so they rely on being transmitted by some vehicle. The most common vehicle that transmits bacteria is the human hand.

A *parasite* is a small living organism that needs a living host to survive. These organisms inhabit the stomachs and intestinal tracts of animals, fish, and poultry, and are passed to people when contaminated food is eaten. One parasite-caused condition is anisakiasis, caused by a roundworm (Anisakis) found in contaminated fish and squid. This parasite lives in the stomach and intestinal tract of fish. The best means of preventing contamination is to completely clean fish immediately after they are caught. Cyclospora is a parasite found in water or foods that have been contaminated with feces. This parasite is often found in undeveloped countries where proper sewage systems have not been constructed. The best means of preventing contamination by this parasite is to avoid drinking or washing with contaminated water.

Unlike bacteria, viruses cannot reproduce in food. They must be in a living cell to reproduce. However, viruses can survive on the surface of food or production equipment and use this surface as a means of transportation between hosts. Unlike bacteria, viruses do not need a protein-rich surface, oxygen, moisture, or a certain pH to survive. Although some viruses can be killed with heat, the only sure way to prevent a foodborne viral illness is to prevent the viral contamination in the first place. Since heat often destroys a virus, foods not often heated, such as salads, desserts, and cold sandwiches, are often the items most likely to transmit viral contaminations to people. Viruses usually spread through contaminated water or unclean hands.

The two most common forms of viral contamination are Hepatitis A and the Norwalk virus. Hepatitis A is most commonly passed from contaminated shellfish and feces-contaminated water. A person who consumes contaminated shellfish may not become sick for many months, but immediately becomes a carrier of the virus. An infected person can infect others through cross-contamination and poor personal hygiene. Unfortunately, a person may infect many people without even knowing that they are carrying the virus.

The Norwalk virus is almost always spread by improper handwashing and poor personal hygiene. This virus stems from contamination of foods that have been in contact with manure and have not been properly cleaned, from contaminated water, or from improper handwashing after using the washroom. The Norwalk virus may be killed by thoroughly cooking foods but has even been found in ice cubes.

*Fungi* (singular *fungus*) are an extremely large group of organisms that range in size from tiny microorganisms to large mushrooms. The most common forms of fungal contamination are from mold and yeast. A *mold* is a cottonlike fungus that is visible to the eye in large clusters. Molds may appear as green or white fuzzy spots on an old loaf of bread or on a block of cheese that has been in the refrigerator too long. ***See Figure 3-2.*** Most food molds are not actually dangerous. They typically only affect the flavor, smell, and overall quality of foods. However, rare types of molds can form toxins such as aflatoxin or produce mycotoxicosis (mold poisoning), resulting in foodborne illnesses, and can even cause cancer in animals. Molds are very different from bacteria in that molds can grow on almost any food at any temperature. Although it is true that molds can be destroyed by heating foods to 140°F for 10 min, the toxins left behind from molds are not destroyed. Freezing prevents the growth of molds but does not destroy them. Foods contaminated with mold must be discarded. One exception to this rule is for hard cheeses, which can be used if 1 in. is removed from around the molded area.

## Molds

3-2　Molds may appear as green or white fuzzy spots on the surface of foods.

A *yeast* is a form of fungi that can be found in foods. Yeast sometimes can be helpful in a controlled state, such as when making bread. Yeast thrives on a diet of water and carbohydrates (sugars or starches). As yeasts feed on moistened carbohydrates, they release alcohol and carbon dioxide gas. This process, called fermentation, is important in making breads and alcoholic beverages. As with molds, most natural yeasts have not been found to be harmful to humans. When uncontrolled, however, they can lead to food spoilage, unpleasant odors, and off tastes. Unwelcome yeasts can be found as a visible slime on the surface of a pickle jar or as a pinkish color on the surface of a container of cottage cheese.

**Chemical Hazards.** The term *chemical hazard* refers to any hazardous substance in chemical form. The most common chemical contaminants include cleaning supplies, pesticides, toxic metals, and other chemical compounds found in a restaurant. For example, the mercury found in some species of fish can be hazardous if too much is consumed. The majority of chemical contaminants can be seen with the human eye, although traces of dangerous chemicals can remain if areas where they are used are not properly cleaned and disinfected.

**Physical Hazards.** The term *physical hazard* refers to any hazard that is not biological or chemical and can lead to a physical injury, such as a cut. Examples of physical hazards include items such as broken glass, staples, sharp pieces of metal, bone, or stones. Such items can be dangerous if eaten, leading to physical harm of teeth or internal digestive organs.

## Cross-Contamination

*Cross-contamination* is contamination that occurs when a biological, chemical, or physical contaminant is transferred from one item to another through an intermediate carrier. For example, a knife used to cut raw poultry is placed on a clean cutting board, and the cutting board is then used to prepare raw vegetables for a salad. Biological contaminants from the raw poultry juices on the knife are passed to the cutting board and subsequently to the raw vegetables. Cross-contamination most commonly occurs when raw foods make contact with ready-to-eat foods, such as when raw chicken and lettuce are cut on the same cutting board. People often cause cross-contamination through careless behavior, unsafe food handling, and poor sanitation practices, such as lack of handwashing.

*Carlisle FoodService Products*

## FOODBORNE ILLNESS

A *foodborne illness* is an illness that is carried or transmitted to two or more people through contact with or consumption of contaminated food. Foodborne illness is often caused by pathogens that are present in food or beverages. Harmful contaminants known as *pathogens* are infectious and toxigenic microorganisms that cause disease. Pathogens include many bacteria, parasites, viruses, and some fungi. If food is not properly handled, these pathogens can contaminate food and cause foodborne illness when the food is consumed. Pathogens reproduce rapidly and, if ingested, can cause vomiting, cramping, headache, sweats, chills, diarrhea, and fever. There are more than 250 known foodborne diseases caused by various pathogens. *See Figure 3-3.*

## Common Pathogens Causing Foodborne Illnesses . . .

| Pathogen | Transmission | Symptons |
|---|---|---|
| Campylobacter jejuni | Contaminated water; unpasteurized milk; raw or under-cooked meat, poultry, or shellfish | Fever, headache, and abdominal cramps followed by diarrhea and nausea that appear 2–5 days after eating; may last 7–10 days |
| Clostridium botulinum | Improperly canned foods, garlic in oil, vacuum-packaged and tightly-wrapped food | Double vision, droopy eyelids, trouble speaking and swallowing, difficulty breathing; symptoms usually appear 18–36 hr after eating, but may appear within 4 hr or after 8 days; fatal if not treated |
| Clostridium perfringens | Food left for long periods in steam tables or at room temperature | Diarrhea and gas pains appear 8–22 hr after eating;symptoms usually last about a day, but less severe symptoms may persist for 1–2 weeks |
| Escherichia coli O157:H7 (a.k.a. E. coli O157:H7) | Contaminated water, unpasteurized milk, raw or rare ground beef, unpasteurized apple juice or cider, uncooked fruits and vegetables, person to person contact | Diarrhea or bloody diarrhea, nausea, abdominal cramps, and malaise can begin 2–5 days after food is eaten, lasting about 8 days; can cause kidney failure |
| Listeria monocytogenes | Ready-to-eat foods such as hot dogs, luncheon meats, cold cuts, fermented or dry sausage, other deli-style meats and poultry, and soft cheeses; unpasteurized milk | Fever, chills, headache, backache, abdominal pain, and diarrhea may occur up to 3 weeks after eating; symptoms may evolve into more serious illness in highly susceptible populations |
| Salmonella (over 2300 types) | Raw or undercooked eggs, poultry and meat, unpasteurized milk and dairy products, seafood, contact by food handlers | Abdominal pain, diarrhea, nausea, chills, fever, and headache usually appear 8–72 hr after eating; symptoms may last 1–2 days |
| Shigella (over 30 types) | Fecal contamination of food and water including food contacted by food handlers with poor personal hygiene | Diarrhea containing blood and mucus, fever, abdominal cramps, chills, and vomiting within 12–50 hr of ingestion; symptoms can last a few days up to 2 weeks |
| Staphylococcus aureus | Person to person through food due to improper food handling; multiplies rapidly at room temperature | Severe nausea, abdominal cramps, vomiting, and diarrhea occur 1–6 hr after eating; symptoms usually disappear within 2–3 days but may last longer if severe dehydration occurs |
| Streptococcus | Milk, ice cream, eggs, steamed lobster, ground ham, potato salad, egg salad, custard, rice pudding, and shrimp salad that has stood at room temperature for several hours before consumption; food contaminated by food handlers with poor personal hygiene; foods made from unpasteurized milk | Sore and red throat, pain on swallowing, tonsillitis, high fever, headache, nausea, vomiting, malaise, rhinorrhea, and occasionally a rash occur within 1–3 days of ingestion |
| Enterococcus | Sausage, evaporated milk, cheese, meat croquettes and pies, pudding, unpasteurized milk, pasteurized milk that is underprocessed or prepared in unsanitary conditions | Diarrhea, abdominal cramps, nausea, vomiting, fever, chills, and dizziness within 2–36 hr following ingestion |
| Vibrio cholerae | Raw and undercooked seafood, contaminated food and water | Symptoms may be absent or mild and occur within 6 hr to 5 days after ingestion; severe illness symptoms include profuse diarrhea, vomiting, leg cramps, loss of body fluids due to dehydration and shock; symptoms typically last 7 days and can result in death if not treated |
| Vibrio vulnificus | Raw fish and shellfish, especially raw oysters | Diarrhea, abdominal pain, nausea, vomiting, fever, and sudden chills within 16 hr of ingestion; some victims develop sores on their legs that resemble blisters; symptoms persist 2–3 days |

**3-3** *Foodborne illness is often caused by bacteria or viruses that may be present in foods or beverages.* (continued on next page)

## . . . Common Pathogens Causing Foodborne Illnesses

| Pathogen | Transmission | Symptons |
|---|---|---|
| Yersinia enterocolitica | Raw meat and seafood, dairy products, produce, and untreated water | Fever, diarrhea, vomiting, and abdominal pain occur 1–2 days after ingestion; symptoms persist for 1–2 days and may be particularly severe in children |
| Cryptosporidium parvum | Food contaminated by food handlers or vegetables fertilized with manure | Severe watery diarrhea, intestinal distress, coughing, and a low-grade fever; symptoms often last 2–4 days; symptoms may last longer in highly susceptible populations |
| Bacillus cereus | Contaminated meats, milk, vegetables, fish and shellfish, rice, potatoes, pasta, cheese products, sauces, puddings, soups, casseroles, pastries, and salads | Diarrheal type: watery diarrhea and abdominal cramps within 6–15 hr after consumption; symptoms usually persist for 24 hr<br>Vomiting type: nausea and vomiting within 0.5 to 6 hr after consumption; duration of symptoms is less than 24 hr |
| Giardia lamblia | Contaminated water, raw vegetables prepared by infected food handlers; cool moist conditions favor its survival | Diarrhea; illness lasts for 1–2 weeks, but chronic cases can last months to years and can be difficult to cure in immune-deficient individuals |
| Noroviruses (Norwalk and Norwalk-like) | Contaminated water, improperly cooked or contaminated shellfish (especially oysters), improperly prepared uncooked fruits and vegetables (especially salads), food or beverages contaminated by an infected food handler | Nausea, vomiting, diarrhea, abdominal cramps, headache and low-grade fever can occur within 24–48 hr after ingestion or contact with the virus; symptoms usually last for 24–60 hr |
| Hepatitis A virus | Contaminated water, fruits, vegetables, iced drinks, shellfish, and salads, including food contacted by food handlers with poor personal hygiene | Fever, malaise, nausea, abdominal discomfort or side pain, vomiting, loss of appetite, dark urine, and pale or white-colored bowel movements usually begin within 2–6 weeks after exposure; symptoms may persist for up to 3 weeks |

**3-3** *(continued from previous page.)*

Bacterial contamination is the leading cause of foodborne illness today. However, not all bacteria are harmful. Yogurt, for example, contains an active bacterial culture that aids in digestion. *Bacteria* are single-celled microorganisms that live in soil, water, organic matter, or the bodies of plants and animals and receive their nourishment by supplying their own food, absorbing dissolved organic matter, or obtaining food from a host. Poor habits such as improper hot and cold holding temperatures, poor personal hygiene habits by food service workers, or allowing food to be contaminated by rodents or insects can all contribute to foodborne illness.

Bacteria divide and double in number once every 20 minutes, and in a matter of hours can multiply into the millions. ***See Figure 3-4.*** Six different environmental conditions combine to make the best environment for bacterial growth. Knowing these six different conditions helps to keep food safe for consumption and guests coming back for more. These conditions are food, acidity, time, temperature, oxygen, and moisture (FAT TOM). FAT TOM is the acronym that identifies the conditions that encourage the growth of foodborne pathogens. Each of these conditions relates to the growth of bacteria in a different way.

## Bacteria

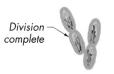

Cells begin to divide

Division complete

**Elapsed Time =** 0 min

**Elapsed Time =** 13 min

**Elapsed Time =** 20 min

**Elapsed Time =** 40 min

**Elapsed Time =** 60 min

**3-4** *Bacteria divide and double in number once every 20 minutes.*

- **Food.** High-protein foods are the type of foods or ingredients that most easily carry bacteria and support bacterial growth. Bacterial microorganisms rely on protein-rich foods for energy to reproduce. Foods high in protein such as meats, seafood, dairy products, and other items containing a protein ingredient, such as chicken pot pie, are considered to be potentially hazardous as they are susceptible to bacteria.

- **Acidity.** All foods fall somewhere between 0 and 14 on a pH scale where 0 is the highest in acidity and 14 is the highest in alkalinity. Right in the middle is 7, which is perfectly neutral or neither acidic nor alkaline. Distilled water has a pH of 7. Most bacteria survive best in acidic conditions that are between 4.6 and 7.5. A protein item such as beef falls around 6.3 and is therefore more susceptible to bacteria and bacterial growth than a lemon, which is highly acidic, having a pH of around 1.8. ***See Figure 3-5.***

## pH Scale

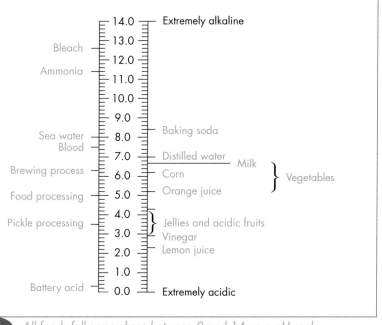

**3-5** *All foods fall somewhere between 0 and 14 on a pH scale.*

## Standard Bacterial Growth Curve

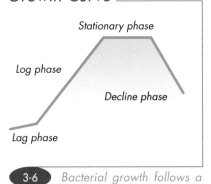

**3-6** Bacterial growth follows a standard curve.

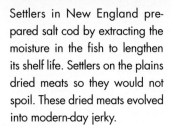

### Historical Note

Settlers in New England prepared salt cod by extracting the moisture in the fish to lengthen its shelf life. Settlers on the plains dried meats so they would not spoil. These dried meats evolved into modern-day jerky.

• **Time.** Every minute that a food is not stored at the proper temperature, bacteria can grow and reproduce. Whenever bacteria are introduced to a new surface or environment, they need time to adjust. The *lag phase* is the adjustment phase where bacteria reproduce slowly. The lag phase can last anywhere from 1–4 hr. Because of this slow growth phase, foods can be left out for relatively short periods of time without a dangerous increase in the bacteria count. After 4 hr, bacteria adjust to their environment and begin to grow and reproduce rapidly. The *log phase* is the rapid growth period for bacteria. The final stage of bacteria development is referred to as the stationary phase. The *stationary phase* is the period of time following the log phase when bacteria reproduce to such an extent that they actually begin to crowd themselves and fight for food, moisture, and space. This overcrowding and fighting for nourishment causes bacteria to begin to die at a relatively quick rate. The *decline stage* is the phase in which the bacteria die and leave behind high levels of toxins. The less time that foods are left in the temperature danger zone, the less opportunity there is for bacteria to multiply and reach dangerous levels. ***See Figure 3-6.***

• **Temperature.** Bacteria reproduce best in a temperature range known as the temperature danger zone. The *temperature danger zone* is a range of temperature between 41°F and 135°F. ***See Figure 3-7.*** Bacteria multiply very slowly at temperatures below 41°F. Bacterial growth is stopped completely at temperatures of 0°F and below, but the bacteria are not killed. In a freezer at 0°F, bacteria enter a state of hibernation, but this state is only temporary. When frozen bacteria return to temperatures between 41°F and 135°F, the bacteria begin to grow again. Bacterial growth is minimal at 135°F, and bacteria are typically destroyed at temperatures of 180°F or above. The food service worker must heat foods to temperatures above 135°F or quickly cool them to temperatures below 41°F to control the growth of harmful bacteria. Potentially hazardous foods left in the temperature danger zone longer than 4 hr must be discarded.

• **Oxygen.** Unlike other living things, not all bacteria need oxygen to live. Some bacteria are aerobic, which means they need oxygen to survive. Other bacteria, known as anaerobic bacteria, can grow and survive without any oxygen at all. *Facultative bacteria* are bacteria that can survive either with or without oxygen. Most kinds of bacteria are facultative bacteria.

• **Moisture.** Bacteria multiply rapidly in moisture. If all moisture is extracted from food, the food can be stored with little chance of bacterial growth. This explains the extended shelf life of powdered eggs, dry milk, and other popular dehydrated foods.

## Temperature Danger Zone

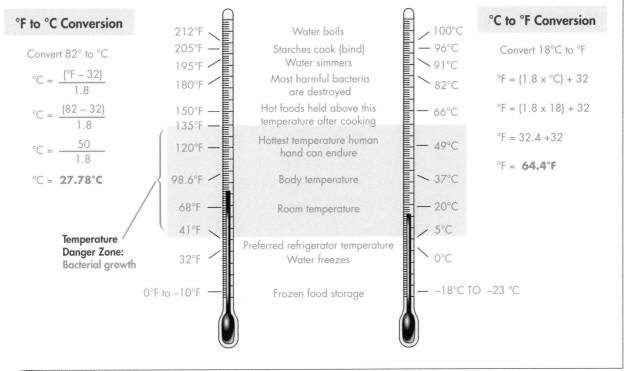

### °F to °C Conversion

Convert 82° to °C

$$°C = \frac{(°F - 32)}{1.8}$$

$$°C = \frac{(82 - 32)}{1.8}$$

$$°C = \frac{50}{1.8}$$

$$°C = \mathbf{27.78°C}$$

**Temperature
Danger Zone:**
Bacterial growth

| | | |
|---|---|---|
| 212°F | Water boils | 100°C |
| 205°F | Starches cook (bind) | 96°C |
| 195°F | Water simmers | 91°C |
| 180°F | Most harmful bacteria are destroyed | 82°C |
| 150°F | Hot foods held above this temperature after cooking | 66°C |
| 135°F | | |
| 120°F | Hottest temperature human hand can endure | 49°C |
| 98.6°F | Body temperature | 37°C |
| 68°F | Room temperature | 20°C |
| 41°F | | 5°C |
| 32°F | Preferred refrigerator temperature Water freezes | 0°C |
| 0°F to –10°F | Frozen food storage | –18°C TO –23 °C |

### °C to °F Conversion

Convert 18°C to °F

$$°F = (1.8 \times °C) + 32$$

$$°F = (1.8 \times 18) + 32$$

$$°F = 32.4 + 32$$

$$°F = \mathbf{64.4°F}$$

**3-7** *The temperature danger zone is between 41°F and 135°F.*

## Potentially Hazardous Food

A *potentially hazardous food* is food that requires temperature control because it is in a form that is capable of supporting the rapid and progressive growth of infectious or toxigenic microorganisms. Potentially hazardous foods must be maintained at a temperature at or below 41°F per FDA *Food Code* § 3-202.11, *Specifications for Receiving—Temperature.* Exceptions to this include milk and shellfish, which may have alternate minimum temperatures specified by state laws. Additionally, eggs are allowed to be maintained at a minimum of 45°F. Potentially hazardous foods that have been cooked must be maintained at temperatures at or above 135°F. Refer to ServSafe® for minimum requirements.

## Highly Susceptible Populations

*Highly susceptible populations* are people more likely to experience food-borne illness than others, such as infants and young children, pregnant women, the elderly, and people who are seriously ill, who take certain medications, or who have life-threatening allergies. People may also be considered highly susceptible if they are receiving food from a facility that provides custodial care. This includes adult day care centers, kidney dialysis centers, hospitals, nursing homes, and senior centers.

### Nutrition Note

Some people have mild, severe, or fatal reactions to certain foods. The following foods account for 90% of all food allergies:

- eggs
- fish
- milk
- peanuts
- shellfish
- soy
- tree nuts
- wheat

Particular care must be taken when serving highly susceptible populations. For example, raw or undercooked eggs, fish, or meat should not be served to people who are considered highly susceptible to foodborne illness. If food contains items that should not be served to highly susceptible populations, it should be noted on the menu. Substitute ingredients such as powdered egg whites or pasteurized eggs can be used to avoid health-hazardous situations.

## FOOD SAFETY GUIDELINES

General food safety guidelines help to ensure that contamination will not occur. The following guidelines should be practiced:

- Get hot foods hot quickly, and keep them at 135°F or above.

- Get cold foods cold quickly, and keep them at 41°F or below.

- Always wash hands with soap and water before starting work, after visiting the restroom, after working with meat, poultry, or seafood, any time they are dirty, and at least once every 4 hr or as directed by local codes.

- Always use sanitized cooking utensils. Always clean and sanitize work areas after they are used.

  - Keep all foods sealed and/or covered as much as possible.

  - Use professional, NSF-certified food handling tools for cutting, cooking, and serving.

  - Purchase USDA-inspected meat only.

  - Thoroughly rinse all fruits and vegetables before using.

  - Use only pasteurized milk.

  - Do not prepare too much food in advance.

*Fluke Corporation*

- Do not allow food to remain between 41°F to 135°F (in the temperature danger zone) for more than 4 hr.

- Do not refreeze thawed meat, fish, or vegetables. Thawing and refreezing causes cellular breakdown and increases susceptibility to decay.

- Do not store canned foods in the opened can.

- Use only pasteurized egg products in items that will not be fully cooked.

- Make sure cans have not been damaged.

- Inspect all foods thoroughly. Contaminated foods do not always have an unusual odor, taste, or appearance.

- Frozen foods should be thawed under controlled conditions, such as in a refrigerator with the temperature set between 34°F and 38°F to prevent cross-contamination.

- Properly dispose of all garbage promptly.

- The refrigerator and freezer are the most important pieces of equipment for controlling bacteria. Check their temperature daily.

- Check all fish, shellfish, and other orders for freshness when delivered.

- Wash the cavity of raw poultry thoroughly.

- Exercise caution with leftovers. Always refrigerate as soon as possible. Reheat quickly to an internal temperature of 165°F.

- Avoid handling food more than necessary. Use disposable plastic gloves when possible and dispose of gloves after contact.

- Clean and sanitize all equipment that has been used on potentially hazardous foods immediately. Equipment such as slicing machines, food shredders and grinders, cutting boards, can openers, and knives are particularly susceptible to contamination.

- Properly cook all foods to their minimum internal cooking temperature or higher and clean and sanitize thermometers after each use.

- Food should not be handled by any person having an acute illness, diarrhea, or open or infected cuts.

- When washing dishes, the wash water temperature should be 160°F and the rinse water should be at least 180°F, and an approved concentration of sanitizer should be used.

- Store glasses, cups, pots, and bowls with their bottoms up.

- Keep dirty dishes, utensils, and towels away from food.

- If ever in doubt about any food, throw it away.

## HAZARD ANALYSIS AND CRITICAL CONTROL POINT (HACCP)

In the 1970s, the FDA developed a food safety program for the NASA space program. This system evolved into what is known as Hazard Analysis and Critical Control Point, or HACCP (pronounced hassip). *HACCP is a systemic approach to the identification, evaluation, and control of food safety hazards.* The HACCP system is used at all stages of food preparation and production.

While some local authorities require use of a HACCP plan for certain high-risk processes, the *Food Code* establishes that the implementation of HACCP is a voluntary effort for retail food service establishments. HACCP plans profile the flow of food through a food service establishment. The regulatory authority must not only approve this plan, but records generated in support of the HACCP plan must be made available for review when requested. An establishment must always consult the local regulatory authority if they are unsure of their requirements, if there are plans to deviate from the requirements, or if there are plans to conduct specialized processes.

When developing a HACCP plan for a food service establishment, the process begins with selection of a food safety team. This team usually consists of managers, kitchen staff, and serving staff, all of whom will be involved with the development, implementation, and execution of the plan. In addition, outside consultants, industry trade associations, university extension services, and local regulatory authorities may be consulted to review or aid in development of the plan. The following steps are taken to develop a HACCP plan:

1. **Develop prerequisite programs.** Prerequisite programs include procedures to address basic operation and sanitation conditions within an establishment or operation. These may include operations such as vendor certification programs, training programs, allergen management, buyer specifications, recipe and process instructions, and supply rotation procedures. These programs function to protect products from contamination by biological, chemical, and physical food safety hazards; control bacterial growth that can result from temperature abuse; and maintain equipment.

2. **Group menu items and food products.** All menu items and food products prepared in an establishment are reviewed and grouped according to the process used for preparation. Process-specific lists identifying all menu items and food products that use a particular process are created to help organize the various groupings. ***See Figure 3-8.***

## Process-Specific Lists

| Process #1<br>Food Preparation with No Cook Step | Process #2<br>Food Preparation for Same Day Service | Process #3<br>Complex Food Preparation |
|---|---|---|
| Salad; house, Caesar | Fried chicken | Soups |
| Fresh vegetables | Broiled fish | Gravies |
| Oysters or clams served raw | Fried oysters | Sauces |
| Tuna salad | Grilled steak | Large roasts |
| Caesar salad dressing | Soup du jour | Chili |
| Cole slaw | Steamed vegetables | Egg rolls |
| Sliced sandwich meats | Cooked eggs | Chicken salad<br>(made from raw chicken) |
| Sliced cheese | Baked potato | |
| Chicken salad<br>(made from canned chicken) | Cranberry-brie appetizer | Beef stew |
| Fruit salad | Risotto | Chicken pot pies |
| Cheese plate | Pancakes and waffles | Twice-baked potatoes |
| Vegetable dip | | Lasagna |
| | | Dim sum |
| | | Roasted turkey |
| | | Stuffed squab |

3-8 *Process-specific lists organize all menu items and food products prepared in a food service establishment according to the preparation processes used.*

3. **Conduct a hazard analysis.** All food safety hazards that exist in a food service establishment are identified. Hazards could be biological, such as microbes; chemical, such as cleaning agents, pesticides, and toxic metal poisoning; or physical, such as ground glass or metal fragments. Hazards could be associated with ingredients or menu items of special concern, such as raw eggs or seafood. Other hazards could be potentially hazardous foods that require specific temperature controls, foods with a history of being associated with an illness, foods that require a great deal of preparation time while in contact with a food service worker, or employees with symptoms of illness.

4. **Implement control measures.** Control measures must be implemented for all prerequisite programs and at critical control points in HACCP plans. Critical control points (CCPs) are various points in food production—from a raw state through processing and shipping to consumption—at which the potential hazard can be controlled or eliminated. Examples are cooking, cooling, packaging, and metal detection. Each control measure is associated with a critical limit. Each control point must have a critical limit and a preventive measure to enforce the critical limit. For a cooked food, for example, a preventive measure might include setting the minimum cooking temperature and time required to ensure the elimination of any harmful microbes. Control measures are designed to prevent, eliminate, or reduce hazards to acceptable levels. ***See Figure 3-9.***

5. **Establish monitoring procedures.** The HACCP plan must identify what is monitored, how it is monitored, the times at which and frequency with which it is monitored, and who is responsible for doing the monitoring. Monitoring procedures should address each critical control point. Such procedures might include determining how and by whom cooking time and temperature should be monitored.

6. **Develop corrective actions.** Corrective actions are actions taken when a critical limit has not been met. When this occurs, the proper corrective action must be executed immediately. For example, a corrective action could be reprocessing or disposing of food if the minimum cooking temperature is not met. Corrective actions must be clearly defined, understood by employees, and easy to implement. Management should document corrective actions to allow for modifications leading to elimination of the problem.

7. **Conduct ongoing verification.** Verification procedures are used to ensure that the HACCP system is working properly. For example, such a procedure could include testing time- and temperature-recording devices that verify that a cooking unit is working properly. Verification should not be confused with routine monitoring. Monitoring is one of the many procedures that must be verified. Verification should be conducted by someone other than the person who is directly responsible for performing the activities specified in the

## Action Lab

How do you know if something is really clean? One way to determine that is to use a product such as Glo Germ™ after cleaning the surface. Try washing your hands following the FDA-recommended handwashing procedure. Then rub about a quarter-size amount of Glo Germ™ gel over both hands like you are putting on hand lotion. Place your hands under a UV lamp and view the glowing germs that remain. Pay close attention to your nails and the areas between your fingers. Try dusting a small amount of Glo Germ™ powder over the entire surface of a stainless steel table. Clean the surface of the table until all the powder disappears. Then pass the UV lamp over the surface to see what areas glow, indicating germs that were left behind. Now try the same technique on a cutting board. Sanitation practices are a critical part of the HACCP plan and keeping food safe in the professional kitchen.

food safety management system. Examples of verification procedures include keeping receiving logs, cooling logs, handwashing logs, and cooking logs. The person conducting verification is responsible for checking logbooks to see that the required number of entries were logged each shift.

8. **Keep records.** Recordkeeping includes keeping records of hazards and their control methods, monitoring safety requirements, and documenting actions to correct potential problems. This also includes documenting activities related to the prerequisite programs.

9. **Conduct periodic validation.** Periodically the entire HACCP program should be reviewed to determine its effectiveness in enforcing food safety standards and proper implementation of the program.

## Hazard Analysis and Critical Control Point (HACCP) Control Measures

| Critical Control Point | Preventive Measure | Critical Limit |
|---|---|---|
| Receiving | Receive food at proper temperatures | 41°F or lower for potentially hazardous foods |
| | Get perishable food into cold storage quickly | 4 hr maximum time in temperature danger zone |
| | Obtain supplies from approved sources | Suppliers must be regulated and inspected by proper authorities |
| Storage | Maintain temperature control | 41°F for perishable foods<br>0°F for frozen foods |
| | Prevent cross-contamination | Ready-to-eat food does not contact raw meat, seafood, or eggs |
| Preparation | Minimize bacterial growth | 4 hr maximum time in the temperature danger zone |
| | Minimize contamination from employees and equipment | Ready-to-eat food does not contact raw meat, seafood, or eggs<br>Employees wash hands when changing tasks and at least once every 4 hr |
| Cooking | Cook foods to proper temperatures | 145°F for eggs, fish, meat<br>165°F for poultry<br>140°F for vegetables |
| Cooling | Use rapid chill refrigeration equipment | Cool to 70°F in 2 hr following cooking |
| | Stir hot foods in an ice water bath | Cool to 41°F in 6 hr following cooking |
| Holding | Keep hot foods at temperatures above the temperature danger zone | Hold at temperature of 135°F or above |
| | Keep cold foods at temperatures below the temperature danger zone | Hold at temperature of 41°F or below |

3-9 *Control measures must be established for each prerequisite program and for each critical control point in an HACCP plan.*

## SANITATION PRACTICES

The food service worker is the most important part of any sanitation program. It has been stated that any cleaning job requires 95% human effort and only 5% mechanical effort. Sanitation efforts are becoming increasingly mechanized, but the food service worker is still responsible for following and maintaining the necessary sanitation standards. *Sanitizing* is the process of reducing the number of microorganisms on a clean surface to safe levels. This is accomplished by using sanitizers. A *sanitizer* is a chemical agent used to sterilize food-contact surfaces.

While the best method used to destroy bacteria is heat, it is not always practical. Bacteria can also be destroyed using sanitizers such as chlorine, iodine, and quaternary ammonium compounds (quats). The number of bacteria destroyed depends on the strength of the chemical used. Chlorine is the least expensive sanitizer. However, it corrodes metal and irritates human skin. Iodine is less corrosive than chlorine, but it stains. Quats are the most expensive sanitizers, yet they are stable at high temperatures and used with warewashing machines.

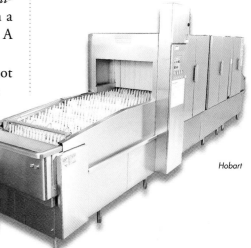

Hobart

Food contamination can be caused by rats, mice, roaches, and flies, which are carriers of disease and bacteria. Most of these pests live in colonies. If one is spotted, it is likely there are more on the premises. To eliminate these pests, ensure that all openings in doors, windows, air vent screens, and other openings are sealed. Check incoming supplies for any indication of pests. Clean all garbage areas before garbage accumulates. If pest control becomes a problem, retain the services of a professional exterminator. *Note:* Pest control poisons and insecticides are dangerous if improperly used.

The Council for Agricultural Science and Technology has estimated that 33,000,000 Americans suffer from foodborne illnesses each year, with 9000 cases resulting in death. The following procedures should be practiced to prepare foods:

1. Sanitize all cutting boards and counter surfaces.
2. Sanitize all utensils. Dishwashers will reach temperatures required to kill bacteria.
3. Wash hands with plenty of soap. Always wash hands after handling raw meat.
4. Sanitize spills from raw meat, fish, seafood, and poultry.
5. Wash fruits and vegetables thoroughly.

### Personal Hygiene

Personal hygiene is the physical care maintained by an individual. It encompasses important areas of health, cleanliness, and outward appearance. Good appearance is essential to all food service workers. People are generally judged by first impressions until others are better acquainted with them. A poor first impression may unfortunately be the end of a relationship. First impressions are made primarily by appearance. If the appearance is neat and clean, the first impression will be a good one. An employer may view their first impression of a job candidate as a reflection of that person's quality of work.

Food service workers must always practice good personal hygiene to help eliminate the spread of bacteria and disease. Personal hygiene requires the following basic rules regarding personal grooming:

- Keep hands and fingernails clean at all times. Use soap and water to clean hands, and wash vigorously. Rinse hands well and dry them with a clean paper towel.

  - Wash hands thoroughly with soap and water after using the restroom, eating, drinking, and touching anything that may contain pathogens or harmful microorganisms.

  - Handle food only as required and with clean hands. Avoid touching the food contact surface of clean utensils with hands after touching food.

  - Never work with open cuts or sores around food. Cuts or sores should be bandaged. Gloves should be worn over injured hands.

  - Do not cough, spit, or sneeze near food or in food preparation areas. Always cover a cough with your arm or sneeze into a handkerchief. Wash hands immediately after using a handkerchief or coughing into your hand.

  - Notify a supervisor and stay home when sick or if experiencing diarrhea, or sore throat and fever.

- Wear clean, proper clothing for the job.

- Control hair by keeping it washed, neatly trimmed, combed, and covered as required.

- Keep body clean by taking a shower or bath daily.

- Keep facial hair clean and trimmed.

- Do not chew gum or smoke while on the job.

- Do not use nail polish, as it may chip off into the food.

- Do not wear artificial fingernails, as they may break off into the food.

- Remove jewelry which may drop into food or cause a safety hazard.

- Do not allow dirty utensils or equipment to touch food.

- Always take the grooming time required to project the best appearance possible and to maintain sanitary conditions.

## Handwashing

All food service employees must keep hands and exposed portions of their arms clean using approved handwashing procedures. ***See Figure 3-10.*** The FDA *Food Code* § 2-301.12, *Personal Cleanliness—Hands and Arms,* specifies the following handwashing procedure for food service employees:

1. Wet hands and arms with hot water (at least 100°F).

2. Work up a lather of soap on fingers, fingertips, areas between the fingers, hands, and arms.

3. Scrub the lathered areas vigorously for at least 15 seconds.

4. Clean under fingernails and between fingers.

5. Rinse hands and arms thoroughly with clean, running warm water.

6. Use a paper towel to turn off faucet.

7. Dry hands with either disposable paper towels or a heated-air hand drying device per FDA *Food Code* § 6-30.12, *Numbers and Capacities—Handwashing Facilities.*

Refer to the CD-ROM for the **Handwashing** Media Clip.

## Handwashing Procedure

1. Wet hands and arms with hot water (at least 100°F).

2. Work up lather of soap on fingers, fingertips, hands, and arms.

3. Scrub vigorously for at least 15 seconds.

4. Clean under fingernails and between fingers.

5. Rinse hands and arms thoroughly with clean, running warm water.

6. Use paper towel to turn off faucet.

7. Dry hands and arms with another paper towel, or use a heated-air hand drying device.

**3-10** *The handwashing procedure specified in the FDA Food Code requires hands to be vigorously scrubbed for at least 15 seconds.*

All handwashing facilities mush be equipped with soap and an approved means for drying hands. All washrooms used by food service employees must display signs notifying food service employees that they must wash their hands before returning to work. Sinks used for food preparation or utensil washing do not need to be equipped with handwashing aids such as soap or hand drying equipment. All hand sanitizers must be FDA-approved for contact with food and food preparation utensils.

Food service employees must wash hands and exposed portions of arms immediately before beginning any food preparation task involving food, clean equipment, or utensils. In addition, hands must be washed under the following circumstances:

- before starting work

- after using the restroom

- after touching the hair, face, or body

- after coughing, sneezing, or using a handkerchief or disposable tissue

- after eating, drinking, chewing gum, and using tobacco products

- before and after touching raw meat, poultry, or fish

- after handling chemicals that may affect food safety

- after touching money

- after touching unsanitized equipment, work surfaces, and cloth towels

- after taking out the garbage

## Warewashing

Many problems in sanitation can be traced to improper dish, silverware, glassware, and pot washing. Dishes should be scraped and rinsed before placing them in the dish racks. Silverware should be soaked before washing. Glassware should be washed in clear water using a compound recommended for glassware. The washing, rinsing, and drying procedures required depend on the type of equipment available, the sanitation program approved by management, and/or the sanitation products used. However, regardless of equipment, program, or product, the temperature of wash water should exceed 160°F. The temperature of rinse water should exceed 180°F. In most cities, wash and rinse water temperature requirements are specified by the local board of health or other authority having jurisdiction. This ensures that all bacteria are destroyed in the washing and rinsing process.

Hot water, friction, and detergent are necessary for proper results when pot washing. Of the three, friction is generally the most neglected. Pots should be scraped clean, placed in hot water containing a good detergent, and scrubbed thoroughly. Care must be taken when using steel wool and other types of scouring pads because pieces will come off as they are used. The pots are then passed through 180°F rinse water. The third step in warewashing is to immerse the pots in a separate sink compartment filled with hot water that contains sanitizing chemicals such as iodine, chlorine, or quats. *See Figure 3-11.*

## Warewashing

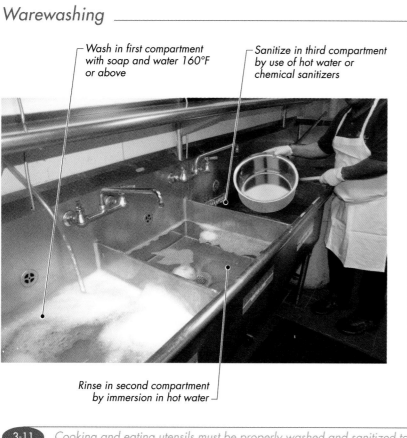

Wash in first compartment with soap and water 160°F or above

Sanitize in third compartment by use of hot water or chemical sanitizers

Rinse in second compartment by immersion in hot water

**3-11** *Cooking and eating utensils must be properly washed and sanitized to prevent contamination.*

Preparation and cooking tools are also a source of bacterial growth. For example, food particles cling to the blade of a knife when cutting certain foods. If the dirty knife is stored and used the following day, it transfers bacteria from the blade to the item cut. This also can occur with forks and other tools. Whether tools are used by the entire kitchen staff or are tools used exclusively by the chef or cook, they must be thoroughly washed after use.

Strict sanitation standards must also be followed in cleaning stationary equipment. All stationary equipment and attachments must be cleaned after use, usually with soap and water. Hard rubber and plastic cutting surfaces should be cleaned as recommended by the manufacturer.

Most states require food service managers to complete a ServSafe® manager certification training course and pass the certification exam. Certification is obtained by taking courses required by the state board of health that pertain to the classification and types of bacteria, foodborne illness, and related safety and sanitation subjects. ServSafe® certification is valid for five years. However, individual states and employers may have more stringent guidelines regarding certification renewal.

## Heating And Cooling Food

Heating foods to their minimum required internal cooking temperature ensures that bacteria are destroyed. The FDA *Food Code* establishes minimum internal cooking temperature standards. ***See Figure 3-12.*** Food must be heated to the specific minimum temperature, and that temperature must be maintained for the required time to effectively destroy bacteria. The higher the internal temperature, the shorter the required period. At some internal temperatures bacteria destruction is instant, so the temperature does not have to be maintained for a minimum time. For example, when ground beef is heated to 168°F, the bacteria are instantly killed. At an internal temperature of 145°F, however, the temperature must be maintained for 3 min to kill all bacteria that may be present.

| FDA Minimum Internal Cooking Temperatures | | | | | |
|---|---|---|---|---|---|
| Food | Temperature* | Time† | Food | Temperature* | Time† |
| Eggs | 145 | 0:15 | Roasts (beef, pork) | 130 | 112 |
| Fish | 145 | 0:15 | | 131 | 89 |
| stuffed | 165 | 0:15 | Minimum oven temp (conventional) 350°F if less than 10 lb | 133 | 56 |
| | | | | 135 | 36 |
| Pork, beef, lamb, or veal | 145 | 0:15 | | 136 | 28 |
| | 145 | 3 | | 138 | 18 |
| | | | | 140 | 12 |
| injected/ground | 150 | 1 | | 142 | 8 |
| | 168 | instant | Minimum oven temp (convection) 325°F if less than 10 lb | 144 | 5 |
| stuffed | 165 | 0:15 | | 145 | 4 |
| Poultry | 165 | 0:15 | | 147 | 2:25 |
| Stuffed pasta | 165 | 0:15 | | 149 | 1:50 |
| Stuffing containing fish, meat, or poultry | 165 | 0:15 | Minimum oven temp (conventional or convection) 250°F if greater than 10 lb | 151 | 1 |
| | | | | 153 | 0:50 |
| Fruits and vegetables | 140 | instant | | 155 | 0:25 |
| Reheated, fully cooked, potentially hazardous foods | 165 | 0:15 | | 151 | 0:14 |
| | | | | 158 | instant |

\* in °F

† in min

Foods must be heated to minimum internal cooking temperatures as established by the FDA Food Code.

Beef may be served raw or undercooked if it is served to persons not included in a highly susceptible population. However, the raw or undercooked beef must be extremely fresh. The beef must meet FDA code requirements for "whole-muscle, intact beef," and all external surfaces must be cooked to a surface temperature of 145°F and achieve a color change. In addition, raw fish or egg products may also be served to persons not included in a highly susceptible population under specific guidelines listed in FDA *Food Code* § 3-401.11(D). Items unacceptable for consumption by highly susceptible populations should be noted on the menu.

Care should be taken when thawing potentially hazardous foods to prevent growth of bacteria. Foods must be thawed either in a refrigerated environment of 41°F or less, or completely submerged under running water per FDA *Food Code* § 3-501.13, *Temperature and Time Control—Thawing*. Frozen food may also be thawed directly as part of the cooking process or thawed in a microwave and immediately cooked.

Cooling foods for storage using the two-stage cooling method also helps prevent bacterial growth. Potentially hazardous food that has been cooked must be cooled to 70°F within 2 hr following cooking. Food must then be cooled to 41°F or less for storage within the next 4 hr, for a total of 6 hr. If the food was prepared with ingredients held at room temperature, it must be cooled to 41°F within 4 hr. Raw eggs must be immediately cooled in refrigerated equipment maintaining a temperature of 45°F or less. Local health requirements may vary. Always follow local health department requirements. The following methods may be used to decrease cooling times:

- Place the food items in shallow pans.

- Separate the food items into small, thin portions.

- Use rapid cooling equipment.

- Place the food in a container held in an ice water bath.

- Add ice to the food.

- Use storage containers that maximize heat transfer, such as aluminum pans.

## Transporting Food

Caterers and larger food service operations transport prepared or partially prepared foods from a central kitchen. In transporting foods, steps must be taken to ensure the food is not contaminated and that bacteria do not make the food potentially hazardous by multiplying while in transit. The following precautions should be followed when transporting foods:

- The transport containers used must be clean, tightly sealed, and designed for efficient cleaning.

*Carlisle FoodService Products*

- The transport containers must have the required refrigeration or heating elements to maintain the proper temperature. Cold foods require temperatures of 41°F or below. Hot foods should be held above 135°F.

- Use the shortest route possible to the area where the food will be served. Minimize loading and unloading time.

- Foods on display for a buffet or salad bar should not be at room temperature for more than 1 hr. Cold food should be kept at a temperature of 41°F or less and hot foods at 135°F or more. Cold foods are kept cold using ice or a refrigeration unit. Hot foods are kept hot using steam tables or chafing dishes.

- Displayed foods must be protected with a sneeze guard (shield that protects food from being contaminated). Leftover display foods that are not individually packaged should be discarded at the end of the meal.

## KITCHEN SAFETY

Safety programs must be in place and safe practices must be employed by all workers on a continuing basis. The Occupational Safety and Health Administration (OSHA) is responsible for setting and enforcing workplace safety standards in the United States. Employers are responsible for safety training and ensuring that employees follow OSHA standards. It is important that food service workers are informed of all safety procedures when they are hired. Employees must report all accidents and safety hazards to the employer or supervisor.

It is important to be aware of situations that may cause injury and to take proper precautions to prevent injuries from occurring. In the event that an accident or injury does occur, a supervisor should be notified and immediate medical attention should be sought. Following an injury, proper precautions must be taken to avoid further injury or wound contamination. Precautions must also be taken to protect food and kitchen equipment from contamination from cuts or personal injuries. The most common injuries resulting from accidents in the professional kitchen are cuts, burns, falls, and strains and sprains.

### Cuts

Cuts can easily occur in the professional kitchen, as knives and other cutting implements are in constant use. The frequency and seriousness of cuts can be reduced by practicing proper cutting procedures and common sense. For instance, tucking fingers away from the knife blade when using a knife reduces the risk of injuring fingers. When cuts occur, they should be treated properly to prevent infection and complications. For serious cuts, emergency medical help should be contacted immediately. The following procedure is for treatment of minor cuts:

1. Wear disposable gloves when treating any person with a bleeding injury.

2. Compress the wound by applying pressure to stop any bleeding. First, remove any foreign objects from the wound. Then, use a clean cloth or bandage to apply light and steady pressure to a wound for 20 min to 30 min, if necessary. Do not repetitively remove pressure to check the status of the wound, as this can hinder clot formation and prolong bleeding time. *See Figure 3-13.*

3. If bleeding is severe or does not stop, have the injured person lie down. Elevate the wound to reduce blood flow to the area.

4. Remove any dirt or debris from the wound. Clean the bleeding area with fresh water. Soap can irritate a wound, so avoid contacting a cut directly with soap.

5. Apply an antibiotic to prevent infection and encourage the body's natural healing process.

6. Apply a bandage or sterile gauze to the wound area.

7. After treating the cut, carefully remove gloves and wash hands thoroughly.

When working in the kitchen, all cuts must be properly covered with a waterproof covering such as a finger cot or disposable glove. A *finger cot* is a protective sleeve placed over the finger to prevent contamination of a cut.

### Burns

Burns are generally more painful and take more time to heal than cuts. Burns that occur in the professional kitchen are classified as either minor or serious. Minor burns may result from popping grease or by handling hot pans with wet or damp towels. Pot holders should be used to protect hands when moving or handling hot cookware. *See Figure 3-14.* Towels should not be used to handle hot cookware as they may not provide adequate insulation from heat and could cause cross-contamination. The following procedure can be used for treating minor burns:

1. Cool the burn under cold running water. Do not put ice on the burn.

2. An aloe vera lotion or antibiotic ointment can be applied to the area to relieve pain after the wound has cooled.

3. Cover the burn with a bandage.

Serious burns may be caused by splashed grease, escaping steam, and gas ignited incorrectly. Serious burns are classified as first-degree, second-degree, or third-degree burns. First-degree burns affect only the top layer of skin and appear red and swollen. Second-degree burns occur when the burn area extends beyond the upper layer of skin into the second layer, known as the dermis. Blisters, intense pain, and swelling result from second-degree burns. Second-degree burns less than 3″ in

## Wound Compression

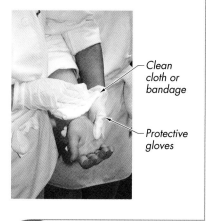

*Clean cloth or bandage*

*Protective gloves*

**3-13** *Compressing a cut can help stop bleeding.*

diameter can be treated as minor burns. Third-degree burns involve damage to body tissue well beyond the first and second layers of skin. Damage may affect fat, muscle, or even bones. If the burn is severe, it should be treated promptly by trained medical personnel. The following steps may be performed on severe burns while waiting for emergency medical personnel to respond to a burn victim:

1. Remove any smoldering materials and prevent any further exposure to smoke and heat.

2. Cover the burn lightly with a cool, moist bandage or clean cloth.

3. Ensure that the victim is breathing. Cardiopulmonary resuscitation (CPR) may be required if the victim stops breathing.

## Hand Protection

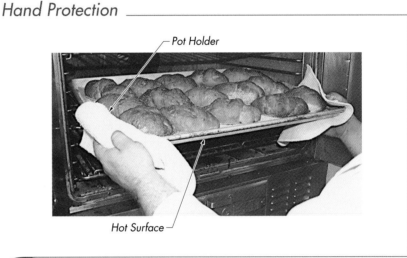

Pot Holder

Hot Surface

**3-14** *Pot holders should be used to protect hands when moving or handling hot cookware.*

### Falls

Falls can cause serious accidents in the professional kitchen. Falls may be caused by wet floors, spilled food, grease, torn mats, and damaged floors. OSHA 29 CFR 1910.22, *Walking-Working Surfaces,* requires all places of employment to be kept clean, orderly, and in a sanitary condition. This includes keeping all floors clean and dry. If a spill occurs, it should be mopped up immediately. Wet floor areas should be designated with warning signs. Nonslip matting can help reduce the risk of falling in areas that tend to be wet. Step-up/step-down areas can be indicated with hazard tape.

The likelihood of falls in the professional kitchen can be minimized by keeping floors clean and dry and by wearing proper footwear. Nonslip shoes should be worn to provide traction in potentially slippery areas. Sandals, open-toe shoes, and high-heeled shoes should not be worn in the professional kitchen.

## Strains and Sprains

Strains and sprains are not as serious as the other types of accidents, but are painful and can result in the loss of working hours. A strain is a bodily injury resulting from excessive tension, effort, or stretching of muscle and ligament tissue. A sprain is an injury resulting specifically from excessive stretching of ligaments. These injuries can be caused by lifting heavy or oversized objects, carrying large numbers of objects at once, or repetitive reaching across tables or counters. Strains and sprains can be prevented by not attempting to carry loads that are too heavy, by using proper lifting technique, and by wearing slip-resistant shoes. Before lifting and carrying objects, it is important to ensure that the path to be used is clear of obstacles and free of hazards. When lifting objects from the ground, apply the following procedure:

1. Bend the knees and grasp the object firmly.

2. Lift the object, straightening the legs and keeping the back as straight as possible.

3. Move forward after the whole body is in the vertical position, keeping the load close to the body and steady.

Use of the proper lifting technique will also help prevent possible back injuries. Assistance should be sought when lifting heavy objects. ***See Figure 3-15.***

## Proper Lifting Procedure

Keep Back Straight

1. Bend knees and grasp object firmly.

2. Lift object by straightening legs.

3. Move forward after whole body is in vertical postion.

**3-15** Using proper lifting technique can help avoid potential sprains, strains, and back injuries.

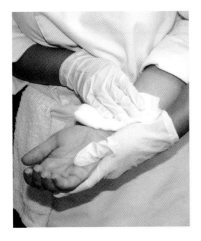

In the event that a sprain or strain occurs, proper first aid treatment should be sought to prevent further injury. Common first aid techniques for treating sprains and strains include the following:

- isolation of the injured limb to prevent further use, using restraining devices such as splints
- rest of the injured area; use of sprained or strained areas should be avoided if possible
- use of elastic wrap or a bandage to compress the injured area
- application of ice to a sprained or strained area immediately after injury to reduce potential swelling; prolonged applications of ice should be avoided, as tissue damage can occur
- elevation of the injury whenever possible to limit swelling

## SUMMARY

Cleanliness is an absolute requirement in the professional kitchen. Foodborne illnesses, triggered by harmful bacteria, can occur as a result of improper food handling, sanitation, and personal hygiene procedures. Highly susceptible populations, including the elderly and young children, are especially at risk. Direct contamination may result from biological, chemical, and physical hazards in the kitchen. Cross-contamination is a danger resulting from poor personal hygiene, unclean equipment, or improper food transportation methods. Knowing the conditions in which bacteria thrive and using appropriate personal hygiene and sanitation procedures will help ensure the safety in a food service establishment. Each food service worker is responsible for maintaining personal cleanliness, both to ensure food safety and to project a professional appearance. Ultimately, management is responsible for maintaining a clean food service establishment for customers.

Refer to the CD-ROM for Quick Quiz® questions related to chapter content.

## Review Questions

1. Describe the five agencies involved in the establishment and regulation of food safety standards.

2. Describe three types of hazards that can cause direct contamination in food.

3. Identify common causes of foodborne illnesses.

4. List the six different environmental conditions that combine to make the best environment for bacterial growth.

5. What precautions should be taken in food preparation for highly susceptible populations?

6. Describe the team involved in developing a Hazard Analysis and Critical Control Point (HACCP) plan for a food service establishment.

7. Contrast the advantages and disadvantages of common sanitizers.

8. Describe the five procedures taken for preparing foods that are free of harmful bacteria.

9. Explain why employees must maintain good personal hygiene.

10. Describe the seven-step handwashing procedure for food service employees as specified in the FDA *Food Code*.

11. Describe the best method to use for washing pots.

12. What six methods can be used to decrease cooling times of food in order to prevent bacterial growth?

13. Describe the temperature requirements for transporting and displaying food.

14. List the most common injuries resulting from accidents in the professional kitchen.

# Food Service Equipment

*A professional kitchen* needs to be clean, organized, well equipped, and laid out in such a way as to allow all of the staff to be in tune with one another. Just as in a band, where each musician has a position on the stage, a professional kitchen is divided into different areas for performing various tasks. In the professional kitchen, these areas are broken down into smaller work sections for performing specific types of tasks, and even smaller work stations for performing a specific task. Just as a musician cannot play without an instrument, each area, section, and station must contain all of the equipment needed to perform required tasks.

## Chapter 4

## Chapter Objectives:
1. Identify the major areas of the professional kitchen.
2. Explain the concept of mise en place.
3. Describe the specifications required for NSF-certified tools and equipment.
4. List the safety guidelines recommended for operating and maintaining commercial equipment.
5. Describe the equipment commonly used in each area of the professional kitchen.

## Key Terms
- delivery area
- storage areas
- preparation and processing areas
- sanitation and safety areas
- mise en place
- NSF international
- receiving area
- wire shelving unit
- walk-in unit
- reach-in unit
- steam table
- proofing cabinet
- work section
- work station
- grill
- flashbake oven
- fire-suppression system
- ventilation system

## THE PROFESSIONAL KITCHEN

The professional kitchen is a fast-paced work environment that is the heart of the food service operation. ***See Figure 4-1.*** The kitchen is where food and other items are received, stored, prepared, cooked, plated, and distributed to the dining room service staff. The dining room staff uses the kitchen not only to pick up the foods to be served, but also to return dirty plates and other used service items en route to the dishwashing area. Because of heavy traffic flow, constant use of equipment, and the required speed and volume of production, the layout of a professional kitchen needs to be carefully planned and the equipment used needs to be commercial grade. The professional kitchen is typically divided into several major areas including receiving, storage, preparation and processing, and sanitation and safety.

4-1  *The fast-paced professional kitchen is the heart of the food service operation.*

The *receiving area* is the entry point where anything ordered by the food service operation, such as food, supplies, or equipment, enters the building. In this area, time and temperature requirements for food items must be monitored.

*Storage areas* are designated areas within the professional kitchen where food items are stored. These areas include dry storage, cold storage, and hot storage. Dry storage is a secure place in the operation that is temperature-controlled so that it stays cool and dry. Here, all nonperishable items such as canned or bottled food products and dry goods such as flour, sugar, and dried pastas are held until needed. Keeping these items stored in one central location also helps with inventory control as they can be easily and quickly counted. Cold storage refers to refrigerated areas where any fresh items such as produce, meats, or seafood that need to be kept cold are stored. These areas can be as small as an under-the-counter refrigerated cabinet or as large as a completely refrigerated warehouse large enough to drive a truck into. Hot storage areas are used to keep foods hot. These areas are typically used to hold foods at the proper and safe temperature until needed, such as when waiting to serve prepared food at a banquet. Hot storage areas are also used to hold hot foods on a buffet or cafeteria serving line. Pizzas are often transported for delivery in insulated bags that are considered hot storage units.

The *preparation and processing areas* are areas in a professional kitchen are where food items are cut, trimmed, prepared, sliced, and cooked. These areas are subdivided into various work sections, each containing several work stations designed for performing a particular kitchen task.

*Sanitation and safety areas* are locations where sanitation or safety equipment is kept. Sanitation areas include any area where items used in the kitchen, the dining room, and anywhere else in the operation are cleaned and sanitized. Areas such as a pot and pan washing station or the dish room are examples of this type of area. Another type of sanitation area is a custodial closet or any place where workers responsible for cleaning floors, tables, or other work and service areas store cleaning supplies and cleaning equipment such as mops and buckets. Safety areas include areas such as first aid stations, fire extinguisher stations, and MSDS (material safety data sheets) information areas. These areas should be checked frequently to be sure that all safety materials are easily accessible and equipment is in good working order.

The equipment used in a professional kitchen must be commercial grade. Professional-grade equipment is made to withstand the wear and tear of the fast-paced professional kitchen, has higher heat output to make cooking faster, and must be able to be easily cleaned. Because the burners on a commercial range are so large and heat to such a high temperature, even the cookware used on the range top must be of a commercial, or heavy-duty, quality. A residential pot or pan could buckle, warp, or even melt if left on a commercial range for even a short period.

No matter how well a kitchen is laid out or equipped, work is only performed efficiently if proper mise en place is practiced. *Mise en place* (meez ahn plahs) is a French phrase meaning "put in place" and refers to having the tools needed to perform tasks properly, having the ingredients for a recipe properly prepared ahead of time, and having stations set up properly so that they are prepared to accommodate orders coming into the kitchen. In the food service industry, mise en place commonly refers to proper organization and preparation prior to service. Professional chefs

**Historical Note**

Marble has the ability to remain cooler than the air around it. Before refrigeration was invented, butchers and chefs used slabs of solid marble to keep foods cold. When animals were butchered outdoors and the cuts of meat sold in open-air markets, they would be held on marble to keep them fresh and cool.

*Chef's Tip:*

Did you know that a surface can be clean but not sanitized? It's true. In the food service industry, the term "clean" means free of visible soil, while "sanitized" means free from harmful microorganisms. A surface that looks clean might not have been sanitized.

know the importance of proper mise en place and stress its significance daily to their staff. Having mise en place means that the kitchen and staff are as ready as possible for tasks to come. Practicing mise en place every day is key to a successful career as a culinary professional.

## Equipment Standards

Common items such as stoves, mixers, pots, and pans can be found in both a restaurant and a home, but there is a difference in how they are made. Equipment designed for home use is not designed to withstand the demands of the professional kitchen and does not meet specifications for commercial equipment. Commercial equipment, the kind used in professional kitchens, is designed according to NSF (formerly known as the National Sanitation Foundation) sanitation standards. *NSF International* is an organization involved with standards development, product certification, education, and risk management for public health and safety. NSF-certified products bear the blue NSF Mark. ***See Figure 4-2.***

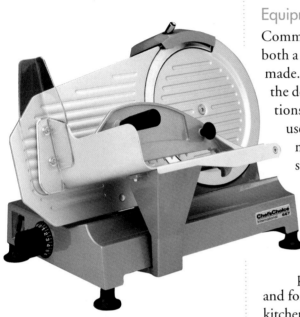

Chef's Choice® by EdgeCraft Corporation

All equipment in the professional kitchen must have the NSF Mark to be acceptable for use. NSF-certified items have passed rigorous inspection for being easily cleaned and maintained and for being able to withstand the daily wear and tear of a professional kitchen. The presence of the NSF Mark on tools and equipment indicates that the item meets the following specifications:

• The item is able to be cleaned easily.
• The item has smooth, nonporous (nonabsorbent), nontoxic, corrosion-resistant surfaces (such as painted enamel or stainless steel).
• Internal corners are sealed and smooth while external edges such as those around the outside edge of a table are rounded and smooth.
• The item is easily taken apart for cleaning and maintenance.
• Waste liquids and other waste are removed easily from the unit.

In general, NSF-certified tools and equipment are made to stand up to the wear and tear of daily use in the professional kitchen. Even items such as beverage glasses and plates used in food service are made from heavier materials than those used in the home. The heavier construction of commercial tools and equipment results in higher cost compared to similar products designed for home use. In turn, commercial tools and equipment last longer and do not need to be replaced as often.

NSF is partnered with local health departments to ensure safe and sanitary conditions in public food service facilities. There are so many different versions of kitchen tools and equipment that it is impractical for NSF to inspect every product on the market. Instead, NSF focuses on commercial items used in the food service industry. Tools and equipment for home use often are not NSF-certified because home tools are not used to the same extent as commercial tools and equipment.

## NSF Mark

The NSF Mark has evolved since its inception in 1944.

NSF-certified products bear the blue NSF Mark.

Edlund Co.

1944

1953

1974

1989

Current

NSF International

4-2　All equipment in the professional kitchen must have the NSF Mark to be acceptable for use.

## Equipment Safety

Many types of equipment are available to make food preparation and processing easy and efficient. One glance around a professional kitchen reveals a variety of commercial equipment that a chef must understand in order to operate it safely. Some of these items are fast machines designed to reduce preparation and cooking times and increase consistency, while other items simply provide a safe and durable work surface for effectively performing a task. Prior to operating any commercial equipment, a few safety guidelines must be understood to protect against personal injury and damage to equipment. Unsafe or careless operation of equipment can lead to serious injury. The following safety guidelines are recommended for operating and maintaining commercial equipment:

- Read all manufacturer instructions for safe and proper operation of all kitchen equipment prior to use.
- Ensure that all available safety features such as blade guards are installed and used.
- Ensure that equipment is securely stationed or anchored to prevent the equipment from falling or slipping.
- Turn off and unplug all equipment before cleaning or disassembling.
- Sanitize equipment after cleaning to prevent foodborne illness.
- Ensure equipment is turned off and unplugged before reassembling after servicing or cleaning. The equipment should remain unplugged until ready for use.
- Notify a supervisor immediately if any malfunction or damage is detected on any equipment. Post a caution notice on the equipment to protect employees from injury.

Refer to the CD-ROM for the
**Tools and Equipment**
Flash Cards.

*Dollies*

*Carlisle FoodService Products*

**4-3** *Dollies are used in the receiving area to transport items.*

Proper cleaning and sanitation of commercial equipment is essential for safe food handling. The presence of bacteria on equipment in the professional kitchen can create cross-contamination and spread foodborne illness. Each piece of equipment must be disassembled and each component thoroughly washed, rinsed, and sanitized to prevent the spread of bacteria and foodborne illness.

## RECEIVING AREAS

The first steps for safe food handling in a professional kitchen occur the moment food enters the building through the receiving area. The receiving area has many more functions than simply being a place where food that has been purchased is dropped off. The *receiving area* is the area of the professional kitchen where all delivered items are checked for freshness, appropriate amounts ordered, temperature, and price. If any problem is detected with items at the time they are received, they can be returned to the supplier immediately. *See Figure 4-3.*

### Receiving Equipment

In order to operate a receiving area adequately, certain pieces of equipment are necessary. A properly outfitted receiving area should contain dollies, inspection tables, measurement equipment, and labeling equipment. *See Figure 4-4.*

**Dollies.** Dollies are used in receiving areas to help the food service worker move heavy items or large boxes from one area to another. In addition, pallet jacks are used to move large amounts of stock.

**Inspection Tables.** Inspection tables in the receiving area allow the receiving clerk to open boxes for inspection without having to bend down. Stainless steel tables are most commonly used as they are easy to clean and maintain.

**Measurement Equipment.** Many frozen and fresh perishable items such as meat, seafood, and dairy products are delivered to the receiving area each day. Thermometers are used to check that frozen and fresh perishable foods are delivered at a low enough temperature according to industry standards. Frozen foods should be delivered at a temperature of 0°F or below. Fresh perishable foods should be delivered at a temperature of 41°F or below. If foods are delivered at temperatures that are above the maximum allowed temperature, they should be rejected and returned to the supplier.

The most common thermometer used in the receiving area is an infrared, or laser point, thermometer. An infrared thermometer allows the user to simply point the unit at the food item and squeeze the trigger. The thermometer instantly measures the surface temperature of the item and indicates the measured temperature on the display panel.

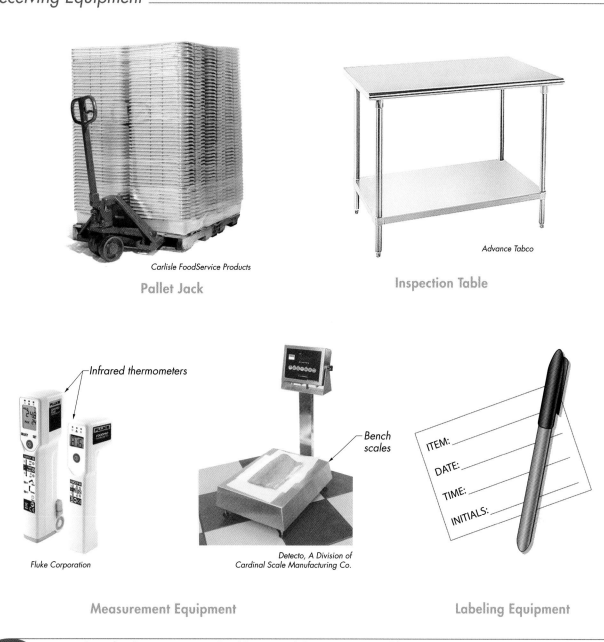

Carlisle FoodService Products

**Pallet Jack**

Advance Tabco

**Inspection Table**

Infrared thermometers

Fluke Corporation

Bench scales

Detecto, A Division of
Cardinal Scale Manufacturing Co.

**Measurement Equipment**

ITEM:
DATE:
TIME:
INITIALS:

**Labeling Equipment**

**4-4** *A properly outfitted receiving area should contain dollies, inspection tables, measurement equipment, and labeling equipment.*

Since many of the items such as meat and seafood that are ordered and delivered are sold and priced by weight, it is important for these items to be weighed upon arrival. Platform, counter, and portion scales are types of scales commonly used in a receiving area. *See Figure 4-5.* Platform and counter scales are used to weigh large or heavy boxes and bags, while portion scales are used to weigh smaller items such as portion-controlled cuts of meat.

## Digital Scales

**Platform**

**Counter**

**Portion**

*Detecto, A Division of Cardinal Scale Manufacturing Co.*

**4-5** *Platform, counter, and portion scales are types of scales commonly used in a receiving area.*

**Labeling Equipment.** All items must be labeled and dated upon arrival so that food service workers know which items to use first. Clearly labeling food items enables the workers to use the oldest items first. Markers are used for labeling boxes. Paper adhesive labels are used if necessary to identify containers. Pens or pencils are used to check the order and invoice for accuracy before moving the items to the appropriate food storage areas.

After an item is received, it is marked with the delivery date prior to being moved into a food storage area. Dating of each received item is essential to ensure older items are used before newly delivered items. If newer items are placed in front of older items frequently, the older items do not move to the front of the storage area or get used. Over time, older items continue to be overlooked at the back of the freezer while the newest ones are always the first ones used. Instead, new items should always be placed at the back of the storage area, and old items are rotated to the front for accessibility using the FIFO stock rotation method. The items at the front of a shelf are always the oldest in storage, and therefore are used before any newer items. The items at the back of a shelf are the newest in storage, and should not be used until all older items are used.

## STORAGE AREAS

All food, beverage, and supply items need to be stored for security, safety, and inventory control. The three types of storage areas in a professional kitchen are dry storage, cold storage, and hot storage. Dry storage is an area used to store dry goods such as flour and canned goods. Cold storage is an area used to store perishable items that need to be kept in a refrigerator or freezer. Hot storage is an area used to keep hot foods hot while serving or until needed for service. If storage areas are not properly maintained, it will affect the ability of a food service operation to practice proper mise en place.

InterMetro Industries Corporation

### Dry Storage

Dry goods refer to any canned items and dry food items such as dried pasta, flour, and sugar that do not need refrigeration. *Dry storage* is a clean and secure area where dry goods are kept before use. Typical dry storage rooms are kept at temperatures between 50°F and 70°F. There are many pieces of equipment in the dry storage area that make it easier to keep food items clean and free of rodent or bug infestation. The equipment in a dry storage area consists of shelving units, security cages, and various different storage containers. ***See Figure 4-6.***

**Shelving Units.** Shelving units are available in a variety of styles, sizes, and shapes for the professional kitchen. They are typically made of stainless steel to maximize durability and cleanliness. All shelves and shelving units must be a minimum of 6″ above the floor to allow access for cleaning underneath the unit. Styles vary depending on the intended use, and include wire shelving, speed racks, overhead shelves, and various specialty styles.

A *wire shelving unit* is a shelving unit made of stainless steel wire. Wire shelving units are available in many different sizes to be used in almost any size room or area. They are commonly available in 4′ and 6′ heights. Wire shelving is primarily used to store boxed food for later use or relocation.

## Storage Equipment

*Advance Tabco*
**Shelving Units**

*InterMetro Industries Corporation*
**Security Cages**

*Carlisle FoodService Products*
**Storage Containers**

**4-6** *A dry storage area is equipped with shelving units, security cages, and various storage containers.*

## Can Rack Shelving

*InterMetro Industries Corporation*

**4-7** *Can rack shelves are stocked from the top so older items rotate to the bottom for use.*

*Can rack shelving* is shelving with rails in which cans of product can be loaded from the top. These rails allow the can to roll toward the bottom for easy removal. ***See Figure 4-7.*** *Dunnage rack shelving* is shelving consisting of reinforced stainless steel platforms that serve to store items at least 6″ above the floor. Dunnage rack shelving allows cleaning beneath the shelf, which reduces the chances of pest or rodent infestation.

An *overhead shelf* is a shelf mounted on the wall or above a work surface. Overhead shelves are commonly located at a work station in the preparation or production area of the kitchen. Overhead shelves are designed to hold items such as spices, kitchen tools, or other items specifically needed at the work station.

A *speed rack,* also called a tallboy, is a tall cart with rails intended to hold entire sheet pans of food. These storage units come equipped with wheels so that they can be moved to various areas of the kitchen as needed. *See Figure 4-8.*

*Pot and pan rack shelving* is overhead storage that suspends pots and pans from a rack. This allows pots and pans to dry quickly after washing and provides more room in the kitchen. *Utensil rack shelving* is overhead storage with suspended hooks to allow kitchen tools to be easily accessible.

**Security Cages.** A *security cage* is a lockable wire cage storage unit used to hold expensive or dangerous items such as fine china or hazardous chemicals. Security cages are constructed similarly to wire shelving but come equipped with cage-style doors that can be locked for added security. Security cages often have caster wheels so they can be moved to various locations if needed.

**Storage Containers.** Storage containers come in many different shapes and sizes and are used throughout the professional kitchen. They are used to store food safely while maximizing the use of shelf space in refrigerators, in freezers, or on transport equipment. Medium and small storage containers usually have flat, tight-fitting, snap-on lids so the container does not spill or leak. These tight-fitting lids will also help preserve product freshness and prevent the product from absorbing odors from other foods. Many containers have graduated measurements on the sides for ease in determining the amount of product stored in the container.

Large storage bins are used most often to store bulk items such as flour, sugar, and rice. They usually have rounded inside corners so that they are easy to clean. Large storage bins may be equipped with wheels and sliding lids so they can be stored beneath a table and pulled out when needed. *See Figure 4-9.*

The most common and practical storage containers are made of strong plastic material such as polyethylene or polyurethane. Glass containers are never used in a professional kitchen because glass breaks too easily and, if broken, can potentially result in glass fragments in or around foods. Aluminum containers should not be used because aluminum reacts with acidic foods, causing them to discolor and develop a metallic taste. All containers should be labeled with a description of the contents and the time and date the item was placed in the container. The label allows workers to quickly determine what a container holds and the freshness of the contents.

## Cold Storage

Having enough cold storage space is essential for the safe handling and storage of foods in food service operations. The three major types of refrigeration and freezer units are walk-in, reach-in, and roll-in units. *See Figure 4-10.*

## Speed Racks

Rails hold sheet pans

Wheels allow movement to various locations

Cres Cor

**4-8** *Speed racks hold entire sheet pans of food and can be rolled to various locations.*

## Storage Bins

Carlisle FoodService Products

**4-9** *Large mobile storage bins are used to store bulk items such as flour, sugar, and rice.*

*Kolpak*

**Walk-in Units**

*True FoodService Equipment, Inc.*

**Reach-in Units**

*True FoodService Equipment, Inc.*

**Roll-in Units**

**4-10** *Common types of cold storage include walk-in units, reach-in units, and roll-in units.*

No matter the style of freezer or refrigeration unit, all freezers must be kept at 0°F or below, while all refrigeration units must be kept at 41°F or below. Refrigeration and freezer units should be cleaned regularly. A solution of baking soda and water can be used to clean interior walls, shelves, and racks. The exterior of the units can be cleaned with a mild soap and water solution.

**Walk-in Units.** A *walk-in unit* is a room-size insulated storage unit used to store whole cases of product as they are delivered. Walk-in units are often outfitted with adjustable shelving. Wire shelving units used in all refrigerators and freezers are either stainless steel or metal dipped into a plastic-like resin so that the moist environment inside the refrigerator does not cause rust or corrosion. True to their name, walk-in units allow the food service worker to walk into the unit to store or retrieve refrigerated products. Carts or racks can be wheeled directly into the unit. This is convenient when moving and storing large quantities of refrigerated products. Walk-in units are capable of storing thousands of pounds of food products until they are ready for use. A unique feature of walk-in units is that they can be made in any size or shape to fit the needs of an operation. Walk-in units become entire rooms of refrigeration or freezer storage. Typically, walk-in freezers and refrigerators are placed side by side so that bulk foods are stored close together.

**Reach-in Units.** A *reach-in unit* is a temperature-controlled cabinet for storing food items. Reach-in units can be individual units or can have up to two or three doors, each with shelves the size of a standard sheet pan. These reach-in units are usually located throughout the kitchen, creating convenient refrigerated or freezer storage near where the items may be needed. For example, there may be a reach-in freezer near a deep fryer used to store frozen French fries or breaded appetizers. Reach-in units are usually more than 6′ tall.

A *lowboy unit* is a type of reach-in unit located beneath a work surface. This allows the chef to prepare foods on the work surface and have adequate refrigerated storage beneath. A *chill drawer* is a refrigerated pull-out drawer that is located beneath a work surface to hold items to be cooked to order. They are typically used to hold steaks and other meat items near a broiler station. ***See Figure 4-11.***

**Roll-in Units.** A *roll-in unit* is an individual refrigeration unit that allows speed racks to be rolled in and out of the unit through a door opening that is just above floor height. Roll-in units are similar to reach-in units in size and dimensions. Roll-in units have a small ramp from the floor surface to the door opening to allow racks to be rolled in or out.

*Undercounter Reach-in Cold Storage*

Lowboys

*True FoodService Equipment, Inc.*

Chill Drawers

**4-11** *Lowboys and chill drawers are variations of reach-in units that provide convenient access to refrigerated storage at a work station.*

Hot storage includes various equipment for holding and serving prepared hot foods. Hot foods must be held at temperatures above 140°F to prevent bacteria growth. Maintaining these temperatures can be challenging for catering and banquet professionals who may have to prepare food and then transport it to another location. The use of proper holding and serving equipment can help maintain the necessary temperatures. It is imperative for the safety of the customer that proper temperatures are maintained during cooking, holding, transporting, and service. When used properly, holding and serving equipment provides a safe and efficient environment for prepared hot foods.

**Bain-Marie.** A *bain-marie* is a water bath used to keep sauces, soups, and other soft foods hot. The term refers to both the water bath and the inserts that hold food above the water bath. A bain-marie uses the same principle as a double-boiler. The difference between the two is that the double-boiler is made up of two vessels, each with a long handle, that nest one inside the other. A bain-marie can be a similar arrangement with two vessels nested inside one another without any handles, or it can be a steam table that is used to hold hot foods. **See Figure 4-12.** Pastry chefs use a bain-marie to melt chocolate as it is gentler than melting the chocolate directly over a flame. Because chocolate melts at a low temperature, the heat of an open burner not only is unnecessary, but it could scorch and burn the chocolate.

*Bain-Maries* _____

A kettle-style warmer is one type of bain-marie.

Bain-marie insets are available in many sizes.

Carlisle FoodService Products

**4-12** *A bain-marie is a hot water bath used to heat foods evenly.*

**Steam Table.** A *steam table* is an open-top table with heated wells that are filled with water. Foods are placed in hotel pans and the pans are placed into the top of the table. A *hotel pan* is a stainless steel pan used to cook, serve, or hold food. The hot water and steam beneath the hotel pans in a steam table keeps food hot for service. *Note:* Foods in the steam table must be kept covered to prevent heat loss. The heated wells are controlled by a thermostat so temperature can be adjusted as needed. A steam table is often also called a bain-marie because it is also filled with hot water and pans of food are placed above the hot water to keep them hot. ***See Figure 4-13.***

*Steam Tables*

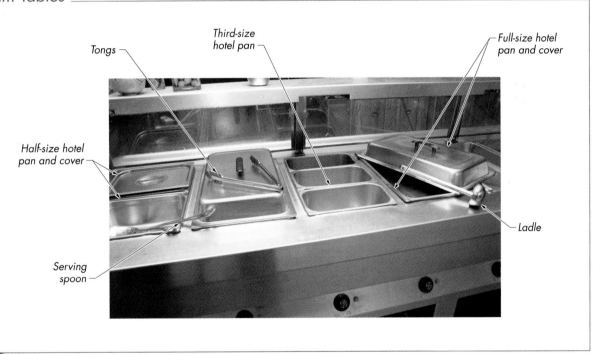

**4-13**  *A steam table is an open-top table with heated wells filled with water to keep foods hot for service.*

**Proofing Cabinet.** A *proofing cabinet* (also called a holding cabinet) is a tall and narrow stainless steel box where temperature and humidity can be adjusted. A proofing cabinet has a temperature dial, a humidity dial, and a thermometer. It has two very different uses. First, as a proofer, it is used for proofing dough. *Proofing* is the process of letting yeast dough rise in a warm (85°F), moist environment until the dough doubles in size. Yeast grows very well in this warm and moist environment. All yeast dough products must be proofed before baking.

## Proofing Cabinets

Cres Cor

**4-14** *Proofing cabinets accommodate standard food service trays and pans and are mounted on wheels for mobility.*

Most proofing cabinets accommodate standard food service trays and pans and are mounted on wheels for easy mobility. Other proofing cabinets are roll-in versions where yeast dough products can be wheeled in on speed racks for proofing. These cabinets can be used as holding cabinets to keep food hot without drying it out. They are also called food warmers. The holding cabinet can be set to a temperature above 135°F to hold items that were previously cooked until they are needed for service or a banquet. ***See Figure 4-14.***

**Overhead Warmers.** An *overhead warmer* is a heat source located above a hot food service area to keep plates of prepared food hot for service to a customer. They are usually electric rod-style elements that heat up similarly to an electric oven. These red-hot rods radiate heat toward the countertop and keep plates beneath them hot. ***See Figure 4-15.***

Overhead warmers can also be bulb-style units such as those used at a carving station or a French fry station. At a carving station, chefs will place large cuts of roasted meat under heat lamps to carve in front of the customer. Heat lamps usually have two 250W bulbs in either red or white. The intense heat from the bulbs keeps foods that are beneath them very warm.

## Overhead Warmers

Cres Cor

**Carving Station**

Carlisle FoodService Products

**French Fry Station**

**4-15** *Overhead warmers are heat sources located above a hot food service area to keep prepared foods hot for service.*

**Chafing Dishes.** A *chafing dish* is a hotel pan warmer with a reservoir for heated water and a heat source below. Chafing dishes are usually made from stainless steel. They can be used as portable bain-maries. Because they are portable, a portable heat source is needed to heat the water in the unit. *Canned fuel* is a gelled, flammable fuel form that is placed beneath a chafing dish; when lit it provides hours of heat. Canned fuel is sold in 2-hr, 4-hr, and 6-hr sizes. ***See Figure 4-16.***

**Utility Carts.** Utility carts make the transporting of food, equipment, or other items from one area to another easier and faster. Most food service utility carts are constructed of stainless steel or plastic and equipped with heavy-duty wheels. They either have two or three shelves to allow transportation of many items at one time. ***See Figure 4-17.***

## Utility Carts

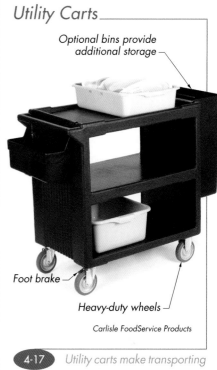

Optional bins provide additional storage

Foot brake

Heavy-duty wheels

Carlisle FoodService Products

**4-17**  *Utility carts make transporting many items at once easy.*

## Chafing Dishes

Cover

Stand

Portable heat source

Hotel pan

American METALCRAFT, Inc.

**4-16**  *A chafing dish is a hotel pan warmer with a portable heat source.*

**Insulated Carriers.** An *insulated carrier* is an insulated container made of heavy polyurethane or other plastic material designed to hold hotel pans of hot or cold foods for transport. Their lids have an insulated gasket sealer and lock securely in place. ***See Figure 4-18.***

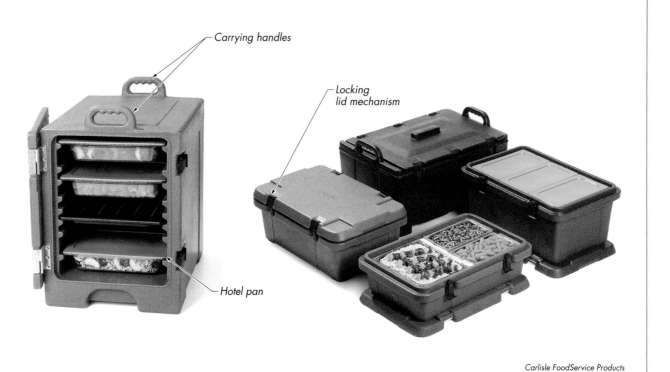

Carrying handles

Locking
lid mechanism

Hotel pan

Carlisle FoodService Products

**4-18** Insulated carriers hold hotel pans of hot or cold food for transport.

**Coffee-Brewing Systems.** Coffee is a popular beverage that is often brewed in large batches and held for service. Common types of coffee-brewing systems include automatic coffee urns, automatic coffee brewers, and thermal server systems. *See Figure 4-19.*

An *automatic coffee urn* is a large, fully automatic coffee-brewing unit that is connected to a water supply line. The direct connection to a water supply line allows an unlimited supply of hot water for making coffee. Some coffee urns require manual agitation of the brewed coffee to ensure uniform quality. Coffee urns brew several gallons of coffee per brewing cycle and are used most often by high-volume food service settings.

An *automatic coffee brewer* is a coffee brewer that dispenses coffee into a glass pot held on a warming plate. It can also be used to make other hot beverages such as hot cider and hot chocolate. The unit may or may not be attached to a water supply. If there is no connected water supply, water must be added to the machine for each brewing. The coffee brewer may have multiple warming plates designed to keep several pots hot at once.

## Coffee-Brewing Systems

**Automatic Coffee Urns**

**Automatic Coffee Brewers**

**Thermal Server Systems**

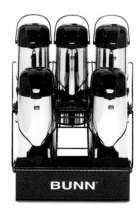

*Bunn Corporation*

**4-19** *Coffee is often brewed in large batches and held for service.*

Thermal server systems make it easy to brew, hold, and transport hot coffee for events. A *thermal server system* is a coffee-brewing system that brews and dispenses coffee into insulated servers that may be stored at convenient locations throughout a dining establishment. Thermal servers can hold coffee at serving temperature for extended periods without a direct heat source.

## PREPARATION AND PROCESSING AREAS

The preparation and processing areas of the kitchen contain many skilled workers operating as parts of different groups. Since many different things can be happening in a large kitchen at one time, it is helpful to group the workers and the areas they are working in into smaller sections. All professional kitchens are divided into work sections.

A *work section* is an area where kitchen professionals are all working for the same purpose and at the same time. For example, in a bakery work section, all of the bakers are there to bake breads and other pastry items. The area is not used for other purposes such as meat fabrication or soup making. It is a work section dedicated solely to the production of baked goods. Typical work sections found in the professional kitchen include prep, hot foods, garde manger, bakery and pastry, short order, banquet, beverage, and sanitation sections. Work sections are further divided into several work stations. ***See Figure 4-20.***

| Professional Kitchen Divisions | |
| --- | --- |
| Work Sections | Work Stations |
| **Prep section** | Butcher/meat fabricator<br>Mise en place for other stations |
| **Hot foods section** | Broiler station<br>Grill station<br>Griddle station<br>Sauté/sauce station<br>Fry station<br>Vegetable station |
| **Garde manger section** | Cold food preparation<br>Salad station (cleaning and preparing)<br>Sandwich station<br>Showpiece/centerpiece station<br>Cold platter preparation (fruit, meat cheese, etc.)<br>Dessert preparation station (may also be in a separate baking and pastry section)<br>Plated dessert station (may also be in a separate baking and pastry station) |
| **Bakery and pastry section** | Dough mixing and proofing station<br>Dough rolling and forming station<br>Baking and cooling station<br>Dessert preparation station<br>Plated dessert station |
| **Short order section** | Fry station<br>Griddle station<br>Grill station<br>Broiler station |
| **Banquet section** | Meat carving station<br>Food holding and plating station<br>Steam cooking station<br>Dry-heat cooking station (roasting, broiling, grilling, etc.) |
| **Beverage section** | Hot beverage station<br>Cold beverage station<br>Alcoholic beverage station |
| **Sanitation section** | Dish machine operator station<br>Warewashing (pot and pan) station<br>Bus station (for used service items) |

**4-20** *The professional kitchen is divided into work sections and workstations.*

A *work station* is an area in a work section for a specific task to be performed by a specific person. For example, multiple line cooks are assigned to each work section. These cooks have individual responsibilities such as working on a broiler or grill. These individuals are then assigned to specific work stations within the work section. Work stations are designed to be relatively self-sufficient and should have the necessary tools, equipment, work space, and power sources needed to function. For example, hamburgers may be prepared on a broiler. The work station where they would be prepared is called the broiler station and should be equipped with a broiler, the gas needed to operate the broiler, utensils needed to flip the hamburgers over, adequate cold storage to hold uncooked food and accompaniments (sliced tomatoes, lettuce, pickles, cheese), seasonings, and plates to hold prepared food. The broiler cook should not need to leave the broiler to find any tools needed to prepare broiled foods. ***See Figure 4-21.***

**Prep Section.** A *prep section* is an area containing tools and equipment used to prepare basic items that may be needed by multiple stations. The title "prep" is short for "preparation." For example, a prep station cook may clean lettuce to be used at a salad station. A prep cook may also make stocks that will be turned over to someone else who will make soups or sauces from them. A prep section usually has basic equipment such as sinks, cutting boards, work tables, slicers, and mixers. Prep stations may include a range where stocks, grains, and sauces are prepared prior to service. All prep work necessary for service is completed in the prep station prior to service.

**Hot Foods Section.** A *hot foods section* is an area containing equipment such as a broiler, deep fryer, and open-burner range as well as the refrigeration and storage needed to operate the section. Typical stations within a hot foods section include broiler stations, grill stations, griddle stations, sauté/sauce stations, fry stations, and vegetable stations. As an order comes into the kitchen, the different stations within the hot foods section will go to work. An order for a pasta dish is sent to the sauté station, while an order for steak is sent to the broiler. The greatest challenge for a hot foods station is in delivering all of the orders (which are probably prepared at different stations) to a given table at the same time. A couple dining together may order sautéed pasta for one person and a medium-well steak for the other person. The couple expects both dishes to be served at the same time, not 10 min apart.

**Garde Manger Section.** A *garde manger section* is an area of the professional kitchen where salads, salad dressings, deli and cheese trays, cold appetizers, cold sandwiches, buffet showpieces or centerpieces, cold platters, and charcuterie are prepared. The garde manger section is also known as the cold kitchen. In a small operation there may be only one

**Broiler Station**

**4-21** *The broiler station should be equipped with all of the tools necessary to prepare broiled foods.*

person in the section whose title is garde manger. In a large operation such as a hotel, this section may be much larger, with multiple garde manger assistants. Typical stations within a garde manger section include cold food preparation, salads, sandwiches, dessert preparation, and plated dessert stations. Typical equipment in the garde manger station includes slicers, stand mixers, blenders, and refrigeration units.

**Bakery and Pastry Section.** A *bakery and pastry section* is an area containing assorted pieces of equipment for producing all of the baked goods and pastries for the operation. Typical equipment found in a bakery and pastry section includes assorted stand mixers, proofing cabinets, dough sheeters, scales, ovens, refrigeration units, and work tables. Work tables in the bakery and pastry section may have tops made of wood, marble, or quartz, as they are preferred by pastry chefs and bakers. Typical stations within a bakery and pastry section include dough mixing and proofing, dough rolling and forming, baking, cooling, dessert preparation, and plated dessert stations.

**Short Order Section.** A *short order section* is an area that makes casual foods that are cooked quickly. This kind of food may be considered "fast food" because it can be prepared quickly and is comparable to the types of offerings at a fast food restaurant. A short order cook works in this section and is responsible for preparing items such as hamburgers, deep-fried appetizers, griddled sandwiches such as a Reuben, and perhaps omelets and pancakes. Short order cooks are usually found in casual-themed operations such as a pub or snack shop, or in a hotel kitchen preparing room service orders. Typical stations in a short order section include fry, griddle, grill, and broiler stations. A separate prep station for mise en place may be included in the short order section.

**Banquet Section.** A *banquet section* is an area where preparation and production for private dining functions for large or small groups takes place. The banquet section will have banquet cooks working together to produce everything needed for a large event. Different from a typical kitchen, the banquet section is planned and organized to efficiently allow precise timing of large-scale food production. A banquet is planned in advance and operates differently from a restaurant where it is unknown how many guests will show up on any given night and what items will be ordered. A banquet section may contain all of the equipment that is dispersed among the various sections of a traditional kitchen, as it functions as a self-sufficient operation independent of the main kitchen area. Typical stations within a banquet section include meat carving, food holding, food plating, steam cooking, and dry-heat cooking (roasting, broiling, grilling, etc.) stations.

**Beverage section.** A *beverage section* is an area containing items needed for the service staff to provide beverages to the guests. Typically, a

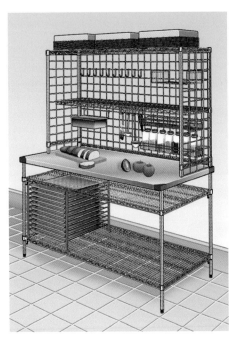

InterMetro Industries Corporation

beverage section contains an ice bin, soft drink dispenser, glass racks, drink blenders, coffee brewers with hot tea, water, and refrigeration units. Stations within a beverage section may include hot beverage, cold beverage, and alcoholic beverage stations. ***See Figure 4-22.***

**Sanitation section.** A *sanitation section* contains cleaning supplies and the large commercial dish machine and dish racks for washing all serviceware items and kitchen smallwares. All of the cleaning supplies and equipment such as mops, buckets, mop wringers, wet floor signs, brooms, and dustpans are stored in the sanitation section. Typical stations within a sanitation section include dish machine, warewashing, maintenance, and bus stations.

Smaller operations such as a snack bar may have only one cook who handles multiple pieces of equipment and cooking methods. Extremely large operations may have many duplicate stations, such as a large steak house with multiple broiler stations. No matter the size of the operation, the various pieces of equipment in each section are placed close together to allow the most efficient production possible.

## Preparation and Processing Equipment

On a television cooking show, the chef prepares a recipe with ingredients that have already been sliced, diced, or prepped in some way to make the process quicker and easier. If this mise en place is not done prior to the start of the show, the chef cannot make the same number of recipes that can be made when prep work is done in advance. Most prep work is processed in a prep kitchen using preparation and processing equipment. Work tables provide a surface with the perfect height for all preparation tasks and safe and efficient operation of prep tools. Some of the common preparation and processing equipment used in the professional kitchen include work tables, can openers, electric slicers, food processors, blenders, mixers, and juicers. ***See Figure 4-23.***

**Work Tables.** A few different styles of tables are used in the professional kitchen. Stainless steel tables are used throughout the kitchen. These tables have very durable surfaces and are constructed for easy cleaning. Butcher block tables are only used in the baking or pastry area. These tables are not used in hot food production as wood surfaces are porous and can breed bacteria. There are too many potentially hazardous foods used in the hot foods kitchen that make a wood surface a poor choice. Marble and granite work tables are sometimes used in baking and pastry areas for sugar and chocolate work, but are being replaced with quartz-surface tops as they are NSF approved and completely food-safe.

Carlisle FoodService Products

4-22 This bar-style beverage station can be moved to different locations.

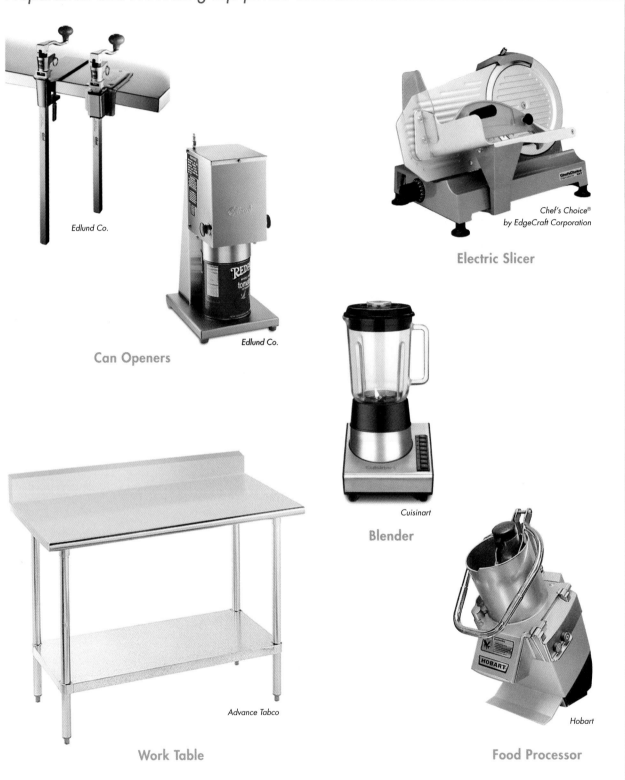

Edlund Co.

Chef's Choice®
by EdgeCraft Corporation

**Electric Slicer**

Edlund Co.

**Can Openers**

Cuisinart

**Blender**

Advance Tabco

**Work Table**

Hobart

**Food Processor**

**4-23** *Preparation and processing equipment used in the professional kitchen includes work tables, can openers, and slicing, cutting, puréeing, blending, and liquefying equipment.*

**Table-Mounted Can Openers.** Professional kitchens use heavy-duty can openers that are mounted to the edge of the table. These can openers open large canned goods quickly and easily. Care should be taken to inspect the blade of these openers frequently as dull blades will leave metal shavings in the cans as they are opened. It is also important to clean and sanitize openers after each use to prevent cross-contamination and foodborne illness.

**Electric Slicers.** An *electric slicer* is a tool used to slice foods such as meats and cheese into uniform slices. It has a regulator for providing a wide range of slice thicknesses up to ¾″ and has a feed grip that grips material firmly on top or serves as a pusher plate for slicing small end pieces. A slicer has a circular blade that rotates at a high speed, slicing items as they are pushed over it. It can be manually operated by pushing the item across the blade with the pusher plate, or it can be switched to automatic to slice a product from one end to the other. All slicers have safety guards built in to help protect the user from the sharp revolving blade. It is extremely important to ensure that these guards are in place at all times when using a slicer. In addition to slicing, a slicer can be used for shredding lettuce and cabbage. *Note:* Exercise caution when using the slicer as the blade is very sharp.

**Food Processors.** A *food processor* is an appliance used to purée, chop, grate, slice, and shred food. It has a removable bowl, an S-shaped blade, and a lid, all sitting on top of a base that houses a powerful motor. A food processor can quickly chop, purée, blend, or emulsify foods, and has additional disk blades for shredding, slicing, and julienning foods and grating cheese.

The *buffalo chopper* (also called a food chopper) is an appliance used to process larger amounts of a product into roughly equal-size pieces. Common uses are grinding stale bread for bread crumbs and grinding large quantities of onions. Food passes under a hoodlike top which houses a large S-shaped blade. The coarseness of the cut is dependent on how long the food is left in the machine. The more times it passes under the hood and blade, the finer the cut. There is a safety switch built in that prevents the machine from operating if the hood-style lid is open.

A *vertical cutter/mixer* (VCM) is an appliance used to cut and mix foods simultaneously for fast volume production. It is usually floor mounted with a 15-80 qt capacity bowl. There are two movable parts within the bowl: the knife blades, which move at high speed, and the mixing baffle, which is operated manually to move the product into the cutting knives. Besides speed of production, another advantage of using a VCM is that the product being processed is never bruised because the knife blades operate at high speed and slice the product in mid-air. VCMs have a safety switch built in that prevents the machine from operating if the lid is open. VCMs can make large amounts of mixed products, such as 60 lb of fresh mayonnaise, in a matter of minutes.

**Chef's Tip:** When using a can opener, always wash and sanitize the small, pointed blade used to puncture the can. The tip of the blade enters the can and usually touches food within. If the blade tip is not maintained, cleaned, and sanitized daily, it could become a source for cross-contamination. Simply running the can opener through the dish machine every day ensures that the blade will be free of harmful microorganisms.

**Blenders.** A blender is an appliance used to chop, blend, purée, or liquefy food. It is similar to a food processor but has a tall and slender canister with a four-toothed blade. It is designed for puréeing drinks, soups, and soft foods as it creates a whirlpool effect while blending. It can also be used to crush ice.

An immersion blender (also known as a stick mixer) is a narrow, handheld blender with a rotary blade at the end. It is similar to the blender in its abilities but is used to purée a product in the container it was prepared in. For example, it can be inserted into a saucepot to purée a soup.

**Juicers.** A *juicer* is an electric or manual device used to extract juice from fruits and vegetables. There are two types of commercial juicers. The first is called a reamer. Reamers can be manual or electric and are used to extract the juice from citrus fruits. To use an electric reamer, fruits are cut in half and placed on the juicer screen. The arm is lowered to squeeze the fruit and extract the juice. The second style of juicer is called a juice extractor. These electrical machines create juice by liquefying raw vegetables, fresh fruits, and even herbs. The fiber and pulp are separated from the juice.

**Mixers.** The mixer is one of the most versatile pieces of equipment in the professional kitchen. It has U-shaped arms that secure a stainless steel mixing bowl and a rotating head that can accommodate three different attachments. The whip is used for whipping volume into products. The paddle is used for general mixing and creaming. The dough hook is used for kneading bread dough. Mixers have multiple speeds for light or aggressive mixing. Bench mixers, also called table-top mixers, come in a variety of sizes ranging from 4½ qt to 20 qt capacities. Floor mixers, also known as volume mixers, range from 30 qt to 140 qt capacities. Most mixers have additional attachments such as grinders and shredders. *See Figure 4-24.*

*Mixers*

Bench Mixer      Floor Mixer    *Hobart*

4-24   *Most mixers have additional attachments for grinding and shredding.*

## Cooking and Baking Equipment

Cooking and baking equipment can be either electric, gas, or steam operated, and is usually placed in a central location in the professional kitchen to be easily accessible. When choosing cooking and baking equipment for a professional kitchen, the establishment's menu should be considered, as the types of items frequently cooked will help to determine what equipment is needed. For example, if the menu is for a steakhouse, there is a greater need for broilers and grills than for griddles and open-flame ranges. Common types of cooking and baking equipment available include deep-fat fryers, ranges, griddles, grills, tilt skillets, ovens, broilers, steamers, combination steamer/ovens, and kettles.

**Deep-Fat Fryers.** A *deep-fat fryer* is a cooking unit used to cook foods by submersion in hot fat. They operate by heating the fat to a temperature between 200°F and 400°F and have a thermostat to regulate the temperature. They are sized by the number of pounds of fat they can hold. For example, a 20 lb fryer holds 20 lb of fat. ***See Figure 4-25.***

*U.S. Department of Agriculture*

## Dry-Heat Cooking Equipment

Grills

Griddles

Ranges

Deep-Fat Fryers

*Vulcan-Hart, a division of the ITW Food Equipment Group LLC*

**4-25** *Common dry-heat cooking equipment includes deep-fat fryers, ranges, and broilers.*

Charlie Trotter's

4-26 Flat tops provide a large, even heating surface.

## Nutrition Note

A grill cooks food more nutritiously than a griddle does because fat is able to drip away from the item being grilled. On a griddle, food sits in the fat on the griddle; therefore, the food retains more fat and calories.

**Ranges.** A *range* (also called a stove top) is a large appliance with surface burners used to cook food. Ranges are most commonly seen with four or six open-flame burners, although some units may have electric burners. The open-flame burner allows for a more intense and direct heat, and it is very easy to regulate its intensity. Some ranges have a heavy steel plate or flat top that covers the burners. This flat top, although it takes much longer to heat, provides even heat across a much larger surface than an individual burner. ***See Figure 4-26.*** Instead of one pot being able to sit on a single burner, a flat top allows multiple cooking vessels to be placed over the heat at one time. Many ranges have both open burners and a flat top cooking surface together on the same unit.

Another type of range that is becoming popular is called an induction range. An *induction range* is an electric range that uses a magnetic coil below the surface of the burner to heat food rapidly. Induction burners interact with cast iron or magnetic stainless cookware, creating heat within the material of the cookware. The surface of the induction range never gets hot, so it is always safe to touch. Instead, the cookware itself becomes hot. An induction range will not heat stainless steel cookware, so cast iron or magnetic cookware must be used.

**Griddles.** A *griddle* is a cooking surface made of metal on which foods are cooked. Griddles are usually self-standing units, but can also be part of a stove top. They have a solid surface and are commonly used for cooking items such as pancakes, hamburgers, eggs, and bacon. The surface temperature is controlled with a thermostat. Griddles should be seasoned after each cleaning to avoid surface corrosion. To season a griddle, heat the surface to 300-350°F and apply about one ounce of cooking oil per square foot of the surface. Spread the oil over the entire surface, wiping off any excess. Repeat the procedure until a slick, mirror-like finish is achieved.

**Grills.** A *grill* is a cooking unit consisting of a large metal grate, also referred to as a grill, placed over a heat source. The heat source may be gas or another burning fuel such as charcoal or wood. Grills are commonly used to cook meats, seafood, and poultry. Food is placed on the preheated metal grate over the heat source and is turned over about halfway through the grilling process to cook the other side.

**Tilting Skillets.** A tilting skillet (also referred to as a tilting braising pan) is a piece of cooking equipment with a large-capacity pan, a thermostat, a tilting mechanism, and a cover. The pan of a tilting skillet can hold between 30 gal. and 40 gal. The tilting skillet is a highly versatile piece of equipment that can be used as an oversize skillet, bain-marie, stockpot, proofing oven, kettle, or evenly heated range top. Although a tilting skillet can be used to pan-fry or sauté, it should not be used for deep-fat frying.

**Ovens.** An oven is an enclosed cabinet where food is cooked by being surrounded by dry, hot air. It can be either gas or electric powered. A *conventional oven* is an enclosed heating cabinet typically located beneath a range or within a wall unit. This is the most common type of oven. The interior cabinet of a standard oven has multiple adjustable wire racks. Other types of ovens found in the professional kitchen include convection ovens, deck ovens, wood-burning ovens, flashbake ovens, and microwave ovens. *See Figure 4-27.*

*Ovens*

Blodgett Oven Company

**Deck Oven**

Amana Commercial Products

**Microwave Oven**

Vulcan-Hart, a division of
the ITW Food Equipment Group LLC

**Convection Oven**

Earthstone Ovens, Inc.

**Wood-Burning Hearth Oven**

Vulcan-Hart, a division of
the ITW Food Equipment Group LLC

**Flashbake Oven**

4-27　Commercial ovens found in the professional kitchen include convection ovens, deck ovens, wood-burning hearth ovens, and flashbake ovens.

To test the theory that microwaves heat water molecules, gather two small identical dishes that are microwavable. Make sure that one is completely dry (not even recently washed). Place 2 tbsp flour in the dry dish. Rinse the other dish under water and dry it with a paper towel. Now add 2 tbsp flour and 1 tbsp water to this dish. Place the first (dry) dish in the microwave, heat for 30 seconds, and remove. Check to see if either the dish or the flour got hot. Next, place the second dish (containing flour and water) in the microwave, heat for 30 seconds, and check it to see if it or its ingredients got hot. Record the results.

A *convection oven* is an oven with an interior fan that circulates the dry, hot air throughout the cabinet. This fan creates more even and intense heating as it keeps the air moving while the item is cooking. Cooking temperatures are about 50°F lower in a convection oven than in a conventional oven because the airflow leads to more efficient cooking. Heat is evenly distributed by the air circulation, allowing the oven to be loaded to capacity and still provide the required heat. Convection ovens are available in gas and electric models and as floor ovens with roll-in dolly, table models, counter models, and stack ovens. A convection oven increases productivity, reduces shrinkage, and cooks more uniformly than a conventional oven. Cooking cycles are completely automatic. In a convection oven, the same quantity of food can be cooked in less space and using less fuel than in a conventional oven.

A *deck oven* is a drawerlike oven that is commonly stacked one on top of another, providing multiple-temperature baking shelves. Food items are placed directly on the floor of a deck oven.

A *wood-burning oven* is an oven with a curved hearth made of masonry that is heated with wood. Once the ashes are swept out, the oven is used for baking or roasting foods. This oven has gained popularity recently due to the smoky flavor the wood imparts to the foods cooked in it.

A *flashbake oven* is an oven that uses both infrared and visible light waves to cook foods quickly and evenly. The light waves cook both from above and below the food. Because the heat from these lights is so intense, there is no loss of flavor or moisture in the foods being cooked.

A microwave oven is not a typical oven at all. A *microwave oven* is a cooking unit using electronically generated microwaves to heat water molecules within foods to cook them. Microwave ovens utilize electromagnetic waves generated by a magnetron. The waves, when striking water molecules in a food, create heat energy. Foods cooked in a microwave oven must contain some amount of water for heat to develop. Foods cooked in the microwave do not brown as they do in a conventional oven. In addition, larger foods are turned to ensure even cooking. More cooking time must be allowed for additional food items cooked at the same time. In the professional kitchen, microwave ovens are used primarily for thawing, heating convenience foods, and reheating cooked foods. A microwave oven requires no preheating. Cooking times must be accurately controlled by a timer. Foods heated in a microwave oven should be placed on china, plastic, or paper containers. Never use metal containers or objects in a microwave oven. Metal surfaces reflect microwaves and can cause damage to the oven.

**Broilers.** A *broiler* is a large piece of cooking equipment in which the heat source is located above the food instead of below it. Food is placed on the broiler and the heat—either gas flame, electric, or ceramic stones—cooks the food from above. Broilers can be stand-alone or combined with an oven. The three basic types of broilers include the standard broiler, salamander, and rotisserie. ***See Figure 4-28.***

*Broilers* _____

Vulcan-Hart, a division of
the ITW Food Equipment Group LLC

**Standard Broiler**

Vulcan-Hart, a division of
the ITW Food Equipment Group LLC

**Salamander**

Henny Penny Corporation

**Rotisserie**

**4-28** *A broiler uses a heat source located above or, in the case of a rotisserie, behind the food.*

A *salamander* is a small overhead broiler that is usually attached to an open burner range. The heat source is not nearly as intense as a standard broiler. A salamander is primarily used to brown, glaze, melt cheese, or finish cooking foods.

A *rotisserie* is a sideways broiler. Instead of the heat being above the food, it is behind it. Foods are placed on a steel rod or spit that revolves past the heat source to ensure even heating. Rotisseries are most often used to cook poultry and meats.

**Steamers.** There are two types of steamers used in the professional kitchen. The first, convection steamers, generate steam with the help of an internal boiler. Here the steam generated circulates around the food, cooking it rapidly. Another kind of steamer is called a pressure steamer. Pressure steamers heat water to above the boiling point in a pressure-controlled, sealed cabinet. The combination of pressure and high temperatures cooks food much quicker than a traditional convection steamer. The steaming cooking method will be discussed further in a future chapter on cooking methods. ***See Figure 4-29.***

### Nutrition Note

Low-fat cooking can be achieved if meat is roasted in a rotisserie or on a raised rack in a roasting pan, allowing the fat to drip away from the meat. If meat is placed on the bottom of the roasting pan, the meat will sit in its own fat while roasting and will retain the fat when served.

## Steam Equipment

*Vulcan-Hart, a division of
the ITW Food Equipment Group LLC*

**Steamer**

*Blodgett Oven Company*

**Combi Oven**

*Vulcan-Hart, a division of
the ITW Food Equipment Group LLC*

**Steam-Jacketed Kettle**

*Vulcan-Hart, a division of
the ITW Food Equipment Group LLC*

**Trunnion Kettle**

**4-29** *Steamers, combi ovens, and kettles are all used to cook food.*

**Combination Ovens.** Combination ovens, commonly called combi ovens, can cook foods in a variety of ways. A *combi oven* is an oven that cooks food by surrounding them with moist, hot air that is circulated using the convection cooking method. With steam present in the oven, shrinkage is reduced, but food still acquires a brown surface. The steam in the oven produces excellent hard rolls or bread and a quality baked potato. Combi ovens are available in several sizes to meet the needs of various food service establishments.

**Kettles.** A *steam-jacketed kettle* is a large cooking kettle that has a hollow lining into which steam is pumped. It cooks foods quickly and evenly while reducing the chance of burning or scorching foods. This is accomplished by heating the food without the steam touching it. There is a large spigot on the bottom of the kettle for easy draining of the kettle.

Another version of the steam-jacketed kettle is a trunnion kettle. The *trunnion kettle* is a small steam-jacketed kettle. Instead of a drain on the bottom of the unit, a trunnion kettle can be tilted to empty it by pulling a lever or turning a wheel.

## SANITATION AND SAFETY AREAS

Sanitation and safety are important to the success of any food service operation. Customers go to a restaurant to enjoy a well-cooked meal prepared by a skilled culinary staff and served by a pleasant and service-minded wait staff. If any of these components is lacking, it could mean a disaster for the overall dining experience. Improper sanitation can be the quickest path to failure even if the food and service are the very best.

A food service establishment should have a designated area where chemicals, cleaning supplies, and other sanitation equipment are stored. Staff should be well trained in proper sanitation procedures to ensure the operation is kept clean and sanitary. Likewise, having specific safety areas in the operation ensures that emergency care items such as first aid kits and fire extinguishers are easily located in an emergency.

*Charlie Trotter's*

### Sanitation and Safety Equipment

Standards are established by local health departments for various sanitation and safety equipment. These standards help food service establishments service customers safely and provide a safe workplace. Properly working, clean sanitation equipment is one key factor to a successful food service operation. It may take only one case of a foodborne illness to completely devastate a business.

Guests coming to a restaurant want to see clean conditions, clean washrooms, and no evidence of pests such as roaches, flies, or rodents. If a restaurant looks as if the staff does not care about the cleanliness of the operation, the customer assumes the staff does not care about the quality or wholesomeness of the food, either.

Although sanitation and safety items used in a food service operation are not actually used in food production, these items are vital to the safety and well-being of the staff and customers. Common sanitation and safety equipment found in the professional kitchen includes three-compartment sinks, commercial dishwashers, food waste disposers, fire extinguishers, fire-suppression systems, ventilation systems, and first aid kits.

**Compartment Sinks.** Commercial sinks are constructed with stainless steel and have rounded corners, making them easier to clean. Typical warewashing sinks have three compartments. The first compartment is for hot soapy wash water, the second is for hot rinse water, and the last is for sanitizing solution. All sinks should be cleaned and sanitized prior to being used for food preparation. Separate hand sinks should be located throughout the kitchen area as food preparation sinks and warewashing sinks should not be used for handwashing. *See Figure 4-30.* Four-compartment sinks are required in some food service establishments. The first compartment is used to rinse off large debris before following the typical warewashing sequence.

*Three-Compartment Sinks*

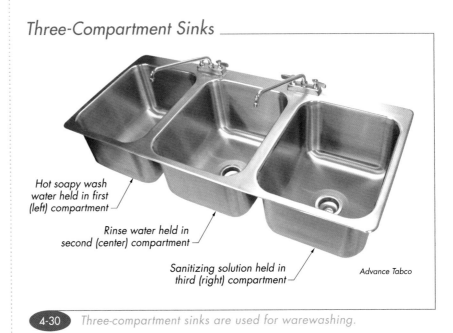

Hot soapy wash water held in first (left) compartment

Rinse water held in second (center) compartment

Sanitizing solution held in third (right) compartment

Advance Tabco

**4-30** *Three-compartment sinks are used for warewashing.*

**Commercial Dishwashers.** The commercial dishwasher makes cleanup of soiled dishes fast and efficient. A single-tank dishwasher has a door that is raised up to load racks of prescraped dirty dishes and glassware. After the rack is loaded, the door is closed and the washing cycle automatically begins. When the cycle is complete, the washing stops and the doors can be lifted to remove the clean service items. A carousel or multitank machine takes racks by conveyor through the cycle of prewash, wash, rinse, sanitize, and dry. Again, prescraped dishes are loaded into racks which are fed into the machine. The racks emerge from the other end of the machine and travel down the conveyor to the end of the run where they can be removed and stored. *See Figure 4-31.*

**Food Waste Disposers.** A *food waste disposer* is a food grinder mounted beneath warewashing sinks to eliminate solid food material. Food waste disposers are commonly called garbage disposers. Solid food material is rinsed from plates into sinks prior to being loaded into a commercial dishwasher. Solid food waste left in pots and pans is also washed into sinks at a warewashing station. The majority of solid foods should be scraped into a garbage can and not rinsed into the disposer. The disposer is only designed to eliminate bits of food that would otherwise be caught in the sink drain. *See Figure 4-32.*

## Commercial Dishwashers

Hobart

4-31    *Commercial dishwashers provide efficient cleanup of soiled dishes and small kitchen equipment.*

**Fire Extinguishers.** Fire extinguishers must be stationed at various areas throughout the kitchen and the entire food service operation. They are metal canisters filled with pressurized dry chemicals, foam, or water. Fire extinguishers are operated manually, meaning that a person must squeeze the trigger to dispense the chemical contained in the extinguisher. The chemical should be directed at the base of a fire to extinguish it. There are three major classes of fire extinguishers used in the professional kitchen: Classes A, C, and K. *See Figure 4-33.* Each class is designed to eliminate a particular type of fire. Class A fire extinguishers are for fires involving common combustible materials such as trash, wood, or paper. Class C fire extinguishers are used to extinguish electrical equipment fires. Class K fire extinguishers are used specifically for cooking-grease fires. It is important to check fire extinguishers frequently to make sure they are adequately charged and ready in case of emergency. Checking them at the end of each month during kitchen inventory is a good way to include a safety check in the operation.

**Fire-Suppression Systems.** A fire-suppression system is required over any open-flame cooking surface or combustible surface such as a deep fryer. A *fire-suppression system* is an automatic fire extinguishing system that is activated by the intense heat generated by a fire. Fire-suppression systems are not manual like fire extinguishers. The system includes discharge nozzles located over each piece of cooking equipment. These spray heads are connected to a highly pressurized tank containing a fire-extinguishing chemical. If a fire occurs under the suppression system, intense heat triggers the spray heads to release a chemical, thus extinguishing the fire. *See Figure 4-34.*

## Food Waste Disposer

In-Sink-Erator

4-32    *Food waste disposers eliminate solid food material rinsed into warewashing or preparation sinks.*

*Chapter 4 — Food Service Equipment*   **135**

## Fire Extinguishers

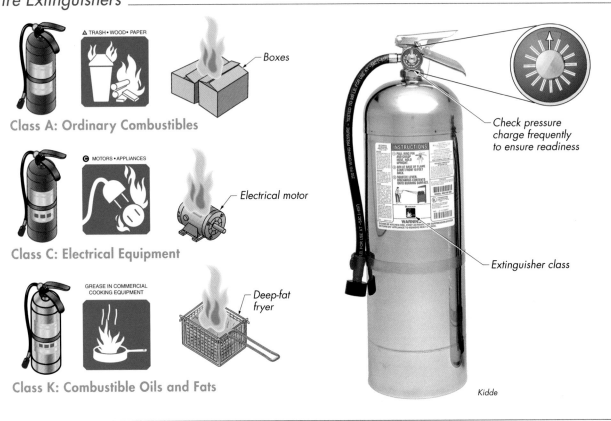

Class A: Ordinary Combustibles

Boxes

Class C: Electrical Equipment

Electrical motor

Class K: Combustible Oils and Fats

Deep-fat fryer

Check pressure charge frequently to ensure readiness

Extinguisher class

Kidde

**4-33** *The three classes of fire extinguishers found in the professional kitchen are Classes A, C, and K.*

## Fire-Suppression Systems

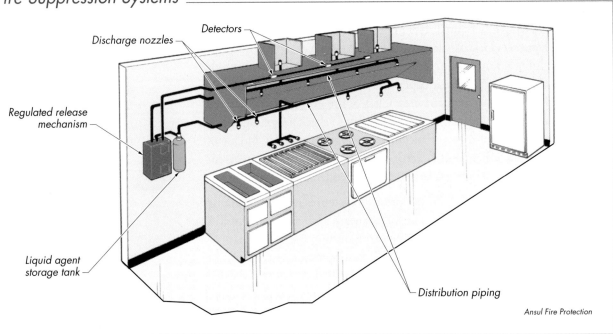

Discharge nozzles

Detectors

Regulated release mechanism

Liquid agent storage tank

Distribution piping

Ansul Fire Protection

**4-34** *A fire-suppression system provides fire protection for commercial cooking equipment.*

**Ventilation Systems.** A *ventilation system* is a large exhaust system that sucks heat, smoke, and fumes out of the kitchen and into the outside air. A properly working ventilation system is essential in any food service operation to remove excess heat, smoke, and fumes that can build up in the kitchen. The exhaust hood of the ventilation system contains the cooking line fire-suppression system. It also has long ductwork that carries air from the exhaust hood out of the building. A large fan located at the end of the ductwork creates a vacuum that draws exhaust air out from the building. As the hot, greasy air is drawn up the hood, grease tends to build up on the interior surface and the filters. This grease must be cleaned off the hood and fire-suppression components regularly as it has the potential to catch fire. ***See Figure 4-35.***

*Ventilation System*

Badger Fire Protection

**4-35** A ventilation system draws hot air away from the cooking area.

**First Aid Kits.** Even when safe knife skills and equipment use are practiced in the kitchen, it is almost inevitable that some injuries will occur. First aid kits contain all the necessary equipment for first-response treatment of kitchen injuries. ***See Figure 4-36.*** A first aid kit should contain an ample supply of the following:

• adhesive strips in various sizes

• gauze bandages

• first aid cloth tape

• finger cots (thimble-like latex sleeves to cover adhesive strips)

## First Aid Kits

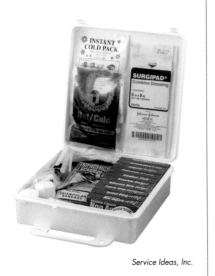

*Service Ideas, Inc.*

**4-36** *A first aid kit should be clearly identified and easily accessible.*

Refer to the CD-ROM for Quick Quiz® questions related to chapter content.

- sterile gauze dressing pads
- cotton swabs
- rubbing alcohol
- burn cream or spray
- peroxide
- antibiotic lotion
- ipecac syrup (to induce vomiting if necessary)
- emergency ice packs
- eyewash (for flushing chemicals or foreign objects from the eye)

First aid kits should be stored in plain view and never locked. In the event that an injury cannot be adequately treated by available first aid equipment, the paramedics should be called as soon as possible.

## SUMMARY

A professional kitchen is composed of many different areas that must function together efficiently to prepare food and perform other related duties. Equipment used in a professional kitchen must be of professional-grade quality to withstand the wear and tear of constant use. To ensure that the equipment is sufficiently durable and easy to keep clean, NSF International certifies all equipment approved for use in a commercial food service operation.

The specific areas found in a professional kitchen include receiving, storage, preparation and processing, and sanitation and safety. The preparation and processing area is further divided into various work sections. A well-run professional kitchen contains a wide variety of equipment for proper inspection, measurement, storage, and preparation of foods. A complete supply of sanitation and safety equipment is equally important to ensure the health and safety of the employees and customers.

In brief, a professional kitchen must not only be clean, but it must also be organized and supplied in such a way to effectively accommodate all necessary kitchen activities.

1. Identify the major areas of the professional kitchen.

2. In reference to a professional kitchen, what is the meaning of mise en place?

3. List the specifications that tools or equipment must meet in order to be NSF certified.

4. Describe the seven safety guidelines recommended for operating and maintaining food service equipment.

5. Describe the various functions of the measurement equipment required in the receiving area.

6. Contrast the variety of shelving units found in the dry storage area of a professional kitchen.

7. List the three types of cold storage units.

8. How does a bain-marie differ from a double-boiler?

9. Describe the functions of a proofing cabinet.

10. Differentiate between overhead warmers and chafing dishes.

11. Describe two methods of transporting food to and from hot and cold storage areas.

12. List the advantages of the three common types of coffee brewing systems.

13. Name the typical work sections found in the preparation and processing area of a professional kitchen.

14. Why are butcher block tables not used in hot food production?

15. What is the difference between a food processor and a blender?

16. Explain why the mixer is one of the most versatile pieces of equipment in the professional kitchen.

17. Identify the distinguishing features of common oven types, including the following: standard, convection, deck, wood-burning, flashbake, microwave, and combination (combi) ovens.

18. List common sanitation and safety equipment found in the professional kitchen.

19. Differentiate between fire extinguishers and fire-suppression systems.

# Hand Tools and Smallwares

## Chapter 5

# Hand tools and smallwares

**Hand tools and smallwares** are small items that can be moved from one station to another in the professional kitchen, and include equipment such as knives, pots and pans, scales, strainers and sifters, turners and tongs, and thermometers. Using appropriate hand tools and smallwares can make a big difference in the ease and efficiency of food preparation. Time spent learning to distinguish between the many handheld tools and pieces of cookware will be evident in the results achieved in the professional kitchen.

## Chapter Objectives:

1. Differentiate between hand tools and smallwares.
2. List the functions of portioning and measuring equipment commonly found in a professional kitchen.
3. Identify the main categories of professional cookware and bakeware.
4. Explain the uses for different kinds of strainers, sifters, and sieves.
5. Describe common types of turning and grabbing tools.
6. Describe various mixing and blending tools.
7. Identify common cutting tools.
8. Contrast the functions of common pastry tools.
9. Describe various types of specialty cookware.

## Key Terms

- **hand tools**
- **smallwares**
- **measurement equivalent**
- **portion scale**
- **baker's scale**
- **digital scale**
- **measuring spoon**
- **ladle**
- **portion control scoop**
- **dry measuring cup**
- **liquid measuring cup**
- **large volume measure**
- **instant-read thermometer**
- **electronic probe thermometer**
- **infrared thermometer**

## HAND TOOLS AND SMALLWARES

*Hand tools* are handheld implements used in cutting, preparing, and serving food. Hand tools such as whisks, tongs, rubber and metal spatulas, and pastry bags all have specific purposes and are designed to make the work of preparing and serving food easy and efficient. Hand tools are "action" tools that cut, mix, blend, strain, or whip food items during preparation. *Smallwares* are items that are used to store or hold foods at some stage of the preparation process. Smallwares such as sheet pans, mixing bowls, and cutting boards are the "nonaction" pieces of small kitchen equipment. The culinary professional must practice the proper use and care of each hand tool and smallware in order to be successful in the professional kitchen.

Hand tools and smallwares have various uses and applications throughout the professional kitchen. Common types of hand tools and smallwares include portioning and measuring equipment; cookware and bakeware; strainers, sifters, and sieves; turning and grabbing tools; mixing and blending tools; cutting tools; pastry tools; and specialty cookware. ***See Figure 5-1.***

## The Professional Kitchen

Turning and grabbing tools

Strainers and sieves

Hotel pans

**5-1** Hand tools and smallwares are located throughout the professional kitchen.

## PORTIONING AND MEASURING EQUIPMENT

Portioning and measuring equipment in the professional kitchen includes equipment for measuring weight, volume, temperature, and time. Knowing the correct piece of equipment to use to accurately measure something requires some practice. Whether measuring ingredients for a recipe in given weight, volume, or serving size, a culinary professional must understand common weights and measures.

Most dry recipe ingredients are given in weight. Common weight measurements include the ounce (oz) and the pound (lb). There are 16 oz in a pound. Some recipe ingredients may be given in volume. Common volume measurements are the teaspoon (tsp), tablespoon (tbsp), cup (c), fluid ounce (fl oz), pint (pt), quart (qt), and gallon (gal.). Liquids are commonly given in volume measure because it is faster and easier to measure. When measuring dry ingredients by volume, consideration must be made for how firmly the ingredients are packed into the measuring device or whether the ingredients should be leveled or slightly heaped. These are not concerns when weighing an ingredient. Many pastry chefs will only use weight measurements even for liquids because weight is a more precise method of measurement. Measurement quantities are usually abbreviated in the recipes. ***See Figure 5-2.***

A *measurement equivalent* is the amount of one form of measure that is equal to another form of measure. ***See Figure 5-3.*** Measurement equivalents can be used to convert from one form of measure to another. For example, cups can be converted to tablespoons using the measurement equivalent 1 cup = 16 tbsp. If the recipe calls for ½ cup of water, multiply the measurement equivalent (16) by the specified measure (½); $16 \times \frac{1}{2} = 8$. A ½ cup of water is equal to 8 tbsp. Occasionally a recipe may call for a "pinch" of some ingredient. This is roughly equal to ⅛ tsp.

### Measurement Abbreviations

| | |
|---|---|
| tsp | teaspoon |
| tbsp | tablespoon |
| c | cup |
| pt | pint |
| qt | quart |
| gal. | gallon |
| oz | ounce |
| fl oz | fluid ounce |
| lb | pound |

**5-2** *Measurements are commonly abbreviated in recipes.*

### Measurement Equivalents

| Measure | Equivalent | Fluid Weight |
|---|---|---|
| 1 dash | ½ pinch | — |
| 1 pinch | approx. ⅛ tsp | — |
| 1 tsp | 60 drops or ⅓ tbsp | — |
| 1 tbsp | 3 tsp or 1⁄16 cup | ½ fl oz |
| 1 jigger | 3 tbsp | 1½ fl oz |
| 1 cup | 16 tbsp or ½ pt | 8 fl oz |
| 1 pt | 2 cups or ½ qt | 16 fl oz or 1 lb |
| 1 qt | 2 pt or ¼ gal. | 32 fl oz or 2 lb |
| 1 gal. | 4 qt or 8 pt | 128 fl oz |

**5-3** *A measurement equivalent is the amount of one form of measure equal to another form of measure.*

## Kitchen Scales

**Portion Scale**

**Baker's Scale**

**Digital Scale**

*Edlund Co.*

**5-4** *The most common types of scales include mechanical portion scales, baker's scales, and digital scales.*

### Weight-Measuring Equipment

Weight, or the heaviness of a substance, is measured in the professional kitchen using various types of scales. The most common scales include portion scales, baker's scales, and digital scales. When using a scale, always check that the scale is set to zero weight before adding any ingredients. When using a container to hold food that is being weighed, remember to place the empty container on the scale first. Then set the scale to zero. Remove the container and fill it with the food. Finally, place the filled container on the scale to obtain the actual weight of the food. Not setting the scale to zero when using a container results in inaccurate measurements.

A *portion scale* (spring scale) is a scale with a spring-loaded platform and a mechanical dial display. Mechanical portion scales are used for measuring food servings and for pre-portioning food. The most common type of portion scale has a large dial on the front of the scale which is graduated from ¼ oz to 32 oz. Scales can also be calibrated in grams, ounces, or pounds. The rotating needle on the dial indicates the weight of the item placed on the platform. This type of scale is used when exact serving portions are required, such as portioning deli meats for a sandwich. Portion scales are also available with a digital readout for more accuracy. **See Figure 5-4.**

A *baker's scale* (balance scale) is a scale with two platforms that use a counterbalance system to measure weight. The food to be weighed is placed on one platform, and a weight is placed on the other platform. Weight is added to or removed from the second platform until the scale balances. The beam between the two platforms has a smaller weight that is used for fine adjustment of the scale. A baker's scale is used to measure most baking ingredients. To weigh 8 oz of egg whites on a baker's scale, a container large enough to hold the egg whites is placed on the first platform of the scale. The scale is balanced by placing the appropriate amount of weight on the second platform. An additional 8 oz of weight is added to the second platform. Then, egg whites are added to the container on the first platform until the scale balances again. With a balanced scale, the correct amount (8 oz) has been measured and added to the container. A baker's scale ensures accuracy in measuring the proper amount of ingredients. Because it does not have a spring-loaded platform with a maximum weight load, the baker's scale can handle much heavier weights than spring scales. The baker's scale can be used to weigh up to 10 lb where most portion scales can handle only up to 2 lb.

A *digital scale* is a scale with a spring-loaded steel platform and a digital display. Digital scales are operated similarly to mechanical portion scales. A digital scale can measure quantities of as little as ⅟₁₀ oz, while a mechanical portion scale and baker's scale cannot. Digital scales are commonly used at salad bars or produce markets where price is determined by weight.

## Volume-Measuring Equipment

In the professional kitchen, volume is measured with various tools that are sized to contain a specific volume. Examples of volume-measuring tools include measuring spoons, ladles, portion control scoops, dry and liquid measuring cups, and large volume measures. Can sizes are another way to indicate the volume of a food item. Some measuring tools are designed to hold a single specific volume. Other measuring tools use indicators that divide the measure into smaller volumes.

A *measuring spoon* is a spoon used to measure a small volume of ingredients. Measuring spoons usually come in a set of four spoons with volume measurements of ¼ tsp, ½ tsp, 1 tsp, and 1 tbsp. Some sets are also stamped with the metric equivalents of these units. Professional sets are made of stainless steel to avoid reacting with foods and to maintain their original shape. A *ladle* is a stainless steel cup-like bowl attached to a long handle and used to stir or serve soups, stocks, dressings, and sauces. It is especially handy when portion control of these items is desired. The long handle of the ladle allows the cup to reach the bottom of large pots. Ladles are available in many sizes, and usually the capacity is stamped on the handle in ounces or milliliters for easy reference. Ladle sizes range from ½ oz to 32 oz. ***See Figure 5-5.***

*Carlisle FoodService Products*

A *portion control scoop* (disher) is a stainless-steel scoop of a specific size with a thumb-operated release lever used to serve food in accurate amounts. Scoops are sized by numbers typically ranging from 6 to 40. The numbers indicate the number of level scoopfuls equal to 1 qt. As the number of the scoop increases, scoop capacity decreases. Each size of scoop has an approximate capacity in ounces as well as an equivalent volume in cups or tbsp. Scoops can be used for portioning batters, meat patties, potatoes, rice, bread dressing, croquette mixtures, bound salads, and, of course, ice cream.

A *dry measuring cup* is a metal cup used to measure dry ingredients. A set of dry measuring cups typically consists of ¼ cup, ⅓ cup, ½ cup, and 1 cup measures. A dry measuring cup does not have a pour lip, as the top edge of the cup is the actual measurement. The cup must be completely filled to the brim and leveled off for the contents to equal the measure of the cup.

A *liquid measuring cup* is a plastic or glass cup used to measure volumes of liquid ingredients. Liquid measuring cups are sold in capacities of 1 cup to 1 gal. These containers are graduated in 1 oz or ½ oz increments as well as in milliliters. They are constructed with a pour lip to pour liquid contents without spilling.

## Measuring and Portioning Tools

### Ladle Capacity

| Ladle Size | Approximate Weight of Portion |
|---|---|
| 1 tbsp | ½ oz |
| ⅛ cup | 1 oz |
| ¼ cup | 2 oz |
| ⅜ cup | 3 oz |
| ½ cup | 4 oz |
| ¾ cup | 6 oz |
| 1 cup | 8 oz |
| 1½ cup | 12 oz |
| 3 cup | 24 oz |
| 4 cup | 32 oz |

*Carlisle FoodService Products*

**Measuring Spoons**

*The Vollrath Company, LLC*

**Dry Measuring Cups**

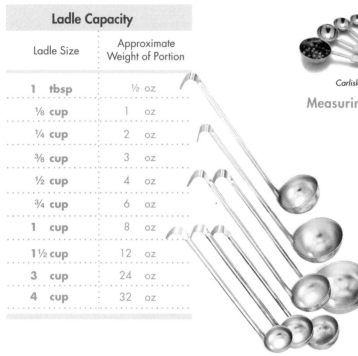

*The Vollrath Company, LLC*

**Ladles**

*Carlisle FoodService Products*

**Liquid Measuring Cups**

### Scoop Capacity

| Scoop No. | Weight | Measure |
|---|---|---|
| 6 | 4¾ oz | ⅔ cup |
| 8 | 4½ oz | ½ cup |
| 10 | 3⅓ oz | ⅖ cup |
| 12 | 2⅞ oz | ⅓ cup |
| 16 | 2½ oz | ¼ cup |
| 20 | 1⅞ oz | 3⅕ tbsp |
| 24 | 1½ oz | 2⅔ tbsp |
| 30 | 1⅛ oz | 2⅕ tbsp |
| 40 | ⅞ oz | 1⅗ tbsp |

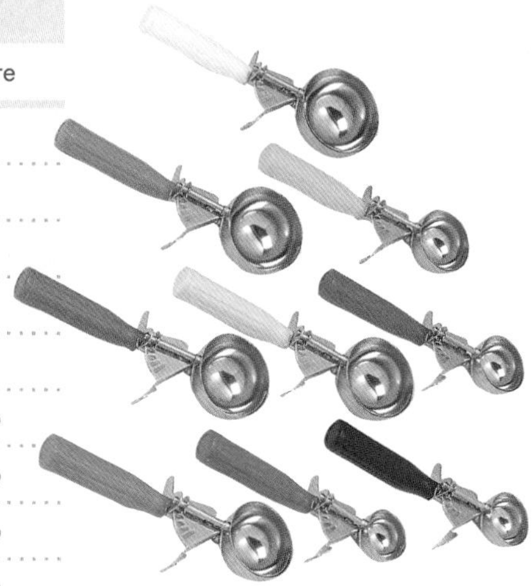

*Carlisle FoodService Products*

**Portion Control Scoops**

*The Vollrath Company, LLC*

**Large Volume Measures**

**5-5** *Volume-measuring tools include measuring spoons, ladles, portion control scoops, dry and liquid measuring cups, and large volume measures.*

A *large volume measure* is a large, graduated aluminum container used to measure volume. Large volume measures are most commonly made of aluminum and are available in 1 gal., ½ gal., 1 qt, and 1 pt capacities as well as metric equivalents. These containers are usually graduated in quarters. For example, each 1 gal. measure is divided into 4 qt measurements. They will usually have pour lips for pouring liquids without spilling.

Some recipes may call for food ingredients by can size. Can sizes are standardized by can manufacturers. To convert a can size to a standard weight or measure, the equivalent weight or measure must be known. **_See Figure 5-6._** The most common can size in commercial cooking is the No. 10 (or #10) can, which holds approximately 3 qt, or between 6 lb 8 oz and 7 lb 5 oz.

InterMetro Industries Corporation

| Standard Can Capacity | | | |
|---|---|---|---|
| Can Size | Average Weight | Average Volume | Average Servings |
| **#1 picnic** | 10½ oz | 1¼ | 2 to 3 |
| **#211** | 12 oz | 1½ | 3 to 4 |
| **#300** | 15 oz | 1¾ | 3 to 4 |
| **#1 tall (or 303)** | 1 lb | 2 | 4 |
| **#2** | 1 lb 4 oz | 2½ | 5 |
| **#2½** | 1 lb 12 oz | 3½ | 7 |
| **#3** | 2 lb 2 oz | 4¼ | 8 |
| **#3 cylinder** | 2 lb 14 oz | 5¾ | 10 to 12 |
| **#5** | 3 lb 8 oz | 7 | 14 |
| **#10** | 6 lb 8 oz | 13 | 25 |

* in cups

 **5-6** *To convert can size to a standard weight or measure, the equivalent weight or measure must be known.*

## Temperature-Measuring Equipment

For sanitation and food safety, it is important that food cooked in the professional kitchen is stored at and cooked to required temperatures. Thermometers are used to measure temperatures of cooked and stored foods. Several types of thermometers are used in the professional

If you turn the nut on the back of the readout head on an instant-read thermometer, you can recalibrate or adjust the accuracy of the thermometer. There are two instances where exact temperature is known without a thermometer: the freezing point of water (32°F) and the boiling point of water (212°F). If you fill a cup to the top with shaved ice and add cold water to fill in any space between the shavings, and then insert a stem thermometer, the thermometer should display 32°F. If the stem thermometer shows a different temperature, turn the nut on the back of the readout head until the display indicates 32°F. Next, if you bring a small saucepot of water to a boil and insert the stem thermometer into the water, the thermometer should display 212°F. If it does not display 212°F, turn the nut again until the correct temperature is displayed. Now you have correctly recalibrated your thermometer.

*Cooper-Atkins Corporation*

kitchen, but the most common are instant-read thermometers, candy and fat thermometers, electronic probe thermometers, and infrared thermometers.

An *instant-read thermometer* (stem thermometer) is a thermometer with a long stainless steel stem attached to either a digital or mechanical display. The instant-read thermometer is small enough to carry in a pocket and has a plastic sleeve case with a pocket clip. The stem of the thermometer is inserted into foods during cooking to determine the internal temperature. However, this thermometer is not intended to be left in food as it cooks because doing so could result in damage to the thermometer. A stainless steel adjustable nut attached to the back of the readout head can be turned to adjust and recalibrate the thermometer if needed. The thermometer should be checked for accuracy on a regular basis. The accuracy can be checked by inserting the stem into a glass of shaved ice and calibrating the display to read 32°F (the freezing point of water) or by placing the stem into a pot of boiling water and calibrating the display to read 212°F (the boiling point of water). Calibrating to 32°F is the preferred method. The accuracy of an instant-read thermometer should be checked if it is dropped. The stem should be sanitized each time the thermometer is used to prevent cross-contamination. ***See Figure 5-7.***

*Candy/deep-fry thermometers* are thermometers used to measure temperatures of hot substances as they are cooking. A candy/deep-fry thermometer has a long stainless steel stem similar to that of an instant-read thermometer, but it has a much larger display and can withstand much higher temperatures. Candy/deep-fry thermometers have a clip for attaching the thermometer to the side of a pot. The clip allows the thermometer to be left in the pot during cooking time without being held. When removing a candy/deep-fry thermometer from a hot substance, care must be taken to avoid placing the hot thermometer immediately into something cold, as the change in temperature may damage the thermometer. This sudden change in temperature from hot to cold is called thermal shock.

An *electronic probe thermometer* (thermocouple thermometer) has a thin stainless steel stem attached by wires to a battery-operated readout device. The stem is placed into a food item, and the internal temperature of the item is displayed on the handheld readout. The stem of an electronic probe thermometer is much thinner than a traditional stem thermometer. Electronic probe thermometers give accurate and immediate temperature readings in both Fahrenheit and Celsius. Although they have excellent accuracy, they are not as small and compact as a stem thermometer.

# Thermometers

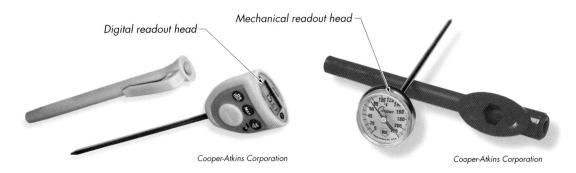

Digital readout head

Mechanical readout head

Cooper-Atkins Corporation

Cooper-Atkins Corporation

**Instant-Read Thermometers**

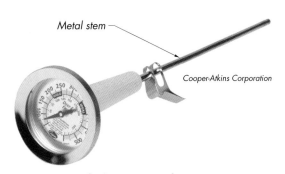

Metal stem

Cooper-Atkins Corporation

**Candy/Deep-Fry Thermometer**

Infrared thermometers measure surface temperature

Cold foods on a salad bar must be kept at 41° or lower

Cooper-Atkins Corporation

Fluke Corporation

**Electronic Probe Thermometer**

**Infrared Thermometer**

**5-7** *Types of thermometers used in the professional kitchen include instant-read (stem), candy/deep-fry, electronic probe, and infrared.*

## Kitchen Timers

**Hanging Timer**

**Countertop Timer**

**Digital Pocket Timer**

*Browne-Halco (NJ)*

**5-8** *A kitchen timer will sound an alarm when a specified period of time has passed.*

An *infrared thermometer* is a thermometer that measures the surface temperature of an item through the use of infrared laser technology. Infrared thermometers are noncontact thermometers, meaning that they do not physically touch the items they are measuring. Rather, the infrared laser is pointed at a food item and the external temperature is taken immediately. The digital display on the unit will indicate the measured temperature. These thermometers are useful for checking temperatures on salad bars and steam tables. They are also used to check the temperature of items as they are delivered to the receiving area of a professional kitchen.

### Time-Measuring Equipment

A *kitchen timer* is a time-measuring tool that indicates the amount of time that has passed, or sounds an alarm when a specified time period has ended. Timers come in a variety of styles including large hanging units, countertop timers, and small digital pocket timers. ***See Figure 5-8.*** Timers allow a chef to work on multiple projects at once. A timer can be set to ring at a critical time in the preparation of a dish, such as when an item needs to be removed from the oven or when a particular ingredient needs to be added. While waiting for the specific time period to pass, the chef is free to work on other tasks.

## PROFESSIONAL COOKWARE AND BAKEWARE

Professional cookware and bakeware are specially constructed to be more durable than cookware and bakeware designed for home use. This added durability is needed to withstand the larger burners on commercial ranges and stoves and the heavy use that occurs in a professional kitchen. Professional cookware is designed to withstand intense heat and flame while protecting food from burning. Cooking and baking equipment are sized to accommodate large quantities of food. Common types of cookware found in the professional kitchen include warming and steaming cookware, oven cookware, and stovetop cookware including various pots and pans.

### Warming and Steaming Cookware

Many types of cookware are used in the professional kitchen to warm and steam food. Warming and steaming cookware includes multipurpose cookware that may be used for storing prepared foods prior to or following cooking. Common types of cookware used for warming and steaming include bain-maries, double boilers, steamers, and hotel pans.

A *bain-marie* is a round stainless steel food storage container with high walls used for holding sauces or soups in a hot or cold water bath or steam table. As discussed earlier, the term bain-marie refers not only to the hot water bath used to gently heat food, but also to the container used to hold or store the food in a hot water bath or on a cold food table such as a salad bar. The round containers are also commonly called bain-marie inserts. Bain-marie inserts have many applications in the professional kitchen such as holding soup in a soup-warming unit, holding salad dressings on a salad bar, and holding hot fudge on a sundae bar. Bain-maries are available in many sizes from 1¼ qt to 11 qt. *See Figure 5-9.*

A *double boiler* is a round stainless steel pot that sits on top of another pot containing simmering water. The simmering water gently heats foods that are placed in the top pot. Double boilers are sometimes called bain-maries because they have an insert which holds food above a hot water bath. However, a double boiler is different from a bain-marie in that a double-boiler is made up of two vessels that "nest" inside one another, each with a long handle. The bottom vessel is filled with water and heated to either hold or cook the food that is placed in the top vessel. A double boiler is often used to avoid scorching the food being heated, as the heat transferred through water is more gentle than the intense heat of an open flame. The open flame of a burner can create such intense heat that it evaporates the moisture in the food being heated, sometimes drying it out. A double boiler will heat foods only as hot as the temperature of the water and steam beneath the upper vessel, or approximately 212°F. An open flame, even on a low setting, can reach temperatures over 400°F. A double boiler is commonly used for melting chocolate and making hollandaise sauce. *See Figure 5-10.*

A *tiered steamer* is a round stainless steel pot with a perforated steamer insert. A tiered steamer is similar to a double boiler except that the steamer insert is perforated. The perforations allow steam from the simmering or boiling water below to rise into the insert. The steam cooks food contained in the steamer insert. *See Figure 5-11.*

A *hotel pan* is a stainless steel pan used to cook, serve, or hold food. It is important to note that hotel pans should only be used to cook foods in an oven, convection steamer, or pressure steamer. They should never be placed directly on an open-burner range. Hotel pans have a lip around the outer edge to support them above the hot water of a steam table or chafing dish. They come in a variety of shapes and sizes to allow many different arrangements in a steam table such as half pans, quarter pans, and third pans. *See Figure 5-12.* Foods can be stored in a hotel pan in the refrigerator or freezer. Hotel pans are available in perforated versions for steaming and draining foods. Hotel pans are also made in a heavy plastic material that can either be used in a steam table or in a refrigerated cold table such as a salad bar. While these heavy plastic pans are not nearly as durable as stainless steel hotel pans, they are much less expensive.

Bain-Maries

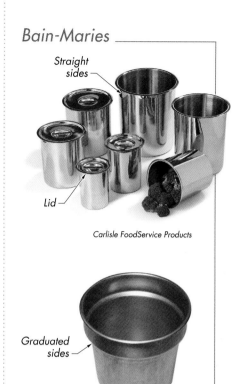

Straight sides

Lid

*Carlisle FoodService Products*

Graduated sides

5-9    A bain-marie is a round stainless steel food storage container with high walls used for holding foods in a hot water bath, steam table, or cold storage unit.

Double Boilers

The top pot of a double boiler is a tiered insert for holding food items

The bottom pot holds simmering water

*Carlisle FoodService Products*

5-10    Double boilers consist of a lower pot nested inside an upper pot.

## Steamers

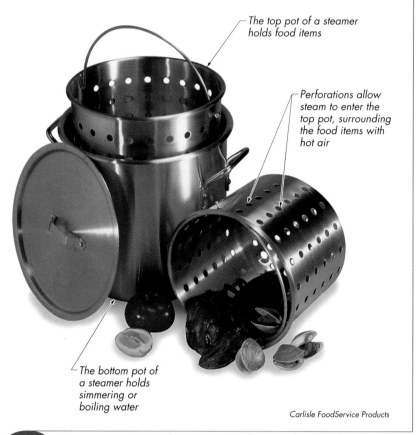

The top pot of a steamer holds food items

Perforations allow steam to enter the top pot, surrounding the food items with hot air

The bottom pot of a steamer holds simmering or boiling water

Carlisle FoodService Products

**5-11** *A steamer consists of a round stainless steel pot with a perforated steamer insert.*

### Oven Cookware

Various types of cookware are used for roasting meats and vegetables in commercial ovens. The most common types include sheet pans, bake pans, and strapped roasting pans. In addition, baking mats provide nonstick benefits and easy cleanup when used with professional oven cookware.

A *sheet pan* is a flat pan with very low sides. Sheet pans come in either full (17¾″ × 25″ × ¾″) or half pan (17¾″ × 12 ⅞″ × ¾″) sizes. They are used for cooking meats such as bacon, sausage links, or roasted chicken in the oven. They are also used for baking items such as cookies, sheet cakes, or rolls. Sheet pans can be purchased with a nonstick surface. *See Figure 5-13.*

A *bake pan* is a rectangular pan with 2″–tall sides and loop handles. Bake pans are the same length and width as a sheet pan. Bake pans are used for roasting or baking items such as vegetables, apples, pasta dishes such as lasagna, casseroles, and some meats.

**Chef's Tip:**
Shallow pans are best for fast cooling, cooking, and flash-freezing. A deep pan takes much longer to safely cool or completely heat foods. Shallow pans help food service professionals follow proper HACCP procedures because they offer less opportunity for food to remain in the temperature danger zone.

## Hotel Pans

| Hotel Pan Capacity | | |
|---|---|---|
| Pan Size | Depth* | Capacity† |
| **Full** | 2½ | 8 |
| | 4 | 13 |
| | 6 | 20 |
| **⅔** | 2½ | 5½ |
| | 4 | 6½ |
| | 6 | 10 |
| **½** | 2½ | 3½ |
| | 4 | 5½ |
| | 6 | 8 |
| **½ long** | 2½ | 3½ |
| | 4 | 5½ |
| | 6 | 8 |
| **⅓** | 2½ | 2½ |
| | 4 | 4 |
| | 6 | 6 |
| **¼** | 2½ | 2 |
| | 4 | 3 |
| | 6 | 4½ |
| **⅙** | 2½ | 1 |
| | 4 | 2 |
| | 6 | 2½ |
| **⅑** | 2½ | ⅝ |
| | 4 | 1⅛ |

* in inches
† in quarts

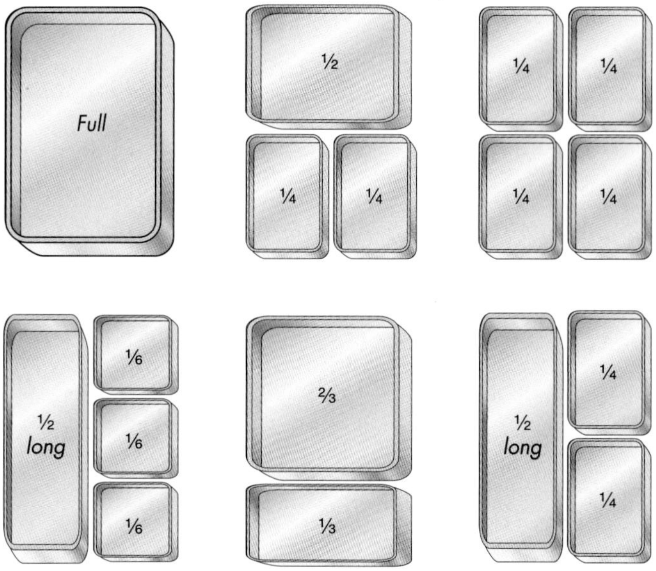

The Vollrath Company, LLC

5-12 Hotel pans are used to cook, serve, and store food, are available in a variety of sizes, and can be used in various configurations in steam tables.

## Oven Pans

*Carlisle FoodService Products*

**Sheet Pans**

*Lincoln Foodservice Products, Inc.*

**Roasting Pans**

*Carlisle FoodService Products*

**Bake Pans**

**5-13** *Sheet pans, bake pans, and roasting pans are common types of oven cookware.*

A *roasting pan* is a rectangular pan with deep (4″ to 5″) sides. Roasting pans can be constructed from stainless steel or aluminum. A roasting pan is similar to a sheet pan in length and width. Strapped roasting pans come with reinforced straps for added strength and durability and are used to roast medium-size to large pieces of meat or poultry. Roasting pans can be purchased with or without covers and are available in various sizes.

Food-grade silicone is a material that has recently become available in the professional kitchen. A *silicone baking mat* is a woven silicone nonstick mat. ***See Figure 5-14.*** These mats do not need greasing as food does not stick to them. They can be used in the freezer, refrigerator, or oven and can withstand temperatures from –40°F to 580°F. Silicone is proving to be such a popular item that various types of baking pans, such as muffin pans and cake pans, are now available in silicone versions. Silicone oven mitts and cooking utensils are also available.

## Stovetop Cookware

Stovetop cookware includes various types of pots and pans used to prepare foods on stovetop ranges. These pots and pans vary in size and shape, to best suit a particular purpose or application. Some pans, such as the saucepan, are very versatile and have many uses and applications. Others, such as the crepe pan, are designed for specific uses.

A *saucepan* is a small, slightly shallow pan with straight or slightly sloped sides. Saucepans are used when a food item is to be cooked in some amount of liquid. Saucepans are commonly used for preparing small amounts of sauce, for shallow poaching, and for reheating small amounts of liquid foods such as soup or marinara sauce. The shallow depth and wide surface area of a saucepan help reduce the risk of scorching thicker liquids such as tomato. The shallow depth and wide surface area also make it easy to retrieve poached items from the pan without breaking or damaging the item. ***See Figure 5-15.***

*Sauté pan* is a general term used to describe round, shallow-walled pans used for sauté applications. This type of pan is also known as a skillet. Common types of sauté pans include sauteuse and sautoir pans. The more commonly recognizable sauté pan is the sauteuse. The *sauteuse* is a round, sloped, shallow-walled pan with a long handle. The top diameter of the sauteuse ranges from 6″ to 16″. A sauteuse is primarily used for sautéing meats and vegetables. The sloped walls of the sauteuse enable the chef to flip foods in the pan without the use of a turner or spatula. An *omelet pan* is the smallest of the sauteuse skillets and often comes with a nonstick finish. A *sautoir* is a round, shallow pan with straight sides. It is also used to sauté meats and vegetables.

A *crepe pan* is a small skillet with very short sloped sides. Crepe pans are usually made from rolled (blue) steel. This type of steel is thinner than that used in other commercial skillets, and it heats very quickly. The fast heating enables the crepe to cook quickly without sticking to the pan. As the name implies, crepe pans are only used to prepare crepes.

A *cast iron skillet* is a shallow-walled pan made of thick, heavy iron to withstand extreme high heat. Cast iron skillets are used for pan-broiling and pan-frying items such as chicken, pork chops, and veal cutlets. These pans have long handles and are available in diameters ranging from 6½″ to 15¼″.

Silicone Baking Mats

5-14  *Silicone baking mats provide nonstick surfaces that can withstand temperatures of 580°F.*

## Stovetop Pans

Saucepans
*Carlisle FoodService Products*

Sauté Pans
Sauteuses
*Carlisle FoodService Products*

Sautoir
*Carlisle FoodService Products*

Crepe Pan
*The Vollrath Company, LLC*

Cast Iron Skillet
*Lodge Manufacturing, Tennessee, U.S.A.*

French Grill
*Lodge Manufacturing, Tennessee, U.S.A.*

**5-15** *Stovetop pans are shallow, wide pieces of cookware used on stovetop ranges.*

A *French grill* is a cast iron plate with raised ridges on one side and a completely flat cooking surface on the other side. It is used to turn an open burner on a range into a grilling surface. The ridged side is used to create crosshatch markings on food. French grills are commonly used to grill fragile foods, such as a delicate fish fillet or a polenta cake, that may fall apart if cooked on a large commercial grill. Some restaurants that do not have a grill use a French grill to mark an item with crosshatch marks before finishing the item in an oven.

A *stockpot* is a large, round, high-walled pot that is taller than it is wide. Like all pots, it is meant for range or stovetop cooking. It has loop-style handles on the sides for easy lifting. A stockpot is used for boiling and simmering items such as bones for soup, stock, and other foods. The tall, narrow shape of the stockpot helps reduce excessive evaporation by leaving a smaller surface area where liquid is exposed to surrounding air. A smaller surface area results in less surface evaporation during cooking. The stockpot can also be fitted with a spigot-style drain at the base to drain off liquids without lifting the large, heavy pot. Stockpots can be made of aluminum or stainless steel. Sizes of stockpots range from 2½ gal. to 40 gal. ***See Figure 5-16.***

## Pots

Stockpots
*Carlisle FoodService Products*

Rondeaus
*Carlisle FoodService Products*

Saucepots
*Carlisle FoodService Products*

Fish Poacher
*World Cuisine, Inc.*

**5-16** *Pots are deep narrow pieces of cookware used on stovetop ranges.*

A *rondeau* is a wide, shallow-walled, round pot used for braising, stewing, and searing meats. It has a heavy metal base, allowing for longer cooking times. Rondeaus are available in sizes from 15 qt to 28 qt and have side loop handles similar to a stockpot. A rondeau is also often called a brazier, although in classical cuisine a brazier was usually square. Today, the two names are used interchangeably.

A *fish poacher* is a long, thin style of pot with loop handles that is specifically designed to poach fish. Its shape is designed to accommodate the body of a fish. The fish poacher has a mesh screen that sits in the bottom of the pan, used to lift the poached fish from the pan without breaking it.

A *saucepot* is a small form of a stockpot. Like the stockpot, it also has loop handles for easy lifting and is used for cooking on the top of the range when stirring, simmering, or boiling is necessary. Saucepots are commonly used to prepare liquid foods such as soups and sauces when the quantity desired is not enough to fill a stockpot. Typically, a stockpot is used for items with a thin consistency such as broth

The first cookware consisted of carved stones that were filled with food and heated on a fire. Eventually clay pottery was developed. This solid but fragile cookware was often coated with plant gum to seal the porous material, resulting in waterproof cookware known as earthenware. Earthenware was commonly suspended over an open flame, rather than placed directly on it, as it was too fragile to handle the intense heat of direct contact with an open flame. Earthenware remains a preferred type of cookware in some parts of the world because of its low production cost.

soups and stocks, whereas a saucepot might be used for thicker items such as chili or cream soups. The shallower depth of a saucepot makes it easier to stir the pot all the way to the bottom, reducing the risk of scorching.

### Bakeware

Bakeware includes metal pans and tins used for baking various breads and pastries, including cakes, pies, and cookies. Professional bakeware is typically made from stainless steel or aluminum. Some professional bakeware is available with a nonstick coating. Silicone bakeware is also becoming available as the material becomes more popular for its non-stick and temperature-resistant qualities. Common types of bakeware include loaf pans, cake pans, pie tins, tart pans, and muffin pans.

A *loaf pan* is a short, deep, rectangular pan typically used for baking loaves of bread. A loaf pan is used to bake foods where a rectangular shape is desired, such as a meatloaf, bread loaf, or pound cake. *See Figure 5-17.*

A *cake pan* is a baking pan used for making cakes and similar baked goods. Cake pans come in almost any shape (round, square, heart-shaped), with straight sides and a medium depth. A typical cake pan set will contain a variety of sizes of the same shape of pan. Cakes can be made using one size of pan and stacking the layers for a taller appearance, as with a double- or triple-layer cake. Tiered cakes are made by using pans of varying sizes and stacking the layers from largest on the bottom to smallest on the top, as with a traditional wedding cake.

Types of cake pans include springform pans and tube pans. A *springform pan* is a round pan with a flat bottom and detachable sides. The shape of a springform pan is similar to a round layer cake pan, except that the round sides can open and close with a clasp-style clip. Springform pans are used to bake thick, wet mixes such as cheesecakes. A *tube pan* is a round baking pan with a second round tube in the center, resulting in a ring-shaped area to hold batter. Tube pans have removable bottoms to help remove the product after baking. They are commonly used to make dessert items such as angel food cake.

A *pie pan* is a round, somewhat shallow pan with sloped sides used for baking pies or pie crusts. Pie pans usually have a depth between 1½″ and 2″. Pie pans are commonly used to make fruit or custard-filled pies. A *tart pan* is a round, shallow baking pan, often equipped with a removable bottom. A tart pan usually has a depth of about 1″. The sides of a tart pan are slightly sloped and can be fluted or smooth. Tart pans are used to bake sweet or savory pastry tarts with a very delicate crust. The removable bottom makes it much easier to remove the tart from the tart pan without breaking the delicate crust.

## Bakeware

*Browne-Halco (NJ)*

**Springform Pan**

*Lincoln Foodservice Products, Inc.*

**Pie Pan**

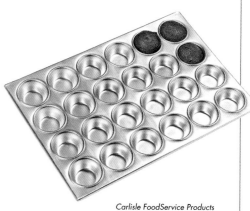

*Carlisle FoodService Products*

**Muffin Pan**

*Carlisle FoodService Products*

**Loaf Pans**

**Cake Pan**

*Lincoln Foodservice Products, Inc.*

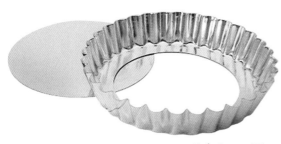

*Matfer Bourgeat USA*

**Tart Pan**

*American METALCRAFT, Inc.*

**Tube Pans**

**5-17** *Bakeware includes various shapes and sizes of pans used for baking breads, cakes, and pastries.*

A *muffin pan* is a large rectangular pan with numerous round inserts. Muffin pans are used to bake muffins and cupcakes. The diameter and depth of the round insets can vary, and are available in sizes from miniature muffin size to jumbo size.

## STRAINERS, SIFTERS, AND SIEVES

Strainers, sifters, and sieves are kitchen tools used to separate various items during the cooking process. Often, they are used to remove solids from liquids or remove large objects from a purée. Strainers are typically used to separate solids from liquids. Sifters are used to aerate and remove any large impurities from powdered or dry ingredients. Sieves are used to purée soft foods while removing any large pieces. Strainers, sifters, and sieves can be used interchangeably in some applications.

In its most basic form, a *strainer* is a bowl-shaped woven mesh screen tool, often with a handle. It is used to strain and drain foods. For example, a strainer may be used to hold grapes under a faucet for washing or to catch small vegetables cooked in broth as the broth is poured out. Forms of strainers include skimmers, spiders, colanders, china caps, chinois, and cheesecloth. ***See Figure 5-18.***

A *skimmer* is a flat, stainless steel perforated disk connected to a long handle, used to skim impurities or food from soups, stocks, and sauces. Skimmers can also be made with a thin wire mesh stretched over a slightly curved frame. A spider is similar in shape to a skimmer, but the open wire design of the spider makes it perfect for removing items from hot fat such as in a deep-fat fryer.

A *colander* is a bowl-shaped perforated metal strainer usually made from stainless steel or aluminum. The bowl of the colander is situated on top of a metal ring base that keeps the colander bowl raised above the surface of the counter or sink. Colanders are commonly used for draining and rinsing foods such as cooked pasta. They are also used for washing items such as fruits and vegetables.

A *china cap* is a perforated cone-shaped metal strainer used to strain gravies, soups, stocks, sauces, and other liquids. A china cap can also be used to purée soft foods by forcing food through the strainer with a pestle (a bat-shaped wooden food plunger). China caps have a clip on the outside rim so they can be hooked onto a pot or bain-marie while straining.

A *chinois* is a china cap that strains liquids through a fine-mesh screen. The fine mesh of a chinois is used to produce extremely smooth sauces or soups. A chinois is often called a bouillon strainer because the tightness of the mesh weave makes it ideal for straining fine sauces and consommé. Particular care should be taken with a chinois as it is fragile compared to most commercial smallwares. If mishandled, the fine mesh screen can become smashed or torn.

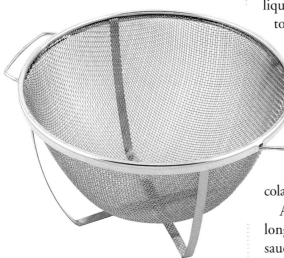

Browne-Halco (NJ)

## Strainers, Sifters, and Sieves

Woven mesh basket

Open wire

Thin wire mesh

Browne-Halco (NJ)

Carlisle FoodService Products

Browne-Halco (NJ)

**Strainer**

**Skimmers**

**Spider**

Perforated metal basket

Perforated metal cone

Carlisle FoodService Products

American METALCRAFT, Inc.

Lincoln Foodservice Products, Inc.

**Colanders**

**China Cap**

**Chinois**

Browne-Halco (NJ)

Browne-Halco (NJ)

All Clad Metalcrafters

Browne-Halco (NJ)

**Sifter**

**Tamis**

**Food Mill**

**Ricer**

5-18 *Strainers, sifters, and sieves are used to separate various items in the cooking process.*

## Cheesecloth

*Cheesecloth is used to strain stocks and fine sauces.*

### Action Lab

Observe how using a sifter can make a big impact on a recipe. Take out a dry cup measure. Measure out four level cups of flour. Next, place the measured flour in a sifter. Sift the flour and then measure the flour again using the same dry cup measure. Note how the volume of flour has increased, but the weight remains the same.

*Cheesecloth* is loosely woven cotton gauze used to strain stocks and fine sauces. Cheesecloth should be rinsed in cold water to remove any loose fibers or strings that could fall into the item being strained. It is often used in place of a chinois. Cheesecloth is used to hold a bouquet garni. **See Figure 5-19.**

A *sifter* is a cylindrical metal container with a hand crank and a woven metal screen stretched across the bottom that is used to aerate and remove lumps from dry ingredients. A crank on the side of the sifter turns a wire paddle inside to work the material being sifted through the screen. Sifting removes lumps from powdered or dry goods such as cocoa, sugar, and flour so that there are no lumps in the final product. A sifter also is used to make products light by incorporating air into the mixture.

A *sieve* is a fine-mesh or perforated strainer used to purée or sift or to remove liquid from a food. A *tamis* (drum sieve) is a flat, round sieve with a wood or aluminum frame and a mesh screen bottom. The tamis is used to sift lumps from dry ingredients and to purée soft foods such as meats or cooked produce. To purée, a rubber spatula is used to force foods through the wire mesh bottom of the tamis.

A *food mill* is a hand-cranked sieve with a bowl-shaped body and is used to purée soft or cooked foods. Food is placed in the body of the food mill. A hand-operated crank controls a blade located at the bottom of the bowl. Turning the crank rotates the blade, which in turn cuts the food into smaller and smaller pieces. Gravity forces the smallest pieces of food through a perforated disk. Disks of various gauges are available to produce very coarse to very fine purées, depending on the size of holes on the disk.

A *ricer* is a sieve with a plunger used to purée food by pushing the food through a perforated metal plate. Soft food, such as cooked potatoes, is placed in a well between two handles. The handles are squeezed together, forcing the food through the perforated plate.

## TURNING AND GRABBING TOOLS

When preparing food in the professional kitchen, tools are often needed to turn or grab hot food items. Using the proper tool for the task makes work easy and safe, and results in a consistent finished product. Turning and grabbing tools include tongs, turners, spatulas, and peels. **See Figure 5-20.**

*Tongs* are a spring-type metal tool consisting of two grippers. Tongs are used to pick up and serve foods without touching the food with hands. Tongs do not poke holes in foods, so foods retain their shape and moisture.

An *offset spatula* (offset turner) is a tool with a wide metal blade that is bent upward and back toward a handle. Often, foods such as pancakes or hamburgers need to be turned over to cook the other side.

The design of the offset spatula keeps hands away from hot cooking surfaces. The offset spatula is used to serve foods or to turn foods over while cooking. It is used to turn foods on a griddle, grill, or broiler. Offset spatulas come in solid, perforated, and slotted versions. Perforated and slotted spatulas allow foods to drain off fats or liquids. Fine-edged slotted spatulas are often used to lift and drain fragile foods such as eggs or fish fillets.

A *peel* is a long, flat, narrow piece of wood or metal shaped like a wide, thin paddle. It is used when baking pizzas or breads to place items into or remove items from the oven.

## Turning and Grabbing Tools

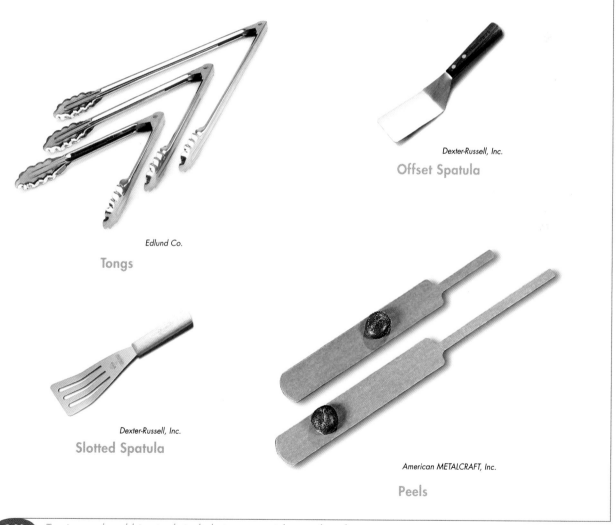

Edlund Co.

**Tongs**

Dexter-Russell, Inc.

**Offset Spatula**

Dexter-Russell, Inc.

**Slotted Spatula**

American METALCRAFT, Inc.

**Peels**

**5-20** *Turning and grabbing tools include tongs, spatulas, and peels.*

## MIXING AND BLENDING TOOLS

There are many times when ingredients for a recipe need to be combined, mixed, or blended. Tools used for mixing and blending include mixing bowls, mixing paddles, spoons, and whisks.

A *mixing bowl* is a large, smooth bowl used for mixing small or large amounts of ingredients. Mixing bowls are available in various sizes from ¾ qt to 45 qt and are made of stainless steel and aluminum. Stainless steel is preferable to aluminum because it does not react with foods that contain acid. Acidic foods chemically react with aluminum bowls, discoloring the bowls and resulting in a metallic taste. ***See Figure 5-21.***

A *mixing paddle* is a long-handled paddle used to stir foods in deep pots or steam kettles. Mixing paddles are typically made of stainless steel. The long handles enable them to reach to the bottom of deep pots or kettles.

## Mixing and Blending Tools

American METALCRAFT, Inc.

**Mixing Bowls**

Carlisle FoodService Products

**Mixing Paddles**

The Vollrath Company, LLC

**Kitchen Spoons**

Carlisle FoodService Products

**Wire Whisks**

5-21  *Mixing and blending tools include mixing bowls, mixing paddles, kitchen spoons, and wire whisks.*

A *kitchen spoon* is a large spoon made of stainless steel. Kitchen spoons are available in solid, perforated, or slotted versions. The solid spoon is used for serving food that is served in a sauce or very soft foods such as creamed spinach. Slotted and perforated spoons are used to drain liquid off the foods to be served.

A *wire whisk* is a mixing tool are made of many stainless steel wires bent into loops and sealed into a stainless steel handle. Common wire whisks are commonly available in two styles, the balloon whisk and the rigid wire whisk. The balloon whisk has very flexible wires, allowing the user to whip a great amount of air into items such as egg whites. It is sometimes called a piano wire whisk. A rigid wire whisk, often called the French whisk, uses heavier gauge wire and is longer than a balloon whisk. It is used more to stir thick substances such as mashed potatoes or a heavy batter.

## SCRAPING TOOLS

Scraping tools are used frequently in the professional kitchen to scrape batter and food from containers, mixing bowls, pots, pans, and prep areas. Scraping tools are commonly made from flexible rubber or plastic, or thin, flexible metal blades. Common scraping tools include the rubber spatula, dough cutter or bench scraper, and plastic bowl scraper.

A *rubber spatula* is a scraping tool consisting of a wide rubber blade attached to a long handle. Rubber spatulas are used to scrape food from bowls, pots, and pans. Spatulas made of rubber come in two forms, a standard (cold temperature) spatula and a silicone spatula. A standard spatula should not be used for cooking because it melts very easily. A silicone spatula, on the other hand, can withstand temperatures up to 650°F. ***See Figure 5-22.***

A *dough cutter* (bench scraper) is a flat stainless steel blade attached to a sturdy handle. It is used to cut dough into portions and scrape all dough fragments from the baker's table.

A *bowl scraper* is curved plastic scraping tool used to scrape food items from curved surfaces. The bowl scraper is handheld and used to completely clean a mix or dough from a bowl. Its flexible structure allows it to curve with the shape of the container being scraped.

## CUTTING BOARDS AND CUTTING TOOLS

A variety of cutting boards and cutting tools are used in the professional kitchen. A *cutting board* is a cutting surface designed to protect work surfaces from cuts and scratches. Only plastic cutting boards are used in a professional kitchen. Plastic is nonporous so it will not absorb liquids. Nonporous material greatly reduces the risk of cross-contamination and foodborne illness if properly cleaned and sanitized. Many companies produce colored cutting boards so that food service operators can use different colors for different foods. ***See Figure 5-23.***

*Scraping Tools*

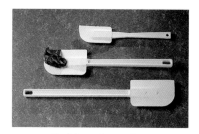

**Rubber Spatulas**

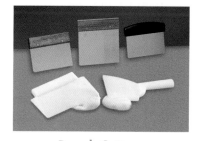

**Dough Cutters**

**Bowl Scrapers**

*American METALCRAFT, Inc.*

**5-22** *Scraping tools are used to scrape foods from containers or off surfaces.*

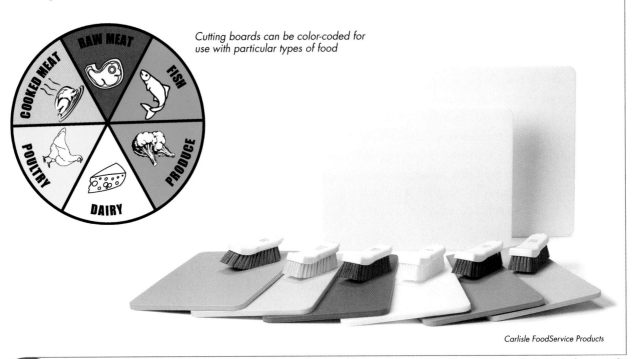

Cutting boards can be color-coded for use with particular types of food

COOKED MEAT • RAW MEAT • FISH • PRODUCE • DAIRY • POULTRY

Carlisle FoodService Products

**5-23** Cutting boards are available in multiple colors, allowing food service operations to color-code boards for specific applications.

Cutting tools range from knives to specialized tools for slicing particular food items or producing garnishes. Some of the most common cutting tools include cutting boards, mandolines, box graters, microplanes, kitchen shears, egg slicers, fruit corers, and cherry or olive pitters. Knives and cutlery will be discussed more thoroughly in Chapter 6.

A *mandoline* is a manual slicing tool with adjustable steel blades used to cut food into consistently thin slices. The mandoline can slice items nearly paper thin and can also produce julienne and waffle cuts. A mandoline must always be used with a hand guard in place or while wearing a cut-resistant glove. ***See Figure 5-24.***

A *box grater* is a stainless steel box with grids of various sizes used to cut food into small pieces. A box grater has fine-gauge grids for grating foods very fine and large-gauge grids for grating foods coarsely. Box graters are used most often for grating cheese and zesting citrus fruits. ***See Figure 5-25.***

A *microplane* is a razor-sharp handheld grater that shaves food into fine or very fine pieces. Microplanes are available in a variety of styles and are often used to shave chocolate, lemon zest, and cheeses. ***See Figure 5-26.***

## Mandoline

Browne-Halco (NJ)

**5-24** A mandoline is a food slicer with available attachments for making slices, julienne cuts, and waffle cuts.

*Kitchen shears* are heavy-gauge scissors used in the preparation of foods. They can be used for various tasks from cutting butcher's twine to trimming artichoke leaves. ***See Figure 5-27.***

Specialty tools such as the egg slicer, fruit corer, and cherry pitter are tools designed for a specific purpose. An *egg slicer* is a slicing tool with a series of tightly pulled wires that, when pressed against a peeled hard-cooked egg, will slice the egg into consistently thick slices. A *fruit corer* is a cylindrical tool that is used to remove the center core from fruits such as apples or pineapples. A *cherry pitter* (or olive pitter) is a specialty tool used to remove the pits from small fruits such as cherries and olives. The cherry is placed on a small ring beneath a metal rod. The user squeezes the two handles together and the rod pierces the fruit, pushing out the pit.

### PASTRY TOOLS

Pastry tools are specifically designed for working with pastry dough. These tools perform various functions such as rolling and flattening dough, cutting dough into shapes, and decorating pastries for presentation. Common pastry tools found in the professional kitchen include rolling pins, dough dockers, pastry wheels, pastry brushes, palette knives, pastry bags and tips, bench brushes, and pie or cake markers. ***See Figure 5-28.***

## Box Grater

Browne-Halco (NJ)

**5-25** A box grater has multiple grids for grating food into fine and coarse pieces.

## Microplane

Browne-Halco (NJ)

**5-26** A microplane shaves ingredients without ripping or shredding them.

## Kitchen Shears

Browne-Halco (NJ)

**5-27** Kitchen shears are used for tasks from cutting twine to trimming artichokes.

## Pastry Tools

World Cuisine, Inc.

**French Rolling Pin**

American METALCRAFT, Inc.

**Dough Dockers**

Messermeister

**Pastry Wheel**

Carlisle FoodService Products

**Pastry Brushes**

Dexter-Russell, Inc.

**Palette Knife**

Messermeister

**Pastry Bag and Tips**

American METALCRAFT, Inc.

**Bench Brush**

American METALCRAFT, Inc.

**Pie Markers**

**5-28** *Pastry tools are used when working with pastry dough or decorating pastries for final presentation.*

A *rolling pin* is a cylinder used to flatten pastry dough, bread crumbs, or other foods. It is often made of wood, but also is available in ceramic, marble, metal, and plastic. Rolling pins range in length from 10½″ to 25″. Handles may be attached on each end of the roller. Many chefs prefer French rolling pins, which do not have handles and may have tapered ends.

A *dough docker* is an aluminum, heavy plastic, or stainless steel roller with pins used to perforate dough. Dough dockers are used when preparing certain yeast dough products to help the dough bake evenly without blistering from the oven heat.

A *pastry wheel* is a dough-cutting tool with a rotating disk attached to a handle. The disk has a cutting edge that is rolled across dough to cut any shape desired. A pizza wheel is a version of a pastry wheel.

A *pastry brush* is a brush used to apply liquids such as egg wash or butter onto baked products. Plastic-bristle brushes should not be used on hot foods, as the heat will cause the plastic to melt onto the food. Brushes with natural bristles (basting brushes) are used for hot food items as the natural bristles will not melt. Brushes are available with silicone bristles that can withstand temperatures up to 650°F.

A *palette knife* (also called a cake spatula) has a long, flat, narrow blade with a rounded edge and is most often used for icing cakes. Palette knives come in lengths from 3½″ to 12″ and are available with blades ranging from semiflexible to highly flexible.

A *pastry bag* is a cone-shaped paper, canvas, or plastic bag with two open ends, the smaller of which is fitted with a pastry tip. A pastry bag is used to pipe soft food into a decorative pattern, such as decorating cakes with icing or piping whipped potatoes. A *pastry tip* is a cone-shaped plastic or metal tip that is fitted into the narrow end of a pastry bag to shape soft foods. Pastry tips fit into pastry bags and allow foods to be squeezed out in an array of decorative patterns or shapes.

A *bench brush* is a brush with long bristles set in vulcanized rubber attached to a wood handle and is used to brush excess flour from the bench (baker's table) when working with pastry or bread dough.

A *pie marker* (or cake marker) is a round, heavy wire tool with guide bars used for marking pies or round cakes for cutting. The guide bars or blades are pressed onto the pie or cake, leaving marks indicating where to cut to create equal portions. Markers are available in various diameters and portion sizes. Markers can also be used to mark pizzas for cutting.

## OTHER TOOLS

Hundreds of additional hand tools are available to food service professionals. In addition to the tools already discussed, meat mallets and funnels are common tools found in the professional kitchen.

The Vollrath Company, LLC

**5-29** Funnels are used for pouring liquids into containers.

A *meat mallet* (meat tenderizer) is a hammerlike tool used to pound and break the connective tissues and muscle fibers in tough cuts of meat. Breaking the muscle fibers in a piece of meat makes the meat tender. The aluminum head is cast with a coarse pattern on one side and a fine pattern on the other.

A *funnel* is a tapered bowl and tube used to pour a liquid from a larger container into a smaller container. Liquid is poured into the bowl and filters through the small tube. Funnels are used to avoid spilling when trying to pour liquid into small openings. ***See Figure 5-29.***

## SPECIALTY COOKWARE

There are many different styles and varieties of cookware in addition to those mentioned in this chapter. Some specialty items such as molds are used to give a specific shape to items as they are cooked or plated. ***See Figure 5-30.*** For example, pâté molds are typically lined with pâté dough, filled with a forcemeat of poultry, meat, game, seafood, or vegetables, and baked slowly in an oven. Timbale molds are mainly used to shape rice, polenta, or other items when plated. Timbale molds can also be used to bake custards and numerous other items. There are many other types of molds such as those used to form petit fours, aspic, brioche, and mousse.

## Specialty Molds

Matfer Bourgeat USA

**5-30** Specialty cookware used in the professional kitchen includes many types of molds, each providing a specific shape for foods such as pâté, rice, polenta, custards, petit fours, aspic, brioche, and mousse.

## SUMMARY

There are hundreds of different hand tools and smallwares available to the professional chef. Knowing the best tool to use in a given situation will result in optimal efficiency and results in the professional kitchen. Hand tools and smallwares include items for portioning and measuring, warming, steaming, cooking, baking, separating or puréeing, turning and grabbing, mixing and blending, scraping, cutting, working with pastry dough, and various specialty applications. Through proper training and practice, aspiring chefs will be able to master the use and care of each of these important tools.

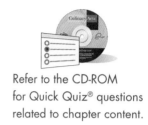

Refer to the CD-ROM for Quick Quiz® questions related to chapter content.

## Review Questions

1. What is the basic difference between hand tools and smallwares?

2. Contrast the main features and uses of a portion scale, a baker's scale, and a digital scale.

3. List the common measuring tools used to determine volume.

4. Compare the main advantages of different types of thermometers used to measure temperatures of cooked and stored foods.

5. Name four types of cookware used for warming and steaming.

6. Describe various uses for a silicone baking mat.

7. Identify the differences and similarities between the sautoir and the sauteuse.

8. Identify two bakeware items that have detachable bottoms.

9. Describe the main uses of five different kinds of strainers.

10. What is the main difference between a sifter and a strainer?

11. List the five main categories of hand tools.

12. What are the differences between a pastry brush and a bench brush?

13. Contrast the functions of terrines and pâté molds with those of timbale molds.

CulinaryArts

PRINCIPLES AND APPLICATIONS

# Knives and Cutlery

*Knife skills* are among the most important skills for an aspiring chef to develop. Mastery of knife skills is achieved only through practice and commitment. The professional chef uses different kinds of knives for a variety of tasks, from dicing vegetables to cutting up poultry. Each knife in the chef's tool kit is used to perform specific tasks. It is important to know which knife to use for each task. Using the proper cutting technique and handling knives safely is paramount in the professional kitchen.

## Chapter 6

## Chapter Objectives:

1. Identify the major parts of a knife.
2. Describe the main uses and distinguishing features of professional knives, cutlery, and garnishing tools.
3. Demonstrate the methods used for safely handling knives.
4. Demonstrate how to sharpen and hone a knife.
5. Describe the basic knife cuts used in the professional kitchen.
6. Demonstrate how to make the following cuts: rondelle, diagonal, oblique, chiffonade, butterfly, batonnet, julienne, brunoise, paysanne, dicing, chopping, and mincing.

## Key Terms

- **blade**
- **tang**
- **bolster**
- **rivets**
- **French knife**
- **cleaver**
- **santoku knife**
- **utility knife**
- **boning knife**
- **butcher's knife**
- **slicer**
- **paring knife**
- **tourné knife**
- **clam knife**
- **oyster knife**
- **parisienne scoop**
- **vegetable peeler**
- **zester**
- **channel knife**
- **whetstone**
- **sharpening steel**

## KNIVES

There are many different tools in the chef's professional tool kit, but the most important of these are knives. The only tools more fundamental to cooking and food production than knives are human hands. ***See Figure 6-1.***

*The Chef's Tool Kit*

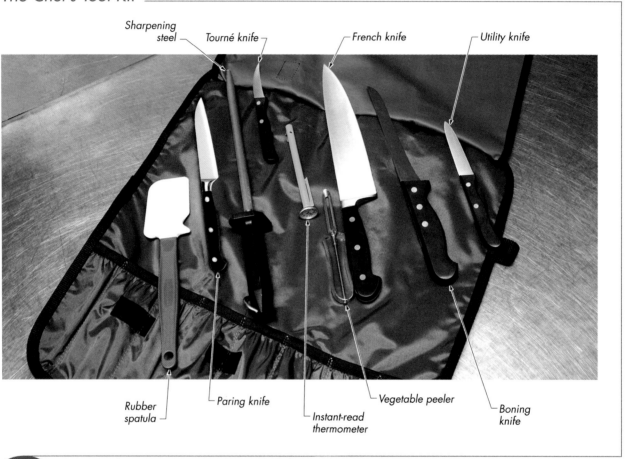

Sharpening steel — Tourné knife — French knife — Utility knife

Rubber spatula — Paring knife — Vegetable peeler — Boning knife

Instant-read thermometer

**6-1** *Knives are the most important tools in a chef's kit.*

Refer to the CD-ROM for the **Tools and Equipment** Flash Cards.

Knives should be selected for quality and comfort and should feel balanced in the hand. There have been many changes in the overall construction of knives over the years, resulting in lighter, more stain-resistant blades and edges that resist dulling and sharpen easily. Like any tool, a high-quality knife may cost more initially but provides superior performance and durability over time. All knives must be kept sharp using a sharpening steel and whetstone. A sharp knife is safer and performs much better than a dull knife. A dull knife requires more pressure or force to cut through food, making it more likely for the knife to slip, which could result in serious injury. A sharp knife requires less force and therefore allows the user to focus on technique and safety.

## Parts of the Knife

Knives come in a wide variety of shapes, sizes, materials, and price ranges. Each shape and variety of knife is designed for a specific purpose, but can possibly be used for additional tasks. It is important to understand the materials, construction, and intended use of each knife to ensure safe and proper handling. Various parts make up a knife, and include the blade, tang, bolster, rivets, and handle. *See Figure 6-2.*

**Blade.** A *blade* is the sharp part of the knife used for cutting. The blade material and overall construction must be considered when choosing a knife. Various materials can be used to manufacture knife blades including carbon steel, stainless steel, high-carbon stainless steel, and ceramic. Most professional cutlery has either a high-carbon stainless steel or ceramic blade. *See Figure 6-3.*

Carbon steel was used widely in the past to make professional knives. Carbon steel was a popular material because the metal was soft, making the knife easy to sharpen. There are a few disadvantages to carbon steel knives. First, the soft metal makes it hard to keep the edge sharp for long periods of use. Carbon steel knives also discolor over time if they come in contact with highly acidic foods such as tomatoes. This discoloration makes the knives look rusted or tarnished. It also causes some foods to oxidize or turn brown. Finally, carbon steel knives can leave a metallic taste on acidic foods. This happens because carbon is a reactive substance. Because of these disadvantages, carbon steel is seldom used today.

Blades constructed from stainless steel became popular because stainless steel, as the name suggests, does not discolor and does not react with acidic foods. Stainless steel knives keep their beautiful shine for life. Stainless steel is also a very hard metal, so stainless steel blades do not lose sharpness as rapidly as carbon steel knives. Some chefs dislike that the metal is so hard because it is difficult to resharpen a dull blade. However, once the blade is resharpened, it keeps its edge much longer than a carbon steel blade.

## Parts of the Knife

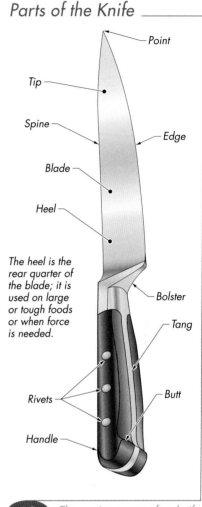

The heel is the rear quarter of the blade; it is used on large or tough foods or when force is needed.

**6-2** *The major parts of a knife include the blade, bolster, tang, rivets, and handle.*

## Blade Materials

Chef's Choice® by EdgeCraft Corporation

Kyocera Tycom Corporation

**High-Carbon Stainless Steel**     **Ceramic**

**6-3** *Blades for professional cutlery are typically manufactured from either high-carbon stainless steel or ceramic.*

High-carbon stainless steel combines the best qualities of carbon steel and stainless steel. High-carbon stainless steel produces a blade that is easy to keep sharp, never changes color, and does not transfer any metal taste to foods. Most knives used in the professional kitchen are made of high-carbon stainless steel.

Ceramic is a blade material that it is not yet widely used in the food service industry. Ceramic blades provide a sharper edge for a longer period than any other material, do not react with acidic foods, and are very easy to keep clean. The only real disadvantage is that the ceramic is somewhat fragile. If a ceramic blade is dropped on the floor, it may chip or break. Ceramic blades are commonly used for sushi and thinly slicing raw meat.

The shape of a steel blade for a professional knife can either be stamped or forged. Stamped blades are cut using machines that work like cookie cutters to cut out a desired shape from a sheet of thin metal. Forged blades are made from red-hot steel hammered into a desired shape, ground to a sharp edge, and polished to a shiny finish. Forged blades are better-quality blades and are much heavier in construction than stamped blades. Stamped-blade knives are considerably less expensive than knives with forged blades.

**Tang.** The *tang* is the unsharpened tail of the blade that extends into the handle of a knife. The highest-quality knives have a tang that extends all the way to the end of the handle. The tang has holes for securing the handle to the blade with rivets. A partial tang that is shorter and uses fewer rivets than a full tang. Partial tang knives are less durable then full-tang knives, but may be acceptable for infrequent or light use. A rat-tail tang is a narrow rod of metal that runs the length of the handle but is not nearly as wide as the handle. Rat-tail tangs are fully enclosed in the handle and are generally less durable than full or partial tangs. *See Figure 6-4.*

**Bolster.** A *bolster* is a thicker band of metal located where the blade joins the handle. The purpose of the bolster is to provide strength to the blade and prevent food from entering the seam between the blade and handle.

**Rivets.** *Rivets* are metal fasteners used to attach the tang of a knife to the handle. All high-quality knives have multiple rivets that pass through the handle material and holes in the tang. The rivets should be flush with the surface of the handle so that food particles cannot become trapped around the rivets.

**Handle.** The handle of the knife can be made from wood, plastic, steel, titanium, or a number of synthetic materials. It is important to keep the handle clean to ensure a good grip. Knives should never be placed in a commercial dishwasher as the intense heat and chemicals may cause the handles to warp and crack.

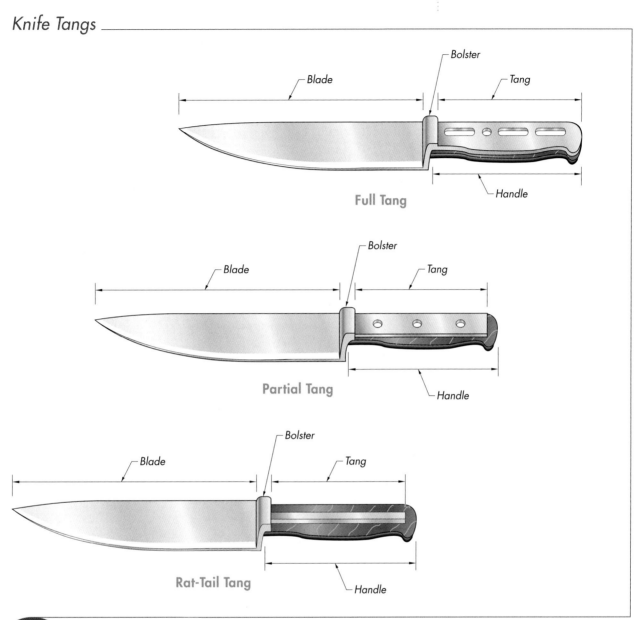

**Full Tang**

**Partial Tang**

**Rat-Tail Tang**

**6-4** *A tang is the tail of the blade as it continues through the handle.*

## Knife Types

A chef uses many different knives in the professional kitchen. Knowing which knife is best to use in a given application makes working with knives safer and more efficient. Some of the most common types of knives used in the professional kitchen are the French knife, slicer, cleaver, santoku knife, utility knife, boning knife, butcher's knife, paring knife, and tourné knife. ***See Figure 6-5.***

A *French knife* (also called a chef's knife) is a large, multipurpose knife with a tapering blade used for slicing, chopping, mincing, and dicing. It is the most versatile knife used in the professional kitchen. Near the handle, the heel of the blade is wide and tapers to a point at the tip. Generally, a bolster separates the blade from the handle. A French knife should be evenly balanced in weight between the blade and the handle to prevent fatigue of the hand or wrist. The most popular blade lengths are 8″, 10″, and 12″.

## Knife Types

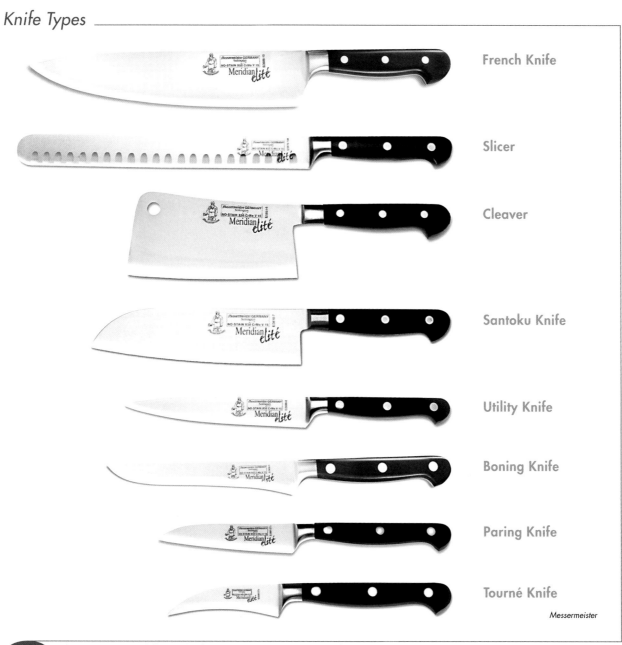

French Knife

Slicer

Cleaver

Santoku Knife

Utility Knife

Boning Knife

Paring Knife

Tourné Knife

*Messermeister*

**6-5** *There are many different knife types available in the professional kitchen.*

A *slicer* is a knife with a narrow, flexible blade usually between 10″ and 14″ long. Slicers are most commonly used at a carving station to slice roasted meats in front of guests. Slicers are available with a serrated or saw-like edge, and can be used to slice breads, cakes, and pastry items.

A *cleaver* is a heavy, rectangular-bladed knife. Cleavers are used to chop through bones. A santoku knife is a relatively new knife in the American kitchen. A *santoku knife* is an Asian-style knife with a razor-sharp edge and a heel that is perpendicular to the spine. In this way, the santoku knife resembles a small cleaver. There are often dimples along the edge that prevent food from sticking to the blade as it slices. The santoku knife is becoming popular with American chefs because of its lightweight, ultra-thin blade. Blades range in size from 5″ to 8″ in length.

A *utility knife* is a multipurpose knife used for anything from cleaning and cutting fruits and vegetables to carving roasted poultry. A utility knife typically has a 6″ to 10″ stiff blade that is similar in shape to the blade of a French knife but much narrower at the heel.

A *boning knife* is a short, thin knife with a 6″ to 8″ pointed blade. It is used to remove raw meat and fish from bones with minimal waste. The blade may be either stiff or flexible. Boning knives with heavier stiff blades are used for larger cuts of meat requiring heavy force, as in boning large food service cuts of beef. Those with thinner, more flexible blades are used for lighter work such as fabricating poultry.

A *butcher's knife* is a heavy knife with a curved, pointed blade that is from 7″ to 14″ in length. The tip of the blade curves about 25° upward and resembles a saber sword. The butcher's knife is commonly used for cutting, sectioning, and portioning raw meat such as steaks. ***See Figure 6-6.***

A *paring knife* (vegetable knife) is a short knife with a 2″ to 4″ very stiff blade. It is used for trimming and peeling fruits and vegetables. A *tourné knife* (bird's beak knife) is a small knife that is similar to a paring knife but has a curved blade. This knife is used to carve vegetables or potatoes into football-shaped pieces by rotating the food while carving.

## Cutlery and Garnishing Tools

In addition to knives, special cutlery and garnishing tools are used for specific applications. ***See Figure 6-7.*** A *clam knife* is a specialty tool with a short, flat, round-tipped blade with a sharp edge used to open clams. An *oyster knife* is a specialty tool with a short, slightly thin, dull-edged blade with a tapered point used to open oysters. A *kitchen fork* (meat fork or chef's fork) is a large fork with two long prongs. Kitchen forks are used for holding meats for carving and are typically used at a chef's carving station.

A *parisienne scoop* (melon baller) is a scoop with a stainless steel blade typically formed into a round half-ball cup attached to a handle. It is most commonly used for cutting various fruits and vegetables into uniform spheres. To use a parisienne scoop, the scoop is pressed firmly into the food item with

**Butcher's Knives**

Dexter-Russell, Inc.

**6-6** Butcher's knives have curved, pointed, heavy blades.

a twisting motion. By making the cuts as close together as possible, waste is minimized.

A *vegetable peeler* is a cutting tool with a metal blade attached to a metal handle used to remove the skin or thin peel from fruits and vegetables. The blade is a curved piece of metal with a slot at the top of the curve housing two sharpened edges, formed over a pin or axis attached to the handle. The double-edged blade can be used in either hand, and allows movement to contour to the shape of the fruit or vegetable. Peelers come in a variety of styles.

A *zester* is a small hand tool that has a stainless steel blade with five or six sharpened holes in its end. The small holes are dragged across the skin of a citrus fruit such as a lemon to yield small strings or "zest" from the peel. Zest can be added to foods as a natural flavoring.

A *channel knife* is a cutting tool with a thin metal blade with one raised channel used to remove a large string from the external surface of an item. A channel knife leaves a decorative pattern on the surface of an item.

## Specialty Tools

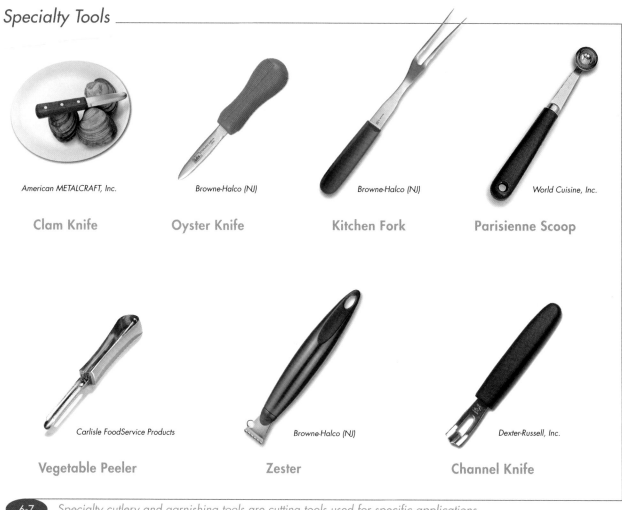

American METALCRAFT, Inc.

**Clam Knife**

Browne-Halco (NJ)

**Oyster Knife**

Browne-Halco (NJ)

**Kitchen Fork**

World Cuisine, Inc.

**Parisienne Scoop**

Carlisle FoodService Products

**Vegetable Peeler**

Browne-Halco (NJ)

**Zester**

Dexter-Russell, Inc.

**Channel Knife**

**6-7** *Specialty cutlery and garnishing tools are cutting tools used for specific applications.*

## KNIFE SAFETY

Because of their sharp edges, knives can be dangerous, and improper use can lead to personal injury. Always follow appropriate safety precautions when holding, carrying, using, washing, and storing knives.

*The following are recommendations for safe knife handling:*

- Always keep knives sharp. Injury is more likely to occur with a dull knife than with a sharp one.
- Never attempt to catch a falling knife. If it is going to fall, move away and let it fall.
- Always wipe a blade with the sharp side facing out.
- Always pass a knife to someone else by laying it on a table and sliding it forward.
- When walking with a knife, always keep the knife at the side of the body with the tip pointing downward and the sharp edge facing behind.
- Never leave knives in a sink as someone could reach in and be injured.
- Never wash knives in a commercial dish machine as the heat and chemicals can ruin the handles.
- Use only clean, sanitized knives on a whetstone or steel to avoid cross-contamination.
- Always wipe the blade after using a whetstone or sharpening steel to remove metal residue.
- Never use a knife to pry a lid off of a container.
- Always cut on a cutting board and never directly on a table.
- Knives should be honed after each use to maintain their edge.
- Always clean and sanitize knives before putting them away.
- Store knives safely in their own sleeves or attach knife guard covers to avoid accidents.

### Sharpening and Honing Tools

Before using any knife, check to make sure the knife edge is sharp and properly maintained. Having a sharp knife helps to prevent injury as the user does not need to exert as much pressure with a sharp knife compared to a dull knife. Whetstones and sharpening steels are commonly used to sharpen and maintain knife blades. ***See Figure 6-8.***

### Sharpening Knives

A *whetstone* is a stone used to grind the edge of the blade to the proper angle for sharpness. The blade of the knife is held at a 20° to 25° angle to the stone and dragged across the stone from tip to heel with equal but light pressure. Two-sided whetstones have a medium-grit or coarse

Refer to the CD-ROM for the **Knife Safety** Media Clip.

*Chef's Tip:*
Knives should never be left in a sink. They should be washed, dried, and immediately stored after use.

*Sharpening and Honing Tools*

Browne-Halco (NJ)
**Whetstone**

Dexter-Russel, Inc.
**Sharpening Steel**

**6-8** *Whetstones and sharpening steels are used to sharpen and hone knives and cutlery.*

Refer to the CD-ROM
for the **Knife Sharpening**
Media Clip.

side and a fine-grit side. A three-sided whetstone sits over a reservoir of either mineral oil or water to keep the stone lubricated during use. This stone has coarse-grit, medium-grit, and fine-grit sides, allowing the user to completely grind a new edge on a very dull knife. After using the whetstone, it is important to use a butcher's steel to align or "hone" the edge of the blade. The butcher's steel straightens the very thin edge of the blade to ensure a razor-sharp finish.

Determining the 20° angle needed to sharpen a knife is not difficult. The counter and stone are at 0°. The knife blade is at 90° when held straight above the stone as if it were cutting the stone in half. Tilting the knife halfway to the counter (to 45°) and halfway again results in the perfect sharpening angle (between 20° and 25°). **See Figure 6-9.**

## Determining Angles for Sharpening

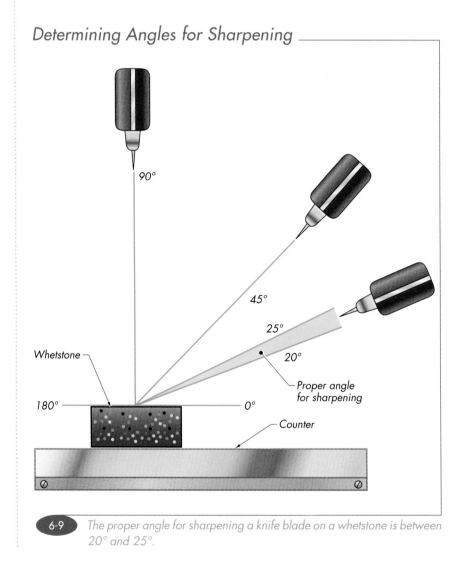

6-9 *The proper angle for sharpening a knife blade on a whetstone is between 20° and 25°.*

## To sharpen a knife, apply the following procedure:

1. Lay the edge of the knife near the tip of the whetstone at a 20° to 25° angle. ***See Figure 6-10.***

2. Use four fingers to guide and stabilize the knife and begin to draw the blade against the surface of the stone.

3. Continue to draw the knife blade against the stone from tip to heel in a smooth and controlled motion, maintaining the approximate 20° to 25° angle.

4. Carefully flip the knife over and begin the same process on the other side.

5. Be sure to use the same number of strokes on each side to create an even and sharp edge. *Note:* Always start on the coarsest side of a whetstone and work toward the finest-grit side.

*Sharpening Knives*

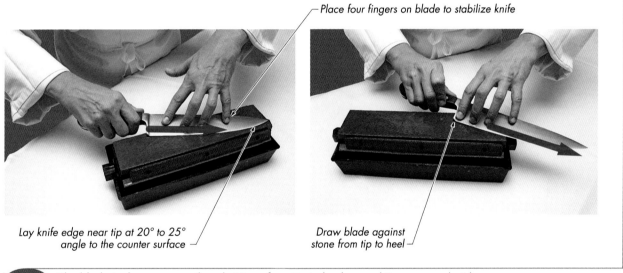

Place four fingers on blade to stabilize knife

Lay knife edge near tip at 20° to 25° angle to the counter surface

Draw blade against stone from tip to heel

**6-10** *The blade is drawn across the whetstone from tip to heel several times on each side.*

### Honing Knives

All professional knife kits come with a sharpening steel. A *sharpening steel* is a steel rod approximately 18″ long with a handle that is used to maintain a sharp edge on a knife. A sharpening steel is usually made of hardened steel, but may also have a surface that is ceramic or diamond-impregnated. The tip of the sharpening steel is usually magnetic so that it can catch metal fragments as they are removed from the blade. The sharpening steel does not sharpen the edge of the blade; it hones, or aligns, the blade. *Honing* (also known as truing) is the process of aligning a blade's edge and removing any burrs or rough spots on the blade. A sharpening steel is used in a manner similar to the whetstone, although it does not sharpen a dull knife. A sharpening steel should be used after sharpening a knife on a whetstone as well as between sharpenings to maintain and straighten the knife's edge.

## To hone a knife with a sharpening steel, apply the following procedure:

1. Hold the sharpening steel perpendicular or pointed toward the floor with the left hand (or right hand if left-handed). ***See Figure 6-11.***

2. Hold the knife blade at a 20° to 25° degree angle in relation to the steel.

3. Pass the blade along the steel from the tip to the heel while maintaining the angle and gentle pressure.

4. After each stroke along the steel, switch the knife to the other side of the steel and draw the blade again against the steel. Repeat three to five times on each side of the knife blade.

5. After ensuring the edge of the knife is straight and sharp, wipe off the blade to remove any metal residue.

### Honing Knives

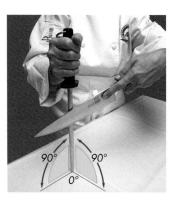

*1. Hold steel 90° from the surface.*

*2. Place knife edge 20° to 25° to the steel.*

*3. Pass blade along the steel from tip to heel; repeat 3 to 5 times on each side.*

*4. Wipe off blade to remove residue.*

**6-11** *Honing is used to align a blade edge and remove any burrs.*

## KNIFE SKILLS

Knife skills are among the most important skills for a culinary student to master. Many different knife cuts are used in the professional kitchen. A chef must know the differences between these cuts and be able to execute them accurately and efficiently. ***See Figure 6-12.***

Before using a knife, it must be properly sharpened and honed with a whetstone and sharpening steel. Proper grip of the knife and position of the guiding hand ensure safety and control when sharpening knives.

### Knife Grips

There are different acceptable methods for gripping a knife, but for consistency, students should first learn the method used most often by culinary professionals. This method provides the best control, stability, safety, and accuracy. To begin, the knife is held by the handle while resting the index finger flat against one side of the blade and the thumb on the other side of the blade. ***See Figure 6-13.*** This grip provides the best control of the knife. For the student, it may feel a little awkward at first, but with practice the knife feels safe and secure in the gripping hand. Placing the index finger on one side of the blade and the thumb on the other is similar to choking up on a baseball bat for better swing control.

*Knife Skills*

6-12   *Mastery of knife skills requires practice.*

*Knife Grip*

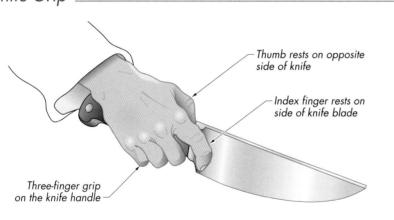

- Thumb rests on opposite side of knife
- Index finger rests on side of knife blade
- Three-finger grip on the knife handle

6-13   *The knife is gripped by the handle with the index finger resting flat against one side of the blade.*

The hand not holding the knife is the guiding hand. The guiding hand is responsible for guiding the item to be cut into the knife. Many students find it difficult to master the positioning of the guiding hand, but with practice it becomes second nature. To position the fingers of the guiding hand, pretend to pick up an egg resting on a table. Note the positioning of the hand and fingers as they reach for the object. The arm is rotated to bring the fingertips to the surface of the table without changing the position of the hand or fingers. This position can also be achieved by trying to imitate the shape and motion of a spider walking on the table. The fingertips should all touch the surface of the table and be slightly tucked back. This hand position is used to guide foods toward the blade of the knife. Food is held between the thumb and fingertips as it is pushed forward toward the knife.

Using the proper knife grip, the tip of the knife is placed on the cutting board. The guiding hand is placed next to the knife blade in the proper position, with fingertips slightly tucked under. The guiding hand should be near the back half of the blade (not the tip), usually close to where the manufacturer brand name or logo is placed on the blade. The side of the blade should rest against the knuckle of the middle finger on the guiding hand. *See Figure 6-14.* Resting the blade of the knife against the knuckle of the middle finger and tucking the fingertips back reduces the chances of cutting fingers.

## Guiding Hand Position

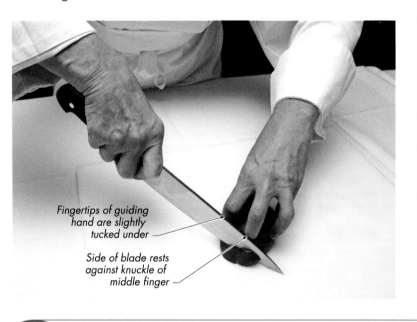

Fingertips of guiding hand are slightly tucked under

Side of blade rests against knuckle of middle finger

**6-14** *The knife is gripped by the handle with the index finger resting flat on one side of the blade.*

## Cutting Movements

Once the proper method of gripping the knife and positioning of the guiding hand is mastered, proper cutting movement can be practiced. When cutting with a French knife, a rocking motion is used. The edge of the knife blade is curved and allows the knife to have a smooth, rocking motion. The tip of the knife slides forward about 1″ when the handle is brought down. Likewise, as the handle is raised up, the tip of the knife slides backward. This rocking movement, coupled with the correct position of the guiding hand, creates a controlled motion that can be used to efficiently chop, dice, and shred.

To achieve a rocking movement, apply the following procedure:

1. Using the proper cutting grip, place the tip of the knife on a cutting board.

2. With the front third of the knife on the cutting board, begin rocking the blade of the knife up and down, along the curve of the blade. The motion is similar to the motion of a rocking chair.

3. While rocking the knife up and down, allow the tip of the knife to slide forward a few inches on the downward motion and pull back on the upward motion. *See Figure 6-15.*

4. Ensure that the blade continually rests against the knuckle of the middle finger when practicing the rocking motion.

*CULINARY PROCEDURES*

## Cutting with a French Knife

1. Place tip of French knife on cutting board and press down on knife handle.

2. Continue pressing down on handle and slide blade slightly forward, following the curve of the blade.

3. When heel of blade is flat on cutting surface, slide blade backward and raise handle to position knife for next slice.

**6-15** *The curved blade of a French knife allows a smooth rocking motion for slicing food.*

## BASIC KNIFE CUTS

Once an understanding of proper knife control is achieved, the basic knife cuts used in the professional kitchen can be executed. Basic cuts are designated with standard measurements that are accepted throughout the food service industry. A uniform cut ensures that items cook evenly and look much more appealing in the finished product. Common cuts used daily in the professional kitchen include slicing, dicing, chopping, and mincing, as well as specialty cuts such as the tourné. Mastery of basic knife cuts requires practice and commitment.

### Slicing Cuts

Slicing involves passing the blade of the knife slowly through an item to make long, thin pieces. In slicing, the knife is either dragged backward or slid forward through the item. This is different from the chopping method where the knife movement is up and down, similar to the movement of a paper cutter. The motion required for slicing is one of the main reasons that knife blades are long and thin. Slicing cuts include the rondelle, diagonal, and oblique. *See Figure 6-16.*

*Slicing Cuts*

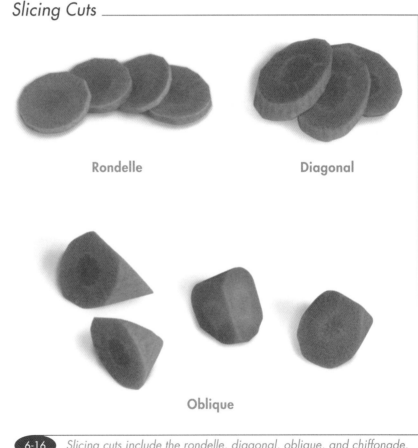

Rondelle

Diagonal

Oblique

**6-16** *Slicing cuts include the rondelle, diagonal, oblique, and chiffonade.*

A *rondelle cut* is a cut that produces disk-shaped slices. Rondelle (pronounced ron-dell) cuts are produced from slicing cylindrical vegetables such as cucumbers and carrots straight through. This cut is also known as the round cut. To make a rondelle cut, the item is placed perpendicular to blade. The item is guided into the knife while slicing to create round disks. ***See Figure 6-17.***

## Rondelle Cuts

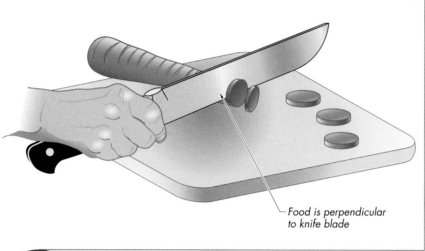

Food is perpendicular to knife blade

**6-17** *Rondelle cuts produce disk-shaped slices.*

A *diagonal cut* is a cut that produces oval slices. Diagonal cuts are made from cylindrical vegetables that are cut on the bias. This cut is used mainly in decorative vegetable side dishes or for ingredients in Asian-style dishes. To make a diagonal cut, place the item at a 45° angle to the knife blade. Guide the item while slicing to create oval slices. ***See Figure 6-18.***

## Diagonal Cuts

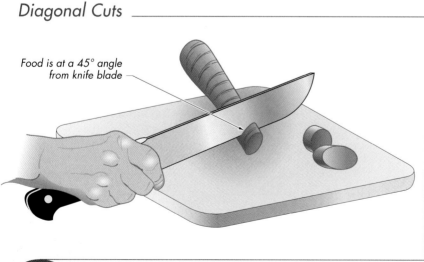

Food is at a 45° angle from knife blade

**6-18** *Diagonal cuts produce elongated oval slices.*

An *oblique cut* is a cut that produces small, rounded pieces with two angled sides. The oblique cut is similar to the diagonal cut in that 45° angle slices are also made on a cylindrical item. However, the oblique cut produces much larger pieces that are more wedge-shaped than a slice.

## To make an oblique cut, apply the following procedure:

1. Guide the item to be cut into the blade at a 45° angle and slice (discard this first cut). ***See Figure 6-19.***
2. Roll the item over halfway (180°) and slice again with knife in same position. This slice produces the first usable piece.
3. Roll the item back to the original position and slice again.
4. Repeat process, rolling the item forward and backward to make rounded, triangular, wedge-shaped pieces.

## Oblique Cuts

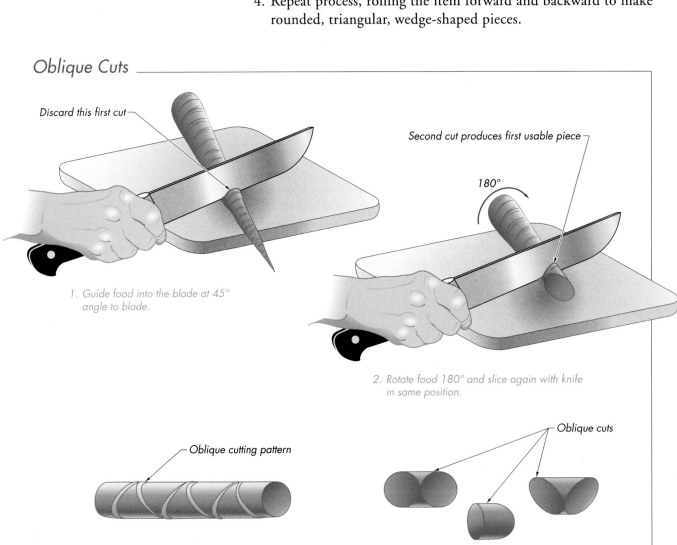

Discard this first cut

1. Guide food into the blade at 45° angle to blade.

Second cut produces first usable piece

180°

2. Rotate food 180° and slice again with knife in same position.

Oblique cutting pattern

Oblique cuts

**6-19** *Oblique cuts produce round pieces with two angled sides.*

A *chiffonade cut* is a cut that produces thinly sliced or shredded leafy vegetables or herbs. Chiffonade (pronounced shif-fo-nahd) cut items can be used as ingredients or as a base under displayed cold foods. To make a chiffonade cut, first wash the leafy items, stack the leaves on top of one another, and roll the stack lengthwise like a cigar. Place the cigar-shaped roll on the cutting board perpendicular to the knife blade. Use a rocking motion to thinly slice the roll as it is fed with the guiding hand into the knife blade. The result is finely shredded leaves or herbs. ***See Figure 6-20.***

## Chiffonade Cuts

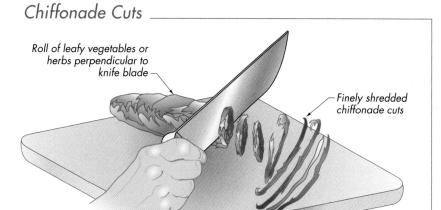

Roll of leafy vegetables or herbs perpendicular to knife blade

Finely shredded chiffonade cuts

**6-20** *The chiffonade cut is used to finely shred leafy vegetables or herbs.*

A *butterfly cut* is a cut used for a thicker meat, fish, or poultry item to separate the item almost completely in half horizontally. The butterfly cut is used to open a thick piece of food and create a thinner overall product that cooks quicker. Butterfly cuts are also used to create a pocket in the food to prepare the item for stuffing. The butterfly cut is different from other slicing cuts because it is a horizontal cut as opposed to a vertical cut.

## To make a butterfly cut, apply the following procedure:

1. Place the item to be cut on a clean work surface.
2. Place the guiding hand flat on top of the item to hold it in place.
3. While holding the knife blade horizontally (parallel to the table), place the edge at the approximate midpoint of the item's thickness.
4. Slice into the item to create a pocket for stuffing. Slice the item almost all the way through, leaving ¼″ to ½″ connected to create a thinner item that cooks faster and more thoroughly.

Refer to the CD-ROM for the **Fine Julienne** Media Clip.

## Stick Cuts

Many cooks who have not attended culinary school have a general understanding of what a dice looks like. However, in the professional kitchen there is an exact dimension for each cut that is accepted and practiced. The terms "batonnet" and "julienne" refer to stick-shaped cuts that are cut to precise measurements. These cuts then can be further cut down into cube-shaped cuts known as dice cuts. Professional cuts are used when uniformity and consistency are necessary for an attractive presentation and to show that the kitchen staff pays attention to detail.

Batonnet (pronounced bah-toh-ney) and julienne (pronounced ju-lee-en) are very similar cuts with only the dimensions of each differing. A *batonnet cut* is a cut that produces a stick-shaped item with dimensions of ¼" × ¼" thick × 2" long. A *julienne cut* is a cut that produces a stick-shaped item with dimensions of ⅛" × ⅛" thick × 2" long. They are both begun by squaring off the item to be cut. ***See Figure 6-21.***

## Batonnet and Julienne Cuts

*Items are trimmed, peeled, and cut into 2"-long pieces.*

1. Square off four sides to make a rectangle.

2. Slice in even thickness; ¼" for batonnet, or ⅛" for julienne, ¹⁄₁₆" for fine julienne.

3. Cut slices into sticks that are the same thickness as original slices.

**6-21** Stick cuts, such as julienne and batonnet, produce long thin strips.

To make a stick cut such as a batonnet or julienne, apply the following procedure:

1. Cut the item into pieces that are 2" in length.
2. Carefully square off four sides of a 2"-long item to make a rectangle. Save scraps for another use.
3. Cut even slices from the rectangle of the desired thickness needed to make the cut (¼" for batonnet, ⅛" for julienne, ¹⁄₁₆" for fine julienne).
4. Stack a few of the slices on top of each other and carefully slice into sticks that are the same thickness as the original slice.

## Dice Cuts

Dice cuts such as large dice, medium dice, small dice, brunoise (pronounced broo-nwahz), and fine brunoise are achieved by further cutting stick cuts into cubes. To produce a dice cut, the stick cut of the appropriate dimension is cut down into cubes with six equal sides. A *brunoise cut* is a cut that produces a cube-shaped item with six equal sides measuring ⅛". A fine brunoise cut measures ¹⁄₁₆". ***See Figure 6-22.***

### To make a brunoise cut, apply the following procedure:

1. Gather a small bundle of julienne-cut sticks.
2. Place the bundle of julienne sticks perpendicular to the knife's blade.
3. With a slicing motion, cut through the bundle at equally spaced ⅛" intervals to produce six-sided cubes ⅛" × ⅛" × ⅛".

CULINARY PROCEDURES

### *From Stick Cuts to Dice Cuts*

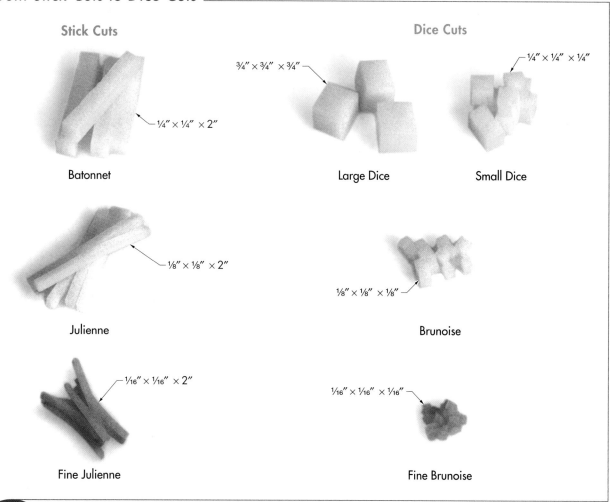

**Stick Cuts**

Batonnet — ¼" × ¼" × 2"

Julienne — ⅛" × ⅛" × 2"

Fine Julienne — ¹⁄₁₆" × ¹⁄₁₆" × 2"

**Dice Cuts**

Large Dice — ¾" × ¾" × ¾"

Small Dice — ¼" × ¼" × ¼"

Brunoise — ⅛" × ⅛" × ⅛"

Fine Brunoise — ¹⁄₁₆" × ¹⁄₁₆" × ¹⁄₁₆"

**6-22** *Dice cuts are created by further cutting stick cuts into cubes.*

## Paysanne Cut

6-23 *A paysanne cut is used to make small tile-shaped slices.*

The same procedure is used to make other dice cuts. A fine julienne is used to produce a fine brunoise, a batonnet is used to produce a small dice, etc. Unlike other dice cuts, mirepoix (pronounced meer-pwah) does not require specific dimensions or particular attention to detail. *Mirepoix* is a rough cut of carrots, onions, and celery measuring approximately ¾″ × ¾″ × ¾″, used to make soups and stocks. A mirepoix is meant to impart flavor but is strained from the finished product and discarded.

A *paysanne cut* is a cut that produces a thin, square, tile-shaped cut with dimensions of ½″ × ½″ × ⅛″ thick. "Paysanne" (pronounced pay-sahn) can also refer to triangular or round pieces with similar dimensions. Cutting paysanne is done in the same manner as the medium dice except that each ½″ × ½″ stick is cut into ⅛″ slices. The "tile-cut" paysanne is usually used to garnish a soup or other dish presented elegantly. ***See Figure 6-23.***

### Dicing an Onion

An onion consists of many layers of membrane that prevent it from being diced in the same manner as a solid item such as a carrot. For this reason, a modified procedure is used to dice onions to any desired size. ***See Figure 6-24.***

CULINARY PROCEDURES

## To dice an onion, apply the following procedure:

1. With a paring knife, cut off the stem end and lightly trim the root end of the onion. *Note:* Do not cut the root end off completely. This end holds all of the layers of the onion in place, preventing the onion from falling apart.

2. Make a thin slice from root end to stem end through the outer peel only and use the tip of the paring knife to pull the outer peel away.

3. Cut the onion in half from the stem end to the root end.

4. Lay the onion on the cutting board with the flat side down.

5. Make vertical slices down through the onion with the tip of the knife, from the stem end to the root end, again leaving the root end intact. The closer the slices, the smaller the finished dice.

6. Make two or three horizontal cuts through the onion, using caution not to slice all the way through, again leaving the root end intact.

7. Turn the onion so that the stem end is against the blade and make slices the thickness of the desired dice and slice all the way through from stem end to root end.

## Dicing an Onion

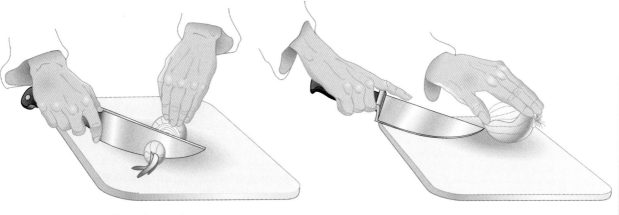

1. Cut off stem leaving the root end intact.

2. Score outer peel from root to stem and peel outer layers of onion.

3. Cut onion in half from stem to root.

4. Lay onion with flat side down, make vertical slices from stem to root, leaving root intact.

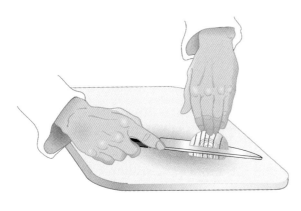

5. Make two or three horizontal slices parallel to cutting surface.

6. With stem against blade, make slices the thickness of desired dice, slicing completely through to the root end.

**6-24**  Leaving the root intact while dicing an onion will hold the many layers together until the final dice cut is made.

## Chopping and Mincing

Chopping is a cut that is not used for many items in the professional kitchen. Many chefs today allow only parsley, hard-cooked eggs, and garlic to be chopped. *Chopping* is rough-cutting an item so that there are relatively same-size small pieces throughout, although there is no uniformity in shape required. *Mincing* is finely chopping an item to yield a very finely cut product. It is commonly used for garlic, shallots, and fresh herbs.

### To chop parsley, apply the following procedure:

1. Wash parsley in cold water, drain well, and remove parsley sprigs from the stems. ***See Figure 6-25.***

2. Compress parsley into a tight pile with the guiding hand and feed the parsley into the knife, making an up-and-down motion with the knife cutting off—almost shaving off—small amounts of parsley from the pile.

3. With the guiding hand opened flat, rock the back of the knife up and down over the parsley. Continue to move the blade up and down while moving the blade back and forth to finely chop the parsley.

4. Place the chopped parsley in a clean towel or double-layered cheesecloth, twist tightly, and rinse under cold water.

5. Squeeze the cloth gently to release the water. Parsley should be fluffy and ready to use.

### To mince shallots, apply the following procedure:

1. Clean and peel the shallot in the same manner as an onion. ***See Figure 6-26.***

2. Make vertical cuts through the shallot, using caution not to cut through the root end.

3. Make one horizontal cut through the shallot, taking care not to cut through the root end.

4. Make slices perpendicular to the other cuts to create a small dice.

5. With the guiding hand resting flat on top of the tip end of the knife, rock the knife up and down to mince the shallot.

## Chopping Parsley

1. Wash parsley in cold water, drain well and remove parsley sprigs from stems.

2. Compress parsley into a tight pile and feed the pile of parsley through the knife, while rocking the knife to finely cut the parsley.

3. With the guiding hand opened flat, rock the blade up and down repeatedly to finely cut the parsley.

4. Place chopped parsley in a clean towel or cheesecloth and rinse with cold water.

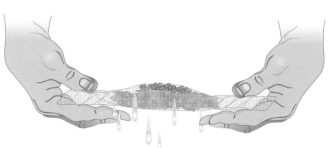

5. Squeeze the cloth gently to release water.

6-25    Parsley is one of the few items that are chopped in the professional kitchen.

## Mincing Shallots

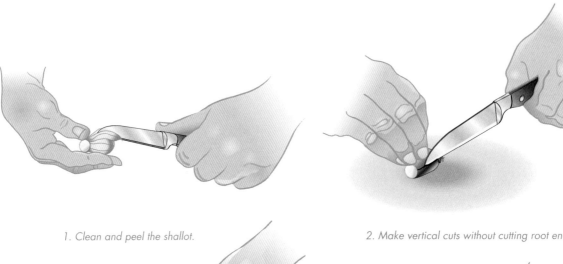

1. Clean and peel the shallot.

2. Make vertical cuts without cutting root end.

3. Make one horizontal cut through shallot without cutting the root end.

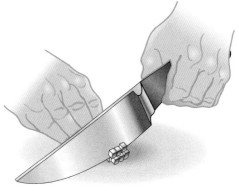

4. Make slices perpendicular to the other cuts to create a small dice.

5. With guiding hand resting flat on top of the knife, rock up and down to mince the shallot.

**6-26** The procedure for mincing shallots requires the root end be left intact to hold the layers together before the dice cut is made.

## Tourné

"Tourné" (pronounced toor-nay) is a French word from the verb "to turn." A tourné is one of the most difficult cuts to master. It is a hand-carved football-shaped item with seven sides and flat ends. It can be used to make a beautiful side dish to accompany an elegant entrée. ***See Figure 6-27.***

*Tourné Cut* _____

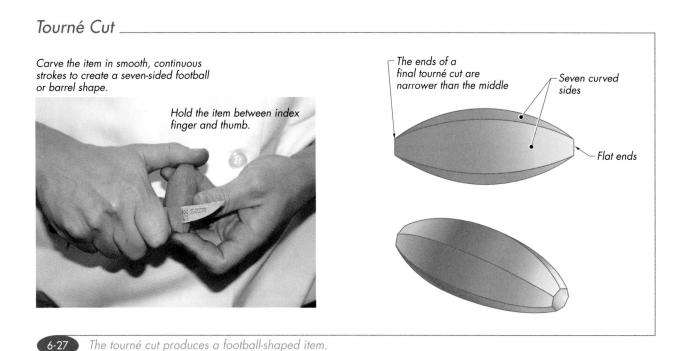

Carve the item in smooth, continuous strokes to create a seven-sided football or barrel shape.

Hold the item between index finger and thumb.

The ends of a final tourné cut are narrower than the middle

Seven curved sides

Flat ends

**6-27** *The tourné cut produces a football-shaped item.*

## To make a tourné cut, apply the following procedure:

1. Wash and peel item, as all trimmings can be used in other ways.
2. Cut a cylindrical item such as a carrot into 2″-long sections as when cutting julienne. If necessary, cut the thickness into 1″ widths. Wide items such as potatoes may be cut into four or six pieces.
3. Hold the item in the guiding hand between the index finger on one end and the thumb on the other.
4. Using a tourné knife, begin carving the item from one end to the other in smooth, continuous strokes, creating a seven-sided football or barrel shape.
5. As each slice is carved, notice that the ends are narrower than the middle.

CULINARY PROCEDURES

## SUMMARY

Excellent knife skills are one of the fundamental skills developed by professional chefs. Students studying to be professionals in the food service industry must practice and master professional knife skills. Safe knife handling and proper sanitation are as important as having the ability to produce accurate and consistent knife cuts. Knives must be kept sharp and stored safely so they are ready when needed. A dull knife is more dangerous to use than a sharp one. To produce basic cuts with skill, speed, and precision requires daily practice. In the professional kitchen, the basic knife cuts such as batonnet, julienne, brunoise, dices, and paysanne cuts all have specific dimensions for consistency.

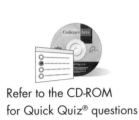

*Browne-Halco (NJ)*

Refer to the CD-ROM for Quick Quiz® questions related to chapter content.

## Review Questions

1. Identify the main parts of a knife and explain the function of each.

2. Describe the advantages of stainless steel blades over carbon steel blades.

3. Which four knives are considered appropriate for cutting vegetables?

4. What's the main function of the cleaver?

5. List six types of specialty tools used to cut foods.

6. Explain why sharp knives are safer than dull knives.

7. What is honing?

8. Describe how to safely carry a knife.

9. Describe how to safely pass a knife to another person.

10. Before placing knives in storage, what steps should be taken to ensure cleanliness?

11. Describe how knives should be stored.

12. Describe the five steps used to sharpen knives.

13. Describe the method that is used most often by culinary professionals for gripping a knife.

14. Identify the differences between slicing and chopping.

15. Identify the differences between a diagonal cut and an oblique cut.

16. Describe reasons for using the butterfly cut.

17. List the steps taken for making a julienne or batonnet cut, including the thickness required for batonnet, julienne, and fine julienne.

18. Describe dice cuts.

19. What type of cut is commonly used for garlic, shallots, and fresh herbs?

20. List the four steps used to make a tourné cut.

# Standardized Recipes and Cost Control

National Pork Producers Council

*Standardized recipes and cost control* are essential to running a profitable food service establishment. For example, if a customer has a favorite dish at a particular restaurant, the restaurant is likely to see repeat business from that customer. However, if the customer returns to the restaurant and receives something that tastes different, the restaurant may lose that customer. A lack of standardization can lead to a loss of business, which translates into a loss of profit. Recipe cost, the menu, portion size, purchasing, receiving, storing, issuing, and waste all factor into cost control.

## Chapter Objectives:

1. Identify the defining elements and benefits of standardized recipes.
2. Convert customary measurements to metric measurements and vice versa.
3. Distinguish among weight, volume, and count.
4. Demonstrate two recipe conversion formulas.
5. List factors to consider when converting a recipe to a larger or smaller yield.
6. Calculate the unit cost, edible portion cost, and total cost of a recipe.
7. Demonstrate how to calculate the yield percentage of a food item.
8. Explain how costing is used in the food service industry.
9. Contrast fixed and variable costs.
10. Explain ways to control costs in a food service establishment.

## Key Terms

- standardized recipe
- yield
- portion size
- weight
- volume
- as-purchased cost
- unit cost
- edible portion
- trim loss
- yield percentage
- shrinkage
- costing
- fixed cost
- variable cost
- par stock
- requisition
- obsolescence

## STANDARDIZED RECIPES

In a professional food service establishment, many menu items are prepared again and again. The key to having each item turn out the same way every time is by following standardized recipes. A *standardized recipe* is a set of directions for preparing a particular menu item to ensure consistent quality, portion size, and cost. A standardized recipe accomplishes the following tasks:

- lists specific ingredients by quantity and in the order they are used

- provides a set of unambiguous step-by-step instructions for preparation or assembly

- provides accurate portion size and total yield

- ensures a consistent, quality product

- ensures consistent cost of producing that item

- eliminates waste due to not overproducing

- decreases errors in food orders by detailing what is included

All standardized recipes must include the recipe name, yield, portion size, quantities of ingredients, step-by-step preparation procedures, cooking temperature, and cooking time. While not required, nutritional information may also be included. ***See Figure 7-1.***

## Standardized Recipes

| Shrimp Creole | |
| --- | --- |
| **Yield:** 25 Servings<br>**Portion Size:** 9 oz | **Cooking Temperature:** 175°F<br>**Cooking Time:** 20 min |
| **Ingredients** | **Procedure** |
| 8 lb shrimp<br>3 qt prepared creole sauce<br>salt and pepper to taste<br>3 lb prepared white rice | 1. Cook shrimp by steaming.<br>2. Bring creole sauce to a simmer in saucepot.<br>3. Add cooked shrimp. Blend gently into sauce with kitchen spoon season to taste.<br>4. Place 2 oz preprared rice on each serving plate.<br>5. Dish shrimp into casserole with 6 oz ladle. |

**7-1** *Standardized recipes include all information necessary to prepare dishes of consistent quality, size, and cost.*

**Recipe Name.** The name of a recipe should match the name of the item as listed on the menu. For example, if a restaurant serves a "Tortilla Soup" but the recipe has the name "Southwestern Tortilla Soup," kitchen staff may be unsure whether the recipe is for the same soup as the menu item. Differences in naming can lead to confusion in the kitchen.

**Yield.** A *yield* is the amount of a food or beverage product that results from a specific recipe after fabrication, preparation, and cooking. Yield may be given as portions such as 16 (3-oz) servings, size of item produced (such as one 8″ pie), or a measured amount of product (such as 2 gal. of soup).

**Portion Size.** *Portion size* is the actual size of an individual serving. For example, a recipe that yields 1 gal. may produce 16 portions of 8 fl oz each. ***See Figure 7-2.***

Portion Size

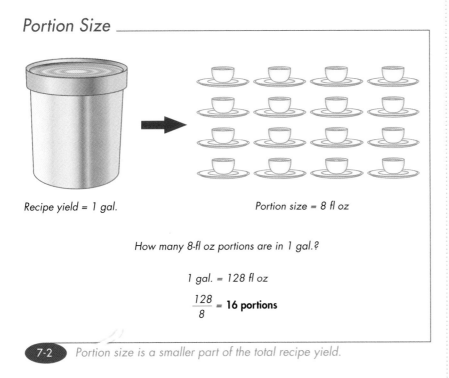

Recipe yield = 1 gal.                     Portion size = 8 fl oz

How many 8-fl oz portions are in 1 gal.?

1 gal. = 128 fl oz

$$\frac{128}{8} = \textbf{16 portions}$$

**7-2**    *Portion size is a smaller part of the total recipe yield.*

**Ingredients.** Standardized recipes list each of the ingredients used in the recipe and the amount needed of each one. The ingredients are listed in the order that they are incorporated into the recipe, helping ensure that none of the ingredients are left out.

**Procedures.** Step-by-step procedures direct when and how ingredients should be added to a recipe. These steps have been determined by thorough testing of the recipe and should always be followed closely. Important preparation steps are often listed just before or after the listed ingredient. For many ingredients the preparation (e.g., slicing, dicing) is done prior to measuring the ingredient. For example, "½ c sliced black olives" requires the olives be sliced before they are measured. In contrast, "½ c black olives, sliced" requires the olives be measured before they are sliced. Therefore, the first example of "½ c sliced black olives" requires more olives for use in the recipe.

**Cooking Temperature.** The temperature at which items are cooked greatly affects the outcome of the product. For example, an item that is pan-fried over high heat may burn on the outside before the inside is cooked through. The cooking temperature for a range is typically listed as low, medium, or high. Other pieces of equipment, such as deep-fat fryers and ovens, have thermostats that can be set to a specific temperature. ***See Figure 7-3.***

*Cooking Temperatures*

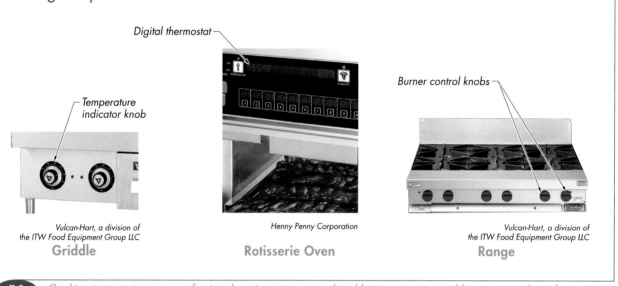

Digital thermostat

Temperature indicator knob

Burner control knobs

*Vulcan-Hart, a division of the ITW Food Equipment Group LLC*
**Griddle**

*Henny Penny Corporation*
**Rotisserie Oven**

*Vulcan-Hart, a division of the ITW Food Equipment Group LLC*
**Range**

**7-3** *Cooking temperatures on professional equipment are regulated by temperature and burner controls or thermostats.*

**Cooking Time.** Cooking time is always specified on a recipe. However, certain variables may have an effect on cooking time, such as the temperature of an item at the start of cooking, accuracy of the thermostat, or the desired degree of doneness (such as a steak being cooked to medium rare or medium well).

**Nutritional Information.** Nutritional information is not required on a standardized recipe, but many customers want to know the nutritional content of a dish, such as information about its fat, sodium, or carbohydrate content. Having this information readily accessible can help employees answer customers' questions.

## STANDARDIZED MEASUREMENTS

In order for a recipe to be successful, the ingredients and the resulting portions must be measured accurately. Too much of one ingredient or too little of another can result in a dish that does not meet customer expectations or simply does not work. Serving the wrong portion size can result in the establishment not having enough servings on hand for a meal service or dissatisfaction from guests who receive a smaller portion than is served to someone else.

Using standardized recipes also makes it much easier to increase or decrease the yield of a recipe on demand. In a professional food service establishment, ingredients are measured by weight, volume, and count.

## Systems of Measurement

Two systems of measurement widely used in culinary arts are the customary system and the metric system. The customary system uses the foot (ft or ′) or inch (in. or ″) to measure length, the pound (lb) and ounce (oz) to measure weight, and the gallon (gal.), quart (qt), pint (pt), cup (c), and fluid ounce (fl oz) to measure volume. The metric system uses the meter (m) to measure length, kilogram (kg) to measure weight, and liter (l) to measure volume.

Prefixes are used in the metric system to represent multipliers. For example, the prefix kilo (k) represents 1000, so 1 kilometer (km) = 1000 m. The most common metric prefixes used in food service are kilo (1000), deca (10), deci (1⁄10), and milli (1⁄10000).

To represent a quantity using another prefix, multiply the quantity by the number of units that equals one of the original metric units. For example, to change 544 g to kilograms (1 g = 0.001 kg), multiply 544 by 0.001 (544 × 0.001 = **0.544 kg**). To change 35 l to centiliters (1 cl = 100 l), multiply 35 by 100 (35 × 100 = **3500 cl**). From this information, we can see that a milliliter is 1/1000 of a liter and a kilogram is 1000 grams.

Customary and metric measurements are converted from one system to the other by multiplying by a conversion factor. A *conversion factor* is the number of units of one measurement that equals one unit of another measurement. Conversions are performed by applying the appropriate conversion factor. Equivalency tables are given to convert between systems.

To convert a customary measurement to a metric measurement, multiply the measurement by the number of metric units that equals one of the customary units (conversion factor). For example, to convert 2 ounces into grams, multiply 2 by 28.35 (2 × 28.35 = **56.7 g**). To convert a metric measurement to a customary measurement, multiply the measurement by the number of customary units that equals one of the metric units (conversion factor). For example, to convert 5 liters into quarts, multiply 5 by 1.057 (5 × 1.057 = **5.285 qt**).

## Weight

*Weight* is a measure of the heaviness or mass of an object. When measuring ingredients for use in a recipe, weight is the most accurate of all measurement techniques. Scales are used to measure the weight of dry ingredients in grams, ounces, or pounds. Common types of scales found in the kitchen are portion scales and baker's scales.

### Action Lab

Select a standardized recipe and label each of the parts. Read the recipe carefully and then explain any tips for preparing the recipe successfully.

*Chef's Tip:*

It does not really matter if ingredients are listed in metric or customary units. What is important is using equipment with the appropriate units to measure ingredients. Most volume measures indicate both customary and metric scales.

When weighing items on a portion scale, use the following procedure to ensure accuracy:

1. Place the empty container that will be used to hold the ingredients on the scale. *See Figure 7-4.*

2. Zero out the scale by adjusting the scale to read zero. This is done so that the weight of the container is not reflected in the final measurement.

3. Place ingredients to be measured in the container until the desired weight is achieved.

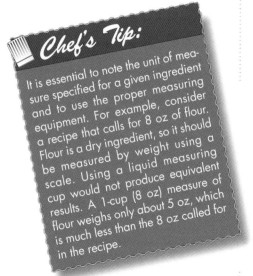

Chef's Tip:

It is essential to note the unit of measure specified for a given ingredient and to use the proper measuring equipment. For example, consider a recipe that calls for 8 oz of flour. Flour is a dry ingredient, so it should be measured by weight using a scale. Using a liquid measuring cup would not produce equivalent results. A 1-cup (8 oz) measure of flour weighs only about 5 oz, which is much less than the 8 oz called for in the recipe.

## Weighing Items

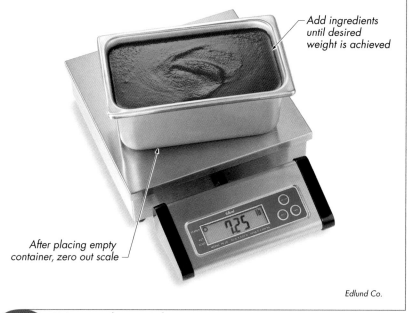

Add ingredients until desired weight is achieved

After placing empty container, zero out scale

Edlund Co.

**7-4**  *Using a scale to weigh items ensures accuracy.*

## Volume

Volume measures are most commonly used to measure liquid ingredients in the professional kitchen. *Volume* is the amount of space that an item occupies. In application, volume is determined with measuring spoons and liquid measures indicating units such as teaspoons (tsp), tablespoons (tbsp), cups (c), pints (pt), quarts (qt), liters (l), gallons (gal.), and bushels (bu). Some types of dry ingredients, such as ground seasonings or spices, may be indicated by volume in a recipe. For example, a recipe may call for 2 tsp oregano. However, most ingredients do not have the same volume and weight. For example, 1 c water weighs 8.3 oz yet has a volume of 8 fl oz.

To avoid making costly errors in the professional kitchen, it is important to understand the difference between ounces and fluid ounces. Using customary measurements, weight is measured in ounces (oz), while volume is measured in fluid ounces (fl oz). One ounce equals ¹⁄₁₆ of a pound. One fluid ounce equals ¹⁄₁₆ of a pint. However, 1 oz does not equal 1 fl oz. For instance, 1 fl oz of buttermilk weighs more than 1 fl oz of water.

The same amount of two different ingredients often does not weigh the same amount. For example, ½ c of fudge is much heavier than ½ c of corn flakes. The ingredients weigh different amounts because they have different densities. Density is the mass (weight) of an object per a given unit of volume.

## Count

*Count* is the measurement of whole items or the actual number of items being used. Count can be used for whole ingredients used in a recipe, such as 2 whole eggs or 1 medium banana. Count can also be used for portion control of a recipe, such as 24 (8-oz) chicken breasts.

Many items are sized by count. For example, a 90-count package of potatoes contains potatoes of a size that average 90 potatoes per 50-lb case. As the count number gets smaller, the size of the item gets larger. For example, a 60-count package of potatoes contains 60 potatoes per 50-lb case. If a case of 90-count potatoes and a case of 60-count potatoes each weigh 50 lb, there are 30 fewer potatoes in the 60-count case. Therefore, a 60-count potato must be larger than a 90-count potato.

## CONVERTING RECIPES

Each recipe produces a specific yield. When a larger or smaller yield is needed, the recipe is converted to produce the desired yield. *Recipe conversion* is the process of increasing or decreasing a recipe to produce a larger or smaller yield than that of the original recipe. The process of recipe conversion involves basic multiplication and division. There are two recipe conversion formulas, one for total yield conversion and one for portion size conversion.

## Total Yield Conversions

A *total yield conversion* is a method of recipe conversion where the total yield of the recipe is converted regardless of portion size. A conversion factor is necessary to perform a total yield conversion. In this case, the conversion factor is the number that results when the desired yield is divided by the original recipe yield.

*Edlund Co.*

To determine the conversion factor for a total yield conversion, apply the following formula:

$$CF = \frac{DY}{OY}$$

where

$CF$ = conversion factor
$DY$ = desired yield
$OY$ = original yield

For example, what is the conversion factor needed to convert a recipe from an original yield of 2 gal. to a desired yield of 5 gal.?

$$CF = \frac{DY}{OY}$$

$$CF = \frac{5\ gal.}{2\ gal.}$$

$$CF = 2.5$$

Thus, 2.5 is the conversion factor needed to convert a recipe yield of 2 gal. to 5 gal.

A total yield conversion is also used to convert a recipe when the number of portions is increasing or decreasing, but the portion size is not changing. For example, a total yield conversion can be used to convert a recipe from an original yield of 16 (8-fl oz) portions to a desired yield of 50 (8-fl oz) portions.

$$CF = \frac{DY}{OY}$$

$$CF = \frac{50}{16}$$

$$CF = 3.125$$

Once the conversion factor is determined, the recipe can be converted to the desired yield. To perform a total yield conversion, the original yield of each ingredient in the recipe is multiplied by the conversion factor. To convert each ingredient to the desired yield, apply the following formula:

$$DY = OY \times CF$$

where

$DY$ = desired yield
$OY$ = original yield
$CF$ = conversion factor

For example, what is the desired yield of a recipe for Butternut Squash and Ginger Bisque that has an original yield of 2 gal. and a conversion factor of 2.5?

$$DY = OY \times CF$$
$$DY = 2 \text{ gal.} \times 2.5$$
$$DY = \textbf{5 gal.}$$

While a conversion factor can be used to increase or decrease the total yield of a standardized recipe, the converted recipe produces portions of the same portion size as the original recipe. ***See Figure 7-5.***

## Total Yield Conversion

Original yield = 2 gal.
Desired yield = 5 gal.
Conversion factor = 2.5

$$CF = \frac{DY}{OY}$$

$$\frac{5}{2} = \textbf{2.5}$$

### Butternut Squash and Ginger Bisque (Yield: 5 gal.)

| Original Quantity | × | CF | = | Desired Quantity |
|---|---|---|---|---|
| 4 butternut squash, roasted | | 2.5 | | 10 |
| 8 oz onion, diced | | 2.5 | | 20 oz |
| 4 oz butter | | 2.5 | | 10 oz |
| 32 oz chicken broth | | 2.5 | | 80 oz |
| 1 tsp cinnamon | | 2.5 | | 2½ tsp |
| ¼ tsp cardamom | | 2.5 | | ⅔ tsp |
| ¼ tsp ginger | | 2.5 | | ⅔ tsp |
| 2 tsp cracked black peppercorns | | 2.5 | | 5 tsp |
| 4 tsp kosher salt | | 2.5 | | 10 tsp |
| 8 oz brown sugar | | 2.5 | | 20 oz |
| 16 oz heavy cream | | 2.5 | | 40 oz |

**7-5** *A total yield conversion is used to alter the recipe yield, regardless of portion size.*

## Portion Size Conversions

A *portion size conversion* is a method of recipe conversion where both the total yield and portion size of the recipe are converted. As when performing a total yield conversion, a conversion factor is needed to perform a portion size conversion. This involves determining both the total original yield and total desired yield of the recipe. These two yields are then used to determine the appropriate conversion factor. To determine the total original yield of a recipe, apply the following formula:

$$TY = NP \times PS$$

where
$TY$ = total original yield
$NP$ = number of original portions
$PS$ = original portion size

For example, what is the total original yield for a recipe that produces 32 (8-fl oz) portions?

$$TY = NP \times PS$$
$$TY = 32 \times 8$$
$$TY = \textbf{256 fl oz}$$

To determine the total desired yield for a recipe, apply the following formula:

$$DY = DP \times DS$$

where
$DY$ = total desired yield
$DP$ = number of desired portions
$DS$ = desired portion size

For example, what is the total desired yield for a recipe to produce 64 (6-fl oz) portions?

$$DY = DP \times DS$$
$$DY = 64 \times 6$$
$$DY = \textbf{384 fl oz}$$

Once the total original yield and total desired yield are determined, the recipe can be converted using a total yield conversion formula. First, determine the conversion factor. For example, what is the conversion factor needed to convert a recipe that produces 32 (8-fl oz) servings to produce 64 (6-fl oz) servings?

$$CF = \frac{DY}{OY}$$
$$CF = \frac{384}{256}$$
$$CF = \textbf{1.5}$$

Thus, 1.5 is the conversion factor needed to convert a recipe that produces 32 (8-fl oz) servings to produce 64 (6-fl oz) servings.

Once the conversion factor is determined, perform a total yield conversion by multiplying the original yield of each ingredient in the recipe by the conversion factor. For example, convert a recipe for Butternut Squash and Ginger Bisque that produces 32 (8-fl oz) portions to produce 64 (6-fl oz) portions. *See Figure 7-6.*

## Portion Size Conversion

Original Yield = 32 (8-fl oz) portions
NP (number of original portions) × PS (portion size) = TY (total yield)
$$32 \times 8 = \mathbf{256 \ fl \ oz}$$

DY (desired yield) = 64 (6-fl oz) portions
DP (desired portions) × DS (desired size) = DY (desired yield)
$$64 \times 6 = \mathbf{384 \ fl \ oz}$$

$$\frac{DY}{OY} = CF$$

$$\frac{384}{256} = \mathbf{1.5}$$

**Butternut Squash and Ginger Bisque** (Yield: 64 [6-fl oz] portions)

| Original Quantity | × CF | = Desired Quantity |
|---|---|---|
| 4 butternut squash, roasted | 1.5 | 6 |
| 8 oz onion, diced | 1.5 | 12 oz |
| 4 oz butter | 1.5 | 6 oz |
| 32 oz chicken broth | 1.5 | 48 oz |
| 1 tsp cinnamon | 1.5 | 1½ tsp |
| ¼ tsp cardamom | 1.5 | ⅓ tsp |
| ¼ tsp ginger | 1.5 | ⅓ tsp |
| 2 tsp cracked black peppercorns | 1.5 | 3 tsp |
| 4 tsp kosher salt | 1.5 | 6 tsp (or 2 tbsp) |
| 8 oz brown sugar | 1.5 | 12 oz |
| 16 oz heavy cream | 1.5 | 24 oz |

**7-6** *A portion size conversion alters the portion size in addition to the total recipe yield.*

## Recipe Conversion Issues

When recipes are increased or decreased by a relatively small amount, the conversion can be performed relatively easily. But when recipes are converted to produce dramatically different yields, such as from 25 portions to 500 portions, problems often occur. These problems stem from not making adjustments to the cooking equipment, timing, or temperature, differences in moisture loss, and recipe error.

**Cooking Equipment.** When converting a recipe to a larger or smaller yield, not only do the quantities of each ingredient increase or decrease, but the size of needed equipment must also increase or decrease respectively. For example, a recipe for preparing 1 lb of rice may call for a 2-qt saucepot. When the recipe is converted to produce 5 lb of rice, a different piece of cooking equipment is required. Similarly, when the recipe is converted to produce 8 oz of rice, a smaller saucepot should be used to ensure successful preparation.

**Timing.** Correct preparation and cooking times are essential for success when following recipes, and remain relatively stable when converting recipes to larger or smaller yields. Preparation and cooking times are not significantly altered when performing a total yield conversion. For example, a single steak on a grill cooks in roughly the same amount of time as 10 steaks on a grill. One tray of cookies bakes in approximately the same amount of time as three trays of cookies.

There are times, however, when a slight modification to preparation or cooking time is necessary when converting a recipe. For example, converting a recipe with a small yield to a produce a very large yield may require a slightly longer mixing time in order to properly incorporate the larger amounts of ingredients. Likewise, if a cake recipe written to produce a 3″-thick cake is prepared as a much thinner sheet cake, the thinner cake will be fully cooked in a much shorter length of time than required for the original recipe.

The temperature of ingredients may also require a change in cooking time. For example, a large roast held at room temperature while it is trimmed and prepped requires less time to roast than a roast that was trimmed and prepped the previous day and held in refrigeration prior to cooking. It takes more time to heat the center of a very cold roast than it takes to heat the center of a roast that has been held at room temperature.

**Temperature.** Cooking temperature does not usually need to be altered when converting a recipe yield. However, changing the density of the main ingredient could require a change in cooking temperature. For example, substituting boneless chicken breasts with bone-in chicken breasts requires an increase in cooking time because bones do not conduct heat well, which slows the cooking process. Because cooking time increases, cooking temperature may need to be lowered 25°F to 50°F to prevent the exterior of the food from overcooking before the center of the food is fully cooked.

A change in cooking temperature may also be necessary to account for the use of different equipment to prepare a recipe. Using a convection oven instead of a conventional oven requires a decrease in cooking temperature by approximately 75°F. This is because convection ovens produce more even cooking at lower temperatures.

**Moisture Loss.** When preparing a converted recipe, a minor loss of moisture may result in the finished item having a drier texture. In some cases, the item may be completely ruined. Moisture loss can occur when a recipe is converted to a smaller yield without adequately converting the equipment to maintain consistent evaporation or surface area. For example, if a recipe that yields 50 servings of rice pilaf is converted to yield 10 servings, the required cooking equipment should be scaled to correspond to the smaller yield. Using a pan with too much surface area would cause excess moisture loss through evaporation, and the rice would not absorb the correct amount of moisture, resulting in a tough, undercooked pilaf.

**Recipe Error.** In some instances a mathematical error can ruin the recipe. Recipe error is more likely to become a problem when a recipe is converted to produce a much larger yield. When adding ingredients that have potential to significantly affect the taste of a recipe, such as herbs, spices, acids, or wine, it is best to add the converted amount slowly and taste the recipe throughout the addition to ensure appropriate flavor.

*Cres-Cor*

## CALCULATING FOOD COSTS

A food service establishment needs to continually monitor both the cost of food purchased and the cost of meals served to ensure the establishment earns more in food sales than it pays in expenses. Calculating food cost involves much more than simply adding up the cost of ingredients on a plate. Most food items are purchased in bulk (large quantities), as it is much less expensive than purchasing food in small amounts. Items sold in bulk often need to be processed further to make them usable as ingredients or portions. When this is done, the purchased price of a bulk item may differ greatly from the actual per serving cost. This is done by calculating unit cost, edible portion cost, and total recipe cost.

### Calculating Unit Cost

To accurately calculate how much a recipe costs to produce, the cost of each ingredient used in the recipe must be determined. Since many items are purchased in large quantity or in bulk form, an as-purchased cost must be converted to an individual unit cost. *See Figure 7-7*. An *as-purchased (AP) cost* is the entire original cost of a bulk item. For example, an as-purchased cost of $14.90 for a case of 80 meatballs means that $14.90 is the price paid for the entire case.

## Item Cost

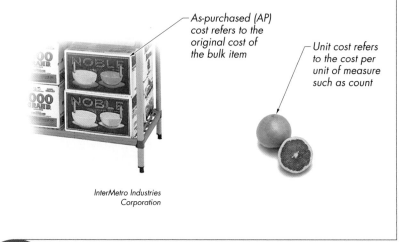

As-purchased (AP) cost refers to the original cost of the bulk item

Unit cost refers to the cost per unit of measure such as count

InterMetro Industries Corporation

**7-7** *Item cost may be expressed as the AP cost of a bulk item or per the unit of measure.*

A *unit cost* is the cost of a bulk item per unit of measure, such as pound, ounce, gallon, cup, tablespoon, or count. The unit of measure is determined by what is called for in a particular recipe. For example, if a recipe calls for 8 oz of flour and the flour was purchased as a 50-lb bag, the as-purchased cost for the 50-lb bag of flour must be converted to a unit cost of price per ounce.

To calculate the unit cost of an item, apply the following formula:

$$UC = \frac{AP}{NU}$$

where

$UC$ = unit cost of item

$AP$ = as-purchased cost

$NU$ = number of units per case

For example, what is the unit cost per frozen meatball in a case of 80 meatballs that costs $14.90?

$$UC = \frac{AP}{NU}$$

$$UC = \frac{14.90}{80}$$

$$UC = \textbf{\$0.19 per meatball}$$

It is more difficult to calculate unit cost for an item measured in small units of weight or volume. A recipe may call for 4 oz of cake flour when the as-purchased cost for cake flour is for a 50-lb bag. To determine unit cost, the weight of the bulk item measured in pounds must first be converted to ounces.

For example, what is the unit cost for 1 oz of cake flour when a 50-lb bag of cake flour costs $34.78?

1.  Determine the weight of the bulk item in ounces. There are 16 oz in a pound, so the number of pounds is multiplied by 16.

    50 × 16 = 800 oz

2.  Determine the unit cost.

$$UC = \frac{AP}{NU}$$
$$UC = \frac{34.78}{800}$$
$$UC = \$0.4 \text{ per oz}$$

The conversion is similar when a volume measure is required. A recipe calls for 3 fl oz of olive oil. A case of six 1-gal. olive oil containers is purchased for $111.00. What is the unit cost for 1 fl oz of olive oil?

1.  Determine the cost per gallon container.

$$UC = \frac{AP}{NU}$$
$$UC = \frac{111.00}{6}$$
$$UC = \$18.50 \text{ per gal.}$$

2.  Determine the weight of the bulk item in fluid ounces. There are 128 fl oz in a gallon, so the number of gallons is multiplied by 128.

    1 × 128 = 128 fl oz

3.  Determine the unit cost.

$$UC = \frac{AP}{NU}$$
$$UC = \frac{18.50}{128}$$
$$UC = \$0.14 \text{ per fl oz}$$

As long as an item is used right out of the package with no waste, such as flour, oil, or mayonnaise, this method of determining unit cost is highly accurate. If an item requires fabrication or preparation that results in any loss or waste from trimming, filleting, skinning, deboning, cooking, or shrinkage, the edible-portion cost must be determined before determining unit cost.

## Calculating Edible-Portion Cost

Many food items lose volume, weight, or count in preparation. An *edible portion (EP)* is the amount of a food item that can be used in a recipe after unusable parts are trimmed away. ***See Figure 7-8.*** For example, bananas are peeled for use in a fruit plate, a fish is filleted prior to sautéing, and stalks are trimmed from broccoli to produce flowerets.

## Edible Portion

**7-8** *An edible portion is the usable part of a food item left after inedible parts are trimmed away.*

The amount of as-purchased product is determined by measuring the item before preparation. The trim loss can be determined by measuring the amount of waste material after preparation. *Trim loss* is unusable volume, weight, or count that is removed from an as-purchased product prior to preparation.

To determine the edible portion of a food item, apply the following formula:

$$EP = PA - TL$$

where

$EP$ = edible portion
$PA$ = as-purchased amount of product
$TL$ = amount of trim loss

When determining the edible portion, the same units must be used for all factors. Therefore, if an as-purchased product is measured in ounces, trim loss and edible portion must also be measured in ounces. For example, what is the edible portion when 4 ears of corn weigh 48 oz and have 20 oz of trim loss from unusable corn cobs?

$$EP = PA - TL$$
$$EP = 48 - 20$$
$$EP = \textbf{28 oz}$$

The *edible-portion cost* is the cost of the usable part of a bulk item. For example, the as-purchased cost of broccoli is $0.49 per lb, but for every 1 lb of broccoli, only ¾ lb is edible. When calculating edible-portion cost, the as-purchased cost is divided by the edible portion. The edible-portion cost accounts for the fact that the cost to serve an item is greater than is reflected by the original as-purchased cost. To calculate edible-portion cost, apply the following formula:

$$EC = \frac{AP}{EP}$$

where
$EC$ = edible-portion cost
$AP$ = as-purchased cost
$EP$ = edible portion

For example, what is the edible-portion cost for broccoli with an as-purchased cost of $0.49 per lb that has a 0.75 lb edible portion per 1 lb?

$$EC = \frac{AP}{EP}$$

$$EC = \frac{0.49}{0.75}$$

$$EC = \textbf{\$0.65 per lb}$$

## Calculating Yield Percentages

A *yield percentage* is the ratio of the edible portion of a food to the as-purchased amount. In other words, it is the ratio of the amount of usable food to the original weight of food purchased. To determine yield percentage, apply the following formula:

$$YP = \frac{EP}{PA} \times 100$$

where
$YP$ = yield percentage
$EP$ = edible portion
$PA$ = as-purchased amount of product
$100$ = constant

For example, what is the yield percentage for iceberg lettuce when 6 heads of lettuce weigh 48 oz and yield 28 oz of edible lettuce?

$$YP = \frac{EP}{AP} \times 100$$

$$YP = \frac{28}{48} \times 100$$

$$YP = 0.58 \times 100$$

$$YP = \textbf{58\%}$$

When yield percentage is known, it can be used to calculate the edible portion of an as-purchased item. To determine the edible portion of an as-purchased item using yield percentage, apply the following formula:

$$EP = PA \times YP$$

where
$EP$ = edible portion
$PA$ = as-purchased amount of product
$YP$ = known yield percentage

For example, what is the edible portion of a 24-lb case of green peppers with a known yield percentage of 44%?

$$EP = PA \times YP$$
$$EP = 24 \times 44\%$$
$$EP = 24 \times 0.44$$
$$EP = \mathbf{10.56\ lb}$$

Likewise, if the edible portion of a food item and the yield percentage are known, the as-purchased amount of the item can be determined. This is useful for determining how much of an item must be ordered. To determine the as-purchased amount of an item when the edible portion and yield percentage are known, apply the following formula:

$$PA = \frac{EP}{YP}$$

where
$PA$ = as-purchased amount of product
$EP$ = edible portion
$YP$ = known yield percentage

For example, how much whole, fresh salmon must be ordered when 25 lb of salmon fillets are needed, with a known yield percentage of 82%?

$$PA = \frac{EP}{YP}$$
$$PA = \frac{25}{82\%}$$
$$PA = \frac{25}{.82}$$
$$PA = \mathbf{30.5\ lb}$$

**Yield Tests.** *Yield tests* are commonly used to determine raw yields, cooking-loss yields, and shrinkage yields. A raw yield test is a procedure used to determine the yield percentage of a food product where the trimmed parts have no other use in the professional kitchen. This is the most basic form of yield test. For example, peeling a banana results in usable fruit (edible portion) and unusable peel (trim loss). When preparing foods, the trim loss must be accounted for to maintain accurate food costs.

In addition to loss from preparation, there can also be loss during the cooking process. A *cooking-loss yield test* is a procedure used to determine the total weight as served of a food product. The *total weight as served (AS)* is the weight of a food product after fabrication and/or loss from cooking. For example, a 13-lb uncooked lamb rib may yield only 10 lb of lamb chops after fabrication and cooking. ***See Figure 7-9.***

## Total Weight as Served

French rack of lamb trimmed for preparation

Prepared lamb chops after loss from cooking

**7-9** Total weight as served is the weight of a food item after cooking.

Performing a cooking-loss yield test can determine how much of a food product must be purchased and prepared to produce the needed amount of servings for a meal service or large function, such as a banquet. To determine the total weight as served of a food product, apply the following formula:

$$AS = NP \times WP$$

where
$AS$ = total weight as served
$NP$ = number of portions prepared
$WP$ = weight per portion

For example, what is the total weight as served for an uncooked 18-lb roast that yields 36 prepared 6-oz portions?

$$AS = NP \times WP$$
$$AS = 36 \times 6$$
$$AS = 216 \text{ oz } (\textbf{13.5 lb,} \text{ } 216 \text{ oz} \div 16 = 13.5 \text{ lb})$$

**Shrinkage.** When the total weight as served is known, the percentage of shrinkage can be determined. *Shrinkage (S)* is weight or volume that is lost from a food item during the cooking process. For example, meat may lose as much as 30% of its weight as fat is rendered out during the roasting process. ***See Figure 7-10.*** To determine the amount of shrinkage, a food is weighed just prior to cooking and again when cooking is complete. The weight as served is then subtracted from the uncooked weight. For example, the shrinkage of a 12-oz ground beef patty that weighs 10 oz after cooking is 2 oz (12 – 10 = 2 oz).

Knowing the percentage of shrinkage enables a chef or manager to identify how shrinkage affects the true cost of a product. To determine the percentage of shrinkage, the amount of shrinkage is divided by the weight of the product prior to cooking. To determine the percentage of shrinkage, apply the following formula:

$$PS = \frac{S}{W} \times 100$$

where
$PS$ = percentage of shrinkage
$S$ = shrinkage
$W$ = weight before cooking
100 = constant

For example, what is the percentage of shrinkage for 1 lb (16 oz) of ground beef that weighs 13 oz after it is cooked and the fat is drained off? First, determine the shrinkage (16 – 13 = **3 oz**).

$$PS = \frac{S}{W} \times 100$$
$$PS = \frac{3}{16} \times 100$$
$$PS = 0.19 \times 100$$
$$PS = \textbf{19\%}$$

Knowing the percentage of shrinkage for a particular food product is necessary to determine how much raw product is needed to produce a desired amount of cooked food. For example, if 100 tacos are needed for a function, and each taco contains 2 oz of cooked ground beef, how much raw ground beef is needed? Simply taking the number of portions (100) and multiplying by the portion size (2 oz) indicates that 200 oz, or 12½ lb of cooked beef is required.

## Shrinkage

**7-10** *Shrinkage is weight or volume lost during the cooking process.*

To determine the amount of raw product to purchase when the percentage of shrinkage is known, apply the following formula:

$$W = \frac{AS}{(100 - PS) \div 100}$$

where
$W$ = weight before cooking
$AS$ = weight as served
$100$ = constant
$PS$ = percentage of shrinkage

For example, how much raw ground beef is needed to produce 12½ lb (200 oz) of cooked ground beef when the shrinkage is known to be 19%?

$$W = \frac{AS}{(100 - PS) \div 100}$$

$$W = \frac{200}{(100 - 19) \div 100}$$

$$W = \frac{200}{81 \div 100}$$

$$W = \frac{200}{0.81}$$

$$W = 246.9 \text{ oz} = 15\frac{3}{8} \text{ lb}$$

## Calculating Total Recipe Cost

A successful food service establishment must practice accurate and effective cost control. A key to cost control is to prepare all items from standardized recipes. *Costing* is the process of determining the total cost of preparing a recipe based on the individual costs of all ingredients. The cost of an individual portion of a recipe can be determined when the recipe has been costed. The cost of ingredients for an individual portion should be used to determine the selling price of that menu item. To calculate total recipe cost, the unit cost of each ingredient must be determined in the amount needed for the recipe, and then the unit cost of each ingredient is added together. A recipe costing form is often used to help determine total recipe cost. ***See Figure 7-11.***

When calculating total recipe cost, it is important to figure unit cost for each item in the same unit of measure that is called for in the recipe. For example, if a recipe calls for 3 oz of an ingredient that was purchased in a 10-lb case for $15, the purchase price must be converted from pounds to ounces by finding the total weight of the case in ounces and then determining the price per ounce (10 lb × 16 oz = 160 oz; $15 ÷ 160 oz = **$0.09 per ounce**).

## Recipe Costing Forms

| Recipe Costing | | | | | | | | |
|---|---|---|---|---|---|---|---|---|

**Recipe:** Shrimp Creole
**Yield:** 25 Servings
**Portion Size:** 9 oz

| Item No. | Ingredient | Quantity | As Purchased (AP) | | Yield Percentage | Edible Portion (EP) | | Total Cost |
|---|---|---|---|---|---|---|---|---|
| | | | Amount | Cost | | Amount | Cost | |
| 1532 | **shrimp** | 8 lb | | | 88% | | | |
| 254 | **prepared creole sauce** | 3 qt | — | — | — | — | — | — |
| 1906 | olive oil | ¼ c | 1 gal. | $25.86 | 100% | | | |
| 1148 | medium onion, chopped | 6 | | | 83% | | | |
| 1112 | celery, chopped | 3 c | | | 75% | | | |
| 1126 | red bell pepper, chopped | 3 | | | 85% | | | |
| 1159 | garlic, minced | 6 cloves | | | 81% | | | |
| 103 | chicken stock | 1.5 qt | 1 gal. | $1.03 | 100% | | | |
| 2106 | canned tomatoes, chopped | 5.5 lb | 6.83 lb | $5.26 | 100% | | | |
| 1874 | bay leaf | 3 | | | 100% | | | |
| 1852 | cajun seasoning | 2 oz | 20 oz | $4.17 | 100% | | | |
| 1886 | **salt** | to taste | | | 100% | | | |
| 1890 | **pepper** | to taste | | | 100% | | | |
| 1216 | **prepared white rice** | 3 lb | | | 100% | | | |

**7-11** *Recipe costing forms are used to calculate the total cost of food items in a recipe.*

## MANAGING FOOD COSTS

There are many costs that must be monitored and managed to successfully run a food service establishment. *Cost* is the actual price an establishment pays for goods (e.g., food, rent, beverages, cleaning supplies) or services (e.g., service staff, cleaning crew, insurance, repairs). There are two types of costs: fixed costs and variable costs. A *fixed cost* is a cost that does not change as sales increase or decrease. Rent, real estate taxes, and employee salaries are examples of fixed costs. A *variable cost* is a cost that increases or decreases in proportion to the volume of production. When sales are up and business is brisk, variable costs increase. When sales are down and business is slower, variable costs decrease. Food and beverage purchases and hourly payroll are examples of variable costs because they increase or decrease depending on the activity of the business.

There are many aspects of a food service operation that require a watchful eye to keep costs in line. The menu, portions, purchasing, receiving, storing, issuing, and waste are all important aspects of controlling costs.

## The Menu

The menu is the roadmap to success in any food service establishment. *See Figure 7-12.* It is the document that markets the establishment to the customer by offering interesting dishes that satisfy customer needs and desires, are profitable, and are competitive with alternatives offered elsewhere. Careful consideration must be taken when creating or changing a menu. Factors such as available equipment and kitchen facilities, ingredients, employee skills, and customer demand must be considered.

## Menus

### Menu

#### Appetizers

baked brie en croute
with sundried tomato pesto and aged balsamic glaze

pan–seared crab cake
with caramelized Vidalia onions, fire-roasted
bell peppers and lemon butter sauce

sautéed black mussels
with garlic, white wine and parsley

#### Soups

chilled English cucumber soup
with charentais melons, yuzu pearls, salmon ceviche and roe

light lobster bisque
with bomba rice and summer vegetables

creamy yellow spring pea soup
with house-smoked trout

leek and potato potage
with Wellfleet clams and fines herbes

cream of Alsace cabbage soup
with home-smoked sturgeon and caviar

#### Entrées

molasses & black pepper
basted pork tenderloin
with ginger-scented mashed sweet potatoes, peach-fig chutney

seared duck breast
with cherry reduction sauce, mascarpone polenta cake,
applewood-smoked bacon braised greens

**7-12** *The menu defines the offerings within a food service establishment.*

In order to prepare a menu item, the kitchen must be equipped with all necessary cooking and preparation equipment. The staff must be able to handle the preparation of all main dishes and side dishes as needed to meet customer demand. If any fabrication is to be performed in-house, space must be allocated for those tasks as well. Seasonal availability and the cost of products must also be considered. For example, a dish with pomegranate seeds may look beautiful, but pomegranates are difficult to acquire out of season and then are very costly if a source is found.

Customers are selective about the foods they are willing to try and how much they are willing to spend. A chef may desire to offer exotic menu items, but in order for the establishment to make a profit on those items, customers must order and pay for them. If there is not enough demand for an exotic item, it can be very costly to an establishment. Regardless of whether the item is ordered, the establishment must keep the ingredients for it on hand. Unused ingredients may spoil and have to be thrown away in the event that it is not ordered.

## Portion Control

A recipe should specify the number of portions it produces and the size of each portion. For example, a recipe that fills a 4″ full-size hotel pan and yields 24 portions should also indicate the size of the portioning or serving utensil needed to produce that number of portions. Using portion control equipment such as ladles, spoons, spatulas, portion control scoops, balances and portion scales, and electric slicers helps ensure accurate portions are served and portion costs are maintained. *See Figure 7-13.*

## Portion Control Equipment

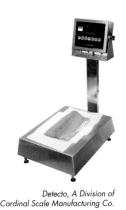

Carlisle FoodService Products

Detecto, A Division of
Cardinal Scale Manufacturing Co.

Carlisle FoodService Products

**7-13** *Portion control equipment such as ladles, spoons, and scoops are used along with scales and slicers to control portion size and costs.*

## Purchasing

Purchasing involves much more than just buying products. Purchasing is actually the selection and the procurement of specific goods. The first step in purchasing is development of purchase specifications. A *purchase specification* is an accurate description of an item, noting the grade, quality, packaging, and unit size required. Purchase specifications are commonly used to establish overall quality standards for desired products. Developing and using these specifications helps ensure that exactly the right products are received. Once a purchase specification is developed, accurate quotes can be requested from multiple purveyors.

In order to determine the appropriate amount of any item to order, a par stock must be established. *Par stock* is the amount of inventory needed on hand to provide adequate supply from one delivery to the next. Establishing a par stock for each item helps ensure that neither too much nor too little product is ordered. Maintaining par stock helps minimize the chance of either running out of items or having to throw away spoiled items. Once the purchase specifications and par stock are determined, a purchase is made using the following procedure:

1. Compile the purchase order based on the specifications and the par stock.

2. Obtain price quotes from multiple vendors based on the purchase order.

3. Choose the best vendor (based on availability, price, and service) and place the order.

4. Receive the order, checking items ordered against the invoice for accuracy.

5. Properly store the order using FIFO and following proper storage requirements.

6. Follow up with the vendor on any pricing errors or discrepancies in quality, product, or quantity.

7. Issue product to specific areas as requested and needed.

Food is classified as perishable or nonperishable. *Perishable items* have a short shelf life and should be purchased frequently in relatively small amounts. Items such as fresh fish, dairy products, fresh meats, and produce are common examples of perishable items. *Nonperishable items* have a much longer shelf life and typically can be kept for six months to a year. Nonperishable items are usually stored at room temperature in their original packing containers in a storeroom. Products such as canned tomatoes, tea bags, and jars of olives are examples of nonperishable food items.

## Receiving

Once purchase specifications are developed and the items are purchased, they are delivered to the establishment. At large food service establishments, the person who receives merchandise is called the receiving clerk. In a smaller establishment, the chef or sous chef may receive the merchandise. All delivered items should be checked against the original invoice and signed for. *See Figure 7-14.* It is important to check deliveries carefully before signing to ensure the following standards are met:

*Receipt of Deliveries*

7-14 *Deliveries are signed for upon receipt and then entered into inventory.*

- all delivered items were ordered

- all items ordered were delivered

- all delivered items match the items listed on the invoice

- all purchase specifications are met

- perishable items such as meat, seafood, poultry, and produce are fresh and of the appropriate weight

- perishable items such as meat, seafood, poultry, and dairy products are delivered at a proper temperature to comply with health department and industry standards; temperatures must be checked by placing a thermometer between packages

- the price charged is the same as the price quoted

- packaging is free of any musty or strange odors

- no items are damaged (e.g., dented cans, crushed boxes, torn packages)

- all packaging is free of water damage or stains from contact with liquids

- all frozen foods are free of large ice crystals (a sign that the item has thawed and been refrozen)

Once a delivery has been received and signed for, it is the property of the establishment, so it is important to check that each of these standards is met prior to signing for the delivery.

## Storage

Storing foods properly is essential for preventing spoilage, waste, pilferage, contamination, and infestation. Immediately after the items have been received and checked in, they should be labeled, dated, and properly entered into inventory using the FIFO system. New items should be stored behind older items so that the oldest items are used first.

All food items in either dry or cold storage areas must be kept 6″ off the floor. Food can never be stored on the floor itself, no matter how clean the floor. Dry storage areas must be well ventilated to prevent foods from molding or obtaining musty odors. A proper pest control plan must be in place to ensure that products do not come in contact with rodents or other pests. Cold storage areas (refrigerators and freezers) need to be kept clean and organized. Thermostats need to be checked daily to ensure that refrigeration units are operating at proper temperatures to keep food safe.

## Issuing

Since theft can be an issue in food service establishments, employee access to storage areas is often limited. Some establishments use an issuing procedure, where items that have been received and stored can be moved into production only after a requisition is issued. A *requisition* is an internally generated invoice that is used to aid in tracking inventory as it moves from storage to production.

## Waste

The goal of food service establishments is to make a profit. If staff is wasteful with products, the profitability of the establishment is reduced. Waste can result from overproduction, spoilage, misuse of trim or byproducts, and poor recordkeeping.

Overproduction occurs when too much product is produced and there is no means of using it. Any unused excess product must be thrown away. Spoilage occurs when food goes bad and must be discarded before it is used. Using the FIFO method of stock rotation is the first step in limiting spoilage of ingredients. Other ways to limit spoilage include ordering the correct amount of an ingredient based on par stock and avoiding obsolescence. *Obsolescence* is the removal of an item from the menu while ingredients for that particular item are still in stock.

Fabrication of products such as meat, poultry, fish, and vegetables often results in usable trim or byproducts. It is wasteful to discard these items. Instead, usable trim or byproducts should be saved for the production of stocks or other food items. Likewise, high-quality leftovers can potentially be used in other dishes. For example, meat from roasted chicken can be pulled, diced, and used in a chicken soup.

A good recordkeeping system can also help reduce potential waste. If an item must be discarded due to overproduction or spoilage, the loss should be recorded. When ordering in the future, records of waste can be useful in determining if particular items should be ordered in smaller amounts.

*Frieda's Inc.*

## SUMMARY

Serving great-tasting food is not enough to guarantee that a food service establishment will be successful. Policies and regulations need to be in place to ensure that a consistent quality of food is served every day by employees demonstrating exceptional customer service. The need for consistent quality is the reason that recipes are standardized in most food service establishments. A standardized recipe helps ensure that each dish turns out the same way and keeps the cost of preparing that dish consistent as well.

From the time a menu is written, to the minute a product is received and stored, to the moment the item is served to the customer, the flow of food must be controlled. The cost of all products purchased by the establishment must be monitored. Portions served must be consistent in size and quality. Standardization and cost control are essential to maintaining a successful and profitable food service establishment.

Refer to the CD-ROM for Quick Quiz® questions related to chapter content.

## Review Questions

1. Identify the seven key elements of standardized recipes.

2. What is the difference between yield and portion size?

3. Define conversion factor.

4. Describe the method for converting a customary measurement to a metric measurement.

5. Contrast the definitions and measuring devices for weight, volume, and count.

6. Describe the formulas used for total yield conversion.

7. Explain the difference between a total yield conversion and a portion size conversion.

8. Describe factors to consider when converting a recipe to a larger or smaller yield.

9. Describe the formula for calculating the unit cost of an item.

10. Describe the relationship between edible-portion and trim loss.

11. Define edible-portion cost.

12. Describe the formula for calculating yield percentage.

13. What are the differences between a raw yield test and a cooking-loss yield test?

14. Describe the formula for determining the total weight as served of a food product.

15. Identify a benefit of knowing the percentage of shrinkage of a food item.

16. Explain how costing is used in the food service industry.

17. Contrast fixed and variable costs.

18. List factors to consider when designing a cost-effective menu.

19. Identify common types of portion control equipment that should be used to ensure portion consistency of menu items.

20. What is a purchase specification?

21. List the standards to check for upon delivery of purchased items.

22. Why is adequate ventilation a requirement for dry storage areas?

23. How can issuing procedures contribute to cost control?

24. Identify four practices that contribute to wastefulness in food service establishments.

25. Define obsolescence.

# Principles and Methods of Cooking

## Chapter 8

*Cooking methods* vary depending upon the type of food item to be cooked. An understanding of how and when to apply various cooking methods helps ensure foods taste good, are easy to digest, and are free of harmful organisms. Cooking occurs when foods are exposed to heat. Heat is transferred to foods through conduction, convection, and radiation. Various dry-heat, moist-heat, and combination cooking methods are used to transfer heat to foods.

## Chapter Objectives:

1. Describe how cooking affects the various characteristics of food.
2. Describe the processes of conduction, convection, and radiation as they relate to cooking.
3. Demonstrate common dry-heat cooking methods and explain the advantages of each.
4. Demonstrate common moist-heat cooking methods and explain the advantages of each.
5. Demonstrate the two main types of combination cooking methods and provide reasons for using these methods.

## Key Terms

- **conduction**
- **convection**
- **radiation**
- **dry-heat cooking**
- **deep-frying**
- **pan frying**
- **sautéing**
- **griddling**
- **roasting**
- **carryover cooking**
- **baking**
- **broiling**
- **moist-heat cooking**
- **steaming**
- **boiling**
- **blanching**
- **simmering**
- **poaching**
- **combination cooking**
- **braising**
- **stewing**

## COOKING EFFECTS ON FOOD

*Cooking* is the process of subjecting foods to heat in order to accomplish three main goals: to make them taste better, to make them easier to digest, and to kill harmful microorganisms that may be present in the food. When heat is introduced to food through the cooking process, characteristics of the food begin to change. All foods are made up of various amounts of proteins, carbohydrates, fats, water, vitamins, and minerals. Each of these nutrients is sensitive to heat, and as nutrients change through heat exposure, the nutrients in turn change some aspect of the food, such as smell, flavor, color, or texture. Understanding how heat affects the characteristics of food helps achieve the desired results when cooking.

### Proteins

A *protein* is a complex organic compound found in living plant and animal cells. When proteins are exposed to heat, they begin to firm and coagulate. *Coagulation* is the process of changing from a high-moisture (liquid) state to a low-moisture (semi-liquid or solid) state. For example, a raw egg coagulates when boiled, resulting in a hard-cooked egg. This can also be seen with chicken breast, which is very high in protein and water. As the chicken cooks, it changes from a raw, watery state to a more solid state; it eventually loses much of its moisture content and becomes very dry if overcooked.

### Carbohydrates

A *carbohydrate* is a complex chemical substance found in food and is the primary energy source for the human body. Carbohydrates are classified as simple and complex. A simple carbohydrate, also known as a simple sugar, is a carbohydrate with little or no nutritional value. When simple carbohydrates are eaten, the body stores them as fats. Simple carbohydrates begin to darken as the temperature of the food rises, changing the flavor and aroma of the food dramatically. This change in simple sugars is referred to as caramelization. To *caramelize* is to heat sugar (or sugars naturally present in foods) until it liquefies and turns golden brown in color. ***See Figure 8-1.*** Caramelization occurs when any food begins to brown during cooking. Caramelization is the crispy brown outside of a steak that was cooked on a grill. It is the wonderful smell given off from a freshly baked, golden brown loaf of bread. It is the sweet taste of grilled onions. Basically, caramelization is responsible for many of the wonderful flavors, aromas, and colors of many cooked foods.

A *complex carbohydrate,* also known as a starch, is a carbohydrate that provides energy to the body. When complex carbohydrates are eaten, the body immediately uses them as an energy source. Unused

## Action Lab

Try this experiment to test the theory of heating proteins. Take two chicken breasts that are the same size and weight. Place one of the raw chicken breasts in the refrigerator. Take the second breast and cook it in a sauté pan until thoroughly cooked. Now remove the first breast from the refrigerator and place the two breasts side by side to examine any change in size or shape. Also, weigh each of the breasts to see if either one has changed in weight. Discuss your findings.

carbohydrates are stored by the body as fats. Complex carbohydrates react differently to heat than simple sugars do. As starches are heated, they begin to absorb water and swell. This is why blueberry muffin batter is very wet in its raw form but expands as it bakes. The starch molecules in the flour absorb moisture from the batter and steam created during the baking process. Starches also absorb moisture when gravy is made. Adding a starch mixture to chicken stock eventually thickens the broth and makes it into chicken gravy because the starch molecules absorb some of the water content of the stock. This is the reason why chicken stock can be changed into chicken gravy simply by incorporating a starch component.

### Fats

*Fats* are organic compounds found in plants and animals that later become energy when consumed as food. Fats can be either solid or liquid at room temperature. Solid fats begin to melt and turn into a liquid as heat is introduced. When fats are liquid at room temperature, they are called oils. In cooking, fat is used as a general term for various substances including butter, margarine, lard, shortening, and various oils. In deep-frying, fats are used to transfer heat to foods. The heat helps simple carbohydrates to caramelize, creating the crispy brown exterior of fried foods.

### Water

Water in foods evaporates in the form of steam as foods are heated. The result is that foods cooked over a long period of time become increasingly tough and dry if an effort is not made to retain the moisture. Moisture can be retained through various moist-heat cooking methods. Keeping food items covered while cooking also prevents steam from escaping.

When a freshly cooked piece of meat is cut, steam is released. This steam is actually water that was naturally in the meat. As meat is heated, internal water begins to change to a gas (steam), and when the meat is cut, gas is released into the air. Food should be left to rest after it is cooked so the food can retain moisture that might otherwise be released as steam. *See Figure 8-2.* For example, if a loaf of bread were to be cut as soon as it came out of the oven, the bread would immediately begin to dry out and lose much of its freshness. If a newly roasted piece of meat is cut, moisture may be released by the juices running out of the hot meat onto the cutting board and by evaporating into the air as steam. If the same roast is allowed to rest for 10 to 20 min after it is removed from the oven, the moisture stays in the meat.

In some instances, it may be desirable to allow water to cook out of a liquid food. Such is the case in preparing sauces, stocks, or soups. By allowing water to escape, the resulting food acquires a more concentrated

*Caramelization*

**Raw Onions**

*Onions begin to sweat*

**Partially Cooked Onions**

**Fully Cooked Onions**

8-1 *Caramelization occurs when the sugars naturally present in food liquefy and turn brown in color.*

## Allowing Food to Rest

National Pork Producers Council

**8-2** *Food should be left to rest after it is cooked so that it can retain moisture that might otherwise be released as steam when the food is cut.*

flavor and thicker consistency. *Reduction* is the process of gently simmering water out of a liquid through evaporation to concentrate flavor and thicken the texture of the food. Sauces, stocks, or soups may be cooked for an extended period of time over low heat to maximize the amount of reduction taking place.

### Vitamins and Minerals

Vitamins are organic substances that are necessary for proper nutrition. Minerals are inorganic substances that are necessary for proper nutrition. Vitamins and minerals are present in all foods in their natural state; however, their composition is very fragile. The retention of vitamins and minerals should be a concern when cooking foods, as overheating food can destroy nutrients. The longer a food cooks, the more nutrients the food loses. For this reason, a cooking method should be chosen that helps to preserve the nutritional value of the food.

### Color and Texture

Color and texture are also affected by heat in foods. The longer a food cooks, the more color loss and texture change occurs. Red meat that is cooked for several hours over low heat becomes very tender. With too long a cooking time, however, red meat cooks to a well-done state and loses most of its color, turning gray. The texture of well-done meat is very tough or chewy. Green vegetables also lose their color, changing from bright green to a yellowish gray color as the green pigment, called chlorophyll, is destroyed by heat. The crisp texture of raw vegetables is lost as heat exposure increases. Properly cooked vegetables are tender and juicy, while overcooked vegetables become soft and mushy. It is important to understand proper cooking methods so the best texture, color, aroma, flavor, and nutritional values are achieved.

### HEAT TRANSFER METHODS

No matter what type of food is being prepared, there are only three methods used to transfer heat to an item: conduction, convection, and radiation. **See Figure 8-3.** Conduction is a method of heat transfer relying on direct physical contact of food with a hot surface, as an egg in an omelet pan. Convection is a method of heat transfer where a substance such as hot air or hot water surrounds the food, as in roasting a turkey or steaming vegetables. Radiation relies on heat transfer through heat waves produced by a heating element, as in a toaster oven or broiler. Each of the heat transfer methods cooks food in a different manner and produces different results. To achieve the desired results, the appropriate heat transfer method must be chosen.

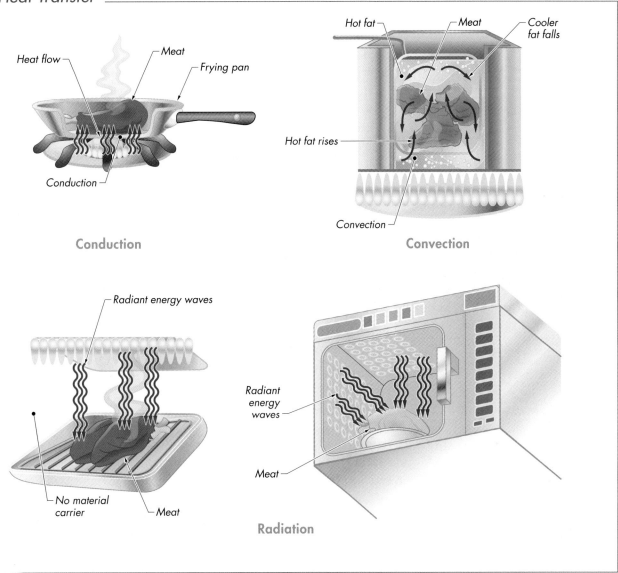

8-3 *Heat transfer methods include conduction, convection, and radiation.*

## Conduction

Conduction is the most basic method of heat transfer. *Conduction* is a method of heat transfer where heat is passed from one object to another through physical contact. When an egg is cracked into an omelet pan and begins to cook, heat transfer occurs as a result of the egg physically coming in contact with the hot metal pan. The egg cooks as a result of the egg's direct contact with the hot pan and the pan's direct contact with the hot flame. All of this direct contact enables the heat to conduct from the heat source (flame) to the pan and finally to the egg in the pan. Common cooking methods using conduction heat transfer include sautéing, stir-frying, griddling, and panbroiling.

## Convection

Convection is a method of heat transfer from a heat source to food through a surrounding substance called a medium. Hot air surrounding a turkey in an oven, hot water surrounding pasta in a pot, and the fat surrounding fried chicken in a deep fryer are examples of different cooking media. When a deep fryer is turned on, the fat inside the fryer heats up through direct contact with the heat source in the bottom of the fryer. Fat near the surface of the fryer circulates downward toward the bottom of the fryer as the hottest fat at the bottom circulates up toward the surface.

This circular motion of the colder medium falling toward the bottom and hotter medium swirling toward the top is an example of convection heat transfer. Convection heat transfer is the reason that the fat is a consistent temperature throughout the fryer after it reaches a set temperature. Pieces of chicken at the upper surface of the fat in a deep fryer cook at the same speed as chicken at the bottom of the fry basket. Likewise, when broccoli is placed in a pot of boiling water, the broccoli near the surface of the water cooks at the same speed as the broccoli at the bottom of the pot, closest to the flame. Through convection, the hotter molecules of water circulate from the bottom of the pot toward the surface, and the colder molecules at the surface circulate toward the bottom of the pot. This constant circulation of warmer and cooler water molecules maintains a steady and even temperature environment.

Atmospheric air provides another medium for convection heat transfer. Even though the air is not a solid touchable surface, it is made up of gas molecules similar to those that make up water. When it is cold outside, even though the coldness can't be touched, it feels cold on the body. This is because the cold air molecules surround the body and transfer heat away from the body, creating the sensation of coldness. The same principles hold true in an oven. The dry, hot air acts as a medium and heats and cooks the food through convection heat transfer. If a roast is put in an oven at 400°F, warmer and cooler air molecules circulate around the meat, cooking and browning the roast.

Many conventional ovens use the principle of convection heat to cook foods by surrounding the food with hot air. The heat source is located at the bottom of the oven, and air circulates naturally. Restaurants use commercial convection ovens equipped with a circulating fan to roast meats or other foods. The fan inside the convection oven helps the warmer and cooler air molecules circulate throughout the oven and provides even heat circulation to evenly cook items. However, conventional ovens do not have a built-in fan to circulate internal air. Therefore, conventional ovens can be hotter near the upper racks and cooler near the lower racks because the air does not circulate as effectively.

Cooking techniques can also play an important part in creating convection heating. For example, when a pot of chili is stirred so it does not burn, the stirring motion creates convection heating. This is because cold molecules are stirred into hot molecules, actually helping the conduction heating at the bottom of the pot turn into convection heating through the circulation of hot and cold molecules.

## Radiation

*Radiation* is a method of heat transfer from a heat source to food through heat waves. Radiation heat transfer can be done in two ways, infrared cooking and microwave cooking. In infrared cooking, an electric or ceramic element is heated until it glows red. This can be seen in a toaster when the electric element glows red as it heats up. The result is the radiant transfer of heat from the glowing element to a slice of bread. Radiant heat transfer also occurs in a broiler where there is either an overhead gas flame, electric heating element, or ceramic tiles that glow red as they become hot. A steak under a broiler cooks on top first, even though it is the bottom of the steak that is in contact with the hot broiler rack. The food beneath the broiler is not cooked from any direct contact with the cooking surface, but rather from the heat waves that are given off by the hot elements.

The second source of radiation heat transfer is microwave cooking. Here, electromagnetic waves generated by a microwave oven strike the food and heat up water molecules in the food at an amazing speed. The result is that the food cooks quickly but does not brown. Because microwave radiation cooking works through the water, fat, and sugar in foods, items that do not contain water, such as an empty glass plate, do not heat up. An empty plate can sit in a microwave operating on high power and it never becomes hot; however, if food is placed on the plate and the microwave is turned on, the plate becomes hot as the food heats up. In this case, the plate becomes hot due to conduction heat transfer. As the food begins to heat up, it conducts heat to the plate through direct contact.

## DRY-HEAT COOKING METHODS

Foods can be cooked in different cooking media, such as fat, water, and air, and heat is transferred to foods in different ways, such as conduction, convection, and radiation. There are two general types of cooking methods, dry-heat cooking and moist-heat cooking, and more specific cooking methods that use dry heat, moist heat, or both. Understanding how to apply these cooking methods and knowing which cooking method creates the desired result for a particular dish is essential.

*Dry-heat cooking* is any cooking method that uses hot air, hot metal, a flame, or hot fat to conduct heat to the food without any moisture. Fat is a dry-heat cooking medium. At first it might seem that, since the fat in a deep fryer is a liquid, it is moist-heat cooking. Remember that a dry-heat cooking method browns foods, while a moist-heat cooking

**Historical Note**

In 1946, while conducting experiments on radar technology, Dr. Percy Spencer made a discovery. He was testing a vacuum tube he had created and he noticed that a candy bar in his pocket began to melt. He wasn't sure if the melting was related to his experiment, so he placed a few popcorn kernels near the vacuum tube opening. Dr. Spencer watched as the kernels began to shake, jump, and eventually pop. He delayed the radar technology testing and began testing the effects of irradiation heat transfer. This in turn led to the development of the microwave oven.

method does not. It is therefore possible to see why fat is a dry-heat method, because foods cooked in hot fat turn brown and crunchy if they are prepared properly.

Dry-heat cooking methods include deep-frying, pan-frying, sautéing, griddling, panbroiling, grilling, roasting, baking, and broiling. *See Figure 8-4.* Meats that are best cooked using dry cooking methods are those that are very tender, with little connective tissue, and that can be served medium rare.

## Dry-Heat Cooking Methods

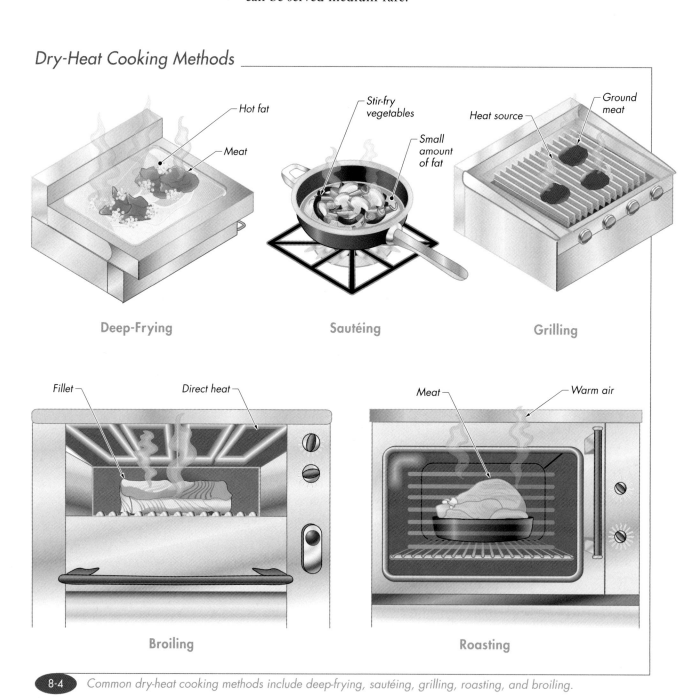

Hot fat

Meat

**Deep-Frying**

Stir-fry vegetables

Small amount of fat

**Sautéing**

Ground meat

Heat source

**Grilling**

Fillet

Direct heat

**Broiling**

Meat

Warm air

**Roasting**

**8-4** *Common dry-heat cooking methods include deep-frying, sautéing, grilling, roasting, and broiling.*

## Deep-Frying

*Deep-frying* is a cooking method that involves submerging foods in very hot fat, usually between 350°F and 375°F. For best results, only a fat designed for use in deep-frying should be used. Most foods are battered or breaded before deep-frying. This helps the food to brown and become crispy and gives foods a coating that prevents them from burning or drying out. Foods that are deep-fried are added to the fryer by either the swimming method or the basket method.

In the swimming method, usually used for battered items, the item is battered and then slowly dropped into the hot fat without a fryer basket. A basket is not used because battered foods stick together as they fry and may also stick to the basket. By slowly adding each battered item to the fryer, the battered item first sinks to the bottom of the fryer and then floats back up to the top. When the side that is facing down in the hot fat turns brown, the item is flipped over to brown the other side.

In the basket method, items (often individually frozen) are added to a fryer basket that is sitting on top of a pan, not over the fryer. It is important to not have the basket over the fryer while filling the basket, as crumbs can fall into the fryer and shorten the usable life of the fat. The basket should never be overfilled, as items near the top of the basket do not cook properly. All of the items in the basket should be submerged when the basket is lowered. Once the basket is filled, it is submerged in the hot fat, and when the items are fully cooked, they are removed from the fryer and placed in a drain pan to allow excess fat to drain off.

## To deep-fry foods, apply the following procedure:

1. Preheat the deep fryer to the desired temperature (foods have a tendency to become too greasy if fried at temperatures lower than 325°F).
2. Clean, cut, and prepare the items to be deep-fried by breading, battering, or lightly dredging the item.
3. Use either the basket method or swimming method to add the items to the fryer.
4. Fry until the items are golden brown. Overdone items are dry and tasteless. Some items float on top of the fat when done.
5. Remove the items from the fryer and place on paper towels or a drain pan to allow excess fat to drain away. This helps retain crispness and makes the item more digestible.
6. Clean/strain fat to remove all food particles that are either floating or that may have settled to the bottom of the fryer. This helps extend the life of the fat in the fryer.

CULINARY PROCEDURES

*Chef's Tip:*
Remember to move fried items away from the fryer before seasoning with salt. Preventing salt from entering the fry fat prolongs the life of the fat.

**Fry Coatings.** Three main types of fry coatings used for deep-frying are dredging, breading, and battering. *Dredging* is the process of lightly dusting an item in seasoned flour or fine bread crumbs for deep-frying. *Breading* is a three-step process used to coat and seal an item in preparation for deep-frying. It is also known as the standard breading procedure. An item is first dipped into flour (dredged), then dipped into a mixture of beaten egg and liquid, and lastly dipped into a bread crumb mixture. The bread crumbs adhere to the flour and egg, coating and sealing the food item. One hand is used exclusively to coat items with wet ingredients and the other hand is used exclusively to coat items in the dry ingredients. By using separate hands for wet and dry ingredients, the hands do not become breaded in the process of moving food from one ingredient to the next. Fried chicken and mozzarella sticks are often breaded using this procedure. *Battering* is the process of dipping an item in a wet mixture of flour, liquid, and fat for deep-frying. Fish fillets and seafood are typically battered before frying. Tempura is an Asian method of battering beef, chicken, shrimp, or vegetables. Once an item has been dredged, breaded, or battered, it is ready for deep-frying.

## To dredge foods, apply the following procedure:

1. Coat items in flour or use very fine ground crumbs for a very light and crispy exterior.
2. Shake off excess coating.
3. Fry coated items immediately until done. Use the basket method for large individual food items and the floating method for multiple food items or small food items.
4. Remove items from the fryer and drain to remove excess fat.

## To bread foods, apply the following procedure:

1. Dredge items in seasoned flour.
2. Shake off excess flour.
3. Dip the items in an egg wash (a mixture of beaten egg and milk or water) and coat thoroughly.
4. Place the items in a crumb mixture, coat thoroughly, and shake off excess crumbs before frying.
5. Fry breaded items until done.
6. Remove items from the fryer and drain to remove excess fat. **See Figure 8-5.**

## Standard Breading Procedure

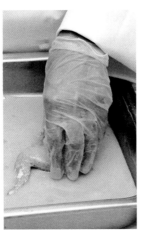

1. Dredge item in flour; shake off excess flour.

2. Dip in egg wash.

3. Coat with bread crumbs.

4. Fry until done.

**8-5** *The standard breading procedure is used to coat and seal food items for deep-frying.*

## To batter foods, apply the following procedure:

1. Prepare desired recipe of batter.
2. If items to be battered are damp, lightly dredge the items in flour before placing in batter. This allows the batter to stick to the item. If the items are dry, they can be dipped directly in the batter without dredging.
3. Dip the items in the prepared batter.
4. Using the swimming method, slowly add the battered items to the fryer.
5. Fry battered items until done
6. Remove items from fryer and drain to remove excess fat.

CULINARY PROCEDURES

### Pan-Frying

*Pan-frying* is a cooking method that involves cooking food in a smaller amount of fat than deep-frying. Eggplant parmigiana and country-fried steak are some foods that are typically pan-fried. Foods that are to be pan-fried are usually breaded using the standard breading procedure and then placed in a pan with hot fat (usually 350°F to 375°F). There should be enough hot fat to cover about one-half to two-thirds the thickness of the food. A common reason for pan-frying something instead of deep-frying it is to allow a thin item, such as a fish fillet, to lay flat in the pan and to prevent it from curling up. After the item being pan-fried is brown on the submerged side, it needs to be flipped over to brown the other side. Other foods that are usually pan-fried include boneless chicken breasts and pork chops.

## Sautéing

**8-6** *Flipping items during sautéing ensures even cooking.*

### Sautéing

Sautéing and stir-frying are two terms that are often used interchangeably. The difference between sautéing and stir-frying is that sautéing is done in a shallow pan with sloped sides and stir frying is done in a Chinese wok. *Sautéing* is cooking food quickly in a small amount of fat. Sautéing is done over high heat in a sauté pan, using caution to not allow the small amount of fat in the pan to burn. Items in the sauté pan can be flipped with a sharp motion of the wrist. Flipping items during cooking helps to ensure even cooking without burning. Vegetables and strips of meat or chicken are some foods that are commonly sautéed or stir-fried. ***See Figure 8-6.***

Stir-frying foods well involves knowing what order to cook various vegetables. Stir-fry vegetables according to their density. Denser vegetables, such as broccoli and carrots, require more cooking time than less dense vegetables, such as bok choy. Always stir-fry meat separately and then add it to the vegetable so each retain their flavors.

### To sauté vegetables, apply the following procedure:

1. Prepare item(s) to be cooked by cleaning and cutting as necessary.

2. Place sauté pan on burner over high heat.

3. After pan is hot, add a small amount of oil, butter, or other fat to the pan. There should be just enough fat to coat the bottom of the pan.

4. When the fat is hot, add the item(s) to be sautéed, using caution to not overfill the pan. If the pan is too full, the items wind up steaming or simmering rather than sautéing.

5. After the pan heats up completely again, use a flicking motion of the wrist to toss and flip the vegetables.

### To sauté meat items, apply the following procedure:

1. Trim meat and season with salt and pepper.

2. Dredge the meat in flour if desired. Dredging thin items such as meat cutlets helps the meat to brown evenly and often results in a moister final product.

3. Heat the pan, and add a small amount of fat to avoid sticking.

4. Add meat to the pan and brown quickly on one side at a moderate temperature. This makes the meat crisper and gives it a more appetizing appearance.

5. Turn the meat with tongs and brown the second side. Meat should be golden brown on both sides when finished.

## Griddling, Panbroiling, and Grilling

Griddling, panbroiling, and grilling are similar in that they are all dry-heat cooking methods that use a heat source from below and do not use fat. *Griddling* is the process of cooking foods on a solid metal cooking surface called a griddle. The heat comes from below the cooking surface but items are not in contact with a flame. Usually a small amount of fat is placed on the hot griddle to prevent foods from sticking to it. The temperature of a griddle is adjustable, which makes it a versatile piece of equipment in the kitchen. It is most often used to cook pancakes, eggs, and some meats. A *grooved griddle* is a griddle with raised ridges that create grill marks on foods. The grill marks are created where food has rested on the raised ridges while cooking. Because it is a solid surface, a grooved griddle does not generate as much smoke as an open grill.

*Panbroiling* is the process of cooking food in a small amount of fat in a skillet or sauté pan. As an item cooks, it is important to pour off excess fat and oil that accumulates. Panbroiling is used in restaurants that do not have a grill to cook steaks. Usually the steak is seared on both sides during panbroiling and then finished in an oven. *Searing* is the process of browning the surface of a food item quickly and with high heat to seal in the juices.

## To panbroil an item, apply the following procedure:

1. Season both sides of item to be panbroiled.

2. Heat the sauté pan. Do not add fat.

3. Place item in the hot sauté pan and brown one side. Do not cover the pan. Covering the pan causes steam to develop.

4. Cook at a moderate temperature to prevent excessive browning or burning. Cooking over moderate heat also helps to keep the item juicy and more tender.

5. Pour off any fat that appears in the sauté pan and keep the pan as dry as possible. If fat is left to build up in the pan, the cooking method turns into sautéing and further progresses to pan-frying.

6. Use tongs to turn the item and brown the other side. This browning helps develop intense flavor in panbroiled foods.

7. Cook until desired doneness. Cooking time depends on the cut of meat, type of meat, thickness, degree to which it is done, and quality of meat.

CULINARY PROCEDURES

*Grilling* is the process of cooking food over a heat source on open metal grates. Meats and vegetables are foods that are commonly grilled. Cooking food on a grill results in crosshatch markings. *Crosshatch markings* are crisscrossed charred lines created when food comes in contact with a hot grill. By rotating the food 60° on the grill, a crisscross pattern results. Crosshatching the presentation side of grilled food enhances the presentation of the dish. ***See Figure 8-7.*** Meats and vegetables may be placed separately on the grill, or strung on a skewer shish kebab-style. Restaurants usually use gas or electric grills. Grills are seasoned before use. *Seasoning* is the process of preparing a cooking surface for use to prevent food from sticking. Grilling foods produces a smoky and charred flavor.

CULINARY PROCEDURES

### To grill an item, apply the following procedure:

1. Use a wire grill brush to scrape the grill clean.
2. Wipe the metal grill grates with a paper towel lightly coated in vegetable oil to season it.
3. Preheat the grill until very hot. This helps prevent foods from sticking.
4. Brush a small amount of oil on the food item(s) to be grilled and season the food as desired.
5. Place the item(s) on the grill with the presentation side facing down. Do not move the item(s) until dark charred lines from the grill grates have developed.
6. Rotate the item(s) 60° with tongs to create crosshatch markings.
7. Use tongs to turn the item(s) and cook the other side.
8. Cook item(s) to desired degree of doneness. *Note:* It is not as important to crosshatch the second side of the grilled item, as it faces down on the plate.

## Roasting

*Roasting* is cooking by surrounding food with dry, indirect heat (heated air). Roasting is usually accomplished in an oven. Roasting is commonly used for large cuts of meats, whole birds, or vegetables. Sometimes the term baking is used interchangeably with roasting, but for the most part, baking refers to breads and pastries. In addition, roasting can also be used to describe rotisserie-style cooking in which items are cooked on a rotating spit over or next to an open flame.

## Crosshatch Markings

1. Place meat on grill at an angle.

2. Without flipping meat, rotate meat 60°.

3. Turn when meat is half done; crisscross pattern is made.

**8-7** Meat is positioned on the grill to obtain the desired crosshatch markings.

The length of time an item should be roasted varies depending upon the size of the item, the temperature of the oven, and the type of item being roasted. Roasted meats are often cooked with a mirepoix (pronounced meer-pwah). A *mirepoix* is a rough-cut vegetable mixture consisting of 25% carrots, 50% onions, and 25% celery cut into approximately 1″ pieces and used to flavor stocks and sauces. ***See Figure 8-8.***

## Mirepoix

25% Celery

50% Onions

25% Carrots

**8-8** A mirepoix is a rough-cut vegetable mixture consisting of 25% carrots, 50% onions, and 25% celery.

Roasting typically requires periodic basting of the item over the duration of the cooking time. *Basting* is the process of continually brushing or ladling juices and fat over an item during the cooking process to aid in moisture retention.

Roasting also results in the occurrence of carryover cooking. *Carryover cooking* is the rise in internal temperature of an item after it is removed from the oven due to residual heat on the surface of the item. The internal temperature may rise between 5°F and 10°F during this time. Allowing roasted items to rest on a counter after being removed from the oven permits carryover cooking to occur. This also helps the item to retain its moisture, which would escape as steam if the item were cut too soon after cooking.

## To roast an item, apply the following procedure:

1. Preheat the oven to the desired temperature. Meats are typically roasted at temperatures between 300°F and 350°F. Roasting meat at lower temperatures prevents shrinkage and loss of moisture.

2. Trim away any excess fat and season item to be roasted. Some items, such as a turkey, may be brushed with oil or butter before roasting to encourage even browning.

3. Place the item on a roasting rack in a roasting pan. The item may be placed on a mirepoix instead of a rack, if preferred. Either method keeps the bottom surface of the item off the pan and out of its own juices.

4. Leave item uncovered while roasting for maximum exposure to the hot, dry air. This enables the item to caramelize. Covering an item while it is roasting creates unwanted steam.

5. Baste item as necessary.

6. Remove roasted item from oven when it is close to the desired degree of doneness or close to a specific internal temperature.

7. Allow the item to rest. Carryover cooking completes the cooking process.

## Baking

Baking is the primary cooking method used in preparing yeast breads, quick breads, cookies, cakes, pies, and pastries. *Baking* is cooking by surrounding the food with dry heat in an oven. The only difference between baking and roasting is the type of food being prepared; the method of preparation is identical. Baking time, like roasting time, varies depending upon the size of the item, the temperature of the oven, the type of item being baked, and the particular ingredients used. Before placing an item in the oven, the oven must be preheated to ensure correct baking of the item. Baking instructions and procedures for all recipes must be followed closely. When baked items are removed from the oven, the items need to rest before being cut. This carryover cooking time also helps the baked items retain their moisture while the baking process is completed .

## Broiling

*Broiling* is cooking with a direct heat source above the item. It is similar to roasting, with the difference that the heat source is above the food instead of below the food. The meat is exposed to an overhead flame in gas cooking and to an overhead heating element in electric cooking. Meats and vegetables can be broiled, and appetizers such as bruschetta are frequently prepared with the broiling method.

Often a gas broiler has ceramic tiles near the flame that heat up and glow red. The purpose of the ceramic tiles is to provide more concentrated heat than the flame can provide by itself. A commercial broiler in a restaurant has open metal grates similar to a grill. The concentrated heat from the heat source above heats up the grates. The hot grates create the distinctive crosshatch markings on broiled items. ***See Figure 8-9.***

### Broiling

*Vulcan-Hart, a division of the ITW Food Equipment Group LLC*

**8-9** *Broiling is a dry-heat cooking method that uses an overhead heat source to cook food.*

## To broil food, apply the following procedure:

1. Use a wire grill brush to scrape the broiler grates clean.
2. Brush a small amount of oil onto the item to be broiled and season with desired seasonings.
3. Preheat the broiler. This helps to keep foods from sticking.
4. Place the item on the broiler with the presentation side facing down. Do not move the item until it has had time to develop dark, charred lines from the broiler grates.
5. Rotate the meat 60° with tongs to create crosshatch markings.
6. Use tongs to turn the item and cook the other side.
7. Cook item to desired degree of doneness. *Note:* It is not as important to crosshatch the second side of the broiled item because it faces down when presented on the plate.

CULINARY PROCEDURES

## MOIST-HEAT COOKING METHODS

*Moist-heat cooking* is any cooking method that uses liquid (including stocks and sauces) or steam as the cooking medium. This involves cooking foods either by submerging them in hot liquid or by exposing them to steam. Because of this, the natural flavor and smell of the food is highlighted. It is important to understand the different types of moist cooking to know which method is best for a given preparation. Moist-heat cooking methods include steaming, boiling, blanching, simmering, and poaching. *See Figure 8-10.*

### Moist-Heat Cooking Methods

| Method | Temperature of Cooking Liquid | Appearance of Cooking Liquid | Common Uses |
|---|---|---|---|
| **Steaming** | Greater than 212°F | Moist hot air emitted rapidly from boiling liquid | Vegetables, eggs in the shell, shellfish, some fish, and poultry |
| **Boiling** | 212°F (at sea level) | Rapid motion in water as large bubbles break the surface | Potatoes, pasta, and vegetables (for blanching) |
| **Blanching** | 212°F (at sea level) | Rapid motion in water as large bubbles break the surface | Vegetables, fruits, removal of impurities from bones for stocks |
| **Simmering** | 181°F to 205°F | Small bubbles have formed and come to the surface | Stews, sauces, soups, meats, poultry |
| **Poaching** | 160°F to 180°F | Very tiny bubbles may have formed on bottom of pot, but they do not break the surface; Liquid shows very little movement and emits slight steam | Delicate fish fillets, eggs out of the shell, soft-fleshed fruits |

**8-10** *Moist-heat cooking methods include steaming, boiling, blanching, simmering, and poaching.*

### Steaming

*Steaming* is a moist-heat cooking method that uses steam as a convection medium to heat food. In steaming, foods are placed over boiling water, usually on a rack inside a covered pot. The movement of the steam around the food cooks the food gently and evenly all around. Many vegetable preparations call for steaming. Broccoli, green beans, carrots, Brussels sprouts, and asparagus are some examples of vegetables that are typically steamed. Other foods such as pot stickers can also be prepared with the steaming method.

Often, other ingredients are placed in the boiling water to add flavor to the food being steamed. An *aromatic* is an ingredient added to a food to enhance its natural flavors and aromas. Aromatics such as wine, herbs, or spices added to the boiling water release flavors and smells into the hot, moist air and are absorbed into the food as it cooks.

In a professional kitchen, foods can be cooked in a commercial convection steamer. *See Figure 8-11.* A convection steamer uses steam in combination with pressure to cook foods more quickly than traditional steaming. Unfortunately, because the process is much quicker and the steam is controlled by the mechanics of the equipment, aromatic ingredients cannot be added to the steaming environment.

## Steaming

Vulcan-Hart, a division of the ITW Food Equipment Group LLC

**8-11** *A commercial convection steamer uses a combination of steam and pressure to reduce cooking time.*

## Boiling

Of the moist-heat cooking methods, boiling is probably the most common by name, but probably one of the least-common methods actually used in a professional kitchen. *Boiling* is a moist-heat cooking method that uses liquid heated to the boiling point as a convection medium to heat food. The only items that are actually boiled in a professional kitchen are pasta, potatoes, and some other starches. To boil food, a large amount of liquid is heated to the boiling point (212°F for water at sea level). The colder molecules of liquid circulate toward the bottom of the pot and the warmer molecules circulate toward the top. This circulation keeps the temperature consistent throughout the pot. Different liquids boil at different temperatures, so depending on the liquid, the boiling point may be less or more than 212°F. As the temperature of a liquid rises, it changes from a normal state to a simmer, and finally reaches full boil. When a liquid reaches full boil, large bubbles form at the bottom surface of the pan and rapidly rise, breaking at the upper surface of the liquid. ***See Figure 8-12.***

## Water Temperature

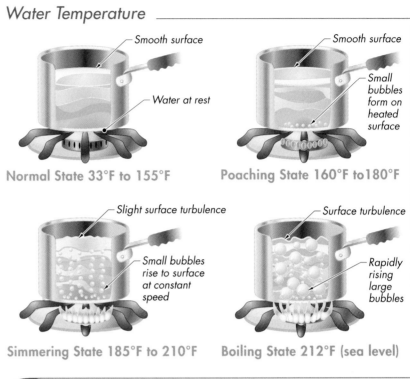

Normal State 33°F to 155°F
- Smooth surface
- Water at rest

Poaching State 160°F to 180°F
- Smooth surface
- Small bubbles form on heated surface

Simmering State 185°F to 210°F
- Slight surface turbulence
- Small bubbles rise to surface at constant speed

Boiling State 212°F (sea level)
- Surface turbulence
- Rapidly rising large bubbles

**8-12** *As the temperature of a liquid rises, the liquid changes from a normal state, to a poaching state, to a simmer, and finally to a full boil.*

At elevations close to sea level, the boiling point of a liquid remains at 212°F. At higher elevations, such as Denver, Colorado, the atmospheric pressure drops. For every 1000′ of elevation above sea level, the boiling point of water drops 2°F. No matter how high the burner is turned on, the temperature of water can never rise above the boiling point. In relation to cooking at higher elevations, this means that the length of cooking time increases to compensate for a lower boiling point.

## Blanching

*Blanching* is a quick moist-heat cooking method used to partially cook an item. There are a few reasons for using a blanching method in the professional kitchen. The first reason is to prepare vegetables and fruits for use. Blanching can make vegetables such as tomatoes easy to peel, partially soften hard vegetables, brighten and set color in produce, and even eliminate bitter or undesirable flavors. Usually when produce items are blanched, they are immediately refreshed or shocked in ice water to stop the cooking process. *Shocking*, or refreshing, is the technique of quickly stopping the cooking process in foods by plunging them into ice water. Blanching in fat is performed in the same manner as blanching in water, but fat is only used for potatoes that are going to be deep-fried.

**Chef's Tip:**

French fries should be blanched in fat before deep-frying them completely. This is because potatoes contain a large amount of water. If freshly cut potatoes are dropped into a deep fryer, they immediately sear on the outside and seal in the water. A few minutes after they are fried, the water in the potatoes begins to steam out, making the French fries very soggy. By blanching the potatoes in 225°F to 250°F fat for a few minutes, the internal water cooks out of the potatoes. Then the temperature of the fat should be raised to 375°F and the potatoes deep-fried until golden brown so they won't be soggy.

## To blanch items in water, apply the following procedure:

1. Clean and prepare items to be blanched.
2. Bring pot of water to boil.
3. Place the items in rapidly boiling water.
4. When desired effect is achieved (peel begins to loosen or color brightens on the vegetable) remove the items from the boiling water and immediately submerge in ice water to stop the cooking process.

The second form of blanching is blanching to remove impurities. Blanching to remove impurities is a method used to rid bones of blood proteins and impurities when making soup or stock from bones. If bones are not blanched prior to use in a soup or stock, the proteins and impurities in the bones ultimately make the soup or stock cloudy.

## To blanch bones to remove impurities, apply the following procedure:

1. Place the bones in a pot of cold water.
2. Turn heat on full and bring water to a boil.
3. Reduce heat to a simmer and cook for a few minutes.
4. Remove the bones and plunge them into cold water.
5. Discard the blanching water.

This blanching method removes loose blood proteins and impurities from the bones that could cause a stock or soup to become cloudy. The bones are then brought to a simmer again in new, clean water to extract their flavor for a stock or soup.

### Simmering

Most foods cooked in liquid are actually simmered, not boiled. *Simmering* is a moist-heat cooking method that uses liquid heated below the boiling point as a convection medium to heat food. To simmer means to heat liquid to a temperature causing very light bubbling but less than a full boil. The liquid should be between 185°F and 205°F to maintain a simmer. Soups and stews are simmered, not boiled, while cooking. Corned beef is an example of meat commonly prepared with the simmering method. Potatoes, whole grains (pilaf and risotto), cereals (polenta), and legumes are also simmered. It is important to maintain a constant and even temperature when cooking foods using the simmering method. Most often, the simmered item is cooked directly in the cooking liquid, allowing it to absorb more moisture. Additionally, it may be desirable to add a mirepoix to the liquid to enhance the flavor of the dish.

### Action Lab

Fill a pan with water and place it on a heat source. Observe the water begin to form tiny bubbles on the bottom surface of the pan before it reaches the boiling point. The bubbles grow and slowly start to rise as the temperature increases. When the bubbles are rising to the surface at a consistent pace, the water is simmering. Reducing the heat at this point helps maintain a steady simmer. Experiment with the temperature of the heat source to achieve a steady simmer without reaching a full boil.

As previously stated, proteins coagulate and become firm when exposed to heat. If a protein-rich food such as a corned beef brisket is cooked using the boiling method, the brisket shrinks and becomes tough. If the same brisket is simmered, it remains very tender after cooking.

**To simmer an item, apply the following procedure:**

1. Clean and prepare the item to be simmered.
2. Add enough liquid to a pan to completely cover the item.
3. Bring the liquid up to the desired temperature between 185°F and 205°F. Note: It is often easier to bring the liquid to a boil first and then lower to the desired temperature.
4. Add the mirepoix if desired.
5. Carefully place the item into the hot liquid and maintain a steady temperature till done. Doneness is usually based on tenderness or cooking time.

## Poaching

To *poach* means to cook foods in a shallow amount of liquid held between 160°F and 180°F. Poaching is used for very delicate foods such as fish fillets or eggs out of their shell. Often some seasonings and a mirepoix are added to the poaching liquid when cooking items with delicate flavors such as fish. The seasonings and mirepoix make the poaching liquid more flavorful and the flavor can be absorbed by the item being cooked. The poaching liquid is often reduced and incorporated into a sauce for the finished dish.

**To poach an item, apply the following procedure:**

1. Clean and prepare item to be poached.
2. Bring the appropriate amount of liquid (usually just enough to cover the item) to a boil and then lower the temperature of the liquid to between 160°F and 180°F.
3. Gently lower the item into the poaching liquid.
4. Cook the item until desired doneness.
5. Cool the poached item directly in the poaching liquid or serve immediately.

## COMBINATION COOKING METHODS

Some cooking methods combine moist and dry cooking methods to achieve the best result. *Combination cooking* includes any cooking method that uses both moist and dry cooking methods. Combination cooking methods are usually used on tougher cuts of meat to make them more tender and flavorful. The two most common combination cooking methods are braising and stewing. Both methods are similar in that they have a two-step process; they first brown the meat using a dry-heat method and then simmer the meat using a moist-heat method. The main difference between these two methods are the size of the meat cooked and the amount of liquid used to cook it.

### Braising

*Braising* is a combination cooking method for larger (roast-size) pieces of meat. In braising, the large piece of meat is only covered halfway with the braising liquid. This allows a braised meat to be subjected to both simmering and steaming of the meat during the cooking process. The meat is seared on all sides in a small amount of fat. ***See Figure 8-13.*** Aromatic vegetables are also added during the searing process. Then a flavorful liquid (usually a stock) is added to the pot. The liquid is brought to a simmer and the meat is cooked slowly until tender. The seared vegetables intensify the flavor of the braising liquid.

### Historical Note

Before slow cookers, many people took a tough cut of meat and roasted it in an oven without any prior preparation. The frequent result was a dry and tough piece of meat for dinner that night. Slow cookers first became popular in the 1970s and 1980s because they allowed homemakers to put a large piece of meat in the cooker with some water and vegetables and leave it on to cook all day. For best results, meat should be seared on all sides before adding it to the cooker. At the end of the day, the "pot roast" is tender and has a mild flavor from the savory vegetables and slow cooking process.

*Braising*

Preparation

Presentation

**8-13** *When braising, meat is seared on all sides before liquid is added. The braising liquid can later be used to make a sauce.*

## To braise meat, apply the following procedure:

1. Clean, trim, and season the meat to be braised.

2. Heat a heavy-bottomed braising pan that is large enough to hold the meat with a cover.

3. Add a small amount of fat to the pan.

4. Add the meat and sear all sides until golden brown. This is also referred to as browning, and it helps to develop a more intense flavor and color.

5. Remove the meat from the pan.

6. Add some aromatic vegetables (such as mirepoix) to the pan and sauté.

7. Add a small amount of tomato product to the vegetables and cook an additional minute.

8. Add some flour to make a roux. Stir until a thick paste-like consistency is achieved.

9. Add some of the cooking liquid to the roux. Stir until completely absorbed.

10. Return the meat to the braising pan and add enough braising liquid to cover the meat halfway.

11. Cover the pan. Let the meat simmer until tender, turning periodically while cooking. Cook at a low temperature either on the range or in the oven (oven temperature should be below 250°F); lower cooking temperatures result in less shrinkage and a better flavor.

12. When the meat is cooked, remove it from the pan and keep warm. The time required for cooking depends on the size, thickness, grade, and type of meat used.

13. If additional thickener is needed, add more roux at this time.

14. Strain and reduce the braising liquid to make a sauce.

### Stewing

*Stewing* is a combination cooking method for bite-sized or slightly larger pieces of meat that typically incorporates a thick and flavorful sauce. Stewing uses more liquid than braising, because the meat is completely covered in liquid. It is important to make sure that the pieces of meat are all similar in size so the pieces cook evenly. Another important step in the stewing method is that after the stew meat has been browned, just enough liquid is added to the meat to completely cover it. The pan is kept on the stove with just enough heat to maintain the simmer without boiling, or covered and placed in a 250°F oven. Usually aromatic vegetables cut into 1″ to 1½″ pieces are added when the meat is about three-fourths done. If added earlier, the vegetables may overcook. Sometimes the vegetables are cooked separately and added to the meat at the end. The liquid, usually a stock or sauce that may be already flavored, develops even more flavor from the browned meat and the slow cooking process. **See Figure 8-14.**

## Stewing

National Chicken Council

**8-14** *Stewing is a combination cooking method for bite-sized or slightly larger pieces of meat that usually incorporates a thick and flavorful sauce.*

## To stew meat, apply the following procedure:

1. Clean and trim the meat to be stewed, and cut into 1½" to 2" cubes.
2. In a hot pan, add a small amount of fat.
3. Add the meat and sear all sides until golden brown.
4. Add some flour to make a roux. Stir until a thick paste-like consistency is achieved.
5. Slowly add the cooking liquid, stirring constantly to avoid lumps, until it just covers the meat.
6. Bring the liquid to a simmer and keep covered to preserve moisture and flavor.
7. Add aromatic vegetables when meat is three-fourths done if desired.
8. The stew is done when the meat become tender.

If the sauce needs to be thicker, the meat and vegetables are strained from the pan and a roux is used to thicken the sauce. The meat and vegetables are added back to the sauce when it has reached the desired consistency.

## SUMMARY

Cooking refers to the process of transferring heat to food through conduction, convection, or radiation. When heat is introduced to food, the properties of the food change in various ways. Nutrients are changed in various ways when heat is introduced to foods. Proteins become firm and coagulate, starches in complex carbohydrates gelatinize and absorb moisture, and the simple sugars in simple carbohydrates caramelize. Cooking methods are classified as dry-heat, moist-heat, or combination methods.

Dry-heat cooking methods use hot air, hot metal, a flame, or hot fat to transfer heat to food. Food is browned, or caramelized, as a result of dry-heat cooking methods. Common dry-heat cooking methods include deep-frying, pan-frying, sautéing, griddling, panbroiling, grilling, roasting, baking, and broiling. Moist-heat cooking methods use liquid or steam as a medium to transfer heat through convection. Common moist-heat cooking methods include steaming, boiling, blanching, poaching, and simmering. Combination cooking methods use both dry and moist heat, typically in an effort to make a tougher cut of meat more tender and flavorful. Braising and stewing are typical combination cooking methods. A thorough understanding of how heat and the use of various dry-heat, moist-heat, and combination methods of cooking affects different foods ensures successful preparation that produces desired results.

Refer to the CD-ROM for Quick Quiz® questions related to chapter content.

## Review Questions

1. List the three main goals of cooking food.

2. What term is used to describe the change that takes place as simple carbohydrates begin to darken as temperatures rise during dry-heat cooking?

3. Why is it recommended that bread or roasted meat be allowed to rest after it is cooked?

4. How does cooking affect the vitamins and minerals in foods?

5. Identify different cooking methods that use conduction.

6. List three types of media that can be used in convection heat transfer.

7. Microwave cooking and infrared cooking are examples of what type of heat transfer cooking method?

8. Why is deep-fat frying considered a dry-heat cooking method instead of a moist-heat cooking method?

9. Describe the differences between the swimming method and the basket method for deep-frying foods.

10. Describe the three main types of fry coatings used for deep-frying.

11. Describe the five-step procedure for sautéing vegetables.

12. Outline the differences among griddling, panbroiling, and grilling.

13. Describe the composition and purpose of mirepoix.

14. What is the difference between baking and roasting?

15. Describe the seven-step procedure for broiling an item on a commercial boiler.

16. List five types of moist-heat cooking methods.

17. Describe the two forms of blanching that are used in the professional kitchen.

18. What is the temperature requirement for simmering?

19. Name the two most common combination cooking methods.

20. What types of food are commonly cooked using a combination cooking method?

CulinaryArts
PRINCIPLES AND APPLICATIONS

# Nutrition and Menu Planning

U.S. Department of Agriculture

## Chapter 9

# Nutrition and menu planning

*Nutrition and menu planning* each play an important role in the food service industry. Healthy eating involves understanding how the carbohydrates, fats, proteins, vitamins, minerals, and water in food impact the body. It also involves choosing nutritious foods from among the choices that a menu presents. Well-planned menus provide choices that deliver a range of nutritious and enticing dishes to satisfy customers of varying ages and differing tastes. The menu is a primary means of advertising the food served in an establishment. Planning menus involves determining the type of food, equipment, skills, and prices that will attract and retain customers.

## Chapter Objectives:

1. List the six major categories of nutrients.
2. Describe the main types of carbohydrates and their effects on the body.
3. Describe the different types of fat and how they impact health.
4. Contrast complete and incomplete proteins and their health effects.
5. Identify essential vitamins and minerals, their sources, and their effects on the body.
6. Describe the role that water plays in maintaining good health.
7. Identify the major factors to consider for healthy menu planning.
8. Describe the dietary guidelines developed to aid consumers in making smart dietary choices.
9. Identify the various functions of a menu.
10. Describe common menu types and classifications.
11. Describe the function of truth-in-menu guidelines.

## Key Terms

- **nutrient**
- **carbohydrate**
- **calorie**
- **insoluble fiber**
- **soluble fiber**
- **monounsaturated fat**
- **polyunsaturated fat**
- **saturated fat**
- **trans fat**
- **hydrogenation**
- **cholesterol**
- **complete proteins**
- **incomplete proteins**
- **water-soluble vitamins**
- **fat-soluble vitamins**
- **minerals**
- **dietary reference intake**

## NUTRIENTS

Nutrition is a key component when planning menus and preparing meals. *Nutrition* is the study of food and nourishment that focuses on how food is taken in and utilized. A *nutrient* is a substance that provides nourishment. Nutrients are divided into six main categories: carbohydrates, fats, proteins, vitamins, minerals, and water. Dietary reference intakes (DRIs), formerly known as recommended dietary allowances (RDAs), have been established to identify how much of each nutrient is necessary for good health.

### Carbohydrates

A *carbohydrate* is a nutrient that provides the body with energy in the form of sugars and starches. Carbohydrates should make up 55% to 60% of a person's daily calories. A *calorie* is a measure of the amount of energy a food item provides to the body. A typical recommended diet for an adult man contains between 2200 and 3000 calories daily. For an adult woman, typical recommendations are between 1800 and 2400 calories daily. Recommendations are based on age, height, weight, and activity level. Based on an average 2000-calorie diet, 1100 calories should come from carbohydrates. When carbohydrates are consumed, the human body breaks them into simple sugars and absorbs them into the bloodstream. There are two major types of carbohydrates, simple and complex. ***See Figure 9-1.***

## Carbohydrates

9-1    Simple and complex carbohydrates provide the body with energy in the form of sugars and starches.

**Simple Carbohydrates.** Simple carbohydrates, also known as simple sugars, are found naturally in fruits and milk. Simple sugars such as glucose and fructose have a sweet flavor, are soluble, and are easily digestible. Simple sugars are also found in the refined sugars that are often added to foods as sweeteners. Eating too many refined sugars can lead to excess weight and can contribute to health problems. Because fruits and dairy products contain other nutrients in addition to simple sugars, they are nutritionally preferred over refined sugars.

**Complex Carbohydrates.** Complex carbohydrates, also known as starches, are found in plant-based foods such as grains, legumes, and vegetables. Common sources of complex carbohydrates include dried beans, peas, lentils, rice, pasta, and whole-grain breads. Foods containing high amounts of complex carbohydrates also provide the body with many vitamins and minerals. For example, whole-grain bread has more complex carbohydrates, vitamins, and minerals than white bread. Complex carbohydrates take longer to digest, leaving the body feeling fuller for a longer period than simple carbohydrates.

*Fiber* is a form of a complex carbohydrate that is nondigestible and nonnutritive, but it is essential in a healthy diet. It is found only in plant-based foods such as whole grains, legumes, and vegetables. There are two kinds of fiber: insoluble and soluble.

*Insoluble fiber* is fiber that will not dissolve in water. It functions like a sponge, absorbing water in the body and expanding. This creates a sense of fullness. Insoluble fiber aids proper digestion by helping to remove harmful wastes from the digestive tract. Insoluble fiber is found in many fruits, vegetables, and whole-grain foods. *Soluble fiber* is fiber that dissolves in water. Soluble fiber has been shown to decrease blood serum cholesterol levels. Sources of soluble fiber include beans, oats, barley, fruits, and vegetables. The USDA *Dietary Guidelines for Americans* recommends consuming 14 g of fiber per 1000 calories consumed.

**Nutrition Note**

The USDA recommends 25 g to 35 g of fiber daily, but many people consume less than 10 g per day. Whole grains, fruits, and vegetables are excellent sources of fiber. By choosing foods that contain complex carbohydrates, people can increase their fiber intake.

## Fats

Fats belong to a large group of chemical compounds called lipids. A *fat* is a nutrient that provides energy, promotes healthy skin, and carries fat-soluble vitamins such as A, D, E, and K throughout the body. Fat also cushions and protects vital organs from physical damage. Types of fats include monounsaturated fats, polyunsaturated fats, saturated fats, and trans fats. Daily fat consumption should equal no more than 30% of total daily calories. Saturated fat should equal no more than 10% of total fat consumed daily.

**Monounsaturated Fats.** A *monounsaturated fat* is a fat found in plant-derived foods and oils such as olives, olive oil, peanuts, peanut oil, and canola oil. These fats have been shown to lower blood cholesterol. For this reason, they are a nutritionally recommended source of dietary fat. Many people in Mediterranean countries, such as Greece and Italy, include olive oil in their daily diets. The fact that these countries have very low rates of cardiac disease is widely attributed to a diet high in monounsaturated fats and low in saturated fats.

**Polyunsaturated Fats.** A *polyunsaturated fat* is a fat found in plant-derived foods such as corn oil, soybean oil, sunflower oil, and sesame oil. Polyunsaturated fats are another nutritionally recommended source of dietary fat. Like monounsaturated fat, polyunsaturated fat has been found to lower blood cholesterol. Because they are better sources of dietary fat, polyunsaturated and monounsaturated fats are sometimes referred to as "good fats."

**Saturated Fats.** A *saturated fat* is a fat found in animal-derived foods such as beef, whole milk, cream, cheese, butter, eggs, poultry skin, and fats such as shortening and lard used in baked goods. Saturated fats are one of the biggest sources of fat in the American diet, but unfortunately are a poor type of fat to consume. Saturated fats are unhealthy because they raise blood cholesterol.

**Trans Fats.** A *trans fat* is a solid fat made from liquid vegetable oil through a process called hydrogenation. *Hydrogenation* is the change that occurs when additional hydrogen atoms are forced to bond with molecules found in plant oils such as corn oil. The presence of additional hydrogen atoms transforms the liquid into a solid. Many margarines and shortenings are created using hydrogenation and contain trans fats. Trans fats are also present in many processed foods.

In 2006 the FDA began requiring that trans fat information be listed on all nutritional labels. ***See Figure 9-2.*** Like saturated fats, trans fats have been found to raise blood cholesterol. Limiting consumption of trans fats is recommended as trans fats may increase the risk of heart disease.

*Trans Fats*

Trans fat information

| **Nutrition Facts**<br>Serv. Size 1 Bar (60g)<br>Servings 1<br>**Calories** 220<br>  Fat Cal. 35<br>* Percent Daily Values (DV) are based on a 2,000 Calorie Diet. | **Amount/serving** | **% DV*** | **Amount/serving** | **% DV*** |
|---|---|---|---|---|
| | **Total Fat** 4g | 6% | **Potassium** 135mg | 4% |
| | Sat. Fat 1g | 4% | **Total Carb** 43g | 14% |
| | Trans Fat 0g | | Fiber 5g | 19% |
| | **Cholest.** 15mg | 6% | Sugars 20g | |
| | **Sodium** 200mg | 10% | **Protein** 4g | |
| | Vitamin A 20% • Vitamin C 0% • Calcium 0% • Iron 20%<br>Vitamin E 10% • Thiamin 20% • Riboflavin 20% • Niacin 20%<br>Vitamin B6 20% • Folic Acid 20% • Phosphorus 10% | | | |

**9-2** *Trans (hydrogenated) fats are identified on nutrition labels.*

*Cholesterol* is an odorless white waxy substance present in all cells of the body. It is a member of the lipids family and is essential in many body processes. The body manufactures all of the cholesterol it needs through the liver. Animal-derived foods, such as beef and poultry, and many processed foods also contain cholesterol.

Cholesterol is transported throughout the body by two main types of proteins in the blood: high-density lipoprotein (HDL) and low-density lipoprotein (LDL). Cholesterol carried by HDL is considered good cholesterol, because HDL transports the cholesterol to the liver, where it is reused or broken down and disposed of. High amounts of HDL are desirable in the body because of evidence suggesting high levels of HDL can reduce the risk of a heart attack.

Cholesterol carried by LDL is considered bad cholesterol, because it has the potential to create deposits in the blood vessels. LDL carries cholesterol to wherever it is needed throughout the body. However, if there is excess cholesterol in the body, the cholesterol carried by LDL begins to stick to arterial walls. The buildup of cholesterol could lead to a blockage in blood flow and could eventually cause a heart attack or stroke.

Daily dietary cholesterol should be limited to 300 mg. Research indicates that consumption of saturated fats and trans fats can increase LDL levels in the blood. Monitoring and limiting total daily fat consumption is necessary for good health.

National Cattlemen's
Beef Association

## Proteins

A *protein* is a complex chain of amino acids that is used by the body to help build and repair body tissues. Hair, eyes, skin, muscles, and bones are made of proteins. Consuming protein as part of a nutritional diet helps keep body tissues in good condition. If carbohydrates fail to provide sufficient energy, the body uses proteins for additional energy instead of using them for building and repairing body tissue.

Proteins are made up of chains of amino acids. The human body requires approximately 22 amino acids. It can produce all but nine of them itself. These nine are called the essential amino acids because they can only be obtained through food sources. The 13 amino acids made by the body are called nonessential amino acids.

*Complete proteins* are proteins that contain all nine essential amino acids and are found in animal-derived foods such as beef, chicken, fish, eggs, cheese, and other dairy products. *Incomplete proteins* are proteins that do not contain all nine essential amino acids, but if combined correctly with other proteins, a complete protein can be obtained. Incomplete proteins are found in vegetables, dried legumes, and nuts. It is important, especially when following a vegetarian diet, that a variety of incomplete proteins are consumed to ensure that the body obtains all nine of the essential amino acids. Protein should make up 12% to 15% of total daily calories. This is approximately equal to 60 g to 75 g of protein daily.

## Vitamins

*Vitamins* are nutrients needed in very small quantities to nourish the body. Vitamins do not provide the body with energy so they do not affect the calorie count of food items. Vitamins are used by various metabolic processes. Eating a balanced diet each day helps ensure the body obtains all of the vitamins required to function properly. Vitamins are categorized as either water-soluble or fat-soluble.

**Water-Soluble Vitamins.** *Water-soluble vitamins* are vitamins that dissolve in water. These vitamins are easily transported in the bloodstream and need to be replenished daily as they are not stored in the body. Water-soluble vitamins include the B-complex vitamins (thiamin, riboflavin, niacin, pantothenic acid, $B_6$, biotin, folate, and $B_{12}$), and vitamin C. ***See Figure 9-3.***

## Water-Soluble Vitamins

| Name | Source | Function |
|---|---|---|
| Thiamin (B$_1$) | Legumes, pork, pecans, spinach, oranges, cantaloupe, milk, eggs | Metabolism |
| Riboflavin (B$_2$) | Milk, cheese, leafy green vegetables, liver, yeast, almonds, soybeans | Metabolism, red blood cell formation, respiration, antibody production, growth regulation |
| Niacin (B$_3$) | Chicken, fish, milk, eggs, yeast, nuts, fruits and vegetables | Metabolism |
| Pantothenic acid (B$_5$) | Whole-grain cereal, legumes, eggs, meat | Metabolism of carbohydrates, proteins and fats |
| Vitamin B$_6$ | Beans, fish, meat, poultry, some fruits and vegetables | Protein and red blood cell metabolism, hemoglobin production, immune system and blood sugar level maintenance |
| Biotin (B$_7$) | Dairy products, seafood, yeast, cauliflower, chicken breast, egg yolk | Cell growth, fatty acid production, metabolism of fats, carbon dioxide transfer, strong hair and nails |
| Folate (B$_9$) | Leafy green vegetables | Cell production and maintenance |
| Vitamin B$_{12}$ | Fish, meat, poultry, eggs, dairy products | Nerve cell and red blood cell maintenance, DNA construction |
| Vitamin C | Fruits and vegetables | Collagen formation, metabolism |

**9-3** *Water-soluble vitamins include the B-complex vitamins and vitamin C.*

*Thiamin* is a B vitamin that aids in metabolism and is found in legumes, pork, pecans, spinach, oranges, cantaloupe, milk, and eggs. *Riboflavin* is a B vitamin that aids in metabolism, red blood cell formation, respiration, antibody production, and growth regulation and is found in dairy products, leafy green vegetables, liver, yeast, almonds, and soybeans. Riboflavin can be destroyed by sunlight so it is important to store foods containing riboflavin in containers and locations that limit exposure to sunlight.

*Niacin* is a B vitamin that aids in metabolism and is found in chicken, fish, milk, eggs, yeast, nuts, fruits, and vegetables. *Pantothenic acid* is a B vitamin that aids in metabolism and is found in whole-grain cereals, legumes, eggs, and meat.

*Vitamin B$_6$* is an essential nutrient for protein and red blood cell metabolism; helps in hemoglobin production, immune system maintenance, and blood sugar level maintenance; and is found in beans, fish, meat, poultry, and in some fruits and vegetables. *Biotin* is a B vitamin that aids in cell growth, the production of fatty acids, metabolism, the transfer of carbon dioxide, and hair and nail growth. It is found in dairy products, seafood, yeast, cauliflower, chicken breast, and egg yolk.

*Folate* is a B vitamin that is used in cell production and maintenance, is particularly important during infancy and pregnancy, and is found in leafy green vegetables. The synthetic form of folate is folic acid. *Vitamin B$_{12}$* is a nutrient that helps maintain healthy nerve cells and red blood cells and is needed in the construction of DNA; it is found in fish, meat, poultry, eggs, and dairy products.

All of the B-complex vitamins are found in fruits and vegetables with the exception of $B_{12}$. Vitamin $B_{12}$ is only found in animal products. Strict vegetarians need to supplement their diets with a $B_{12}$ vitamin.

*Vitamin C* (ascorbic acid) is a vitamin that aids in the formation of collagen, ligaments, joints, bones, and teeth and is found in fruits and vegetables. Vitamin C acts as an antioxidant in the body, protecting cells from damage from oxygen. Vitamin C is also important in maintaining a healthy and strong immune system. The greatest sources of vitamin C are citrus fruits, such as oranges, grapefruit, lemons, and limes.

**Fat-Soluble Vitamins.** *Fat-soluble vitamins* are vitamins that dissolve in fat. Unlike water-soluble vitamins that must be replenished daily, fat-soluble vitamins can be stored in the liver for later use. Fat-soluble vitamins include vitamin A, vitamin D, vitamin E, and vitamin K. ***See Figure 9-4.***

Scurvy, a potentially fatal disease caused by a vitamin C deficiency, results in bleeding gums, loose teeth, and broken capillaries under the skin. In the early 1900s, scurvy was very common among sailors. At that time, naval diets included an abundant amount of fruits and vegetables when a ship initially set sail. However, if the ship was on a long voyage, the galley would run out of fresh produce and signs of scurvy would develop among crew members. Eventually, it was discovered that citrus fruits prevented scurvy, so sailors were given a daily dose of lemon or lime juice.

## Fat-Soluble Vitamins

| Name | Source | Function |
|---|---|---|
| **Vitamin A** | Liver, eggs, dark green vegetables | Vision, bone growth, cell division, reproduction |
| **Vitamin D** | Fortified milk and cereal, dairy products, UV rays | Calcium and phosphorus absorption, bone mineralization |
| **Vitamin E** | Vegetable oils, nuts, leafy green vegetables | Immune system, DNA repair, metabolism |
| **Vitamin K** | Leafy green vegetables | Blood clotting |

9-4　*Fat-soluble vitamins include vitamin A, vitamin D, vitamin E, and vitamin K.*

*Vitamin A* is a fat-soluble vitamin that is crucial for good eye health, and cell division and reproduction. Vitamin A is found in the form of retinol or carotenoids in various foods. Retinol is found in animal-derived foods and is converted to vitamin A in the body. The red, yellow, and orange pigments (known as carotenoids) in fruits and vegetables also can be converted to vitamin A in the body. Beta-carotene, which is the most abundant carotenoid, is the orange pigment in vegetables such as carrots. Of all carotenoids, beta-carotene is most easily converted to vitamin A.

*Vitamin D* is a fat-soluble vitamin responsible for the formation of healthy bones and teeth and helps the body use calcium and phosphorus properly. A deficiency in vitamin D can lead to rickets. Rickets is a disease in which bones grow abnormally, resulting in soft bones and bowed legs. Vitamin D is unique because it is produced by exposure to ultraviolet (UV) rays in sunlight. The body produces vitamin D from cholesterol in the body after being exposed to sunlight. People who have a limited amount of exposure to sunlight need to ensure they eat foods such as milk and cereal that are fortified with vitamin D. Egg yolks also contain vitamin D.

*Vitamin E* is a fat-soluble vitamin that protects cells from oxygen damage. It is used in the immune system, for DNA repair, and in metabolism. It is found in vegetable oils, nuts, and leafy green vegetables. Vitamin E is considered an antioxidant.

*Vitamin K* is a fat-soluble vitamin that has an essential role in the formation of blood clots. Blood clotting is necessary when the skin is cut or a wound is received. The clotting action of the blood prevents major blood loss by sealing the wound. Vitamin K is found in dark green leafy vegetables, fruits, dairy products, and egg yolks.

## Minerals

*Minerals* are inorganic substances that are used in various processes throughout the body. The need for minerals in the human body is quite small, but it is important to good health that these small needs are met. Minerals are divided into two categories, macrominerals and trace minerals. Macrominerals, including calcium, phosphorus, and magnesium, are required in larger amounts than trace minerals, such as iron, iodine, and zinc. Regardless of the amount needed, it is crucial that the body receives all minerals necessary for good health. Eating a well-balanced diet and a wide variety of foods each day ensures the body receives sufficient amounts of needed minerals. *See Figure 9-5.*

| Minerals | | |
|---|---|---|
| Name | Source | Function |
| **Calcium** | Milk, dairy products, broccoli, dark leafy greens | Muscle and blood vessel contraction and expansion, secretion of hormones, sending messages in the nervous system, development of bones and teeth |
| **Phosphorus** | Milk, dairy products, legumes, nuts, peanut butter | Construction and replication of DNA, transport of energy, cell structure |
| **Magnesium** | Green vegetables, legumes, nuts, whole grains | Muscle and nerve function, immune system maintenance, blood pressure and blood sugar regulation, metabolism, protein synthesis |
| **Sodium** | Table salt | Maintaining blood pressure and fluid levels |
| **Potassium** | Fruits, vegetables, dairy products, seafood | Maintaining fluid levels |
| **Iron** | Red meat, fish, poultry, legumes, whole grains, spinach | Oxygen transport, cell growth, cell differentiation |
| **Iodine** | Iodized table salt, seafood, fish | Energy consumption and metabolic rate |
| **Zinc** | Oysters, red meat, poultry | Immune system, DNA synthesis |
| **Selenium** | Fruits, vegetables, meat, seafood, nuts | Antioxidant, heart function, immune system |
| **Copper** | Seafood, organ meats, legumes, nuts, seeds | Red blood cell formation, collagen formation |

**9-5** *While minerals are needed in only small amounts, they are important for the function of various processes throughout the body.*

**Macrominerals.** Calcium and phosphorus are the most abundant minerals in the body, making up approximately 90% of the nutrients needed for development of strong bones and teeth. Deficiencies in these minerals may cause stunted growth and early development of osteoporosis. *Calcium* is a macromineral that is used in muscle contraction, blood vessel contraction and expansion, secretion of hormones and enzymes, and sending messages to the nervous system. It is the most abundant mineral in the body. Good sources of calcium include dairy products, broccoli, and dark leafy greens.

*Phosphorus* is a macromineral that is used in the structure of DNA and cellular membranes. Phosphorus is needed for cellular growth and replication of DNA. It is also needed to aid in the transport of energy at the cellular level in the body. However, too much phosphorus in the body can lead to calcium depletion. Food sources of phosphorus include milk, dairy products, legumes, nuts, and peanut butter.

*Magnesium* is a macromineral that is essential for muscle contraction, normal bowel functions, bone and tooth structure, and nerve transmission. It is also important in the metabolism of carbohydrates, and it influences the release of insulin, the hormone that regulates blood sugar levels. Magnesium is found in dark green vegetables, nuts, legumes, and whole grains.

Wisconsin Milk Marketing Board

*Sodium* is a mineral that the body uses to help maintain blood pressure levels and fluid balance. It is found most often in common table salt. Too much sodium in the body causes high blood pressure (hypertension), which can lead to heart attack and stroke. For this reason, the USDA recommends limiting sodium intake.

*Potassium* is a mineral that works with sodium to maintain blood pressure and fluid levels in the body. It is found in fruits, vegetables, dairy products, and seafood.

**Trace Minerals.** *Iron* is a trace mineral that is necessary to hemoglobin production, the substance in blood that carries oxygen throughout the body. Iron also plays a role in cell growth and differentiation. Foods high in iron include red meat, poultry, legumes, whole grains, and green leafy vegetables such as spinach.

*Iodine* is a trace mineral that helps control energy use in the body and helps regulate the thyroid gland, which controls metabolism. An iodine deficiency may lead to the development of a goiter, which is a condition where the thyroid gland becomes enlarged. The most common sources of iodine are fish, seafood, and iodized salt.

*Zinc* is a trace mineral that helps in the growth and maintenance of all tissues and helps the body manufacture fats, carbohydrates, and proteins. It is found in almost every cell in the body. Zinc also contributes to the process of healing wounds. It helps maintain the senses of taste and smell as well as contributing to growth and development during pregnancy, childhood, and adolescence. Protein-containing foods, whole grains, legumes, and fortified cereals are good sources of zinc.

*Selenium* is a trace mineral that is needed in small amounts to make selenoproteins, which are antioxidants. The heart, the immune system, and the thyroid gland require small amounts of selenium. Selenium can be found in fruits and vegetables grown in areas with a high selenium content in the soil. It can also be found in some meats, seafood, and nuts.

*Copper* is a trace mineral that is used with iron in the formation of hemoglobin to make red blood cells. Copper also aids in the formation of collagen, which gives strength to bones, teeth, muscles, cartilage, and blood vessels. Sources of copper include seafood, organ meats, legumes, nuts, and seeds.

### Water

Water is essential to life. It contains no calories and is present in all cells of the body. It is used by the body for virtually all functions involving transportation, including absorption, digestion, circulation, and excretion, and is necessary in maintaining body temperature. The human body loses about 1 qt of water daily through normal body processes, so water must be replenished daily to ensure a healthy system. Many foods contain water, including fruits, vegetables, and fruit juices.

Carlisle FoodService Products

## MENU PLANNING

Menus are one of the most important aspects of successful food service. Menus communicate the food and beverage items offered by the food service establishment. When planning menus, it is also vital to consider health, evaluate the needs of potential customers, develop recipes that satisfy those needs, and review and modify recipes appropriately.

### Health Concerns

Many people are increasingly concerned about eating healthy due to rising obesity rates, allergies to certain foods, and specific dietary concerns. Menus that emphasize nutritious foods or contain health-conscious items portray an establishment that cares about its customers' concerns.

**Obesity Rates.** Obesity rates in industrialized nations have risen as lifestyles have become more sedentary. Advances in technology have led to people spending more time in front of computer, and television and less time on physical activities for both recreation and work. Also, food has become more plentiful, and portion sizes have increased. The average person is not even aware of what a healthy portion size looks like. A healthy portion of meat should be no more than 3 oz, which is about the size of a deck of cards. *See Figure 9-6.* Many restaurants often offer double or even triple the recommended portion size.

**Food Allergies.** Food-related allergies are also a growing concern in the food service industry. Many people have allergic reactions to specific foods. Allergic reactions to food can be severe or even fatal. It is estimated that

**Nutrition Note**

Portion sizes offered at restaurants are often two to three times larger than USDA recommendations for healthy eating. Healthy menus offer smaller portions or more fruits and vegetables in relation to the entrée.

2% of adults and about 5% of infants and young children in the United States suffer from food allergies.

The USDA requires all food labels to clearly state if products contain any of the eight major allergens: milk, eggs, fish, shellfish, tree nuts, peanuts, wheat, or soybeans. These allergens account for 90% of all documented food allergy-related cases. Food service staff must be educated on ingredients in all menu items offered in order to ensure customers do not suffer an allergic reaction.

**Dietary Concerns.** Common dietary concerns include heart disease and diabetes. Heart-healthy menu items are becoming mainstream as more people try to lower their cholesterol levels. Diabetics look for foods lower in fat and sugar. Another dietary concern that is gaining the attention of many professional chefs is celiac disease. Celiac disease is a disease that involves an inability to digest gluten and gluten-related proteins. The disease causes damage to the interior lining of the small intestine and can destroy the organ's ability to absorb nutrients. Gluten is found in all foods containing wheat flour. If a person with celiac disease consumes food containing gluten, the person may become very ill. An increase in the availability of gluten-free foods is a result of increased awareness of celiac disease.

## Evaluating Customer Needs

Several factors are taken into consideration when developing menu items for potential customers. Some of the more important factors to consider are customer age, location, and type of clientele.

Considering the age of the customers is important. Infants, preschool-aged children, and older adults are considered highly susceptible populations and are more likely to contract a foodborne illness. In addition, sense of taste can differ between age groups. Young children may not care for strong flavors, for example. Older adults may prefer softer and easily digestible foods. Healthy and tasty menu items can be carefully chosen to satisfy customers of all ages. For example, young children may prefer macaroni and cheese to potatoes au gratin. Likewise, older adults may prefer a bound salad to a spinach salad.

The location of an establishment can factor into the kind of menu choices that will likely prove successful. For example, the establishment may be located in the center of a metropolitan area where there are a lot of business people who frequently eat lunch out. In such a location, there is probably a large base of people who seek healthy menu options.

The type of clientele ultimately must be considered when planning menus. A successful establishment will offer menu choices that cater to the needs and desires of customers. For example, if the local population has an interest in vegetarian choices, an establishment could be successful by offering some meatless entrées. If the establishment is targeting families with children, the menu may offer a variety of choices for children and additional choices for adults.

## Healthy Portions

*Meat and Meat Alternatives*
¼ of the plate

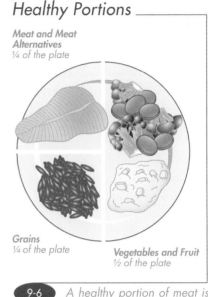

*Grains*
¼ of the plate

*Vegetables and Fruit*
½ of the plate

**9-6** A healthy portion of meat is no more than 3 oz.

### Recipe Development

Once the needs of customers have been identified, one of the easiest ways to begin recipe development is to look at existing menu choices. If fried chicken is on the menu, it could be replaced by a broiled or grilled chicken breast entrée. When creating new items, a chef should consider using the most nutritious ingredients and cooking methods. Including a variety of fruits and vegetables will create balanced meals that are appealing in flavor and color. Limiting ingredients that are high in fat, sugar, or empty calories will also help create more health-conscious recipes.

### Recipe Modification

Substituting ingredients in menu offerings can create new choices that offer more nutritional balance. Key areas to consider when modifying recipes are starches and fats. Replacing processed grains and starches with whole-grain options provides more complex carbohydrates. Excellent sources of whole-grains include wheat berries, wild rice, red rice, bulgur, kasha, barley, and quinoa.

When modifying a recipe, a chef should also evaluate the existing fat content. There may be alternatives to the types of fat used in the recipe. For example, it may be possible to replace a saturated fat such as butter with a monounsaturated fat such as olive oil.

## GOVERNMENT GUIDELINES

Many people do not know how much or which types of foods to consume on a daily basis to maintain a healthy body weight. It can be hard to know which foods are healthy and which foods are unhealthy. The federal government has developed laws and guidelines to help the average consumer understand nutritional information and make smart dietary choices. Standardized nutritional labeling makes it easy for consumers to see if a product meets their specific dietary needs. Dietary guidelines can be used to determine reasonable portion sizes. The USDA MyPyramid Food Guidance System recommends the daily intake of food from the five major food groups and emphasizes the need for daily exercise. Having a thorough understanding of these government guidelines can help a chef create menus that support a healthy lifestyle.

### Nutrition Facts Labels

The Nutrition Labeling and Education Act of 1990 requires food manufacturers to display a Nutrition Facts label on all foods. The Nutrition Facts label is designed to inform consumers of the nutritional value of prepackaged foods and provides a standard for comparing various foods based on nutritional components. A Nutrition Facts label includes information on serving size as well as nutritional values including number of calories and amounts of fat, cholesterol, sodium, carbohydrate, protein, and nutrients per serving. Finally, the Nutrition Facts label breaks each of these values into the equivalent percentage of recommended daily consumption based on a 2000-calorie diet. *See Figure 9-7.*

**Servings.** Serving size and the number of servings per container are listed first on the Nutrition Facts label. This establishes the recommended serving size on which all nutritional values for the product are based. Most serving sizes are provided in units such as cups, ounces, or pieces, followed by the metric unit in grams.

**Calories.** The second item on the Nutrition Facts label identifies the number of calories per serving and the number of those calories that come from fat. A healthy number of fat calories should be not more than 30% of the total calories. To determine the percentage of fat calories, divide the number of calories from fat by the total number of calories per serving, and multiply the result by 100. For example, if a serving contains 220 calories, and 40 calories are from fat, fat calories are 18.2% of total calories (40 ÷ 220 = 0.182; 0.182 × 100 = 18.2%). Because the fat calories are less than 30% of total calories, this could be considered a healthy choice.

However, if a serving contains the same 220 calories and 120 calories are from fat, fat calories are 54.5% of total calories (120 ÷ 220 = 0.545; 0.545 × 100 = 54.5%). Because the fat calories are more than 30% of total calories in this choice, this food item would be considered unhealthy.

**Nutrition.** The next section of the Nutrition Facts label lists the nutritional breakdown of fat, cholesterol, sodium, carbohydrates, and protein. Each nutrient is listed in grams per serving.

The total fat portion of the Nutrition Facts label lists total fat per serving as well as a breakdown of the total fat content into saturated fat and trans fat. Since January 1, 2006, all Nutrition Facts labels have included the amount of trans fat in addition to total fat and saturated fat.

Cholesterol, sodium, and carbohydrates are listed next. Carbohydrates are important to note, because the label can help determine if the item contains simple or complex carbohydrates. If there is a high amount of dietary fiber listed underneath carbohydrates, this is a good indication that the food has a large amount of complex carbohydrates. If there is a high amount of sugar listed, such as in a can of soda, this is a good indication that the food is high in simple carbohydrates. Protein is the last item listed in the nutritional breakdown.

**Percent Daily Values.** The Percent Daily Values listed on the Nutrition Facts label are the amounts of nutrients a healthy person needs each day based on a 2000-calorie diet. The Percent Daily Value is given for each nutrient. The Percent Daily Value is the percentage of each nutrient compared to the total dietary reference intake (DRI) for that nutrient. For example, if the DRI for sodium is 2250 mg, and the item contains 45 mg of sodium, it contains 2% of the total daily

*Nutrition Facts Label*

| **Nutrition Facts** | |
|---|---|
| Serving Size | 29 crackers (30g) |
| Servings Per Container | about 12 |

| **Amount Per Serving** | |
|---|---|
| **Calories** 130 | Calories from Fat 40 |

| | % Daily Value* |
|---|---|
| **Total Fat** 4.5g | **7%** |
| Sat. Fat 1.5g | **8%** |
| Trans Fat 0g | |
| **Cholesterol** 0mg | **0%** |
| **Sodium** 360mg | **15%** |
| **Total Carbohydrate** 20g | **7%** |
| Dietary Fiber less than 1g | **3%** |
| Sugars 0g | |
| **Protein** 4g | |

| | | | |
|---|---|---|---|
| Vitamin A 2% | • | Vitamin C | 0% |
| Calcium 4% | • | Iron | 10% |

*Percent Daily Values (DV) are based on a 2,000 calorie diet. Your daily values may be higher or lower depending on your calorie needs:

| | Calories: | 2,000 | 2,500 |
|---|---|---|---|
| Total Fat | Less than | 65g | 80g |
| Sat. Fat | Less than | 20g | 25g |
| Cholesterol | Less than | 300mg | 300mg |
| Sodium | Less than | 2,400mg | 2,400mg |
| Total Carbohydrate | | 300g | 375g |
| Dietary Fiber | | 25g | 30g |

**9-7** *The Nutrition Facts label is designed to inform consumers of the nutritional value of prepackaged foods.*

allowance for sodium. As people may have different dietary needs, the bottom portion of the label lists the maximum total fat, saturated fat, cholesterol, sodium, total carbohydrate, and fiber a healthy person should eat daily based on a 2000- and 2500-calorie diet.

## Dietary Guidelines

The U.S. Department of Agriculture (USDA) has published the *Dietary Guidelines for Americans* for almost 100 years. The publication is revised every five years. Early recommendations by the government were aimed at preventing nutritional deficiencies and diseases. Today, nutritional deficiencies are much rarer. Instead, key recommendations focus on disease prevention and healthy eating. Excessive intake of the wrong types of food can lead to the development of obesity and chronic diseases such as heart disease.

In recent revisions to the *Dietary Guidelines,* the USDA has significantly increased recommended daily consumption of fruits, vegetables, whole grains, and fat-free or low-fat milk and milk products. They have reduced recommended daily consumption of meats, total grains, oils, solid fats, and sugars. Instead, they recommend eating lean meats, poultry, fish, beans, eggs, and nuts. Emphasis is placed on a diet that is low in saturated fats, trans fats, cholesterol, salt (sodium), and added sugars.

In order to ensure that a person meets daily nutrition requirements, dietary reference intakes (DRIs) have been established. A *dietary reference intake (DRI)* is a quantity of a particular nutrient that is sufficient to meet the nutrient requirements of 97% to 98% of all healthy individuals in a group. An *adequate intake (AI)* is the dietary intake value that is used when no DRI has been established, and there is no estimated average requirement calculated for a nutrient. For example, there is no DRI for calcium due to an insufficient amount of research to date. Instead, there is an AI value for calcium that serves as a guideline until a DRI is established.

## MyPyramid Food Guidance System

In 2005 the USDA released the MyPyramid Food Guidance System. ***See Figure 9-8.*** The system emphasizes daily exercise and suggests daily servings for each of the five food groups: grains, vegetables, fruits, low-fat milk and dairy products, and lean proteins. MyPyramid differs from previous recommendations in that it does not suggest a number of servings. Instead, it provides a more accurate measure of the total amount of food from each group to consume daily. For example, the food guide recommended 2 to 3 servings of protein per day. MyPyramid changed that recommendation to 5½ oz of protein per day. MyPyramid also suggests eating across the pyramid to ensure that a diet contains all necessary nutrients needed by the body. The stairs on the left side of MyPyramid represent the need for daily exercise.

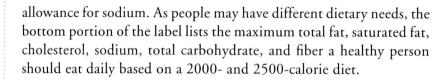

**MyPyramid**
STEPS TO A HEALTHIER YOU
MyPyramid.gov

| GRAINS | VEGETABLES | FRUITS | MILK | MEAT & BEANS |
|---|---|---|---|---|
| Make half your grains whole | Vary your veggies | Focus on fruits | Get your calcium-rich foods | Go lean with protein |
| Eat at least 3 oz. of whole-grain cereals, breads, crackers, rice, or pasta every day | Eat more dark-green veggies like broccoli, spinach, and other dark leafy greens | Eat a variety of fruit | Go low-fat or fat-free when you choose milk, yogurt, and other milk products | Choose low-fat or lean meats and poultry |
| | Eat more orange vegetables like carrots and sweetpotatoes | Choose fresh, frozen, canned, or dried fruit | | Bake it, broil it, or grill it |
| 1 oz. is about 1 slice of bread, about 1 cup of breakfast cereal, or ½ cup of cooked rice, cereal, or pasta | Eat more dry beans and peas like pinto beans, kidney beans, and lentils | Go easy on fruit juices | If you don't or can't consume milk, choose lactose-free products or other calcium sources such as fortified foods and beverages | Vary your protein routine — choose more fish, beans, peas, nuts, and seeds |

For a 2,000-calorie diet, you need the amounts below from each food group. To find the amounts that are right for you, go to MyPyramid.gov.

| Eat 6 oz. every day | Eat 2½ cups every day | Eat 2 cups every day | Get 3 cups every day; for kids aged 2 to 8, it's 2 | Eat 5½ oz. every day |

**Find your balance between food and physical activity**
- Be sure to stay within your daily calorie needs.
- Be physically active for at least 30 minutes most days of the week.
- About 60 minutes a day of physical activity may be needed to prevent weight gain.
- For sustaining weight loss, at least 60 to 90 minutes a day of physical activity may be required.
- Children and teenagers should be physically active for 60 minutes every day, or most days.

**Know the limits on fats, sugars, and salt (sodium)**
- Make most of your fat sources from fish, nuts, and vegetable oils.
- Limit solid fats like butter, margarine, shortening, and lard, as well as foods that contain these.
- Check the Nutrition Facts label to keep saturated fats, *trans* fats, and sodium low.
- Choose food and beverages low in added sugars. Added sugars contribute calories with few, if any, nutrients.

*Information reported by the U.S. Department of Agriculture*

**9-8** *The MyPyramid Food Guidance System emphasizes exercise in addition to eating healthy foods from the five major food groups.*

## Whole Grain

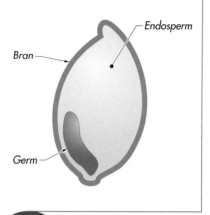

**9-9** *The three key parts of a whole grain are the bran, germ, and endosperm.*

**Grains.** Whole grains are an important source of nutrients. They are a main source of dietary fiber. Whole grains contain three key parts: bran, endosperm, and germ, ***See Figure 9-9.*** The bran provides fiber, B vitamins, and trace minerals. The endosperm provides energy in the form of carbohydrates, protein, and B vitamins. The germ provides vitamin E, B vitamins, trace minerals, and phytonutrients. Refined grains lose most of the bran and germ from the grain kernel as they are processed. The recommended daily serving of grains is 6 oz.

**Vegetables.** A person eating a 2000 calorie daily diet should be eating 2 cups to 3 cups of vegetables per day. Adding vegetables to a diet can help reduce the risk of chronic diseases such as stroke, type 2 diabetes, and some forms of cancer. The USDA specifically recommends eating more dark green and orange vegetables and more dried beans and peas.

**Fruits.** A person eating a 2000-calorie daily diet should consume 1½ cups to 2 cups of fruits per day. Eating a variety of fruits is best, and fresh fruit is preferred over fruit juices, which often contain refined sugars.

**Milk and Dairy Products.** Milk and dairy products are an important source of calcium, which helps to help maintain bone strength. The recommended daily serving of milk and dairy products is 3 cups per day. In addition to various forms of milk and cream available, this group also contains butter, cheese, cottage cheese, sour cream, and yogurt. The USDA recommends choosing low-fat or fat-free dairy products. Lactose-free products and other sources of calcium should be consumed if a person does not or cannot consume milk products.

**Meats, Fish, and Nuts.** Meat, poultry, fish, dried beans or peas, eggs, nuts, and seeds are part of this food group. Fish, nuts, and seeds contain monounsaturated and polyunsaturated oils and are a healthy choice. It is important to take care when selecting fish, as some varieties may contain high levels of mercury that can be dangerous when the fish is eaten in large quantities. Lean or low-fat meats are preferred over more fatty cuts. The recommended daily intake of foods from this group is between 5 oz and 6 oz for adults.

**Sugars and Sweetening Agents.** Sugars and sweetening agents are considered discretionary calories. Discretionary calories are the calories left over after all recommended daily allowances of the major food groups have been met. The average person is allowed between 100 and 350 discretionary calories per day.

**Fats and Oils.** Solid fats and oils are also considered discretionary calories, yet in small amounts are important for proper health and nutrition. The best sources of nutritious fats and oils are nut and vegetable oils. Intake of solid fats such as butter, shortening, or lard should be as limited as possible, as these fats are high in saturated fat. The maximum amount of saturated and trans fats consumed in a day should be under 24 g for people on a 2000-calorie diet.

**Exercise.** Daily exercise is important to staying healthy. *The Dietary Guidelines for Americans* emphasizes the need to include regular physical activity as a way to improve one's health and sense of well-being, and to maintain a healthy body weight. *Physical activity* is any bodily movement produced by skeletal muscles that results in energy expenditure. It is recommended that adults engage in at least 30 min of moderate-intensity physical activity daily to reduce the risk of chronic diseases, such as heart disease, type 2 diabetes, colon cancer, and osteoporosis. To prevent the gradual accumulation of excess weight in adulthood, up to 30 additional min per day may be required. Walking for 20 min to 30 min daily will help improve overall health.

## PLANNING MENUS

Many individuals may be involved in the planning and writing of menus. In a family-run restaurant, the chef and the owner may plan the menu collaboratively. This helps ensure that the foods being offered and prepared satisfy the cost and profit needs of the owners. The menu offerings, portion sizes, selling prices, and food costs must meet the needs of both the operation and the customers.

In a large hotel, the executive chef often works with top management to create the menu. In a franchise or chain operation, the menu, portion sizes, recipes, and prices are determined by a central office. In hospitals, healthcare settings, and nursing facilities, the menu and recipes are written by a registered dietician, sometimes with the help of a chef. In an institutional food service setting, a food service manager and kitchen manager or head cook usually work together on the menu.

*Charlie Trotter's*

### The Function of the Menu

The menu is the primary way that an operation informs customers of the types of food offered and how much each dish costs. The menu serves many management functions within the operation as well. It also dictates the following:

- the type of customer that will come to the restaurant, which is also referred to as the target market
- the type and amount of each product needed on hand
- the knowledge, experience, and skill the staff must have
- the equipment needed
- the selling price of an item, based on food cost
- the percentage of profit each item should generate, based on the food cost subtracted from the sales price, which is also known as the profit margin

There are many styles of menus, with the style of menu offered depending on the type of operation. Some restaurants offer multiple courses for a set price, while others charge separately for each dish. A menu can be offered on a large overhead sign, a single sheet of paper, or even in the style of a book, with pages offering foods from appetizers to desserts. The physical form the menu takes is often influenced by the menu type. Menu types include à la carte, semi-à la carte, prix fixe, table d'hôte, California, and meal specific.

**À la Carte.** An *à la carte menu* is a menu that prices all food and beverage items separately. À la carte menus can be found in restaurants from casual quick-service restaurants, where a sandwich is ordered and charged separately from a side of fries, to steak houses, where a side of vegetables is ordered and charged separately from a steak entrée.

**Semi-à la Carte.** A *semi-à la carte menu* is a menu that has the entrée accompanied by and priced to include the starch and vegetable and usually a choice of soup or salad, while the appetizer and dessert courses are presented and charged separately. Some quick-service restaurants use both à la carte and semi-à la carte menus when they offer their "value meal" combinations, where a diner can order a main dish, side dish, and beverage for a set price.

**Prix Fixe.** A *prix fixe menu* is a menu that offers a complete meal with all courses included at a set price. There are typically a few options within each course. For example, a prix fixe menu may offer an appetizer, soup, salad, sorbet, entrée, and dessert for a set price, with three or four choices in each category, such as a choice from three different entrées. The customer is able to choose an item from each category. Prix fixe menus are often used in upscale restaurant settings.

**Table d'Hôte.** A *table d'hôte menu* is a menu that identifies the item that will be served for each course as determined by the chef. It similar to a prix fixe menu in that everything is included in the cost of the meal. The difference is that a table d'hôte menu does not offer choices to the customer. A plated banquet where the customer has no choices is an example of a table d'hôte menu.

**California.** A *California menu* is a menu that offers diners all of the food and beverage selections for any meal served throughout the day. A California menu usually resembles a book made with heavy laminated pages. A diner has the option of ordering breakfast, lunch, or dinner items at any time. Restaurants that offer a California menu are typically open 24 hours a day.

**Meal Specific.** A *meal-specific menu* is a menu written to be inclusive of a specific meal only, such as breakfast, lunch, or dinner. Meal-specific menus feature only dishes that are typical for that time of day, such as eggs, pancakes, and waffles on a breakfast menu and sandwiches and salads on a lunch menu.

## Menu Classifications

Once the menu type is determined, the decision needs to be made as to how often the selections will change. Some menus change with seasonal offerings, while others remain fairly unchanged throughout the year. Developing a menu that represents the operation well, is pleasing to the customers, and that accomplishes the cost and profit goals of the operation is the right menu to have. Classifications for menus include cycle menus, fixed menus, and market menus.

A *cycle menu* is a menu is written for a specific period and repeats after that period ends. Cycle menus are common in institutional food service settings such as hospitals and school cafeterias. A cafeteria menu with 30 days of predetermined menus that repeats each month is an example of a cycle menu. In a setting where there are new customers daily, such as in a hospital cafeteria, a cycle menu may repeat every five days.

A *fixed menu* is a menu that is developed and then rarely changes. Fixed menus are commonly found in quick-service restaurants where the entire organization has the same menu and it rarely changes. They can also be found at entertainment arenas, such as ballparks and concert venues, where the same menu is available throughout the season and does not change.

A *market menu* is a menu that changes often to coincide with changes in available products. Market menus allow the chef to be creative and to use seasonal food items at peak freshness. Market menus change frequently, as they revolve around seasonal items that are at their peak for only a short time.

## Truth-in-Menu Guidelines

When planning menus, it is important for the manager or chef to also make sure that the menu provides accurate information. In the United States, the federal government has enacted truth-in-menu guidelines requiring accuracy in statements made on menus. Truth-in-menu guidelines prohibit the following:

- misrepresenting the portion size of an item, such as advertising a 12-oz steak but serving a 10-oz steak
- misrepresenting the quality or grade of an item, such as listing USDA Prime beef when serving USDA Select beef
- misrepresenting a brand of product by serving another brand of product in its place, such as by advertising Coca-Cola® but serving another brand of cola
- misrepresenting dietary or nutritional information, such as listing a product as low fat, low sodium, or gluten free when it is does not meet criteria to be labeled as such
- misrepresenting preservation methods, such as advertising fresh fish but serving fish that was previously frozen
- misrepresenting the type of product served, such as advertising an expensive variety of fish but serving a less expensive variety in its place

Truth-in-menu guidelines were developed to protect customers and to ensure that customers get what they pay for. Food service establishments must be able to support and prove statements related to the nutritional content of items served. For example, if a menu item is listed as "heart healthy," the operation must have factual documentation showing that components of the dish are, in fact, heart healthy.

## SUMMARY

The human body needs a wide variety of nutrients to function properly. The six major types of nutrients include carbohydrates, fats, proteins, vitamins, minerals, and water. A well-balanced diet provides the right amount of each essential nutrient. Just as consuming too little of a particular nutrient can affect a person's health, eating too much of certain nutrients can also have a negative impact.

Understanding the nutritional value of a food and how the body uses various nutrients helps customers make wise food choices. This same understanding helps food service professionals better serve customers by offering menu selections that are both nutritious and tasty. The federal government has developed laws and guidelines to help consumers understand nutritional information and make smart dietary choices. Likewise, food service professionals should be conscious of the most recent USDA guidelines when planning a menu. A chef should also be aware of truth-in-menu laws and guidelines.

The type of food service establishment often influences the type and classification of menu chosen. Types of menus include à la carte, semi-à la carte, prix fixe, table d'hôte, California, and meal specific. Once the type of menu is determined, a classification such as cycle, fixed, or market is used to indicate how often food items are changed. Menu development also involves the use of the truth-in-menu guidelines that ensure customers will be served the food that is advertised on the menu.

Refer to the CD-ROM for Quick Quiz® questions related to chapter content.

1. Identify the six major categories of nutrients.

2. What is the difference between simple and complex carbohydrates?

3. Give examples of the general health advantages that fats provide to the human body.

4. Explain the effects of trans fats on the body.

5. Explain why cholesterol carried by high-density lipoprotein (HDL) is considered good cholesterol.

6. Describe the differences between complete and incomplete proteins.

7. Identify the general category of vitamin that dissolves in water and needs to be replenished daily, and provide examples of this type of vitamin.

8. List the benefits and common sources of vitamin E.

9. What are the two main types of minerals?

10. What condition might result from an iodine deficiency?

11. Describe the importance of water for maintaining good health.

12. What is considered a healthy portion size of meat?

13. Why is the age of clientele an important consideration when planning a menu?

14. Explain possible ways to modify starches in existing recipes to make menu items more nutritious.

15. Define dietary reference intake (DRI).

16. How does the MyPyramid Food Guidance System differ from previous recommendations by the USDA?

17. How many minutes of daily walking does the *Dietary Guidelines for Americans* recommend to improve one's overall health?

18. In addition to informing customers of offerings and prices, what management functions are served by a menu?

19. What are six types of inaccuracies prohibited by truth-in-menu guidelines?

20. What is the difference between an à la carte menu and a semi-à la carte menu?

21. What is the difference between a prix fixe menu and a table d'hôte menu?

22. What is the difference between a California menu and a meal-specific menu?

23. Compare a cycle menu, a fixed menu, and a market menu.

# Seasonings and Flavorings

*Seasonings and flavorings* can intensify or improve the natural flavors in foods. Every type of food has a distinct and unique flavor. Seasonings are ingredients that intensify the flavor of a food without changing its natural flavor. Flavorings add to or change the natural flavor of a food. Successfully using seasonings and flavorings requires an understanding of sensory perception. How customers perceive food determines their opinion of a food service establishment.

## Chapter Objectives:

1. Describe the three facets of sensory perception and their affect on how we experience food.
2. Explain the difference between seasonings and flavorings and list specific categories and examples of each.
3. Describe the major guidelines for cooking with herbs and spices.
4. Identify the general rules for purchasing and storing herbs and spices.
5. Compare and contrast the types of nuts and seeds commonly used in cooking.
6. Discuss the quality, smoke point, and function of common types of cooking oils.
7. Identify the quality and characteristics of different types of vinegar.
8. Describe common condiments and their uses.

## Key Terms

- sensory perception
- seasonings
- flavorings
- lemon zest
- herbs
- spices
- fines herbes
- herbes de Provence
- nuts
- oils
- vinegars
- condiments

## SENSORY PERCEPTION

There is a saying that "you eat with your eyes." This means that if food does not look appetizing a person is likely to think that it does not taste very good. If a food looks appetizing, however, a person often can hardly wait to taste it. Smell is another sense that can influence a person's judgment of a food. For example, some people are hesitant to taste certain varieties of cheese because of strong or pungent smells. The smell of cheese intensifies as it ages and this smell, or aroma, may keep some people from wanting to try aged cheeses.

*Sensory perception* is the ability of the eyes, nose, mouth, and skin to gather information and evaluate the environment. When eating, a person uses four of the five sense organs—the eyes, nose, taste buds, and skin—to evaluate food. These four sense organs evaluate appearance, smell, flavor, and texture, enabling a person to describe and identify foods.

### Presentation

The presentation of food almost immediately prompts a judgment on whether it will taste good or bad. It is important for a chef to plan the way a plated course is presented to the customer. Careful thought should be given to making every dish look appealing to the customer. Foods should be properly cooked (not overcooked or undercooked) when they are served. Meals should have a good mix of color and be neatly plated. A beautifully plated meal is first eaten with the eyes while the mouth is anxiously waiting to take a taste. ***See Figure 10-1.*** Cooked vegetables, for example, look more appetizing when they are vibrant shades of green, red, yellow, or orange. Vegetables that are overcooked lose much of their natural color, resulting in a grayish appearance. If vegetables have lost color, chances are they have also lost a lot of flavor, and people may hesitate to eat them.

### Flavor

What does it mean when someone tastes an orange and says it "tastes like an orange"? What words could be used to describe the flavor of an orange to someone who had never tasted one? The easiest way to describe a flavor is to focus on the three sensory experiences that help the brain to determine a sense of flavor. These three sensory experiences are aroma, touch, and taste. When food is eaten, signals about these three sensory experiences are sent to the brain. The brain can then come to a conclusion about the flavor of the food.

## Presentation

MacArthur Place Hotel, Sonoma

10-1   *A beautiful presentation will cause the customer to anticipate the taste of the food.*

**Aroma.** The nose can detect thousands of aromas. All of the flavors associated with foods, such as the difference in flavor between a red grapefruit and an orange, are actually determined in the nose. This is because the nose has many more sensory cells than the tongue. Therefore, the nose can differentiate among foods that are very similar in taste. A description such as "a vanilla-flavored coffee" really describes the overall aroma of the item rather than its taste. The tongue may taste the sweetness and bitterness of the coffee, but it is the nose that senses the vanilla and the roasted smell of the coffee beans.

**Touch.** Touch sensations are physical feelings sensed by nerve endings. Touch sensations that contribute to flavor are detected by nerves in the skin of the mouth, nose, and throat. These feelings also play a role in the detection of flavors, as the nerves "feel" the flavors. For example, one flavor with a strong physical sensation is pepper. Certain aspects of the flavor of pepper are sensed in the back of the throat, which explains why pepper can sometimes make a person cough. Other nerves sense the feel of pepper in the nose and mouth, while the heat of pepper is sensed by nerves in the nose and under the tongue.

**Taste.** There are countless flavors available to taste. Does this mean that the tongue can actually identify each flavor? The truth is that the tongue can clearly identify only four major types of flavor: salty, sweet,

**Action Lab**

Gather five cocktail straws, 1 tsp lime juice, 1 tsp lemon juice, 1 tsp orange juice, 1 tsp apple juice and a glass of water. Blindfold a partner and have your partner hold his or her nose. Now, using a different cocktail straw for each liquid, place a drop of juice on your partner's tongue and have your partner describe the flavor while holding his or her nose. Have your partner drink water between each flavor. Repeat with each liquid and describe your findings.

bitter, and sour. The tongue is covered in small bumps called *papillae*. Each of these papillae is covered with hundreds of minute cells called *taste buds.* These taste buds can detect different flavor characteristics. ***See Figure 10-2.*** As food enters the mouth, molecules relating to the four types of flavor (salty, sweet, bitter, and sour) pass over the tongue and are sensed by the taste buds. The saliva in the mouth also plays a very important role in the function of taste buds. Without saliva, foods would be too thick to be absorbed by the tiny taste bud cells and could not be identified.

### Taste

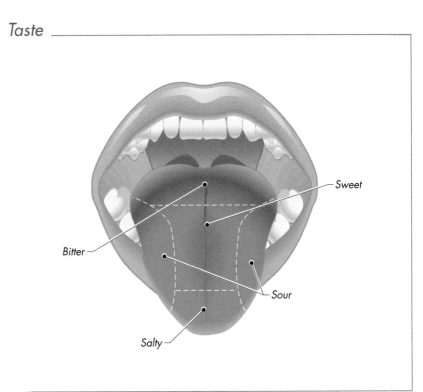

10-2 Taste buds on different parts of the tongue identify different flavor characteristics.

### Texture

Food texture can be a determining factor in a person's eating decisions. Even if a food looks and smells appealing, an unappetizing texture can cause a person to not want to eat that item. For instance, foods such as bananas and avocados may taste great to most people, but some people do not eat these items because they do not like the mushy texture. The best meals are made up of foods with complementary textures. A soft food should be served with a crisp counterpart. A chewy food is paired well with a crunchy or creamy item. Imagine a plate of moist roast turkey, crisp broccoli spears, and creamy mashed potatoes. A variety of textures makes a meal interesting and provides contrast between the items on the plate. ***See Figure 10-3.***

*Texture*

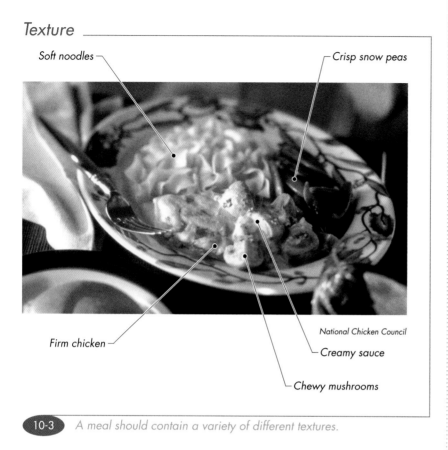

Soft noodles

Crisp snow peas

Firm chicken

National Chicken Council

Creamy sauce

Chewy mushrooms

**10-3** *A meal should contain a variety of different textures.*

## SEASONINGS

Although each food has a particular flavor, it is often desirable to enhance the flavor of food through the use of seasonings. *Seasonings* are items added to enhance the natural flavor of foods. When seasonings are used correctly, the flavor of each particular seasoning should not be noticeable. If a seasoning is used incorrectly or in too great a quantity, it can make the item being prepared inedible or unpleasant to eat. The most common seasonings used are pepper, salt, and lemon zest.

## Pepper

Pepper is the most widely used seasoning. It comes in many forms and colors. Black, white, and green peppercorns are the berries of the same climbing vine known as the pepper plant. This plant is native to the tropics and never grows farther than 20° from the equator. Although they are the berries of the same plant, the varieties of peppercorns result from the berries being processed differently. The most common kinds of pepper are black pepper and white pepper. Black pepper is made from dried, immature berries. The peppercorns are picked when still slightly green, dried by the sun, and sold as whole peppercorns, cracked pepper, or ground pepper. ***See Figure 10-4.***

## Peppercorns

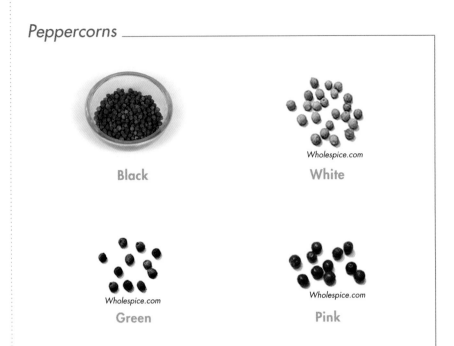

Black

White
*Wholespice.com*

Green
*Wholespice.com*

Pink
*Wholespice.com*

**10-4** *Black peppercorns are the most common variety of peppercorns.*

*White pepper* is made from mature berries of the pepper plant. Ripe berries are fermented and the outer covering containing most of the spiciness is removed. White pepper is much milder in flavor than black pepper and is sold in whole or ground form. White pepper is generally more expensive than black pepper because it requires more labor to cultivate and process. Some manufacturers can produce a white-colored pepper by mechanically removing the outer skin from black peppercorns rather than going through the ripening and fermenting process. The product is not considered to be equivalent to white pepper, as its flavor remains closer to that of black pepper. White-colored pepper produced using this mechanical process is labeled "decorticated."

Green peppercorns are the same immature berries as the black peppercorns, but rather than being dried by the sun they are placed in a salty brine or vinegar or are freeze-dried. Pickled green peppercorns are soft and have a taste and texture similar to capers. The freeze-dried version is often cracked or ground as a seasoning.

Pink peppercorns do not come from the plant that produces white, black, and green peppercorns but rather are the berries of an evergreen tree native to South America. Pink peppercorns do not have the same heat as other varieties of pepper. The flavor of pink peppercorns is similar to that of rosemary or pine.

## Salt

Salt (sodium chloride) is used to enhance the flavor of many foods. Salt has also been used throughout history as a means of preserving foods, as it prevents or at least greatly slows the growth of harmful microorganisms on food. Salt is sold in many forms including table salt, sea salt, kosher salt, and pickling salt. ***See Figure 10-5.***

## Salts

Table

Sea

*The Spice House.com*

Kosher

Pickling

**10-5** *Salts vary in grain size and weight.*

The correct amount of salt to add depends on the food. Salt has a more intense taste and flavor on cold foods than on warm or hot foods. A good rule of thumb is that more salt can always be added to an item, but it cannot be taken away. Therefore, it is important to always taste food before adding salt.

Salt is typically produced by pumping purified water through underground salt deposits. The resulting water, or "brine," that reaches the surface is then allowed to evaporate, leaving behind salt crystals. These crystals are finely ground and mixed with anticaking chemicals to prevent the salt from absorbing moisture and clumping. In the United States, iodine is added to salt as a preventive against swelling of the thyroid, commonly known as a goiter. Salt that has added iodine is labeled as iodized salt. This is the salt known as table salt.

**Sea Salt.** As the name suggests, *sea salt* is salt produced through the evaporation of seawater. Large, shallow reservoirs are allowed to fill with seawater, which is then prevented from flowing back into the ocean. As the shallow pool of seawater evaporates, a thick layer of crystallized sea salt is formed. Sea salt also contains the minerals calcium, potassium, and magnesium. The high concentrations of these additional minerals give sea salt a more intense flavor. Sea salt is available in fine and coarse crystals and is used in both cooking and preserving food. Many chefs consider sea salt to be the most highly prized variety of salt.

**Kosher Salt.** *Kosher salt* is a type of salt used for curing, seasoning, and preparing kosher foods. Kosher salt contains no iodine or anticaking additives. It is lighter in weight than table salt and sticks to food very well. Cuts of raw red meat, such as beef, are packed in kosher salt for a period of time so that the salt draws blood and moisture from the meat. The meat is considered purified after the excess blood has been removed.

**Pickling Salt.** *Pickling salt* (canning salt) is a very pure form of salt that has no residual dust, iodine, or other additives. The purity of pickling salt makes it a perfect seasoning to use when pickling foods as it does not cloud the pickling brine or settle to the bottom of the jar.

## Lemon Zest

*Lemon zest* is the grated peel and pith of a lemon that is often used as a flavor enhancer. ***See Figure 10-6.*** Only a small amount is needed in any dish. Lemon zest is added to various sauces, baked goods, and sautéed or grilled meats. The purpose of using lemon zest is not to make things taste lemony, but rather to add a fresh aroma and taste to the food.

## FLAVORINGS

*Flavorings* are items that add a new flavor to food or alter the natural flavor of food. Vanilla and raspberry are examples of flavorings with such intense flavors that they overtake the flavor of other ingredients in a recipe and emerge as the dominant characteristic of the recipe. All extracts used in baking are flavorings, as they add an intense flavor to whatever they are used in. Flavorings also include herbs, spices, vinegars, and other condiments.

## Herbs

*Herbs* are a group of aromatic plants whose leaves, stems, or flowers are used to add flavor and aroma to food. The majority of herbs are grown in temperate climates. Although dried herbs are available throughout the year, the flavors and aromas of herbs change slightly when dried. For this reason, chefs usually prefer to use fresh herbs. ***See Figure 10-7.***

## Lemon Zest

— Lemon

Zest —

**10-6** *Lemon zest enhances flavor by giving a fresh aroma and taste to food.*

When preparing a recipe, it is important to add fresh herbs at the appropriate time to ensure maximum flavor and aroma. When making hot foods, herbs should be added near or at the end of the cooking process. For maximum flavor when preparing cold dishes, it is better to add fresh herbs a few hours before the dish is served. Dried herbs, on the other hand, can be incorporated into a dish near the beginning of the cooking time since they are not as delicate and it takes longer to extract their flavor. When substituting dried herbs in a recipe that calls for fresh herbs, always use a smaller amount of dried herbs than the amount of fresh herbs called for in the recipe.

Dried herbs should be stored in airtight containers out of direct light to help preserve their flavors and aromas. Freshly cut herbs should be wrapped loosely in damp paper or cloth and placed in a plastic bag to prevent them from drying out and wilting. Properly wrapped fresh herbs should be refrigerated between 35°F and 45°F.

Florida Tomato Committee

**Basil.** *Basil* is the pointy green leaf from a plant of the same name. This savory herb blends well with many different foods. Varieties of basil have hints of other flavors, including lemon, garlic, cinnamon, clove—even chocolate.

Sweet basil is the most common type of basil and is used to flavor tomato sauces, pesto, pasta, fish or chicken dishes, vegetables, and egg dishes, as well as some vinaigrettes and infused oils. Opal basil is very similar to sweet basil in flavor but is slightly milder. The major difference is in the color of the plant itself. Opal basil leaves are slightly more firm, crinkled, and have a vibrant plum color.

**Bay Leaves.** Bay (sweet laurel) is an herb rich in flavor. A *bay leaf* is the thick, aromatic leaf of the evergreen bay laurel tree grown in the Mediterranean. Bay leaves are used for flavoring soups, roasts, stews, gravies, and meats, and for pickling. Because bay leaves are very tough and can be sharp along the edges, they are always removed and discarded before serving.

**Chervil.** *Chervil* is an herb with dark green, curly leaves that is native to Russia. It is grown in England, northern Europe, and the United States. Chervil has a flavor is similar to parsley but is more delicate with a hint of licorice. Chervil can be used in salads, soups, and egg and cheese dishes and is also commonly used as a delicate garnish.

**Chives.** A *chive* is a very delicate herb that has a mild onion flavor. Chives are available year-round and can be purchased fresh, flash frozen, or dried. Chives grow in small bunches with hollow, grass-shaped, green sprouts. If left unpruned, the chive sprout blossoms with purple flowers that can also be used to flavor a salad or other dish. Chives add color and flavor to cream cheese, egg dishes, soups, salads, and are a common garnish on top of a baked potato with sour cream.

**Historical Note**

Bay (sweet laurel) has historical symbolism. Roman emperors wore a laurel wreath because it was considered a symbol of a triumphant leader or champion. Celebrated scholars and athletes throughout history have been crowned with laurel wreaths.

## Herbs

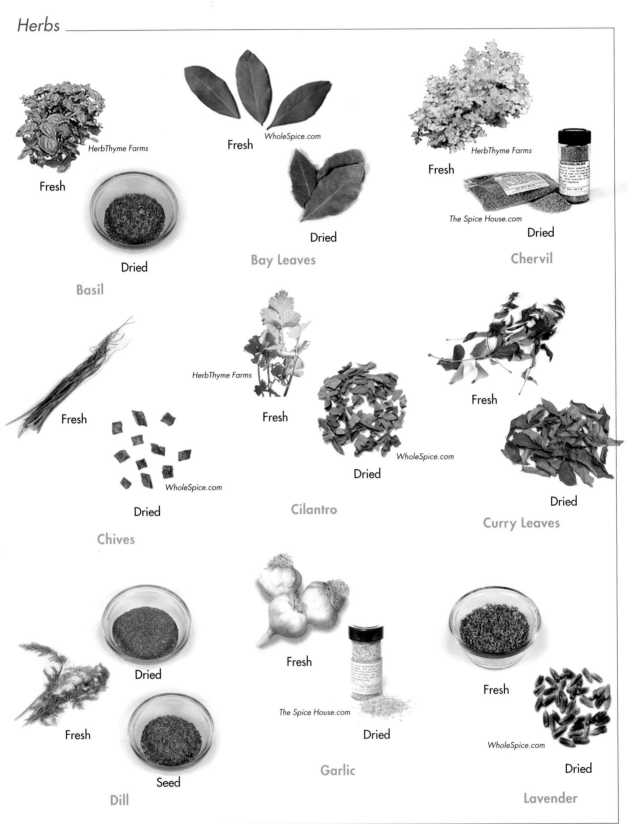

Fresh *HerbThyme Farms*

Fresh

Dried

**Basil**

Fresh *WholeSpice.com*

Dried

**Bay Leaves**

Fresh *HerbThyme Farms*

*The Spice House.com*

Dried

**Chervil**

Fresh

Dried *WholeSpice.com*

**Chives**

Fresh *HerbThyme Farms*

Fresh

Dried *WholeSpice.com*

**Cilantro**

Fresh

Dried

**Curry Leaves**

Dried

Fresh

Seed

**Dill**

Fresh

*The Spice House.com*

Dried

**Garlic**

Fresh

Dried *WholeSpice.com*

Dried

**Lavender**

**10-7** Herbs are aromatic plants whose leaves, stems, or flowers are used to add flavor and aroma to food. (continued on next page)

# Herbs

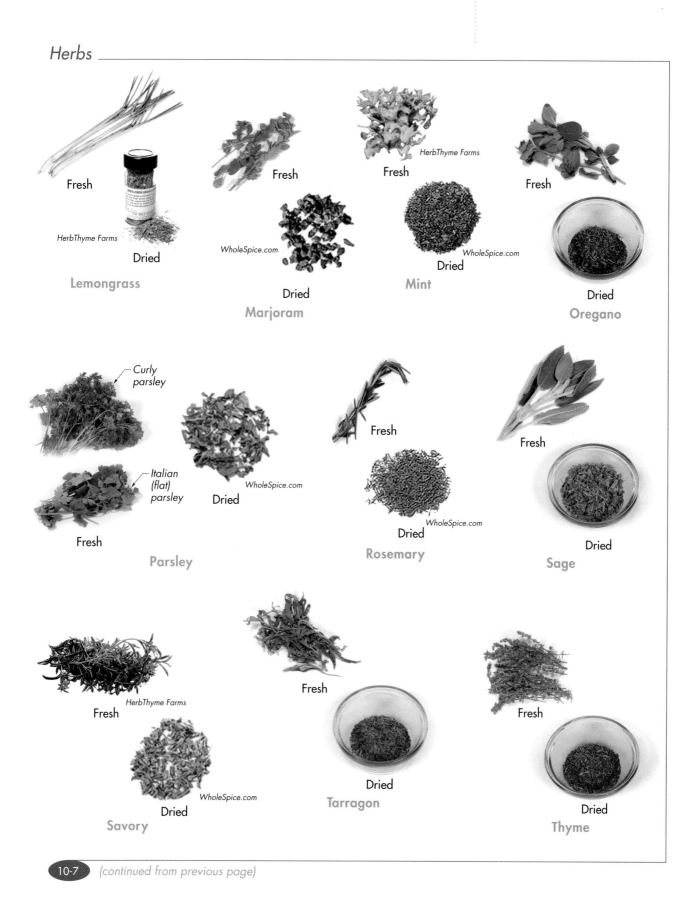

Fresh

HerbThyme Farms

Dried

**Lemongrass**

Fresh

WholeSpice.com

Dried

**Marjoram**

Fresh

HerbThyme Farms

Dried

**Mint**

Fresh

Dried

**Oregano**

Curly parsley

Italian (flat) parsley

Fresh

WholeSpice.com

Dried

**Parsley**

Fresh

WholeSpice.com

Dried

**Rosemary**

Fresh

Dried

**Sage**

Fresh

HerbThyme Farms

WholeSpice.com

Dried

**Savory**

Fresh

Dried

**Tarragon**

Fresh

Dried

**Thyme**

**10-7** *(continued from previous page)*

**Cilantro.** *Cilantro* (also known as Chinese parsley) is the stem and leaves of the cilantro plant. Cilantro has a distinct flavor that is sharp and slightly lemony. It is usually added just before serving because heating cilantro nearly destroys its flavor. It is most commonly used in Mexican, Asian, and South American dishes.

**Curry Leaves.** A *curry leaf* (neem leaf) is a shiny green leaf from a small tree native to southern India. These bay-leaf-shaped leaves have a pungent curry aroma and slightly lemony flavor. They are commonly used in Indian and Southeast Asian cuisine.

**Dill.** *Dill* is a member of the parsley family with feathery, blue-green-colored leaves. The leaves have a slightly herbal and aniselike flavor and can be purchased fresh or dried. The plant also yields small, flat, aromatic seeds, which are preferred over the dried leaves. Dill is a welcome addition to green beans, potato salad, poached fish, marinated cucumbers, cauliflower, and some lamb dishes.

**Garlic.** Garlic is not an herb in the true sense of the word but is used in the same manner as an herb. *Garlic* is a member of the lily or onion family and has a strongly flavored and aromatic bulb, commonly called a head. The head has a dry, papery covering that separates individual sections known as cloves. Garlic chives come from an entirely different species of plant than regular chives. Also called Chinese chives, garlic chives are solid, flat, grass-shaped sprouts that have a mild garlic flavor.

**Lavender.** *Lavender* is a member of the evergreen family with bluish-green, spiky stems and purple flowering tops. Lavender is an herb used in classical Provençal cuisine. It is most commonly known for its aromatic qualities. The flowering top has a mild lemony flavor.

**Lemongrass.** *Lemongrass* (citronella) is the fibrous stalk of a tropical grass. Only the white-colored part of the stalk is commonly used. Lemongrass adds a fresh lemony flavor without adding the acidity of a lemon. It is commonly used in Asian dishes.

**Marjoram.** *Marjoram* is an herb with short, oval leaves from a plant in the mint family. Marjoram has a flavor that is similar to a cross between fresh oregano and thyme. Marjoram is used to flavor soups, stews, sausage, cheese dishes, and lamb dishes.

**Mint.** The term "mint" can be applied to members of a large family of similar herbs, the most common of which are spearmint and peppermint. Mint is widely known for its cool, refreshing flavor. *Mint* is a versatile herb with soft, bright green leaves that can be used both in sweet and savory dishes as well as in teas and condiments. Mint is used in the preparation of lamb dishes, poached fish, vegetables, fruit salads, iced tea, and fruit drinks.

**Oregano.** *Oregano* is an herb with dark green leaves and a pungent, peppery flavor. It is native to the Mediterranean and is grown extensively in Greece and Italy. Oregano leaves are small, slightly curly, and oval in shape. Oregano so closely resembles the flavor of marjoram that it is

## Nutrition Note

Garlic not only makes some foods taste better, but also is good for your health. Long ago, garlic was used to treat illnesses from the common cold to the plague. Garlic has been called an "herbal wonder drug" because of its many health benefits. Tests have shown that garlic may help lower cholesterol and reduce the risk of heart disease. Scientists are also investigating whether allyl sulfur, a compound found naturally in garlic, may slow or prevent the growth of cancer cells in the human body.

sometimes called "wild marjoram." Oregano is often used in Italian and Greek cuisine.

**Parsley.** *Parsley* is an herb with dark green leaves that is used as both a flavoring and a garnish. The most popular types are curly parsley and Italian (flat) parsley. Parsley has a tangy flavor that is most prominent in the stems. The stems can be used in a classic bouquet garni. Parsley is used in soups, salads, stuffing, stews, sauces, potatoes, and vegetable dishes.

**Rosemary.** *Rosemary* is an herb that is a member of the evergreen family with needlelike leaves that have an aroma of fresh pine and a hint of mint. It is available either dried or fresh, although the dried form is quite hard and not very pleasant to eat. It is considered to be one of the two strongest herbs, the other being sage. Rosemary is fragrant and sweet-tasting and pairs beautifully with grilled or roasted meats, especially lamb, chicken, pork, and duck. Because its needles are secured to a branch, it can be easily removed when the correct flavor has been achieved.

*Messermeister*

**Sage.** *Sage* is a fragrant herb with narrow, velvety, greenish-grey leaves. Sage is a member of the mint family that grows in the wild and is also used as an ornamental shrub. It can be purchased as whole leaves, rubbed (crushed), or ground. Sage is often used to flavor stuffing, stews, and sausages, and in bean or tomato preparations.

**Savory.** *Savory* is a member of the mint family and has smooth, slightly narrow leaves. Most savory is grown in France and Spain. Summer savory is considered to be the best variety because it is harvested during summer when the leaves are at peak quality, producing a delicate, spicy-sweet flavor. Savory is used in flavoring beans, meat, fish, egg dishes, stuffing, and some baked goods.

**Tarragon.** *Tarragon* is an herb native to Siberia. It is a small perennial plant, with smooth, slightly elongated leaves. It is best known as the flavoring in béarnaise sauce. Fresh tarragon leaves are often used to decorate aspic and chaud-froid pieces. Like anise, tarragon has a flavor that suggests a touch of licorice. Tarragon blends well with seafood and tomato dishes and is a delightful addition to salads and salad dressings. A favorite among French chefs, tarragon is used in French cuisine more often than any other herb.

**Thyme.** *Thyme* is an herb whose leaves and tender stems are picked just before its blossoms start to bloom. This member of the mint family has been popular in the United States for many years as a fragrant but not overpowering herb used with fish and shellfish. Thyme is available in fresh or dried form. It is a welcome addition to beef stew, clam chowder, oyster stew, meat loaf, poultry, and vegetable preparations, and is also a key ingredient in a classic bouquet garni. Thyme is grown principally in southern France and Spain. There are nearly a hundred varieties of thyme, including some that have slight tastes of nutmeg, lemon, mint, and sage.

## Spices

*The Spice House.com*

*Spices* are derived from the bark, seeds, roots, buds, or berries of very aromatic plants. Unlike herbs, most spices come from humid, tropical climates and are typically available only in dried form. The majority of spices available in the marketplace today can be purchased either whole or ground. ***See Figure 10-8.*** The flavors of spices range from sweet to savory and from mild to hot. Some aromatic plants have certain parts that can be used as an herb and others that can be used as a spice. Coriander is a good example of this type of aromatic plant, as its leaves are used as an herb (cilantro) and its seeds are used as a spice (coriander).

**Allspice.** *Allspice,* also know as Jamaican pepper, is the dried, unripened fruit of a small pimiento tree that flourishes in Jamaica. It is because of its complex flavor, suggesting a combination of cinnamon, nutmeg, and cloves, that this pea-shaped spice is called "allspice." Allspice is used in both whole and ground form in the preparations of pies, cakes, puddings, stews, soups, preserved fruit, curries, relishes, and gravies.

**Anise.** *Anise* is a small annual plant from the parsley family that produces a comma-shaped seed known as the anise seed. Fresh anise seed has a greenish color. A tan or brown color indicates that anise is old and should be discarded. This licorice-flavored spice is used in coffee cake, sweet rolls, cookies, sweet pickles, licorice products, candies, and cough syrups.

Star anise, also known as Chinese anise, is not related to the anise spice. It is actually a star-shaped seed fruit from the Chinese magnolia tree. Although similar in flavor to anise, the flavor of star anise is more intense and slightly bitter. It is one of the main components in a Chinese five-spice blend.

**Achiote Seeds.** *Achiote (annatto) seeds* are red, corn-kernel-shaped seeds of the annatto tree, native to South America. When ground or used in cooking, they give off a yellowish-orange color that is used as a natural food coloring for butter, cheese, and smoked fish. Achiote seeds are also used in many South American and Mexican dishes.

**Capers.** *Capers* are the unopened flower buds of a shrub that grows wild in the Mediterranean. The buds are used only after being cured in strongly salted white vinegar. They develop a sharp, salty flavor. Large, medium, and small caper varieties are available for purchase. Capers should be stored in their original brine. Adding water or additional vinegar to the brine will cause the capers to spoil. Capers are commonly used in fish and poultry dishes and are one of the key ingredients in tartar sauce.

**Caraway Seeds.** *Caraway seeds* are the small crescent-shaped brown seeds of the caraway plant. The caraway plant grows in Holland, Germany, England, and Poland. Caraway seed is used to flavor bread, sauerkraut, pork, and cabbage. Although caraway seed is best known in the United States as the seed found in rye bread, it can be used in cheese, potatoes, stews, and soups with excellent results.

**Cardamom.** *Cardamom* is the dried, immature fruit of a tropical bush in the ginger family. The fruit consists of a yellowish pod about the size of a small grape, which holds the dark, aromatic cardamom seeds. It is the second most expensive spice; only saffron is more costly. Cardamom has a lemony flavor and a pleasant aroma. It is available in whole or ground form. Cardamom is used in pickling, coffee cake, curries, and Danish pastries. It is commonly used in Middle Eastern cuisine.

**Cayenne Pepper.** *Cayenne pepper* is a pungent powder ground from the small pods of certain varieties of hot peppers in the capsicum family. The peppers are grown mainly in South America, the West Indies, and Africa. Cayenne pepper is also known as red pepper because of its bright red-orange color. Cayenne pepper is used in soups, cream dishes, meat, fish, cheese, and egg dishes. It should be used in moderation, as it is very hot and strong.

**Celery Seeds.** *Celery seeds* are the tiny brown seeds from the lovage herb. Celery seeds can be purchased whole or in ground form. This spice is grown in France, India, and the United States. It is used to flavor cole-slaw, potato salad, sauces, soups, dressings, fish, and certain meats. A blend of celery seed and salt, called celery salt, is often sprinkled over a Chicago-style hot dog.

**Cinnamon.** *Cinnamon* is the dried, thin, inner bark of a small evergreen tree native to India and Sri Lanka. After the orange-brown bark has been removed from the tree, it is rolled and cut into 3″ long quills known as cinnamon sticks. Cinnamon has a distinctive flavor and aroma. It can be purchased whole or in ground form. It is used in baking, pickling, preserving, pies, cakes, puddings, stewed fruits, custards, and sweet doughs.

**Cassia.** *Cassia* (Chinese cinnamon) is the bark of a small evergreen tree that is thicker and darker in color than cinnamon. Cassia bark is processed in the same manner as cinnamon bark. The flavor of cassia is much stronger and is not as smooth as that of cinnamon. Cassia is also less expensive than cinnamon. It is commonly used as a substitute for cinnamon. In fact, most spices labeled as cinnamon sold in the United States are actually cassia, as labeling laws do not require companies to distinguish between the two.

**Cloves.** A *clove* is the dried, unopened bud of a tropical evergreen tree primarily found in the East Indies and on islands off the coast of Africa. The cloves grow in clusters and are ready to be picked when the buds turn red. Dried cloves are dark brown. The clove is sometimes referred to as the nail-shaped spice because of its resemblance to a builder's nail. It is thought to possess the most pungent flavor of all the spices. Cloves are used in pickling and in the preparation of roast pork, corned beef, baked ham, soups, applesauce, pumpkin pie, and cakes.

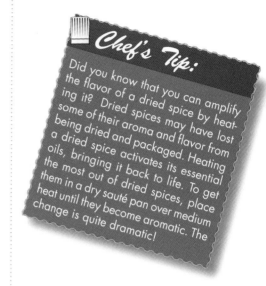

*Chef's Tip:*

Did you know that you can amplify the flavor of a dried spice by heating it? Dried spices may have lost some of their aroma and flavor from being dried and packaged. Heating a dried spice activates its essential oils, bringing it back to life. To get the most out of dried spices, place them in a dry sauté pan over medium heat until they become aromatic. The change is quite dramatic!

## Spices

Allspice

Anise

Achiote (Annatto) Seeds

Capers

Caraway Seeds

Cardamom

Cayenne Pepper

Celery Seeds

Ground    Stick

Cinnamon

Cassia

Cloves

Coriander

Crushed Red Pepper

Cumin

Fennel Seed

*WholeSpice.com*

*WholeSpice.com*

*WholeSpice.com*

*WholeSpice.com*

*The Spice House.com*

*WholeSpice.com*

*WholeSpice.com*

*The Spice House.com*

*WholeSpice.com*

*WholeSpice.com*

**10-8** *Spices are derived from the bark, seeds, roots, buds, or berries of aromatic plants. (continued on next page)*

# Spices

**Fenugreek** — *WholeSpice.com*

**Filé Powder**

Ground — Root **Ginger** — *WholeSpice.com*

**Horseradish** — *The Spice House.com*

**Juniper Berries**

**Mace** — *The Spice House.com*

**Mustard Seeds**

Ground — Whole Kernal **Nutmeg**

**Paprika** — *WholeSpice.com*

**Poppy Seeds**

**Saffron**

**Sesame Seeds** — *WholeSpice.com*

**Szechwan Pepper** — *The Spice House.com*

**Tamarind**

**Turmeric** — *WholeSpice.com*

**Wasabi** — *The Spice House.com*

**10-8** *(continued from previous page)*

**Coriander.** *Coriander* is the ridged seeds of the coriander plant. Although this seed comes from the same plant as the herb cilantro, the flavor of coriander is completely different from that of cilantro and the two should never be substituted for each other. Coriander is a small, round, light-brown seed, similar in appearance to whole white pepper. It has a pleasant taste that suggests the combined flavors of sage and lemon peel. Coriander is grown in Morocco and many Mediterranean countries. It is used in candies, pickles, baked goods, hot dogs, sausages, and curry dishes.

**Crushed Red Pepper.** *Crushed red pepper,* also known as crushed chilies or chili flakes, is a blend of crushed hot chile peppers. These hot flakes are used typically as a flavoring for soups or as a condiment on pizza.

**Cumin.** *Cumin* is the dried, aromatic seeds of a plant that is a member of the parsley family. Cumin is native to Egypt. The spice has a slightly bitter, warm flavor and is commonly used in Middle Eastern, Indian, and Mexican cuisines. Cumin is grown in Morocco, India, Egypt, and South America. It is an essential ingredient in curry powder and chili powder. Cumin is used to flavor chili, soup, tamales, and rice and cheese dishes.

**Fennel Seed.** *Fennel seed* is the small seedlike fruit of the fennel plant. Fennel seed is light brown with a greenish tint and resembles anise in flavor and aroma. Fennel is grown chiefly in India and Eastern Europe and is commonly used in Scandinavian cooking and Italian baking. Fennel seed enhances the flavor of roast duck, some chicken and pork dishes, and is often used in sausage preparation.

**Fenugreek.** *Fenugreek* is a spice that is the pepple-shaped seed of a plant in the pea family. Its flavor is similar to burnt sugar and leaves a bitter aftertaste. The orange color of fenugreek seeps into dishes in which the spice is used. Fenugreek is mainly used in Indian curry dishes, chutneys, and vegetable preparations.

**Filé Powder.** *Filé powder* is made from the ground leaves of the sassafras plant. It is a common ingredient in Creole and Cajun cuisine. Filé powder is a standard ingredient in gumbo preparation. The ground leaves act as a thickener and a seasoning.

**Ginger.** *Ginger* is a spice that comes from the bumpy root of a tropical plant grown in China, India, and Jamaica. The large root is commonly called a hand, as the thick offshoots resemble fingers. When the plant is about a year old, the roots are dug up and sold fresh, dried, ground, crystallized, candied, or pickled. Ginger has a warm, pungent, spicy flavor that adds zest to cakes, cookies, pies, fruits, puddings, and some meat preparations. Its dried form is an essential ingredient in the preparation of gingerbread.

The Spice House.com

**Horseradish.** Horseradish comes from a large, brown-skinned, white root of a perennial shrub related to the radish. It grows mainly in cool climates and has a biting, spicy flavor. *Horseradish* is a spice obtained by peeling and grating horseradish root. Grated horseradish can be served as a condiment on roasted grilled or boiled meats, as well as fish and shellfish dishes. Horseradish is also commonly incorporated into cream sauces or compound butters.

**Juniper Berries.** Juniper berries come from an evergreen bush called a juniper. *Juniper berries* are small, purple berries with an aromatic flavor similar to that of rosemary. Juniper berries are slightly tart. They are most commonly crushed and used in the preparation of wild game dishes. They are also the key ingredient from which the alcohol gin is made.

The Spice House.com

**Mace.** *Mace* is the lacy red-orange covering of the nutmeg kernel. Mace turns dark yellow when dried. Its flavor somewhat resembles that of the nutmeg, although it is not as pungent. Mace is a traditional spice in pound cake and sweet doughs. It is also used in chocolate dishes, oyster stew, spinach, and pickling.

**Mustard Seeds.** *Mustard seeds* are the extremely tiny seeds of the mustard plant, available in three different forms: yellow, brown, and black. Yellow seeds are the mildest in flavor, while black are the most intense. Brown mustard seeds have an intensity that falls in between those of the yellow and black varieties. Dark brown mustard seed is most often used in Chinese cuisine. Mustard seeds have virtually no aroma but release a very hot, pungent flavor when ground and mixed with water. Ground mustard reaches maximum flavor after being allowed to stand in liquid for about 10 min. Mustard seed enhances the flavor of pickles, cabbage, beets, sauerkraut, sauces, salad dressings, ham, frankfurters, and cheese.

**Nutmeg.** *Nutmeg* is the kernel from a large, yellow, nectarine-shaped fruit of a large tropical evergreen. When the nutmeg fruit is ripe, the outer hull splits open, exposing the sister spice mace, which partially covers the nutmeg kernel. The sweet, warm, spicy flavor of nutmeg can be used to enhance cream soups, custards and puddings, baked goods, potato dishes, and sauces.

**Paprika.** *Paprika* is a spice made by grinding paprika peppers after the seeds and stems have been removed. There are two kinds of paprika used in the professional kitchen, Spanish and Hungarian. Spanish paprika is mild in flavor and has a bright red color. Hungarian paprika is darker in color and has a more pungent flavor. Paprika is grown in Spain, central Europe, and the United States. It is used as a colorful garnish for many foods. It is also sometimes used to brown food. Paprika is a necessary ingredient in Hungarian goulash, chicken paprika, Newburg sauce, French dressing, and veal paprika.

**Poppy Seeds.** *A poppy seed* is the very small, blue-gray seed of the cultivated poppy plant grown chiefly in Holland, India, and the Middle East. Poppy seeds have a mild, nutty flavor. Poppy seeds are best known for garnishing rolls and bread. They can also be used in butter sauces for fish, vegetables, and noodles.

**Saffron.** Saffron is the most expensive spice in the world, but a little will go a long way. *Saffron* is a spice made from the dried, bright red stigmas of the purple flower of the saffron plant. It takes about 225,000 stigmas, or over 75,000 flowers, to produce 1 lb of this desirable spice. The three stigmas of each flower are removed by hand and dried, producing delicate yet potent red-orange strands.

Saffron is grown in Spain and the Mediterranean region. It imparts a perfumelike aroma and deep yellow color that is desired in preparations of certain rice dishes and fine baked goods. The use of saffron is common in Scandinavian and Spanish cuisines. Saffron strands should be steeped in warm or hot water before use to release their color, aroma, and flavor.

American METALCRAFT, Inc

**Sesame Seeds.** *Sesame seeds* are small, flat, honey-colored seeds with a nutty flavor that come from pods on the sesame plant. The sesame plant is native to India and is also grown in Central America, Egypt, and the United States. Sesame seeds are baked on rolls, bread, and buns to supply a nutty flavor to a product. The seeds are also toasted and stirred into butter, and served over fish, noodles, and vegetables. White sesame seeds are common in Indian and Asian cuisines, while black sesame seeds are common in Japanese dishes. Sesame seeds are ground to form tahini, a traditional Indian spread.

**Szechwan Pepper.** *Szechwan pepper* is a spice made from the dried, ground berries of an ash tree native to China. The result is a fiery-flavored powder used in many Chinese dishes. It is a also component of Chinese five-spice powder.

**Tamarind.** *Tamarind* is a spice that comes from long pods that grow on the tamarind tree. The long pods contain seeds and pulp that are dried. Tamarind is native to Africa and is commonly used in Indian curries, as well as African and Mediterranean cuisines. Tamarind is also the main ingredient in Worcestershire sauce.

**Turmeric.** *Turmeric* is a spice made from the root of a lily-like plant in the ginger family. Sometimes referred to as "Indian saffron," it is native to Asia and is used not only in food preparation but also in medicine as a dye. Turmeric is associated with saffron because both possess a deep yellow color and are used in much the same way in food preparation. When turmeric roots are ground, a bright yellow powder is produced that has a taste similar to mustard. Turmeric is an important ingredient in curry powder and is also used in making some prepared mustards.

**Wasabi.** *Wasabi* is the light-green-colored root of an Asian plant with a hot and tangy flavor similar to horseradish. It is commonly sold in the

United States in a dried powder form or as a paste. When it is mixed with water, a fiery green-colored condiment is produced. Wasabi paste frequently accompanies sushi or sashimi.

## Blends

Spice or herb blends decrease production time because the chef only needs to reach for one item. Blends can be created to achieve specific results, as with Cajun spice. *See Figure 10-9.* Purpose-specific blends may also find use in other dishes, such as when pickling spices are used to add flavor to a pot roast. There are many types of blends. Some of the commonly used blends include Cajun spice, chili powder, Chinese five-spice powder, curry powder, fines herbes, Herbes de Provence, pickling spice, poultry seasoning, and seasoning salt.

*Cajun spice* is a spice blend that may consist of red and black pepper, oregano, salt, thyme, garlic, ground fennel seeds, and paprika. Each manufacturer uses its own unique blend. Cajun spice is used in preparing blackened foods including seafood, steaks, pork chops, and chicken. It adds a spicy but complementary flavor.

*Chili powder* is a spice blend consisting of Mexican peppers, oregano, cumin, garlic, and other spices. It is used in preparing chili, tamales, stews, Spanish rice, gravies, and appetizers.

*Chinese five-spice powder* is a combination of equal proportions of ground Szechwan pepper, cloves, star anise, cinnamon, and fennel seeds. It is commonly used in Chinese cuisine.

## Blends

The Spice House.com
**Cajun Spice**

The Spice House.com
**Chili Powder**

The Spice House.com
**Chinese Five-Spice Powder**

WholeSpice.com
**Curry Powder**

The Spice House.com
**Fines Herbes**

WholeSpice.com
**Herbes de Provence**

The Spice House.com
**Pickling Spice**

**Poultry Seasoning**

WholeSpice.com
**Seasoning Salt**

10-9 *Blends add a lot of flavor to many dishes. Many chefs also create their own signature blends.*

*Curry powder* is a blend of a number of spices, typically including combinations of cloves, cumin, coriander, black pepper, red pepper, mustard, cinnamon, nutmeg, ginger, cardamom, fenugreek, and turmeric. These and additional spices are blended to create the flavor and color of curry. The color and, to some degree, flavor of curry varies depending on the manufacturer. There are also many different forms of curry. For example, the curry dishes in some regions of India, such as Madras, are hot because of the inclusion of generous amounts of red pepper. In other regions of India , such as Bombay, a milder, sweeter curry is more common. Curry powder is used to make curries of meat, fish, chicken, and eggs. It is also used to season rice, soups, and some shellfish dishes.

*Fines herbes* is the combination of parsley, chives, tarragon, and chervil used in classical French cuisine. The combination may also include thyme. *Herbes de Provence* is another classical mixture of herbs used in the cuisine of southern France. A combination of dried thyme, bay leaves, basil, fennel, rosemary, and lavender can be used as a rub for roasted or grilled meats and poultry as well as in various savory yeast breads.

*Pickling spice* is a blend of several whole spices such as cloves, cinnamon, nutmeg, peppercorns, allspice, bay leaves, ginger, mace, crushed red pepper, dill seed, fennel, and mustard seed, and is primarily used in pickling. It is also an excellent addition to stocks, soups, relishes, sauces, and some meat preparations, such as pot roast and sauerbraten. When pickling spice is used in cooking, it is usually added in a sachet d'épices. The spice is tied inside a piece of cheesecloth, added to the preparation, and removed when the desired flavor is achieved.

*Poultry seasoning* is a ground mixture often containing sage, thyme, marjoram, savory, pepper, onion powder, and celery salt. It is most often used to season chicken, turkey, pork, veal, lamb, and white meat game birds, but has other uses as well. For example, it can be used to add flavor to stuffing or bread dressing, meat loaf or dumplings.

*Seasoned salt* is a generic term for a blend of salt and other savory ingredients. Various types of seasoned salts are used as flavor enhancers in all sorts of dishes. However, seasoned salt is most commonly used on meat, poultry, and seafood.

## COOKING WITH HERBS AND SPICES

Cooking with herbs and spices requires patience, practice, and the ability to follow some basic rules. General guidelines for using herbs and spices in cuisine include the following:

- Herbs and spices should be used to enhance the natural flavor of food.

- Herbs and spices should not disguise the flavor of any food.

- A specific herb or spice should not dominate the flavor of a food. Exceptions to this rule include curry or chili dishes, where the defining character of the dish relies on the use of specific spices or herbs.

*Chef's Tip:*

Not all seasoned salt is the same. Always read the label on seasoned salts and tenderizers to see if they contain monosodium glutamate (MSG). Monosodium glutamate is a chemical substance used as a flavor enhancer and tenderizer. It is important to remember that some individuals are allergic to MSG and can have mild to severe reactions to foods that contain it. Studies have also shown that MSG causes a chemical reaction in the brain that may trigger anxiety attacks in some people.

- Always season in moderation. More seasoning can always be added, but removal of seasonings is impossible.

- If a dish has a long cooking time, add spices early in the cooking process to extract the most flavor and add herbs near the end to protect them from losing their flavor.

- If a dish has a short cooking time, add herbs at the beginning.

- To release the flavor of dried herbs, rub the herbs between the palms before adding the herbs to a preparation.

- When using fresh herbs in soups, sauces, or gravies, tie the herbs in a sachet d'épices or bouquet garni for easy removal after use.

- In uncooked dishes, herbs or spices should be added several hours before serving to allow time for their flavor to develop. This is especially important in salad dressings, fruit juices, and marinades.

- Bay leaves used in food preparation must be removed before a dish is served because they are hard, sometimes sharp, and could pose a choking problem.

## PURCHASING HERBS AND SPICES

Herbs and spices should always be bought in small quantities as their flavor can deteriorate quickly even when they are stored under proper conditions. The best storage location for dried herbs and spices is a cool, dry place, where heat cannot remove any flavor and dampness cannot cause caking. As soon as a spice is ground it starts to lose flavor. However, storage in a tightly closed container helps to prevent deterioration. Dried herbs lose their flavor even faster than ground spices. Good color, strong flavor, and aroma are important points to consider when buying dried herbs. However, most professionals prefer to purchase herbs whole if possible because whole herbs retain flavor and freshness longer. A large bunch of fresh herbs should be stored in a manner similar to storage of freshly cut flowers, with the leaves upright and the stems in cool, clean water. Smaller bunches can be wrapped in a clean, damp towel to prevent drying out. They should be stored in a refrigerator kept between 35°F and 40°F.

Herbs and spices should be tested for freshness when delivered. A good test is to examine the herb or spice for a bright, rich color and fresh appearance. A small amount of an herb can be rubbed in the palm of the hand and smelled. The scent should be fresh and fairly strong. A spice can be tested by placing a small amount in the palm and bringing the hand up to the nostrils. A spice should have a powerful, fresh aroma.

Imitation spices are less expensive than natural spices and are made by spraying the oil of the true spice on a carrier such as ground soy, buckwheat, or cottonseed hulls. However, imitation spices do not possess the same strength and quality as natural spices. The most common imitation spices are imitation pepper, imitation nutmeg, imitation cinnamon, and imitation mace.

## NUTS AND SEEDS

*Nuts* are hard-shelled, dry fruits or seeds that contain an inner kernel. Some nuts, such as walnuts and pecans, contain two kernels inside instead of one. Some foods that are not technically nuts are considered to be nuts because of their similar use in culinary preparations. For example, peanuts are actually legumes that are grown underground, but are considered and used as nuts. Nuts are often added to foods to add texture or a nutty flavor. For example, fish is often coated in chopped nuts before being sautéed to make a crispy exterior. Salads are often tossed with toasted nuts for added crunch and texture as well. Most types of nuts are available both shelled and in the shell. Shelled nuts are often available chopped, sliced, or whole. ***See Figure 10-10.***

*Nuts*

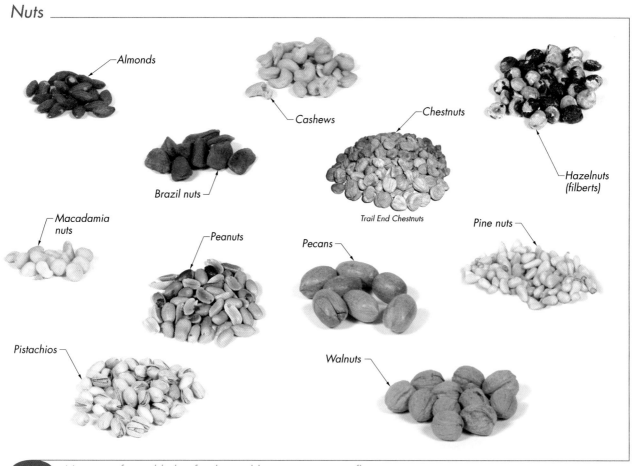

Almonds

Cashews

Chestnuts

Hazelnuts
(filberts)

Brazil nuts

Trail End Chestnuts

Macadamia nuts

Peanuts

Pecans

Pine nuts

Pistachios

Walnuts

**10-10** *Nuts are often added to foods to add texture or a nutty flavor.*

An *almond* is the edible kernel of a small, lozenge-shaped fruit that grows on small trees native to the Mediterranean region. Today, almonds are a chief crop of California. Almonds are marketed in many forms including whole, chopped, slivered, sliced, ground, or as a sweetened paste (also known as marzipan) used in baking.

*Brazil nuts* are contained in the seeds of very large fruit grown on trees native to the rainforests of South America. These huge trees bear plentiful amounts of fruit, with each piece of fruit containing between 10 and 25 large seeds. The seeds have a hard, thick shell, containing a large, white, richly flavored kernel.

A *cashew nut* is the butter-flavored kidney-shaped kernel of the fruit of the cashew tree, grown in India and Africa. The tree is related to the poison ivy plant. Like poison ivy, the cashew plant produces a toxin. The toxin present in cashews is in the shell and for this reason cashews are never sold with the shell on. Cashews are used in many Asian dishes and are a favorite in many baked goods.

Chestnuts differ from any other nut in that they are very sweet, higher in starch, and lower in fat. *Chestnuts* are the fruit of the chestnut tree. They must be cooked before they are used. Most chestnuts are grown in Italy. They are available whole in the shell, dried, canned, or as a sweetened paste to be used in baking.

A *hazelnut* (filbert) is a marble-sized nut from the hazel tree. These rich nuts grow in large clusters. They grow in the wild in the northeastern and upper midwestern regions of the United States. Once shelled, these nuts have a bitter, brown, papery skin that covers the surface. The skin may be removed by roasting the nuts in a 250°F oven for about 15 min. The nuts should be removed from the oven and, while still hot, rubbed with a dry towel to remove the skin.

*Macadamia nuts* are marble-sized nuts from an evergreen tree. Although native to Australia, macadamia nuts are almost exclusively grown in Hawaii. These nuts have a very high fat content and a sweet, creamy taste. Macadamia nuts are almost always shelled at a processing plant because the shell is extremely hard and very difficult to remove. Macadamia nuts are wonderful in baked goods and with chocolate.

As previously mentioned, *peanuts* are not true nuts but are actually legumes that grow underground. The two most common types of peanuts are the Spanish peanut and the Virginia peanut, with the Virginia being the larger and more flavorful of the two types. Peanuts are used in many Asian cuisines, salads, ice cream sundaes, and desserts. Peanut butter is a popular nutty spread made from ground peanuts and a small amount of oil.

A *pecan* is the nut from the pecan tree, which is native to the Mississippi River valley. Pecans have a maple-like flavor and pair well with sweets such as chocolate, cookies, and cakes.

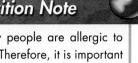

## Nutrition Note

Many people are allergic to nuts. Therefore, it is important to remember that certain foods contain nuts that do not appear to. The following are a few examples:

- Some pestos may contain nuts.
- Some salad dressings may contain nut oils.
- Cakes and desserts may contain marzipan or frangipane (made from almonds) or praline (made from hazelnuts).
- Cheesecake bases may contain nuts to make them crunchier.
- Some Indian dishes are thickened with almonds or peanut flour.
- Some Greek and Turkish dishes contain tahini (made from sesame seeds).
- Some Chinese and vegetarian dishes contain tofu (made from soybeans).

A *pine nut* is the nut of various types of pine trees. Pine nuts are very tender and small and are shaped liked kernels of corn. They are used in salads, baked goods, and many Italian and Spanish dishes, and are one of the main ingredients in pesto sauce.

A *pistachio* is the pale green nut from the pistachio tree. Although the pistachio is native to Asia, most pistachios available in the United States are grown in California. The nut is shaped like a bean inside a shell that is often dyed red or blanched until it is white. Pistachios are used in many baked goods and in some classical food preparations such as galantines and pâtés.

A *walnut* is the fruit of the walnut tree. The black walnut is cultivated in Appalachia and has a much stronger flavor and aroma than that of the English walnut, which is grown in California. Walnuts are used in baked goods, salads, and sweets.

A *coconut* is the large fruit of the coconut palm tree. The coconut is green when young and turns brown with stringy fibers when ripe. Its outer shell is as hard as wood while the inside is made up of moist white flesh and a liquid known as coconut water. (*Note:* Coconut water is not another term for coconut milk, which is a product made with puréed coconut flesh.) Fresh coconuts should feel heavy and the water inside should be audible when the coconut is shaken. Coconut is also available shredded or flaked, sweetened or unsweetened. ***See Figure 10-11.***

## Coconuts

White flesh

Hard shell

**10-11** *Coconuts have a hard outer shell and moist white flesh.*

## Seeds

Pumpkin seeds

Sunflower seeds

**10-12** *Both shelled pumpkin seeds and sunflower seeds can be used in salads.*

A *pumpkin seed* is a seed that comes from many varieties of pumpkins. They are removed from the pumpkin and then slowly dried in an oven. Pumpkin seeds may be soaked in a salt brine prior to being dried. They are often used in salads. ***See Figure 10-12.***

A *sunflower seed* is the seed from the sunflower plant. Sunflower seeds can be eaten either raw or cooked, and are often shelled and tossed into salads.

## OILS, VINEGARS, AND CONDIMENTS

Many ingredients can be added to enhance the flavors of foods. Whether it is a salad dressing that enhances the flavor of a salad, the salsa that accompanies a nacho chip, or the mustard on a ballpark hot dog, condiments and the ingredients they are made from add richness, flavor, moisture, and taste to foods. Oils are available in various colors and flavors. Vinegars can add tartness to a dish. Condiments include many kinds of flavoring ingredients in prepared forms.

### Oils

*Oils* are fats that remain in a liquid state at room temperature. Cooking oils are produced from various types of seeds, plants, vegetables, and nuts. Different oils are better suited for different tasks. For example, the flavor of olive oil is destroyed under intense heat; therefore, it is not typically used for frying. Alternately, canola oil has almost no flavor and has a very high smoke point, making it a very good choice for frying. *Smoke point* is the temperature at which fats begin to break down under heat and begin to smoke or burn.

Many oils are prepared from expensive ingredients and are difficult to produce. Just like the vegetables or seeds that they are produced from, oils can turn rancid and spoil. Oils should be kept in a cool, dry place out of direct sunlight to ensure their shelf life. Infused oils are any variety of oil with added herbs or spices that increase the flavor. Many infused oils are available on the market today, but they are also very easy to produce in the kitchen. ***See Figure 10-13.***

*Oils*

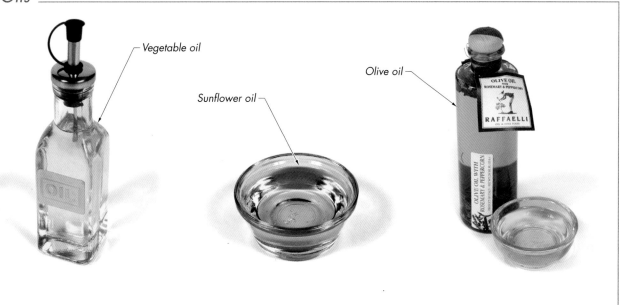

 **10-13** *Oils are fats that remain in a liquid state at room temperature.*

*Vegetable oils* are a group of oils that are extracted from vegetables, cottonseed, corn, grape seeds, peanuts, sesame seeds, or soybeans. Since these oils are not made from animal products, they are cholesterol-free. Vegetable oils are almost completely free of odor and flavor, with the exception of corn oil, which has a slight flavor of cornmeal. The term "vegetable oil" may also refer to a blend of a few of these oils. The term "pure" on the label means that the oil was extracted from only one source. Canola oil, also called rapeseed oil, is extracted from rapeseeds and is high in monounsaturated fats, which are preferred to unsaturated or saturated fats. It has one of the highest smoke points of commercial vegetable oils.

*Nut oils* are relatively expensive forms of vegetable oil that are named for the nuts from which the oil is extracted. For example, walnut oil is extracted from walnuts. Nut oils do not perform very well under heat. They are most often used in finer salad dressings. Nut oils have a short shelf life and turn rancid quickly compared to other vegetable oils. Nut oils should be purchased in small containers that can be used within a few months.

Of all the oils discussed, olive oil is the only oil that is extracted from a fruit rather than a seed. Olive oil is produced in many parts of the world, including Spain, Italy, Greece, France, northern Africa, and the United States. Each region's variety of olive oil differs, if only slightly, just as wines from various regions throughout the world differ in taste, color, and consistency. The differences in quality in olive oils from the same producer are indicated on the label as extra-virgin, virgin, pure, and olive-pomace. Extra-virgin olive oil refers to the first pressing of the olives without the use of heat or chemicals and with a resulting acidic level less than 1%. Virgin olive oil is also oil from the first pressing without the use of heat or chemicals, but with an acid content of as much as 3%. Pure olive oil is produced using heat, and often chemicals, to extract additional oils from the olive pulp after the first pressing. Olive-pomace oil is similar to olive oil that is classified as pure, but the pits of the olive are also pressed in the production of olive-pomace oil.

Many people are under the false impression that oils will last almost indefinitely. Actually, oils stay fresh for approximately six months after being opened and then run the risk of turning rancid or spoiling. Many commercial salad dressings are made with oils that have been winterized. Winterizing is a process that allows the oil to remain clear under refrigeration. Oils that have not been winterized will turn thick and cloudy in cold temperatures.

## Vinegars

*Vinegars* are very sour, clear liquids used in cooking, marinades, and salad dressings. Although most vinegar is produced through the fermentation of wine, a few other alcoholic products can also be turned into vinegars. Vinegars are produced by adding specific bacteria to an alcohol. The bacteria aggressively feed on the alcohol, turning it into

---

### Nutrition Note

Olive oil is rich in monounsaturated fatty acids. Fats supply energy to the body and help transport fat-soluble vitamins such as A and D. However, fats should add up to no more than 20% of daily calories, so use oils sparingly.

acetic acid. Most vinegars used in food preparation have between 5% and 7% acidity. The quality and characteristics of vinegar depend on the wine or alcohol from which it was produced. A higher-quality wine or alcohol will yield a higher-quality vinegar. *See Figure 10-14.*

*Vinegars*

White Wine     Red Wine     Cider

Distilled     Balsamic

**10-14** *Vinegars are very sour, clear liquids and are available in many varieties.*

No matter what the quality level, vinegar should be clear, not cloudy or opaque, and have a consistency similar to water. It should not appear cloudy or murky and should never have any signs of mold. Vinegar exhibiting any of these characteristics should be discarded. It is important to note that commercial vinegars are pasteurized when they are produced. The pasteurization process enables an unopened container of vinegar to remain fresh almost indefinitely. However, once a container has been opened and introduced to air, vinegar should be kept for only about three months.

The most common vinegars used in the professional kitchen include wine vinegars, cider vinegar, distilled vinegar, and balsamic vinegar. *Wine vinegar* is a vinegar produced from red and white wines, champagne, and sherry. A wine vinegar should exhibit the same color characteristics as the wine from which it was produced. Red wine vinegars are aged longer than white wine vinegars. *Cider vinegar* is vinegar made by fermenting unpasteurized apple juice or cider until the sugars are converted into alcohol. It has a soft honey color, mild acidity, and a subtle apple taste. Cider vinegar is often

used as a flavoring on salads and in salad dressings. *Distilled vinegar* is vinegar made by fermenting diluted, distilled grain alcohol. It is most often used in pickling and preserving foods because it has a more acidic quality and more pungent aroma than most other types of vinegar.

*Balsamic vinegar* is a vinegar made by aging red wine vinegar in wooden casks for many years. Balsamic vinegar has been produced in Italy for many centuries. In recent years, it has become popular for use in salad dressings. All balsamic vinegars are made from red wine vinegar. The red wine vinegar picks up characteristics, such as color and aroma, from the type of wood in which it is stored. The highest grades of balsamic vinegar have aged for decades. After aging, the vinegar has darkened intensely and become sweeter. Although balsamic vinegar has a higher acid content than its original red wine vinegar, the sweet characteristics of balsamic vinegar can cover up or mellow the acidic qualities. Because the traditional aging process is labor-intensive and results in a very expensive product, many balsamic vinegars sold in the United States undergo a mechanically produced flavoring and caramelization process.

Other vinegars also used in the professional kitchen include malt vinegar, rice vinegar, and flavored vinegars. *Malt vinegar* is vinegar made from malted barley. It has a lower acidity than other vinegars and is commonly used as a condiment. Many places serve malt vinegar alongside fried foods such as fish and chips. *Rice vinegar* is vinegar made from rice wine. Rice vinegar has a slightly sweet character and is only mildly acidic. It is most often used in Asian cooking and in dipping sauces. *Flavored vinegars* are vinegars made from any traditional vinegar in which other items such as herbs, spices, fruits, vegetables, or flowers are added. Flavored vinegar varieties, such as raspberry vinegar or tarragon vinegar, have been around for centuries and remain popular today.

## Condiments

*Condiments* are savory, spicy, or salty accompaniments to food, such as a relish or a sauce. Condiments are used to enhance the flavor of a dish. The most common condiments found in the professional kitchen include ketchup, mustard, mayonnaise, salad dressing, soy sauce, Tabasco® sauce, and Worcestershire sauce. ***See Figure 10-15.***

*Ketchup* (catsup) is a thick, tomato-based sauce. Because vinegar and sugar are two of the main ingredients in ketchup, it has a slightly sweet-and-sour acidic flavor. Once opened, ketchup will retain its freshness longer if it is stored under refrigeration, although it is not necessary. It loses much of its bright red color when its quality deteriorates with age. It is commonly used on French fries, hamburgers, and hot dogs.

## Condiments

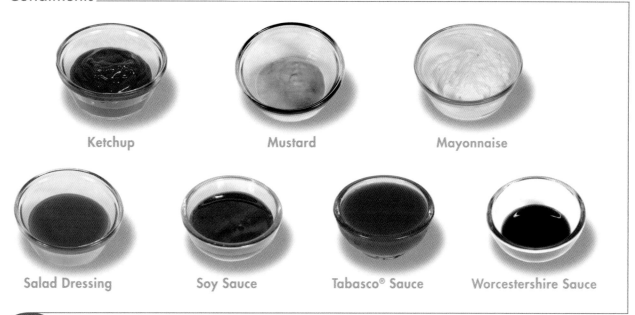

Ketchup   Mustard   Mayonnaise

Salad Dressing   Soy Sauce   Tabasco® Sauce   Worcestershire Sauce

**10-15** *A condiment can be anything used to enhance the flavor of food.*

Mustard comes in many different varieties. *Mustard* is a pungent powder or paste made from the seeds of the mustard plant for use as a seasoning or condiment. It is made from mustard seeds, vinegar or wine, salt, and spices. Some types of mustard are made from ground seeds, resulting in a very smooth product, while others use ground, cracked, and whole seeds for a more rustic appearance. Regardless of the color of seeds used, the unique flavor of mustard is derived from the essential oils that are released when the seeds are crushed. The flavor of mustard can range from mild and tangy to very hot and spicy.

Prepared yellow mustard is made of completely puréed seeds that, when mixed with water, result in a tangy flavor. Prepared yellow mustard can be either used as a topping, such as on a hot dog, or as an ingredient, such as in mustard rub on a rack of lamb. Because it has a high acid content, prepared yellow mustard does not spoil. However, it can dry out and darken on the surface, and the flavor can deteriorate with age.

Dijon mustard originated in Dijon, France. Dijon mustard is light in color, but fairly strong in flavor. Dijon is preferred by many chefs and customers due to its less acidic taste. This is due to the use of verjuice, a sour juice made from unripe grapes, being used instead of vinegar in the making of Dijon mustard. Today Dijon mustard comes in many different flavors, such as walnut, champagne, raspberry, and blue cheese.

English and Chinese mustards are made from ground mustard flour mixed with water. Once water is added to the flour, the mixture becomes intensely hot and spicy. A little bit of these mustards goes a long way. Besides being spicy to the tongue, they can make the eyes water and can interact with the sinuses.

*Mayonnaise* is the thick, uncooked emulsion formed by combining salad oil with egg yolks, vinegar, and seasonings. Mayonnaise may be made in the professional kitchen or purchased in conveniently sized jars. It is used in salad dressings, bound salads, and as sandwich spreads.

Salad dressing of the sandwich spread variety is also known as a cooked dressing. It is similar in texture and appearance to mayonnaise and may be used in place of mayonnaise on sandwiches. Although salad dressing may be mistaken in appearance for mayonnaise, they are not the same. A *salad dressing* is a cooked, mayonnaise-like product usually made from distilled vinegar, vegetable oil, water, sugar, mustard, salt, modified cornflour, xanthan gum and guar gum (as stabilizers), and riboflavin (for coloring).

Salad dressing is much sweeter than mayonnaise and differs from mayonnaise in two main ways. First, a salad dressing contains no egg yolks to act as emulsifiers. Instead chemical emulsifiers are used to make the dressing appear as creamy and rich as a true mayonnaise. Secondly, the FDA requires that all mayonnaise brands contain at least 65% oil in order to be called a mayonnaise. A salad dressing needs only to contain a minimum of 30% oil. Less oil and no egg yolks result in a less expensive product, so most salad dressings are less expensive than most brands of mayonnaise. Salad dressings are used in bound salads, such as tuna salad, and as sandwich spreads.

Salad dressing is also a general term for a sauce served on salad. Prepared salad dressings are used to add fat, acid, flavor, and moisture to lettuce greens. Salad dressings are available in many flavors and styles. Low-fat and fat-free versions of popular salad dressing varieties are readily available. Major categories of salad dressings include mayonnaise-based; buttermilk-, sour cream-, or yogurt-based; emulsified vinegar and oil-based; and broken dressings:

- Mayonnaise-based dressings are very creamy and include thousand island and creamy garlic.

- Buttermilk-, sour cream-, or yogurt-based dressings are creamy dressings that do not contain mayonnaise and include ranch and poppy seed dressings.

- Emulsified vinegar and oil-based dressings include French dressing.

- Broken or separated dressings include vinaigrettes.

*Soy sauce* is a dark brown sauce that is made from mashed soybeans, roasted barley, salt, and water. A culture is added and the mixture fermented in vats for 6 months to 18 months. At the end of the fermentation period, the mixture is pressed and strained to produce soy sauce. Soy sauce is commonly used in Asian food preparations and as a dipping sauce.

*Tabasco®* sauce is a very hot, vinegar-based sauce made from red peppers, vinegar, and salt. Tabasco® sauce is usually packaged in small, shaker-top bottles and is used in flavoring meat sauces, salads, and soups, and on eggs.

*Worcestershire sauce* is a pungent, dark-colored sauce. It is used in cooking and for seasoning meat. Although Worcestershire sauce is made from many spices, its key ingredient is tamarind. It is most commonly served on grilled steak or as an ingredient in sauces, stews, and soups.

## SUMMARY

Food has a way of awakening all of the senses. From the sizzle of a red-hot fajita, to the aroma of fresh sautéed garlic, to the tart taste of lemon in a lemon meringue pie, the senses of sight, taste, smell, and touch come alive with food. The senses identify different qualities of foods and, through sensory perception, people can enjoy many distinct aspects of a dish.

Seasonings and flavorings used in food preparation have the power to make a good dish taste incredible. Pepper is the most widely used seasoning and comes in many forms and colors. Salt, another common seasoning, is sold in many forms including table salt, sea salt, and kosher salt. Lemon zest adds a fresh aroma and taste to the food. Examples of flavorings include a wide assortment of herbs, spices, and herb or spice blends. Herbs and spices should enhance, not disguise, the flavors of food and should be used in moderation. Similarly, adhering to specific guidelines for purchasing and storing herbs and spices is important for ensuring a high level of quality.

A wide variety of nuts and seeds are often added to foods to add texture or a nutty flavor. For instance, fish can be coated in nuts before being sautéed. Common nuts used in cooking include almonds, Brazil nuts, and pecans. Walnuts, pumpkin seeds and sunflower seeds are often used in salads.

Cooking oils add a variety of flavors to a dish, while vinegars add a tart flavor. Finally, condiments such as ketchup, mustard, mayonnaise, salad dressing, and soy and Worcestershire sauces are used to enhance the flavor of various dishes. All in all, a chef must have a thorough understanding of a wide variety of seasonings, flavorings, and common condiments in order to best complement the natural flavors of food.

Refer to the CD-ROM for Quick Quiz® questions related to chapter content.

1. Identify the three sensory experiences that determine the flavor of a food.

2. List the four most common seasonings used for cooking or finishing a dish.

3. Contrast the storage methods for both dried and fresh herbs.

4. Identify the most common type of basil and the products it is often used in.

5. Describe common uses for lemongrass.

6. What are the differences between mint and savory?

7. Describe common uses for capers.

8. Describe the difference between cinnamon and cassia.

9. List the main features and uses for fenugreek.

10. Describe the common forms of mustard seed found in nature and in the market.

11. List the four colors that peppercorns come in.

12. What is Cajun spice usually composed of?

13. What is the difference between fines herbes and herbes de Provence?

14. Give two examples of dishes in which it is acceptable for the herb or spice to dominate the flavor of the food.

15. What step should be taken to release the flavor of dried herbs before adding the herbs to a preparation?

16. For uncooked dishes, at what point should herbs or spices be added to the dish?

17. What are the main differences between imitation spices and true spices?

18. What are the main features that distinguish chestnuts from other nuts?

19. Describe the main characteristics and uses for pine nuts.

20. How do nut oils differ from other types of vegetable oils?

21. Identify the main qualities of virgin olive oil.

22. In general, how long do oils stay fresh after being opened?

23. List the four most common types of vinegar used in professional kitchens.

24. Describe the distinguishing characteristics of yellow mustard, Dijon mustard, and English and Chinese mustards

25. Describe the main features that distinguish salad dressing from mayonnaise.

26. What is the key ingredient in Worcestershire sauce?

# Eggs and Breakfast Foods

## Chapter 11

*Eggs and breakfast foods* are the first dishes that many restaurants use to develop a new employee's organizational skills, timing, and speed. A breakfast cook must be able to handle many orders at one time and must be able to work quickly. Many nutrition experts have stated that breakfast is the most important meal of the day because it helps the body operate at maximum efficiency. It can be thought of as the fuel to get a body's engine running for the day.

## Chapter Objectives:

1. Explain the basic composition and major classifications of eggs.
2. Identify storage, safety, and sanitation requirements to follow when using eggs.
3. Demonstrate common dry-heat and moist-heat preparation methods used to cook eggs.
4. Describe common techniques for preparing and serving pancakes, waffles, and French toast.
5. Describe common breakfast meats and how they are prepared.
6. Identify types of breakfast cereals, juices, and side dishes.
7. Contrast continental breakfasts and breakfast buffets.

## Key Terms

- yolk
- emulsifier
- white
- chalazae
- salmonella
- sunny-side up eggs
- over-easy eggs
- basted eggs
- scrambled eggs
- omelet
- folded omelet
- rolled omelet
- frittata
- shirred eggs

## BREAKFAST FOODS

Breakfast foods consist of various egg dishes, meats, fruits, breads, and cereals. ***See Figure 11-1.*** Of these, the most popular and most common breakfast food is the egg. Simple preparations such as scrambled eggs are quick and relatively easy, while more complex preparations such as eggs Benedict or a breakfast quiche require more time and practice.

### Breakfast Foods

MacArthur Place Hotel, Sonoma

**11-1** *A breakfast buffet can offer a wide variety of breakfast foods.*

## EGGS

Eggs are used in many preparations in the professional kitchen. Knowledge of egg composition, various egg uses, and proper cooking methods will help the cook produce better results. Eggs are a complete protein food. They are high in vitamin content and are easy to digest when cooked properly. Besides their importance in breakfast preparations, eggs can also be featured as luncheon and dinner entrées. In fact, eggs are a food that may be used in any part of the menu, from appetizer to dessert. Eggs are relatively inexpensive; therefore, serving eggs or egg dishes as a meal can help keep food costs down.

Eggs are so versatile that they can be used in a variety of ways and in different preparations. Some common applications of eggs include the following:

- a thickening or binding agent, such as in meat loaf, custard, pudding, and pie filling

- an adhesive agent, such as in breading and coatings

- an emulsifying agent, such as in mayonnaise, salad dressing, and hollandaise sauce

- a clarifying agent, such as in consommé and aspic

- a lightening agent (incorporating air), such as in soufflés, sponge cakes, and meringues

- an entrée, such as in breakfast preparations including omelets and prepared egg dishes

## Egg Composition

An egg is composed of four main parts: the shell, shell membrane, yolk, and white. *See Figure 11-2.* The *shell* of the egg is the thin hard covering composed of calcium carbonate. Although the shell protects the fragile yolk and white, it is fragile as well. While the shell feels like a hard surface, it is porous. This means that odors can be absorbed through the shell. If eggs are stored next to strong smelling foods such as garlic or onions, the egg will absorb some of the odors. The fact that the egg is porous also means that the egg will evaporate some of its moisture through its shell over time.

In addition to the shell, the fragile yolk and white are also protected by a shell membrane. The *shell membrane* is a thin, skin-like material located directly under the egg shell.

The *yolk* is the yellow portion of the egg where all of the fat as well as much of the protein, vitamins, and minerals are found. Although the yolk is only about one-third of the weight of the egg, it contains more than three-fourths of the calories and all of the cholesterol found in the egg. For many years it was thought that eggs contained so much cholesterol that they were unhealthy to eat. The American Heart Association has now determined that eggs do not have as much cholesterol as was originally thought, and that eggs are packed with vitamins and other nutrients. Another product found in the yolk is lecithin, which acts as an emulsifier in products such as mayonnaise and hollandaise. An *emulsifier* is a substance that enables two substances that would usually not mix well, such as oil and water, to blend together into a smooth substance.

The *albumen* is the clear portion of the raw egg, which makes up two-thirds of the egg and consists mostly of ovalbumin protein. It is the primary food source for the developing embryo and provides a cushion from the hard shell covering. The albumen contains the majority of protein found in the egg as well as a high level of riboflavin. When cooked, the albumen turns white.

As it is cooked, the albumin begins to turn white and change into a solid, which is why people refer to it as the egg white. Another compo-

Chef's Tip:

The color of an egg's shell represents the particular breed of hen that laid the egg. The color of the shell has no effect on the flavor, quality, or nutritional composition of an egg.

## Egg Composition

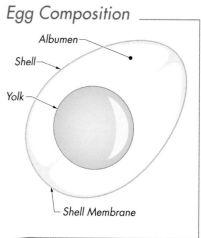

11-2 An egg is composed of four main parts: the shell, shell membrane, yolk, and white.

nent of the egg found in the albumin is two cords called the chalazae. The *chalazae* are the two small stringy parts of the egg white that anchor the yolk to the white. People often misunderstand it to be the start of the chick embryo but it is not. It is a safety feature that nature gives to eggs to keep the internal components anchored together.

## Grading and Size of Eggs

The size of an egg and the color of its shell have no bearing on its quality. The most important factors are the appearance and quality of the interior, the condition of the shell, the size of the air pocket within the egg, and the overall cleanliness of the shell. The United States Department of Agriculture (USDA) is responsible for the grading standards and inspections of eggs when they come to market. The grades are AA, A, and B. The highest quality egg, Grade AA, has a firm yolk and white, with both standing tall when broken onto a flat surface. As the egg ages or is of a lower grade, the yolk and the white will begin to lay flatter and spread out further on the flat surface. *See Figure 11-3.* The white will also be looser and more watery, and the natural air pocket inside the egg will grow larger. However, it is important to note that this grading has nothing to do with any nutritional characteristics.

## USDA Egg Grades

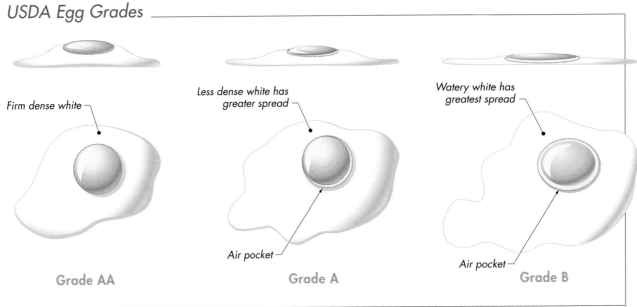

Firm dense white

Less dense white has greater spread

Watery white has greatest spread

Air pocket

Air pocket

Grade AA

Grade A

Grade B

11-3 *The yolk and the white of lower-grade eggs lay flatter and spread out farther on the flat surface.*

Eggs are also classified according to size. The three grades of eggs are sorted into various sizes based on the minimum weight per dozen eggs. Each named size category weighs three ounces per dozen more than the previous size. *See Figure 11-4.* The following classifications of eggs are listed according to size from smallest to largest, where size represents the minimum weight per dozen:

* pee wee—15 oz to 18 oz
* small—18 oz to 21 oz
* medium—21 oz to 24 oz
* large—24 oz to 27 oz
* extra large—27 oz to 30 oz
* jumbo—30 oz or more

The most common size of egg used in food service establishments is large. Most recipes are written using large eggs as the standard.

## Egg Sizes

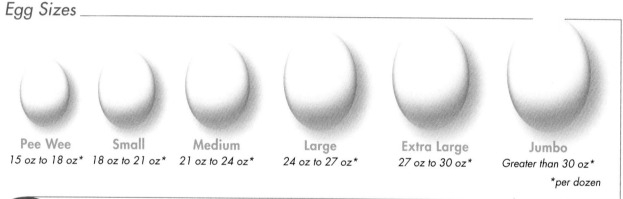

| Pee Wee | Small | Medium | Large | Extra Large | Jumbo |
|---|---|---|---|---|---|
| 15 oz to 18 oz* | 18 oz to 21 oz* | 21 oz to 24 oz* | 24 oz to 27 oz* | 27 oz to 30 oz* | Greater than 30 oz* |

*per dozen

**11-4** *Eggs are classified according to size, from pee wee to jumbo.*

## Storing of Eggs

Eggs should be stored under refrigeration at a temperature around 36°F. Storing eggs even for a day at room temperature will age them more than they would have aged in a week under refrigeration. Older eggs are best for making hard-cooked eggs as the shells are easier to peel. Once hard-cooked, eggs can last in the refrigerator for up to one week. Fresh eggs in the shell can be held for about one month past the packaged date.

## Egg Safety and Sanitation

Like all protein-rich foods, eggs are a potentially hazardous food if not handled and stored properly. Protein-rich foods provide excellent breeding grounds for harmful bacteria. The form of bacteria most commonly associated with eggs is salmonella. *Salmonella* is a bacteria found in the intestinal tracts of all chickens and can cause

### Action Lab

A fresh egg, when placed in a pot of cold water, will sink to the bottom—but, conversely, an old egg will float. As the egg's moisture evaporates through the shell, it is replaced with air from the outside. As more egg is replaced with air, the egg gains the ability to float in water rather than sink. Gather a fresh egg and an egg that is two or more weeks old. Place each egg in a pot of tap water. If the egg is old enough it will float.

food poisoning. Even though eggs are cleaned and sanitized at packing houses, some traces of the harmful bacteria may still remain. If you have an egg with a dirty shell, it is best to discard it. It is also important to ensure that when you crack an egg, the raw egg doesn't come in contact with the outer shell. This high presence of bacteria is the reason that it is best to thoroughly cook eggs prior to eating them. Many operations will use pasteurized egg products that have been heated to a specific temperature for a brief period of time to kill harmful bacteria before they are packaged. This pasteurization process does not cook the egg.

In the professional kitchen, fresh eggs are used more than any other form of egg. This is because fresh eggs are the most versatile. Eggs can also be purchased in 5 lb frozen cartons or in 30 lb cans of eggs, egg yolks, or egg whites. These forms and quantities are convenient only if the eggs are to be used in bulk food preparations and baked products. Dried eggs are also available, but are not often used in food service establishments. Dried eggs are occasionally used in bake shops.

Cooked eggs perish rapidly, mainly because of their delicate nature. Whether frying, poaching, scrambling, shirring, or basting, best results can always be obtained by cooking eggs in small quantities and as close to serving time as possible. Cooked eggs held for even as short a time as 10 min will show signs of loss of quality. Eggs that are scrambled or poached can be held using special holding techniques.

*The following tips are helpful for proper egg storage:*

- Always store eggs in the refrigerator. If eggs are left in a warm place, they lose freshness rapidly. Remember that an egg left out for a day at room temperature will age almost as much as it would have after a full week of refrigeration.
- When storing leftover egg yolks in the refrigerator, cover them with a small amount of water and plastic wrap. If left uncovered, they will form a surface crust and dry out rapidly.
- Leftover egg whites will keep for up to a week in the refrigerator if they are placed in a tightly covered and sanitized container.
- Frozen egg products should be thawed gradually in the refrigerator. After they have thawed, they should be stirred thoroughly before use. One quart of frozen eggs equals approximately 24 fresh large eggs.

## METHODS OF EGG PREPARATION

Both dry-heat and moist-heat cooking methods can be used to prepare eggs. It is important to remember that the white of the egg cooks at a lower temperature than the yolk. This is why the white of an over-easy egg is cooked but the yolk is still runny. The most common mistake people make when cooking eggs is either overcooking or undercooking them. Too high heat or too long a cooking time will cause eggs to toughen and become rubbery. It can also change their color and flavor. The yolk of the egg may even form a green outer ring when overcooked due to the natural sulfur found in the white that reacts with the natural iron found in the yolk. This can often be seen in an overcooked hard-cooked egg or scrambled eggs that sit in a steam table on a buffet for a long time.

A hard-cooked egg, when cooked correctly, has a distinctive appearance. *See Figure 11-5.* The white is bright white and firm but does not feel rubbery. The yolk, which is the true indicator, looks almost fluffy. The color is a pale yellow all the way through, and there is no difference in color near where the yolk meets the white.

## Hard-Cooked Eggs

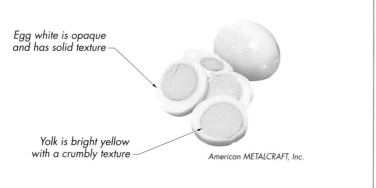

Egg white is opaque and has solid texture

Yolk is bright yellow with a crumbly texture

American METALCRAFT, Inc.

**11-5** *A correctly hard-cooked egg has a distinct visual appearance.*

Eggs are a very fragile food and therefore need to be cooked appropriately. Eggs are worked using various dry-heat and moist-heat cooking methods.

### Dry-Heat Cooking Methods for Eggs

Eggs can be cooked using various dry-heat cooking methods such as pan-frying, sautéing, and baking. Common egg preparations using dry-heat cooking methods include fried eggs, sautéed or scrambled eggs, omelets, and baked or shirred eggs.

**Fried Eggs.** Pan-fried, or fried eggs as they are more commonly called, include sunny-side up, over-easy, over-medium, and over-hard. The main difference between variations is in cooking time and proper technique. *Sunny-side up eggs* are lightly cooked fried eggs with unbroken yolks and are never flipped over to cook the other side. *Over-easy eggs* are fried eggs with unbroken yolks that are flipped over to the other side and cooked easy (lightly), medium, or hard (well). *Basted eggs* are fried eggs with unbroken yolks that are cooked the same as sunny-side up eggs, but the tops are slightly cooked by tilting the pan and basting the eggs with hot butter from the pan.

Eggs should always be fried to order, and fresh Grade AA eggs should be used. The fresher the egg, the more the yolk stands up and the less runny the white is. For best results, eggs should be fried in clarified butter, shortening, or margarine.

Selecting the correct size sauté pan is necessary for best results. For a single egg, the sauté pan should be 4″ in diameter at the bottom. For an order of eggs (two eggs), the pan should be 6″ in diameter. The pan should have sloped, shallow walls and a long handle. Usually a nonstick pan is used to help prevent the egg from sticking and potentially breaking the yolk. To begin, the eggs are cracked into a small dish so that the raw eggs can be inspected for quality, as well as to help ensure that the yolks do not break when added to the pan. The pan is heated over moderate heat and about 1 tsp of clarified butter, oil, or margarine is placed in the bottom of the pan. Next, the eggs are taken from the dish and placed in the pan. The hot oil in the pan will solidify the eggs immediately so the whites will not spread. The eggs are cooked to the desired doneness.

Chef's Tip:
Whole butter is not used to cook eggs as it burns very easily and at a very low temperature.

## To make a fried egg, apply the following procedure:

1. Choose the appropriate size sauté pan and heat the pan over medium heat.
2. Crack eggs into a small dish or bowl.
3. Place a small amount of clarified butter, margarine, or oil into the hot pan.
4. Slide the eggs into the pan.
5. When whites are firm, gently flip eggs over to cook the other side (remember that sunny-side up eggs and basted eggs are never turned over).
6. Fry eggs to desired degree of doneness (over-easy, over-medium, or over-hard).
7. When eggs are done, gently flip them back over so that the first side is up.
8. Slide the eggs out onto a plate and serve, taking care to wipe away any excess butter or oil.

*The following tips will help prevent mistakes when frying eggs:*

- Do not use a poorly conditioned pan or griddle. Eggs can stick to a poorly conditioned pan or griddle, causing them to burn and break.
- Do not fry with too much oil. When eggs are slid out of a pan and onto a plate, they will be very greasy. There is also danger of being burnt when the eggs are flipped over as the oil can splash out of the pan.
- Do not fry with too little oil. Eggs may stick to the pan and will then burn and probably break when they are removed from the pan if too little oil is used.
- Fry eggs at the correct temperature. Eggs burn and are usually over-cooked on the outside but may be too runny inside when cooked at too high a temperature, and egg whites spread too rapidly when cooked at too low a temperature.

American METALCRAFT, Inc.

**Sautéed Eggs.** Sautéing covers some popular styles for cooking and serving eggs. Scrambled eggs, omelets, and frittatas all are prepared using the sautéing method.

*Scrambled eggs* are eggs that are whisked while raw to combine the yolk and white and then sautéed to produce fluffy curds. Scrambling is the easiest method to choose when preparing eggs in quantity. Scrambled eggs may be prepared in several ways. Eggs can be scrambled in a steam jacket kettle or tilt braiser, on a griddle, in a double boiler, or in a sauté pan on the range. The best method for scrambling smaller amounts is in a sauté pan.

To prepare scrambled or sautéed eggs, eggs are broken into a stainless steel, glass, or china bowl. An aluminum bowl is never used because it will discolor the eggs as they are whisked. The eggs are seasoned with salt and pepper and whisked slightly with a wire whisk or kitchen fork. A small amount of milk or cream can be added if desired (about 2 oz to 3 oz of liquid for each pint of eggs). Too much milk will cause weeping (giving off water) after the eggs are cooked.

The beaten eggs are poured into a heated, oiled, or buttered sauté pan, where they start to coagulate immediately. After the eggs are added to the pan, the heat is reduced and the egg mixture is carefully stirred from the bottom with a high-heat spatula. Scrambled eggs are properly cooked when they are soft and fluffy, a little moist, but not runny. Scrambled eggs should always be slightly undercooked, as they will become firm as they are held for service. If the eggs are browned or overcooked, they will become dry, hard, and unpalatable. If scrambled eggs are to be held more than 5 min before serving, a medium béchamel sauce or a little heavy whipping cream can be added. This will extend the holding time and prevent the eggs from drying out and discoloring.

To make scrambled eggs, apply the following procedure:

1. Break the eggs into a stainless steel mixing bowl, season with salt and pepper, and add a small amount of milk or cream.
2. Beat the eggs with a wire whisk or kitchen fork until well-mixed and fluffy.
3. Heat a sauté pan with a small amount of clarified butter or oil.
4. Pour the eggs into the pan, reduce the heat, and stir slowly with a high-heat spatula until the eggs are set but still moist.
5. Serve immediately.

*The following tips will help prevent mistakes when making scrambled eggs:*

- Cook eggs at the correct temperature. Eggs will usually burn and be overcooked, dry, and brown when cooked at too high a temperature.
- Do not stir eggs excessively. Egg particles become too fine, giving a poor appearance, if eggs are excessively stirred during cooking.
- Do not hold cooked eggs too long. The eggs can develop unappetizing colors and will have an unpleasant flavor if held too long in a steam table.
- Scramble with the correct amount of oil in the pan. Eggs will become greasy if too much oil is used, and they will stick, become tough, dry out, and burn if too little oil is used.

**Omelets.** An *omelet* is an egg dish made with beaten eggs and cooked into a solid form. Omelets are one of the most-ordered egg dishes in restaurants. They can be filled with a wide variety of ingredients. The number of combinations of ingredients is endless. When an omelet is served with another item such as bacon or mushrooms, it often takes the name of that accompanying item—for example, bacon omelet or a mushroom omelet.

Omelets are almost always prepared to order and usually consist of two or three eggs per order. If they are held for even a short period of time, they lose their fluffiness and become tough and rubbery. There are three varieties of omelets: folded omelets, rolled (French style) omelets, and frittatas.

A *folded omelet* is an omelet that is cooked until nearly done and then folded before serving. *See Figure 11-6.* An omelet pan, which usually has a nonstick surface, is the tool of choice for omelets, as it has sloped sides that make the omelet easier to flip. Any meat or vegetable ingredients that will be included in a folded omelet will need to be cooked, or at least heated, prior to adding them to the beaten eggs. The exception to the rule is cheese, which is added as the omelet is folded near the end of the cooking process.

## To make a folded omelet, apply the following procedure:

1. Heat an omelet pan over medium-high heat and add clarified butter, margarine, or oil.
2. In the hot omelet pan, cook any meats or vegetables to be used in the omelet.
3. In a bowl, whisk the eggs until well-blended and season with salt and pepper.
4. Pour the beaten eggs into the hot pan with the cooked ingredients and stir with a high-heat rubber spatula until somewhat set.
5. With the spatula, gently lift up the side of the cooked egg and let any uncooked or liquid egg run under the cooked portion.
6. Add cheese if desired.
7. Fold one-half or one-third of the omelet into the center using the spatula.
8. Slide the unfolded end of the omelet gently onto a plate and flip the omelet over it so that a folded omelet is served. *Note:* If a tri-fold is made, the seam should be underneath.

*The following tips will help prevent mistakes made when making folded omelets:*

- Do not use a poorly conditioned pan or a pan that is not nonstick. Eggs stick and the omelet will break when rolled or folded if the pan does not have a nonstick surface or is poorly greased.
- Use the correct amount of oil. If there is too little oil in the pan, the eggs will stick and burn. If there is too much oil, the hot grease will splatter when the eggs are added, and excess grease may spill out onto the plate when rolling or folding the omelet.
- Do not overcook the omelet. An omelet will become too brown and will crack when rolled or folded if it is overcooked.
- Do not cook the omelet ahead of service. An omelet loses its fluffiness and becomes tough and rubbery if it is prepared too far in advance.

## Folded Omelets

1. Heat pan and fat.
2. Cook any fillings and ingredients.

3. Whisk eggs and season to taste.

4. Stir eggs and filling ingredients over heat until somewhat set.

5. Lift gently to allow undercooked egg to run underneath.

6. Add cheese if desired.

7. Fold half of the omelet into the center using the spatula.

8. Roll remaining unfolded side on top and serve.

**11-6** Making a folded omelet involves cooking the filling and egg mixture without overstirring and carefully folding the omelet to serve.

A *rolled omelet,* or French omelet, is an egg mixture that is cooked and folded, and cooked filling ingredients are added through a slit cut in the omelet prior to serving. ***See Figure 11-7.*** Like a folded omelet, any meat or vegetable ingredients are precooked, but the difference is that these additional ingredients are not added into the raw egg mixture and cooked with it. When the omelet is slightly set but still in a very moist condition, the pan should be tilted to about a 60° angle and a kitchen fork or high-heat spatula should be used to roll the eggs. The egg should remain pale yellow and be browned as little as possible. Once the egg portion of the rolled omelet is cooked, rolled tightly, and plated, any additional cooked ingredients and cheese may be placed into a slit cut into the top of the omelet.

## Making Rolled Omelets

1. Heat pan and fat.
2. Whisk eggs and season to taste.

3. Stir eggs over heat until somewhat set.
4. Lift gently to allow undercooked egg to run underneath.

5. Roll one quarter of omelet toward center.

6. Roll omelet onto plate.

 11-7 A rolled omelet is similar to a folded omelet, except that any filling ingredients are added into a slit after the omelet has been cooked and folded.

## To make a rolled omelet, apply the following procedure:

1. Heat an omelet pan over medium-high heat and add clarified butter, margarine, or oil.

2. In a separate bowl, whisk eggs until well-blended and season with salt and pepper.

3. Pour the egg mixture into the omelet pan and gently shake the pan while stirring the egg mixture gently with a high-heat spatula.

4. When the egg mixture is almost set, stop stirring and lift up one side of the omelet and allow any liquid egg mixture to run underneath to cook.

CULINARY PROCEDURES

5. Take the side of the omelet farthest away from the plate it is to be served on and fold that side over about one-fourth of the way to the center.

6. Place the omelet pan over the serving plate with the unfolded end toward the plate, and use the spatula to roll the omelet out of the pan and onto the plate with the seam facing down.

*Note:* If filling is desired, cut a slit into the top of the omelet and gently spread the slit open. Place the precooked filling ingredients or cheese into the slit and serve.

Very light and fluffy omelets have become popular in some areas, but the folded or rolled omelet is featured on most menus. A fluffy omelet can be prepared in two ways. The first way is by separating the yolk from the white and beating each to a soft foam. The beaten white and yolk are then folded gently together until well mixed. The eggs are then poured into a hot, greased pan and cooked in the same manner as a folded or rolled omelet. This type of omelet can also be finished in a 325°F oven. When finished, the fluffy omelet should be served at once. The second way to make a fluffy omelet is to use an immersion blender (or electric stick mixer) to whip air and volume into the egg mixture. When the egg mixture is added to the hot pan and oil, it will cook much lighter than a mixture that was simply whisked by hand.

A *frittata* is basically a traditional folded omelet, but it is served open-faced after being browned under a broiler or in a hot oven. ***See Figure 11-8.*** The ingredients are all previously cooked and mixed with the egg, but instead of folding the omelet, it is left open and cheese (if desired) is added on top. The omelet is then placed under a broiler or in a very hot oven to melt and brown the cheese.

*Frittatas* _____

1. Cook the frittatta similarly to a folded or rolled omelet.

2. Top with cheese if desired.

American Egg Board

3. Serve immediately.

**11-8** A frittata is an open-faced flat omelet finished in a broiler or hot oven.

## To make a frittata, apply the following procedure:

1. In an omelet pan over medium high-heat, add clarified butter, margarine, or oil.
2. Add any meat or vegetable ingredients to the omelet pan and cook until done.
3. In a separate bowl, whisk eggs and season with salt and pepper.
4. Add the egg mixture to the pan with the other cooked ingredients and stir with a high-heat spatula until almost set.
5. Gently lift the edge of the frittata so that any uncooked egg can run under and be cooked.
6. Top with cheese if desired.
7. Place under a broiler or in a hot oven to melt and brown the cheese.
8. Serve immediately.

CULINARY PROCEDURES

**Shirred Eggs.** *Shirred eggs* (baked eggs) are prepared using an oven. Shirred eggs are cooked in a small, shallow, oven-proof casserole dish that has been coated with butter. The casserole dish is placed on a range and the eggs are cooked at a medium heat until the whites are set. The dish is then transferred to an oven or broiler to lightly cook the tops of the eggs. In either case, it is important to never overcook the eggs. Shirred eggs may be served with a variety of foods. Ham, sausage, and Canadian bacon are the most popular accompaniments. ***See Figure 11-9.***

HELPFUL TIPS

*The following tips will help prevent mistakes when making shirred eggs:*

- Do not cook at too high a temperature. Shirred eggs become hard and tough and are usually burned if cooked at too high a temperature.

- Do not cook at too low a temperature. Shirred eggs spread and the yolk has a tendency to break if cooked at too low a temperature.

## *Shirred Eggs*

11-9  Shirred eggs are prepared in a casserole dish and finished in an oven.

## Moist-Heat Cooking Methods for Eggs

With moist-heat cooking, eggs, whether in the shell or removed from the shell, are cooked in water. The hot water cooks the egg more gently than dry-heat cooking. Common egg preparations using moist-heat cooking methods include soft-cooked, medium-cooked, and hard-cooked, as well as poached eggs.

**Cooking Eggs in the Shell.** Although the common term for cooking eggs in the shell is "boiled eggs," the fact is that eggs in the shell should be simmered, never boiled. Boiling tends to toughen the texture of the egg and can create a green coating around the outside of the yolk. Simmering at a temperature of approximately 195°F is recommended.

Eggs should be room temperature before they are placed in the hot water or they may crack. If the eggs are left in the refrigerator until it is time to cook them, warm water should be run over the eggs before placing them in the hot water.

To prepare eggs in the shell, apply the following procedure:

1. Place the eggs in a pot and cover them with cold water.
2. Place the pot on the range and bring the water to a boil.
3. Reduce the heat to allow the pot to simmer and begin timing.
4. Cook to desired doneness.
   - Soft-cooked (also called soft boiled) — 3 min to 5 min
   - Medium-cooked — 6 min to 8 min
   - Hard-cooked — 9 min to 12 min

 *The following tips will help prevent mistakes when cooking eggs in the shell:*

- Do not cook at too high a temperature. Eggs become tough and rubbery when they are cooked at too high a temperature. A green ring may appear around the yolk.

- Do not cook at too low a temperature. Eggs are often undercooked when prepared at too low a temperature.

*Chef's Tip:*
*An exposed yolk will actually absorb water and become mushy. If the yolk of an egg is exposed, do not place the peeled egg in water; instead, place it in a bowl and cover it with plastic wrap or a damp cloth.*

If eggs are to be served in the shell, they should be plunged into slightly cold water and served immediately. If the eggs are to be held for use in preparation, such as sandwiches, egg salad, deviled eggs, or garnish on a salad, they should be cooled in ice-cold water for about 5 min immediately after cooking and then placed in a bain-marie, covered with water, and stored in the refrigerator.

A hard-cooked egg is peeled by cracking the shell gently on a hard surface or by rolling it on a hard surface. Peeling is started at the large end of the egg, and the shell is removed by moving downward, toward the narrow end. Keeping the egg submerged in cold water will help to loosen the shell.

To coddle eggs for such preparations as Caesar salad dressing, the eggs are brought to room temperature and placed in a pot. Boiling water is added to the pot until the eggs are covered. The pot is then covered with a lid for 1 min to 2 min and left to stand, without heating. Next, the eggs are immediately removed from the hot water and run under cold water for a few minutes until thoroughly cooled. The resulting yolk is coddled and ready to use as an emulsifier.

**Poached Eggs.** Poached eggs are removed from the shell prior to cooking. Poached eggs are very popular on breakfast menus for dishes such as eggs à la Florentine and eggs Benedict. Salt and vinegar are used in the preparation of poached eggs. The salt and vinegar cause the white to set firmly around the yolk when the egg is placed in the water, keeping the white from spreading. The acetic acid from the vinegar toughens the ovalbumin protein contained in the egg white, and when the white is set firmly around the yolk, a more appetizing look is obtained. ***See Figure 11-10.*** This acid will not affect the flavor of the egg when used in diluted quantities. Although poached eggs are very easy to prepare, the challenge is in determining doneness as well as transporting eggs from the poaching water to the plate without breaking the yolks.

*Poached Eggs* _____

*1. Add vinegar and salt to help keep the white intact.*

*2. Cook poached eggs 3 min to 5 min in simmering water.*

*3. Serve eggs as an entrée.*

**11-10** *Poached eggs are simmered eggs that have been removed from the shell prior to cooking.*

About 12 eggs at a time can be poached in each gallon of liquid. The water may be used for three different batches before being discarded.

To prepare poached eggs in quantity, the eggs are slightly undercooked. The eggs are then placed immediately in cold water to stop further cooking and held until ready to serve. To serve, the precooked eggs are reheated in hot salted water. The quantity method is usually for banquet or buffet preparations. For individual breakfast preparations, eggs should be poached to order.

To prepare poached eggs, apply the following procedure:

1. Fill a pan with enough water to cover the eggs (about 2½").
2. Add 1 tbsp of salt and 2 tbsp of distilled white vinegar for each gallon of water.
3. Bring the liquid to a boil and then reduce the liquid to a simmer (about 195°F to 200°F).
4. Break the eggs into a bowl and slide them into the simmering liquid; the eggs should slide gently down the side of the dish so the yolks remain in the center of the white and do not break.
5. Cook as desired (usually 3 min to 5 min is sufficient as most people prefer the yolk to still be somewhat soft).
6. Remove with a skimmer, perforated ladle, or slotted spoon; drain well and serve on buttered toast.

 *The following tips will help prevent mistakes made when poaching eggs:*

- Do not use too much vinegar. Too much vinegar toughens the eggs and affects flavor and smell.
- Do not cook at too low a temperature. Eggs cooked at too low a temperature become too tender and difficult to handle when they are served.
- Do not cook at too high a temperature. Eggs cooked at too high a temperature become tough and are usually overcooked.

## BREAKFAST FOODS AND SIDE DISHES

Breakfast is the most important meal of the day as it is the first nourishment that the body receives after the period of fasting that takes place during sleep. Breakfast foods mean many different things to different people. A restaurant breakfast menu will include egg dishes and breakfast meats such as ham, bacon, and sausage as well as varieties of breakfast potatoes. Pancakes come stacked high while waffles can be topped with everything from blueberries to whipped cream and nuts.

### Pancakes and Waffles

Popular breakfast preparations are pancakes (also known as hotcakes, griddle cakes, or flapjacks) and waffles. Both are popular because they are easy to digest, can be served in a variety of ways with a variety of toppings, and usually have a low menu price. Waffles and pancakes are examples of quickbreads.

Pancakes and waffles must always be cooked to order and served piping hot on a hot plate or platter. They are typically served with butter and toppings such as syrup, jam, jelly, and fruit. Maple syrup is the most popular topping; however, fruit or fruit-flavored syrups are also popular. Pancakes and waffles cost very little to make even when featured with a high-cost accompaniment such as strawberries, cherries, blueberries, or apples. Both pancakes and waffles are complemented well by sausage, ham, or bacon.

The basic recipe for pancakes consists of flour, sugar, salt, baking powder, baking soda, milk or buttermilk, eggs, and butter or oil. The dry ingredients are sifted together, and the milk, eggs, and butter are whisked together in a separate bowl. Then the liquid mixture is slowly combined with the dry mixture. The batter is spooned onto a hot oiled griddle or pan. The surface of the pancake will begin to bubble, and when the bubbles pop, the pancake is ready to be flipped over. *See Figure 11-11.* The surface should be golden brown. When the second side is cooked to a light golden brown, the pancakes may be served.

## Pancakes and Waffles

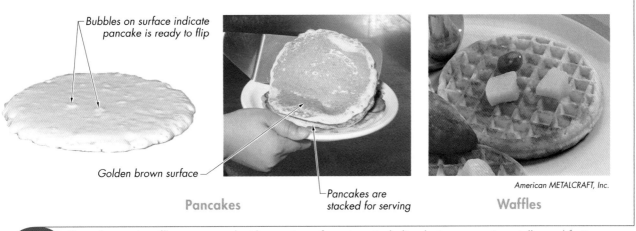

Bubbles on surface indicate pancake is ready to flip

Golden brown surface

**Pancakes**

Pancakes are stacked for serving

American METALCRAFT, Inc.

**Waffles**

**11-11** *Pancakes and waffles are served with a variety of toppings including butter, syrup, jam, jelly, and fruit.*

### Crepes

*Crêpes* are French pancakes that are lighter and thinner than basic pancakes. The surface is cooked to a pale cream color with almost no browning. Typically, crêpes are filled with fruit, rolled up, and topped with confectioners' sugar. They may be served either for breakfast or as a dessert. *See Figure 11-12.*

The procedure for making waffles is very similar to pancakes except that a waffle iron is used in place of a griddle or pan. Batter is poured onto a well-oiled waffle iron; the lid is closed, and batter is cooked for a few minutes on both sides simultaneously. Waffles are most often served with a fruit topping and whipped cream.

## Preparing Crêpes

1. Use just enough batter to lightly coat a small omelet pan.

2. Spread the batter across the diameter of the pan and cook over medium-low heat.

3. Before the crêpe begins to turn golden, flip with a rubber spatula.

4. When both sides are lightly golden brown, add fruit filling if desired.

5. Fold an edge of crêpe toward center and roll onto plate.

6. Dust with confectioners' sugar if desired and serve warm.

**11-12** Crêpes are French pancakes typically filled with fruit and topped with confectioners' sugar.

## French Toast

American Egg Board

**11-13** French toast is also a popular breakfast item.

### French Toast

Another alternative to pancakes is French toast. Although heavier than a pancake or waffle, French toast is also a popular breakfast item. ***See Figure 11-13.*** Slices of thick-cut bread are dipped into a batter primarily consisting of eggs and milk. The batter-dipped bread is then placed onto a heated and lightly oiled griddle or pan and cooked on each side until golden brown. French toast can be served with any of the accompaniments served with pancakes or waffles.

### Breakfast Meats

Breakfast meats commonly include sausage, bacon, Canadian bacon, and ham. All of these breakfast meats come from pork. Bacon, Canadian bacon, and ham are cured by smoking, whereas most sausage is fresh. Breakfast meats are available precut. These meats are usually precooked and then reheated for service when the breakfast demand is large. When the demand is smaller, breakfast meats are cooked to order. Precooking is done to speed service, since breakfast must be served quickly.

Breakfast sausage is available in the form of patties or links. Patties may be purchased in bulk or prepared in the kitchen from ground fresh pork and spices. Sausage patties are usually portioned and formed into 1 oz or 2 oz servings. Sausage patties can be precooked by baking on sheet pans in the oven at 350°F, cooking in a skillet on the range, broiling under the broiler, or grilling on a griddle.

Sausage links typically range in size from about 1 oz each (16 per lb) to 2 oz each (8 per lb). Sausage links are also made from ground pork that has been mixed with seasonings and spices. The portion served for breakfast is generally three or four links if they are smaller and one or two links if they are larger. Sausage links are cooked by separating the links, placing them on sheet pans, and baking at 350°F until done. *See Figure 11-14.* After cooking, the links are placed in a hotel pan and held for service. Sausage links should be cooked in small amounts as any that are left over will be dry the next day.

## Breakfast Sausages

**11-14**  *Sausage links are typically placed on sheet pans and baked at 350°F.*

Bacon comes from the belly of the pig. Sliced bacon is packaged in a few different forms. Slab-packed bacon is presliced but the user has to pull the slices apart. With shingled bacon, the sliced strips of bacon are slightly overlapped. Laid-out or layer-packed bacon is packaged with each slice separated and spread out on a sheet of baking or parchment paper. *See Figure 11-15.*

## Sliced Bacon

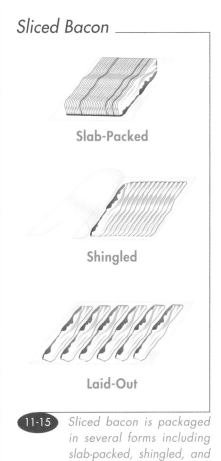

**Slab-Packed**

**Shingled**

**Laid-Out**

**11-15**  *Sliced bacon is packaged in several forms including slab-packed, shingled, and laid-out.*

Laid-out bacon is the style most often used in a food service establishment as it is convenient and does not require someone to spend time taking the slices apart. Most commercial establishments are willing to pay a few more cents per pound for the convenience. The most common thickness of sliced bacon is an 18–22 count, meaning there will be between 18 and 22 slices per pound.

Bacon slices should be placed on a sheet pan side by side but not touching. The bacon is baked at 350°F until three-quarters done. The bacon is then removed from the oven. To drain the bacon it is placed in a hotel pan with slices of bread lining the bottom of the pan. The bread drains the bacon of any excess grease and keeps the slices from lying in the grease until service. This method of cooking is recommended because it reduces shrinkage and curling, improves appearance, and makes the cooking more uniform. Bacon may also be cooked in a pan on the range, under the broiler, or on a griddle.

Canadian bacon, the boneless, smoked, pressed loin of a pork, is popular on breakfast menus. It is cut from the center of a pork chop. The perfect round shape of Canadian bacon lends itself well to using it as a base for poached eggs, as in an eggs Benedict. It is also the perfect shape for a breakfast sandwich. Since this meat is also already cooked, it is simply heated under a broiler, on a griddle, or in a skillet just prior to serving.

Ham comes from the hip and thigh of the pig. It is usually purchased cooked in a form that is boneless or boned and rolled. This form provides portions of 3 oz or 4 oz when sliced. Since the meat is already cooked, it is simply heated under a broiler, on a griddle, or in a pan before serving.

## Potatoes

Potatoes are usually served fried or griddled at breakfast. Potatoes may be served à la carte or included in featured breakfast combinations such as "two eggs with ham, hashed brown potatoes, toast, and beverage." Popular breakfast potato preparations are hashed brown (grated potatoes shaped into a patty and pan-fried or deep-fried), home-fried (large diced red potatoes pan-fried in a skillet or on a griddle with seasonings), and lyonnaise (either diced or sliced potatoes cooked with grated or diced onions). *See Figure 11-16.*

## Cereals

The two types of cereal commonly served in food service establishments are cold (ready-to-eat) cereal and hot cereal. Cold cereal has been popular in the United States for many years. Cold cereals come in a variety of textures and grains and range from sweeter varieties marketed to children to the whole grain and fiber varieties marketed to adults.

*Idaho Potato Commission*

## Breakfast Potatoes

Hash browns

Home-fried (skillet) potatoes

*Idaho Potato Commission*

**11-16** *Breakfast potatoes include options such as hashbrowns and home-fried (skillet) potatoes.*

Hot cereals are also popular as a breakfast item, especially during colder months. Hot cereals are made up of whole, cracked, ground, or flaked grains. They need to be cooked on the stove and are usually served with milk or cream, nuts, cinnamon, sugar, butter, or fruit. Some examples of hot cereals are oatmeal, Cream of Wheat®, and grits.

### Pastries

Assorted pastries are often featured on the breakfast menu. The most popular types of pastries are sweet rolls, doughnuts, coffee cakes, and Danish pastries. If possible, sweet rolls and Danish pastries should be served warm. ***See Figure 11-17.*** Some food service establishments have their own bakeshops and take great pride in producing pastries in their own baking facilities. However, many food service establishments purchase these items from commercial bakeries to reduce labor costs.

### Fruits and Juices

Any kind of fruit can be served for breakfast. Most often, fruit is served as an accompaniment or garnish, rather than the main item. Ripe fruit that is in season is the best choice to serve at breakfast. Typically a melon such as honeydew or cantaloupe is sliced or served in chunks. Grapes can be washed and left in small bunches, or fresh strawberries can be sliced and fanned as a garnish on a breakfast plate.

### Nutrition Note

Fruits are loaded with fiber, vitamins, and nutrients. They are generally low in calories, sodium, and fat. So give your body a nice treat today, eat some fresh fruit!

## Danish Pastries

*Danish pastries are fruit-filled and should be served warm.*

Both fruit and vegetable juices are common breakfast menu items. Juices may be purchased fresh, frozen, or canned. Frozen juices should be allowed to stand for a time after mixing. Breakfast juices commonly include orange juice, grapefruit juice, cranberry juice, pineapple juice, tomato juice, prune juice, and mixed or blended juices such as cranapple. A standard serving of juice is a 4 oz glass.

## CONTINENTAL BREAKFAST

The continental breakfast, made popular in European countries, is a light breakfast consisting of fruit or juice, toast or pastries, and coffee or tea. No heavy cooking is required. Continental breakfasts do not provide the balanced nutrition of more complete breakfasts, but they conserve labor costs.

## BREAKFAST BUFFETS

Like a traditional buffet, the breakfast buffet is designed to display foods in a manner that stimulates the appetite. Many popular breakfast preparations, such as scrambled eggs, hashed brown potatoes, ham, pancakes, assorted Danish pastries, fresh fruit, and French toast, contain colors that can be very attractive when arranged properly. The secret of a successful breakfast buffet is to offer a variety of foods that are well-prepared, arranged to take full advantage of the natural colors, and attractively priced. *See Figure 11-18.*

## Breakfast Buffet

Carlisle FoodService Products

**11-18** *Breakfast buffets offer a variety of popular breakfast preparations displayed in a manner that stimulates the appetite.*

### SUMMARY

Breakfast is considered by many to be the most important meal of the day. It can be thought of as fuel for the body's engine. Without it, a body gets off to a slow start and may not have a great deal of energy. The position of breakfast cook is demanding and fast paced. It is one of the best positions in a restaurant to allow a new cook to develop speed, organization, and timing. Although many items can be served for breakfast, the most popular and most common dish is the egg.

Eggs are used for many food preparations because of their unique properties. For example, they are used to incorporate air, assist as an adhesive for coating items prior to frying, and act as an emulsifier in items such as mayonnaise. As a breakfast dish, eggs have long been a popular choice. This is due to many factors such as reasonable cost, versatility, and availability any time of the year. Interesting items made with eggs such as frittatas, omelets, and eggs Benedict make egg dishes desirable. Other items that are served for breakfast include quickbread dishes such as waffles and pancakes, French toast, breakfast meats, potato side dishes, hot and cold cereals, fresh fruits and juices, and assorted pastries.

Refer to the CD-ROM for Quick Quiz® questions related to chapter content.

1. In addition to being used as entrées, list other common applications of eggs.

2. Identify the portion of the egg where all of the fat, as well as much of the protein, vitamins, and minerals, are found.

3. Describe the USDA's six classifications of eggs according to size.

4. Why is it recommended to cook eggs in small quantities and as close to serving time as possible?

5. Describe three ways to fry eggs.

6. Explain the five-step procedure for making scrambled eggs.

7. Explain the eight-step procedure for making a folded omelet.

8. How is a frittata different from a traditional folded omelet?

9. Describe the preferred method for peeling the shell of a simmered egg.

10. Describe the procedure for poaching eggs.

11. List the basic ingredients for pancakes.

12. Describe how to cook sausage links.

13. Identify the difference between home-fried and lyonnaise potatoes.

14. Give examples of common types of hot cereal.

15. What is the standard serving size of breakfast juice?

16. Compare the advantages of continental breakfasts with the advantages of breakfast buffets.

# Recipes

## Buttermilk Pancakes

*yield: 10 pancakes (2 per serving)*

| | |
|---|---|
| 2 c | all-purpose flour |
| 2 tbsp | granulated sugar |
| 1 tsp | salt |
| 2 tsp | baking powder |
| 1 tsp | baking soda |
| 16 oz | buttermilk |
| 2 | whole eggs (beaten) |
| 2 tbsp | melted unsalted butter |
| 2 tbsp | corn or vegetable oil |

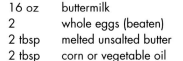

1. Sift all dry ingredients together in large bowl.
2. In a separate bowl, whisk together buttermilk (whole milk can also be used), eggs, and butter until well blended.
3. Add the milk mixture to the flour mixture and mix gently. Do not overmix, as the batter may become tough. *Note:* The mixture should still be fairly lumpy after it is properly mixed.
4. With a clean paper towel, wipe some oil onto a hot griddle or sauté pan set over medium-high heat.
5. Pour the batter onto the hot surface using a portion control scoop or ladle for size consistency.
6. Cook until the surface of each pancake has many small bubbles on it.
7. When the bubbles begin to pop, flip each pancake over to cook the other side.
8. Serve the pancakes immediately.

### Nutrition Facts

**Per serving:** 361.3 calories; 33% calories from fat; 13.6 g total fat; 100.7 mg cholesterol; 1044.3 mg sodium; 230.1 mg potassium; 48.5 g carbohydrates; 1.4 g fiber; 10 g sugar; 47.2 g net carbs; 11 g protein

## Crêpes

*yield: 16 crêpes (2 per serving)*

| | |
|---|---|
| 1 qt | milk |
| 10 oz | flour |
| 6 | eggs |
| ½ tsp | salt |
| 3 oz | melted butter |
| 3 oz | sugar |
| | jelly, jam, preserves, marmalade, or applesauce |
| 4 oz | confectioners' sugar |

1. Place the eggs in a mixing bowl and mix by hand using a wire whisk, or mix in the mixing machine at medium speed using the paddle for approximately 1 min.
2. Add the milk and continue to mix until the milk is blended with the eggs.
3. Combine the dry ingredients, sift, and add gradually to the liquid mixture. Mix for approximately 1 min or until all the dry ingredients are blended with the liquid mixture.
4. Add the melted butter and mix until well blended.
5. Pour the batter into a bain-marie and place in the refrigerator until ready for use.
6. Using a well-conditioned egg skillet or omelet pan, coat with melted butter or shortening and heat slightly.
7. Using a 2 oz ladle, coat the bottom of the skillet with the crêpe batter. While pouring the batter into the skillet, rotate the skillet in a clockwise direction so the batter will spread uniformly over the bottom of the skillet and the coating will remain very thin.
8. Place the skillet on the range and cook one side, then flip or turn by hand and cook second side. Brown the crêpes very lightly.
9. Remove from the skillet and place on sheet pans covered with wax paper. If crêpes are stacked, place a sheet of wax paper between each crêpe.
10. Spread each crêpe with jelly, jam, preserves, marmalade, or applesauce and roll up.
11. Dust with confectioners' sugar and serve three crêpes to each order.

### Nutrition Facts

**Per serving:** 741.3 calories; 19% calories from fat; 16.8 g total fat; 193.7 mg cholesterol; 284.5 mg sodium; 351.9 mg potassium; 134.7 g carbohydrates; 2.2 g fiber; 85.6 g sugar; 132.5 g net carbs; 12.8 g protein

## Waffles

*yield: 5 waffles (1 per serving)*

| | |
|---|---|
| 2 | eggs |
| 12¾ oz | milk |
| 8½ oz | cake flour |
| 1 tbsp | baking powder |
| 3½ tbsp | sugar |
| 3¼ oz | melted butter |

*Chef's Choice® by EdgeCraft Corporation*

1. Place the eggs in a mixing bowl and mix in the mixing machine at medium speed using the paddle for approximately 1 min.
2. Add the milk and continue to mix until the milk is blended with the eggs.
3. Combine the dry ingredients, sift three times, and add to the egg-milk mixture. Mix for 1 min.
4. Add the melted butter and mix until well blended.
5. Pour the batter into a bain-marie and place in the refrigerator until ready for use.
6. Brush the top and bottom of the waffle iron with salad oil. Heat to approximately 375°F.
7. Pour enough batter on the waffle grid to barely cover it. The amount used will depend on the size and shape of the waffle iron.
8. Let the waffle cook for about 1 min before lowering the top of the iron. When the top is lowered, cook about 1 min to 2 min longer. Exercise caution while cooking because the top grid is usually hotter than the bottom grid.
9. Serve two waffles per serving with jam, jelly, syrup, marmalade, or fruit.

**Nutrition Facts**

**Per serving:** 422.9 calories; 42% calories from fat; 20.2 g total fat; 145.5 mg cholesterol; 359 mg sodium; 197.7 mg potassium; 50.9 g carbohydrates; 0.8 g fiber; 13.2 g sugar; 50.1 g net carbs; 9.5 g protein

## French Toast

*yield: 8 pieces (2 per serving)*

| | |
|---|---|
| 8 slices | white bread |
| 3 | eggs |
| ⅔ c | milk |
| 3¼ tsp | sugar |
| ½ tsp | vanilla |
| 4 oz | confectioners' sugar |

1. Break the eggs into a stainless steel bowl and beat with a wire whisk.
2. Add the milk, sugar, and vanilla, and beat until well blended.
3. Pour the mixture into a hotel pan. Dip each slice of bread into the batter and coat both sides of the bread.
4. Remove the bread from the batter. Let drain slightly.
5. Brown the bread on both sides by placing it on a lightly oiled skillet or griddle.
6. Serve two pieces to each order with desired syrup, fruit topping, or butter, and then sprinkle with confectioners' sugar.

**Nutrition Facts**

**Per serving:** 227.4 calories; 26% calories from fat; 6.7 g total fat; 162.7 mg cholesterol; 409.4 mg sodium; 159.5 mg potassium; 30.9 g carbohydrates; 1.2 g fiber; 8.1g sugar; 29.7 g net carbs; 9.9 g protein

**Recipe Variation:**

**5** **Cinnamon:** To make cinnamon French toast, add the cinnamon to the vanilla first and mix well. Then add the milk, sugar, and finally the eggs. This reverse procedure prevents lumps and evenly distributes the cinnamon in the batter.

**6** **Whole Grain:** Substitute whole grain bread.

## Spanish Omelet

*yield: 1 omelet*

| | |
|---|---|
| 1 tsp | onion, diced |
| 1 tsp | red pepper, diced |
| 1 tsp | green pepper, diced |
| 2 | eggs, beaten |
| 1 tsp | salad oil |
| | salt and pepper to taste |
| 1 tbsp | cheddar cheese |
| | (or other desired cheese) |
| 1 tbsp | salsa |
| 1 tsp | sour cream |

1. In a nonstick omelet pan over medium heat, add the salad oil.
2. When the oil is hot, add the vegetables and sauté for 1 min.
3. Add the beaten eggs and stir with a high-heat rubber spatula until almost set.
4. Flip omelet over gently to cook other side.
5. Place cheese in center of omelet and fold one side over.
6. Slide omelet onto plate.
7. Garnish top of omelet with salsa and sour cream.

### Nutrition Facts
**Per serving:** 234.2 calories; 67% calories from fat; 17.8 g total fat; 433.6 mg cholesterol; 444.7 mg sodium; 214.2 mg potassium; 2.8 g carbohydrates; 0.4 g fiber; 1.7 g sugar; 2.4 g net carbs; 15.1 g protein

## Quiche Lorraine

*yield: 2 pies (twelve 5-oz servings)*

*Pie Crust*

| | |
|---|---|
| 1 c | flour |
| 3 oz | cream cheese |
| 4 oz | butter |
| 1 | egg yolk |
| ½ tsp | salt |
| 1 c | dried beans (any variety) |

*Filling*

| | |
|---|---|
| 5 oz | onions |
| 5 oz | small diced bacon |
| 5 oz | parmesan or Romano cheese, grated |
| 5 oz | swiss cheese, grated |

*Custard*

| | |
|---|---|
| 4 | eggs |
| 1 pt | whole milk |
| ¼ tsp | white pepper |
| ¼ tsp | salt |

1. Gently mix all ingredients for crust, taking care not to overmix. Dough should still be crumbly when mixed.
2. Roll dough out to fit greased pie tin.
3. Cover dough with aluminum foil; fill with dried beans and prebake crust until set.
4. While crust is prebaking, sauté diced bacon until crisp, add diced onions and sweat.
5. Take the crust out of the oven and remove the dried beans and the aluminum foil.
6. Mix bacon/onion mixture, add cheese mixture, mix well, and fill crust.
7. Whisk eggs, add milk, salt, and pepper, and whisk well again.
8. Pour custard mixture over onion mixture in crust.
9. Bake at 375°F for 25 min or until a toothpick inserted in center comes out clean.
10. Let rest out of oven for 10 min before cutting.

### Nutrition Facts
**Per serving:** 351.9 calories; 64% calories from fat; 25.5 g total fat; 151.4 mg cholesterol; 543.8 mg sodium; 219.6 mg potassium; 15.9 g carbohydrates; 1.8 g fiber; 3.1 g sugar; 14 g net carbs; 14.9 g protein

# Garde Manger

Daniel NYC

*Garde manger* refers to anything that is prepared and served cold in the professional kitchen. The term is also used to describe a chef who is responsible for preparing cold foods. A garde manger (pantry chef) prepares items such as salads, salad dressings, dips, assorted cheeses, fruit and vegetable platters, deli meats, cold charcuterie, carved centerpieces, cold sandwiches, and cold hors d'oeuvres. A garde manger must use as much skill when preparing cold dishes as other chefs use when preparing recipes that require complex cooking.

## Chapter Objectives:

1. Describe the five basic types of salad.
2. Identity common varieties of lettuce, chicory, and other greens.
3. Identify common herbs and edible flowers often added to salads.
4. Demonstrate the procedure for washing salad greens, for removing the core from a head of lettuce, for preparing romaine lettuce, and for removing the rib from loosely packed greens.
5. Describe common guidelines for storing greens.
6. Demonstrate the procedure for preparing a basic French vinaigrette, an emulsified vinaigrette, and a mayonnaise.
7. Identify the six presentation categories of salads.
8. Demonstrate how to prepare a tossed salad.
9. Contrast the six different categories of cheese.
10. Describe the guidelines for preparing, cooking with, and storing cheese.
11. Demonstrate how to glaze food with aspic.
12. Describe methods of presenting cold foods.

## Key Terms

- **salad greens**
- **emulsion**
- **tossed salad**
- **composed salad**
- **bound salad**
- **gelatin salad**
- **cheese**
- **soft cheese**
- **semi-soft cheese**
- **blue-veined cheese**
- **hard cheese**
- **crudites**
- **chaud froid platter**
- **aspic**
- **charcuterie**
- **pâté**
- **terrine**
- **galantine**
- **ballotine**

## SALADS

Salads are a popular menu item, with limitless possible combinations of flavors and ingredients. Salads can contain raw or partially cooked ingredients and can be served hot or cold. They do not necessarily contain lettuce, although many do. Salads typically are served with a salad dressing, which can be sweet, savory, or a combination of both. Salads can be served as an appetizer, a main course, a side salad, a separate course, or a dessert, depending on the ingredients used. ***See Figure 12-1.***

## Salads

Appetizer Salad

Main Course Salad

*Entourge*

**12-1** *Salads can be categorized according to when they are served in a meal.*

An *appetizer salad* is a salad that is served as a starter to a meal. Appetizer salads are the most basic type of salad, often featuring common varieties of lettuce such as iceberg or romaine, tomato wedges, cucumber slices, croutons, and dressing. Upscale restaurants often serve more sophisticated salads as appetizers.

A *main course salad* is a fairly complete and balanced dish containing protein (such as meat, seafood, eggs, or legumes) and an assortment of vegetables. The portion of a main course salad should be as large as that of an entrée so that the salad adequately satisfies the appetite. A Caesar salad with grilled chicken is a common main course salad. Main course salads are popular in North America but are less common in Europe.

A *side salad* is a small salad served to accompany a main course. If a main course is hearty or heavy, an accompanying side salad is typically light and refreshing. If a main course is light, a heavier side salad such as a pasta, grain, or potato salad is typically served. Sometimes a very small amount of salad is served as a garnish for an entrée.

A *separate course salad* is a salad served as a course of its own. Separate course salads are common in Europe, where a salad course may follow a main course. The purpose of a separate course salad is to refresh and cleanse the palate prior to dessert. These salads are usually light and are often served with a vinaigrette dressing.

A *dessert salad* is a sweet salad usually consisting of nuts, fruits, and sweeter vegetables such as carrots, and may be bound by gelatin. Dessert salads are usually served with yogurt, whipped cream, or a citrus-flavored dressing.

## Salad Nutrition

Salad greens are healthy, low-calorie, low-fat ingredients that are high in vitamin A, vitamin C, iron, and fiber. However, when meat, cheese, croutons, salad dressing, or other garnishes are added, a low-fat salad quickly becomes a high-calorie, high-fat course. To prepare a low-calorie, low-fat salad, the amounts of fat-containing ingredients used must be limited. Choosing a fat-free or low-fat salad dressing can help limit the amount of fat and calories in a salad.

## SALAD GREENS

*Salad greens* are a category of vegetables consisting of edible leaves, so called because they are used almost exclusively in raw salads or as a garnish. Although the name suggests that these are green vegetables, salad greens can be red, yellow, white, and even rusty brown. Two major categories of salad greens are lettuce and chicory, in addition to several other types of edible greens.

## Lettuce Varieties

Lettuce varieties of salad greens are tender and fragile, and should be kept covered when refrigerated. The most common varieties of lettuce are butterhead, iceberg, romaine, and leaf. ***See Figure 12-2.*** Blends of different lettuce varieties are also popular and available premixed and prewashed. Blends of young field greens such as mesclun and micro-greens are also popular.

**Butterhead Lettuce.** Butterhead lettuce has pale green leaves with a sweet, buttery flavor, tender texture, and delicate structure. Before use, the leaves must be thoroughly washed, and blemished leaves or spots must be removed by clipping with salad shears or scissors. Butterhead lettuce may be mixed with other greens or it may be served alone with an appropriate dressing. The natural shape of the leaf forms a shallow bowl that can be used as an edible serving dish for individual portion-size salads or single-portion foods. Butterhead lettuce can also be used as a lettuce wrap in place of a tortilla or slice of bread for a low-carbohydrate meal.

Chef's Tip:

No matter what type of salad is served with a meal, it should not contain the main ingredient used in the entrée. For example, if an entrée is a chicken dish, any salad served during the same meal should not contain chicken.

## Lettuce Varieties

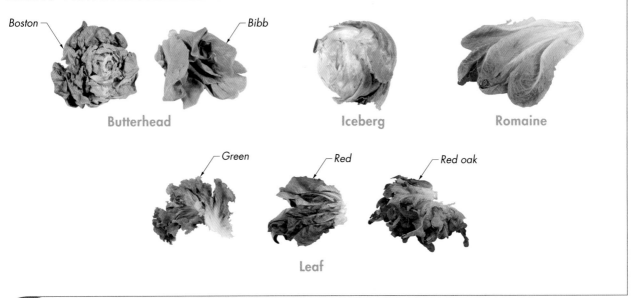

Boston

Bibb

**Butterhead**

**Iceberg**

**Romaine**

Green

Red

Red oak

**Leaf**

**12-2** *Common types of lettuce include butterhead, iceberg, romaine, and leaf.*

Varieties of butterhead lettuce include Boston lettuce and Bibb lettuce. They are similar to each other in appearance and flavor, with the main difference being the size of the leaf. Boston lettuce (buttercup lettuce) grows in a loosely packed, round head. The outer leaves of Boston lettuce are light green and the inner leaves are light yellow. Care must be taken when cleaning and cutting Boston lettuce because the leaves are fragile and bruise easily. For the same reason, it does not ship well and must be grown close to market. Bibb lettuce (limestone lettuce) is similar to Boston lettuce, in that it has a loosely packed, round head of a similar size, but the leaves are darker green, crisper, and smaller than those of Boston lettuce.

**Iceberg Lettuce.** Iceberg lettuce is the most common salad green. It has a round, compact head, pale green leaves, crisp texture, mild flavor, and excellent keeping qualities. It retains crispness even after being cut or processed. It usually only requires a light washing, because the head is so compact that dirt is unable to penetrate the leaves. After washing, only the core and a few of the darker, limp outside leaves are removed before use.

**Romaine Lettuce.** Romaine lettuce has long, fairly dark green leaves that grow in a loosely packed, elongated head. The leaves are darker in color near the edges and lighter in color near the thick center rib. Romaine lettuce has a mild, sweet flavor and blends extremely well with other greens. Because of its loose head, dirt collects in the ridges of the leaves during growth, so it must be washed thoroughly before use. Romaine lettuce does not bruise easily when cut.

**Leaf Lettuce.** Leaf lettuce has a rich color, soft leaves, and a very mild flavor. It grows in a loose bunch, which makes it easy to wash and clean but also makes it prone to damage and bruising. Red and green varieties of leaf lettuce are readily available. It is often used on a sandwich to add color and separate the meat or cheese from the bread. It should be kept covered in the refrigerator until ready to use.

**Mesclun.** *Mesclun* (spring mix) is a blend of young salad greens. These greens are baby versions of the varieties just mentioned and are more tender and milder tasting. The petite leaves are perfect for delicate, individually plated salads.

**Microgreens.** *Microgreens* are very small immature greens that are tender and flavorful. Microgreens are the first sprouting leaves of an edible plant. In addition to use in salads, microgreens are commonly used in hors d'oeuvres. Common varieties of microgreens include greens from beets, spinach, kale, and turnips. Chefs often use them as a garnish on elegant dishes.

## Chicory

*Chicory* is a hearty, flavorful, and slightly bitter variety of leafy vegetable. It comes in many colors from various shades of green to yellow, red, or rusty brown. Chicory also comes in different shapes from thick, torpedo-shaped bulbs to long, thin, weedlike leaves. Common varieties of chicory include Belgian endive, curly endive, escarole, radicchio, and Chinese cabbage. ***See Figure 12-3.***

*Chicory*

Belgian Endive    Curly Endive

Escarole    Radicchio

12-3    *Chicory is a hearty, flavorful, and slightly bitter variety of leafy vegetable that comes in many colors and shapes.*

**Nutrition Note**

Chicory is a rich source of the flavonoid kaempferol. Research has shown that flavonoids, which are antioxidants, have profound effects of the body's ability to prevent cancer and even have antiaging effects on the skin.

**Endive.** Some forms of endive have weedlike leaves, while others have oblong, tightly packed heads. Endive has a slightly bitter flavor that is considered desirable by many chefs and is especially pleasant when paired with a flavorful vinaigrette.

Belgian endive has a slender, tightly packed, elongated head that forms a point. It is the sprouting head of a chicory root and is usually about 4″ to 6″ in length. The leaves are creamy white in the center and yellow, slightly green, or purple along the edges. Often the leaves are removed one by one to keep the shape of the leaf intact. Whole leaves can be filled with a spread or dip as hors d'oeuvres. The head can also be split in half lengthwise before it is cleaned and can be braised or grilled. A half head portion set on another contrasting green as a base makes an attractive salad. As the name suggests, the Belgian endive originated in Belgium, but it is now commonly grown in California as well.

Curly endive has curly, twisted, thin leaves that grow into a loose bunch. The leaves vary in color from dark green on the outer leaves to pale green or white in the center and base. Sometimes curly endive is simply called chicory. It is also called frisée (pronounced free-ZAY). Curly endive has a strong bitter flavor pairs well with strongly flavored cheese and acidic vinaigrettes. It can also be used effectively as a garnish.

Escarole (broad leaf endive) is dark green in color and has thick, broad leaves with an irregular shape that grow in a loose, fan-shaped bunch. It is similar in taste to curly endive but not quite as bitter. The base of the bunch can be bitter and tough, so it is often discarded. Escarole is almost always blended with sweeter, milder greens. It can also be braised or sautéed as a vegetable.

**Radicchio.** Radicchio has a small, compact head of red leaves, similar to a small head of red cabbage. Radicchio is usually mixed with other salad greens to add flavor and color. The leaves form a bowl shape, similar to those of butterhead lettuce to hold individual-size salads. Radicchio is also commonly sautéed or braised and served as a side vegetable.

**Chinese Cabbage.** Chinese cabbage, also know as celery cabbage or Napa cabbage, has an elongated head of coarse, tightly packed leaves, approximately 12″ long. It is similar in shape to romaine lettuce. The outer leaves are light green with a mild cabbage taste, while the inner leaves are almost entirely white with much less flavor. Its crisp texture enhances any mixed greens preparation. Chinese cabbage should be used in moderation when mixed with other greens.

## Other Greens

In addition to lettuce and chicory, other green leafy vegetables such as spinach, watercress, and arugula are commonly used in salads to add flavor, texture, and color. *See Figure 12-4.* These other greens are more intense in flavor than the traditional lettuce varieties so they are used sparingly and only added in small quantities to salads.

**Nutrition Note**

Chinese cabbage is an excellent source of calcium. Many people who are lactose intolerant suffer from calcium deficiency. Having a diet that contains Chinese cabbage can help a person, especially in one of these populations, have enough calcium for healthy bones and to fight off osteoporosis.

## Other Greens

Spinach

Watercress

Arugula

**12-4** *Greens used in salads also include leafy vegetables such as spinach, watercress, and arugula.*

**Spinach.** Spinach is a dark green leafy vegetable that can be served alone or blended with other greens. In preparation, the long tough stem at the base of each leaf must be removed. Then each leaf must be washed two or three times to remove dirt and grit that can collect in the ridge of the stem. Spinach leaves should be firm and crisp; yellowed or limp leaves should be discarded. Spinach salad is topped with hot bacon dressing, diced bacon, sliced mushrooms, and chopped hard-cooked eggs.

**Watercress.** Watercress has small, disk-shaped, dark green leaves that grow on thin stalks. The leaves are tender and fragile and should be carefully removed from the stem for use. Watercress has a slightly peppery taste, similar to that of a turnip, and has a tendency to stimulate the appetite. While it can be added to other mixed greens, it is most popular as a garnish for salad, steak, fruit, or soup preparations. Watercress also lends itself to puréeing in cream sauces and soups.

**Arugula.** Arugula has flat, oval leaves with frilled edges. It has a strong peppery flavor and is seldom served by itself, but can add zesty flavor when used in a salad or puréed into a cream sauce. Larger leaves are not used as they are tough and have an overly strong flavor.

**Mâche.** Mâche (lamb's lettuce) has small, dark green, slightly curled leaves. It is tender and mild in flavor and pairs well with mild-flavored vinaigrettes.

**Dandelion Greens.** Dandelion greens grow wild, but due to their popularity they are now also cultivated. The cultivated greens are more tender and milder in taste than ones grown in the wild. The greens are fairly smooth but have a slightly rough, irregular edge. Wild dandelions are tasty until the yellow flower blooms, and then they become bitter and tough. This green is an early spring favorite that can now be found year-round. Dandelion greens can be blended with other salad greens or can be served alone.

**Kale.** Kale has curly leaves found in a loose bunch and is a member of the cabbage family. There are several varieties of kale, with leaves in shades of pale yellow, green, pink, and purple. All varieties of kale are edible but the green varieties are best suited for cooking and salads, while the other colors are best for garnishing.

## Herbs and Specialty Items

Other items such as fresh herbs and edible flowers can be added to salads for an interesting flavor and colorful appearance. ***See Figure 12-5.*** The leaves of fresh herbs such as basil, cilantro, chives, dill, mint, oregano, parsley, sage, and savory can be included in fresh salads to add flavor, but should not be overpowering. Small leaves can be added whole, but larger leaves should be torn or cut chiffonade. Stems should be discarded.

## Herbs and Specialty Items

Oregano
Sage
Rosemary
Rose

**12-5** *Herbs and edible flowers can be added to salads for additional flavor, aroma, and color.*

Edible flowers can add flavor and aroma as well as a colorful visual appeal to salads. Only flowers grown without the use of pesticide by growers that specialize in edible flowers should be used. The most common varieties of edible flowers include roses, nasturtiums, pansies, primroses, and violets. Flowering herbs such as chive blossoms, oregano flowers, and thyme flowers can also be used.

Sprouts have become a popular leafy vegetable on sandwiches, in salads, and as a garnish on prepared foods. Sprouts have a fresh flavor that can range from slightly nutty to slightly spicy. Common types of sprouts include chickpea, daikon, green lentil, and mustard.

## PREPARING AND STORING SALAD GREENS

Lettuce and other salad greens grow close to the ground and have a tendency to be dirty or gritty from soil. Even if salad greens look clean, it is important to keep in mind that insects, dust, dirt, pesticides, and fertilizers may be hidden between the leaves. Therefore, all salad greens must be washed, drained, trimmed, and properly stored before use. If prepared or stored incorrectly, greens can bruise, brown (rust), and lose crispness.

Proper washing of salad greens is only one part of safe salad production. A person preparing salads must also practice proper sanitation habits. Because greens are not cooked when making a salad, it is especially important that hands be washed thoroughly. Most local health department codes require employees to wear single-use gloves when preparing salads and to change the gloves frequently during the course of production. Many documented cases of mass contamination have occurred because an employee did not sufficiently wash hands or effectively use single-use gloves when preparing salads.

### Washing Salad Greens

To wash salad greens, they should first be cut or torn to desired size, and then completely submerged in a sink of clean, cold water. The greens are gently stirred to rinse away dirt and then removed from the dirty water. The sink is then rinsed out completely and refilled with clean, cold water. The process is repeated two or three times until the water in the sink remains clean, with no visible dirt or sediment. Loosely packed greens should be separated and each leaf washed if the leaves are to be kept whole. The task must be done gently because tender greens can bruise easily. It may even be necessary to cut some of the elongated heads in half lengthwise in order to remove all dirt, grit, and sand.

### To wash salad greens, apply the following procedure:

1. Cut or tear greens to appropriate size. **See Figure 12-6.**
2. Fill a clean and sanitized sink with cold water.
3. Submerge greens completely in the cold water.
4. Gently stir greens to rinse. Remove greens from the dirty water prior to draining the sink.
5. Rinse sink out completely and refill with cold water. Repeat steps 3 through 5 until no dirt is felt on the bottom of the sink.
6. Remove greens from water and spin in a salad spinner to dry.
7. Store in a pan with a perforated pan insert and cover with a damp paper towel or plastic wrap until needed.

CULINARY PROCEDURES

---

### Action Lab

Take a head of lettuce and cut it in half. With one half of the head, use a knife to cut the lettuce into bite-size pieces. With the other half, tear the lettuce into pieces of similar size. Let the lettuce sit at room temperature for 30 min, then observe the difference between the cut and torn lettuce. Is the cut lettuce less bruised than the torn lettuce? Many chefs believe that cutting greens bruises them. However, tearing greens takes longer, costing an establishment time and money. Whole leaves from greens such as butterhead lettuce or radicchio are usually torn free from the head by hand. All other greens are prepared more efficiently with a sharp knife.

## Washing Salad Greens

1–2. Cut greens to desired size and fill a clean and sanitized sink with cold water.

3–5. Submerge in cold water and gently stir to remove dirt; drain, rinse, and refill sink until no dirt remains at bottom of sink.

6–7. Dry in a salad spinner; store in a pan with perforated inset and cover.

**12-6** Salad greens are rinsed in clean, cold water until dirt no longer collects at the bottom of the sink.

*Chef's Tip:*

Salad greens should not be washed until the day they are needed for service. Washing salad greens speeds the rate of spoilage and wilting. Washing greens too early can result in wilting before preparation.

All salad greens must be dried completely after washing. Wet greens become limp in a short amount of time, whereas dried greens remain crisp for a longer period of time. Also, oil-based vinaigrettes or mayonnaise-based dressings do not stick to wet leaves. Salad dressing clings best to the surface of dry leaves. Using a salad spinner is the best way to dry lettuce, as the spinning of the internal basket (centrifugal force) throws water from the leaves without damaging them. ***See Figure 12-7.***

## Salad Spinner

**12-7** A salad spinner dries leafy greens using centrifugal force.

## Preparation

Solidly packed head lettuces such as iceberg lettuce must have the core removed and be cut to appropriate size before washing and draining the leaves. If the leaves are not cut to appropriate size, it is nearly impossible for soaking water to reach the center of the tightly packed heads to effectively remove pests, dirt, or chemicals.

### To remove the core from a solidly packed head of lettuce, apply the following procedure:

1. Hold the head of lettuce in the palm of one hand with the core facing down. ***See Figure 12-8.***
2. Carefully hit the core end of the head against a clean work surface to free the core from the rest of the head.
3. Grab the core and pull to remove it from the head.

### Removing the Core from Head Lettuce

1–2. Hold lettuce with core facing down; hit core against clean work surface.

3. Grab core and pull to remove.

**12-8** *The core is removed from head lettuce before washing and cutting the lettuce.*

### To prepare romaine lettuce, apply the following procedure:

1. Trim any damaged leaf tips with a stainless steel French knife and remove any discolored or bruised outer leaves. ***See Figure 12-9.***
2. Make two or three lengthwise cuts along the head with a French knife, leaving the base intact to hold the leaves.
3. Cut crosswise at 1″ to 1½″ intervals to produce bite-size pieces.
4. Wash, dry, and store leaves according to standard procedure.

## Preparing Romaine Lettuce

1. Trim damaged tips and remove any bruised outer leaves.

2. Make 2 or 3 lengthwise cuts.

3–4. Make crosswise cuts at 1" to 1½" intervals; clean in cold water; spin dry; store in a covered container.

**12-9**  A head of romaine lettuce is cut lengthwise and crosswise into pieces of desired size.

Loosely packed greens such as spinach collect dirt in the center rib of the stem. The rib of loosely packed greens must be removed prior to preparation.

### To remove the rib from loosely packed greens, apply the following procedure:

1. Fold each leaf in half so that the two sides of the top surface meet, exposing the bottom of the leaf. **See Figure 12-10.**
2. With the other hand, pinch the thick stem and rib and carefully pull from the leaf to remove and discard.
3. Cut, wash, dry, and store leaves according to standard procedures.

*CULINARY PROCEDURES*

## Cleaning Greens

Remove rib

Fold leaf in half

Remove stem

**12-10**  Remove the center rib of loose-packed greens before washing.

### Storage

All greens should be stored in their original packaging until needed. Delicate greens such as butterhead lettuces deteriorate faster than heartier or firmer varieties such as romaine. All washed greens should be drained thoroughly and stored in a perforated stainless steel pan with a second solid pan as an underliner to hold the drippings. Washed greens should be covered with a damp paper towel or plastic wrap to retain crispness. Crispness may be improved by adding slices of lemon to cold water and letting the greens soak for a short period. When removed from this solution, the greens should be drained again thoroughly.

Greens and other delicate produce must be stored away from tomatoes, apples, pears, and other fruits that emit ethylene gas. Storing salad greens near these items can cause the greens to wilt and deteriorate rapidly. Greens are best stored at a temperature between 35°F and 38°F, which is colder than the storage temperature of other produce (40°F and 48°F). *Note:* If any part of a green is frozen, the appearance and texture are damaged and the green must be discarded.

## ADDITIONAL SALAD INGREDIENTS

The combination of ingredients that can be used in a salad is bound only by a chef's imagination. However, it is important to remember that a salad should complement other courses. The most common ingredients used in salads are vegetables, proteins, starches, fruits, and nuts. ***See Figure 12-11.***

*Salad Ingredients*

Peaches — Beets — Black beans — Rice — Orzo pasta

Walnuts — Daniel NYC — Mandarin orange — Pine nuts — Crawfish

**12-11** *In addition to leafy greens, salads may include vegetables, proteins, starches, fruits, and nuts.*

## Vegetables

Vegetables used in a salad can be either raw or cooked depending on their original texture. Harder vegetables such as cauliflower need to be blanched prior to use in a salad, while softer vegetables such as tomatoes can be added raw. An entire salad can be made from a single type of vegetable such as a coleslaw made from cabbage or pickled beet salad made from blanched beets. Some common examples of cooked vegetable salads (served cold) include potato salad, beet salad, and roasted pepper and artichoke salad. Some common raw vegetable salads include cucumber salad and tomato and red onion salad.

## Proteins

Any sort of protein can be used in a salad as a main ingredient or a garnish. Common types of protein used in salads include cooked poultry, meats, seafood, nuts, and cheese. A protein ingredient can be used as the main body of a salad, such as chicken salad. Protein ingredients can also be used to make a salad hearty and more flavorful. When adding a protein to a salad, it is important to practice food safety and sanitation.

## Starches

Starches such as breads, grains, legumes, and pasta can add variety, texture, and nutritional value to many salads. Breads are often used as a salad garnish, such as toasted croutons. Bread can also be an ingredient in a tossed salad such as in panzanella, or bread salad. In panzanella, crusty bread that is toasted and torn is tossed with tomatoes, basil, other desired ingredients, and a red wine vinaigrette.

Grain and legumes are nutritious additions to a salad and can also serve as a main ingredient. With people becoming more health conscious, salads with whole grains, low-fat proteins, and legumes are gaining in popularity. Grains to be used in a salad should be cooked slightly less than al dente so that they retain shape and have a slight bite. Overcooked grains become mushy as they absorb dressing. Grains such as wheat berries, barley, and all forms of whole grain are well suited for salad preparations. Because they do not soften with the addition of dressing, beans and other legumes to be used in salad must be cooked until fairly soft.

Pasta salads are a popular starch-based salad. They can be as traditional as a rotini pasta with basic French vinaigrette to as contemporary as a couscous salad with pomegranate, mint, and curry. Pasta used in a pasta salad should be slightly undercooked so that it does not become soggy or break apart when stirring. Pasta is absorbent after it is cooked, so it is important to ensure that the flavor of the dressing is still present after the salad has been allowed to rest.

*Wisconsin Milk Marketing Board*

## Fruits and Nuts

When used in a salad, fruits can be fresh, dried, or roasted, and nuts can be salted, glazed, or seasoned and toasted. Fruits and nuts add textures, flavors, and aromas to a salad. Depending on the fruit that is added and how it was prepared, it can add acidic qualities, sweetness, chewiness, or crispness to a salad. Nuts, depending on how they are seasoned and prepared, can also add sweetness, saltiness, spiciness, or cultural flavors to a salad. A complementary mixture of fruit and nuts can take a basic green salad to a new level, such as by adding dried cherries, fresh pears, and maple-glazed pecans to a bleu cheese and mesclun salad with a balsamic vinaigrette.

## SALAD DRESSINGS

A salad dressing is considered the sauce for a salad. A well-prepared salad can be ruined by a low-quality dressing, because in most cases the flavor of the dressing is the first flavor that is tasted. Therefore, a dressing must be prepared with the finest ingredients available and the utmost care.

Many different salad dressings can be prepared from scratch or purchased prepared. The cost of a prepared salad dressing may be higher than preparing it from scratch. Most dressings are prepared from a vinaigrette or mayonnaise base.

An *emulsion* is a mixture of two typically unmixable liquids (such as oil and water) that are forced to bond with each other to result in a creamy, smooth product with a uniform appearance. ***See Figure 12-12.*** A *temporary emulsion* is a temporary mixture of two typically unmixable liquids that eventually separate into their original state when allowed to rest. Whisking oil and vinegar together creates a temporary emulsion in a basic French vinaigrette. As the oil and vinegar are whisked together rapidly, the oil is broken into tiny droplets that are then surrounded with a coating of vinegar. If the mixture of oil and vinegar is allowed to rest, it eventually separates back to a pool of oil floating on a pool of vinegar.

A *permanent emulsion* is a mixture of two typically unmixable liquids forced to bond permanently with the aid of an emulsifier, such as egg yolk. Mayonnaise is an example of a permanent emulsion, formed by dripping oil into a small amount of vinegar and yolks while whisking rapidly and continuously. The oil naturally repels from water in the vinegar, but the egg yolk acts as an emulsifier, preventing the separation of the two opposing liquids. Each yolk can emulsify up to a maximum of 7 fl oz of oil. If more oil is added, the excess oil separates out from the vinegar, breaking the emulsion. If the mixture is not whisked rapidly and continuously, or if the oil is not added very slowly to the vinegar, the emulsion will break, resulting in separation of the vinegar and oil.

### Vinegar and Oil Selection

No matter what the principal ingredients, flavorings, or seasonings, all dressings are produced in the same manner. It is important to match a vinegar to an appropriate oil. Some oils, such as olive oil, have stronger flavors and are better paired with a more intense vinegar, such as a balsamic vinegar. Other oils, such as canola or safflower, are lighter or more neutral in flavor and are better paired with a sweeter, more delicate vinegar, such as an herbal or fruit vinegar. Nut oils work best with sherry or champagne vinegar because the crispness of the vinegar brings out the earthiness of the nut oil.

Emulsion

*Oil*

*Vinegar and egg yolk mixture*

**12-12** *An emulsion is a uniform mixture of two typically unmixable liquids, such as oil and vinegar.*

*Chef's Tip:*

When preparing an emulsion, ingredients should sit at room temperature for 1 hr prior to preparation. Ingredients at room temperature are much easier to emulsify than cold ingredients.

### Vinaigrettes

A basic French vinaigrette is commonly referred to as simply "basic French" in the food service industry, and should not be confused with the orange-red dressing known as French or Russian. A basic French vinaigrette is a combination of two main ingredients, oil and vinegar, with the addition of flavorings and seasonings.

In a basic French vinaigrette, the ratio of oil to vinegar is 3:1, or three parts oil to one part vinegar. When using intensely flavored oil, less than three parts oil is recommended to balance the flavors of vinegar and oil. When preparing a citrus vinaigrette, where half or more of the vinegar is replaced with a citrus juice, less than three parts oil is needed. Also, if mildly flavored vinegar such as an aged balsamic is used, less than three parts oil is needed. Because a basic French vinaigrette is a temporary emulsion, it must be stirred thoroughly or shaken again before dressing a salad or being served separately. The acid in a vinaigrette can balance the fattiness of some dishes and can liven up the overall flavor of a dish. A vinaigrette can be served over broiled or grilled fish, seafood, vegetables, and poultry. *See Figure 12-13.*

### Vinaigrette Dressing

Vinegar

Oil

Alaska Seafood Marketing Institute

**12-13** *A basic French vinaigrette is a temporary emulsion of oil and vinegar combined with seasonings and flavorings that can be used to season entrées as well as salads.*

Flavors, seasonings, and other ingredients can be added to a basic French vinaigrette or emulsified vinaigrette to produce flavored dressings. Minced garlic, shallots, Dijon mustard, and sugar are common additions to vinaigrettes. Some ingredients, such as mustard and paprika, help to stabilize the emulsion. Garlic and shallots are used to add a savory flavor, mustard adds a touch of zest, and sugar mellows out the sharpness of vinegar. Additionally, ingredients such as fresh herbs, unique spices, or acidic fruit juices can be added to further intensify flavor. *Note:* Fresh herbs intensify the flavor of a dressing quickly. Dried herbs should be warmed slightly in a sauté pan to reactivate essential oils before use in a dressing.

## To prepare a basic French vinaigrette, apply the following procedure:

1. Carefully select a vinegar (or other acid) and an oil that will pair well with each other and that complement the item they are intended to dress. ***See Figure 12-14.***
2. Place seasonings, any flavoring ingredients, and vinegar in a mixing bowl and whisk to incorporate.
3. Slowly add the oil in a fine stream while whisking continuously to form a temporary emulsion.
4. Let vinaigrette rest an hour or two to marry the flavors.
5. Whisk as needed to incorporate oil and vinegar prior to use.

## Preparing Basic French Vinaigrettes

1–2. Select vinegar (or other acid) and oil; whisk in seasonings and flavorings.

3–5. Slowly add oil while whisking continuously; let rest; whisk again prior to serving.

**12-14** *Seasonings and flavorings are added to the vinegar (or other acid) before oil is incorporated.*

To prepare an emulsified vinaigrette, apply the following procedure:

1. Collect all ingredients and allow to sit at room temperature for 1 hr prior to preparation. **See Figure 12-15.**

2. Place pasteurized yolks or whole eggs in a mixing bowl and whisk until foamy and light.

3. Add all dry ingredients and any flavoring ingredients to egg and whisk to incorporate.

4. Add about one-fourth of the liquid ingredients and whisk well to incorporate.

5. While whisking rapidly, carefully begin adding the oil very slowly in a fine stream until a smooth emulsion forms.

6. Continue to add the oil slowly in a fine stream until it begins to get thicker. Then begin adding a little more of the liquid, alternating with the oil until all of the ingredients are incorporated and emulsified. *Note:* It is important that the dressing is whisked rapidly the entire time so that the oil and vinegar are not allowed to separate.

7. Check the consistency and seasonings. It should be smooth and well seasoned. If it is too thick, a little water can be whisked in to correct the consistency.

*Preparing Emulsified Vinaigrettes*

1–3. Bring all ingredients to room temperature; whisk eggs or yolks until foamy; add all dry ingredients and flavoring ingredients; whisk well.

4–5. Add ¼ of liquid and whisk well; slowly add oil while whisking until thick.

6–7. Alternate adding more liquid and oil until incorporated; check consistency and seasonings.

**12-15** Egg yolks are used to create a permanent emulsion in an emulsified vinaigrette.

## Mayonnaise

Mayonnaise is an important dressing because it is the foundation for many popular dressings, such as thousand island, green goddess, and bleu cheese. Mayonnaise is a prepared by forming a permanent emulsion between oil and egg yolks. The quality of a mayonnaise depends on the quality of oil used. It is usually best to prepare mayonnaise with oil that is fairly neutral in flavor, such as corn, canola, or cottonseed oil. Olive oil is not recommended because it has a stronger flavor than many other oils and might limit the use of the mayonnaise as a base for preparing other dressings. If olive oil is used, it is best to blend it with another oil to reduce the strong flavor.

### To prepare a mayonnaise, apply the following procedure:

1. Collect all ingredients and allow to sit at room temperature for 1 hr prior to preparation. *See Figure 12-16.*
2. In a stainless steel or other nonreactive mixing bowl, whisk seasonings and half of the vinegar into the yolks until light-colored, airy, and foamy.
3. Add oil very slowly (a drop at a time), until the emulsion begins to form and the mixture thickens slightly.
4. Once about ¼ of the oil has been added, increase addition of the oil, adding the oil in a fine stream while whisking rapidly and continuously to incorporate.
5. As the mixture becomes thick and hard to whisk, add a small amount of vinegar to thin slightly.
6. Continue alternating vinegar and oil while whisking until all the oil and vinegar have been added.
7. Check and adjust seasonings as necessary.
8. Transfer the mayonnaise into a chilled storage container and refrigerate until needed.

CULINARY PROCEDURES

## SALAD VARIETIES

The term "leafy green salad" can refer to any salad made with a base of greens. Tossed salads, composed salads, and bound salads can all be types of leafy green salads. Leafy green salads can be as basic as iceberg lettuce, a wedge of tomato, and grated carrots topped with a choice of salad dressing, or as complex as a mixed baby field greens with diced pears, Gorgonzola cheese, glazed pecans, and a balsamic vinaigrette.

## Preparing Mayonnaise

1–2. Bring all ingredients to room temperature; whisk together seasonings and half the vinegar into yolks until foamy.

3. While whisking, begin adding oil a drop at a time until mixture begins to thicken.

4. When half the oil has been added, increase to a fine stream.

5–8. If mixture gets too thick, add a bit of vinegar; alternate vinegar and oil until incorporated; adjust seasonings; refrigerate.

**12-16** *Mayonnaise is the foundation for many popular salad dressings.*

All salads fall into one of six presentation categories. The style of a salad varies depending on how the salad is to be plated or served. The six presentation categories are tossed, composed, bound, vegetable, fruit, and gelatin.

### Tossed Salads

A *tossed salad* is a mixture of leafy greens such as lettuce, chicory, spinach, or fresh herbs other items such as meats, vegetables, fruits, nuts, cheese, and/or croutons, and served with a dressing. There are endless combinations of ingredients that can be used in constructing tossed salads, but care should be taken to use ingredients with complementary flavors and textures. A dressing should be chosen that does not overpower or conflict with the ingredients. ***See Figure 12-17.***

## Tossed Salad

*Florida Tomato Committee*

**12-17** *In a tossed salad, leafy greens are tossed with vegetables and a dressing.*

To make a tossed salad, apply the following procedure:

1. Select a few varieties of leafy greens so the salad will be colorful and have a variety of textures and flavors. *See Figure 12-18.*

2. Cut, wash, dry, and store greens according to the standard procedure until ready for use.

3. Prepare any garnishes or additional ingredients that will be added to the salad. Store properly until needed.

4. Prepare the appropriate salad dressing, remembering that hearty greens can pair well with lighter or heavier dressings, while delicate greens require a lighter dressing.

5. Carefully combine the greens, additional ingredients, and dressing in a mixing bowl and toss gently to coat distribute all components evenly.

6. Serve the appropriate portion size as quickly as possible, as dressing speeds up wilting of the greens.

CULINARY PROCEDURES

## Preparing Tossed Salads

*1–2. Select greens; cut, wash, and dry properly.*

*3. Prepare garnish or other ingredients.*

*4. Prepare salad dressing.*

*5–6. Toss all ingredients to gently coat with dressing; serve immediately.*

**Chef's Tip:**

When pairing leafy greens with a dressing, a common rule of thumb is that delicate or fragile leaves should be combined with a lighter, thinner dressing. A heavy dressing can crush delicate greens and mask delicate flavors and textures. Heartier, crisper greens such as romaine or iceberg lettuce work well with any type of dressing.

**12-18**  *Tossed salads should be prepared just prior to serving.*

## Composed Salads

A *composed salad* is a salad consisting of a base, body, garnish, and dressing carefully composed and arranged attractively on a plate. ***See Figure 12-19.*** The base of a composed salad serves as the foundation on which a salad is built. The base is not only an edible component of the salad, but a colorful liner between the rest of the salad and the plate. A base typically is a bed of salad greens, fruit, or vegetables. Often a base is a cup-shaped leaf in which the rest of the ingredients can be placed, resulting in a beautiful presentation.

## Composed Salads

Coconut garnish

Watermelon coulis garnish

Baby greens and watercress body

Dressing

Watermelon base

Charlie Trotter's

**12-19** Composed salads have a base, body, garnish, and dressing.

The body of a composed salad consists of the main ingredients, such as tossed salad greens, pasta, cooked grains, vegetables, a fruit medley, or precooked proteins such as chicken salad. The garnish for a composed salad is similar to the garnish for a hot food plate. The purpose of a garnish is to add color, flavor, and texture to a dish. It serves as an edible decorative ingredient to a salad. It may be as simple as a tomato wedge or a fanned strawberry or as substantial as a sliced hard-cooked egg or slices of grilled meat, poultry, or seafood.

A dressing should work to bring all the flavors, textures, and components of the salad together. It should complement the ingredients used and not be so heavy or overpowering in flavor that it masks other flavors in the salad. Dressing can be applied to a composed salad by ladling it on top of the salad, by tossing the salad in the dressing, or by spraying the dressing on the salad, depending on the ingredients and the desired presentation.

Applying dressing to a salad with a ladle after it is plated is common in family-style restaurants and at some banquets. When applying dressing with a ladle, the dressing is not equally distributed throughout the salad. Some parts of the salad may have too much dressing, while the ingredients in the center and base of the body of the salad may not have any dressing.

Salads are sometimes sprayed with a dressing in large-scale banquet operations. Using this method, salads are composed on sheet pans prior to the function instead of directly on plates, as plated salads take up a great deal of room. Then a light dressing such as a vinaigrette is placed in spray bottles. The bottles are shaken to mix the dressing, and the salads are sprayed with a mist of vinaigrette before being lifted from the sheet pan onto individual plates.

## Bound Salads

A *bound salad* is a combination of a main item, flavoring ingredients, and seasonings held together with a binding agent. ***See Figure 12-20.*** Many types of ingredients can be used for the main item in a bound salad including cooked meats, seafood, poultry, hard-cooked eggs, potatoes, pastas, grains, vegetables, fruits, or legumes. Minced chives, onions, and celery are often used as flavoring ingredients, and mayonnaise is used most often as the binding agent. Other binding agents include vinaigrettes, yogurt, and creamy dressings. Popular bound salads include chicken salad, tuna salad, potato salad, and kidney bean salad. Bound salads can be served in many forms including the following:

*Bound Salads*

*National Chicken Council*

**12-20** *A bound salad is composed of a main ingredient, flavoring ingredients, seasonings, and a binding agent.*

- the body of a composed salad, such as a tomato half stuffed with crab salad
- a sandwich filling, such as tuna salad served on a toasted bun
- a side dish such as potato salad

Some types of bound salads are primarily served as a side dish, while other types are primarily served as sandwich filling. For example, chicken salad is often served as a sandwich filling, but is not typically served as a side dish. A kidney bean salad is often served as side dish but is not served as a sandwich filling.

*The following tips are helpful when preparing bound salads:*

- Items in a bound salad should complement each other while letting the main ingredient stand out.
- Chilled ingredients should be used so that a salad is not in the temperature danger zone during or after assembly.
- Cooked items should be prepared in advance and chilled completely before use (unless the item is to be served hot, such as hot German potato salad).
- Ingredients should be cut into consistent sizes for a pleasing finished appearance.
- Ingredients should be evenly combined with dressing to bind the salad together and distribute flavor evenly throughout.
- All ingredients should be mixed with the dressing as close to the time of service as needed to ensure safe food handling.

## Vegetable Salads

**12-21** *Vegetable salads can be made with raw or cooked vegetables.*

## Fruit Salads

**12-22** *Fresh fruit salads should be prepared just prior to service to ensure freshness.*

## Vegetable Salads

Vegetable salads are salads that are primarily composed of vegetables. Vegetables used in vegetable salads may be either precooked or raw. Beet salad, cauliflower salad, and green bean salads are examples of cooked vegetable salads where the vegetables are blanched in heavily salted water, refreshed in ice water, and drained well prior to use. Raw vegetable salads are popular because of the freshness of the ingredients. Coleslaw, cucumber and onion, and carrot and raisin salads are examples of raw vegetable salads. ***See Figure 12-21.***

When making vegetable salads, the structure of the vegetable is important for overall quality. Green vegetables can turn yellow and look unappealing if left in an acidic solution such as a vinaigrette for too long a time. Therefore, when making a salad with mushrooms, zucchini, red peppers, and asparagus served with a vinaigrette, for example, it is best to leave asparagus out of the salad until just prior to serving.

Vegetables to be used in a vegetable salad must be washed and trimmed to the desired size. If the vegetables are to be blanched, grilled, or cooked in any other manner, it should be done prior to assembling the salad so the vegetables are completely cooled prior to use. For presentation, the vegetables can be combined with dressing before or after plating, or the dressing may be served on the side.

## Fruit Salads

Fruit salads can be prepared from fresh, canned, or frozen fruits. ***See Figure 12-22.*** Some fruit salads can be mixed and tossed with success, but most depend on arranging the fruit in an attractive manner, as in a composed salad. Fruit salads are fragile, can discolor rapidly, and become soft when fruit is cut and exposed to air too far in advance. They should be prepared as close to serving time as possible and served well chilled.

Some fresh fruits, such as apples, avocados, bananas, and pears discolor when cut and exposed to air. To prevent rapid discoloration and to improve flavor, these fruits should be dipped or lightly tossed in liquids that contain citric acid, such as lemon, orange, pineapple, or lime juice. Using a nonreactive stainless steel knife helps to prevent discoloration.

Fresh fruits, like fresh vegetables, are superior to canned or frozen products in taste and texture. However, professional kitchens may rely on the convenience of canned or frozen products in order to meet production demand. Canned fruit should always be stored in a cool, dry place. For best results, canned fruit should be refrigerated overnight prior to use in a salad. Once a can is open, unused contents should be stored in an approved plastic storage container.

If a dressing is to be used with a fruit salad, it is typically a lighter dressing made with yogurt, whipped cream with a fruit purée, or honey. In place of a dressing, fruit salads can be tossed with a slight amount of fruit juice or fruit purée, a chiffonade of fresh mint, or a splash of a sweet liquor.

## Gelatin Salads

Another salad variety is the gelatin salad. A *gelatin salad* is a salad made from flavored gelatin formed in a mold. Gelatin salads can be presented in many different forms, offer a variety of colors, and are easy and inexpensive to prepare. ***See Figure 12-23.*** Cut-up fruit or vegetables or grated cheese can be mixed in and suspended in the gelatin before it sets. Fresh fruit floats and canned fruit sinks in gelatin.

In any gelatin preparation, the correct ratio of gelatin powder to liquid must be used. Fruit-flavored gelatin powder packaged for commercial use contains 1 lb 8 oz of gelatin powder. It is recommended that a total of 1 gal. of water or fruit juice be added: ½ gal. of boiling liquid to dissolve the gelatin, and ½ gal. of cold liquid to cool the mixture.

*Gelatin Salads*

*The following tips are helpful when preparing gelatin salads:*

- Mix the gelatin in a stainless steel container.
- Be sure all the gelatin is thoroughly dissolved in the hot liquid before the cold liquid is added.
- If speed is required in forming the gel, place ice or frozen fruit or juice in the cold liquid to accelerate the action.
- Gelatin sets more quickly at a cold temperature, so placing it in the coldest area of the refrigerator (lower area) produces a gel more quickly.
- Never attempt to speed the gel by placing it in the freezer. Freezing would cause the mixture to crystallize and upon defrosting the gelatin would become liquid.

When preparing a gelatin mold for display on a buffet, the volume of the selected mold must be able to hold the amount of prepared gelatin. To determine if a mold can hold the proper amount, the mold is filled with water. The water is then poured into a liquid measure to determine the volume of the mold. If the amount of water held by the mold equals no more or less than 1 cup from the recipe yield, the mold is the correct size for the recipe.

Many gelatin recipes call for whipped cream or beaten egg whites to be folded in. This procedure is used to give a fluffy, spongy texture to the gelatin preparation. Whipped cream or egg whites should be

**12-23** *A gelatin salad is made from flavored gelatin formed in a mold.*

*Chef's Tip:*

*Whipped products should be whipped just before use in a gelatin salad, never in advance. Cream whips best when utensils and cream are cold. Egg whites whip best when they are held at room temperature. All whipping utensils must be clean and free of oil or grease for best results.*

whipped at high speed using a wire whip or an electric mixer until soft peaks form and are stiff enough to hold their shape. If the whipped ingredient is under- or over-whipped, the texture of the finished product may be affected.

If mayonnaise, sour cream, whipped cream, beaten egg whites, fruits, or vegetables are to be added to the gelatin, the gelatin mixture must be chilled until slightly thickened before the additions are made. This ensures even distribution of the added ingredients.

When layering two or more gelatin mixtures in the same pan or mold, each layer poured into the pan or mold must be chilled until slightly firm before adding the next layer. This is done to prevent one layer from running into another. If one layer is set too firm, the layer placed on top may slip off or separate from the completed mold when the entire salad is unmolded. Before adding more gelatin to a layer that has already set, be sure the gelatin to be added is completely cool. If the gelatin being added is too warm, it can melt the set layer and the two mixtures will run together.

To unmold large molds, such as those used on a buffet service, a large sink is filled with about 6″ of warm water. The mold is dipped into the warm water up to about ½″ from the top of the mold, and tilted slightly from side to side until the gelatin mixture separates from the sides of the mold. A serving dish or tray is then placed upside-down over the top of the mold. The dish or tray and mold are then turned over and shaken slightly until the gelatin releases from the mold.

To unmold a small individual mold, the mold is held in the palm of one hand with the open end resting in the palm. The mold is then placed under warm running water for 1 sec to 2 sec. As soon as the mold is removed from the water, the top of the mold is tapped with the palm of the free hand. The gelatin preparation should drop out of the mold and into the palm of the hand. If the gelatin preparation does not drop out easily, it can be placed under warm running water a second time.

Chef's Tip:
If a gelatin layer becomes too firm, remove it from the refrigerator and let it set at room temperature for approximately 15 min or until it loses some of its firmness before adding the next layer.

## CHEESE

*Cheese* is a dairy product consisting of the coagulated (thickened), compressed, and usually ripened curd of milk that has been separated from the whey. However, some cheeses are made from the whey. *Curd* is the thick, casein-rich part of coagulated milk. *Whey* is the watery part of milk. There are many varieties of cheese made from cow's milk as well as milk from sheep, goats, and other animals. Most types of cheese used in the professional kitchen are made from cow's milk.

Cheese is most often made from milk that has been coagulated or curdled using either rennet (a substance that contains acid-producing enzymes) or an acid. After the milk curdles, it is separated into the cheese curd and whey. The flavors and textures of each cheese differ depending on what type of animal the milk came from, the diet of the animal, the percentage of butterfat in the cheese, and how long it has been aged. Cheeses are generally classified into six different categories: fresh, soft, semisoft, blue-veined, hard, and processed.

## Fresh Cheese

Fresh cheeses are usually not aged or allowed to ripen, as they spoil easily. Cheese varieties in this category include feta, mozzarella, Mascarpone, cottage cheese, ricotta, cream cheese, Neufchâtel, chèvre frais, and baker's cheese. These cheeses should be used soon after they are purchased. *See Figure 12-24.*

*Fresh Cheeses*

Feta

Mozzarella

Mascarpone

*Wisconsin Milk Marketing Board*

**12-24** *Fresh cheeses such as feta, mozzarella, and Mascarpone do not age well and should be used immediately.*

*Feta* is a fresh cheese of Greek origin made from sheep's or goat's milk. The cheese is slightly cured for a period that can range from a few days to four weeks. It has a salty taste and, when aged for a long period, becomes very salty and dry. Because of this condition, it is always wise to taste feta, if possible, before making a purchase. When aged properly, feta has a creamy texture, pleasant saltiness, and a soft to semisoft consistency. The smell is similar to cider vinegar and the taste has a faint trace of olives. Feta can be used in snacks, salads, sandwiches, and cooked dishes, such as lasagna and omelets.

*Mozzarella* is a very tender cheese with a soft, plastic-like curd. Mozzarella is primarily made from cow's milk. In the process of making mozzarella cheese, the whey is drained from the curd and reserved for making ricotta. When mozzarella is melted it has an elastic or rubbery consistency. It is commonly used in pizza and lasagna. *Mozzarella di bufala* (buffalo mozzarella) is an unripened fresh cheese made from water buffalo's milk, or a combination of cow's milk and water buffalo's milk. It is prized for its rich, slightly sour and acidic flavor.

*Mascarpone cheese* is a cream cheese of Italian origin. It has a smooth texture, a white or pale yellow color, and a buttery, subtly sweet flavor. It can be used with good results in savory dishes, but is most often seen in desserts, such as tiramisu. Mascarpone works well as a spread or can be served on its own with fruit, liqueur, or a dusting of cocoa.

*Cottage cheese* is a fresh cheese and is perhaps the simplest of all cheeses. It has a fine, mildly sour taste. It is known by many names, including Dutch cheese, pot cheese, smearcase, and, in some localities, popcorn cheese (because of its large curds). It is marketed in about five different varieties: small curd, large curd, flake curd, home-style, and whipped. It is available either plain or creamed. Cottage cheese is used in appetizers, salads, cheesecakes, and cooked dishes. It is highly perishable and should always be stored at a low temperature.

*Ricotta* is a white and creamy cheese that is somewhat similar to cottage cheese, yet it is made from the whey of other cheeses instead of milk. It has a bland, yet sweet flavor. Ricotta made in Europe is made from the whey of other cheeses, while ricotta made in North America is made using a mixture of whey and whole or skim milk. Ricotta blends well with the flavors and textures of other foods. It is an important ingredient in pasta dishes such as lasagna and manicotti.

*Cream cheese* is a soft, fresh cheese with a rich, mild flavor. It is an uncured cheese made from cream or a mixture of cream and milk. Gum arabic may be used as a stabilizer to extend the product life of the cheese. Cream cheese that does not contain gum arabic has a lighter, more natural texture but does not keep as well. Cream cheese is used extensively in the professional kitchen in the preparation of items such as canapé spreads, sandwiches, salads, salad dressings, and desserts such as cheesecake.

*Neufchâtel* is a soft, fresh cheese that originated in the Normandy region of France. It is somewhat similar to cream cheese, but has more moisture and less fat. When young, Neufchâtel has a mild, slightly salty flavor, becoming more pungent as the cheese ripens. It is made from whole or skim milk or a mixture of milk and cream. Although Neufchâtel is most often sold as a fresh cheese, it is also available cured. Because of the smooth texture of this cheese, it spreads and blends well and is used in canapé spreads, salads, salad dressing, and many dessert items. Neufchâtel that is made in the United States has a smoother texture and is more similar to cream cheese than the French product.

*Chèvre frais* is fresh goat's milk cheese. The word *chèvre* is French for "goat," and is used for any variety of goat's milk cheese. While chèvres are available in textures ranging from soft to firm, the soft, fresh varieties are most popular. Chèvre frais has a pure white color and a soft, spreadable texture that is slightly dry. It has a mild, slightly peppery flavor and is often blended with herbs or spices. It can be used

Chef's Tip: Cheese labeled "pur chèvre" is made of exclusively goat's milk. Other cheeses labeled as "chèvre" may be made from a blend of cow's milk and goat's milk.

in cooking with good results and is also used as a spread. Montrachet is among the most popular varieties.

*Baker's cheese* is a fresh skim milk cheese that is much like cottage cheese but is softer and finer grained. In making baker's cheese, the curd is drained in bags rather than in vats. Baker's cheese is used in making cheesecakes, pies, and certain pastries.

### Soft Cheese

A *soft cheese* (rind-ripened cheese) is a cheese that has been sprayed with a harmless mold to produce a thin skin. The living mold then works to ripen the soft cheese by reacting with the rind (the exterior skin of the cheese), resulting in a soft, suedelike coating on the cheese and a soft interior. A soft cheese will become extremely soft and somewhat runny once it is fully ripened. Soft cheese varieties include Brie and Camembert. ***See Figure 12-25.***

*Soft Cheeses* _____

Brie

Camembert

*Wisconsin Milk Marketing Board*

**12-25** *Soft cheeses have a rind produced by a harmless mold.*

*Brie* is a soft cheese with a strong odor, a sharp taste, and a creamy white color. It originated in France but is made in many countries including the United States. Brie is very similar to Camembert cheese, another of the popular French cheeses, but due to variations in manufacturing and ripening, there are differences in flavor and aroma. Brie is usually produced in small disks and is known chiefly as a buffet or dessert cheese. It can also be used in hors d'oeuvres, salads, sandwiches, or melted in hot entrées.

*Camembert* is a soft cheese made from cow's milk with a yellow color and a waxy, creamy consistency. The thin rind has the appearance of felt. Like Brie, Camembert is made in many countries, including the United States. It is served most often as a dessert accompanied by crackers and fruit. For optimal eating quality, the cheese should be left at room temperature prior to serving.

## Dry-Rind Cheeses

Monterey Jack

**12-26** *A dry-rind cheese is allowed to ripen with the external rind exposed to the air producing a dry, almost woody exterior.*

## Semisoft Cheese

A *semisoft cheese* is firmer than soft cheese but is not as hard as hard cheese. Semisoft cheeses are produced using one of three different ripening processes that result in a dry-rind, washed-rind, or waxed-rind cheese.

**Dry-Rind Cheeses.** A *dry-rind cheese* is a cheese that is allowed to ripen with its exterior exposed to air. The circulation of air dries out the exterior, producing a dry, almost woody rind. Although the rind becomes hard and dry, the interior of the cheese remains tender and smooth. Common examples of dry-rind semisoft cheeses are Monterey Jack, bel paese, and havarti. *See Figure 12-26.*

*Monterey Jack* is a dry-rind semisoft cheese that displays a smooth texture, a creamy white color, and a mild taste. It is sometimes described as having a taste similar to that of American Muenster. Monterey Jack that is aged for a longer-than-average period becomes harder in texture and zestier in flavor. It is used in the professional kitchen for sandwiches, salads, and certain entrée dishes, especially in Mexican or Southwestern cuisine.

*Bel paese* is a semisoft cheese with a buttery flavor and light color and melts easily. It originated in Italy but is now produced in both Italy and the United States. Sometimes substituted for mozzarella cheese, it is allowed to mature for nearly eight weeks while dry-rind ripening.

*Havarti* is a Danish dry-rind semisoft cheese made from cow's milk with a buttery, somewhat sharp flavor. It ages for approximately three months, and during the process develops small "eyes," or openings, similar to those of Swiss cheese but much smaller.

**Washed-Rind Cheeses.** A *washed-rind cheese* is a semisoft cheese that has an exterior rind that is washed, or rinsed, with a fluid such as brine, wine, olive or nut oil, or fruit juice. Washing the rind with these products generates the growth of harmless bacteria that ripen the cheese through the rind to the interior. Common examples of washed-rind semisoft cheeses are Limburger, Muenster, brick, and Port Salut. *See Figure 12-27.*

*Limburger* is a washed-rind semisoft cheese with a characteristic strong aroma and flavor. Limburger cheese was first marketed in Limburg, Belgium. Much of this cheese is made in Germany and the United States. Limburger cheese is either made from whole or skim cow's milk. It has a creamy texture that is developed through ripening in a damp atmosphere for a period of two months. Limburger cheese is served most often with crackers and onions as a dessert or buffet cheese.

*Muenster* is a washed-rind semisoft cheese with a flavor between that of brick cheese and Limburger. It was first produced in the vicinity of Munster, Germany. European Muenster ranges from mild to sharp in taste and may have a strong aroma due to aging. A version of this cheese is produced in North America and is comparatively mild. Muenster is marketed in cylindrical form and is used as a buffet or sandwich cheese.

## Washed-Rind Cheeses

Limburger

Muenster
*Wisconsin Milk Marketing Board*

**12-27** *A washed-rind cheese is washed, or rinsed, with liquids such as wines, oils, or fruit juices.*

*Brick cheese* is a washed-rind semisoft cheese made from cow's milk with a mild, sweet flavor and a texture that is firm yet elastic, with many small holes. It slices and melts well. The name "brick cheese" may stem from the bricks that were once used to weight presses used in making the cheese. The cheese is also sold in a brick shape. Brick cheese is used most often in the professional kitchen for cheese platters and sandwiches.

*Port Salut* is a washed-rind semisoft cheese that has a soft, smooth, orange-colored rind and a glossy ivory-colored interior. Its flavor may range from mellow to robust, depending on the age of the cheese, and has been compared to that of Gouda cheese. In some instances, the aroma of the cheese is like a mild Limburger. The cheese is used as a dessert, for appetizers, and served with apple pie.

**Waxed-Rind Cheeses.** *Waxed-rind cheese* is a cheese produced by dipping a wheel of freshly made cheese into a liquid wax and allowing the wax to harden. The cheese ripens while encased in the wax. Although the wax coating is hard, the cheese inside stays soft. Common waxed-rind semisoft cheeses include Edam, Gouda, and fontina. *See Figure 12-28.*

*Edam* is a waxed-rind semisoft cheese made from cow's milk with a firm, crumbly texture and is usually shaped like a ball with a slightly flattened top and bottom and a red wax coating. The red coating is one of the identifying characteristics of the cheese. Edam is named after its birthplace, Edam, which is in the Netherlands. Edam cheese

**Historical Note**

Port Salut was created around 1865 by Trappist monks at an abbey in Port du Salut, France. It is now manufactured in abbeys in various parts of Europe as well as one monastery in Kentucky. The Trappists have kept the exact process a secret, but a similar cheese is made outside the monasteries in Europe and North America.

made for consumption within the Netherlands is rubbed with oil but is not colored. The cheese for export is colored red on the outside, rubbed with oil, wrapped, and shipped. Edam cheese made in the United States is covered with a thin coating of red paraffin to give it its characteristic color. It is used most often as a dessert cheese on platters and on buffet tables.

## Waxed-Rind Cheeses

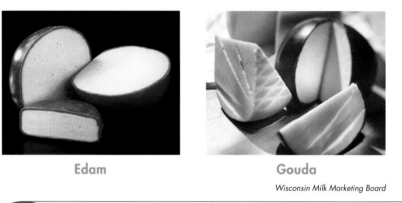

**Edam**

**Gouda**

Wisconsin Milk Marketing Board

**12-28** *Waxed-rind cheeses ripen while encased in wax.*

*Gouda* is a semisoft waxed-rind cheese that is similar to Edam cheese, but contains more fat. It also originated in Holland, in the Dutch province of Gouda and, as is the custom, was named after its place of origin. Gouda is usually shaped like a flattened ball or formed into a loaf. Neither Edam nor Gouda is recommended for cooking. They are generally served as dessert or buffet cheeses.

*Fontina* is a waxed-rind semisoft cheese made from cow's milk. It has been produced in the Alps since the 12th century. Young fontina cheese is somewhat soft and becomes harder as it ages. It is perfect for use in fondues and has a nutty, herbal flavor.

### Blue-Veined Cheese

A *blue-veined cheese* is a cheese produced by inserting harmless live mold spores into the center of ripening cheese with a needle. The blue vein that runs through these cheeses indicates where a needle was inserted and mold spores were released. The mold is safe to consume and decreases the time required for a cheese to ripen. It also adds a distinctive flavor to the cheese. After the mold spores are injected, the exterior of the cheese wheel is salted to help keep the surface dry and prevent the mold from overtaking the exterior. Common blue-veined cheeses include bleu, Gorgonzola, Roquefort, and Stilton. *See Figure 12-29.*

## Blue-Veined Cheeses

**Bleu Cheese**

**Gorgonzola**

*Wisconsin Milk Marketing Board*

**12-29** *Blue-veined cheeses contain a harmless blue-green mold.*

*Bleu cheese* is a semisoft blue-veined cheese made from cow's milk and is characterized by the presence of green-blue mold. It is made in the United States in the style of European-made Gorgonzola, Roquefort, and Stilton. Bleu cheese is generally produced in wheels weighing about 7 lb. In the professional kitchen, bleu cheese is used in bleu cheese dressing, salads, sandwiches, and cheese platters.

*Gorgonzola* is a blue-veined cheese that is mottled with characteristic blue-green veins produced by a mold known as *penicillium glaucum*. The surface of the cheese is protected with aluminum foil. Gorgonzola cheese is generally cured for a period of six months to a year. It originated in the village of Gorgonzola, which is near Milan, Italy. Today it is made chiefly in the regions of Lombardy and Piedmont. Gorgonzola is used in salads and salad dressings and as a dessert and buffet cheese.

*Roquefort* is a blue-veined cheese characterized by a sharp, tangy flavor and by blue-green veins that flow through the white curd. The blue-green veins are created by spreading a powdered bread mold over the curd before ripening. The cheese is cured for a period of two to five months, depending on the sharpness desired. It was first made in the village of Roquefort, France. Roquefort is made from ewe's milk. Although a variation is made from cow's milk in other countries, the word Roquefort can only be used for traditional Roquefort-made cheese due to a French regulation. Roquefort is used principally as a dessert cheese but is also used in salads.

*Stilton* is a blue-veined cheese made from cow's milk with a flavor that is milder than Roquefort or Gorgonzola. It was first made in the village of Stilton, England. Stilton, like the other blue-veined cheeses, has a crumbly texture and veins of blue-green mold running through the curd and wrinkled rind. Although some Stilton is available into North America, it is not imported in great quantities. It is used chiefly as a dessert and buffet cheese.

### Hard Cheese

A *hard cheese* is a firm, somewhat pliable and supple cheese with a slightly dry texture and buttery flavor. Because hard cheeses have a firmer overall texture than softer varieties, they grate and slice well, making them perfectly suited for sandwich preparations. Common examples of hard cheeses are Cheddar, Swiss, provolone, Gruyère, Cheshire, and Manchego. ***See Figure 12-30.***

## Hard Cheeses

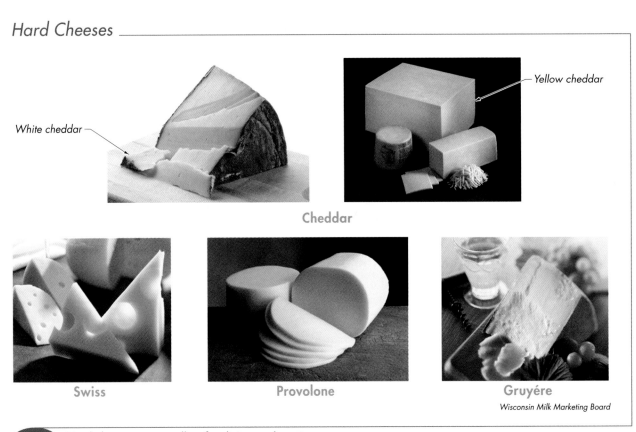

White cheddar

Yellow cheddar

Cheddar

Swiss

Provolone

Gruyére

Wisconsin Milk Marketing Board

**12-30** *Hard cheeses are excellent for slicing and grating.*

*Cheddar* is a hard, mild to sharp-tasting aged cheese. It originated in the village of Cheddar in Somersetshire, England, and was introduced to North America by British colonists. Cheddar cheese accounts for a significant portion of all cheese made in the United States. The sharpness of Cheddar cheese varies based on the length of the aging period. Longer aging produces a sharper taste. Cheddar is used in the preparation of Welsh rarebit, fondue, soufflés, canapé spreads, and sandwiches. It is also eaten on crackers or used to top chili, and is sometimes served with apple pie. Colby cheese is a U.S. product similar to mild Cheddar.

*Swiss cheese* is a term for various varieties of hard cheese with large holes (eyes). Emmenthaler is the quintessential Swiss cheese and is a product of Switzerland. Other varieties of cheese labeled as Swiss cheese may be U.S. products of a similar style. Swiss cheese has an elastic body and a mild, sweet flavor. The holes in Swiss cheese are developed by gas-producing bacteria that release carbon dioxide during the ripening period. The carbon dioxide forms bubbles that result in the characteristic holes. The curing period for Swiss cheese varies depending on where the cheese is made. In the United States, it is typically cured three to four months, while in Switzerland it is cured for six to ten months. The longer curing period used in Switzerland produces cheese with larger holes and a more pronounced flavor. Swiss cheese is used in many preparations, from sandwiches to stuffed veal chops. It is a traditional cheese for fondue.

*Provolone* is a hard cheese with a mild to sharp taste (depending on age) and an elastic texture. It originated in southern Italy but is now produced in other parts of Italy and in North America. It is light in color and can be cut without crumbling. A distinguishing characteristic of provolone is that it is often formed into the shape of a pear, sausage, or cone and corded for hanging. Provolone is used in the preparation of many Italian dishes, especially pizza.

*Gruyère* is a hard cheese made in Switzerland that is similar in many ways to Emmenthaler, but has smaller holes and a sharper taste. Gruyère is an excellent cheese for use in cooking. In the professional kitchen, Gruyère is used in fondue, veal cordon bleu, and sautéed veal chops.

*Cheshire cheese* is the oldest of the named English cheeses and is classified as a hard cheese. It originated in Cheshire County, and is similar to Cheddar cheese but has a texture that is more crumbly and less dense. It is available in a natural white color or a deep yellow. The yellow is the result of adding annatto (a vegetable dye) to the curd. Cheshire cheese derives its unique, mildly salty taste from the salty soil in that area of England. The cows that graze on the grass grown there produce salty milk, which is used in making the cheese. Cheshire cheese is used in the same ways as Cheddar and makes an excellent Welsh rarebit or fondue.

*Manchego* is a hard, sheep's milk cheese produced in the La Mancha region of Spain. It is aged for at least 3 months and has a rich golden color. Manchego has a tangy, slightly salty flavor. It is the most well-known cheese from Spain.

**Grating Cheeses.** A *grating cheese* is a hard, crumbly, dry cheese used grated or shaved onto food prior to serving. The crumbly texture makes it quite difficult to slice a grating cheese smoothly. Grating cheeses are usually produced as large wheels, many weighing close to 100 pounds. The most common grating cheeses are Parmesan, Romano, and asiago. ***See Figure 12-31.***

**Nutrition Note**

Swiss cheese has been shown to reduce the risk of dental problems. Researchers believe that Swiss cheese stimulates the production of saliva in the mouth, which helps to wash food off of the teeth. The milk proteins in Swiss cheese also act to neutralize the acid component found in plaque. This results in stronger tooth enamel and reduced tooth decay.

## Grating Cheeses

Wisconsin Milk Marketing Board
**Parmesan**

**Romano**

Wisconsin Milk Marketing Board
**Asiago**

**12-31** *Grating cheeses are the hardest cheeses produced.*

*Parmesan* is a grating cheese that originated in Parma, Italy. It is extremely hard, has a granular texture, and can keep indefinitely when properly cured. Because of its texture, Parmesan is classified with a group of Italian cheeses known as *grana* (meaning "grain"). Parmesan cheese is widely produced in the United States and Argentina, but Parmesan from Italy is superior to Parmesan produced elsewhere. The cheese is rubbed with oil and dark coloring from time to time through the aging period. It is available in grated form or whole, and can be used as a table cheese when still slightly moist. Parmesan is considered a seasoning cheese because it is used to season preparations such as onion soup or pasta dishes.

*Romano* is a grating cheese that is similar to Parmesan but softer in texture. In Italy, Romano is used both grated and as a table cheese. It was first made in the vicinity of Rome from ewe's milk. Today it is made in other parts of Italy and other countries, and is also made from cow's milk and goat's milk. Romano has a granular texture, a sharp flavor, and a brittle, black rind. It is aged for five to eight months if it is to be used as a table cheese. If it is to be grated, it is aged for about a year. A longer aging period sharpens the flavor. In the professional kitchen, Romano is used as a seasoning in pasta dishes or in other applications, such as for topping au gratin vegetable dishes.

*Asiago* is a grating cheese with a nutty, toastlike flavor. High-quality asiago often contains a crunchy material resulting from an amino acid found in milk that crystallizes during the aging process. American-made asiago has a taste that is different from that of imported Italian asiago, mainly due to the diets of the cows that produce milk for the cheese. Italian asiago is produced from the milk of cows that graze in fields where herbs, medicinal plants, and flowers grow. This diet yields a more complex and flavorful cheese than American-made asiago.

## Processed Cheese

A *processed cheese* is a mixture of any variety of finely ground cheese scraps that are mixed with nondairy ingredients and melted together. The result is a cheese product that is uniform in flavor and texture and can be packaged in just about any shape or size. For the best processed cheese, care must be taken to select cheeses that are fully cured and sharp in flavor. ***See Figure 12-32.***

While a processed cheese is made up almost entirely of cheese, a *processed cheese food* is a cheese-based product that may contain as little as 51% cheese. The other remaining 49% can be made up of dairy or nondairy products, such as emulsifiers that allow the product to be heated and meltable. Traditional hard cheeses such as Cheddar become stringy when melted, but processed cheese and cheese food melt to a consistently smooth, silky texture. Common types of processed cheese food include American cheese and cold-pack cheese.

Processed cheese has certain advantages over other cheeses. It is economical, melts fairly easily and evenly, and can be kept for a longer time than other types of cheese. Numerous varieties of processed cheese are found in the supermarket. It is used extensively in the professional kitchen.

*American cheese* is processed cheese produced by mixing shredded varieties of various cheeses with some dairy or nondairy products, melting it together, and pouring it into molds to shape it. In the United States, the term American cheese is commonly used to describe any type of processed cheese. American cheese melts more smoothly than most other cheese and has a texture that is softer, smoother, and more uniform than most cheeses.

*Cold-pack cheese* is a processed or blended cheese made from pasteurized milk without the aid of heat. Cold-pack cheese is creamy and ranges from white to orange in color. It is available in many different flavors and is often mixed with a variety of spices and seasonings.

## CHEESE PREPARATION AND STORAGE

Cheese is a perishable food item that must be stored and served properly. Because cheese is a dairy product that contains a considerable amount of butterfat, absorbs odors, spoils relatively quickly, and can dry out if left exposed to air. In order to serve cheese at its best, it is important to understand proper storage, handling, and serving requirements. These requirements are relatively consistent for all cheese varieties with the exception of fresh cheese, such as cream cheese and cottage cheese, which should always be served chilled. Cheese can be used in fondue, salads, on sandwiches, in soups, or simply enjoyed with crackers, fruit, or bread. ***See Figure 12-33.***

### Processed Cheese

**12-32** *Processed cheese is made from a mixture of finely ground cheese scraps.*

### Historical Note

Processed cheese was made in Germany and Switzerland as early as 1895, but the first patent for processed cheese in the United States was not issued until 1916. Approximately one-third of all cheese made in the United States is marketed as processed cheese.

## Cheese Fondue

Wisconsin Milk Marketing Board

**12-33** Cheese fondue is a hot cheese preparation served with items for dipping.

### Serving Cheese

Cheese is often served as a snack food or as an ingredient in a dish such as a salad, sandwich, vegetable, or entrée in North America, while in Europe cheese is traditionally served with a continental breakfast or as a separate course at the end of a meal. When preparing a cheese platter, it is important to include a variety of cheeses with varying textures, degrees of ripeness, and flavors, from a variety of milks if possible. Cheese should ideally be served at room temperature. If cheese has been refrigerated, it should be removed about 1 hr prior to service. Room-temperature cheese exhibits the best flavor and texture. Cheese should not be left at room temperature for more than a few hours or it becomes oily and may dry out.

### Cooking with Cheese

Cheese preparations are generally cooked at the lowest temperature possible to prevent proteins in the cheese from hardening. This allows the cheese to maintain an even texture and smooth consistency. Cheese becomes tough and stringy if overheated. Therefore, when preparing cheese sauces, fondues, or adding cheese to a soup, the cheese should be added at the end of the cooking process so that it melts smoothly and incorporates evenly into the product. For example, when making smooth cheese sauce for nachos, cheese should be added after a béchamel sauce has thickened. If cheese is added directly to the milk during the béchamel preparation, it separates, settles to the bottom, and is not incorporated.

*The following tips are helpful when cooking with cheese:*

- When baking a dish containing cheese, use an oven temperature of 325°F to 350°F.
- When making mornay sauce (cheese sauce) or a cheese soup, the liquid should be thickened prior to adding the grated cheese. If the sauce or liquid is thin, the cheese will separate and become stringy and will settle to the bottom of the pot.
- Pasteurized processed cheese products can withstand heat better than other cheese products, but should still be cooked using a low temperature to avoid overcooking.
- When broiling a cheese preparation, keep the cheese several inches below the fire and broil only until the cheese melts. Excessive heat toughens the cheese.
- For a rich cheese flavor in breads and quick breads, use a sharp-flavored cheese such as Parmesan, Cheddar, or sharp processed cheese.
- Use a low or moderate temperature and minimum cooking time, because excessive heat and prolonged cooking toughens the product.
- When adding Parmesan to a sauce or soup, sift the cheese and stir while adding to avoid lumps. Add the cheese just before the item is removed from heat.

## Storing Cheese

Cheese is perishable and should be kept tightly wrapped in plastic wrap or plastic bags in the refrigerator. Wrapping cheese helps keep it moist and keeps out air and odors from other foods. There are a few exceptions, such as processed cheese in aerosol cans and squeeze packs, that are stored at room temperature. Baker's cheese can be kept moist by placing it in a container, smoothing the surface until even, and covering it with a thin layer of fine sugar. When refrigerated properly, cheese should retain freshness as follows:

- Fresh cheeses keep for a week to 10 days.
- Soft cheeses keep for about two weeks.
- Semisoft cheeses keep for two to three weeks.
- Hard cheeses keep for about one month.
- Grating cheeses keep for a couple of months.

## COLD FOODS PRESENTATION

Platters of cold foods can be relatively simple to prepare. Some of the most popular cold foods presentations involve platters, charcuterie, canapés or other cold hors d'oeuvres, and raw bars. Centerpieces and garnishes can be used to add visual appeal to a cold foods presentation.

### Cold Platters

Cold foods can be displayed on various-size platters. A cold foods platter is a garde manger's opportunity to provide a sample of various foods as well as to highlight the chef's expertise. Typical kinds of platters include raw foods platters, chaud froid platters, and platters of cured or smoked foods. *See Figure 12-34.*

*Cold Platters*

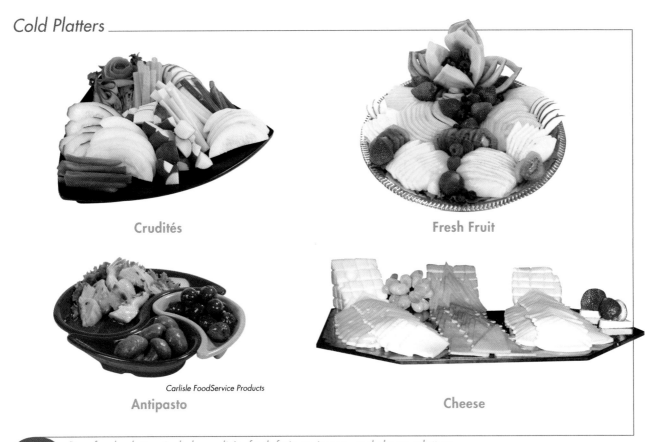

Crudités

Fresh Fruit

*Carlisle FoodService Products*

Antipasto

Cheese

**12-34** *Raw foods platters include crudités, fresh fruit, antipasto, and cheese platters.*

Food on raw foods platters should be arranged to show the freshness and quality of the ingredients used. Fresh vegetables can be cut into sticks for dipping into flavorful dips. A raw vegetable platter is often called a crudités (pronounced crew-dee-tay) platter, as *crudités* is the French word for "raw." For a fresh fruit platter, slices of fresh fruit should be sliced and shingled for a uniform, neat presentation. Fruits should not be cut too far in advance because many fruits oxidize or deteriorate rapidly when cut. Antipasto platters provide an arrangement of cured meats and cheeses as well as some marinated or pickled vegetables. Cheese platters should have a nice variety of cheeses on them as well as some fresh fruits, breads, and crackers. Regardless of the type of platter being made, it is best to have an attractive presentation with an assortment of colors and textures when possible.

A *chaud froid platter* is a platter of hot food items presented cold. Chaud froid comes from the French words *chaud* (pronounced show), meaning "hot," and *froid* (pronounced fwa), meaning "cold." A chaud froid platter contains items such as meats, seafood, poultry, and vegetables that have been cooked, chilled, and glazed with either clear or white aspic. *Aspic* is a savory jelly made from clarified meat, fish, or vegetable stock and gelatin that is used to glaze foods. Aspic is used to preserve the color of foods as well as prevent them from drying out due to exposure to air. Food with an aspic glaze has the appearance of artificial food because it has such a beautiful finish. Clear aspic produces a beautiful clear finish, while white aspic produces a decorative, shiny surface.

Aspic-coated foods can be lifted with an offset spatula and arranged on a serving platter. *Note:* It is important that aspic-glazed items are not handled with bare hands or fingers as the shiny aspic shows fingerprints.

**Chef's Tip:**

Some chefs prefer use a dipping fork to dip items into aspic glaze, as doing so results in a much smoother coating. A disadvantage of this method, however, is that the bowl of aspic becomes contaminated with the food and cannot be used for other applications or stored for any length of time.

## To glaze food with aspic, apply the following procedure:

1. Lay items to be glazed on a glazing screen or cooling rack over a sheet pan, presentation-side up. Cover tightly and refrigerate until needed.

2. Prepare instant clarified aspic according to package directions.

3. Place hot aspic in a nonreactive bowl over an ice bath and, while stirring constantly, cool until aspic becomes slightly thicker in consistency.

4. Remove items to be glazed from the refrigerator. Using a spoon or ladle, carefully and consistently ladle just enough aspic to completely cover the item to be glazed. It is important that the correct amount of aspic is poured over the surface of each item to ensure the entire item is coated and to avoid any drips or unevenness in the glaze.

5. Place glazed items in the refrigerator to allow the aspic to set.

6. Repeat steps 4 and 5 at least three times to completely and uniformly glaze each item.

CULINARY PROCEDURES

## Smoked Salmon

**12-35** Cured or smoked meats and seafood make an attractive platter when sliced thin and arranged with complementary accompaniments.

Cured and smoked items are very popular in cold foods presentations. Smoked and cured seafood and meats are standard brunch items and are commonly found at upscale receptions and weddings. As with other cold foods platters, smoked items should be arranged attractively on the platter with accompaniments that complement the foods. For example, cured salmon can be served with capers, diced red onions, and chopped hard-cooked eggs. Cured and smoked meats can be served with gherkins (miniature pickles), assorted mustards, and spreads. Cured and smoked items should be sliced fairly thin to ensure they are tender and not overpowering in flavor. ***See Figure 12-35.***

### Charcuterie

*Charcuterie* (pronounced shar-coo-ta-REE) refers to the production of pâtés, terrines, galantines, sausages, and other products traditionally made from pork. Today these products are also made from other sources, including other meats, poultry, seafood, and vegetables.

A *pâté* (pronounced pah-tay) is a preparation made from a finely ground meat layered with additional garnishes or ingredients and baked in a loaf-shaped pan. The traditional form of pâté, called *pâté en croûte* (pronounced pah-tay ahn crewt), is made with pork forcemeat (finely ground meat) that is wrapped in pastry dough and baked in a loaf-shaped pan. Pâté en croûte is French for "pâté in crust." ***See Figure 12-36.***

## Pâtés

Crust

Finely ground meat

**12-36** Pâtés are made from finely ground meat layered with garnishes or other ingredients.

A *terrine* is a preparation made from either coarsely or finely ground meat, seafood, poultry, or vegetables and is baked in a loaf-shaped pan. Although the terms pâté and terrine are often used interchangeably, traditionally they are different preparations. A traditional terrine is made with a coarsely ground pork forcemeat baked in an earthenware casserole without a crust. The word terrine is derived from the French word *terre* (pronounced tair), meaning "earth." ***See Figure 12-37.***

## Terrines

**12-37** *Terrines are made from coarsely or finely ground meat.*

A *galantine* is a boned-out cut of meat, poultry, or seafood that is stuffed with forcemeat, poached in stock, cooled, sliced, and glazed with aspic. A traditional galantine is produced from boned-out poultry stuffed with a poultry forcemeat. When preparing a galantine from fish or shellfish, the prepared forcemeat is wrapped tightly in a layer of plastic wrap and covered with aluminum foil to secure the shape before poaching. Similar to the galantine is a ballotine. A *ballotine* is a poultry leg that is boned out, stuffed with forcemeat, poached or braised until tender, sliced, and served hot.

### Raw Bars

A *raw bar* is a section of countertop or buffet line devoted to raw, or cooked and chilled, fish and seafood. ***See Figure 12-38.*** Common items on a raw bar include sushi, sashimi, raw oysters and clams on the half shell, caviar, cooked shrimp, scallops, and lobster meat.

*On the half shell* is a phrase referring to bivalves such as oysters and clams that are opened and served on one half of the shell. Shrimp and crab can be served raw but are typically served cooked. Raw bars commonly serve seafood seasonings and garnishes including lemon wedges, cocktail sauces, and hot sauces such as Tabasco®.

*Raw Bars*

Shrimp

Oysters

Carlisle FoodService Products

12-38  *Raw or chilled seafood is served on raw bars.*

USDC ™
Processed
Under
Federal
Inspection

A garde manger must use extra caution when inspecting seafood, especially seafood that is to be served raw. It is essential to ensure that seafood is from suppliers that guarantee the food is from uncontaminated waters and has been safely handled. All seafood should arrive from the supplier with a packed under federal inspection (PUFI) tag that states the place of origin, the date it was harvested, the wholesale grower, and the seller. The purpose of this information is so that, in the event of an outbreak of foodborne illness, the health department can trace the seafood that was sold to the restaurant. Eating raw meat, seafood, or eggs always comes with some risk. However, a reputable restaurant with a skilled and conscientious staff greatly reduces the opportunity for an outbreak of foodborne illness.

## Centerpieces and Garnishes

Table centerpieces and creative garnishes complete a cold foods presentation. Because many garde manger items are served on platters, there is not a great deal of height to entertain the eye. Most of the action, color, and beauty of garde manger is focused just above the height of the tabletop. Because of this, garde mangers often create centerpieces and garnishes to impress customers and guests.

One of the most dramatic centerpieces is an ice carving. An ice carving is typically about 2′ to 3′ tall and can be carved in any shape or design imaginable by a skilled ice carver. An ice carving is placed on a table in a drain pan to catch any water as the ice melts. Common sculptures found on buffets include vases, swans, baskets, and hearts. These carvings are for decorative purposes only. *See Figure 12-39.* Other carvings can be both decorative and practical. For example, a carving of a large clamshell or treasure chest can hold shrimp cocktail or oysters on the half shell.

*Ice Carving*

**12-39** *Ice carvings can be attractive sculptures or functional serving pieces.*

Refer to the CD-ROM for the **Fluted Mushrooms** Media Clip.

There are many different garnishes that a garde manger can create for a cold buffet as well. Honeydew and cantaloupe can be carved into small swans or beautiful flowers while watermelons can be carved to form a whale or a large basket to hold cut-up fruit. Vegetables can also be carved to garnish a cold foods buffet. A butternut squash, for example, can be carved into a vase and filled with assorted root vegetables such as turnips, daikon radishes, carrots, and beets carved into a beautiful bouquet of flowers. An edible vegetable centerpiece can impress customers and guests as they browse an assortment of cold foods. *See Figure 12-40.*

Garnishes

Tomato rose

Honeydew flower

Butternut squash vase

**12-40** *Fruits and vegetables can be cut into decorative garnishes.*

## SUMMARY

The garde manger is responsible for all cold foods produced in the professional kitchen, including salads, dressings and dips, cheese courses, cold foods platters, charcuterie, and carved centerpieces.

Salads are versatile dishes that can be used in many different ways throughout the course of a meal, as appetizers, side salads, main courses, separate courses, or even desserts. Many salads are made using salad greens, which include lettuces, chicory, and other edible greens such as spinach, watercress, or kale. Care must be taken to properly prepare, wash, and store salad greens. Vegetables, proteins, starches such as grains or pasta, fruits, and nuts are also common salad ingredients.

Salads are often served with a dressing. The most common types of dressing include temporary emulsions, such as vinaigrettes, and permanent emulsions, such as mayonnaise. These dressings form the foundation from which countless varieties of dressings can be prepared.

There are many varieties of salads, and the presentation often depends on how the salad will be plated or served. Tossed salads are mixtures of leafy greens and raw vegetables. Composed salads have a base, body, garnish, and dressing carefully arranged on an individual plate for each serving. A bound salad consists of a main item and other ingredients held together with a binding agent such as mayonnaise. Vegetable salads can be made from raw or cooked vegetables. Fruit salads must be prepared just prior to serving and may be dressed with a light yogurt, whipped cream, fruit purée, or honey dressing. Gelatin salads are made from flavored gelatins formed in a mold.

Cheese comes in varieties ranging from soft fresh cheeses to hard grating cheeses. Cheese can be served as a course or used as an ingredient in a hot or cold dish. When selecting cheese for a cheese platter, a garde manger should consider varieties with varying textures, degrees of ripeness, and flavors.

Cold foods platters produced in the garde manger kitchen include crudités, fresh fruit, antipasto, cheese, chaud froid, and cured or smoked foods. Platters should be arranged attractively to show the freshness and quality of the ingredients used. Charcuterie involves preparation of items such as pâtés, terrines, and galantines, which feature forcemeats that are cooked, cooled, and sliced. Raw bars include presentations of raw, or cooked and chilled, fish and seafood.

In addition to cold foods, a garde manger is responsible for preparation of any centerpieces or garnishes, such as ice carvings and creations made from raw fruits and vegetables. Centerpieces and garnishes allow a chef to display talent and creativity while adding interest to a presentation of cold foods.

Refer to the CD-ROM for Quick Quiz® questions related to chapter content.

## Review Questions

1. List the five basic types of salad.

2. Identify the vitamins and minerals commonly found in salad greens.

3. Contrast the main qualities of butterhead lettuce with those of romaine lettuce.

4. What is the difference between mesclun and microgreens?

5. List common varieties of chicory.

6. Identify three green leafy vegetables used sparingly in salads to add texture, flavor, and color.

7. What are the most common varieties of edible flowers?

8. Why is it important to cut solidly packed head lettuces to appropriate size before washing and draining the leaves?

9. At what temperature should greens be stored?

10. Illustrate the differences between a temporary emulsion and a permanent emulsion.

11. List the six presentation categories of salads.

12. What is the difference between a tossed salad and a composed salad?

13. List common binding agents used for bound salads.

14. Provide examples of common types of vegetable salads.

15. Explain common ways to prevent rapid discoloration in fresh fruit salads.

16. List five tips to follow when preparing gelatin salads.

17. Name the distinguishing characteristics and common types of fresh cheese.

18. Contrast the main qualities and production processes of soft cheeses with those of semisoft cheeses.

19. List four types of blue-veined cheese.

20. Contrast the various qualities of common hard cheeses.

21. Explain why cheese preparations are generally cooked at the lowest temperature possible.

22. What is effect of glazing a chaud froid platter with aspic jelly?

23. Explain the difference between a pâté and a terrine.

24. Identify items commonly included on a raw bar.

25. Explain the function of table centerpieces and creative garnishes.

## Basic French Vinaigrette

9

*yield: 1 pt (sixteen 1-oz servings)*

| | |
|---|---|
| 4 oz | red wine vinegar |
| 1½ tsp | kosher salt |
| ¼ tsp | fresh ground black pepper |
| pinch | granulated sugar |
| 12 oz | canola oil |

1. In a stainless steel mixing bowl, combine the vinegar, salt, pepper, and sugar and whisk to incorporate.
2. Add oil in a steady stream while whisking continuously to form a temporary emulsion.
3. Whisk as necessary before use.

### Nutrition Facts
**Per serving:** 189.2 calories; 98% calories from fat; 21.3 g total fat; 0 mg cholesterol; 176.3 mg sodium; 7.9 mg potassium; 0.5 g carbohydrates; 0 g fiber; 0 g sugar; 0.5 g net carbs; 0 g protein

### Recipe Variation:

**10** **Red Wine Oregano Vinaigrette:** Add 2 tsp dried oregano and 1 tbsp diced scallions (green onions) to the basic French Vinaigrette.

## Mayonnaise

11

*yield: 1 pt*

| | |
|---|---|
| 2 | pasteurized egg yolks |
| ½ tsp | kosher salt |
| pinch | white pepper |
| ½ tsp | dry mustard powder |
| 1½ tbsp | white wine vinegar |
| 14 oz | canola oil |
| ½ tsp | lemon juice |
| 1 shake | Tabasco® sauce |
| 1 shake | Worcestershire sauce |

1. Collect all ingredients needed to make the mayonnaise and allow to sit at room temperature for about one hour prior to attempting the emulsification.
2. In a nonreactive mixing bowl, whisk the yolks with the salt, pepper, mustard, and half of the vinegar until lighter colored and foamy.
3. Begin adding the oil very slowly, while whisking rapidly and continuously, a drop at a time until the emulsion begins to form and the mixture thickens slightly.
4. Once about ¼ of the oil has been added, the oil can be added in a fine stream while whisking rapidly and continuously to incorporate.
5. As the mixture becomes difficult to whisk, add a small amount of vinegar to thin slightly.
6. Continue alternating vinegar and oil while whisking until all the oil and vinegar have been added.
7. Add lemon juice, Tabasco®, and Worcestershire and adjust seasonings if necessary.
8. Place in a chilled, sanitized storage container and refrigerate until needed.

### Nutrition Facts
**Entire recipe:** 3629.9 calories; 98% calories from fat; 406.2 g total fat; 419.6 mg cholesterol; 964.6mg sodium; 75.4 mg potassium; 4 g carbohydrates; 0.4 g fiber; 0.3 g sugar; 3.6 g net carbs; 5.8 g protein

# Recipes

## 12

## Thousand Island Dressing

*yield: 1 pt (sixteen 1-oz servings)*

| | |
|---|---|
| 1 pt | prepared mayonnaise |
| 2¼ tbsp | ketchup |
| 2½ oz | chili sauce |
| 1 | hard-cooked egg, chopped |
| 1 | small shallot, minced |
| ½ tsp | garlic, crushed |
| 2 tsp | pimentos, chopped |
| 1 tbsp | gherkins, chopped (or sweet pickle relish) |
| 2 tsp | capers, chopped |
| 2 tsp | green peppers, minced |
| 1 tsp | fresh chives, minced |
| ½ tsp | fresh parsley, minced |
| 1 tsp | sugar |
| 1 dash | Worcestershire sauce |
| 1 dash | Tabasco® sauce |
| 1 tsp | lemon juice |
| to taste | salt and pepper |

1. Cut all ingredients as specified.
2. Place all ingredients in nonreactive bowl and mix well.

**Nutrition Facts**

**Per serving:** 143.6 calories; 63% calories from fat; 10.3 g total fat; 22.7 mg cholesterol; 277.7 mg sodium; 125.4 mg potassium; 12.6 g carbohydrates; 0.1 g fiber; 2.8 g sugar; 12.6 g net carbs; 1.4 g protein

## 13

## Creamy Garlic Dressing

*yield: 1 pt (sixteen 1-oz servings)*

| | |
|---|---|
| 3 oz | basic French dressing |
| 6 oz | sour cream |
| 6 oz | mayonnaise |
| 2 cloves | garlic, minced |
| ½ tsp | garlic powder |
| dash | Worcestershire sauce |
| dash | Tabasco® sauce |
| to taste | salt and pepper |

1. Add all ingredients except basic French dressing to nonreactive mixing bowl and mix well.
2. Add basic French dressing and mix well.
3. Adjust seasonings to taste.

**Nutrition Facts**

**Per serving:** 89.5 calories; 80% calories from fat; 8.2 g total fat; 7.4 mg cholesterol; 100 mg sodium; 22.6 mg potassium; 4 g carbohydrates; 0 g fiber; 1.6 g sugar; 4 g net carbs; 0.5 g protein

## 14

## Blue Cheese Dressing

*yield: 1 pt (sixteen 1-oz servings)*

| | |
|---|---|
| 3 oz | basic French dressing |
| 6 oz | sour cream |
| 6 oz | mayonnaise |
| ½ tsp | garlic, minced |
| 3 oz | blue cheese, crumbled |
| dash | Worcestershire sauce |
| dash | Tabasco® sauce |
| to taste | salt and pepper |

1. Add all ingredients except basic French dressing to nonreactive mixing bowl and mix well.
2. Add basic French dressing and mix well to incorporate.
3. Adjust seasonings to taste

**Nutrition Facts**

**Per serving:** 107.5 calories; 79% calories from fat; 9.7 g total fat; 11.4 mg cholesterol; 174.1 mg sodium; 34.1 mg potassium; 4 g carbohydrates; 0 g fiber; 1.6 g sugar; 4 g net carbs; 1.6 g protein

## Green Goddess Dressing

*yield: 1 pt (sixteen 1-oz servings)*

| | |
|---|---|
| 1 tbsp | fresh chives, minced |
| 2 tbsp | fresh parsley, chopped |
| 1 clove | fresh garlic, minced |
| 2 tbsp | lemon juice, fresh |
| ½ oz | tarragon vinegar |
| 1¼ c | mayonnaise |
| ½ c | sour cream |
| dash | Worcestershire sauce |
| dash | Tabasco® sauce |
| to taste | salt and pepper |

1. Place all herbs in blender with lemon juice and vinegar and pulse until finely chopped.
2. Add mayonnaise, sour cream, Worcestershire, and Tabasco® and blend well.
3. Adjust seasonings to taste.

---
**Nutrition Facts**

**Per serving:** 91.8 calories; 75% calories from fat; 7.8 g total fat; 9.4 mg cholesterol; 218.3 mg sodium; 29.1 mg potassium; 5 g carbohydrates; 0 g fiber; 1.2 g sugar; 5 g net carbs; 0.9 g protein

---

## Buttermilk Ranch Dressing

*yield: 1 pt (sixteen 1-oz servings)*

| | |
|---|---|
| ½ tsp | kosher salt |
| 3 cloves | garlic, minced |
| 6 oz | buttermilk |
| 9 oz | mayonnaise |
| 2 tsp | white vinegar |
| 3 tbsp | parsley, chopped and rinsed |
| 3 tbsp | minced chives |
| to taste | salt and black pepper |

1. Place salt on minced garlic and cream with side of a French chef's knife in a smashing motion.
2. Whisk together garlic and buttermilk and add mayonnaise, vinegar, parsley, and chives.
3. Adjust seasonings with salt and black pepper.

---
**Nutrition Facts**

**Per serving:** 68.1 calories; 70% calories from fat; 5.4 g total fat; 4.6 mg cholesterol; 202.9 mg sodium; 27.3 mg potassium; 4.7 g carbohydrates; 0 g fiber; 1.6 g sugar; 4.6 g net carbs; 0.6 g protein

---

**Recipe Variation:**

**17** **Caesar Dressing:** Combine 13 oz prepared buttermilk ranch dressing, 3 oz grated Parmesan cheese, 1 tbsp red wine vinegar, ½ tsp lemon juice, 1 tsp garlic powder, and 1 tsp cracked black pepper. Season with salt to taste.

# Recipes

## Sweet Balsamic Vinaigrette

*yield: 1 pt (sixteen 1-oz servings)*

| | |
|---|---|
| 5 oz | balsamic vinegar |
| 2 tbsp | granulated sugar |
| 10 oz | olive oil |
| to taste | salt and pepper |

1. Place vinegar in mixing bowl.
2. Add sugar and whisk until sugar is almost dissolved.
3. Add olive oil while whisking.
4. Season with salt and pepper.

### Nutrition Facts
**Per serving:** 162.2 calories; 94% calories from fat; 17.5 g total fat; 0 mg cholesterol; 18.9 mg sodium; 10.6 mg potassium; 2.2 g carbohydrates; 0 g fiber; 1.6 g sugar; 2.2 g net carbs; 0 g protein

## Croutons

*yield: 10 servings*

| | |
|---|---|
| 10 slices | white bread, crust removed and diced |
| 2½ oz | melted butter |
| ½ tsp | salt |
| ¼ tsp | pepper |

1. Toss all ingredients together to coat well.
2. Place buttered bread on sheet pan and bake in a 350°F oven until lightly browned on all sides, approximately 15 min. *Note:* Croutons can be sprinkled with Parmesan cheese immediately after being removed from the oven, if desired.

### Nutrition Facts
**Per serving:** 82.9 calories; 65% calories from fat; 6.2 g total fat; 15.3 mg cholesterol; 198.8 mg sodium; 14.4 mg potassium; 6.1 g carbohydrates; 0.3 g fiber; 0.5 g sugar; 5.8 g net carbs; 1 g protein

**Recipe Variation:**

## 20

**Whole-Grain Croutons:** Substitute whole-grain bread for the white bread.

## Grilled Chicken Caesar Salad

*yield: 4 entrée servings*

| | |
|---|---|
| 1 lb | romaine lettuce |
| 6 oz | Caesar dressing |
| 4 oz | seasoned croutons |
| 2 oz | Parmesan cheese, grated |
| ½ lb | grilled chicken breasts, sliced into thin strips and chilled |

1. Cut, wash, and dry lettuce.
2. Toss lettuce with Caesar dressing and portion on plate.
3. Top dressed lettuce with chicken breast, croutons, and Parmesan cheese.

### Nutrition Facts
**Per serving:** 336.3 calories; 26% calories from fat; 10.2 g total fat; 61.5 mg cholesterol; 924.4 mg sodium; 490.5 mg potassium; 33.1 g carbohydrates; 3.9 g fiber; 8.4 g sugar; 29.2 g net carbs; 27.9g protein

## Waldorf Salad

*yield: 10 side servings*

| | |
|---|---|
| 3 | red apples |
| 2 | green apples |
| ½ tsp | lemon juice |
| 6 oz | mayonnaise |
| ½ c | whipped cream, whipped stiff (or whipped topping) |
| 1 tsp | vanilla extract |
| 2 tsp | sugar |
| to taste | salt and pepper |
| 4 oz | walnuts, chopped |
| 1 head | leaf lettuce, leaves washed and separated |
| 5 | maraschino cherries, halved |
| 10 sprigs | parsley |

1. Peel and cut apples into small dice and toss with ½ tsp lemon juice to prevent browning.
2. Mix mayonnaise, whipped cream, vanilla, sugar, salt, and pepper well to make dressing.
3. Add apples and walnuts and mix well to incorporate.
4. Serve scoop of salad on lettuce leaf and garnish with a half cherry and parsley.

---

### Nutrition Facts

**Per serving:** 211.9 calories; 62% calories from fat; 15.5 g total fat; 12.6 mg cholesterol; 152.3 mg sodium; 211.6 mg potassium; 18.5 g carbohydrates; 3 g fiber; 9.7 g sugar; 15.5 g net carbs; 2.7 g protein

---

## Praline Pecan and Pear Salad with Sweet Balsamic Vinaigrette and Crumbled Blue Cheese

*yield: 4 servings*

*Praline pecans*

| | |
|---|---|
| ½ c | shelled pecans |
| ½ tbsp | brown sugar |
| 2 tbsp | water |
| to taste | salt and pepper |

*Salad*

| | |
|---|---|
| 4 oz | mesclun greens |
| ½ | ripe pear, medium dice |
| ½ c | praline pecans |
| 1 oz | blue cheese, crumbled |
| 2 oz | sweet balsamic vinaigrette |

1. Place shelled pecans in hot sauté pan and dry pan-roast until edges of pecans are brown. *Note:* Pecans can also be browned in the oven.
2. When pecans are slightly browned, add brown sugar and let melt slightly.
3. Add water and season with salt and pepper to taste. Stir continuously until all sugar is melted and water is completely evaporated.
4. Pour nuts onto buttered aluminum foil to cool.
5. Toss all salad ingredients gently to coat and distribute thoroughly.

---

### Nutrition Facts

**Per serving:** 241.7 calories; 76% calories from fat; 22.2 g total fat; 5.3 mg cholesterol; 175.6 mg sodium; 220 mg potassium; 10.3 g carbohydrates; 3.6 g fiber; 4.8 g sugar; 6.6 g net carbs; 4.4 g protein

---

# Recipes

## 24

## Salata Kalamata

*yield: 4 entrée servings*

| | |
|---|---|
| 4 oz | mesclun greens |
| 2 oz | kalamata olives |
| 2 oz | crumbled feta |
| 4 oz | grape tomatoes, sliced in half |
| 2 oz | sun-dried tomatoes |
| 2 oz | cucumbers, medium dice |
| 4 oz | red wine oregano vinaigrette |

1. Toss all ingredients together gently and serve.

**Nutrition Facts**

**Per serving:** 126.3 calories; 50% calories from fat; 7.3 g total fat; 12.6 mg cholesterol; 684.9 mg sodium; 657.4 mg potassium; 13.4 g carbohydrates; 2.5 g fiber; 6.9 g sugar; 10.8 g net carbs; 4.6 g protein

## 25

## Greek Vinaigrette

*yield: 1 pt (Sixteen 1 oz servings)*

| | |
|---|---|
| ½ c | red wine vinegar |
| 1 tsp | sugar |
| 1½ c | olive oil |
| 2 tsp | lemon juice |
| 2 tsp | oregano |
| 2 tbsp | green onion, minced dice |
| to taste | salt and pepper |

1. Place vinegar and sugar in mixing bowl and whisk until most of the sugar is dissolved.
2. Slowly add the olive oil while whisking.
3. Add lemon juice, oregano, and onion and mix well.
4. Season to taste.

**Nutrition Facts**

**Per serving:** 25.3 calories; 91% calories from fat; 2.6 g total fat; 0 mg cholesterol; 18.4 mg sodium; 7 mg potassium; 0.5 g carbohydrates; 0.1 g fiber; 0.3 g sugar; 0.4 g net carbs; 0 g protein

## 26

## Spinach and Roasted Potato Salad with Caraway Bacon Vinaigrette

*yield: 6 entrée servings*

| | |
|---|---|
| 6 oz | fresh leaf spinach, ribs removed |
| 1 lb | roasted potato cubes |
| 4 oz | bacon, diced |
| 1 | shallot, minced |
| 1 clove | garlic, minced |
| 1 tsp | caraway seed, pan-toasted |
| 1 oz | brown sugar |
| 3 oz | cider vinegar |
| 6 oz | vegetable oil |
| 1 tbsp | whole-grain mustard |
| 1½ tsp | fresh thyme, minced |
| to taste | salt and black pepper |

1. Render diced bacon. Remove bacon from heat when cooked and reserve.
2. Add shallots, garlic, and caraway seed to the bacon fat and sweat until tender.
3. Blend in brown sugar and melt it, using caution to not burn.
4. Remove pan from heat and whisk in vinegar, oil, mustard, and thyme. Whisk until emulsified. Season with salt and pepper to taste.
5. Place spinach, potatoes, bacon, and vinaigrette in a large bowl, toss to mix well, and serve.

**Nutrition Facts**

**Per serving:** 440.4 calories; 74% calories from fat; 37.4 g total fat; 12.9 mg cholesterol; 249.8 mg sodium; 651.9 mg potassium; 23.6 g carbohydrates; 2.6 g fiber; 5.6 g sugar; 21.1 g net carbs; 5.3 g protein

## Green Bean, Tomato, and Prosciutto Salad with Warm Bacon Vinaigrette

*yield: 10 (4-oz servings)*

| | |
|---|---|
| 2 lb | green beans, cleaned, blanched, and shocked |
| 2 | tomatoes, blanched, peeled, and cut into wedges |
| 4 oz | red onion, julienne |
| 4 oz | prosciutto, sliced thin and cut julienne |
| 4 oz | croutons |

*Dressing*

| | |
|---|---|
| 5 oz | bacon, diced |
| 2 tbsp | shallot, minced |
| 1 clove | garlic, minced |
| 1 tbsp | sugar |
| 4 oz | cider vinegar |
| 1 tsp | dijon mustard |
| 1½ tbsp | vegetable oil |
| to taste | salt and black pepper |

1. Render diced bacon until crisp and brown. Remove from pan and drain on paper towel.
2. Reserve 2 tbsp of the rendered bacon fat in the pan and discard remainder.
3. Add shallots and garlic to the bacon fat and sweat over medium heat for 1–2 min.
4. Add sugar and vinegar and heat to dissolve sugar.
5. Remove from heat and whisk in mustard and oil.
6. Season with salt and pepper to taste.
7. Combine all salad ingredients in mixing bowl. Serve with dressing.

### Nutrition Facts
**Per serving:** 200.7 calories; 46% calories from fat; 10.5 g total fat; 17.6 mg cholesterol; 541.2 mg sodium; 312.7 mg potassium; 19.9 g carbohydrates; 4 g fiber; 3.3 g sugar; 15.9 g net carbs; 8.3 g protein

## Thai Noodle Salad

*yield: 8 (4-oz servings)*

| | |
|---|---|
| 1 lb | rice noodles (vermicelli), cooked and cooled |
| 3 oz | bean sprouts |
| 4 | green onions, bias cut in 1" lengths |
| 4 oz | carrot, julienne |
| 1 | red pepper, julienne |
| 4 leaves | napa cabbage, julienne |

*Dressing*

| | |
|---|---|
| 2 cloves | garlic, minced |
| 3 | green onions, minced |
| 2 tsp | chili garlic sauce |
| 2 tbsp | sugar |
| 1 tbsp | lime juice |
| 2 oz | rice wine vinegar |
| 1 tbsp | fish sauce |
| 6 oz | salad oil |
| 1 tbsp | sesame oil |
| 2 tbsp | cilantro |

1. Combine all dressing ingredients and mix well to incorporate.
2. Place salad ingredients and dressing in mixing bowl and toss well to incorporate.

### Nutrition Facts
**Per serving:** 295.5 calories; 66% calories from fat; 22.9 g total fat; 0 mg cholesterol; 227.8 mg sodium; 210.4 mg potassium; 24.8 g carbohydrates; 1.7 g fiber; 4.7 g sugar; 23.1 g net carbs; 1.6 g protein

# Recipes

## 29

## Cucumber Salad

*yield: 6 side servings*

| | |
|---|---|
| 1½ oz | white wine vinegar |
| 1 clove | garlic, minced |
| 1 tbsp | dill |
| 3 oz | vegetable oil |
| 1 tsp | sugar |
| to taste | kosher salt and white pepper |
| 1 tbsp | pickled ginger, julienne (optional) |
| 1 | cucumber, peeled, seeded, cut in half moons |
| ½ | red onion, julienne |

1. Mix all ingredients except cucumber and onion in bowl.
2. Add cucumber and onion and mix well.

---

**Nutrition Facts**

**Per serving:** 137.6 calories; 91% calories from fat; 14.3 g total fat; 0 mg cholesterol; 49.2 mg sodium; 75.9 mg potassium; 2.9 g carbohydrates; 0.4 g fiber; 1.2 g sugar; 2.4 g net carbs; 0.4 g protein

---

## 30

## Asparagus and Tomato Salad with Shaved Fennel and Gorgonzola

*yield: 4 side servings*

| | |
|---|---|
| ¼ head | fennel, shaved |
| 2 oz | red onion, julienne |
| 2 | tomatoes, peeled and cut into wedges |
| ½ lb | asparagus, blanched, cooled, and cut into 1½" sections |
| 2 oz | gorgonzola cheese |

*Dressing*

| | |
|---|---|
| 3 tbsp | olive oil |
| ½ tbsp | red wine vinegar |
| ½ tbsp | balsamic vinegar |
| ½ tsp | garlic, minced |
| 2 tsp | shallots, minced |
| ¼ tsp | thyme, minced |
| ¼ tsp | tarragon, minced |

1. Combine all dressing ingredients in a bowl and whisk to mix well.
2. Toss all remaining ingredients into dressing and mix well. *Note:* Asparagus can be left whole for a composed salad presentation where other ingredients are placed carefully on top of the dressed asparagus.

---

**Nutrition Facts**

**Per serving:** 327.4 calories; 40% calories from fat; 15.2 g total fat; 12.7 mg cholesterol; 235.1 mg sodium; 1105.9 mg potassium; 43.2 g carbohydrates; 3.8 g fiber; 3.8 g sugar; 39.3 g net carbs; 11.4 g protein

---

## Roasted Corn and Black Bean Salad

*yield: 8 side servings*

| | |
|---|---|
| 2 ears | corn, shucked, kernels oiled and roasted (or substitute 1 c frozen corn for fresh shucked) |
| 3 tbsp | red onion, brunoise, rinsed |
| 3 tbsp | red pepper, brunoise |
| 3 tbsp | green pepper, brunoise |
| ¾ c | black beans, cooked and rinsed |

*Dressing*

| | |
|---|---|
| 1 clove | garlic, creamed |
| 1 | shallot, minced |
| 2 tsp | honey |
| 3 tbsp | lime juice |
| 1 tsp | white vinegar |
| 3 oz | vegetable oil |
| ¼ tsp | cumin |
| ½ | chipotle pepper, roasted and minced |
| 3 tbsp | cilantro, chopped |
| to taste | salt and pepper |

1. Roast corn until charred slightly and cool completely.
2. Mix all dressing ingredients well.
3. Add salad ingredients and mix well.

### Nutrition Facts

**Per serving:** 183.7 calories; 53% calories from fat; 11.1 g total fat; 0 mg cholesterol; 85.9 mg sodium; 316.9 mg potassium; 19.9 g carbohydrates; 2.4 g fiber; 2.5 g sugar; 17.5 g net carbs; 3.7 g protein

## Zesty Potato Salad with Dill

*yield: 4 side servings*

| | |
|---|---|
| 12 oz | new red potatoes |
| 2 tbsp | mayonnaise |
| 1½ tbsp | sour cream |
| 1 tsp | garlic, minced |
| 2 tsp | dill, minced |
| 2 tsp | whole-grain mustard |
| 1 tbsp | red pepper, small dice |
| 1 tbsp | green pepper, small dice |
| 1 tbsp | celery, small dice |
| 2 tbsp | red onion, small dice |
| ½ tsp | celery seed |
| 1 tsp | fresh lemon juice |
| to taste | salt and black pepper |

1. Cook potatoes in heavily salted water until al dente. Remove and place in refrigeration to cool completely.
2. Mix remaining ingredients and whisk together.
3. When potatoes are cool, cut into quarters.
4. Add potatoes to dressing and mix gently to incorporate.

### Nutrition Facts

**Per serving:** 118.1 calories; 29% calories from fat; 4 g total fat; 3.9 mg cholesterol; 170.1 mg sodium; 532.4 mg potassium; 19.4 g carbohydrates; 2 g fiber; 0.7 g sugar; 17.4 g net carbs; 2.6 g protein

# Recipes

## 33

## Tuna Niçoise Salad

*yield: 4 servings*

| | |
|---|---|
| 4 | red potatoes, cooked |
| 4 | tuna steaks, grilled and chilled |
| 2 | eggs, hard cooked |
| 4 oz | mixed greens |
| 8 leaves | Boston bibb lettuce |
| ¾ lb | green beans, blanched |
| 1 | tomato, cut into wedges |
| 4 oz | roasted red peppers |
| 12 | Niçoise or Kalamata olives, pitted |
| 4 oz | croutons |

*Dressing*

| | |
|---|---|
| 2 oz | red wine vinegar |
| 2 tsp | dijon mustard |
| 2 tsp | shallot, finely minced |
| 6 oz | vegetable oil |
| ½ tsp | sugar |
| 2 tbsp | mixed herbs (basil, parsley, tarragon) |
| to taste | salt and black pepper |

1. Combine vinegar, dijon mustard, and shallots.
2. Slowly whisk in oil.
3. Season with sugar, herbs, salt, and pepper.
4. Place Boston lettuce on plate as a liner.
5. Mix a small amount of dressing into mixed greens and toss well. Place on top of Boston lettuce.
6. Carefully arrange green beans, tomatoes, roasted red peppers, olives, and croutons on salad greens and top with tuna fillets. *Note:* Canned tuna may be substituted for tuna steaks.

### Nutrition Facts
**Per serving:** 832.0 calories; 58% calories from fat; 55 g total fat; 152.9 mg cholesterol; 533.3 mg sodium; 1324.8 mg potassium; 54.7 g carbohydrates; 7.6 g fiber; 3 g sugar; 47.1 g net carbs; 31.9 g protein

## 34

## Blue Cheese and Walnut Terrine

*yield: 1 terrine*

| | |
|---|---|
| ½ lb | blue cheese |
| ½ lb | cream cheese (room temperature) |
| ½ stick | butter (near room temperature) |
| ½ cup | toasted walnuts |
| 2 oz | mesclun greens |

1. Pulse all together in food processor until fairly well distributed.
2. Line terrine mold with plastic wrap.
3. Remove mixture from food processor and press firmly into plastic-lined terrine mold.
4. Cover mixture completely and refrigerate overnight to set completely.
5. Unmold terrine and slice as needed.
6. Serve terrine with a side of mesclun greens.

### Nutrition Facts
**Entire recipe:** 2386.6 calories; 83% calories from fat; 228.3 g total fat; 541.1 mg cholesterol; 3847.7 mg sodium; 1211.6 mg potassium; 20.5 g carbohydrates; 4.7 g fiber; 3.2 g sugar; 15.8 g net carbs; 75.6 g protein

## Swiss Cheese and Roasted Garlic Fondue

*yield: 1 pt*

| | |
|---|---|
| 2 cloves | garlic |
| 1 tsp | olive oil |
| 6 oz | grated Swiss cheese |
| 12 oz | béchamel |

1. Place garlic in aluminum foil with olive oil.
2. Place in a 300°F oven and roast until soft and lightly browned, about 20 min.
3. Heat béchamel sauce.
4. Using a whisk, slowly add grated Swiss cheese to béchamel. Continue to whisk until all cheese is incorporated and fondue is smooth.
5. Add roasted garlic and whisk well to incorporate.

### Nutrition Facts
**Entire recipe:** 5464.7 calories; 60% calories from fat; 376.6 g total fat; 1044.7 mg cholesterol; 2141.3 mg sodium; 4372.8 mg potassium; 370.9 g carbohydrates; 7.3 g fiber; 143.2 g sugar; 363.6 g net carbs; 155.9 g protein

## Prosciutto-Wrapped Baked Brie en Croûte

*yield: 1 wheel baked brie*

| | |
|---|---|
| ¼ lb | prosciutto |
| 10 oz | Brie cheese (1 wheel) |
| 1 | egg, beaten |
| 1 oz | milk |
| 1 sheet | puff pastry |

1. Lay out sheets of thinly sliced prosciutto and wrap around wheel of Brie.
2. Mix beaten egg and milk to make an egg wash.
3. Lay prosciutto-wrapped Brie on top of a sheet of puff pastry. Egg wash edges of pastry.
4. Wrap edges of pastry around Brie to completely encase the brie.
5. A half hour before baking, place Brie in freezer for 30 min to firm. *Note:* This will help the Brie to retain its shape in the oven.
6. Egg wash the top and sides of the puff pastry and bake at 350°F for 25–30 min until golden brown in color. Serve warm.

### Nutrition Facts
**Entire recipe:** 4847.1 calories; 70% calories from fat; 387.6 g total fat; 1573.9 mg cholesterol; 11307.2 mg sodium; 2663.2 mg potassium; 29.1 g carbohydrates; 0.7 g fiber; 8.1 g sugar; 28.4 g net carbs; 307.8 g protein

# Sandwiches and Hors d'Oeuvres

## Chapter 13

# Sandwiches and hors d'oeuvres

*Sandwiches and hors d'oeuvres* offer opportunities to demonstrate creativity. The sandwich requires the basic assembly of bread, spread, filling, and garnishes. A skilled chef is able to prepare many forms of hot and cold sandwiches including simple closed, multidecker, open-faced, grilled, deep-fried, tea, and wraps. Hors d'oeuvres are foods served before a meal to whet the appetite.

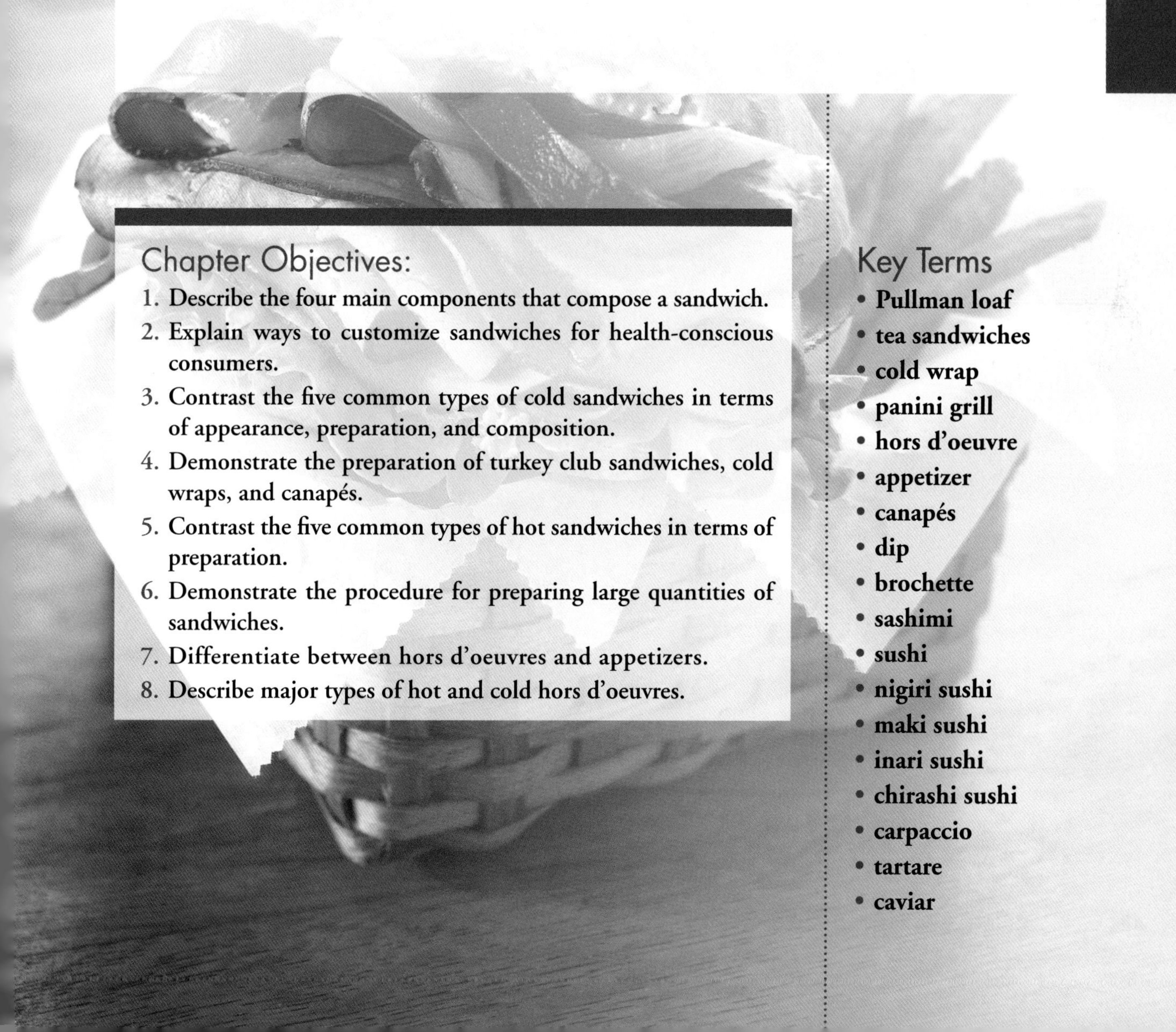

## Chapter Objectives:

1. Describe the four main components that compose a sandwich.
2. Explain ways to customize sandwiches for health-conscious consumers.
3. Contrast the five common types of cold sandwiches in terms of appearance, preparation, and composition.
4. Demonstrate the preparation of turkey club sandwiches, cold wraps, and canapés.
5. Contrast the five common types of hot sandwiches in terms of preparation.
6. Demonstrate the procedure for preparing large quantities of sandwiches.
7. Differentiate between hors d'oeuvres and appetizers.
8. Describe major types of hot and cold hors d'oeuvres.

## Key Terms

- **Pullman loaf**
- **tea sandwiches**
- **cold wrap**
- **panini grill**
- **hors d'oeuvre**
- **appetizer**
- **canapés**
- **dip**
- **brochette**
- **sashimi**
- **sushi**
- **nigiri sushi**
- **maki sushi**
- **inari sushi**
- **chirashi sushi**
- **carpaccio**
- **tartare**
- **caviar**

Stories handed down through culinary history suggest that John Montagu, the fourth Earl of Sandwich, can be credited as the inventor of the modern-day sandwich. The name "sandwich" was derived from his title. Montagu served as the First Lord of the Admirality and as a colonel in the British army. When Captain Cook discovered a small group of islands in the South Pacific, he named them the "Sandwich Islands" after his sponsor, Montagu. The Sandwich Islands are known today as the Hawaiian Islands.

Two legends offer explanations of how the sandwich came to be named after Montagu. The first legend derives from Montagu's reputation as an avid gambler who hated to leave his seat at a card table. He would ask his servants to serve him cured or dried meat between two slices of bread to keep his hands somewhat clean so that he could continue playing cards while eating. The second legend contends that Montagu was a very busy man who would have his servants put some meat between two slices of bread so that he could continue to work while eating.

## SANDWICHES

From a hot dog at a ballpark, to a hamburger from a popular restaurant, to a grilled-cheese sandwich served with a hot bowl of tomato soup, sandwiches are a staple of the American diet. Sandwiches are a convenient food because they are self-contained, which adds to their portability. They are a popular lunch item, as they are easy to transport in a lunch sack or box.

In the professional kitchen, a new employee's first responsibility is often the sandwich station, because sandwich making is a skill that can be quickly learned. This does not necessarily mean that it is easy work. The sandwich station can be one of the busiest and fastest-paced stations in the entire kitchen. Although it is common for the sandwich station to be located in the pantry area of a restaurant, away from the heat of the main kitchen, hot sandwiches such as the hamburger are served from the hot line.

Most sandwiches consist of four main components: the bread, the spread, the filling, and the garnish. *See Figure 13-1.* Although some sandwiches may have all four of these components, it is not mandatory for a sandwich to include every one of them. Due to the number of different breads, spreads, and fillings available, the number of potential sandwich combinations is endless.

## Sandwich Components

National Chicken Council

**13-1** *The four main components of a sandwich are the bread, spread, filling, and garnish.*

It is important to choose sandwich components that complement each other in flavor and texture. Imagine the following as a menu item: *Hickory-Smoked Turkey Sandwich, served on hearty multigrain bread with crisp bacon, Swiss cheese, and a tangy honey-mustard mayonnaise.*

Many people would find this sandwich appetizing, because it offers a balance of sweet, smoky, savory, and bitter or sharp flavors. The tangy honey mustard offers sweet and sour flavors that pull together the range of flavors in the sandwich and create a well-balanced overall flavor. Understanding each of the main components of a sandwich helps a chef to choose interesting combinations that deliver satisfying flavors.

## Breads

Bread has four main purposes in a sandwich; it serves as the packaging, is an economical filler, provides nutrition, and adds flavor and color. Bread serves as easy-to-handle edible packaging that holds the filling, spread, and garnish in place while eliminating the mess that would otherwise occur from eating the sandwich ingredients from hand. Without bread, holding onto the ingredients of a sandwich would be messy and perhaps difficult. Bread is an economical filler, in that it is relatively inexpensive compared with meats, cheeses, or vegetables, yet it can quickly satisfy the appetite.

Nutritional value varies, depending on the bread. Whole-grain and multigrain breads provide more nutrients than refined white or even whole wheat bread. Finally, bread adds flavor and color to a sandwich. Bread can range from the common white, wheat, and rye varieties to more exciting flavors such as onion rye, tomato herb, dried cherry, and walnut breads. Some varieties of bread have a very soft texture, while others, such as sourdough, can be chewy or firm. ***See Figure 13-2.***

## *Breads*

**13-2** *The many varieties of bread offer a wide range of flavors and textures.*

When choosing bread for a sandwich, there are a few important points to consider. The bread should not overpower the sandwich by having a strong flavor that hides the flavor of the filling. Also, bread should not be so soft that it tears when a spread is applied, but should not be so tough that it is difficult to bite or chew. Examples of commonly used sandwich bread varieties in professional kitchens include croissants, various rolls (such as kaiser, onion, French, and Italian), pitas, tortilla wraps, sourdough, foccacia, traditional white bread, rye bread, savory breads (such as onion or herb), and whole-grain breads. The most common form of bread used in professional cooking is the Pullman loaf.

A *Pullman loaf* is a long, rectangular loaf of bread that has a flat top, uniform slices, and a consistently fine texture from one end to the other. Each slice of the Pullman loaf is almost perfectly square. A Pullman loaf is much longer than loaves of bread commonly found in a grocery store or bakery. The Pullman loaf is widely used because its consistent slices make it easy to prepare sandwiches that look the same. It also has a dry texture that helps it resist becoming soggy after a spread such as mayonnaise is applied. Flat loaves of white or wheat bread, such as the Pullman loaf, are also the traditional breads used as a base for finger sandwiches, canapés, and various hors d'oeuvres. **See Figure 13-3.**

## Pullman Loaf

Alpha Baking Co., Inc.

**13-3** *The most common form of bread used in professional cooking is the Pullman loaf.*

Today more than ever before, people are seeking out healthy, high-quality foods, even when it comes to fast food. To satisfy customers, breads and rolls should be fresh, and never served stale or dried out.

## Spreads

Spreads serve a variety of purposes in a sandwich. Some spreads are used to add flavor and moisture, some to add richness from fat they contain, and some are used as binders for certain filling combinations. Spreads may be used to create a moisture barrier for the bread, preventing the bread from becoming soggy. There are few things less appetizing than a soggy sandwich. Since fats repel liquids, fat-based spreads can help keep the moisture of wet fillings from soaking into the bread. Conversely, mayonnaise and salad dressing can make bread soggy if they are added to the sandwich too far in advance.

The main types of spreads fall into one of four categories: mayonnaise, salad dressing, butter, and variety. It is important when choosing a spread to make sure that the spread will not overpower the sandwich, but complement the flavor and taste of the main ingredient.

National Cancer Institute

**Mayonnaise.** Mayonnaise is the most commonly used spread. Because it is made from egg yolks and oil, it adds moisture, flavor, and richness to a sandwich. Mayonnaise is also used as a binding agent for sandwich fillings such as chicken salad or tuna salad. As an oil-based spread, it acts as a moisture blocker, helping to prevent bread from becoming soggy. Mayonnaise can be flavored simply through the addition of other condiments, such as mustard.

**Salad Dressing.** Salad dressing refers not to the kind served over a bed of salad greens, but rather to cooked sandwich spread, such as Miracle Whip®. Salad dressing serves the same purpose as mayonnaise and has a similar flavor and texture. Like mayonnaise, salad dressing spreads can also be flavored with other ingredients. The primary difference between mayonnaise and salad dressing is that salad dressing does not contain any egg yolks.

**Butter.** Butter is one of the most common spreads and can add richness and flavor to a sandwich. Compound butters are often used to add a more complex and intense flavor to a sandwich. An avocado butter could be a delicious spread on a Southwest-style grilled chicken breast sandwich. Compound butters are easy to prepare and are a great way to enhance the flavor and richness of a sandwich.

*Chef's Tip:*

Mayonnaise needs to be kept refrigerated after it is opened, as it is highly susceptible to contamination. It is also extremely important to keep salads, sandwiches, and other items that are made with mayonnaise at safe temperatures.

All butters, whether compound or plain, should be either somewhat soft or should be whipped before use so that they spread evenly and gently without tearing the bread. Often chefs whip a small amount of water into butter to lighten its texture and density, as well as to stretch their supply of butter. Margarine is often used as a substitute for butter to reduce cost or because butter is higher in calories, saturated fats, and cholesterol. However, some varieties of margarine contain trans fats. Scientific evidence suggests that trans fats can raise levels of LDL cholesterol, which can increase the risk of heart disease. Some cities have banned the use of trans fats in professional cooking.

**Variety Spreads.** A variety spread is any other mixture that is of a spreadable consistency and can be added to a sandwich to complement or increase flavor and moisture. For example, flavored cream cheese, spreadable cheese, pesto, guacamole, refried beans, hummus (chick pea purée), baba ghanoush (roasted eggplant purée), tapenade (black olive and caper spread), spreadable pâté, and spicy green olive jardiniere purée are all great spreads that can dramatically enhance the overall flavor of a sandwich. Even some prepared condiments can be considered spreads, including vinegar and oil, vinaigrettes, mustard, ketchup, and fruit preserves of any kind. A chef should be aware that, while these spreads may be flavorful, they do not provide the same moisture barrier to prevent soggy bread that mayonnaise, salad dressing, or butter provide.

## Fillings

A *filling* is the main ingredient in a sandwich and is often the reason that people order it. Types of sandwich fillings include hot or cold meats, poultry, seafood, cheeses, bound salads, or vegetables, or a combination of any of these. *See Figure 13-4.* Many sandwiches are named for their filling, such as a hot dog, a hamburger, or an egg salad sandwich. *See Figure 13-5.*

| | Sandwich Fillings | | |
|---|---|---|---|
| Category | Examples | Category | Examples |
| Beef | Thinly sliced roasted beef (hot or cold)<br>Hamburger patties<br>Corned beef<br>Pastrami<br>Thin steaks such as butt steaks | Seafood | Tuna (fresh steaks or canned)<br>Fish fillets<br>Lox or cured salmon<br>Shrimp<br>Lobster<br>Crab<br>Surimi<br>Sardines |
| Pork | Bacon<br>Barbequed pork<br>Roasted and sliced pork loin<br>Grilled pork tenderloin<br>Canadian bacon<br>Ham | Cheese | Cheddar<br>Swiss<br>American (processed)<br>Provolone<br>Mozzarella<br>Cream cheese |
| Sausage | Italian<br>Polish<br>Frankfurter (hot dog)<br>Bologna<br>Bratwurst<br>Various luncheon meats | Bound salad | Chicken<br>Seafood<br>Tuna<br>Turkey<br>Ham<br>Egg |
| Poultry | Chicken breast<br>Turkey breast | Vegetables and fruits | Lettuce<br>Tomato<br>Red onion<br>Any vegetable suitable for grilling or roasting<br>Bean sprouts<br>Peanut butter<br>Fruit jellies, jams, compotes, or preserves |

**13-4** *Common sandwich fillings include various types of meat, poultry, seafood, cheeses, bound salads, vegetables, and fruits.*

## Fillings

Egg salad

Pullman loaf

13-5 *Fillings such as egg salad give a sandwich its name.*

Sandwiches based on a combination of fillings include the Monte Cristo sandwich, which has ham, turkey and Swiss cheese, and the Reuben sandwich, which has corned beef, sauerkraut, and Swiss cheese. Since the filling is the main attraction of the sandwich, it essential that the filling be prepared and served properly. For example, when serving a hot fish sandwich, the fish should be hot, moist, flaky, and not overcooked (dry, tough, charred, or burnt) or undercooked (wet, cold, or mushy). The temperature of the filling should be appropriately hot or cold, depending on the type of sandwich. The filling should never be at room temperature when a sandwich is served.

### Garnishes

Garnishes increase the visual appeal of a sandwich. Typical sandwich garnishes include olives, raw vegetables (crudités), fruit, and side salads such as potato salad, pasta salad, and coleslaw. Vegetable toppings such as lettuce, tomato slices, slices of raw onion, or pickle spears are also appropriate sandwich garnishes. ***See Figure 13-6.***

### Sandwich Nutrition

Sandwiches can be nutritious, balanced meals or they can be simply great-tasting compact meals where health and nutrition are not factors in their selection. It is important that food service establishments offer healthy options for customers who are health-conscious or who have special dietary needs. For example, many people enjoy a big hamburger with melted cheese, caramelized onions, mayonnaise, and other condiments on a sesame-seed white-bread bun, served with a heaping pile of hot French fries. Likewise, someone watching calories, cholesterol, or fiber may prefer a turkey burger on a whole-grain roll with lettuce and tomatoes, accompanied by a side of fresh fruit.

Combining the ingredients of a sandwich with some thought can yield healthy and nutritious combinations. A well-planned sandwich can contain each of the food groups in the correct proportions, creating a meal that includes proteins, carbohydrates, fats, vitamins, and minerals. Although many sandwiches are not considered healthy because they are high in fat, changing the cooking method or the spread can change the nutritional composition. For example, replacing a deep-fried chicken breast with a grilled chicken breast would provide a healthier sandwich. Simple changes, such as making a chicken salad sandwich with light or fat-free mayonnaise rather than heavy mayonnaise, can have an impact on the nutrition of a sandwich.

## Garnishes

Hot peppers — Carrots — Coleslaw — Lettuce — Fresh tomatoes

Florida Department of Agriculture and Consumer Services, Bureau of Seafood and Aquaculture Marketing

**13-6** Garnishes are added to increase the visual appeal of a sandwich or to complement the flavors of a sandwich.

## SANDWICH STYLES

Many varieties of sandwiches can be served either hot or cold. For example, a cold tuna salad sandwich served hot with a slice of melted cheese becomes a tuna melt. A cold corned beef on rye sandwich can be made into a hot Reuben sandwich by adding some sauerkraut and Swiss cheese and griddling it until hot. Although some sandwich varieties are great served either hot or cold, some sandwiches, such as hamburgers or egg salad, are best served only as designed.

## Cold Sandwiches

Cold sandwiches are so called because their fillings are served below room temperature. Often the fillings are cooked ahead of time and then refrigerated, such as roast beef, turkey, or ham luncheon meats. Cold sandwiches can be separated in five distinct categories: simple cold closed, multidecker, open-faced cold, tea, and cold wraps. Although each type has some or all of the four basic sandwich components, they are quite different in presentation.

**Simple Cold Closed Sandwiches.** Simple cold closed sandwiches are the easiest to recognize, the quickest to prepare, and the most commonly served. Simple cold sandwiches typically have two pieces of bread (or the top and bottom of a bun or roll) surrounding a spread, a filling, and a garnish. A cold ham and cheese on white; a turkey, lettuce, tomato, and mayo on wheat; and a classic Italian submarine sandwich are examples of simple cold closed sandwiches. ***See Figure 13-7.***

### Historical Note

It wasn't until World War II that peanut butter and jelly formed their great-tasting relationship. Soldiers received individual packets of peanut butter and jelly in their K-rations to be used as condiments. Some creative soldiers came up with the idea of spreading peanut butter on one slice of bread and jelly on another and sticking the two together to form a peanut butter and jelly sandwich. The combination of salty, creamy peanut butter and sweet fruit jelly soon became a favorite sandwich, enjoyed by both children and adults.

## Simple Cold Closed Sandwiches

**13-7** *Simple cold closed sandwiches are the most commonly served variety of cold sandwiches.*

Refer to the CD-ROM for the **Multidecker Sandwiches** Media Clip.

**Multidecker Sandwiches.** Multidecker sandwiches are very similar to simple cold closed sandwiches. The main difference is that a multidecker sandwich has three or more pieces of bread, at least two layers of filling, and at least one spread (which is spread on the bread between each layer). The most common multidecker sandwich is the turkey club sandwich.

To prepare a multidecker (turkey club) sandwich, apply the following procedure:

1. Coat one side of three toasted slices of bread with mayonnaise. **See Figure 13-8.**
2. Layer lettuce, tomato, turkey breast, and cheese on one slice of bread. Top with a second slice of bread
3. Add lettuce, tomato, and bacon and top with a third slice of bread.
4. Cut into triangles.
5. Arrange triangles on a wooden skewer or plate in an appealing format.

**Open-Faced Cold Sandwiches.** Open-faced cold sandwiches have a more upscale appearance than simple cold closed sandwiches. When preparing a typical open-faced cold sandwich, a single slice of bread is toasted or grilled to make it firmer. The bread is then lightly coated with a flavorful spread and topped with thinly sliced meat, poultry, seafood, partially cooked or raw vegetables, or a thin layer of a bound salad. The sandwich is then artfully garnished, resulting in an appetizing presentation. A half of a bagel with lox and cream cheese is an example of a common open-faced cold sandwich. An open-faced sandwich is in essence a larger form of the open-faced hors d'oeuvre known as a canapé. **See Figure 13-9.**

**Tea Sandwiches.** A *tea sandwich* is a petite and delicate sandwich that originated in England as a snack served with afternoon tea. The breads used for tea sandwiches are light and soft in texture. The crusts are trimmed off to make them easy to eat without creating a lot of crumbs. The spreads used in tea sandwiches can be very flavorful, and the fillings are usually delicate in flavor, such as cucumber or watercress. Tea sandwiches are usually cut into fancy shapes such as diamonds, circles, rectangles, or squares, or can be rolled and sliced into pinwheels. **See Figure 13-10.**

**Cold Wraps.** A *cold wrap* is a variety of cold sandwich in which a flat bread or tortilla is coated with a spread, topped with one or more fillings, and rolled tightly. It can be served whole or can be cut in half to reveal the filling. Cold wraps are often served wrapped in parchment or waxed paper. When wrapped in paper, a wrap sandwich is one of the cleanest and most convenient sandwiches to eat, because the person eating it can hold the paper rather than the sandwich.

## Preparing Multidecker Sandwiches

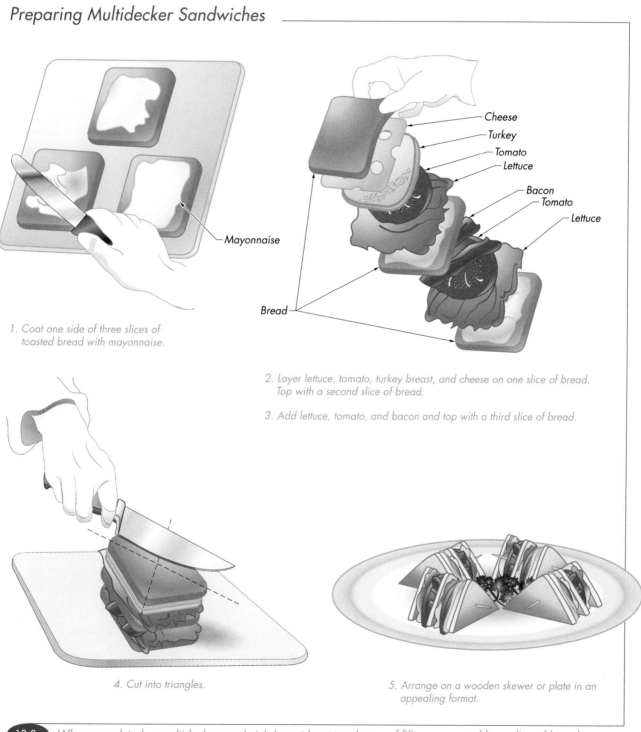

1. Coat one side of three slices of toasted bread with mayonnaise.

Mayonnaise

Cheese
Turkey
Tomato
Lettuce
Bacon
Tomato
Lettuce
Bread

2. Layer lettuce, tomato, turkey breast, and cheese on one slice of bread. Top with a second slice of bread.

3. Add lettuce, tomato, and bacon and top with a third slice of bread.

4. Cut into triangles.

5. Arrange on a wooden skewer or plate in an appealing format.

13-8    When completed, a multidecker sandwich has at least two layers of filling separated by a slice of bread.

## Open-Faced Cold Sandwiches

Toasted bread

Goat cheese

Grilled portobello mushrooms

**13-9** Open-faced cold sandwiches can be more elegant than simple cold closed sandwiches.

## Tea Sandwiches

**13-10** Tea sandwiches are delicate and petite finger foods.

To prepare a cold wrap sandwich, apply the following procedure:

1. Coat a piece of flatbread (such as a tortilla) with a spread. *See Figure 13-11.*
2. Arrange desired fillings on the spread.
3. Fold in the sides of the flatbread to seal the ends of the wrap.
4. Roll the flatbread tightly around the filling.
5. Cut the completed wrap in half on a bias.
6. Garnish as desired.

CULINARY PROCEDURES

*Preparing Wrap Sandwiches*

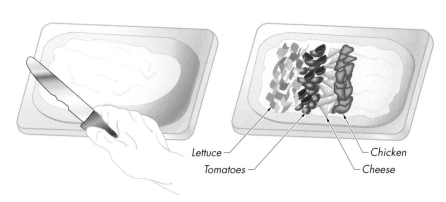

Lettuce
Tomatoes
Chicken
Cheese

*1. Coat flatbread with spread.*

*2. Arrange filling on spread.*

*3. Fold in sides of flatbread so the filling will not fall out when rolling the wrap.*

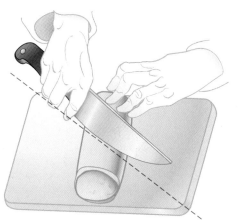

Florida Department of Citrus

*4. Roll flatbread tightly around filling.*

*5. Cut on bias.*

*6. Garnish as desired.*

**13-11** *Wrap sandwiches are prepared by rolling flatbread around a filling.*

## Hot Sandwiches

Hot sandwiches feature cooked fillings and are served hot. Most hot sandwiches are served on warm toasted or grilled bread. Care must be taken when preparing hot sandwiches with uncooked garnishes, such as lettuce, tomatoes, or pickle slices, to avoid wilting the garnish. Fresh garnishes are placed on the sandwich after the hot filling and bread are plated. Common types of hot sandwiches include simple hot closed, open-faced hot, grilled or griddled, deep-fried, and hot wraps.

**Simple Hot Closed Sandwiches.** Simple hot closed sandwiches are made in the same style as simple cold closed sandwiches, with the difference being that the filling is served hot between two pieces of bread or a split roll or bun. Hot sandwiches often contain additional fillings or garnishes that are not heated, such as raw lettuce and tomato. The most common simple hot closed sandwich is the hamburger. Other simple hot closed sandwiches include the fish sandwich, grilled chicken sandwich, hot roast beef sandwich, steak sandwich, and veggie burger. *See Figure 13-12.*

*Simple Hot Closed Sandwiches*

Filling — Bread — Spread — Garnish

13-12 *Simple hot closed sandwiches consist of a top and bottom piece of bread, or a split roll, surrounding a hot filling, spread, and a garnish.*

**Open-Faced Hot Sandwiches.** Open-faced hot sandwiches are sandwiches consisting of a slice or two of either fresh, toasted, or grilled bread placed on a serving plate, topped with hot meat or other hot filling, and covered with gravy, sauce, melted cheese, or another type of topping. Often the open-faced hot sandwich is browned under a broiler just prior to serving. Due to the fact that this variety of sandwich is covered with a sauce, it is usually eaten with a knife and fork rather than by hand. Popular types of open-faced hot sandwiches include hot turkey, hot roast beef, and various kinds of melts such as the patty melt, tuna melt, and turkey melt. ***See Figure 13-13.***

*Open-Faced Hot Sandwich*

Spread

Garnish

Bread

Filling

**13-13** *Open-faced hot sandwiches have a hot filling topped with a sauce or melted cheese.*

The most popular open-faced hot sandwich is one that is usually not thought of as a sandwich: the pizza. A pizza consists of bread dough topped with a spread (tomato sauce), a filling (sausage, vegetables, etc.), and a garnish (cheese or spices) all baked until golden brown. In the case of a plain cheese pizza, cheese is the filling and there is typically no garnish. Although once considered exclusively a casual food, gourmet pizzas can now be found in restaurants around the world. ***See Figure 13-14.***

*Pizza*

National Cattleman's Beef Association

**13-14** *Pizza is a type of open-faced hot sandwich.*

**Grilled or Griddled Sandwiches.** Grilled or griddled sandwiches are those that are usually buttered on the outside and cooked on a griddle, in a sauté pan, or on a panini grill. They sometimes are also referred to as toasted sandwiches because the bread turns a toasty golden brown color during cooking. Grilled or griddled sandwiches typically include cheese as a filling ingredient, as it melts during cooking and holds the bread to the filling. It is important to understand that the filling in a griddled or grilled sandwich is only heated, rather than being fully cooked, during the assembly process. All items that must be fully cooked, such as bacon, chicken, or beef, must be thoroughly cooked prior to use as a filling on a griddled or grilled sandwich. The two most common grilled sandwiches are the grilled cheese sandwich and the Reuben sandwich.

A *panini grill* is an Italian clamshell-style grill made specifically to cook grilled sandwiches. To cook a sandwich on a panini grill, the sandwich bread is first buttered on both exterior sides. Next, the sandwich is placed in the grill and the hinged lid is closed over the sandwich. The hinged lid holds the sandwich in place while it cooks, and grills the sandwich on both top and bottom sides without having to flip the sandwich. A sandwich cooked on a panini grill is often called a panini sandwich. ***See Figure 13-15.***

*Panini Sandwiches*

13-15  *A panini grill is used to cook panini sandwiches.*

**Hot Wraps.** Hot wraps are very similar in appearance to cold wraps. They can be made using one of two methods. In the first method, flatbread is covered with a spread, topped with a cold precooked filling, and rolled up. The resulting rolled wrap sandwich is then cooked in any variety of ways such as by deep-frying, baking, griddling, or grilling it on a panini grill. An alternate method is placing flatbread on a heated griddle, covering it with a spread, topping it with a hot precooked filling, rolling it tightly, and serving. Common examples of hot wrap sandwiches are burritos, fajitas, tacos, and enchiladas. ***See Figure 13-16.***

*Hot Wraps*

*USA Rice Federation*

**13-16** *Hot wraps can be cooked after assembling cold ingredients, or assembled with hot ingredients after cooking.*

**Deep-Fried Sandwiches.** Deep-fried sandwiches are made by dipping a simple hot closed sandwich into beaten eggs (and sometimes additionally into bread crumbs) and gently placing it in a deep-fat fryer. *Note:* When an egg-dipped sandwich is cooked, it must reach a safe internal temperature to properly cook the raw egg. Alternate cooking methods for this variety of sandwich are cooking it on a sheet pan in the oven or on a griddle, as deep-fat frying can make the sandwich somewhat greasy. The most popular sandwich in this category is the Monte Cristo: two pieces of bread filled with cooked ham, turkey, and Swiss cheese and soaked in beaten eggs before it is pan-fried. It is often served with fruit, syrup, or preserves on the side and dusted with powdered sugar.

## SANDWICH STATION MISE EN PLACE

Preparing sandwiches is a skill that relies on organization, consistency, and speed. Whether preparing sandwiches in large quantities, such as for a banquet, or individually to order, the sandwich station must be well organized to reduce the amount of movement needed and maximize efficiency.

Practicing proper sanitation is extremely important throughout the kitchen. Because many sandwiches are served cold and undergo quite a bit of handling by the sandwich maker (slicing the meat, tomatoes, cheese, buns, etc.), it is important to make sure that the ingredients are properly refrigerated and are handled correctly at all times. Sandwich makers should wear gloves when preparing or handling sandwiches to prevent cross-contamination. Gloves should be changed frequently to aid in food safety and sanitation. *See Figure 13-17.*

### Sandwich Sanitation

**13-17** *Gloves should be worn when preparing sandwiches. It is important to change gloves often to avoid contamination.*

### Ingredients

The ingredients needed to prepare sandwiches should all be in place prior to receiving the first order. The primary responsibility of a sandwich cook (also called a prep, pantry, or garde manger cook) is to assemble sandwiches quickly, neatly, and efficiently. This means that spreads should be prepared, meats and cheeses should be sliced and portioned, and garnishes should be washed and ready to use. Garnishes such as lettuce leaves should be separated from the head, and tomatoes should be sliced. When service time comes, nothing should be left to prepare except the sandwich itself. *See Figure 13-18.*

## Sandwich Station

Barker Company

13-18    *Everything should be in place at the sandwich station before the first order is received.*

All items or ingredients in a sandwich station should be within arm's reach to maximize efficiency and speed. Because much of a sandwich maker's work is assembly work, for fast production it is important that the individual making the sandwiches use both hands. For example, one hand can be placing the filling on the bread while the other is reaching for the garnishes.

**Portion Control.** Practicing proper portion control is important in all aspects of food production. It is also important for maintaining an accurate food cost, as the selling price of an item is based on that item's standard portion size. If the portions are not consistent, the actual cost of making each sandwich is not either. Proper portion control also ensures that customers receive the same value for their money. For example, if two customers at the same table order corned beef sandwiches, and one customer receives a sandwich with meat piled high while the other customer receives a skimpy portion, the person who received less meat is sure to complain or at least not be very happy with the sandwich. ***See Figure 13-19.***

All items for sandwich making should be portioned prior to service. All meats should be sliced and portioned by weight. The portions can either be individually bagged in small plastic baggies or can be placed in small piles separated by individual squares of waxed paper and stacked or layered in a storage container. Items such as cheese that are served by the slice should be sliced to an exact thickness to ensure a consistent portion.

## Portion Control

Hobart

**13-19** *Practicing proper portion control is important to maintaining accurate cost control and providing customers with consistent value.*

### Equipment

The equipment needs of a sandwich station are determined by what is on the menu. For example, if deli meats and cheeses are bought sliced, there is no need for a slicer at the sandwich station. If there are hot items, such as hot roast beef with au jus, the station needs to have a steam table to keep the items hot. Equipment found at a sandwich station includes storage equipment, portioning equipment, cooking equipment, and hand tools.

Storage equipment at a sandwich station typically includes refrigerators and steam tables. Refrigerators at sandwich stations are often equipped with condiment tops that can hold garnishes and spreads within easy reach. A steam table or soup warmer is used to hold roasted meats and hot soups.

Portioning equipment is essential to produce consistent portions at the sandwich station. Meats and other fillings must be portioned consistently to avoid customer dissatisfaction. Standard portioning equipment includes scales and portion control scoops.

Cooking equipment may be necessary to prepare hot sandwiches such as hot roast beef, grilled chicken, or grilled cheese sandwiches. Typical cooking equipment for a sandwich station can include a toaster, griddle, grill, broiler, deep fryer, oven, or a Panini grill.

Hand tools such as a variety of knives (paring knife, French knife, and bread slicer), spatulas, spreaders, and cutting boards are basic necessities for a sandwich station. If all mise en place is completed in advance of service, these may be the only tools necessary during production.

## PREPARING LARGE QUANTITIES OF SANDWICHES

Although an establishment may have a different procedure for making each sandwich on the menu, it is possible to make large quantities of sandwiches at one time following a simple assembly procedure. The basic assembly procedure for making any sandwich includes grilling or toasting the bread (if desired), coating one side of the bread with spread, placing a filling (hot or cold) on the spread, topping the filling with a garnish, and adding the top piece of bread.

The most important factor for an efficient sandwich station is to have the necessary ingredients available with the proper mise en place. For example, spreads should be made in advance, garnishes should be cut, washed, and ready for use, and meats should be cooked, trimmed, and presliced or ready to slice and serve as needed.

All ingredients should be arranged within easy reach. For example, right-handed workers should place the bread supply to their left, and left-handed workers should place the bread supply to their right. All spreads and filling ingredients should be directly in front of the worker. Having an organized workspace enables the sandwich maker to produce a large quantity of sandwiches relatively quickly, while reducing risks associated with having food in the temperature danger zone for long periods.

**Action Lab**

Work in teams to prepare at least 20 sandwiches. One team might be assigned tuna or chicken salad sandwiches, while another prepares meat and cheese sandwiches. A third team might be assigned triple-decker sandwiches, while the fourth team prepares tea sandwiches. Discuss the experience and what tips might be used to prepare quantities of sandwiches on the job.

## To prepare large quantities of sandwiches, apply the following procedure:

1. Arrange bread or toast slices in rows directly in front of worker. ***See Figure 13-20.***

2. Using a spatula, coat the slices of bread or toast with desired spread.

3. Place filling on alternate rows of bread. If four rows of bread are used, place filling on two center rows. Spread soft filling evenly, bringing filling to the edge of the bread. Arrange meat or cheese slices so the bread is well covered. Avoid extending beyond the edge of the bread.

4. Arrange the desired garnish on the filling (if preferred).

5. Place the remaining slices of bread on the top of slices containing the filling.

   *Note:* If preparing tea sandwiches, trim the crust edges of the bread using a chef's knife. Cut the sandwich in halves, thirds, or fourths. Sandwiches may be stacked so that several may be cut at the same time.

CULINARY PROCEDURES

## Preparing Large Quantities of Sandwiches

1–2. Arrange bread or toast in rows; coat bread with spread.

3–4. Place filling on alternate rows of bread; arrange garnish on filling.

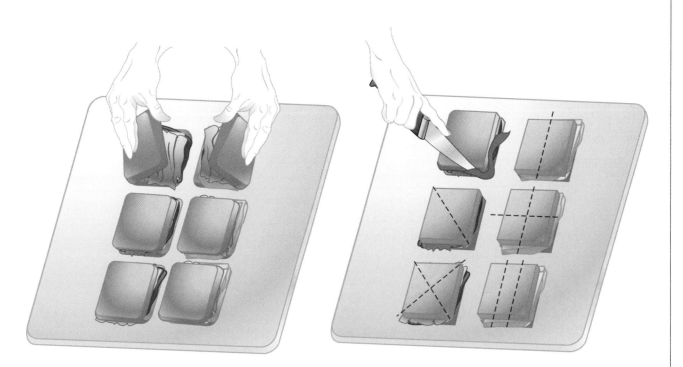

5. Place remaining bread on slices with filling.

**Note:** For tea sandwiches, trim crusts and slice into halves, thirds, or fourths.

**13-20** Large quantities of sandwiches can be prepared at the same time if the sandwich station is well organized.

## HORS D'OEUVRES

Similar to sandwiches, hors d'oeuvres (pronounced oar derves) can be made from a countless number of fillings, spreads, and garnishes, and may be served hot or cold. They can be served closed or open-faced, and their form and ingredients are limited only by the imagination of the chef who is preparing them.

An *hors d'oeuvre* is an elegant, bite-size portion of food, creatively presented and meant to be served apart from a seated meal. Hors d'oeuvres are usually more elegant and often demonstrate more culinary skill than traditional appetizers. Hors d'oeuvre is a French term meaning "outside the work." The translation is appropriate, as hors d'oeuvres were not originally prepared by culinary staff. Rather, these tiny portions of food were prepared by service staff in the homes of royal or wealthy French families. Because the extravagant meals at these palaces and castles took all day to prepare, the service staff would prepare petite portions of food to tide the guests over until the food came out of the kitchen. It was not uncommon for an extravagant meal to be served much later than the time stated on the invitation and to last for eight or more hours.

Although they are very small portions, hors d'oeuvres may substitute for an entire meal when many different varieties are served. This type of meal is referred to as a standing meal. When making hors d'oeuvres, it is important that each portion be small enough to be consumed in a single bite. A diner should never need a knife to eat an hors d'oeuvre. ***See Figure 13-21.***

**Chef's Tip:**

When serving a dipping sauce with food, it is a good idea to provide a serving spoon for the sauce. Serving spoons will help guests who are sharing an item to avoid having to dip partially eaten food into the sauce.

*Hors d'Oeuvres* _____

**13-21** *Hors d'oeuvres are elegant, bite-size portions of food served apart from a seated meal.*

The terms "hors d'oeuvre" and "appetizer" are often used interchangeably. However, there is a difference. An *appetizer* is food typically served as the first course of a seated meal. Appetizers are usually plated and are often large enough for a few people to share. They are usually more casual foods rather than elegant or formal ones. Most appetizers are larger than a single bite and are frequently served with a dipping sauce. Common appetizers include fried mozzarella sticks, battered onion rings, nachos, and Buffalo wings. Fancier appetizers such as beef carpaccio, spring rolls, or bruschetta may be served as the first course in upscale restaurants. ***See Figure 13-22.***

## Appetizers

Jalapeño Poppers

Pizza Bites
Anchor Food Products, Inc.

 Appetizers are typically served as the first of several courses in a seated meal.

### Serving Hors d'Oeuvres

The practice of serving small portions of foods as either a starter course or a complete meal is common all over the world. Spain calls their small portions *tapas,* Greece has *mezzes,* Italy has *antipastos* and the Chinese have *dim sum.* In all of these cultures, both hot and cold starter dishes are common. Common hors d'oeuvres include miniature egg rolls, miniature meatballs, rumaki, stuffed wontons, stuffed phyllo triangles, and various types of sushi.

Cold hors d'oeuvres are often designed to be picked up with the fingers and generally require little more than a napkin. Hot hors d'oeuvres often require an eating utensil, whether a fork or a toothpick, because they are hot and sometimes glazed or coated in a sauce, making them a little messy to eat with the fingers. While cold hors d'oeuvres can usually be prepared well in advance, hot hors d'oeuvres need to be cooked closer to the actual time they are needed and therefore require more last-minute preparation.

*Tru, Chicago*

In professional kitchens, hors d'oeuvres are often prepared as part of the garde manger department. Many chefs at upscale restaurants serve a very petite hors d'oeuvre as a complimentary first course called an *amuse bouche* (pronounced ah-muse boosh). Amuse bouche is a French term and translates literally to "entertain the mouth." Talented chefs put this translation to work by creating amazing combinations of flavors, colors, and textures that get guests excited about the food to follow.

## Canapés

A *canapé* is an hors d'oeuvre that looks like a miniature open-faced sandwich. Canapés have a base, a spread, and an edible garnish. Although most canapés are made on a base of bread, toast, or crackers, canapés can also be made on firm vegetables such as sliced cucumber or yellow squash, Belgian endive leaves, celery stalks, or even cherry tomatoes. Additionally, pastry shells called barquettes, tartlets, and phyllo shells can be topped with any number of fillings, while petite pâte à choux shells known as profiteroles can be filled using a pastry bag. *See Figure 13-23.*

Bread is the most common base for a canapé. Toasted bread works better than plain bread because it can handle more moisture and holds shape better. If freshly sliced bread is to be used untoasted, it must be firm enough to hold the topping without becoming soggy or fragile. Often fresh bread is compressed with a rolling pin to make it denser and thinner.

## *Canapés*

**13-23** *Canapés are petite sandwiches that can be constructed on a base of bread, toast, crackers, crisp vegetables, barquettes, tartlets, phyllo shells, or pâte à choux shells.*

To make canapés using fresh (untoasted) bread, apply the following procedure:

1. Slice bread lengthwise into long, evenly sliced pieces. *See Figure 13-24.* A Pullman loaf is ideal for making canapés.
2. Remove the crusts from the slices and reserve for making croutons or bread crumbs.
3. Pressing firmly with a rolling pin, roll the bread slices thinner.
4. Cover the flattened bread with a thin layer of spread.
5. Cut into squares, triangles, or rectangles, utilizing the entire slice.
6. Garnish as desired.
7. Plate for service.

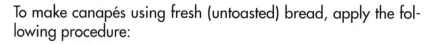

## Preparing Fresh (Untoasted) Canapés

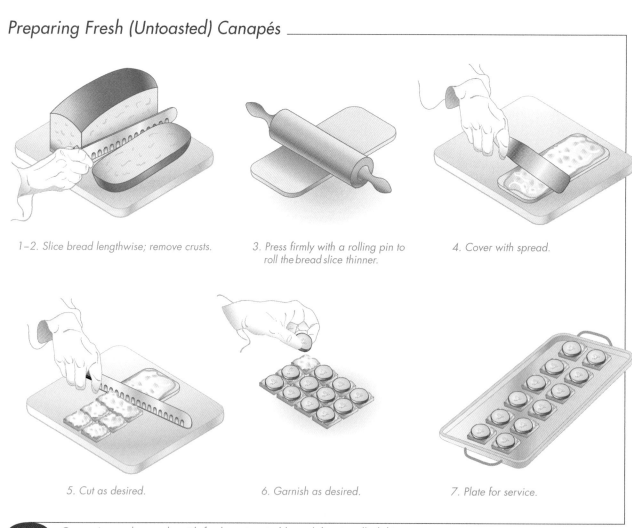

*1–2. Slice bread lengthwise; remove crusts.*

*3. Press firmly with a rolling pin to roll the bread slice thinner.*

*4. Cover with spread.*

*5. Cut as desired.*

*6. Garnish as desired.*

*7. Plate for service.*

13-24    *Canapés can be made with fresh, untoasted bread that is rolled thin.*

To make canapés using toasted bread, apply the following procedure:

1. Slice bread lengthwise into long, evenly sliced pieces. ***See Figure 13-25.*** A Pullman loaf is ideal for making canapés.

2. Remove the crusts from the slices and reserve for making croutons or bread crumbs.

3. Using a broiler or a large toaster, toast the slices on both sides.

4. When evenly golden, remove bread slices and allow to cool completely.

5. Cover the toasted bread with a thin layer of spread and cut into squares, triangles, or rectangles, utilizing the entire slice.

6. Garnish as desired.

7. Plate for service.

*Preparing Toasted Canapés*

*1–2. Slice bread lengthwise; remove crusts.*

*3–4. Toast under broiler; allow to cool completely.*

*5. Cover with spread and cut as desired.*

*6. Garnish as desired.*

*7. Plate for service.*

**13-25** *Canapés are also made on toasted bread.*

To make canapés using the oven method, apply the following procedure:

1. Slice bread lengthwise into long, evenly sliced pieces. *See Figure 13-26.* A Pullman loaf is ideal for making canapés.

2. Remove the crusts from the slices and reserve for making croutons or bread crumbs.

3. Cut the bread into desired shapes and reserve scraps for making croutons or bread crumbs.

4. Brush both sides of the bread cutouts with melted butter and place in a 450°F oven for 8 min to 10 min until golden brown.

5. When evenly golden, remove bread slices from oven and allow to cool completely.

6. Cover the toasted bread with a thin layer of spread.

7. Garnish as desired.

8. Plate for service.

**Spreads.** The spreads for canapés can be a flavorful compound butter or cream cheese spread, a puréed meat or seafood spread, or a bound salad. Flavorful butter and cream cheese spreads are produced by blending seasonings or other flavoring ingredients with softened butter or cream cheese until smooth. Typical flavoring ingredients used in these spreads include lemon, herbs, Dijon mustard, pimiento, blue cheese, smoked salmon, capers, scallions, roasted garlic, and olives. It is important to note that solid items such as pimientos need to be puréed prior to folding into the spread to ensure a smooth, spreadable texture.

Cooked meat, seafood, vegetables, and bound salads are great spreads with texture. Bound salads such as tuna salad, chicken salad, or a grilled vegetable salad not only taste great but have a pleasing appearance when gently piled on a canapé base. These salads should have finely chopped ingredients or be slightly puréed. This can be accomplished using a sieve or a blender. Other items such as spreadable pâtés can add a touch of elegance to a canapé tray. Regardless of the type of spread used, it should be smooth enough to be piped through a pastry bag or spread with a palette knife to yield an attractive appearance.

*Alpha Baking Co., Inc.*

## Preparing Canapés—Oven Method

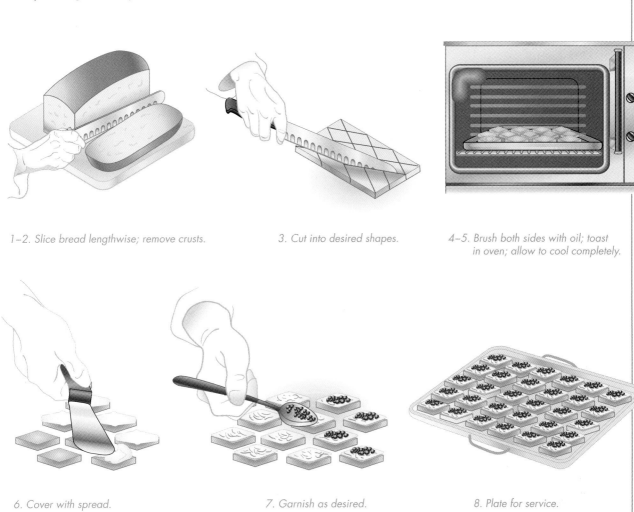

1–2. Slice bread lengthwise; remove crusts.

3. Cut into desired shapes.

4–5. Brush both sides with oil; toast in oven; allow to cool completely.

6. Cover with spread.

7. Garnish as desired.

8. Plate for service.

**13-26** Bread for canapés may also be cut into decorative shapes and toasted in the oven.

**Garnishes.** The garnish for a canapé should be delicate but should finish the canapé, making it look elegant and complete. The garnish can be anything from a small slice of radish or smoked salmon, to a dab of caviar or the leaf of a delicate herb such as chervil.

**Sliced-Vegetable Canapés.** Canapés can be made on slices of fresh, firm vegetables instead of on a bread base. In this type of preparation, firm vegetables such as cucumber, yellow or green squash, or carrots are sliced into rounds or diagonals and topped with a spread. Also, vegetables such as Belgium endive leaves, celery stalks, and pea pods can have any variety of fillings piped into them from a pastry bag. ***See Figure 13-27.***

## Sliced-Vegetable Canapés

Cherry tomato bases

Cucumber bases

**13-27** *Raw vegetables can be used as a canapé base in place of fresh or toasted bread.*

## Cocktails

*Florida Tomato Committee*

**13-28** *Cocktails include mixtures of bite-size fresh fruit and cold cooked or raw seafood.*

### Cocktails

The term "cocktail" is not only used to describe an alcoholic beverage, but also to describe a chilled appetizer. A cocktail may be something as simple as diced mixed fruit or cold cooked or raw seafood. Juice cocktails are made from fruit or vegetable juices and are tangy in taste. They are bright in appearance and are served in chilled glasses. Fruit cocktails and seafood cocktails are attractively arranged bite-sized pieces of food that can be eaten with a cocktail fork. ***See Figure 13-28.***

Fresh seafood such as oysters on the half shell, shrimp, crab, and lobster make excellent cocktails and are often served with either a spicy cocktail sauce or with a vinaigrette-style dressing called a *mignonette*. Seafood and fruit cocktails should be served as fresh and as cold as possible.

### Barquettes, Tartlets, Phyllo Shells, and Bouchées

Barquettes and tartlets are petite pastry shells made from a delicate pâte (dough) that is similar to a thin pie dough. The pâte is pressed into miniature boat-shaped tins for barquettes, or round tins for tartlets, and then baked in the oven. Phyllo cups are made by layering buttered sheets of phyllo dough in miniature muffin tins and then baking. Bouchées are miniature pâte à choux puffs that become hollow inside when they are baked. Each of these pastry items can be filled with savory or sweet spreads and garnished to be beautiful and appetizing. Barquettes, tartlets, phyllo shells, and bouchées can be used to hold hot or cold hors d'oeuvres. ***See Figure 13-29.***

## Phyllo Cups

Cheese and sour cream garnish

Phyllo formed into cups and baked

Hot filling

**13-29** *Phyllo cups can be used to hold hot or cold fillings.*

## Crudités

Raw vegetables are known as *crudités,* a French term meaning "raw things". Crudités compose the body of what is commonly called a vegetable tray. Any vegetable that is acceptable to eat raw (or slightly blanched) can be cut into sticks, bite-size pieces, or any variety of shapes, and arranged beautifully on a serving platter as crudités. A crudités platter usually consists of vegetables such as carrots, celery, broccoli, cauliflower, cherry tomatoes, bell peppers, radishes, zucchini, yellow squash, and cucumbers. ***See Figure 13-30.*** Pickled relish items such as pickles, olives, and assorted hot, sweet, and roasted peppers may also be included.

## Crudités

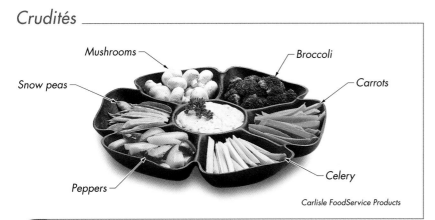

Mushrooms

Snow peas

Broccoli

Carrots

Peppers

Celery

*Carlisle FoodService Products*

**13-30** *A crudités platter may include raw vegetables such as broccoli, carrots, celery, peppers, snow peas, and mushrooms.*

## Dips

A *dip* is a creamy sauce or condiment that is served with hard food such as bread, vegetables, or chips. Dips cover the spectrum from hot to cold, sweet to savory, and smooth to chunky. There are really no specific rules as to how dips can be made or what ingredients can be used. From chips served with salsa to pita bread served with hummus, a wide variety of dips can be prepared.

Cold creamy dips usually have mayonnaise, cream cheese, sour cream, or yogurt as the base ingredient. Creamy dips can be savory or sweet. Virtually any ingredient can be added to the base, such as crab meat, roasted garlic, cooked spinach, or Parmesan cheese. A dip should complement the food item that is dipped into it and should not be overpowering in flavor or leave a bad aftertaste.

Dips that are made using sour cream or mayonnaise are made in the same manner as creamy salad dressings, but a dip is thicker in texture so that it can adhere to the item being dipped into it. Cream cheese is typically whipped with cream, milk, or buttermilk to soften it. Other cold dips, such as guacamole or curried lentil dip, have a purée of vegetables, fruit, or legumes as the main ingredient. Cold dips can be served in a small dish or very creatively, in a hollowed bread bowl, carved squash, or a hollowed head of cabbage or lettuce, for an attractive presentation. ***See Figure 13-31.***

Hot dips usually have a Mornay (cheese sauce), béchamel, or tomato sauce as the base ingredient. A great variety of items can be added as a primary flavoring ingredient. For example, jalapeños are a great accompaniment to a cheddar-based Mornay sauce or goat cheese base, and pesto is a perfect addition to a ripe tomato sauce. Hot dips need to be served from a bain marie or chafing dish so that they are at the appropriate temperature.

## Brochettes

A *brochette* is an hors d'oeuvre consisting of food that is speared onto wooden, metal, or natural skewers and then broiled or grilled. ***See Figure 13-32.*** Examples of natural items that can be used as skewers include branches of rosemary, twigs of lemon grass, or pieces of sugar cane. Brochettes are also called shish kebabs, or kebabs for short. While brochette is a French term, the words "shish kebab" reflect the Middle Eastern origins of this skewered presentation. Almost any sort of solid food can be cooked and served on skewers. Examples of brochettes include a sweet brochette of fresh fruit, a seafood brochette of shrimp and scallops, and a vegetarian brochette of mushrooms, zucchini, cherry tomatoes, peppers, and onions.

Brochettes are a versatile hors d'oeuvre and, if the portion size is enlarged, can be served as an entrée. Savory brochettes are commonly marinated to increase their flavor.

**Dip**

Pita slices

Creamy dip

*Wisconsin Milk Marketing Board*

**13-31** *Cold dip is often served in a small dish and garnished.*

*The following are helpful tips to remember when making brochettes:*

- Soak wooden skewers in water for at least 30 min prior to inserting into food to be grilled. Soaking allows the wooden skewer to absorb water so that it will not burn or catch fire.
- Spray metal skewers with a nonstick cooking spray prior to inserting into foods. This helps prevent foods from sticking to the skewer, which can later make the food hard to remove and eat.
- Foods such as thick pineapple slices should be skewered through the slice from side to side rather than through the center of the piece. This helps prevent the slice from spinning as well as allows it to sit flat on the grill.
- If wooden skewers are very long, they can be cut shorter with a French knife to an appropriate size for an hors d'oeuvre.
- Leave a portion of the skewer exposed at each end so that the guest has a place to hold it.

## Brochettes

Grilled chicken

Scallions

National Chicken Council

**13-32** *Foods served on wooden skewers are referred to as brochettes.*

## Sushi

Sushi is a traditional Asian dish that is commonly served as an hors d'oeuvre but can also be served as a light main course or a full entrée. Japanese-style sushi is the most common style of sushi in North America. The terms sashimi and sushi are often misunderstood. *Sashimi* is raw fish eaten without rice. In a sashimi preparation, thick slices or wedges of very fresh raw fish are served and eaten with traditional garnishes and condiments. *Sushi* is a vinegar-seasoned rice dish garnished with items such as raw fish, cooked seafood, egg, or vegetables. The term sushi is derived from *zushi*, the Japanese term for cold, vinegar-seasoned, cooked rice.

Sushi rice is made by steaming or simmering a sticky, short-grain Asian rice until it is slightly firm and chewy. After the rice is cooked, it is placed in a wooden bowl and gently mixed with a mixture of rice wine vinegar and sugar syrup. A cutting motion is used to not only mix the liquid well into the cooked rice, but also to help cool the rice, preventing it from becoming too soft. After the sushi rice cools, it can be formed by hand into various shapes.

There are many types of sushi available. The most common forms of sushi are nigiri, maki, inari, and chirashi. ***See Figure 13-33.*** *Nigiri sushi* is small hand-formed mounds of vinegar-seasoned cooked rice garnished with a topping. *Maki sushi* is vinegar-seasoned cooked rice layered with ingredients, rolled in a dried seaweed paper called *nori*. Maki rolls are made by layering sushi rice with other ingredients and rolling them tightly in a sheet of nori to form a cylinder shape. The cylinder is then sliced into bite-size portions. Many people in Western culture find maki rolls to be

**Chef's Tip:**
Care should always be taken when serving raw or undercooked food. Highly susceptible populations including young children, pregnant women, and seniors should avoid eating raw or undercooked meat, poultry, seafood, or eggs. If a food contains these items, it should be noted on the menu.

more palatable than other types of sushi. *Inari sushi* is vinegar-seasoned rice and toppings stuffed into a small purse of fried tofu. *Chirashi sushi* (scattered sushi) is sushi consisting of assorted toppings that are either scattered over or mixed into vinegar-seasoned rice.

*Sushi*

**Nigiri Sushi**

Nigiri salmon

Nigiri toro (fatty tuna)

Nori

Salmon-avocado filling

Sushi rice

**Maki Rolls**

13-33 *In North America, the most common forms of sushi include nigiri sushi and maki sushi.*

### Smoked Seafood, Meats, and Poultry

Smoked food items can add a unique flavor to hors d'oeuvres. For example, a piece of smoked bacon wrapped around a scallop and broiled can add a sweet and tangy flavor to a classic rumaki. Almost any food item can be smoked to add a unique flavor. While seafood, meats, and poultry are the foods that are most commonly smoked, other types of food such as cheeses or vegetables can also be smoked. Food can be cold smoked or hot smoked. The process of cold smoking exposes food to smoke but does not actually cook the food. The process of hot smoking adds heat to the smoking chamber so that the food cooks while it is being smoked.

Smoking is achieved by placing food items in a smoker containing soaked wood chips. The wood chips are soaked prior to smoking to prevent them from burning rapidly. The wet wood chips do not ignite, but rather smolder and emit smoke into the smoker cabinet. As food items sit in the smoke-filled cabinet, they absorb the flavor of the wood smoke. The length of time that an item is exposed to the smoke determines the intensity of the smoke flavor in the finished item.

Mildly smoked foods are most popular because the true flavor of the item is not overpowered by the flavor of smoke. Hors d'oeuvres and appetizers with smoked ingredients range from a smoked gouda fondue to a smoked salmon and cream cheese canapé. Smoked pulled pork is a savory ingredient in a hot hors d'oeuvre.

## Cured and Raw Meats

Cured and raw meats are popular items to serve as appetizers or hors d'oeuvres. Cured pork, beef, veal, venison, goat, and sheep meat may be salted, smoked, or air-dried. Proscuitto is a popular example of a cured meat that is often used as an appetizer or hors d'oeuvre. Other types of cured meats and seafood used often in appetizers or hors d'oeuvres include ham, bacon, and kippered herring.

*Carpaccio* is a term used to describe meats or seafood that are sliced thin and served raw. Raw meats such as beef, veal, or lamb can be seared on the exterior, chilled well, and sliced very thin. The thinly sliced meat is then carefully arranged on a serving plate, seasoned with salt and pepper, and often drizzled with a small amount of extra virgin olive oil before serving. *Tartare* is a preparation of freshly ground or chopped raw meat that is seasoned and then served as a small mound. Steak tartare is the most common tartare preparation. Only meats that are very fresh and that have been handled with extreme care can be served raw.

*Charlie Trotter's*

## Grilled and Roasted Vegetables

Grilled and roasted vegetables make a tasty and colorful addition to an hors d'oeuvre platter. Grilled or roasted vegetables are usually marinated briefly in an oil-based marinade, seasoned, and then either grilled over an open flame or roasted in an oven at a very high heat. Roasting or grilling vegetables causes the natural sugars in the vegetables to caramelize and intensify in flavor. Very hard vegetables such as carrots should be first blanched until al dente before grilling so that the result is more tender. A platter of grilled, marinated vegetables that are chilled prior to serving is a common presentation of hors d'oeuvres.

### Caviar

One of the most extravagant and luxurious of hors d'oeuvres, *caviar* is the harvested roe (eggs) of sturgeon fish. Although eggs are harvested from other varieties of fish, such as salmon and whitefish, only sturgeon roe can be labeled simply as "caviar" and does not require the type of fish be listed anywhere on the package. Other fish roe must be labeled as to the type of fish they are from, such as salmon caviar or whitefish caviar. ***See Figure 13-34.***

## Caviar

Salmon roe

Caviar

Whitefish roe

Collins Caviar Company

**13-34** *Only the roe of sturgeon fish can be labeled simply as "caviar," although roe from other fish, such as salmon and whitefish, are also served as caviar.*

Caviar from the Caspian Sea is considered to be the best in the world and is imported from Russia and Iran. Once harvested, a caviar master gently washes the roe, separates the individual eggs (called berries), and carefully blends the eggs with a small amount of salt to preserve them.

Caviar is traditionally harvested from three varieties of sturgeon: Beluga, Osetra, and Sevruga. Although other varieties of sturgeon roe are also caviar, these three varieties are considered to be of the highest quality. Beluga sturgeon (not to be confused with Beluga whales) are the largest of the sturgeons and can weigh over 1500 lb. Beluga produce the largest roe of any fish. Beluga caviar is also the most expensive variety of caviar, although many caviar lovers consider it only the second-best variety of caviar. Osetra caviar is considered the best-tasting caviar. Osetra sturgeon are smaller than Beluga sturgeon and have smaller roe. The roe are medium-sized, with what is considered perfect taste and texture. Sevruga is the smallest of the three varieties of sturgeon and has the tiniest roe. Pressed caviar is a form of caviar made from drained and pressed Osetra and Sevruga caviar. It is drained and compressed into a form resembling a cake of caviar.

There are more than 20 varieties of sturgeon. In the United States, American sturgeon caviar and other fish roe varieties have gained popularity due to availability and a much lower cost. The U.S. government has ruled that the American paddlefish is a sturgeon for food purposes. For this reason, American paddlefish roe are labeled as American sturgeon caviar. Other popular types of fish roe in the United States include salmon, whitefish, and lumpfish caviars.

Caviar is sold either fresh or pasteurized. Fresh caviar is packed and sold in special tins or small glass jars. It has a short shelf life of only up to two weeks unopened and only three to four days once opened. Pasteurized caviar does not have to be stored in refrigeration until it has been opened. Once pasteurized caviar is opened, it has a storage life less than a week.

Caviar should only be handled with a mother of pearl, china, or wooden spoon. Handling caviar with a metal spoon causes a chemical reaction that results in the caviar having a metallic taste. Caviar is classically served on toast points, blinis (small savory pancakes), chilled cooked fingerlings, or new potatoes with a dollop of sour cream. Standard accompaniments for caviar are finely minced cooked egg white, egg yolk, and red onion. *See Figure 13-35.*

### Nutrition Note

Caviar is high in protein, fat, sodium, and cholesterol. Caviar also contains a significant amount of phosphorus and calcium as well as several vitamins.

## Assorted Caviars

Tru, Chicago

13-35  *An assortment of caviar can be an elegant hors d'oeuvre presentation.*

American METALCRAFT, Inc.

Refer to the CD-ROM
for Quick Quiz® questions
related to chapter content.

## SUMMARY

The sandwich is one of the most basic and easy-to-prepare food items. Making sandwiches requires organizational skills, speed, and coordination, as it is a very busy station in the professional kitchen.

A sandwich is typically composed of bread, spread, filling, and garnish. The filling of a sandwich is the main component. Sandwiches are often identified by or named for their fillings. Sandwiches are divided into hot and cold categories, depending on the serving temperature of the filling. The possible combinations of ingredients and even varieties of sandwiches are limited only by the chef's imagination.

Hors d'oeuvres are petite portions of food served separately from the main meal. They are often served at cocktail parties, receptions, or other social gatherings, as they are easy to eat while standing up and mingling with other guests. They can be a meal in themselves when many different varieties are served, but they are typically not served as a course of a seated meal. Although the terms hors d'oeuvre and appetizer are sometimes used interchangeably, an appetizer is often the first course of a seated meal. It can be served as a dish to be shared by a few or several people at a table. The portion size for an appetizer is usually much larger than for an hors d'oeuvre.

Hors d'oeuvres can be as different as the imagination of the chef permits. They can be passed on platters or served plated. The most common varieties of hors d'oeuvres are canapés, barquettes, cocktails, crudités, dips, brochettes, sushi, smoked or cured meats, grilled or roasted vegetables, and caviar. A chef should be creative not only with the ingredients used in an hors d'oeuvre, but also with the flavors and presentation. Each hors d'oeuvre should be as beautiful as it is delicious.

## Review Questions

1. Describe the four general purposes that bread serves in a sandwich.

2. Identify the four main categories of spreads that can be used in sandwiches.

3. List three sandwiches that are named for their filling.

4. Explain how to modify the main ingredients of a hamburger with melted cheese and mayonnaise on a sesame-seed white-bread bun for a customer who is health-conscious.

5. Contrast the presentation styles of the five categories of cold sandwiches.

6. Provide the five-step procedure for making a multidecker sandwich.

7. Provide the six-step procedure for preparing a cold wrap sandwich.

8. Contrast the preparation methods for five common types of hot sandwiches.

9. Describe common ways to portion meats and cheeses for a sandwich station prior to service.

10. Provide the six-step procedure for preparing large quantities of sandwiches, including how to treat tea sandwiches in the final step.

11. Distinguish hors d'oeuvres from appetizers in terms of preparation, presentation, and purpose.

12. List some advantages of serving cold hors d'oeuvres instead of hot hors d'oeuvres.

13. Explain the procedure for making canapés using fresh (untoasted) bread.

14. Describe crudités.

15. Describe different types of sushi, including nigiri, maki, inari, and chirashi.

16. Identify common foods that can be hot or cold smoked to add a unique flavor to hors d'oeuvres.

17. Describe typical ways to serve caviar.

# Recipes

## 37

## Classic Italian Submarine Sandwiches

*yield: 4 servings*

| | |
|---|---|
| 4 | Italian rolls |
| 2 tbsp | extra virgin olive oil |
| 1 tbsp | red wine vinegar |
| 1 clove | garlic, minced |
| 1 tsp | fresh oregano, finely chopped |
| to taste | pepper |
| ½ lb | capicola, thinly sliced |
| ½ lb | Genoa salami, thinly sliced |
| ¼ lb | provolone cheese, sliced |
| ¼ lb | mozzarella cheese, sliced |
| 2 c | romaine lettuce, shredded |
| 1 | medium tomato, sliced |
| ½ | medium red onion, thinly sliced |
| 1 | red pepper, roasted and julienned |
| ½ c | black olives, sliced |

1. Slice each roll in half lengthwise, leaving the two sides attached.
2. Combine olive oil, vinegar, garlic, oregano, and pepper in a small bowl. Mix well.
3. Brush a small amount of the vinaigrette on the bottom cut side of each roll.
4. Layer the meats, cheeses, and vegetables on each roll, placing the meats on the bottom and vegetables on top.
5. Drizzle the remaining vinaigrette on the vegetables.
6. Serve immediately, or wrap tightly in plastic wrap and refrigerate.

---
**Nutrition Facts**

**Per serving:** 905.9 calories; 65% calories from fat; 66.7 g total fat; 132.6 mg cholesterol; 2374.1 mg sodium; 584.4 mg potassium; 37.7 g carbohydrates; 3.4 g fiber; 2.4 g sugar; 34.3 g net carbs; 38.8 g protein
---

## Herbed Chicken Salad Sandwiches

*yield: 6 servings*

| | |
|---|---|
| 12 slices | sandwich bread |
| 4 | boneless, skinless chicken breasts |
| ½ c | mayonnaise |
| 2 tbsp | celery, small dice |
| 1 tbsp | onion, minced |
| to taste | salt and pepper |
| 1 tsp | fresh tarragon, minced |
| 1 tsp | fresh chives, minced |

1. Steam or lightly sauté chicken breasts until cooked through. Cool.
2. When cool, dice chicken into medium dice and place in mixing bowl.
3. Add eggs, mayonnaise, celery, onions, salt, pepper, and herbs. Mix well.
4. Refrigerate for several hours or overnight.
5. Place ½ cup of chicken salad on a slice of sandwich bread. Top with a second slice of bread.
6. Serve immediately, or wrap tightly in plastic wrap and refrigerate.

---
**Nutrition Facts**

**Per serving:** 384.3 calories; 23% calories from fat; 10.2 g total fat; 96.3 mg cholesterol; 632.7 mg sodium; 470.4 mg potassium; 30.4 g carbohydrates; 1.3 g fiber; 3.5 g sugar; 29.1 g net carbs; 40.4 g protein
---

## Triple-Decker Turkey Club Sandwiches

39

*yield: 2 servings*

| | |
|---|---|
| 6 slices | sandwich bread |
| 3 tbsp | mayonnaise |
| 6 leaves | romaine lettuce |
| 1 | medium tomato, sliced into 4 or 8 slices |
| 4 slices | cooked turkey breast |
| 4 strips | bacon, cooked, cut into 4" lengths |

1. Lightly toast bread.
2. Coat one side of six slices of bread with mayonnaise. Set aside two coated slices.
3. Top each of the remaining four mayonnaise-coated slices with lettuce, turkey, tomato, and bacon.
4. Place the two reserved slices of break, spread side down, on top of two of the layered portions to form sandwiches.
5. Top each sandwich with one of the two remaining layered portions to create tripledeckers.
6. Cut each sandwich into four triangle portions and plate.

### Nutrition Facts

**Per serving:** 739.2 calories; 55% calories from fat; 45.5 g total fat; 81.5 mg cholesterol; 1882.7 mg sodium; 1031.7 mg potassium; 56.3 g carbohydrates; 5.4 g fiber; 13.3 g sugar; 50.9 g net carbs; 26.7 g protein

## Smoked Salmon–Wasabi Tea Sandwiches

40

*yield: 12 servings*

| | |
|---|---|
| 6 thin slices | sandwich bread |
| ½ tbsp | wasabi paste |
| 4 oz | cream cheese |
| 4 oz | smoked salmon, thinly sliced |

1. Place bread slices on a cutting board. If slices are taller than ¼" thick, use a rolling pin to gently flatten bread.
2. Using a whisk, beat wasabi paste and cream cheese until well blended.
3. Spread each slice of bread with wasabi cream cheese.
4. Evenly divide salmon and place on three of the six slices of bread.
5. Top salmon with remaining three slices of bread, spread-side down.
6. Trim crusts and cut each sandwich into quarters.
7. Serve immediately, or wrap individually and refrigerate.

### Nutrition Facts

**Per serving:** 71 calories; 50% calories from fat; 4 g total fat; 12.6 mg cholesterol; 285.1 mg sodium; 39.7 mg potassium; 5.4 g carbohydrates; 0.3 g fiber; 0.4 g sugar; 5.1 g net carbs; 3.2 g protein

# Recipes

## Chipotle Turkey Wraps

*yield: 4 servings*

| | |
|---|---|
| ½ c | mayonnaise |
| 3 tbsp | fresh cilantro, chopped |
| ¼ | red onion, minced |
| 2 tsp | chipotle hot sauce |
| 1 | lime |
| to taste | salt |
| 4 | flour tortillas, 8"-10" diameter |
| ½ lb | smoked turkey breast, thinly sliced |
| 2 leaves | romaine lettuce, shredded |
| ½ | tomato, diced |
| 4 oz | cheddar cheese, shredded |

1. In a bowl, combine mayonnaise, cilantro, onion, hot sauce, and juice from half of a lime. Season with salt to taste.
2. Spread each tortilla with chipotle mayonnaise.
3. Layer turkey, lettuce, tomato, and cheese on each tortilla.
4. Roll up tightly. Cut each wrap in half.
5. Serve immediately.

### Nutrition Facts
**Per serving:** 514.3 calories; 44% calories from fat; 25.7 g total fat; 61.8 mg cholesterol; 1544.1 mg sodium; 393.9 mg potassium; 47.6 g carbohydrates; 3 g fiber; 5.9 g sugar; 44.6 g net carbs; 23.2 g protein

## Bistro Hamburgers

*yield: 4 servings*

| | |
|---|---|
| 1 lb | ground beef |
| 1 clove | garlic, minced |
| 2 tbsp | fresh parsley, chopped |
| to taste | pepper |
| 4 | sesame rolls |
| 4 leaves | Boston lettuce |
| ½ | onion, sliced thin |
| ½ | tomato, sliced |
| 2 oz | dill pickle slices |
| to taste | ketchup, mustard, mayonnaise |

1. Add ground beef, garlic, parsley, and pepper. Blend well.
2. Divide into four equal balls. Flatten each ball into a 6"-diameter patty.
3. Grill patties to desired doneness.
4. Slice sesame rolls in half.
5. Place cooked patty on roll. Layer with lettuce, onion, tomato, and dill pickles.
6. Serve with ketchup, mustard, and mayonnaise to taste.

### Nutrition Facts
**Per serving:** 427.3 calories; 41% calories from fat; 19.6 g total fat; 77.1 mg cholesterol; 569.6 mg sodium; 509.9 mg potassium; 33.8 g carbohydrates; 2.1 g fiber; 2.8 g sugar; 31.6 g net carbs; 27.3 g protein

## Philly Beef Sandwiches

*yield: 4 servings*

| | |
|---|---|
| 1 tbsp | olive oil |
| 2 | medium onions, sliced |
| 2 | sweet peppers (red or green), julienned |
| 1 clove | garlic, minced |
| to taste | pepper |
| 1½ lb | beef tenderloin, sliced very thin |
| 4 | Italian rolls |
| 4 slices | mozzarella cheese |

1. In a sauté pan, cook onions, peppers, and garlic until tender. Season with pepper and set aside.
2. Pan-fry beef until brown, but not crispy.
3. Slice each roll in half lengthwise, leaving the two sides attached.
4. Layer beef, onions, and peppers in each roll. Top with a slice of mozzarella cheese.
5. Wrap in foil paper to hold in heat and allow cheese to melt.
6. Serve immediately.

### Nutrition Facts

**Per serving:** 745.4 calories; 51% calories from fat; 42.7 g total fat; 133 mg cholesterol; 577.2 mg sodium; 816.2 mg potassium; 41.4 g carbohydrates; 3.4 g fiber; 5.9 g sugar; 37.9 g net carbs; 46.9 g protein

## Roasted Tomato, Prosciutto, and Garlic Pizza

*yield: 4 servings*

| | |
|---|---|
| 1 batch | pizza dough (see chapter 22) |
| ½ c | all-purpose flour |
| 2 tbsp | corn meal |
| 1 c | tomato sauce (see chapter 14) |
| 1 tbsp | olive oil |
| 1 c | shredded mozzarella |
| 1 tbsp | fresh basil, chopped |
| 1 | medium tomato, sliced |
| 5 cloves | roasted garlic |
| 6 oz | prosciutto, thinly sliced and cut into small pieces |

1. On a lightly floured surface, roll out pizza dough to desired size.
2. Spread corn meal on appropriately sized pizza pan, and place dough on top.
3. Spread tomato sauce evenly over pizza dough, leaving a ½" border around the edge for crust. Top tomato sauce with fresh basil.
4. Brush the edge (crust) with olive oil.
5. Layer with shredded mozzarella, sliced tomatoes, roasted garlic cloves, and prosciutto.
6. Bake in 475°F oven for 10 min, or until cheese is melted and crust is golden brown.
7. Cut into squares or triangles.

### Nutrition Facts

**Per serving:** 643 calories; 33% calories from fat; 24.2 g total fat; 80.2 mg cholesterol; 1709.5 mg sodium; 683.5 mg potassium; 71.3g carbohydrates; 2.9 g fiber; 3.7 g sugar; 68.4 g net carbs; 34.3 g protein

# Recipes

## 45

## Reuben Sandwiches

*yield: 4 servings*

| | |
|---|---|
| 8 slices | dark pumpernickel rye bread |
| 2 tbsp | butter |
| 1 lb | shredded corned beef |
| ½ lb | sauerkraut |
| 4 slices | Swiss cheese |

*Wiscon Milk Marketing Board*

1. Spread one side of each slice of bread with butter.
2. Place the bread butter-side down on a cutting board.
3. Layer corned beef, sauerkraut, and Swiss cheese on four slices.
4. Top each sandwich with remaining buttered slices of bread, butter side up.
5. Grill sandwiches on both sides until browned and cheese has melted. Serve immediately.

### Nutrition Facts

**Per serving:** 522.6 calories; 54% calories from fat; 32.1 g total fat; 102.3 mg cholesterol; 2158.3 mg sodium; 564.6 mg potassium; 28.8 g carbohydrates; 4.8 g fiber; 1.7 g sugar; 24 g net carbs; 29.3 g protein

## 46

## Monte Cristo Sandwiches

*yield: 2 servings*

| | |
|---|---|
| 4 slices | white sandwich bread |
| 3 tsp | Dijon mustard |
| 4 slices | Gruyère cheese |
| 4 slices | ham, sliced thin |
| 4 slices | roasted turkey, sliced thin |
| 2 | eggs |
| 2 tbsp | water |
| 2 tbsp | butter |

1. Arrange bread on cutting board. Spread each slice with mustard.
2. Place two slices of cheese, two slices of ham, and two slices of turkey on each of two slices of bread. Top with additional bread.
3. Whisk eggs and water together until well blended in a shallow bowl.
4. Melt butter in a frying pan or skillet over medium-low heat.
5. Dip each bread side of the sandwiches in the egg batter.
6. Cook in hot butter until cheese melts and bread is golden, flipping as needed, about 4 min per side.
7. Cut sandwiches in half diagonally.

### Nutrition Facts

**Per serving:** 665.1 calories; 51% calories from fat; 39.1 g total fat; 354.6 mg cholesterol; 1882.2 mg sodium; 540.6 mg potassium; 30.2 g carbohydrates; 1.7 g fiber; 4.7 g sugar; 28.5 g net carbs; 46.5 g protein

## Steak and Cheese Quesadillas

*yield: 4 servings*

| | |
|---|---|
| 2 tbsp | vegetable or corn oil |
| ½ lb | skirt or flank steak, thinly sliced |
| ½ tsp | ground cumin |
| ¼ c | chunky salsa |
| 4 | flour tortillas, 7" diameter |
| ½ lb | queso chihuahua (Mexican melting cheese), crumbled |
| 1 tbsp | fresh cilantro |

1. In a sauté pan, heat 1 tbsp of oil. Add beef and cumin. Cook until browned.
2. Add salsa and cook until heated through.
3. Arrange tortillas on a cutting board.
4. Divide cheese evenly between two tortillas. Layer with cooked beef and cilantro.
5. Top with a second tortilla.
6. Heat 1 tbsp of oil in a skillet or sauté pan.
7. Cook stuffed tortillas in hot oil until cheese is melted and tortillas are golden on both sides.
8. Serve with sour cream, salsa, and guacamole for dipping.

### Nutrition Facts

**Per serving:** 514.9 calories; 55% calories from fat; 32.4 g total fat; 87.3 mg cholesterol; 780.3 mg sodium; 378.5 mg potassium; 28 g carbohydrates; 1.7 g fiber; 4.5 g sugar; 26. 2g net carbs; 27.5 g protein

## Onion Rings

*yield: 4 servings*

| | |
|---|---|
| 4 | medium onions, sliced and separated into rings |
| 2 | eggs |
| 8 oz | cake flour |
| ¼ tsp | salt |
| 1 c | milk |
| | vegetable oil for deep-fat frying |

*National Onion Association*

1. Submerge the onion rings in ice cold water. *Note:* This prevents loss of flavor and juice from the onion during cooking.
2. Break the eggs into a stainless steel bowl and beat slightly with a wire whisk.
3. Sift the flour, baking powder, and salt in a separate bowl.
4. Add the dry ingredients slowly to the liquid mixture while whisking briskly with a wire whisk. Whip until a smooth batter forms.
5. Coat the onions with batter.
6. Deep-fry using the floating method in 350°F to 375°F fat until golden brown.
7. Remove from oil and place on paper towels to drain excess fat.

### Nutrition Facts

**Per serving:** 582.5 calories; 49% calories from fat; 32.3 g total fat; 111.8 mg cholesterol; 210.4 mg sodium; 396.3 mg potassium; 62.4 g carbohydrates; 3.1 g fiber; 10 g sugar; 59.3 g net carbs; 11.1 g protein

# Recipes

## 49

## Buffalo Wings

*yield: 4 servings*

| | |
|---|---|
| 12 | chicken wings |
| 1¾ c | all-purpose flour |
| 2 tsp | salt |
| 1 tsp | pepper |
| 2 | eggs |
| ½ c | milk |
| 2½ c | bread crumbs |
| 4 tbsp | Tabasco® sauce |
| 1 tbsp | vinegar |
| 1 tsp | cayenne pepper |
| 2 tbsp | margarine |

*National Chicken Council*

1. Cut chicken wings in half at the joint. Trim tips of wings.
2. Mix flour, salt, and pepper. Coat chicken pieces in seasoned flour.
3. Beat eggs with a wire whisk. Add milk and mix well. Dip floured chicken pieces in egg wash.
4. Coat chicken with dried bread crumbs.
5. Fry in deep fat at 350°F to 375°F until golden brown.
6. Combine Tabasco® sauce, vinegar, cayenne pepper, and margarine in a small bowl.
7. Drizzle sauce over cooked chicken and toss to coat. Serve hot. *Note:* Buffalo wings are named for the city of Buffalo, New York, where the recipe originated. They are often served with celery sticks and blue cheese dip which helps to cut the heat of these spicy chicken wings.

### Nutrition Facts
**Per serving:** 902.7 calories; 36% calories from fat; 36.9 g total fat; 222 mg cholesterol; 1968.6 mg sodium; 534.8 mg potassium; 92.9 g carbohydrates; 4.9 g fiber; 6.4 g sugar; 88 g net carbs; 46.1 g protein

## Scallop Rumaki

*yield: 4 servings*

| | |
|---|---|
| 4 slices | bacon |
| 8 | sea scallops |
| to taste | salt and pepper |

1. Stretch bacon strips to make thinner. Cut each strip to yield 4″ pieces.
2. Wrap one 4″ piece of bacon around each scallop and secure with a toothpick.
3. Place scallops in oven at 350° and cook until bacon is done and scallop is opaque.
4. Season with salt and pepper to taste.

### Nutrition Facts
**Per serving:** 156.2 calories; 74% calories from fat; 13 g total fat; 29.2 mg cholesterol; 357.1 mg sodium; 155.6 mg potassium; 0.9 g carbohydrates; 0 g fiber; 0 g sugar; 0.9 g net carbs; 8.3 g protein

## Crab Rangoon

*yield: 8 servings*

*Wontons*

| | |
|---|---|
| 1 | egg |
| ½ c | water |
| 1 tsp | salt |
| 2 c | all-purpose flour |

*Filling*

| | |
|---|---|
| 4 oz | cream cheese, softened |
| 4 oz | crab meat, cooked and flaked |
| ½ clove | garlic, minced |
| 1 | shallot, finely chopped |
| to taste | black pepper |

1. To prepare wrappers, beat the egg with a wire whisk. Add ¼ cup water and salt.
2. Sift flour into a medium bowl. Form a well in the center of the flour, and pour wet mixture into well. Mix well. Add water as necessary to form dough.
3. Knead dough for 5 min, and allow to rest for 30 min.
4. Roll dough very thin on a floured surface. Cut into twenty-four 3½" squares to form wonton wrappers.
5. Layer cut squares with wax paper and store in refrigerator until needed.
6. To prepare the filling, combine all filling ingredients until well blended.
7. Lay each wonton wrapper on a flat working surface. Wet all four edges of the wrapper with water.
8. Place one teaspoon of filling in the center of each wrapper.
9. Fold each wrapper in half, carefully matching and sealing wet edges. *Note:* Keep the filling in the center of the wrapper and away from the sealed edges.
10. Fold each wrapper in half lengthwise again.
11. With the side containing the filling facing up, grab the top corner from each side and crisscross diagonally so they overlap across the center of the wonton. Cover prepared wontons with a wet paper towel to prevent them from drying out.
12. Fry in deep fat at 350°F to 375°F until golden brown.
13. Drain on paper towels to remove excess fat.

> **Nutrition Facts**
> **Per serving:** 222.3 calories; 24% calories from fat; 6.1 g total fat; 49.5 mg cholesterol; 499.8 mg sodium; 263.1 mg potassium; 32.7 g carbohydrates; 0.9 g fiber; 0.2 g sugar; 31.8 g net carbs; 9.1 g protein

## Potstickers

*yield: 4 servings*

*Filling*

| | |
|---|---|
| 2 oz | Napa cabbage, shredded |
| 4 oz | ground pork |
| 2 | green onions, chopped |
| 1 tsp | white wine |
| ¼ tsp | cornstarch |
| ¼ tsp | sesame oil |
| to taste | salt and pepper |

*Dumplings (or use wonton skins)*

| | |
|---|---|
| ½ c | all-purpose flour |
| ¼ c | boiling water |
| 2 tbsp | vegetable oil |
| ½ c | water |

*Sauce*

| | |
|---|---|
| 1 tbsp | soy sauce |
| 1 tsp | sesame oil |
| ½ clove | garlic, minced |

1. Combine all filling ingredients in a bowl and set aside.
2. In a separate bowl, combine flour and boiling water. Knead until smooth.
3. Roll dough into a 6" long roll and cut crosswise into ½" pieces.
4. Roll each piece into a 3" circle.
5. Place 1 tbsp filling in the center of each dough circle.
6. Bring the edges of the circle up, making 5 pleats, and pinch the edges together at the top to make a pouch.
7. Heat oil in a large skillet. Cook dumplings in hot oil until the bottoms are lightly browned.
8. Add water to pan, cover, and cook 7 min or until all water is absorbed.
9. For sauce, combine soy sauce, sesame oil, and garlic. Serve dumplings with sauce in a separate bowl for dipping.

> **Nutrition Facts**
> **Per serving:** 215.9 calories; 60% calories from fat; 14.6 g total fat; 20.4 mg cholesterol; 243.4 mg sodium; 155.5 mg potassium; 13.8 g carbohydrates; 0.8 g fiber; 0.4 g sugar; 13 g net carbs; 7 g protein

# Recipes

## 53

## Seafood Canapés

*yield: 16 pieces*

| | |
|---|---|
| 4 slices | white bread, crust removed |
| ⅓ c | crab meat, flaked |
| ⅓ c | medium or tiny shrimp |
| 2 tbsp | celery, finely chopped (½ stick) |
| 1 tbsp | green onion, finely chopped |
| ¼ c | mayonnaise |
| 1 tsp | sea salt |
| 2 sprigs | fresh dill |

1. Arrange the bread on a cutting board and slightly flatten with a rolling pin, using even pressure.
2. In a small bowl, combine crab, shrimp, celery, onion, and mayonnaise.
3. Divide the seafood spread evenly among the bread slices, coating each slice evenly.
4. Cut each slice into quarters.
5. Season each canapé with sea salt and garnish with a bit of fresh dill. Serve chilled.

### Nutrition Facts

**Per piece:** 33.2 calories; 38% calories from fat; 1.5 g total fat; 11.3 mg cholesterol; 185.3 mg sodium; 31.6 mg potassium; 2.6 g carbohydrates; 0.1 g fiber; 0.4 g sugar; 2.5 g net carbs; 2.4 g protein

## Tomato and Mozzarella Canapés with Basil Pesto

*yield: 16 pieces*

| | |
|---|---|
| ½ | French baguette (approximately 16″ long, 1″ diameter) |
| ½ c | olive oil |
| 2 c | fresh basil |
| 1½ oz | Parmesan cheese |
| 2 c | toasted pine nuts |
| 1 clove | garlic, minced |
| ½ lb | fresh mozzarella |
| 1 pint | grape tomatoes (approximately 16 tomatoes) |
| to taste | salt and pepper |

1. Slice the baguette on the diagonal into ½″ thick pieces.
2. Lightly brush each slice with olive oil. Place on a baking sheet.
3. Bake in a 350°F oven for 8 min to 10 min or until lightly golden.
4. Blanch basil leaves in boiling water for 2 sec. Transfer immediately to ice water to shock.
5. Finely grate the Parmesan cheese.
6. In a food processor, purée basil, Parmesan cheese, pine nuts, and garlic.
7. Add remaining olive oil, and mix well.
8. Spread basil pesto on toasted bread.
9. Thinly slice mozzarella. Place one slice on each canapé.
10. Take grape tomatoes, thinly slice, and arrange slices on canapé. Use one tomato per canapé.
11. Season with salt and pepper to taste.

### Nutrition Facts

**Per piece:** 278.7 calories; 66% calories from fat; 21.6 g total fat; 9.9 mg cholesterol; 211.1 mg sodium; 481 mg potassium; 16.4 g carbohydrates; 4.9 g fiber; 0.9 g sugar; 11.5 g net carbs; 8.8 g protein

## 55

## Cucumber and Smoked Salmon Canapés with Chive Cream Cheese Spread

*yield: 32 pieces*

| 1 lb | smoked salmon, sliced thin into ½ oz pieces |
|------|---------|
| 1 | medium cucumber |
| ½ c | sour cream |
| 8 oz | cream cheese, softened |
| 1 tbsp | chives, finely chopped |
| ¾ pint | grape tomatoes |
| 1 sprig | parsley |

1. Peel the skin of the cucumber if desired. Slice the cucumber into ¼" thick slices.
2. In a mixing bowl, combine cream cheese, sour cream, and chives until well blended.
3. Spread 2 tsp of chive cream cheese on each cucumber slice.
4. Slice grape tomatoes in half.
5. Garnish each cucumber slice with half a grape tomato, a piece of smoked salmon, and a bit of fresh parsley.
6. Serve cold.

**Nutrition Facts**

**Per piece:** 43 calories; 72% calories from fat; 3.6 g total fat; 11 mg cholesterol; 79.3 mg sodium; 51.2 mg potassium; 0.8 g carbohydrates; 0.1 g fiber; 0.1 g sugar; 0.7 g net carbs; 2 g protein

## Tomato Cream Cheese Bouchées

*yield: 4 servings*

| 1 tbsp | olive oil |
|--------|-----------|
| 1 clove | garlic, minced |
| 1 | medium tomato, diced |
| ½ tbsp | fresh basil |
| to taste | salt and pepper |
| 8 oz | cream cheese, softened |
| 1 batch | puff pastry dough |

1. In a small skillet, heat olive oil. Sauté garlic and tomato 3 min.
2. Add basil, salt, and pepper. Cook 1 min.
3. Purée tomato mixture. Cool to room temperature.
4. In a mixing bowl, combine cream cheese and tomato purée until well blended.
5. On a lightly floured surface, roll out puff pastry dough. Cut into 3" squares.
6. Place a small spoonful of tomato cream cheese mixture on each pastry square.
7. Fold pastry diagonally and pinch edges to contain the filling.
8. Brush each side of pastry lightly with olive oil and arrange on a baking sheet.
9. Bake in 375°F oven for 10 min to 12 min, or until pastries are golden brown.
10. Allow to cool slightly on wire racks. Serve hot.

**Nutrition Facts**

**Per serving:** 83.3 calories; 82% calories from fat; 7.8 g total fat; 20.8 mg cholesterol; 88.5 mg sodium; 55.4 mg potassium; 1.9 g carbohydrates; 0.2 g fiber; 0.3 g sugar; 1.7 g net carbs; 1.7 g protein

# Recipes

## Mushroom Barquettes

*yield: 12 pieces*

*Mushroom filling*

| | |
|---|---|
| ¼ c | fresh mushrooms, sliced |
| ¼ c | sour cream |
| ¼ c | cream cheese, softened |
| ½ c | mozzarella cheese, shredded |
| 1 sprig | fresh rosemary, finely chopped |

1. Follow the recipe for the pâte dough.
2. Combine all filling ingredients in a small bowl until well blended.
3. On a lightly floured surface, roll out chilled dough in a 13" square.
4. Arrange 12 barquette molds closely together. Drape dough over molds. With a rolling pin, roll over molds to cut dough. Press dough into each mold. Prick dough with a fork and chill molds for 20 min.
5. Arrange molds on a baking sheet and cover with parchment paper. Bake at 375°F for 8 min.
6. Cool shells in molds for 10 min. Remove shells and cool completely on a wire rack.
7. Sauté mushrooms and let them cool.
8. Blend all the filling ingredients together.
9. Fill each shell with a spoonful of mushroom filling. Place filled barquettes on a baking sheet and place under the broiler for 3 min.
10. Serve hot.

**Nutrition Facts**

**Per piece:** 134 calories; 74% calories from fat; 11.3 g total fat; 31.3 mg cholesterol; 100.8 mg sodium; 30.5 mg potassium; 6.1 g carbohydrates; 0.1 g fiber; 1.2 g sugar; 5.9 g net carbs; 2.4 g protein

## 58

## Spinach and Bacon Tartlets

*yield: 12 pieces*

*Spinach and bacon filling*

| | |
|---|---|
| 4 oz | bacon |
| 1 | small onion, diced |
| ½ c | fresh spinach |
| 2 oz | Monterey Jack cheese, grated |
| 2 oz | feta cheese, crumbled |
| 3 | eggs |
| ⅓ c | cottage cheese |
| to taste | salt and pepper |

1. Follow the recipe for making pâte dough.
2. In a frying pan, cook bacon over medium-high heat until crispy. Reserve fat. Place bacon on paper towels to drain excess fat. Finely chop cooked bacon.
3. In the reserved bacon fat, sauté diced onion until translucent.
4. Add fresh spinach. Sauté 3 min, stirring frequently. Add chopped bacon and mix well.
5. Arrange tartlet shells on a baking sheet. Evenly divide spinach mixture among tartlet shells.
6. Top spinach with grated Monterey Jack and feta cheeses.
7. In a small bowl, beat eggs with a wire whisk.
8. Add cottage cheese, salt, and pepper.
9. Pour egg mixture over spinach and cheese.
10. Bake at 375°F about 25 minutes, or until filling is cooked through.
11. Serve slightly cooled.

**Nutrition Facts**

**Per piece:** 192.3 calories; 72% calories from fat; 15.8 g total fat; 88.5 mg cholesterol; 274.9 mg sodium; 71.6 mg potassium; 6.7 g carbohydrates; 0.2 g fiber; 1.6 g sugar; 6.5 g net carbs; 6 g protein

## Shrimp Cocktail with Mignonette

**59**

*yield: 4 servings*

| | |
|---|---|
| 12 | fully cooked jumbo shrimp |
| 1 | lemon |

*Mignonette*

| | |
|---|---|
| 1 tsp | black pepper, freshly cracked |
| ¼ c | red wine vinegar |
| 1 | shallot, minced |
| to taste | salt |

1. Arrange shrimp around the rim of a cocktail glass filled with crushed ice.
2. In a small bowl, combine all mignonette ingredients until well blended.
3. Place a small dish in the center of the cocktail glass, and fill with mignonette.
4. Serve cold. Garnish with lemon wedges.

### Nutrition Facts

**Per serving:** 94.7 calories; 2% calories from fat; 0.3 g total fat; 32.2 mg cholesterol; 122 mg sodium; 400.6 mg potassium; 19.1 g carbohydrates; 0.2 g fiber; 0.4 g sugar; 18.9 g net carbs; 6 g protein

**Recipe Variation:**

**60** **Shrimp Cocktail with Cocktail Sauce:** Substitute the mignonette with a cocktail sauce made from ¼ cup ketchup, 1 tbsp chili sauce, 1 tbsp tomato purée, 1 tsp lemon juice, ½ tbsp ground horseradish, and salt to taste.

## Crudités Platter with Vegetable-Dill Dip

**61**

*yield: 4 servings*

*Crudités*

| | |
|---|---|
| 3 | carrots, batonnet cut |
| 3 stalks | celery, batonnet cut |
| 1 c | fresh broccoli florets |
| 1 pint | cherry tomatoes |

*Vegetable-dill dip*

| | |
|---|---|
| ½ c | mayonnaise |
| ½ c | sour cream |
| 1 | shallot, minced |
| 1 tbsp | carrot, grated |
| 1 tbsp | fresh dill, finely chopped |
| 1 tsp | fines herbes |
| to taste | salt |

1. Arrange all raw vegetables on a platter.
2. For dip, in a small bowl, combine mayonnaise, sour cream, minced shallots, grated carrots, dill, and seasonings to taste. Mix well.
3. Place dip in a serving bowl with a small serving spoon.

### Nutrition Facts

**Per serving:** 224.9 calories; 63% calories from fat; 16.3 g total fat; 20.3 mg cholesterol; 367.4 mg sodium; 539 mg potassium; 19.1 g carbohydrates; 2.8 g fiber; 4.6 g sugar; 16.3 g net carbs; 3.2 g protein

## Bleu Cheese Dip

**62**

*yield: 4 servings*

| | |
|---|---|
| 6 oz | cream cheese |
| 2 oz | bleu cheese, crumbled |
| 3 tbsp | sour cream |
| ¼ | medium onion |
| to taste | salt |

1. To make onion juice, place onion in a blender and grind to a fine consistency. Place ground onion in a paper towel to squeeze out the juice. Reserve 1 tsp of onion juice for dip and store extra for later use.
2. Place all ingredients in a mixer and blend thoroughly at a low speed.
3. Use extra cream cheese or sour cream as necessary to reach desired consistency.
4. Chill until ready to serve.

### Nutrition Facts

**Per serving:** 221.6 calories; 82% calories from fat; 20.8 g total fat; 61.4 mg cholesterol; 401.3 mg sodium; 113.4 mg potassium; 2.8 g carbohydrates; 0.1 g fiber; 0.6 g sugar; 2.7 g net carbs; 6.6 g protein

## Nacho Cheese Sauce

*yield: 1 pint*

63

| | |
|---|---|
| 1 pt | béchamel |
| ½ lb | sharp cheddar, grated |
| 2 tbsp | canned jalapeño slices |
| 2 tbsp | canned jalapeño juice |
| ½ tsp | garlic powder |
| ½ tsp | ground cumin |
| to taste | salt and pepper |

*Wisconsin Milk
Marketing Board*

1. Heat béchamel until hot.
2. Slowly add grated cheese to the hot béchamel, whisking constantly.
3. Once all the cheese has been incorporated, add the remaining ingredients and mix well.
4. Serve with tortilla chips.

**Nutrition Facts**

**Entire recipe:** 1843.2 calories; 65% calories from fat; 137.9 g total fat; 409 mg cholesterol; 2050.1 mg sodium; 1091.8 mg potassium; 74.2 g carbohydrates; 1.9 g fiber; 29 g sugar; 72.3 g net carbs; 78.2 g protein

## Chicken Satay with Teriyaki Sauce

*yield: 4 servings*

64

| | |
|---|---|
| ¼ c | peanut butter |
| ¼ c | fresh cilantro, chopped |
| 3 tbsp | lime juice |
| 3 tbsp | water |
| 1 tbsp | fresh ginger, peeled and grated |
| 1 tsp | hot chile sauce |
| to taste | salt |
| 2 lb | chicken breast, cut into 2″ pieces |
| | wooden skewers |

*Teriyaki sauce*

| | |
|---|---|
| ⅔ c | soy sauce |
| ¼ c | mirin (sweet rice wine) |
| ⅓ c | cider vinegar |
| 2 cloves | garlic, minced |
| 3 tbsp | light brown sugar, packed |
| 3 tbsp | fresh ginger, peeled and grated |

1. In a large mixing bowl, combine peanut butter, cilantro, lime juice, water, ginger, hot chile sauce, and salt.
2. Add chicken pieces and toss to coat.
3. Soak wooden skewers in water for 10 min. Skewer chicken on wooden skewers and grill or griddle for 15 min to 20 min or until cooked through.
4. In a small saucepan, bring soy sauce, mirin, and cider vinegar to a simmer. *Note:* Sweet sherry can be substituted for mirin.
5. Add garlic, brown sugar, and ginger. Stir until sugar is completely dissolved.
6. Transfer sauce to a small bowl and place in an ice bath or cool with a cooling wand to room temperature.
7. Serve chicken hot with dipping sauce at room temperature.

**Nutrition Facts**

**Per serving:** 422.5 calories; 22% calories from fat; 11.1 g total fat; 131.5 mg cholesterol; 1718.7 mg sodium; 887 mg potassium; 20.8 g carbohydrates; 1.6 g fiber; 12.6 g sugar; 19.2 g net carbs; 59 g protein

## Tropical Fruit Kebabs

*yield: 4 servings*

| | |
|---|---|
| 2 c | fresh pineapple, cut in 1″ cubes |
| 1 c | red seedless grapes |
| 2 c | pears, cut in 1″ pieces |
| 2 c | mangoes, cut in 1″ cubes |
| 4 | kiwifruits, peeled and quartered |
| 2 tbsp | lemon juice |
| | wooden skewers |

1. Place all fruit in a large mixing bowl.
2. Drizzle lemon juice over fruit and toss to coat.
3. Arrange fruit on wooden skewers, layering pineapple, grape, pear, mango, and kiwifruit, and repeating.
4. Serve chilled.

### Nutrition Facts

**Per serving:** 214.5 calories; 3% calories from fat; 0.9 g total fat; 0 mg cholesterol; 6.4 mg sodium; 639 mg potassium; 55.6 g carbohydrates; 7.8 g fiber; 40.7 g sugar; 47.8 g net carbs; 2.3 g protein

## Prosciutto with Melon

*yield: 4 servings*

| | |
|---|---|
| ½ | cantaloupe, cut into 4 wedges |
| 12 slices | prosciutto, cut paper-thin |

1. Place a wedge of melon on each plate.
2. Arrange three slices of prosciutto on each plate, along side of or wrapped around melon.
3. Serve chilled.

### Nutrition Facts

**Per serving:** 168.2 calories; 39% calories from fat; 7.1 g total fat; 59.5 mg cholesterol; 2293.7 mg sodium; 452.3 mg potassium; 0.8 g carbohydrates; 0.1 g fiber; 0.5 g sugar; 0.8 g net carbs; 23.7 g protein

## California Rolls

*yield: 4 servings*

| | |
|---|---|
| ¼ c | rice vinegar |
| 1 tbsp | sugar |
| 1 tsp | salt |
| 1 c | cooked sushi rice |
| 2 sheets | nori, 8″ × 7″ sheets |
| 1 tbsp | avocado, julienne cut |
| 1 tbsp | cucumber, peeled, julienne cut |
| ½ c | Alaskan king crab meat |
| 1 tbsp | wasabi paste |
| 1 tbsp | pickled ginger |

1. Combine rice vinegar, sugar, and salt over medium-low heat, until sugar and salt are dissolved.
2. Slowly add the rice vinegar mixture to the cooked sushi rice. Let stand until cool. Refrigerate 2 hr or longer.
3. Dry-roast each sheet of nori, moving quickly over direct heat for 30 sec, or until it turns bright green.
4. Place a piece of nori on a sudare (bamboo sushi mat). With wet hands, spread ½ cup sushi rice evenly on the nori, leaving a 1″ border on the top edge.
5. Arrange sticks of avocado and cucumber in a horizontal line across the center of the rice. Top with half of the crab meat.
6. From the bottom edge, lift the nori and mat slightly and roll tightly and evenly, pressing down with each quarter turn.
7. Seal the nori by placing a bit of water on the top 1″ border. Press the seam closed.
8. Repeat procedure with second sheet of nori.
10. With a sharp knife, trim the two ends of the rolls. Cut each roll into six 1″ pieces.
11. Garnish plate with wasabi paste and pickled ginger. Serve cold with soy sauce.

### Nutrition Facts

**Per serving:** 129.3 calories; 6% calories from fat; 1.1 g total fat; 26.8 mg cholesterol; 1117.3 mg sodium; 294.6 mg potassium; 21.8 g carbohydrates; 0.6 g fiber; 3.2 g sugar; 21.2 g net carbs; 13 g protein

# Fruits

Daniel NYC

*Fruits* are an important part of a healthy diet. Improved methods of transportation have made more varieties of fruit available today than ever before. A fruit is most flavorful and of the highest quality at the peak of its growing season. A chef must possess a strong understanding of available varieties of fruit, how to judge freshness and quality, and when the different fruits are in season. Having this information helps a chef choose the best fruits to offer on menus at different times of the year.

## Chapter Objectives:

1. Identify the major categories of fruit and common examples of each category.
2. Demonstrate the following procedures: removing the peel from citrus fruit, segmenting citrus fruit, coring an apple using a paring knife, coring a pineapple, and dicing a mango.
3. Explain common factors to consider when purchasing and storing fruit.
4. Contrast the various methods used for cooking fruit.

## Key Terms

- **fruits**
- **citrus**
- **peel**
- **pith**
- **zest**
- **ceviche**
- **berry**
- **grape**
- **pome**
- **pectin**
- **drupe**
- **melon**
- **tropical fruit**
- **ethylene gas**
- **fruit-vegetable**

## FRUIT CLASSIFICATIONS

*Fruits* are the edible, ripened ovaries of flowering plants that contain one or more seeds. Fruits can add color, texture, and many different flavors to a meal. Many fruits are high in water, fiber, certain vitamins, antioxidants, and phytochemicals, so incorporating fruit into daily meal planning is good for overall health. Fruit typically contains seeds, which are often inedible and must be removed. Fruits are classified into several major categories, including citrus, berries, grapes, pomes, drupes, melons, tropical fruits, and fruit-vegetables.

### Citrus

*Citrus* is a type of fruit that grows on thorny trees or shrubs in tropical regions and has a thick rind and pulpy meat. Citrus fruits are harvested fully ripe, as they do not continue to ripen after they are removed from the plant. Citrus fruits are excellent sources of vitamin C and are usually quite acidic. Fresh citrus fruits should be kept refrigerated to extend their storage life. Citrus fruits include oranges, grapefruits, lemons, and limes. ***See Figure 14-1.***

A citrus fruit commonly has a brightly colored, thick outer rind called a *peel.* The white layer just beneath the peel of a citrus fruit is called the *pith.* The pith is usually removed during preparation because of its bitter taste.

*Citrus Fruits*

Oranges

Grapefruits

Lemons

Limes

14-1 Citrus fruits have brightly colored peels, white piths, and wedge-shaped segments of meat.

To remove the peel from a citrus fruit, apply the following procedure:

1. Slice off the peel and pith from the top and bottom of the fruit. The top of the fruit can be identified by a small circle where the stem has been removed. **See Figure 14-2.**

2. Place one cut end of the fruit on the cutting board. Cut off even strips of the peel, starting at the top edge and moving the blade down while following the contour of the fruit.

*Peeling Citrus Fruit*

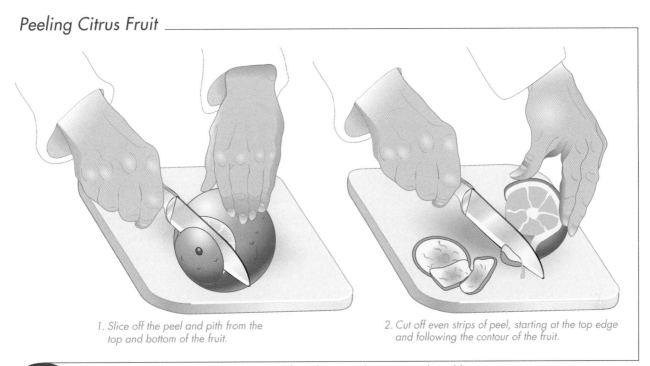

1. Slice off the peel and pith from the top and bottom of the fruit.

2. Cut off even strips of peel, starting at the top edge and following the contour of the fruit.

**14-2** *The pith and peel are removed from citrus fruits because they are tough and bitter.*

The meat of a citrus fruit is naturally divided into wedge-shaped segments. These segments are often separated when preparing citrus fruits for use in a dish or as a garnish.

To segment a citrus fruit, apply the following procedure:

1. Remove the peel and pith by carving around the fruit with a paring knife.

2. Carefully cut along each side of the thin membrane, separating each section of meat from the membrane. When finished, the membrane is left intact while the segments are removed. **See Figure 14-3.**

## Segmenting Citrus Fruit

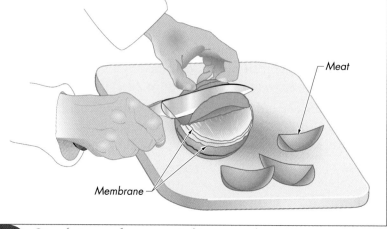

Meat

Membrane

14-3 Citrus fruits are often segmented to remove the meat.

Zesting is accomplished by using a zester, a fine grater, or a thin, sharp blade to scrape the outermost peel of a citrus fruit into fine shreds. *See Figure 14-4.* *Zest* is the colored outermost layer of the peel of a citrus fruit such as a lemon or lime. Zest is typically used to add flavor to a dish, as it contains a high concentration of flavored essential oils. Thin strips of zest can be obtained by running a five-hole zester along the surface of the rind. The white pith can be bitter, so care should be taken not to peel or cut too deeply when zesting. Large strips of zest can be obtained by using a peeler to peel off the colored surface of the rind. The large strips of zest can then be cut into julienne strips. Zest can also be candied for use as a confection or a decoration.

## Zesting

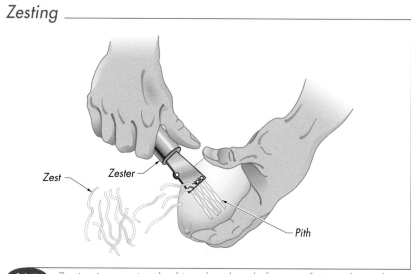

Zest

Zester

Pith

14-4 Zesting is removing the thin colored peel of a citrus fruit, such as a lemon or lime.

**Oranges.** An *orange* is a sweet, round, orange-colored citrus fruit. Valencia and navel oranges are the two most popular varieties of oranges and can be used in many ways. Sauces, juices, jams, jellies, and desserts are just a few of the ways these oranges can be incorporated into cooking. Valencia and navel oranges are available year-round, but their peak season is from December to April. Blood oranges are another, less common variety of sweet orange. Blood oranges are smaller than the Valencia and navel varieties and have a deep red flesh. They are used to make beautifully colored sauces and to add color to salads.

**Grapefruits.** The grapefruit is a hybrid of an orange and a pummelo. A *pummelo* is a pale green to yellow citrus fruit that is larger than a grapefruit, with a thick, spongy rind. A *grapefruit* is a round fruit with a thick, yellow outer rind and rather tart flesh. The two most common varieties of grapefruit are white grapefruit and pink grapefruit. White grapefruit has pale yellow flesh, and pink grapefruit has pink-colored flesh.

**Lemons.** A *lemon* is a tart yellow citrus fruit with high acidity. Lemon juice is used in desserts, as well as many types of sauces that flavor poultry, fish, and shellfish. *Ceviche* is a dish originating from Peru, made from raw fish or shellfish, lemons or limes, chopped onion, and minced chilies. The fish mixture is allowed to rest briefly at room temperature, giving the acid time to "cook" the fish or shellfish. Lemon juice can also be used as a salad dressing.

**Limes.** A *lime* is a small citrus fruit that can range in color from dark green to yellow-green. Limes are less acidic than lemons. Limes are available year-round and lime juice is used in many sauces and desserts. Lime juice adds a fresh, crisp flavor to many ethnic dishes.

**Mandarins.** A *mandarin* is a small, dark-orange-colored citrus fruit that is closely related to the orange but is more fragrant, whose varieties include tangerines, clementines, satsumas, tangors, and tangelos. The peel of a mandarin is thinner than that of an orange and is more easily separated from the meat. Mandarins are often eaten fresh because of their intense sweetness. Descriptions of mandarin varieties are as follows:

- tangerines—a cross between a mandarin and a bitter variety of orange, with a slightly reddish orange peel

- clementines—also a cross between a mandarin and a bitter orange, but with a rougher skin

- satsumas—a small, seedless variety of mandarin

- tangors—a cross between a tangerine and a sweet orange

- tangelos—a cross between a tangerine and a grapefruit

The main difference between tangerines and clementines is where they are grown. Most tangerines are grown in the southeastern United States, while clementines are predominantly grown in Europe, North Africa, and Israel.

## Berries

A *berry* is a type of fruit that is small and has many tiny seeds. Berries include strawberries, blueberries, raspberries, and blackberries. Berries grow on bushes and vines and are harvested ripe, as they do not continue to ripen after harvest. Quality berries are sweet and evenly colored. ***See Figure 14-5.***

*Berries*

Strawberries

Raspberries

Blackberries

Blueberries

**14-5** Berries are small fruits with many tiny seeds.

**Strawberries.** A *strawberry* is a bright red, heart-shaped fruit covered with tiny black seeds. A ripe strawberry should be evenly red. When selecting strawberries, it is important to select only berries that are free of brown or soft spots and that are not near any berries that are molding. A moldy berry can quickly spread mold and spoil other berries. Strawberries are widely used in desserts, sauces, and salads.

**Blueberries.** A *blueberry* is a small, dark-blue berry that grows on a shrub. Blueberries are in peak season from mid-June to August. With such a short season, blueberries can be rather expensive. The high cost of blueberries means that a chef needs to be creative when using them. Blueberries are often used in jams, jellies, desserts, quick breads such as muffins, and salads.

**Raspberries.** A *raspberry* is a subtly tart red fruit that grows on vines. Botanically, raspberries are not considered berries, but for culinary purposes they are categorized as berries. Raspberries are an aggregate fruit, meaning that each berry is a cluster of very tiny fruits. The velvety soft texture of raspberries makes them one of the most fragile members of the berry group. Most raspberries are marketed frozen because of their susceptibility to mold. Raspberries are used in a variety of desserts and can be used to make bright red sauces to complement any dessert.

### Nutrition Note

Blueberries and cranberries contain more antioxidants than most fruits. These antioxidants have been found to reduce the risk of certain types of cancer.

**Blackberries.** A *blackberry* is a sweet, dark purple to black, aggregate fruit that grows on a bramble bush. The peak season for blackberries is from mid-June to August. Blackberries are slightly larger than raspberries and have a shiny outer layer. Blackberries are used in many desserts and sauces because of their sweetness and strikingly dark color.

**Cranberries.** A *cranberry* is a small round fruit that is red in color and sour tasting. The skin of a cranberry is white when young and turns a deep red when ripe. Cranberries grow in cool climates on low vines in cultivated bogs or wetlands. Cranberries should be examined before use and any soft or discolored cranberries should be discarded. Because of their strong, sour flavor, they are most often used in desserts, breads, sauces, and jellies.

**Currants.** A *currant* is a small red or black berry that grows on bushes native to western Europe. Currants grow in grapelike clusters, peaking in late summer. Currants are either red currants or black currants. Currants are used in a variety of ways, including jams, jellies, sauces, and fruit soups. Dried currants used in baking are most often Zante currants. The Zante currant is not a currant at all, but is actually a small, sweet variety of seedless grape. Zante currants are native to the Greek islands.

## Grapes

A *grape* is a type of fruit that has a smooth skin and grows on woody vines in large clusters. Grapes have the longest documented history of any type of fruit and are the most widely grown fruit internationally because of their use in winemaking. Table grapes are varieties suitable for eating, as opposed to varieties grown specifically for winemaking.

The two main classifications of table grapes are white grapes, which are usually green, and black grapes, which are usually red to dark blue in color. Most of the flavor of a grape comes from its skin. It is important to choose grapes that are firm with no discoloration. *See Figure 14-6.*

**Thompson Grapes.** A *Thompson grape* is a seedless grape that is pale to light green in color. Thompson grapes are the most common type of seedless grapes found in the market. These grapes are available year-round, but their peak season is from June to November. Thompson grapes make a great accompaniment to a salad or a cheese platter.

**Red Flame Grapes.** A *red flame grape* is a seedless grape that ranges from a light purple-red color to a darker purple color. Red flame grapes are also commonly available in the market. They are typically sweeter and crisper in texture than Thompson grapes. Red flame grapes are also a wonderful accompaniment to a salad or cheese platter.

*Grapes*

Thompson

Concord          Red flame

**14-6** *Common varieties of table grapes include Thompson, red flame, and Concord.*

**Concord Grapes.** A *Concord grape* is a seeded grape with a deep black color. It is sometimes called a slipskin grape, as its skin is very easily separated from the fruit. Concord grapes are available in the market as a table grape but, because they have seeds, they are used less often than the Thompson or red flame varieties. Concord grapes are commonly used in the making of jams, jellies, and grape juice due to their high sugar content and distinguishable flavor. The majority of grape jam and jelly produced in the United States is made with Concord grapes.

## Pomes

A *pome* is a type of fruit that contains a core of seeds and an edible peel. Examples of pomes are apples, pears, and quinces. Pomes have thin skin and grow on trees or bushes. Quality pomes are free of blemishes and bruises and have no soft spots. ***See Figure 14-7.***

### Pomes

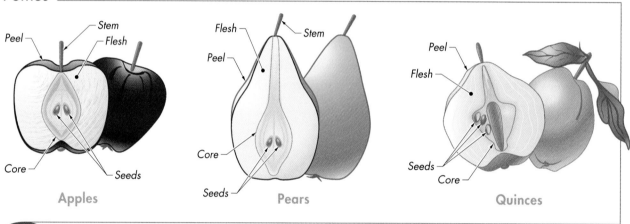

**14-7** *A pome has a core of seeds and an edible peel.*

**Apples.** An *apple* is a hard round pome that can range in flavor from sweet to tart and in color from pale yellow to dark red. Apple trees grow best in temperate zones and can withstand temperatures to −40°F. Apples are available year-round, but their peak season is between late August and early November. Countries that produce the largest amounts of apples include Russia, Germany, France, the United States, and China. There are more than 7500 cultivated varieties of apples. Some varieties of apples are not good for cooking purposes, and others become bitter while baking in the oven. ***See Figure 14-8.***

When preparing an apple for use in a dish, the inner core must be removed. The core contains inedible seeds and tough, inedible fibers. A paring knife or apple corer can be used to remove the core. An apple corer is inserted into the stem end of the apple to cut out a cylindrical portion of apple containing the core. The cylinder is removed and the remaining edible fruit is prepared as desired.

## Apple Varieties . . .

| | Name | Description | Uses | Availability |
|---|---|---|---|---|
| | **Acey Mac** | Sweet, tart, juicy, tender flesh, cooks quickly | Sauces, raw | Sept. to June |
| | **Braeburn** | Sweet, tangy, crisp, aromatic, juicy, yellow flesh | Cooking, sauces, salads, raw | Oct. to July |
| | **Cameo** | Sweet, crunchy | Baking, sauces, salads | Sept. to June |
| | **Cortland** | Sweet, hint of tartness, juicy, tender, white flesh | Baking, salads, sauces, raw | Sept. to April |
| | **Crispin** | Sweet, juicy, crisp | Baking, sauces, salads, raw | Oct. to Sept. |
| | **Empire** | Sweet, tart, juicy, creamy-white flesh | Baking, sauces, salads, raw | Sept. to July |
| | **Fortune** | Crisp and spicy | Baking, sauces, raw | Late Oct. to June |
| | **Fuji** | Spicy sweet, juicy, firm cream-colored flesh, tender skin | Sauces, salads, raw | Oct. to June |
| | **Gala** | Yellow to red; sweet, juicy, crisp yellow flesh | Salads, raw | Sept. to June |
| | **Ginger Gold** | Sweet, tart, fine texture, crisp cream-colored flesh | Salads, raw | Aug. to Nov. |
| | **Golden Delicious** | Yellow; sweet, crisp, light yellow flesh | Baking, salads, raw | Sept. to June |
| | **Granny Smith** | Green; tart, crisp, juicy | Cooking, salads, raw | Sept to June |
| | **Honey Crisp** | Sweet, tart, juicy, extremely crisp, yellow flesh | Baking, sauces, salads, raw | Sept. to Feb. |

**14-8** *Some varieties of apples are best for cooking and baking, while others are more suitable for use in sauces or salads.*
*(continued on next page)*

| | Name | Description | Uses | Availability |
|---|---|---|---|---|
| | **Idared** | Sweet, tart, juicy, firm, pale yellow-green flesh, sometimes rosy pink flesh | Baking, cooking, sauces, salads | Oct. to Aug. |
| | **Jonagold** | Tangy, sweet | Cooking, raw | Oct. to May |
| | **Jonamac** | Sweet, tart | Sauces, raw | Sept. to Nov. |
| | **Jonathan** | Tart | Baking, sauces, salads | Sept. to April |
| | **Macoun** | Honey sweet, tart, juicy, crisp, creamy, yellow flesh | Baking, cooking, sauces, salads, raw | Oct. to Nov. |
| | **McIntosh** | Sweet, tangy, juicy, tender, white flesh | Baking, sauces, salads, raw | Sept. to June |
| | **Northern Spy** | Tart | Baking, sauces, raw | Oct. to Dec. |
| | **Paula Red** | Tart, juicy, crisp, white flesh | Sauces, raw | Aug. to Oct. |
| | **Pink Lady (Cripps Pink)** | Sweet, tangy, firm flesh | Baking, sauces, salads | Nov. to Feb. |
| | **Red Delicious** | Bright to dark red, sometimes striped; mildly sweet, juicy, crisp, yellow flesh | Salads, raw | Oct. to Dec. |
| | **Red Rome** | Mildly tart, firm, greenish-white flesh | Baking, sauces, salads | Oct. to Sept. |
| | **Winesap** | Dark red; spicy, slightly tart | Cider, cooking, raw | Oct. to Jan. |

**. . . Apple Varieties**

14-8 *(continued from previous page)*

To core an apple using a paring knife, apply the following procedure:

1. Slice the apple in half lengthwise, cutting through the stem end to the bottom. ***See Figure 14-9.***

2. Cut each half of the apple in half lengthwise again to quarter the apple.

3. With a paring knife, trim away the center core and seeds and discard.

CULINARY PROCEDURES

## Coring Apples

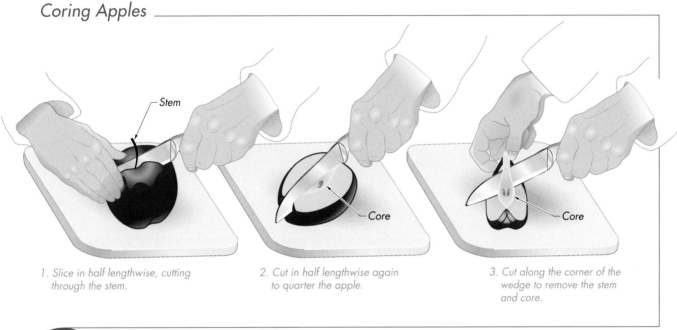

Stem

Core

Core

1. Slice in half lengthwise, cutting through the stem.

2. Cut in half lengthwise again to quarter the apple.

3. Cut along the corner of the wedge to remove the stem and core.

**14-9** *The inedible core must be removed when preparing an apple.*

**Pears.** A *pear* is a bell-shaped pome with a thin peel and sweet flesh. Pears should be harvested before they are ripe, but if picked too early they will not develop their full flavor. Conversely, pears that are picked too late turn brown and become watery inside. If left to ripen on the tree, pears develop concentrations of cellulose, resulting in a grainy texture. There are thousands of different kinds of pears. Commonly available varieties of pears include the Anjou, Seckel, Bosc, and Bartlett. ***See Figure 14-10.***

Anjou pears are generally plump, lopsided, and green in color. The Anjou pear is one of the standard commercially used pears because of its low cost and almost year-round availability. The Anjou pear is not as sweet as some other varieties of pears, but its texture is smooth, with juicy, firm flesh. Anjou pears are best eaten as a snack or in salads. Anjou pears that are not fully ripe can be baked, poached, or roasted.

*Chef's Tip:*

Red pears are cultivated primarily because of their cosmetic appeal. The inside flesh is similar to other pears, with pale color and a similar flavor and texture. Red pears have a tough skin that should be peeled before the pears are cooked.

| Common Pear Varieties | | | |
| --- | --- | --- | --- |
| Name | Description | Uses | Availability |
| Anjou | Green when ripe; sweet and juicy | Raw, salads, baking, poaching, or roasting when underripe | Sept. to July |
| Seckel | Tiny, maroon and olive green; extremely sweet | Desserts, pickling, raw | Sept. to Feb. |
| Bosc | Brown and russeted; dense flesh, highly aromatic, spicy sweet | Baking, cooking | Sept. to April |
| Bartlett | Bright yellow when ripe; aromatic, sweet, and juicy | Raw, cooking, canning | Aug. to Jan. |

**14-10** *Commonly available types of pears include Anjou, Seckel, Bosc, and Bartlett.*

The Seckel is a small pear that is sometimes called a honey pear or sugar pear because of its syrupy fine-grained flesh and complex sweetness. The Seckel pear matures in late summer and early fall. It is superb for use in desserts or can be eaten as a snack.

The Bosc pear has a gourdlike shape, with a brown to bronze-colored peel. The peel of a quality Bosc pear has a yellow rather than a green undertone. The flesh of a Bosc pear is rich, syrupy, and sweet. Because of their firm texture, Bosc pears hold their shape when cooked. Bosc pears are winter pears and need cold storage to ripen fully.

The Bartlett pear is the most widely grown pear worldwide. It is large and golden, with a bell shape. The Bartlett pear has a musky flavor and aroma and creamy, juicy flesh. Bartlett pears are delicious as a snack, can be used in sauces or sorbets, or can be baked in pies or tarts.

**Quinces.** A *quince* is a hard yellow fruit that grows in warm climates. Quinces are believed to be native to Iran. Quinces are not eaten raw because they have a bitter taste. These characteristics disappear during cooking. Quinces are typically cooked in a sugar syrup, which turns the fruit slightly darker with a pinkish tint. Quinces are often used in jams and jellies because they contain a high percentage of pectin. *Pectin* is a chemical found in plants that, when combined with an acid, forms a clear gel used as an edible thickening agent.

## Drupes

A *drupe* is a type of fruit that contains only one seed or pit. Drupes are sometimes referred to as stone fruits. Examples of drupes are peaches, nectarines, apricots, plums, and cherries. ***See Figure 14-11.*** Drupes grow on shrubs and trees and are usually harvested before they are ripe. High-quality drupes are free of blemishes or bruises.

**Peaches.** A *peach* is a sweet, orange to yellowish fruit with downy or fuzzy skin. Native to China, the flavor of the peach lends it to being eaten fresh or used in recipes. The flesh of a peach is juicy, yet firm enough to hold shape. The skin is edible, but the large oval pit contained in a peach cannot be eaten. Peaches ripen and spoil very quickly.

**Nectarines.** A *nectarine* is a sweet, slightly tart, orange to yellowish fruit. Also native to China, nectarines share many characteristics with peaches. In fact, the nectarine is a result of a peach mutation. The firm yellow flesh of a nectarine is juicy and contains a large oval pit.

**Apricots.** Apricots, also native to China, are smaller than peaches or nectarines. An *apricot* is a fruit that has pale orange-yellow skin with a fine, downy texture, and a sweet and aromatic flesh. Apricots are quite delicate and must be harvested before they are ripe to avoid damage in shipping.

**Plums.** A *plum* is an oval-shaped fruit that grows on trees in warm temperate climates with a flesh that can vary from reddish orange to yellow or greenish yellow. The skin can vary in color from a blue-purple to red, yellow, or green. There are more than 2000 varieties of plums. Some types of plums are sweet, juicy, and fragrant, while others can be sour, crisp, or mealy.

### Nutrition Note

Apricots are rich in vitamin A, vitamin C, and potassium.

## Drupes

Peaches

Nectarines

California Fresh Apricot Council

Apricots

U.S. Department of Agriculture

Plums

National Cherry Growers and Industries Foundation

Cherries

**14-11** *Drupes are fruits that contain a single pit or seed, such as peaches, nectarines, apricots, plums, and cherries.*

Damson plums are a blue-skinned variety of plum with an acidic, tart flavor. Damson plums make wonderful jams, jellies, and preserves. Japanese plums are slightly heart-shaped, with pale green or golden yellow flesh that is sweet and juicy. The skin of the Japanese plum varies from crimson or purple to a greenish yellow color.

**Cherries.** A *cherry* is a small smooth-skinned fruit that grows in a cluster on a cherry tree. Cherries have a long, thin stem that holds them on the tree. The skin of cherries typically ranges in color from a bright red to a deep red that is nearly black. There are also golden-skinned varieties. The meat is pulpy and juicy, and ranges in color from a dark yellow-orange to a deep reddish black. The small pit of a cherry is easily removed with a cherry pitter. ***See Figure 14-12.*** The three categories of cherries are sweet, sour, and wild.

## Pitting Cherries

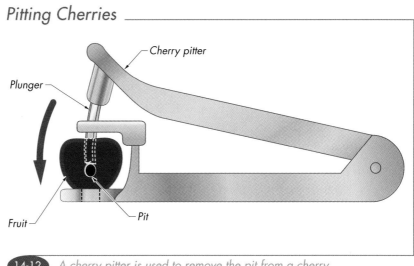

*Cherry pitter*

*Plunger*

*Fruit*

*Pit*

**14-12** *A cherry pitter is used to remove the pit from a cherry.*

Sweet cherries include Bing cherries, bigarreau cherries, and Gean cherries. The Bing cherry is the most common type of cherry in North America and is known for its luxurious sweetness. Bing cherries have a thin skin with a dark red hue. The French use the heart-shaped bigarreau cherry for its sweetness. Gean cherries are red or black, soft-fleshed cherries used in the making of Kirsch, a cherry-flavored brandy.

Sour cherries include Montmorency cherries and Morello cherries. Sour cherries have a dark red hue and, because of their sourness, are more often cooked than eaten raw. These cherries are found in many jams, jellies, pies, and liqueurs.

Wild cherries are quite small, nearly black in color, and not as fleshy as other cherry varieties. These cherries have a pungent, bitter flavor.

## Melons

A *melon* is a type of fruit that has a hard-skin and a soft inner flesh that contains many seeds. Melons typically grow on vines. The hard outer skin, or rind, can be netted or smooth in texture. The seeds contained inside the fruit are removed before serving fresh melon or using it to make dishes such as soups or sorbets. Most melons are picked just before they are ripe. Characteristics of a good melon include firmness and a good aroma. Examples of melons include cantaloupe, muskmelon, honeydew, and watermelon. *See Figure 14-13.*

## Melons

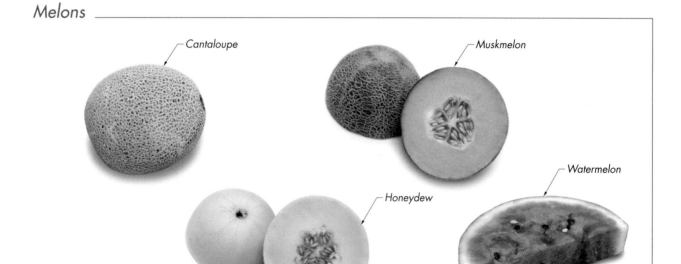

**14-13** *Cantaloupe, muskmelon, honeydew, and watermelon are varieties of hard-skinned melons.*

**Cantaloupes.** A *cantaloupe* is an orange-fleshed melon with rough, deeply grooved skin. The grooves run lengthwise from the stem, much like the grooves on a pumpkin. The most widely cultivated variety of the cantaloupe is the Charentais cantaloupe, which has pale green, lightly ridged skin. Charentais cantaloupe is grown almost exclusively in France. True cantaloupe is rarely found in the United States; American cantaloupe is actually a type of muskmelon.

**Muskmelons.** A *muskmelon* is a round, orange-fleshed melon with beige or brown, netted skin. The inside flesh can range in color from salmon to orange-yellow. American cantaloupe is a type of muskmelon, but not all muskmelons are cantaloupe. There are many hybrids derived from muskmelon, which often causes confusion in the classification.

**Honeydew.** A *honeydew* is a melon with a smooth outer skin that changes from a pale green color to a creamy yellow color as it ripens and has a mild, sweet flavor. Honeydew is available almost year-round, but its peak season is from June through October.

**Watermelons.** A *watermelon* is a sweet, extremely juicy melon that is round or oblong in shape, with pink or red flesh and green skin. The watermelon is named for its high water content—watermelons are 92% to 95% water. The weight of a watermelon varies by variety, but some can weigh up to 30 lb. Watermelon vines grow best in warm climates.

Watermelons are either seedless or filled with a large amount of seeds. Immature watermelon seeds are thin and white. The seeds thicken and turn black when mature. The thick rind of a watermelon ranges from light green to dark green in color, and is often striped or spotted. In some cultures, watermelon rind is eaten as a vegetable. In China, for example, the rind is stir fried, and in Russia the rind is pickled. Watermelon is most often sliced and enjoyed fresh, or puréed into sauces and cold fruit soups.

## Tropical Fruits

A *tropical fruit* is a type of fruit that comes from a hot humid location. Tropical fruits can range in flavor from sweet to tangy and in texture from soft to crisp. Rapid transportation has greatly increased the variety of tropical fruits available. Banana, pineapple, and kiwifruit are tropical fruits that have been available in the United States for many years. Some tropical fruits, such as pomegranates and figs, are native to the Mediterranean and Middle East where they have been enjoyed since the times of ancient Greece and Rome. Mango, papaya, passion fruit, and guava are flavorful tropical fruits that are readily available when in season. Other, more exotic, tropical fruits, such as mangosteens, lychees, and rambutans, have a more limited availability. ***See Figure 14-14.***

**Bananas.** A *banana* is an elongated yellow fruit that grows in a hanging bunch on a banana plant. Banana bunches grow in tiers called hands, with each hand having up to twenty bananas. There are typically five to twenty hands of bananas in a bunch. Two of the most common varieties of bananas consumed in the United States are the sweet banana and the plantain.

The most common variety of sweet banana is the Cavendish banana. This variety is particularly hearty, making it well suited for shipping from tropical farms. Bananas are picked and shipped while still green. *Ethylene gas* is an odorless gas that a fruit emits as it ripens and that encourages other surrounding fruits to ripen. At their destination, bananas are sometimes placed in airtight rooms filled with ethylene gas, which accelerates their ripening. If purchased green and allowed to ripen naturally, their flavor is notably richer. Bananas at peak ripeness have a bright yellow skin that is speckled with brown spots. Bananas can be eaten fresh, added to salads, cooked, or baked into quickbreads such as banana bread.

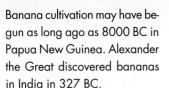

**Historical Note**

Banana cultivation may have begun as long ago as 8000 BC in Papua New Guinea. Alexander the Great discovered bananas in India in 327 BC.

**14-14** *Tropical fruits have a wide variety of flavors and textures.*

A *plantain* is a close relative of the banana but is larger and has a dark brown skin. Unripe plantains are firm and starchy, similar to potatoes. Plantains are usually fried, not peeled and eaten like a sweet banana. They are sometimes served with a sweet sauce over them to reduce their starchy flavor. When extremely ripe, the skin of a plantain turns black, with inner fruit that is soft, deep yellow, and sweet. Ripe plantains can be used for sweet dishes. Plantains can also be dried and ground into banana meal.

**Pineapples.** A *pineapple* is a sweet, acidic tropical fruit with a prickly, pine-cone-like exterior and juicy, yellow flesh. Pineapples grow on a shrublike perennial plant that grows to be about 3′ tall. The pineapple plant bears hundreds of small purple flowers that grow in a spiral pattern around a central axis; these flowers join to form a single fruit, the pineapple. The leaves at the top of the fruit are called the crown. Pineapples take 18 to 20 months to fully mature, and must be fully ripe when harvested because they will not ripen after harvesting. Pineapples are eaten fresh, in salads, baked with ham or in desserts, grilled as an accompaniment, or made into a juice. Before pineapple can be used in a dish or served fresh, the spiny outer covering and tough inner core must be removed.

## To core a pineapple, apply the following procedure:

1. Cut off the top and bottom ends of the pineapple. ***See Figure 14-15.***
2. Place the bottom cut end of the fruit on the cutting board. Cut off even strips of peel, starting at the top and moving the blade down while following the contour of the fruit.
3. Cut the fruit in half lengthwise.
4. Cut each half of the fruit in half lengthwise again to quarter the pineapple.
5. Make a final lengthwise cut along the corner of the wedge to remove the core. The peeled, cored fruit can be further cut and prepared as desired.

CULINARY PROCEDURES

## Coring Pineapple

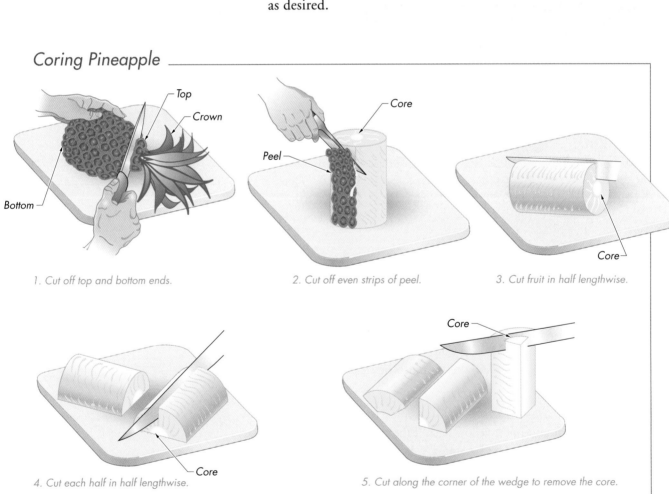

1. Cut off top and bottom ends.

2. Cut off even strips of peel.

3. Cut fruit in half lengthwise.

4. Cut each half in half lengthwise.

5. Cut along the corner of the wedge to remove the core.

**14-15**  The spiny peel and tough core are removed from a pineapple.

**Kiwifruit.** A *kiwifruit* is a small, barrel-shaped fruit, approximately 3″ long and weighing between 2 oz and 4 oz. Kiwifruit has a very thin, brown, fuzzy skin. Its bright green flesh has a white core that is surrounded by hundreds of tiny black seeds. Kiwifruit is native to China and is also known as the Chinese gooseberry. Kiwifruit is harvested when ripe but still firm. It can be eaten fresh or added to fruit salads, ice cream, and cereal. It also makes a colorful garnish when added to a dessert plate. Kiwifruit becomes sweeter as it ripens at room temperature.

**Pomegranates.** The *pomegranate* is a round, bright red fruit with a hard, thick outer skin, and measures about 3″ in diameter. A pomegranate contains a thick white membrane that encloses hundreds of red seeds that are the actual fruit of the pomegranate. The seeds are juicy and sweet, and are used in salads, soups, and sauces. Pomegranate juice is becoming more popular because of its antioxidant qualities.

**Figs.** The *fig* is the small pear-shaped fruit of the fig tree. There are more than 150 varieties of figs. The most common varieties include black figs, green figs, and purple figs. Because figs are highly perishable, they are most often available dried. Figs have a sweet, rich flavor. The texture is somewhat tough and gritty due to the massive amount of tiny seeds inside. Fresh figs are an excellent accompaniment to a cheese platter and are also delicious when made into compotes or jams.

**Mangoes.** A *mango* is an oval or kidney-shaped fruit with orange to orange-yellow flesh. The thin outer skin of the mango can include shades of light yellow, red, green, and pink. There are over 1000 varieties of mangoes grown in tropical regions around the world. Peak mango season is from May through August. The flesh of a mango clings to a large flat stone in the center of the fruit. The peel and stone are removed when preparing mangoes for use in a dish. Mangoes can be used in a variety of ways, from sauces and desserts, to drinks, salads, and salsas.

## To dice a mango, apply the following procedure:

1. Cut the mango lengthwise on one side of the stone. ***See Figure 14-16.***
2. Cut along the other side of the stone to remove it. Discard the stone.
3. Make several cuts vertically and horizontally in the flesh of each section, taking care not to cut through the skin on the other side.
4. Bend the peel backward to push the fruit outward, revealing the cubes.
5. Cut along the contour of the peel to remove the cubes.

## Preparing Mangoes

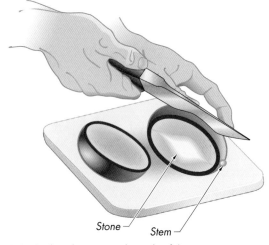

Stone — Stem —

1–2. Cut lengthwise on either side of the stone.

3. Make several cuts vertically and horizontally.

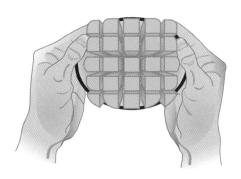

4. Bend the peel backward to push the fruit forward.

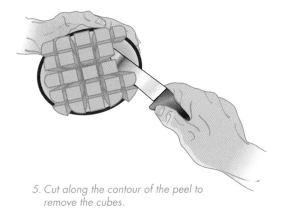

5. Cut along the contour of the peel to remove the cubes.

**14-16** A mango is prepared by scoring the flesh, bending back the peel, and slicing the individual cubes from the peel.

**Papayas.** A *papaya* is a pear- or cylinder-shaped tropical fruit weighing between 1 lb and 2 lb, with inside flesh that ranges in color from orange to reddish yellow. The center cavity is filled with numerous edible seeds that resemble peppercorns and have a sharp taste similar to pepper. Peak papaya season is from April through June. Papayas are ready to be harvested as soon as they are streaked with yellow. Green papayas should not be purchased as they were harvested too early and will not ripen. Papayas contain the enzyme papain, which is used as a meat tenderizer.

**Guavas.** A *guava* is a small oval-shaped fruit, usually 2″ to 3″ in diameter, with thin edible skin that can be yellow, red, or green. The flesh can be white, yellow, or pink in color. Guava juice is often blended into tropical fruit drinks. Guava is also excellent in jams, jellies, and chutneys.

**Passion Fruit.** The *passion fruit* is a small oval-shaped fruit, typically weighing 2 oz to 3 oz, with firm, inedible skin that can be either yellow or purple. As the fruit ripens, the skin becomes thinner and wrinkles. The orange-yellow or pink-green pulp of passion fruit is sweet and juicy, with a gelatin-like texture and a citrus flavor. Its small black seeds are edible. Passion fruit can be used in ice cream, jams, jellies, and a variety of desserts.

**Star Fruit.** The *star fruit* (carambola) is a tart tropical fruit that is shaped like a star when cut crosswise. It is native to Asia and can be 2″ to 5″ long, between 1″ and 1½″ in diameter, and weigh 3 oz to 4 oz. It has a thin edible skin enclosing translucent flesh that is crisp and juicy. Star fruit can be added to a salad or used as a garnish. It has a flavor that ranges from very tart and acidic to fresh and somewhat sweet. Star fruit is ripe when the tips of the "star" begin to turn brown.

**Persimmons.** A *persimmon* is a bright-orange fruit that is similar in shape to a tomato and very sweet when ripe. Unripe persimmons are tannic in flavor and have a chalky aftertaste. Persimmons grow on trees that are native to China. They are a winter fruit that remains on the tree even after the leaves have fallen, with peak season occurring from October through January. Persimmons are eaten raw or baked into breads, cakes, and desserts.

**Mangosteens.** Mangosteens are native to the Philippines, Malaysia, and Indonesia. The *mangosteen* is a round , sweet, juicy fruit, about the size of an orange, with a hard, thick, dark purple rind that is inedible. Underneath the hard rind is a reddish membrane, also inedible, that encloses the flesh. The flesh is white and segmented, similar to that of a tangerine. Mangosteens are best eaten raw. They must be ripe before being harvested, and they spoil quickly. Mangosteens are prized as the most luscious fruit of Asia.

**Prickly Pears.** The *prickly pear* is a fruit with protruding prickly fibers that is a member of the cactus family. Prickly pears are native to the tropical regions of the Americas. The prickly pear is 2″ to 4″ in length and has a very thick, coarse outer skin that can range in shades from green, yellow, and orange, to red or deep purple, depending on the variety. The flesh is sweet and juicy and contains sweet, crisp, edible seeds.

**Lychees.** A *lychee* is a fruit covered with a thin, red inedible shell and has light pink to whitish flesh that is refreshing, juicy, and sweet. *See Figure 14-17.* The lychee also contains an inedible stone. Lychees are native to China and are grown on evergreen trees that can reach heights of 50′ to 60′ and bear an average of 5000 pieces of fruit (300 lb) annually. Lychees must be harvested ripe, as they do not continue to ripen after being harvested.

**Rambutans.** A *rambutan* is sweet, fragrant fruit covered on the outside with soft spikes. The rambutan is native to Malaysia and is related to the lychee. Rambutans are about 2″ in diameter and grow in clusters on an evergreen tree. They are easily opened, revealing fruit similar to that of a lychee, with flesh that surrounds a single inedible seed.

Lychees

*U.S. Department of Agriculture*

**14-17** *A lychee has an inedible red shell and light pink to whitish flesh.*

## Tomatoes

Florida Tomato Committee

 Tomatoes are considered a fruit-vegetable.

## Fruit Packaging

14-19 States have diverse regulations on the weight that containers in a flat can hold.

## Fruit-Vegetables

A *fruit-vegetable* is a fruit that is served as a vegetable. A tomato, for example, is a fruit-vegetable because the tomato itself is the edible seed-bearing part of a plant. ***See Figure 14-18.*** Other fruit-vegetables include peppers, avocadoes, cucumbers, zucchini, and eggplant. Fruit-vegetables contain a lot of water.

## PURCHASING FRUIT

Fresh fruit is packed in cartons, lugs, flats, crates, or bushels, and is sold by weight or count. Lugs can hold between 25 lb and 50 lb of produce, whereas flats are usually used to ship pint- or quart-size containers of produce such as blueberries or strawberries. States have diverse regulations on the weight the containers in a flat can hold. Fruit size may also need to be specified when ordering from a vendor, because a 25-lb case of apples could contain 90, 110, or 125 apples, depending on the size of the fruit. ***See Figure 14-19.***

Some fruits can be purchased in a prepared form. For example, melons can be purchased cleaned, peeled, and cut. This is convenient but costs much more than purchasing the same fruit whole. Prepared fruit also may not taste as fresh as fruit bought whole due to processing.

## Fresh Fruit Grades

The USDA has a voluntary grading program for fresh fruits. Grades are based on a variety of characteristics including size, uniformity of shape, color, texture, and the absence of defects. The following are USDA grades for fresh fruit:

- U.S. Fancy (high quality)
- U.S. No. 1 (overall good quality)
- U.S. No. 2 (medium quality)
- U.S. No. 3 (lowest quality)

Each variety of fruit has individual standards for grading. Some fruit varieties may have additional grades, such as U.S. Extra Fancy, U.S. Utility, or U.S. Commercial, that are specific to the particular variety. Most fruits used in restaurants are of U.S. No. 1 or U.S. Fancy. Lower grades are often processed into sauces, jams, or preserves.

## Ripening Fresh Fruit

As a fruit begins to ripen, it changes in color and size and its flesh becomes soft and succulent. Left on the plant, fruit does not stop ripening when it reaches full maturity. Rather, it continues to ripen, breaking down in texture and flavor and eventually spoiling. Some fruits can continue to ripen and mature after harvest, while other fruits cannot. It is important for a chef to understand when fresh fruit should be purchased in order to have the best flavor and quality. Fruits that continue to ripen after being harvested, such as bananas and pears, are often purchased before they are fully ripe. Other fruits, such as pineapples, have to be harvested fully ripe because they do not continue to mature after they are picked. Fruits of this type should never be purchased unripe.

Ripening can be delayed by chilling. The ripening process can also be accelerated by storing fruit at room temperature or with other fruits that emit a large amount of ethylene gas. Apples, melons, and bananas give off ethylene gas and should be stored away from delicate fruits that could quickly ripen and spoil.

*Chef's Tip:*
Medium-sized fruits have better flavor than smaller- or larger-sized fruits.

## Canned Fruits

Almost any type of fruit can be canned. Peaches, pears, and pineapples are three of the most commonly canned fruits. When being canned, raw fruit is cleaned, canned, and subjected to a very high temperature to ensure that all bacteria that could otherwise cause spoilage or sickness are destroyed. Canned fruit can be stored for indefinite periods as long as it is kept in a cool, dry place. Cans that are dented or bulging should be disposed of as they may contain harmful bacteria.

Fruits are canned in water, light syrup, medium syrup, or heavy syrup. The packing method should be considered when using canned fruit in recipes, as the canning liquid can affect the flavor, texture, and nutritional value of the fruit. After opening canned fruit, any unused portions should be transferred to a storage container and refrigerated.

## Frozen Fruits

Freezing fruit inhibits the growth of microorganisms without affecting the nutritional value of the fruit. Many fruits are individually quick-frozen. This method prevents ice crystals from forming around the fruit as a result of the water in the cell walls bursting. Liquid nitrogen is commonly used to produce a quick chill, which speeds the freezing process.

Fruits such as berries are frozen whole, while other fruits such as peaches are cleaned, sliced, and frozen. Individually quick-frozen fruits are convenient because the entire content need not be thawed to use a small portion. Some frozen fruits also come stored in heavy syrup in plastic containers. Frozen fruits are graded according to the following USDA grades:

- U.S. Grade A or U.S. Fancy
- U.S. Grade B or U.S. Choice
- U.S. Grade C or U.S. Standard

## Dried Cranberries

Frieda's Inc.

 **14-20** *Cranberries are a commonly dried fruit.*

### Dried Fruits

Drying is a food preparation technique that has been around for thousands of years. Dried fruit has a much sweeter taste than fresh fruit due to the sugars being more concentrated. Dried fruit can be added to cakes, pies, biscuits, or muffins, or eaten as a snack. It is also widely used in chutneys and compotes. For maximum shelf life, dried fruit should be stored in an airtight container in a cool, dry place. Commonly dried fruits include apricots, plums, grapes, bananas, apples, figs, and cranberries. ***See Figure 14-20.***

## COOKING FRUIT

Although many dishes involve raw fruit, fruit can also be prepared using various cooking methods. When cooking fruit, it is important to remember that fruit is delicate and can become soft or mushy very quickly. Adding sugar to fruit can help prevent it from becoming mushy. The sugar is absorbed by the cells of the fruit, helping to plump them up and stay firm. Adding lemon juice or an acid works in the same way. Any sort of an alkali, such as baking soda, quickly breaks down cells in the fruit, turning the fruit to mush. Some common cooking methods for fruit include deep-frying, sautéing, grilling, broiling, baking, simmering, and poaching.

### Deep-Frying

Some fruits are suitable for deep-frying because they do not break down when exposed to very high temperatures. Bananas, apples, pears, and peaches are a few fruits that do well when deep-fried. These fruits can be coated with a heavy batter, fried, and sprinkled with powdered sugar for a delightful treat.

Before deep-frying, the fruit is cleaned, and sliced or diced into uniformly sized pieces so that it cooks evenly. The fruit is patted dry with a paper towel to help the batter adhere to the fruit. If the fruit is wet, the batter will not stick. The fruit is dipped in the batter and then submerged in hot fat. When the batter turns golden brown, the fruit is skimmed out of the hot fat. Cooling fruit on a rack or on paper towels allows excess fat to drain. Fried fruits are often garnished with powdered sugar or melted chocolate.

### Sautéing

Chefs sauté fruits in butter, sugar, spices, or liqueur. The fruit develops a sweet, rich flavor and a syrupy, caramelized glaze. Sautéed fruit can be used in many different dessert dishes, such as in crêpe fillings or as a topping for ice cream. It can also be incorporated into savory mixtures that include garlic, onions, or shallots. Savory fruit mixtures pair well with entrées such as pork or poultry.

## Grilling and Broiling

When grilling or broiling fruit, the sugars must be allowed time to caramelize. This happens quickly, as broiling and grilling occurs at very high temperatures. Fruits that are good to broil or grill include pineapples, peaches, grapefruits, bananas, and apples. These fruits can be cut into slices or chunks and soaked in liquor or coated with sugar, honey, or liqueur for extra flavor before cooking.

When broiling fruit, the fruit should be placed on a sheet pan lined with parchment paper. Any grilling of fruit should be done on a clean grill without any residue from previously grilled foods. Fruit can be placed directly on the grill or cooked on skewers to make fruit kabobs. *See Figure 14-21.* Grilled or broiled fruit can be eaten alone or added as an accompaniment.

## Baking and Roasting

Most pomes, drupes (stone fruits), and tropical fruits are well suited for baking. Pears and apples are excellent baked, as their sturdy skin keeps them from falling apart. The inner cavity of an apple or a pear can be stuffed with a flavorful filling to add a special touch. Double-crust or single-crust pies are typically filled with fruits such as apples, peaches, or blueberries. Some baked fruit desserts are called cobblers, which are simply fruit topped with a crust. *See Figure 14-22.* Strudels (crisps) are thin, flaky pastries that are commonly filled with fruits such as apples or cherries. Turnovers are a smaller variation of strudels.

Fruit can also be added to meats that are being roasted. For example, ham is often covered with pineapple rings while roasting to add extra sweetness to the meat.

## Simmering and Poaching

The simmering method is often used to make fruit compotes and stewed fruits. Fresh, frozen, canned, or dried fruit can be simmered. Simmering tenderizes and sweetens fruit. Simmered fruit can be served hot or cold and can accompany a dessert or entrée.

Apples, pears, peaches, and plums are often poached. Fruits are poached in various liquids, such as water, liquor, wine, or syrup. Poaching temperature is 185°F. This low temperature ensures that the fruit retains its shape while cooking.

## SUMMARY

Fruits are essential to the human diet. Fruits are packed with vitamins, minerals, and fiber and provide a sense of fullness without fat or excess calories. There are many varieties of fruit, including citrus fruits, berries, grapes, pomes, drupes, melons, and tropical fruits. Fruit is always most flavorful at the peak of its season. Fresh fruit should be purchased when in season and used at peak ripeness. Fruit can be prepared by deep-frying, sautéing, grilling, broiling, baking, roasting, simmering, or poaching.

## Grilling Fruit

Chicken  Banana  Pineapple

*National Chicken Council*

**14-21** Fruit can be placed on skewers and grilled.

## Apple Cobbler

**14-22** An apple cobbler is a type of baked fruit dessert.

Refer to the CD-ROM for Quick Quiz® questions related to chapter content.

1. Describe the main characteristics of citrus fruits.

2. Demonstrate the procedure for segmenting a citrus fruit.

3. List the varieties of mandarins.

4. Identify common types of berries.

5. Why are Concord grapes commonly used in the making of jams, jellies, and grape juice more than other types of grapes?

6. Explain how to core an apple using a corer.

7. Demonstrate how to core an apple using a paring knife.

8. Contrast the main characteristics of Anjou, Seckel, Bosc, and Bartlett pears.

9. Describe the main characteristics of a drupe.

10. Describe the differences between cantaloupes and muskmelons.

11. List the major differences between sweet bananas and plantains.

12. Demonstrate how to core a pineapple.

13. List the three most common varieties of figs.

14. Demonstrate how to dice a mango.

15. Describe the main characteristics of lychees.

16. Identify the USDA grades for fresh fruit.

17. Describe ways to accelerate and delay the ripening of fruits that continue to ripen after harvesting.

18. Identify common benefits of quick-freezing fruit.

19. List common cooking methods for fruit.

20. Describe how to prepare fruit prior to deep-frying.

21. Describe how to prepare fruit prior to grilling or boiling.

## Recipes

## 68

### Apple Fritters

*yield: 16 fritters*

| | |
|---|---|
| 2 | apples, peeled and sliced |
| 1 c | sifted all-purpose flour |
| 1 tbsp | granulated sugar |
| 1½ tsp | baking powder |
| ¼ tsp | salt |
| ½ c | milk |
| 1 | egg, beaten |
| 1 tbsp | melted butter |
| ½ c | confectioners' sugar |

1. Mix flour, sugar, baking powder, and salt together in a large bowl.
2. Mix milk, egg, and butter together in a small bowl while whisking rapidly. *Note:* Butter that is too hot can cause the egg to curdle.
3. Add the wet ingredients to the dry ingredients and slowly mix until incorporated.
4. Dip the apple slices in the batter to coat.
5. Drop the battered apples into hot fat. Fry at 350°F until golden brown.
6. Remove from fat and allow the cooked fritters to drain.
7. Dust the hot fritters with confectioners' sugar just prior to serving.

**Nutrition Facts**

**Per fritter:** 67.1 calories; 18% calories from fat; 1.4 g total fat; 15.9 mg cholesterol; 89.8 mg sodium; 38.2 mg potassium; 12.4 g carbohydrates; 0.4 g fiber; 5.9 g sugar; 12 g net carbs; 1.5 g protein

### Grilled Fruit Kebabs

*yield: 4 servings*

| | |
|---|---|
| 4 oz | pineapple, cut into 1" cubes |
| 4 oz | honeydew, cut into 1" cubes |
| 4 oz | cantaloupe, cut into 1" cubes |
| 4 | strawberries |
| 2 tsp | powdered sugar |
| 8 | wooden skewers |

1. Soak wooden skewers in water for 10 min.
2. Run 2 skewers through each piece of fruit about ¼" apart. This will prevent the fruit from spinning around on a single skewer.
3. Dust the fruit skewers lightly with powdered sugar and grill on each side for 2-3 min or until marked.

**Nutrition Facts**

**Per serving:** 50.2 calories; 3% calories from fat; 0.2 g total fat; 0 mg cholesterol; 10.2 mg sodium; 214.3 mg potassium; 12.6 g carbohydrates; 1.4 g fiber; 10.5 g sugar; 11.2 g net carbs; 0.7 g protein.

# Recipes

## Baked Apples

*yield: 4 servings*

| | |
|---|---|
| 4 | apples, peeled and cored |
| 4 tbsp | butter |
| ½ c | brown sugar |
| to taste | cinnamon and nutmeg |

1. Place the cored apples upright in a roasting pan so that the hole from the core can be filled.
2. Place 1 tbsp of butter in the center of each apple.
3. Fill the center hole of each apple with 2 tbsp brown sugar.
4. Lightly dust apples with cinnamon and nutmeg to taste.
5. Bake in a 400°F oven, basting often with melted butter, for 35 min to 40 min or until the apples are soft.

### Nutrition Facts
**Per serving:** 266.9 calories; 38% calories from fat; 11.7 g total fat; 30.5 mg cholesterol; 12.3 mg sodium; 213.8 mg potassium; 43.1 g carbohydrates; 1.7 g fiber; 39.4 g sugar; 41.4 g net carbs; 0.5 g protein

## 71

## Poached Pears

*yield: 4 servings*

*Wisconsin Milk Marketing Board*

| | |
|---|---|
| 4 | ripe Anjou pears, peeled |
| 3 | juice oranges |
| 1 | lemon |
| ½ c | red wine (optional) |
| ½ c | pear nectar |
| ⅛ tsp | cinnamon |
| 4 drops | vanilla extract |
| ½ tsp | cornstarch |

1. Peel pears and remove core from the pear, leaving the stem intact and the fruit otherwise whole.
2. Juice the oranges and lemon. Combine juice, red wine (optional), pear nectar, cinnamon and vanilla in a sauce-pot. *Note:* The pot should be just large enough to hold the four pears standing upright.
3. Add pears to the poaching liquid. The liquid should cover the pears halfway. Bring to a simmer and cook over low heat until tender, turning pears halfway through the cooking time.
4. When done, remove pears from liquid and cool completely.
5. Use a few tablespoons of the poaching liquid to make a slurry with the cornstarch.
6. Slowly add the cornstarch mixture to the poaching liquid to create a sauce. Bring to a simmer and stir until thickened.
7. Serve sauce with pears.

### Nutrition Facts
**Per serving:** 152.7 calories; 1% calories from fat; 0.3 g total fat; 0 mg cholesterol; 4 mg sodium; 375.6 mg potassium; 40.4 g carbohydrates; 5.6 g fiber; 22 g sugar; 34.9 g net carbs; 1.2 g protein

## Mango-Peach Cobbler

*yield: 6 servings*

**Filling**

| | |
|---|---|
| 2 | mangoes, diced |
| 4 | peaches, sliced thin |
| ¼ c | granulated sugar |
| 1 tbsp | lemon juice |
| 1 tsp | cornstarch |

**Biscuit Topping**

| | |
|---|---|
| 1 c | all-purpose flour |
| ½ c | sugar |
| 1 tsp | baking powder |
| ½ tsp | nutmeg |
| ½ tsp | salt |
| 6 tbsp | unsalted butter, cold |
| ¼ c | boiling water |

1. Mix mangoes, peaches, sugar, lemon juice, and cornstarch in a 2-qt baking dish.
2. Heat in a 400°F oven for 8 min.
3. Mix flour, sugar, baking powder, nutmeg, and salt in a large bowl.
4. Using a pastry blender, blend butter into the dry mixture until butter is pea-sized.
5. Add boiling water and stir until just combined.
6. Spoon the biscuit topping over the heated fruit.
7. Bake in 400°F oven until topping is golden.

### Nutrition Facts
**Per serving:** 348.4 calories; 30% calories from fat; 12.1 g total fat; 30.5 mg cholesterol; 278.7 mg sodium; 262 mg potassium; 59.8 g carbohydrates; 2.8 g fiber; 40.8 g sugar; 57 g net carbs; 3.2 g protein

**Recipe Variations:**

**73** **Peach Cobbler:** Substitute mangoes with an additional 2 peaches, sliced thin.

**74** **Apricot Cobbler:** Substitute filling with 12 sliced apricots, ½ cup granulated sugar, 1 tbsp flour, 1 tsp lemon juice, and ½ tsp almond extract.

**75** **Blueberry Cobbler:** Substitute filling with 6 cups of blueberries, ½ cups granulated sugar, 2 tbsp cornstarch, and 1½ tbsp lemon juice. Filling does not need to be heated.

## Raspberry Compote

*yield: 4 servings*

| | |
|---|---|
| 4 c | red or black raspberries |
| 2 | tart apples, medium-diced |
| 1 c | brandy |
| ¼ c | sugar |
| 1 tbsp | butter, unsalted |
| 1 tsp | fresh mint, chiffonade cut |

1. Add apples, brandy, and sugar to a saucepan and bring to simmer. Reduce liquid to ¼ cup.
2. Stir in berries.
3. Gently stir in butter and mint until the butter is melted and thoroughly incorporated.
4. Serve warm over an entrée or side dish.

### Nutrition Facts
**Per serving:** 310.9 calories; 10% calories from fat; 3.8 g total fat; 7.6 mg cholesterol; 3.3 mg sodium; 269.4 mg potassium; 36.9 g carbohydrates; 9.8 g fiber; 25.1g sugar; 27.2 g net carbs; 1.7 g protein

# Vegetables

*Vegetables* are an essential part of a healthy diet. As more people are becoming health conscious, chefs are offering even more nutritious menu options. A chef must be knowledgeable about the types of vegetables that are available, how to judge their freshness and quality, and when different vegetables are in season. Like fruits, vegetables are at their best at the peak of their growing season. Vegetables can be prepared in many ways, with different effects on flavor, color, and texture. Having a firm grasp on the selection and preparation of vegetables helps a chef offer delicious, healthy dishes year-round.

## Chapter 15

## Chapter Objectives:

1. Describe the nine major classifications of vegetables.
2. Identify examples within each vegetable classification.
3. Demonstrate how to prepare leeks, beans, artichokes, tomato concassé, bell peppers, and avocados.
4. Describe nine methods for cooking vegetables.
5. Describe the chemical reactions of various vegetables to acidic and alkaline ingredients.
6. Roast a pepper and a head of garlic.

## Key Terms

- **vegetables**
- **cabbages**
- **greens**
- **bulbs**
- **pod and seed vegetables**
- **fungi**
- **roots**
- **tubers**
- **gourds**
- **squashes**
- **stalks**
- **fruit-vegetables**
- **acidic**
- **alkaline**
- **chlorophyll**
- **carotenoid**
- **flavonoid**

## VEGETABLE CLASSIFICATIONS

*Vegetables* are the edible leaves, stalks, roots, bulbs, seeds, and flowers of non-woody plants. Like fruits, vegetables are an excellent source of vitamins, minerals, and fiber. Also like fruits, vegetables are classified by specific characteristics. Some vegetables, such as tomatoes, eggplants, and peppers, are botanically considered to be fruits but are prepared as vegetables. The nine major categories of vegetables include cabbages, greens, bulbs, pod and seed vegetables, roots and tubers, gourds and squashes, stalks, and fruit-vegetables.

### Cabbages

*Cabbages* are a variety of vegetables with edible flowers, leaves, or heads. For the most part, members of the cabbage family grow quickly in cool climates. They can be eaten raw or cooked and are easy to prepare. **See Figure 15-1.**

## Cabbages

**15-1** *The cabbage family includes a wide variety of vegetables with edible flowers, leaves, or heads.*

**Head Cabbages.** A *head cabbage* is a member of the cabbage family and consists of many layers of thick leaves that form a head. The leaves can be smooth or curled, and can be various colors, including green, purple, red, and white. The inner leaves are usually lighter in color than the outer leaves because they are less exposed to sunlight. A dense area known as the heart forms where the leaves attach to the stalk. The inedible heart is removed during preparation.

Head cabbages usually range in weight from 2 lb to 8 lb and in diameter from 4″ to 10″. Head cabbage is available year-round. The best heads are heavy and compact, with shiny, crisp, unblemished leaves.

Head cabbage can be prepared in a variety of ways, such as by steaming, braising, stir-frying, or stuffing. It is often added to soups and stews and pairs well with onions, potatoes, and carrots.

**Brussels Sprouts.** A *Brussels sprout* is a member of the cabbage family that consists of very small round heads of tightly packed leaves that grow along an upright stalk. Brussels sprouts are ready to be harvested when they reach a diameter of about 1″. When choosing, the best sprouts are bright green and have no yellowing leaves. Their peak season is from September through February. Brussels sprouts can be steamed, sautéed, broiled, or grilled.

**Broccoli.** *Broccoli* is a member of the cabbage family with tight clusters of dark green florets on top of a pale green stalk with dark green leaves. When buying broccoli, it is important to choose broccoli that is firm and evenly colored. Broccoli can be eaten raw or can be steamed, sautéed, stir-fried, broiled, or blanched. Broccoli is a great accompaniment to any meal and is available year-round.

**Cauliflower.** *Cauliflower* is a member of the cabbage family with a head of tightly packed white florets on a short, white-green stalk with large, pale green leaves. Some varieties of cauliflower have a purple or greenish tinge to the florets. Cauliflower grows covered with numerous layers of leaves attached to the stalk and surrounding the head. The leaves protect the head of cauliflower from sunlight and preserve its white color. When purchasing cauliflower, it is important to choose heads that are firm and compact. Cauliflower can be eaten raw or can be steamed, sautéed, stir-fried, broiled, or blanched.

Refer to the CD-ROM for the **Preparing Fresh Cauliflower** Media Clip.

**Kohlrabi.** *Kohlrabi* is a hybrid vegetable created by crossbreeding cabbages and turnips, that has a pale green or purple bulbous stem and dark green leaves. Although the entire kohlrabi plant is edible, it is primarily the bulbous stem that is used in commercial cooking. Kohlrabi is available year-round, but its peak season is from June through September. It is sweet and crisp and can be eaten raw or cooked. It is often prepared by sautéing, stir-frying, or blanching. The inner part of the base is sometimes removed, producing a cavity that can be stuffed.

**Bok Choy.** *Bok choy* (pak choi) is a member of the cabbage family that has tender white ribs and bright green leaves. It has a subtler flavor than head cabbage. There are many varieties of bok choy, some having short ribs and others having long ribs. It is readily available year-round. Bok choy can be eaten raw, but it is usually added to soups or stir-fried.

**Napa Cabbage.** *Napa cabbage* (celery cabbage) is a member of the cabbage family that has an elongated head of tightly packed, crinkly, yellow-green leaves with a thick, white center vein. Its flavor is quite subtle due to its high water content. The high water content also results in leaves that are much more tender than those from head cabbage. Napa cabbage can be added to a stir-fry, soup, or stew or used raw in a fresh salad.

**Kale.** *Kale* is a member of the cabbage family that has large, curly leaves that vary in color. Leaf colors can range from green and white to shades of blue or purple. Although all varieties are edible, the green varieties are better for cooking and others are often used as garnishes. Because of its bitterness, kale is rarely eaten raw. It is often added to soups and stews or sautéed in flavorful oil and served as a side dish.

## Greens

There are many different types of greens, or edible leaves. Many greens can be eaten raw, but they are usually cooked to bring out more flavor. Greens are often bitter. Cooking them can decrease their bitterness and make them much more appealing to the palate. *See Figure 15-2.*

**Spinach.** *Spinach* is a dark green, leafy vegetable that is rich in vitamin A, folate, potassium, and magnesium. It grows well in temperate climates. The stems are usually removed before it is cooked. Spinach can be sautéed or added to soups or creamed dishes. It is available quite readily year-round.

**Swiss Chard.** *Swiss chard* is a green with large, dark green leaves that grow on white or reddish stalks. Swiss chard is a relative of the beet, but is grown for its stalks and leaves. It tastes somewhat like spinach and can be prepared in the same manner.

**Sorrel.** *Sorrel* is a green with large leaves that are acidic in flavor. Sorrel was introduced to North America by the Pilgrims, who grew it in their gardens. Sorrel's bright green leaves can be eaten raw and are sometimes used in salads to add a refreshing taste. Sorrel can be cooked in soups and sauces, or sautéed and served as a side dish.

**Collard Greens.** *Collard greens* are greens in the kale family that have large, green leaves with a thick, white vein. Their flavor is similar to kale. Collard greens must be washed thoroughly before being prepared. They can be added to soups and stews, or served as a side dish. Collard greens are available year-round.

**Turnip Greens.** *Turnip greens* are greens of the turnip plant. Young turnip greens have a sweet flavor. As the plant ages, the leaves become bitter. Turnip greens must be thoroughly washed before they are cooked. The root of the turnip plant is considered a separate vegetable.

**Mustard Greens.** *Mustard greens* are the large, dark green leaves of the mustard plant. They have a strong peppery flavor and are often served as a side dish with pork. Mustard greens must be washed thoroughly before cooking.

## Greens

Spinach

Swiss chard

**15-2** *Greens include leafy vegetables such as spinach and Swiss chard.*

**Dandelion Greens.** A *dandelion green* is a green from the dandelion plant, usually considered a weed, with yellow flowers and long, dark green, serrated leaves. Dandelion greens are quite bitter. Often considered a salad green, young dandelion greens are added to salads with a vinaigrette dressing that cuts the harsh bitterness of the green. Dandelion greens can also be served sautéed as a side dish.

**Lettuce.** *Lettuce* is a member of the category of greens known as "salad greens." Lettuce is almost exclusively used in salads or as a garnish. Common types of lettuce include iceberg, leaf, romaine, Belgian endive, Bibb (limestone), Boston, and escarole.

## Bulbs

*Bulbs* are strongly flavored vegetables that grow underground. They are used to flavor many different types of dishes. They are also very fragrant and are used for their aromatic qualities. Examples of bulbs include garlic, onions, shallots, scallions, and leeks. ***See Figure 15-3.***

**Garlic.** *Garlic* is a bulb vegetable made up of several small cloves that are enclosed in a thin, pale, husklike skin. There are several varieties of garlic, each adding great flavor to food. The most common variety is white garlic, which is also the most pungent in flavor. Pink garlic, named for its pinkish outer covering, is another strongly flavored variety. Elephant garlic contains much larger cloves but is milder in flavor than smaller varieties. The flavor of garlic is released when a clove is cut, crushed, or chopped, and increases as a clove is chopped finer. Crushing garlic with the side of a French knife is an easy way to remove the peel. Garlic bulbs should be stored in a cool, dry place.

Refer to the CD-ROM for the **Creaming Garlic** Media Clip.

## Bulbs

**15-3** Bulbs such as garlic, onions, shallots, scallions, and leeks tend to be strongly flavored.

**Onions.** An *onion* is a bulb vegetable made up of many concentric layers of fleshy, juicy leaves. Onions grow in a variety of sizes and in colors including red, yellow, and white. The variety of onion and the climate where it was grown determine how strong its flavor is. Yellow onions (Spanish onions) are mild while white onions are sweet. Red onions are the sweetest variety of onion and are commonly used in salads for their color. Onions are a key flavoring ingredient in many dishes and are an essential part of mirepoix. Onions can be eaten raw, or can be deep-fried, sautéed, stir-fried, roasted, or grilled.

**Shallots.** A *shallot* is a small bulb vegetable that is similar in shape to a bulb of garlic with two or three cloves inside. The outer covering can be bronze colored, rose colored, or pale gray. Shallots have a pink-tinged ivory flesh and a flavor that is more subtle than an onion. When purchasing shallots, it is important to look for those that are firm and dry-skinned and to avoid any that are sprouting.

**Scallions.** A *scallion* is a bulb vegetable with a slightly swollen base and long, bright green leaves that are slender and hollow. Scallions are mildly flavored compared to onions. The best scallions have a pleasant aroma and fresh, brightly colored leaves. Scallions are often added to salads or used as a garnish.

**Leeks.** A *leek* is a long, white vegetable with a cylindrical bulb, similar in appearance to a scallion but much larger. Leeks should be firm and have bright green leaves. The white part of the leek is the portion used most often in recipes. The green leaves are most often used to flavor soups and stocks. Leeks are milder and sweeter than onions. The leaves of a leek are long, wide, and rather flat. It is important to clean leeks well, as soil and grit often become trapped between the layers of the bulb.

## To clean a leek, apply the following procedure:

1. Cut off the root end of the leek just above the roots. ***See Figure 15-4.***

2. Split the leek lengthwise down the center from top to bottom.

3. Rinse thoroughly under cool water to remove any soil or grit that may have settled between the layers.

4. Cut off the top portion of the darker green end of the leek and remove any white portions that look old or not fresh. Cut into desired pieces.

*CULINARY PROCEDURES*

### Preparing Leeks

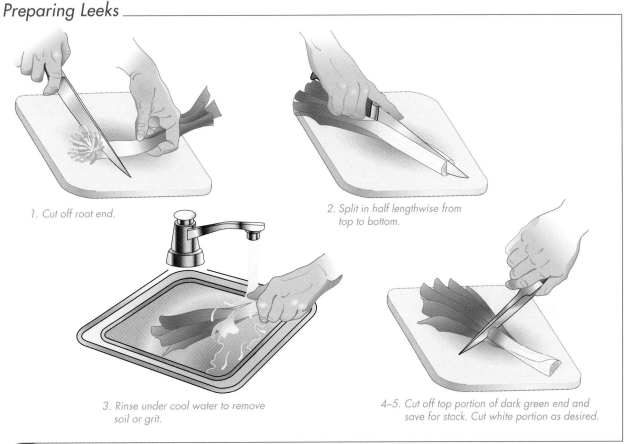

*1. Cut off root end.*

*2. Split in half lengthwise from top to bottom.*

*3. Rinse under cool water to remove soil or grit.*

*4–5. Cut off top portion of dark green end and save for stock. Cut white portion as desired.*

**15-4** *Leeks are split in half lengthwise and rinsed under cool water to remove any soil, grit, or other impurities that may have settled between their layers while growing.*

### Pod and Seed Vegetables

*Pod and seed vegetables* are all varieties of legumes (such as beans and peas), corn, and okra. Some pod and seed vegetables are eaten fresh, some are dried, and some can be used in either fresh or dried form. For example, corn can be served fresh on the cob, as separated kernels in canned corn, or dried as popping corn. The pod and seed vegetable category includes some of the oldest recorded forms of food.

## Dried Legumes

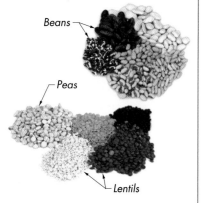

Beans

Peas

Lentils

**15-5** Legumes are plants that produce edible seeds, such as beans, peas, and lentils.

**Legumes.** A *legume* is a plant that produces pods that contain a row of edible seeds. There are thousands of varieties of legumes, but the most popular are beans, peas, and lentils. ***See Figure 15-5.*** In some varieties, the pods are eaten along with the seeds. Legumes are highly nutritious, as they are rich in fiber and protein and contain little or no fat.

Fresh green beans and fresh wax beans are actually immature beans. These legumes have pods that are not fully developed and are therefore completely edible. Fresh beans can be prepared in a variety of ways, including sautéing, broiling, grilling, steaming, or simmering.

There are thousands of varieties of dried beans. Dried beans are either kidney shaped or round and usually measure less than ¼″ long. Some popular dried beans include kidney beans, great northern beans, pinto beans, black beans, cannellini beans, black-eyed peas, and flageolets. Dried beans can be eaten hot or cold and are popular in hearty soups. Some beans are popular puréed, such as pinto beans in a refried bean dish. Dried beans are harvested just as they are beginning to dry. They are then shelled and left to dry. When dried, they become rock hard and must be rehydrated before cooking. Soaking the beans decreases the overall cooking time and results in an even texture throughout. Changing the water once or twice while soaking the beans can remove impurities that could cause gas during digestion.

## To soak dried beans overnight, apply the following procedure:

1. Sort through the dried beans, removing any cracked beans as well as any debris.
2. Rinse the beans several times in cold water.
3. Transfer the beans to a bowl that is large enough to hold 3 parts water to 1 part beans.
4. Let the beans soak overnight in a cool place.

## To quick-soak dried beans, apply the following procedure:

1. Sort through the dried beans, removing any cracked beans as well as any debris.
2. Rinse the beans several times in cold water.
3. In a large stockpot, add 4 parts water to 1 part beans.
4. Slowly bring the beans to a boil. Reduce heat and simmer for 2 min.
5. Remove the beans from heat. Cover and let stand for 1 hr to 2 hr or until beans swell.
6. Discard the water and proceed with the recipe.

Lentils, like dried beans, contain very little fat and are high in protein and fiber. Unlike beans, lentils do not have to be soaked, because they are smaller and are already split when purchased. There are many varieties of lentils, with green, red, and brown being the most common. When preparing a recipe with lentils, the lentils must be thoroughly washed because they often contain small stones. Lentils are used in soups and added to salads. Lentils turn mushy if overcooked.

There are over a thousand varieties of peas, but the most common are garden peas and snow peas. Garden peas are green seeds removed from the pod and are very versatile. Snow peas are sweeter than garden peas and have a flat pod that is entirely edible. Snow peas are commonly found in Asian cuisine.

The majority of peas are sold frozen or canned. Fresh peas can be rather expensive. When purchasing fresh peas, it is important to choose smooth, shiny, bright green pods. Peas also come in a dried form known as split peas, which are harvested fully mature and left to dry. Split peas make a wonderful soup or addition to a stew and are prepared similarly to lentils.

Edamame (fresh soy beans) is a type of fresh shelling pea that has a fibrous, inedible shell. Edamame is typically steamed, chilled, shelled, and then eaten as a snack or as a garnish in salads. Edamame may also be served as a marinated accompaniment to Asian cuisine.

**Corn.** Corn is a hybrid seed grain variety of the wild grass called maize. Corn is the most produced cereal crop worldwide, followed by wheat and rice. The edible seeds of corn, called kernels, grow on a woody cob near the top of the plant. Some varieties of corn are grown solely for seed, others for popping corn, and still others for sweet corn. Sweet corn is commonly eaten as a vegetable, rather than as a grain. Sweet corn is in peak season during the months of July and August. *See Figure 15-6.*

The top half of the cob is surrounded with a fine, hairlike material called silk. The silk is removed prior to cooking. The entire cob is further surrounded by thin leaves called husk. When grilling or roasting sweet corn, the outer leaves of the husk are removed, while the freshest inner leaves are left on the cob. The husk helps the cob to retain moisture on the hot grill and prevents the corn kernels from burning or drying out.

Sweet corn can also be boiled, steamed, or cooked in the microwave. When these methods are used, the entire husk is removed prior to cooking. Cooked sweet corn can be eaten right off the cob (as "corn on the cob") or the kernels can be cut off the cob with a knife.

**Okra.** *Okra* is a fibrous pod vegetable containing round, white seeds. Okra pods have a gelatinous, somewhat slimy liquid inside. Washing okra prior to cooking reduces the presence of this substance.

*Sweet Corn* _____

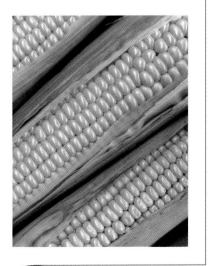

15-6  Sweet corn consists of rows of seeds, known as kernels, that grow on a woody cob.

When preparing okra, the stem end is trimmed off and the pod is thoroughly rinsed in cold water. Okra should be cooked only in stainless steel cookware. Aluminum cookware causes okra to become discolored and dark. Okra is best prepared by steaming, boiling, deep frying, or pickling.

### Fungi

*Fungi* are plantlike organisms such as mushrooms and truffles. Fungi are not actually vegetables, as they lack chlorophyll, flowers, and roots, but they are prepared and used in the kitchen in the same manner as vegetables. There are a tremendous number of cultivated varieties of mushrooms. Cultivated varieties include the common (white), button (cremini), enoki, wood ear, shiitake, portobello, oyster, and chanterelle. Button mushrooms are the smallest of the cultivated varieties, and portobello mushrooms are the largest, with caps measuring up to 6″ in diameter. Mushrooms can be used to add flavor to many dishes and can be sautéed, deep-fried, or eaten raw.

Mushrooms can be purchased fresh, canned, or dried. When purchasing fresh mushrooms, it is important to choose firm mushrooms that are not wrinkled, spotted, or slimy. Fresh mushrooms must be cleaned before being used in a recipe, but should not be soaked in water, as their bodies act like sponges and quickly become saturated. Rather, fresh mushrooms should be lightly rinsed and cleaned with a damp paper towel or soft brush. Dried mushrooms must be rehydrated before they are used in a recipe or they will be too tough. Mushrooms need to be stored in a cool, dry place and require air circulation to stay fresh, so storing them in a paper bag works best. If mushrooms are stored in a plastic container, it should have holes to allow air circulation. ***See Figure 15-7.***

*Chef's Tip:*
Always purchase mushrooms from a certified source. Some poisonous varieties of mushrooms look very similar to safe varieties.

Mushrooms

Common

Oyster

Shiitake

Portobello

Enoki

**15-7** *Mushrooms are edible fungi that exist in many varieties, such as common, oyster, shiitake, portobello, and enoki.*

A *truffle* is an edible fungus with a distinctive taste. Truffles develop underground, making them hard to locate, and take at least five years to form. Highly trained dogs are used to sniff out their distinct odor and dig them up. Truffles are extremely expensive, costing hundreds of dollars per pound, and are found in premier fine-dining restaurants. A dish that includes truffles is expensive.

The two main varieties of truffles are the black and the white. Black truffles are bulbous and have dark flesh with white veins and black warts. Black truffles are highly odorous and are often used to add flavor to pâtés and terrines. White truffles are elongated in shape and can grow quite large. Some are over 4″ in diameter and weigh up to 1 lb. White truffles have a distinct, strong aroma and are excellent in pastas and sauces. Because their flavor is so strong, a small amount of truffle is quite sufficient in any recipe. Black or white truffles are often served thinly shaved over an entrée. Truffle oil is a less expensive way to use truffle flavor in a dish.

### Roots and Tubers

*Roots* are vegetables that extend deep into the soil to reach water and nutrients. *Tubers* are short, thick, oblong vegetables that are part of a plant stem. The potato is a well-known example of a tuber. Roots and tubers such as carrots, parsnips, rutabagas, turnips, beets, radishes, potatoes, jícamas, and sweet potatoes are prepared and used in much the same way. ***See Figure 15-8.***

## Roots and Tubers

Carrots

Rutabaga

Beets

Potatoes

Jícama

Parsnips

Turnips

Radishes

Sweet Potato

Roots

Tubers

15-8 *Roots and tubers grow underground and include carrots and potatoes.*

**Carrots.** A *carrot* is an orange-colored root vegetable that is rich in vitamin A. Carrots are inexpensive and are available year-round. When purchasing carrots, it is important to choose those that are firm and intact with a bright orange color. Carrots can be eaten raw, put into salads, and cooked in a variety of ways. Sautéing, broiling, blanching, or steaming are a few common preparation methods used with carrots.

**Radishes.** A *radish* is a small, round root vegetable that has a peppery taste. The leaves of the radish plant are edible, but the root is more commonly eaten. There are many varieties of the radish, including the red radish, black radish, and daikon (also called Asian radish). The red radish is round and about 1″ in diameter, with a red or pinkish red exterior and crisp, juicy, white flesh that has a sharp, peppery flavor. The daikon is a white-fleshed radish with an elongated shape, similar to that of a carrot. Daikons are usually 8″ to 12″ in length, with a flavor that is much milder than that of a red radish. Radishes can be eaten raw and are often added to salads. Cooked radishes are added to soups and stews and have a much less pungent flavor.

**Beets.** A *beet* is a round root vegetable with a deep reddish purple color. Beets are rich in nutrients, including vitamin A, vitamin C, and potassium. They are often associated with the cuisine of northern European countries such as Russia, Germany, and Ireland. When purchasing beets, those of the highest quality are firm with smooth skin and no spots or bruising. Beets can be eaten raw, cooked, or pickled. Borscht is a famous soup that gets its deep red color from beets.

**Parsnips.** A *parsnip* is an off-white root vegetable, similar in shape to a carrot. Parsnips range from 5″ to 10″ in length and are used in much the same way as carrots. Peak season for parsnips is from late fall through winter, after the first frost. Parsnips that are harvested later in season are sweeter, as the cold converts some of the starches into sugar. Parsnips are most often used in soups or stews, or served alongside a roast. They can be prepared by broiling, roasting, blanching, simmering, or steaming or can be eaten raw.

**Turnips.** A *turnip* is a round, fleshy root vegetable that is purple and white in color. Turnips have a peppery flavor, similar to that of a radish, and are a good source of vitamin C and potassium. They need to be washed and peeled prior to being cooked. Turnips can be simmered and then puréed or mashed like potatoes, or diced and then sautéed, blanched, or simmered. When roasted, they develop a buttery taste.

**Celery Root.** *Celery root* (celeriac) is the knobby root of a type of celery grown for its root rather than its stalk. The hollow stalk of the plant is not commonly eaten. Celery root measures about 4″ to 5″ in diameter and can weigh between 2 lb and 4 lb. Celery root should be washed thoroughly and peeled. It can be diced, simmered, and puréed into a wonderful soup. Celery root oxidizes quickly, so a little lemon juice or vinegar should be added prior to cooking to avoid discoloration. Celery salt is made from ground, dried celery root.

**Rutabagas.** A *rutabaga* is a round root vegetable derived from a cross between a savoy cabbage and a turnip. Rutabagas are actually in the cabbage family. They are often confused with turnips, but rutabagas are longer and rounder in comparison. The flesh also has a yellow tint and the flavor is more distinct than that of the turnip. Rutabagas are high in potassium and vitamin C. They are prepared in much the same way as turnips, and are added to soups or stews or puréed like potatoes.

**Potatoes.** A *potato* is a round, oval-shaped, or elongated tuber that grows underground. The tuber is the only edible part of the plant. The color of potato skin differs among varieties and can be brown, red, yellow, white, orange, or blue. Potato flesh is creamy white to yellow in color. When purchasing potatoes, it is important to choose firm, undamaged potatoes with no signs of sprouting. Potatoes must be stored in a dry, cool, dark place that allows them to breathe. If potatoes are stored without ventilation, they rot quickly.

Potatoes oxidize when peeled or cut. Placing cut potatoes in cold water prevents them from discoloring. Potatoes can be sautéed, broiled, grilled, baked, simmered, or can be deep-fried. Potatoes are added to many soups, stews, and casseroles. They should not be eaten raw due to an inedible starch found in uncooked potatoes. This starch is converted to an edible sugar when the potato is cooked.

**Sweet Potatoes.** A *sweet potato* is a tuber that grows on a vine. Sweet potatoes are an excellent source of vitamin A and potassium. The flesh of a sweet potato is dark orange and becomes moist and sweet when cooked. The skin is edible, although it is often removed before cooking. Peeled or cut sweet potatoes oxidize, so it is important to place them in cold water until they are used. Sweet potatoes can be prepared in a variety of ways and can be incorporated into cakes, pies, breads, and cookies. They are often puréed with cinnamon, butter, nutmeg, or brown sugar to enhance their sweetness. Sweet potatoes are a staple in many Latin American and Asian countries and are popular in the southern United States.

**Jícamas.** A *jícama* is a round tuber between 6″ and 8″ in length that is native to Mexico and Central America. Jícamas are covered with thin, light brown skin that is removed before cooking. Their crisp, white flesh has a delicate, sweet flavor. Smaller jícamas are more flavorful than larger ones. Jícamas should be purchased free of bruises. They can be eaten raw in salads or dips, or cooked and puréed similarly to potatoes.

**Jerusalem Artichokes.** A *Jerusalem artichoke* (sunchoke) is a tuber with a thin, brown, knobby-looking skin, similar to that of ginger root. Jerusalem artichokes are related to sunflowers. The skin is edible but is often removed before cooking. The white flesh is crisp and sweet and can be used in salads. Jerusalem artichokes can be simmered, blanched, or steamed, added to soups, or made into a purée.

## Gourds and Squashes

There are hundreds of different varieties of gourds and squashes. Gourds grow on quick-growing plants with trailing vines, large leaves, and complex root systems. Only a few varieties of gourds are eaten. Squashes also grow on vines and can be cooked or eaten raw. Squashes are classified as either summer or winter squash, based on their storage life. *See Figure 15-9.*

*Winter Squashes*

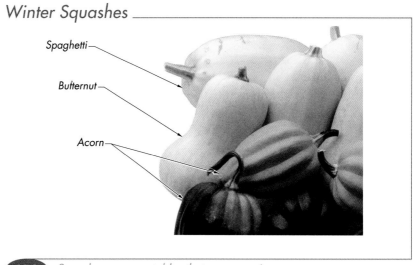

Spaghetti

Butternut

Acorn

15-9 *Squashes are vegetables that grow on vines.*

**Chayote.** *Chayote* is a gourd native to Mexico and Central America. A chayote is similar in appearance to a pear and can grow to be between 2″ and 8″ in length. The edible skin is light green, and the inner flesh is pale green to white in color. A chayote contains a single seed that can be eaten. The flesh and skin can be eaten raw or can be stuffed, added to soups or stews, or stir-fried.

**Summer Squash.** Summer squashes are harvested only 2 days to 8 days after flowering. Varieties of summer squash include zucchini, crookneck (gooseneck) squash, and pattypan. Zucchini is an elongated squash, resembling a cucumber, and is available in green or yellow varieties. Crookneck squash is a yellow squash that resembles a bowling pin with a bent neck. Pattypan is a round, shallow squash with scalloped edges that is best harvested when it is no larger than 2″ to 3″ in diameter. The outer skin, flesh, and seeds of summer squash are entirely edible. Summer squashes are highly perishable, so they should be used soon after they are purchased. They are wonderful sautéed, stir-fried, or grilled.

**Winter Squash.** Winter squash varieties differ from the summer squash in that they are harvested fully ripe. They come in a variety of shapes and sizes. Varieties of winter squash include butternut, acorn, turban, pumpkin, autumn, and spaghetti. Winter squashes have a hollow inner cavity full of seeds that can be cleaned, roasted, and eaten. The inside flesh varies in color and taste by variety. The outer skin is rarely eaten raw due to its hardness. Because winter squashes have such a thick outer shell, they can be stored for months before spoilage occurs. Most winter squashes are roasted and made into soups or puréed and served as a side dish.

*Spaghetti squash* is a winter squash with flesh that can be separated into spaghetti-like strands after it is cooked. Spaghetti squash can be used as a substitute for regular spaghetti noodles and served just like spaghetti. The flesh is pale yellow and tastes very similar to that of a summer squash.

## Stalks

Stalks are the stems of vegetables that contain a high amount of cellulose. As a stem continues to develop, it becomes tougher. Stems are usually harvested while tender to avoid too much cellulose developing. ***See Figure 15-10.***

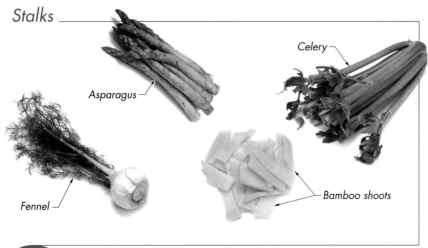

Stalks

Celery

Asparagus

Fennel

Bamboo shoots

15-10 *Stalk vegetables, such as asparagus or celery, are edible stems.*

Refer to the CD-ROM for the **Preparing Fresh Asparagus** Media Clip.

*Chef's Tip:*

Cut artichokes turn brown from oxidation. A little lemon juice, or even a slice of lemon, should be placed on the heart after cutting or removing the choke to keep it from discoloring.

**Asparagus.** *Asparagus* is a vegetable consisting of an edible stalk that is most often referred to as a spear. Asparagus is harvested in the spring while it is young. The longer the asparagus grows, the woodier it becomes, making it unpalatable. The two varieties of asparagus are the green and the white. Green asparagus is more common, and obtains its green color through photosynthesis. White asparagus is grown covered in soil to prevent photosynthesis from taking place. The spears are harvested as soon as they begin to emerge. White asparagus is more tender than green asparagus but does not have as much flavor.

Both varieties of asparagus can be broiled, grilled, simmered, or steamed. Asparagus is often puréed and made into soups, or served whole alongside a main entrée. Raw asparagus is excellent in salads and is a popular ingredient in omelets, quiches, and pasta dishes.

**Celery.** *Celery* is a stalk vegetable that is 12″ to 20″ in length. The outer stalks are green, while the inner stalks are white in color. The inner stalks are sweeter and more tender than the outer stalks. Celery seeds are harvested after the plant flowers. Celery should be shiny, firm, and crisp when purchased. Stalks that are brown or damaged or that have brown or yellow leaves should be avoided. Celery can be eaten raw. Cooked celery is often incorporated into recipes for its savory flavor. Common methods of preparing celery include sautéing, stir-frying, roasting with a meat or poultry dish, or simmering in stock, soup, or stew. Celery is a main component of mirepoix.

**Fennel.** *Fennel* is a celery-like stalk with overlapping leaves that grow out of a large bulb at its base. Fennel has a mild, sweet flavor that is associated with licorice or anise. When purchasing fennel, it is important to choose stalks that are firm and unblemished with healthy-looking, bright green leaves. Fennel can be eaten raw but is usually cooked. It can be diced or sliced, and sautéed, broiled, blanched, or steamed. It can also be puréed into a soup or side dish, or made into an au gratin similar to potatoes.

**Bamboo Shoots.** A *bamboo shoot* is an immature shoot of the bamboo plant, harvested shortly after it emerges from the soil and reaches approximately 6″ in length. Bamboo shoots cannot be eaten raw because they contain a toxic substance that is removed through cooking. Bamboo shoots are mainly available for use canned or dried. Some specialty Asian stores may have fresh bamboo shoots, although they can be quite expensive. They are an excellent addition to a stir-fry or salad.

**Artichokes.** An *artichoke* is the unopened flower bud of a thistle plant. Artichokes that were harvested early and are small can be cooked and served whole. The petals, called leaves, are the edible part of the bud. The tips of the larger leaves can become very sharp, developing a spiny thorn. The dense area where the leaves attach to the stem is known as the heart. As the flower matures, a fuzzy, sometimes thorny, center called the choke develops just above the center of the heart. The choke must be removed before the artichoke can be eaten.

## To prepare an artichoke, apply the following procedure:

1. With kitchen shears, cut off the top third (the thorn) of each of the larger exposed leaves, from the middle of the artichoke down to the base, and discard the removed tips. ***See Figure 15-11.***

2. With a French knife, remove the entire top half of the artichoke, exposing the lighter-colored choke in the center, and discard the top half.

3. With a spoon, thoroughly scrape the choke from the solid heart, and discard the choke.

4. Cut off any extra stem.

5. Squeeze a small amount of lemon onto the exposed heart.

*CULINARY PROCEDURES*

*Preparing Artichokes* _____

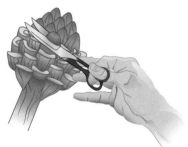

*1. Cut off top third of each leaf to remove the sharp thorn.*

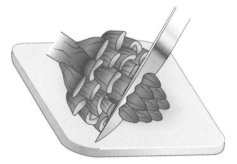

*2. Cut off top half.*

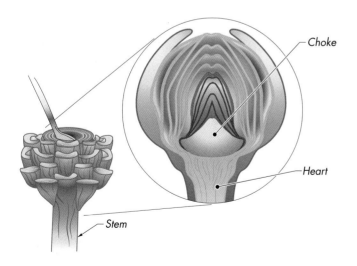

Choke

Heart

Stem

*3–4. Scrape out choke with spoon; cut off excess stem.*

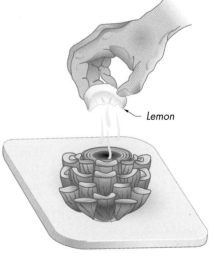

Lemon

*5. Squeeze lemon juice on exposed heart and stem.*

**15-11** *The thorns and choke are removed when preparing an artichoke.*

Once the choke is removed, the artichoke can be simmered, steamed, or baked until the heart becomes tender. Baked artichokes are commonly filled with a mixture of lemon, garlic, and bread crumbs and baked until the filling is golden and the heart is tender.

### Fruit-Vegetables

Fruit-vegetables are botanically considered fruits, but they are prepared as vegetables. Fruit-vegetables are more similar in flavor to vegetables than to fruits, as they are typically more tart or bitter than they are sweet. Cucumbers, tomatoes, peppers, eggplants, and avocados are just a few examples of fruit-vegetables. *See Figure 15-12.*

## Fruit-Vegetables

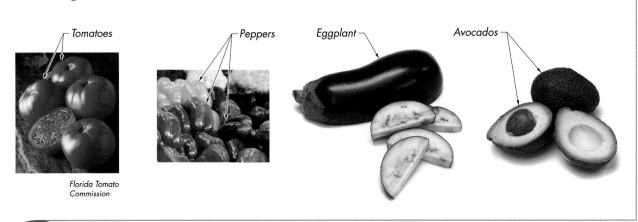

Tomatoes

Florida Tomato Commission

Peppers

Eggplant

Avocados

**15-12** Fruit-vegetables are botanically classified as fruits but are prepared as vegetables.

Refer to the CD-ROM for the **Cucumber Fans** Media Clip.

**Cucumbers.** *Cucumbers* are fruit-vegetables that are often eaten raw or pickled. The cucumber is a widely cultivated member of the gourd family. The most common varieties of cucumbers are English (also called "burpless"), Japanese, Mediterranean, and dosakai, which are yellow, round cucumbers available in parts of India. Cucumbers used for pickling are usually smaller and thicker than cucumbers that are eaten fresh and have bumpy, light-green skin.

**Tomatoes.** A *tomato* is a juicy fruit-vegetable that contains many edible seeds. Tomatoes are available in many different colors, shapes, and sizes. Cherry, plum, beefsteak, yellow, and pear tomatoes are just a few of the more than a thousand cultivated varieties. Tomatoes are highly perishable and should be eaten or used within a few days of purchase. The number of ways tomatoes can be incorporated into a recipe is virtually endless. Tomatoes can be eaten raw, added to salads or sandwiches, or made into soups, sauces, or juice. Tomatoes are also used in many prepared dishes. Concassé (pronounced con-kah-say) is a preparation method where items are peeled, seeded, and then chopped or diced.

To prepare tomato concassé, apply the following procedure:

1. With a paring knife, make a ½″-wide X in the bottom of a tomato, just slightly deeper than the surface of the skin. ***See Figure 15-13.***

2. Place the tomato in a pot of boiling water to blanch for 20 sec to 30 sec or until the skin near the X begins to wrinkle or come free from the tomato.

3. Remove the tomato with a strainer or slotted spoon and shock in ice water until cold.

4. Remove the tomato from the ice water. With a paring knife, make a circular cut around the core, and remove the core.

5. Using the tip of a paring knife, grab the loose tomato skin and peel it away.

6. Cut the tomato in half horizontally and gently squeeze each half to remove the seeds and juice.

7. Slice, chop, or dice the peeled and seeded tomato as desired.

CULINARY PROCEDURES

*Chef's Tip:*

Use caution to not overblanch a tomato. If the tomato is cooked in the boiling water for too long, it becomes soft and cannot be diced cleanly.

## Tomato Concassé

*1. Make a ½″ wide X in the bottom of the tomato.*

*2. Blanch in boiling water for 20 sec to 30 sec.*

*3–4. Shock in ice water; remove the core.*

*5. Peel away skin with a paring knife.*

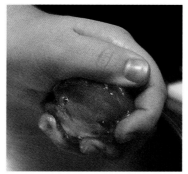

*6. Slice in half and squeeze gently to remove seeds.*

*7. Slice, chop, or dice as desired.*

**15-13**   *Tomato concassé consists of chopped or diced tomatoes that have been peeled and seeded.*

**Bell Peppers.** A *bell pepper* (sweet pepper) is a fruit-vegetable that contains hundreds of seeds in its inner cavity. They are readily available year-round. Bell peppers are plump with a squarish shape created by four lobes. Some varieties of peppers only have three lobes, giving them a heart-shaped appearance. The flesh of a bell pepper is quite crisp. Green peppers turn yellow and ultimately red if left to ripen on the vine. The longer the pepper stays on the vine to ripen, the sweeter it becomes. Red peppers are the sweetest peppers because they are the ripest. Bell peppers are often julienned or diced before use. An alternate preparation technique for peppers is the rolling method.

To julienne or dice a bell pepper, apply the following procedure:

1. Wash the pepper thoroughly.
2. Cut off the top of the pepper just below the stem. Remove the stem and reserve the top of the pepper. ***See Figure 15-14.***
3. Cut off the bottom of the pepper, so that the pepper sits flat on the cutting board, and reserve the bottom piece.
4. Slice down each side vertically, working around the pepper to remove the four sides, leaving the center seed portion intact. Discard the center portion.
5. Lay the pepper sides flat. Julienne or dice to desired size.

*Preparing Peppers*

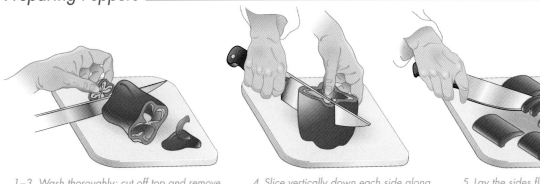

1–3. Wash thoroughly; cut off top and remove stem; cut off bottom.

4. Slice vertically down each side along the seedpod. Dispose of seeds.

5. Lay the sides flat and julienne or dice to desired size.

**15-14** Peppers are often julienned or diced before use.

To remove the seed portion from a bell pepper using the rolling method, apply the following procedure:

1. Cut off the top and bottom of the pepper. Remove and discard the stem. ***See Figure 15-15.***

2. Slice the pepper from top to bottom to open the pepper.

3. Begin to unroll the pepper by cutting the seeds and rib sections from the flesh in a rolling motion.

## Rolling Pepper Method

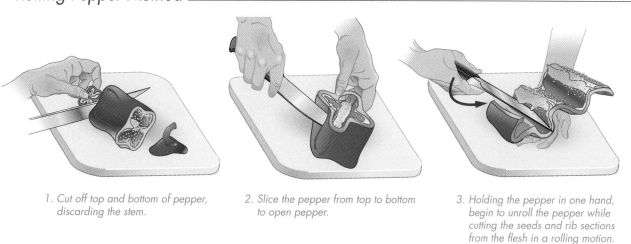

1. Cut off top and bottom of pepper, discarding the stem.

2. Slice the pepper from top to bottom to open pepper.

3. Holding the pepper in one hand, begin to unroll the pepper while cutting the seeds and rib sections from the flesh in a rolling motion.

**15-15** The rolling method is an alternative way to prepare bell peppers.

**Chiles.** A *chile* (hot pepper) is a pepper with a very distinct mild to hot flavor. Varieties include the jalapeño, habañero, poblano, and serrano. Larger chiles, such as the poblano, are much milder than smaller chiles, such as the serrano. Chiles are native to Mexican and Central American cuisines. Chiles are used in a variety of ways. Most of them are used to flavor sauces or soups. Some, like the poblano, can be stuffed and made into *chile relleno.* Most chiles are readily available year-round.

*Chef's Tip:*
Always wear disposable rubber gloves when cleaning chiles such as habañeros or jalapeños. This prevents oils present in the pepper seeds from being absorbed by skin or fingernails and causing burning.

**Tomatillos.** A *tomatillo* (Mexican husk tomato) is a bright green fruit-vegetable the size of a small tomato. They are native to Mexico, where it has been cultivated since the time of the Aztecs. Tomatillos are covered with a thin, papery husk that is removed before they are cooked. The flavor of a tomatillo is rather acidic and tart. Tomatillos are added to soups, made into salsas, or eaten raw in salads.

**Eggplant.** An *eggplant* is a fruit-vegetable with deep purple, edible skin and yellow to white, spongy flesh. Although there are many varieties of eggplant, the most common variety in North America is an oval-shaped eggplant that resembles a large pear. The flesh contains small, brown, edible seeds. Eggplant begins to discolor as soon as it is cut, so it is important to either cook the eggplant immediately after

Refer to the CD-ROM for the **Preparing Fresh Avocados** Media Clip.

it is sliced or to sprinkle a little lemon juice on it. Sliced raw eggplant can be lightly salted and left on paper towels to drain some of the moisture from the eggplant before cooking, which prevents it from later becoming soggy. Eggplant can be roasted and puréed, coated and deep-fried, or used in casseroles. Eggplant is most commonly found in Mediterranean and East Indian cuisines. Moussaka is a famous Mediterranean dish made with eggplant.

**Avocados.** An *avocado* (alligator pear) is a pear-shaped fruit that grows on trees native to Central and South America. Avocados have a rich, buttery flavor. The most common avocado is the Hass variety. Avocados have a very short period of peak ripeness. The skin and pit of an avocado are inedible and must be removed before use. Avocados are most commonly eaten raw or made into guacamole, a popular Mexican dish made with puréed avocados, chiles, onions, garlic, lime juice, and salt. Avocados are used in tacos, sandwiches, and salads.

To remove the pit and skin from an avocado, apply the following procedure:

1. Cut the avocado in half lengthwise from stem end to bottom end, cutting all the way around the pit. ***See Figure 15-16.***
2. With the avocado in both hands, use a twisting motion (like when removing a lid from a jar) to separate the two halves.
3. Place the half containing the pit on a cutting board (pit side up) and gently hit a French knife into the pit.
4. Gently twist the French knife to work the pit free.
5. Using a large spoon, gently scoop the avocado flesh from the skin.

*Pitting Avocados*

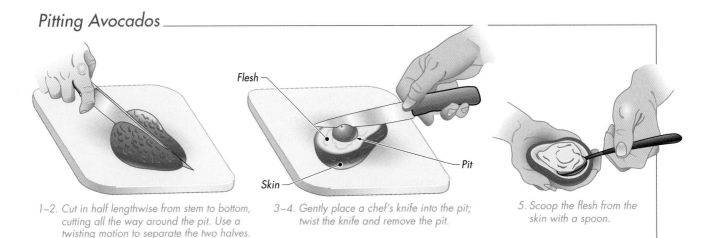

Flesh

Pit

Skin

*1–2. Cut in half lengthwise from stem to bottom, cutting all the way around the pit. Use a twisting motion to separate the two halves.*

*3–4. Gently place a chef's knife into the pit; twist the knife and remove the pit.*

*5. Scoop the flesh from the skin with a spoon.*

15-16  *The pit must be removed from an avocado before use.*

## PURCHASING VEGETABLES

Understanding how to purchase and store vegetables is very important in maintaining the highest-quality produce. Because most fresh vegetables have a short shelf life, it is important to know how to maximize their use. Preparing vegetables during their peak season also costs less, so it is smart for a chef to use seasonal vegetables in a menu.

### Fresh Vegetables

The USDA has a voluntary grading system for vegetables, which includes the following:
- U.S. Extra Fancy
- U.S. Fancy
- U.S. Extra No. 1
- U.S. No. 1
- U.S. No. 2
- U.S. No. 3

The decision about which grade of vegetables to purchase depends on how the vegetables will be used. Some recipes using fresh or slightly cooked vegetables may require premium ingredients, while lesser grades could be acceptable in a soup recipe.

Fresh vegetables are packed in cartons that are referred to as bushels, cases, flats, lugs, or crates and are sold by weight or count. The weight or count of a packed container depends on the size and type of the vegetable.

Vegetables should be stored in a produce cooler, away from meat, poultry, seafood, or dairy products. Potatoes, onions, garlic, and squash should be stored in a cool, dry location that is between 60°F and 70°F. Storing these items in a conventional refrigerator causes their starches to convert to sugar, changing their texture and flavor in an undesirable way. Other vegetables do well stored in a refrigerator at 41°F or cooler. It is important to store vegetables away from fruits that emit ethylene gas, such as apples and bananas, as the fruit could cause the vegetables to overripen and spoil.

### Canned Vegetables

Canned vegetables are a staple in the professional kitchen. They are sold in a variety of commercial sizes, packed by weight. ***See Figure 15-17.*** There are many advantages to using canned vegetables. They are already cleaned, cut, peeled, and cooked, and they have been treated with heat to kill any harmful microorganisms. However, canning often softens vegetables and can sometimes cause nutrient loss. Canned vegetables are graded by the USDA as U.S. Grade A or U.S. Fancy, U.S. Grade B, and U.S. Grade C or U.S. Standard.

### Frozen Vegetables

Frozen vegetables offer the same convenience as canned vegetables, with additional advantages. Frozen vegetables often retain their color and nutrients better than canned vegetables. Like fruits, some vegetables are individually

Kolpak

*Chef's Tip:*
Onions, carrots, and potatoes are graded using an alphabetical system, with Grade A being the best quality.

quick-frozen to preserve their texture and appearance. Some vegetables are blanched before being frozen, which reduces overall cooking time. Other frozen vegetables are already fully cooked and need only to be heated for service. The grading system used for canned vegetables also applies to frozen vegetables. Frozen vegetables are usually packed in 1-lb to 2-lb bags.

| Vegetable Can Sizes | | |
|---|---|---|
| Can Size | Weight | Cans Per Case |
| No. 2 | 20 oz | 24 |
| No. 2½ | 28 oz | 24 |
| No. 300 | 14–15 oz | 36 |
| No. 303 | 16–17 oz | 36 |
| No. 5 | 46–51 oz | 12 |
| No. 10 | 6 lb 10 oz | 6 |

 Vegetables can be purchased in cans of various sizes.

## COOKING VEGETABLES

Common methods for cooking vegetables include sautéing, stir-frying, grilling, broiling, roasting, steaming, blanching, simmering, and deep-frying. Vegetables should be cooked until they are just tender enough to be easily digested. At this stage of cooking, most vegetables retain the majority of their nutritional value, flavor, and color. Overcooked vegetables often lose their bright natural colors, may become mushy in texture, and lose nutrients, as vitamins and minerals are destroyed by excess heat.

### Reactions to Acids and Alkalies

Using acidic or alkaline ingredients when cooking can cause chemical reactions in vegetables, affecting their color and texture. Often an acidic ingredient, such as lemon juice or vinegar, is added to a cooking liquid to contribute flavor to the dish. An alkali, such as baking soda, is often added to tough ingredients, such as dried beans, to speed up the softening process. The problem with adding acidic or alkaline ingredients is that they negatively affect the natural pigments present in vegetables. *See Figure 15-18.* Types of natural pigments in vegetables include chlorophyll, carotenoids, and flavonoids.

**Chlorophyll.** *Chlorophyll* is an organic pigment found in green vegetables, such as spinach, broccoli, and asparagus. With an acidic ingredient in the cooking liquid, these vegetables turn a drab olive green color but retain their naturally firm texture. If an alkali is added to their cooking liquid, green vegetables become brighter in color but mushy in texture.

National Cancer Institute

| Acid and Alkali Reactions | | | | |
|---|---|---|---|---|
| Pigment | Cooked Vegetables* | Acid Added | Alkali Added | |
| **Chlorophyll** | Broccoli | Color loss | Mushy texture | |
| **Carotenoids** | Carrots | Little or no effect | Mushy texture | |
| **Flavonoids** | Beets | Brighter red | Turns blue; mushy texture | |

\* No acidic or alkaline ingredients used

**15-18** *Acidic and alkaline ingredients can affect the color and texture of vegetables.*

**Carotenoids.** A *carotenoid* is an organic pigment found in orange or yellow vegetables such as carrots, yellow squash, tomatoes, or red peppers. Acids have little to no affect on carotenoids. Alkaline ingredients do not affect the color of carotenoids, but do cause these vegetables to become very mushy.

**Flavonoids.** A *flavonoid* is an organic pigment found in purple, dark red, and white vegetables such as red cabbage, beets, parsnips, and cauliflower. Acidic ingredients cause purple or dark red vegetables to turn bright red. Alkaline ingredients cause purple or dark red vegetables to turn blue, and white vegetables to turn yellow. Alkaline ingredients also cause a mushy texture in these vegetables.

## Sautéing and Stir-Frying

Sautéing and stir-frying vegetables is done very quickly in a hot pan with a small amount of oil or butter. Vegetables prepared for sautéing or stir-frying are usually diced small or thinly sliced. Green vegetables turn a lovely bright green color when cooked this way. The finished vegetables should be firm; wilted vegetables are a result of sautéing or stir-frying for too long.

## Grilling and Broiling

Broiling and grilling are two fast and easy ways to prepare vegetables. Grilling or broiling caramelizes the sugars in a vegetable, giving it an appealing flavor. Vegetables should be seasoned as desired, drizzled with a little oil, and placed directly on the grill or under the broiler. The size of the vegetables determines how long they need to cook.

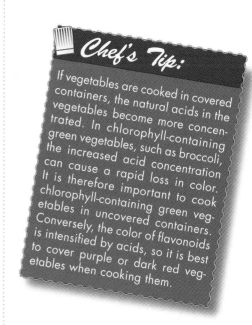

*Chef's Tip:*

If vegetables are cooked in covered containers, the natural acids in the vegetables become more concentrated. In chlorophyll-containing green vegetables, such as broccoli, the increased acid concentration can cause a rapid loss in color. It is therefore important to cook chlorophyll-containing green vegetables in uncovered containers. Conversely, the color of flavonoids is intensified by acids, so it is best to cover purple or dark red vegetables when cooking them.

## Roasting

Vegetables are roasted at a lower temperature and for a longer period than is required for direct-heat cooking methods. Some vegetables, like peppers, onions, carrots, turnips, and parsnips are roasted alongside a large piece of meat to enhance the flavor of the meat. Vegetables to be roasted should be cut into the same shape and size to ensure uniform doneness.

Fire-roasting vegetables over an open flame produces yet another result. This technique enables the vegetable to be cooked whole. Peppers are often fire-roasted as it gives them a distinctive flavor, yet can be done quickly.

**To fire-roast a pepper, apply the following procedure:**

1. Place the washed pepper over an open flame, either directly on a burner or grill or on a mesh grate. An alternate method is to place the pepper in a broiler, directly under the flame. ***See Figure 15-19.***

2. Roast the pepper until charred on all sides, turning continuously. The pepper should not directly touch the flame.

3. When the pepper is completely charred on all sides, remove and place in a paper bag or wrap in plastic and allow to rest for 10 min. This allows the pepper to steam itself and loosens its skin. *Note:* If a plastic bag is used, check the pepper for doneness after 5 min.

4. Remove the pepper and peel away the loosened skin.

5. Rinse the pepper under cold water to remove any remaining pieces of charred skin.

### Fire-Roasting Peppers

1–2. *Place the washed pepper over flame; turn pepper until surface is charred evenly on all sides.*

3. *Let charred peppers rest in a paper bag or wrapped in plastic for 10 min.*

4–5. *Peel away charred skin; rinse in cold water.*

**15-19** *Both bell peppers and chiles are commonly roasted for use in dishes ranging from salads, sandwiches, and soups to main course entrées.*

Roasted garlic is a popular ingredient in many dishes. When garlic is roasted, its flavor becomes rich and sweet.

## To roast garlic, apply the following procedure:

1. With a knife, slice off the top of a head of garlic, just above the tips, to expose the cloves.
2. Place the head of garlic on a sheet of aluminum foil or in a shallow pan, and drizzle with a small amount of olive oil.
3. Place in a 350°F conventional oven for about 20 min to 30 min, until the cloves are soft and golden brown.
4. Gently squeeze the garlic to release the soft meat from the skin

## Steaming

Vegetables are steamed by being placed in a perforated container above a pot of boiling water. Professional kitchens may also have automatic steamers that can accommodate hotel pans, allowing large quantities of food to be steamed in a short amount of time. Steamed vegetables often lack flavor and must be seasoned properly when finished. They can also be finished by sautéing for added flavor.

## Blanching

Blanching is done very quickly in boiling water. The vegetables are then removed and placed into an ice bath to stop the cooking. Some vegetables, such as tomatoes, are blanched to make it easier to remove their skin. Others, like carrots, are blanched and then finished using a broiler or grill to ensure doneness throughout.

## Simmering

Simmering is a common preparation method used for vegetables. Simmered vegetables can be eaten as they are or can be puréed into dishes such as mashed potatoes. Simmering results in bland-tasting vegetables, so it is important to season vegetables well when using this cooking method. Some vegetables are simmered and then finished by sautéing to add flavor.

## Deep-Frying

Deep-fried vegetables are popular appetizers in many restaurants. The vegetables are coated in a batter and fried until crisp. French fries or other forms of deep-fried potatoes are a popular item in many restaurants. Other popular vegetables for deep-frying include onions, mushrooms, cauliflower, zucchini, and eggplant.

## SUMMARY

Vegetables are a vital component of a healthy diet. They are rich in vitamins, minerals, and fiber and low in calories and fat. A chef needs to monitor cooking methods and accompaniments, since they can greatly affect the nutritional value of a vegetable dish. Purchasing, preparing, and serving high-quality vegetables tells customers that a chef cares about their health.

There are many varieties of vegetables, including cabbages, greens, bulbs, pods, seeds, fungi, roots, tubers, gourds, squashes, stalks, and fruit-vegetables. Vegetables are freshest and have the best flavor and color when purchased in season. The following tips help a chef to prepare and cook vegetables with the best taste, color, texture, and nutritional value:

- Cut vegetables into uniform sizes to prevent uneven cooking times.

- Cook vegetables only as long as needed for them to become tender, to avoid loss of nutrients and quality.

- Many vegetables can be blanched prior to service, shocked in ice water to stop the cooking process, and then heated as needed using a variety of methods, so that they stay at peak quality.

Common methods for preparing vegetables include sautéing, stir-frying, grilling, broiling, roasting, steaming, blanching, simmering, and deep-frying. Vegetables can be served as appetizers, salads, in soups, as side dishes, or as entrées.

Refer to the CD-ROM for Quick Quiz® questions related to chapter content.

## Review Questions

1. List the nine major categories of vegetables.

2. Describe common uses for kale.

3. Identify the main characteristics of kohlrabi.

4. Illustrate the nutritional benefits of spinach.

5. Describe common uses for dandelion greens.

6. List common types of bulbs.

7. Describe the procedure for cleaning a leek.

8. What are the nutritional benefits of legumes?

9. Explain how to quick-soak dried beans.

10. Contrast the main characteristics of garden peas and snow peas.

11. Describe the two main types of truffles.

12. List the main characteristics of a jícama.

13. How does spaghetti squash differ from other varieties of squash?

14. Describe the procedure for preparing an artichoke.

15. List three popular varieties of bell peppers.

16. Explain how to remove the pit from an avocado.

17. Describe what results from cooking vegetables in an acid such as vinegar.

18. Describe what results when vegetables are cooked in an alkali such as baking soda.

19. Identify nine ways to cook vegetables.

20. Describe the procedure for roasting a red pepper.

# Recipes

## 77

## Beer-Battered Zucchini

*yield: 8 (5-oz servings)*

| | |
|---|---|
| 1¼ c | all-purpose flour |
| ½ tsp | baking powder |
| 1 tsp | salt |
| 1 c | beer |
| 2 | medium zucchini, cut on the bias into ⅓"-thick slices |
| to taste | salt |
| | vegetable oil, for deep-frying |

1. Place 1 c of flour, baking powder, and salt into a large mixing bowl. Make a well in the center of the flour.
2. Pour the beer into the well and whisk until the mixture is just combined.
3. Strain the batter into a clean bowl and let it rest, covered, for 1 hr.
4. Dust the zucchini slices with the remaining flour. Shake off any excess flour.
5. Dip the floured slices into the batter and coat completely.
6. Working in batches, fry the zucchini slices in 370°F oil using the swimming method until golden.
7. Using a slotted spoon, transfer finished slices to paper towels to drain. Sprinkle with salt to taste.

### Nutrition Facts
**Per serving:** 211.1 calories; 58% calories from fat; 13.9 g total fat; 0 mg cholesterol; 364 mg sodium; 157.4 mg potassium; 17.5 g carbohydrates; 1.1 g fiber; 0.9 g sugar; 16.5 g net carbs; 2.7 g protein.

## 78

## Tempura-Battered Zucchini

*yield: 8 (5-oz servings)*

| | |
|---|---|
| 1¼ to 1½ c | club soda, cold |
| 1¼ c | all-purpose flour |
| ¼ c | cornstarch |
| 2 | medium zucchini, cut on the bias into ⅓"-thick slices |
| 1 tsp | salt |
| | vegetable oil, for deep-frying |

1. Place club soda in a medium-sized mixing bowl and sift in 1 c flour and the cornstarch.
2. Add salt, and whisk until evenly blended.
3. Cover, and refrigerate for 45 min to 1 hr.
4. Dust the zucchini slices with the remaining flour. Shake off any excess flour.
5. Dip the floured slices into the batter and coat completely.
6. Working in batches, fry the zucchini slices in 375°F oil using the swimming method until golden.
7. Using a slotted spoon, transfer finished slices to paper towels to drain. Sprinkle with salt to taste.

### Nutrition Facts
**Per serving:** 214.6 calories; 57% calories from fat; 13.9 g total fat; 0 mg cholesterol; 305.7 mg sodium; 150.3 mg potassium; 20.2 g carbohydrates; 1.1 g fiber; 0.9 g sugar; 19.1 g net carbs; 2.6 g protein.

## Stir-Fried Vegetables

*yield: 6 (5-oz servings)*

| | |
|---|---|
| 3 tbsp | peanut oil or vegetable oil |
| 1 tbsp | fresh ginger, minced |
| 1 tbsp | fresh garlic, minced |
| ½ | red onion, quartered with layers separated |
| 1 head | bok choy, trimmed, cut lengthwise into quarters and then into 1" segments |
| ¼ lb | Chinese broccoli, cut into 1" pieces |
| ½ lb | Chinese long beans, trimmed and cut into 1" pieces (or use fresh green beans) |
| 1 | red bell pepper, cut into 1" pieces |
| 5 | scallions, cut diagonally into 1" pieces |
| ¼ head | napa cabbage, cut crosswise into 1" strips |
| ⅔ c | chicken stock, chicken broth, vegetable broth, or water, heated |
| 1 tbsp | soy sauce |
| 1 tbsp | cornstarch, dissolved in 1 tbsp water |
| 1 | scallion, thinly sliced |
| 2 tsp | toasted sesame seeds |

1. Place a large wok or sauté pan over high heat and add 2 tbsp of oil.
2. When the oil is hot, add the ginger and garlic and stir-fry just until they are aromatic, about 30 seconds. Scoop out the aromatics and set them aside.
3. Add the remaining oil. When it is hot, add the onion pieces and stir-fry until they turn glossy and bright, approximately 1 min to 2 min.
4. Add the bok choy and Chinese broccoli. Stir-fry 1 min to 2 min.
5. Add the beans, bell peppers, and the 1" scallion pieces. Continue stir-frying until they are bright green and glossy, approximately 1 min to 2 min.
6. Add the napa cabbage, ⅓ c of the hot stock, and the reserved aromatics.
7. Continue stir-frying until all the vegetables are tender-crisp, approximately 2 min.
8. Add the remaining stock, soy sauce, and cornstarch mixture, and stir-fry until the vegetables are lightly glazed with sauce, approximately 1 min.
9. Transfer the vegetables to a heated serving dish. Garnish with scallions and sesame seeds. Serve immediately.

---

**Nutrition Facts**

**Per serving:** 151 calories; 48% calories from fat; 8.3 g total fat; 0.8 mg cholesterol; 249.2 mg sodium; 733.9 mg potassium; 16.1 g carbohydrates; 4.2 g fiber; 5 g sugar; 11.9 g net carbs; 6.2 g protein

---

## Grilled Asparagus with Garlic, Rosemary, and Lemon

*yield: 6 (5-oz servings)*

| | |
|---|---|
| 1 lb | asparagus |
| 1½ tbsp | olive oil |
| 2 cloves | garlic, finely minced |
| 1 tsp | lemon zest |
| 2 tsp | fresh rosemary |
| to taste | salt and black pepper, freshly ground |

*Florida Department of Agriculture and Consumer Services, Bureau of Seafood and Aquaculture Marketing*

1. Trim asparagus and discard woody ends.
2. In a small bowl, combine oil, garlic, rosemary, and lemon zest. Mix well.
3. Place asparagus on a heated grill (preferably on a vegetable grill pan, so that the stalks do not fall through the grill grate).
4. Brush mixture onto the asparagus spears.
5. Cook asparagus to desired tenderness. Season with salt and pepper to taste.

---

**Nutrition Facts**

**Per serving:** 64.7 calories; 72% calories from fat; 5.3 g total fat; 0 mg cholesterol; 395.3 mg sodium; 204.3 mg potassium; 3.4 g carbohydrates; 1.3 g fiber; 0 g sugar; 2.2 g net carbs; 2.2 g protein

---

# Recipes

## 81

## Roasted Potatoes and Turnips

*yield: 6 (4-oz servings)*

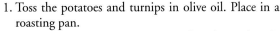

| | |
|---|---|
| 4 | medium potatoes, diced |
| 2 | medium turnips, diced |
| 2 tbsp | extra virgin olive oil |
| 2 sprigs | fresh rosemary |
| 1 sprig | fresh tarragon, minced |
| 1 tbsp | fresh parsley or chervil, minced |
| to taste | salt and pepper |

1. Toss the potatoes and turnips in olive oil. Place in a roasting pan.
2. Garnish with rosemary, tarragon, and parsley or chervil.
3. Roast at 350°F for 45 min to 1 hr, or until vegetables are tender.
4. Season with salt and pepper to taste.

**Nutrition Facts**

**Per serving:** 161.3 calories; 25% calories from fat; 4.7 g total fat; 0 mg cholesterol; 84.8 mg sodium; 683.6 mg potassium; 27.6 g carbohydrates; 3.9 g fiber; 2.7 g sugar; 23.6 g net carbs; 3.3 g protein.

## Stuffed Artichokes

*yield: 4 servings*

| | |
|---|---|
| 4 | large artichokes, cleaned, with chokes removed |
| ½ c | onion, minced |
| 2 cloves | garlic, minced |
| ½ c | olive oil |
| 1½ c | bread crumbs |
| ½ c | scallion greens, thinly sliced |
| ⅔ c | Parmesan cheese, grated |
| 3 oz | lemon juice |
| to taste | salt |
| to taste | pepper |
| 2 oz | white wine |
| 1 tbsp | chopped parsley |

1. Pour 1″ to 2″ of water and 1 tbsp lemon juice in a sauté pan. Place prepared artichokes in pan, top side up.
2. Pour ½ tsp of lemon juice in the center of each artichoke to coat the heart.
3. Bring to a boil over medium-high heat, then lower to a simmer.
4. Cover, and gently steam artichokes for 8 min or until stem end is tender when pierced with a paring knife. Remove from heat and allow to rest, covered, for 5 min.
5. While the artichokes are cooking, sauté the minced onions and garlic in 1 tbsp olive oil until soft.
6. Deglaze onion and garlic with white wine and reduce by half.
7. Remove onion mixture from heat and add parsley, bread crumbs, scallions, Parmesan cheese, remaining lemon juice, salt, and pepper. Mix well. Place the artichokes in a small baking dish or hotel pan just large enough to hold them.
8. Fill each artichoke with the bread crumb mixture. Fill both the center and any spaces between the leaves.
9. Add a small amount of water to just cover the bottom of the pan. Drizzle the remaining olive oil over the stuffed artichokes.
10. Bake at 375°F for 12 min or until golden brown.

**Nutrition Facts**

**Per serving:** 577.4 calories; 52% calories from fat; 34.2 g total fat; 14.7 mg cholesterol; 781.4 mg sodium; 814.6 mg potassium; 52.4 g carbohydrates; 11.3 g fiber; 4.4 g sugar; 41.1 g net carbs; 17.8 g protein.

## Steamed Vegetables with Asian Dressing

*yield: 5 (4-oz servings)*

| ¼ lb | broccoli florets |
|------|------------------|
| ¼ lb | cauliflower florets |
| 1 | carrot, cut into thick diagonal slices |
| ½ | summer squash, cut in thick rounds |
| ¼ lb | asparagus |
| ½ c | scallions |

*Asian Dressing*

| 1½ tbsp | yellow miso |
|---------|-------------|
| 1 tbsp | water |
| ½ tbsp | rice vinegar |
| ½ tsp | soy sauce |
| ½ tbsp | fresh ginger, grated |
| ½ tsp | red chile paste |
| 2 tbsp | peanut oil |
| ½ tsp | sesame oil |

1. Pour 1″ or more of water in a wok or sauté pan and bring to a boil over medium-high heat.
2. Place the vegetables in a bamboo or collapsible steamer.
3. Set the steamer over the boiling water, cover, and cook for 4 min to 6 min, or just until the vegetables are al dente.
4. While the vegetables are cooking, combine the miso, water, vinegar, soy sauce, ginger, and chile to taste, in a bowl, and whisk.
5. Gradually whisk in the peanut oil, starting with a few drops and then adding the rest in a steady stream to make a smooth, slightly thick dressing.
6. Gradually whisk in the sesame oil, starting with a few drops and then adding the rest in a steady stream.
7. Spoon dressing over steamed vegetables. Serve warm or at room temperature.

### Nutrition Facts

**Per serving:** 90.6 calories; 60% calories from fat; 6.4 g total fat; 0 mg cholesterol; 235.2 mg sodium; 329.6 mg potassium; 7.9 g carbohydrates; 2.2 g fiber; 2.5 g sugar; 5.8 g net carbs; 2.8 g protein.

## 84

## Cabbage and White Bean Soup

*yield: ½ gal. (eleven 6-oz servings)*

| ½ c | dried white beans (great northern, navy, or cannellini), picked over and rinsed |
|-----|------------------|
| ½ | onion, diced |
| 1 oz | unsalted butter |
| 1 pinch | ground cloves |
| 1 clove | garlic, finely chopped |
| 1¼ lb | smoked ham hocks |
| 1½ qt | light stock |
| 1 | bay leaf |
| ¾ tsp | fresh thyme |
| ½ lb | Yukon Gold potatoes, peeled and diced into 1″ cubes and stored in water |
| ½ lb | cabbage, cored and cut into ½″ pieces |
| 1 tbsp | fresh parsley, chopped |

1. Place beans in container and cover with cold water at least 2″ over beans and let soak overnight. Drain in a colander the next day. *Note:* The beans can be quick-soaked by placing them in a pot with cold water, bringing to a boil, and simmering for 2 min. After 2 min, remove the beans from heat and allow to rest, uncovered, for 1 hr.
2. In medium stockpot, sweat onion in butter over medium heat. Add garlic and sweat an additional 1 min.
3. Add ham hocks, stock, bay leaf, and thyme. Bring to a simmer, cover, and cook 1 hr, skimming off any impurities as they rise to the surface.
4. Add beans and simmer uncovered, stirring occasionally for 40 min to 50 min or until beans are almost tender.
5. When beans are almost done, add potatoes and cabbage. Return to a simmer and cook uncovered for 20 min to 25 min or until vegetables and potatoes are tender.
6. Remove ham hocks. When they are cool enough to handle, discard skin and bones and then cut the meat into bite-size pieces.
7. Add ham to soup. Season with salt and pepper to taste. Discard bay leaf.
8. Serve in bowls and garnish with croutons.

### Nutrition Facts

**Per serving:** 150.7 calories; 34% calories from fat; 5.7 g total fat; 15.2 mg cholesterol; 327.2 mg sodium; 540.8 mg potassium; 17 g carbohydrates; 2.4 g fiber; 3.2 g sugar; 14.6 g net carbs; 8.3 g protein.

# Stocks and Sauces

*Stocks and sauces* are fundamentally important items produced in the professional kitchen. The French word for stock is fond, which translates as "base" or "foundation." A well-made stock serves as the foundation for many sauces and soups, and stocks are used in various cooking techniques. A sauce, often made with some sort of stock, adds flavor, richness, and eye appeal to a dish. The best sauces complement a dish and never cover it up. Most sauces are slightly thicker than stocks due to either reduction taking place over a long cooking time, or the addition of a thickening agent or puréed ingredients. To prepare stocks or sauces, a chef must have knowledge of the necessary ingredients, preparation procedures, and required storage methods.

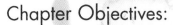

## Chapter 16

## Chapter Objectives:
1. Describe the basic composition of stocks.
2. Describe common methods and guidelines for making stocks.
3. Contrast the two most common methods for cooling stocks.
4. Demonstrate how to make the following items: white stock, brown stock, glace, fish stock, fumet, and roux.
5. Contrast fish stocks and fumets.
6. Explain the process of reduction.
7. Describe the four most common thickening agents used in stocks and sauces.
8. Demonstrate how to add a liaison to a liquid.
9. Describe each of the five mother sauces.
10. Demonstrate how to make hollandaise sauce.
11. Identify three types of butter sauces.
12. Demonstrate how to make beurre blanc sauce and compound butter.
13. Contrast common contemporary sauces.
14. Demonstrate how to make flavored oils.

## Key Terms
- stock
- mirepoix
- remouillage
- bouillon
- glace
- fumet
- gelatinization
- roux
- beurre manié
- slurry
- liaison
- mother sauce
- demi-glace
- jus lié
- lecithin
- beurre
- coulis
- nage

## STOCKS

A *stock* is the strained liquid that results from cooking seasonings and vegetables or the bones of meat, poultry, or fish. Stocks are used to make sauces and soups. Stocks are made primarily by simmering bones, vegetables, and seasonings in water for a length of time to extract their colors, flavors, and nutrients. There are also many concentrated commercial stock bases available from food service distributors. These concentrated bases come in either granulated or paste forms. Usually the granulated or dried forms are predominantly made of salt, while the paste forms are made from the meat, bones, and juices of the product they represent. These commercial bases are often used simply to add additional flavor to a fresh stock. Often fresh vegetables and even some bones are added to a stock made from a commercial base to make the stock taste fresher or like it was made from scratch.

On page 1 of Auguste Escoffier's *Cookbook and Guide to the Fine Art of Cookery*, he says the following in regard to stocks:

> Indeed, stock is everything in cooking … If one's stock is good, what remains of the work is easy; if, on the other hand, it is bad or merely mediocre, it is quite hopeless to expect anything approaching a satisfactory result. The cook mindful of success, therefore, will naturally direct his attention to the faultless preparation of his stock….

The emphasis Escoffier places on producing a good stock reflects the importance of stock as the foundation for numerous dishes.

### Stock Composition

Stocks are composed of water, a flavoring component (such as bones), mirepoix, and aromatics. The composition of stock can vary, but is generally around ten parts water, five parts flavoring component, and one part mirepoix, plus aromatics. ***See Figure 16-1.*** Stock should be relatively clear, not cloudy. Care should be taken to remove any impurities that could possibly cloud or discolor the stock. Overcooked vegetables can also result in a cloudy stock.

**Water.** Cold water is always used as it helps loosen impurities and blood from the bones. These impurities may then float to the surface, allowing them to be skimmed off. If hot water is used to start a stock, the hot water makes the impurities stick to the bones and causes the stock to become cloudy as it simmers.

**Flavoring Components.** Flavoring ingredients such as meat trimmings or bones are used for a meat stock, and vegetables for a vegetable stock. These flavoring ingredients also add to the nutritional makeup of a stock. Bones should be cut into 3″ long pieces. Larger bones should be split in half lengthwise for quicker and more complete drawing out of the collagen (which converts into gelatin), flavor, and nutrients. ***See Figure 16-2.***

## Stock Composition

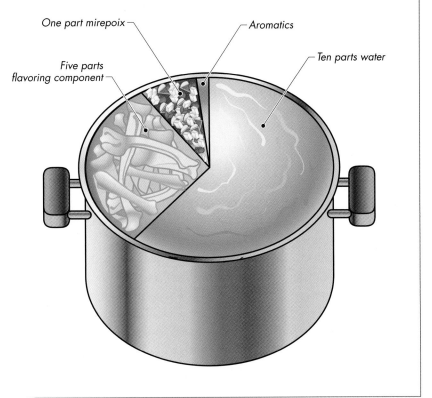

One part mirepoix

Aromatics

Five parts flavoring component

Ten parts water

**16-1** *The composition of stock is approximately ten parts water, five parts flavoring component, and one part mirepoix, plus aromatics.*

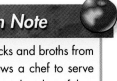

### Nutrition Note

Making stocks and broths from scratch allows a chef to serve a healthier product than if they were purchased prepared. Prepared products are most often loaded with sodium, and many have artificial colors and flavors added. Making stock from scratch allows the chef to fortify it with fresh vegetables, so the resulting stock contains more nutrients. Stocks made from scratch generally contain little to no sodium and no preservatives or artificial flavors or colors.

## Stock Bones

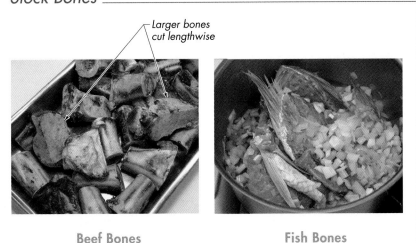

Larger bones cut lengthwise

**Beef Bones**

**Fish Bones**

**16-2** *Large bones used for stock are split in half lengthwise. Smaller bones are cut into 3" pieces.*

**Mirepoix.** Mirepoix is added at a proportion of 1 oz mirepoix for every 8 oz of bones (or a ratio of 1:8). *Mirepoix* is a mixture of 50% onions, 25% celery, and 25% carrots, roughly cut into the appropriate size for the stock being produced. Because the mirepoix is strained from the finished stock, a perfect dice is not required. ***See Figure 16-3.*** A white mirepoix is used for a white stock. A *white mirepoix* is a mirepoix with onions and celery, but the carrots are replaced with parsnips and leeks.

A stock made from large bones, such as beef stock, requires large cuts of mirepoix (approximately 1″ to 2″ square pieces). Fish and chicken stocks that cook for a short time and contain small, thin bones require small cuts of mirepoix (approximately ½″ square pieces). The smaller the cut, the quicker the mirepoix cooks. If a small-cut mirepoix is added to a beef stock, the vegetables become overcooked well before the stock is done. Overcooked mirepoix can lead to the vegetables dissolving, making the stock cloudy and discolored.

Vegetables used in a stock must be thoroughly washed, cleaned, and trimmed prior to using. Although it is not necessary to peel carrots for a mirepoix, most chefs remove the peelings from carrots as the thin peels will dissolve and discolor the stock. Celery leaves are not recommended for use in stock as the leaves are often bitter. Onion peels are sometimes used in brown stocks, but never in white stocks as the peels may darken the stock.

Sometimes a matignon is used in place of mirepoix. A *matignon* is a uniformly cut mixture of onions, carrots, and celery, and may also contain smoked bacon or ham. Matignon is often referred to as "edible mirepoix," because it is cut uniformly and left in the completed dish.

*Mirepoix* _____

16-3 *A mirepoix is a rough-cut vegetable mixture consisting of 25% carrots, 50% onions, and 25% celery.*

**Aromatics.** An *aromatic* is an ingredient such as an herb, spice, or vegetable added to a food to enhance its natural flavors and aromas. There are two standard methods for adding aromatics to a stock: a bouquet garni (pronounced boo-KAY gar-NEE) or a sachet d'épices (pronounced sah-shay day-piece). ***See Figure 16-4.*** A *bouquet garni* is a mixture of fresh vegetables or herbs that are tied into a small bundle with butcher's twine. The twine is then tied to the handle of the pot so that the bundle is easily retrieved, and the bundle is added to the soup, stock, or sauce. Typically, a stalk of celery or a piece of leek acts as the main structure for the bouquet.

A *sachet d'épices* (French for a "sack of spices") is a piece of cheesecloth filled with aromatic ingredients such as herbs, spices, mushroom stems, leeks, and other aromatic vegetables. As with the bouquet garni, the sachet d'épices (or sachet, as it is commonly referred to) is tied with twine to the handle of the pot for easy retrieval. It is used to easily remove aromatics from a stock in the same way a tea bag is used to easily remove tea leaves from a cup of tea.

A stock is usually reduced to concentrate the flavors of the main ingredients, vegetable mirepoix, and aromatics. Salt is never added to aromatics or stock in the early stages, because after reduction the resulting flavor is too intense. Instead, salt is added at the final stage of preparation, if desired.

Aromatics

Bouquet Garni

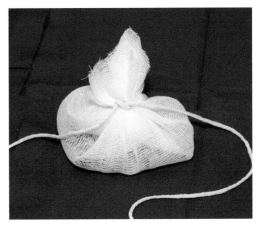

Sachet d'Épices

16-4 *A bouquet garni and a sachet d'épices are used to add aromatics to a stock.*

## Cooking Times for Stocks

| Type | Cooking Time* |
|------|---------------|
| Beef | 8 hr |
| Chicken | 4 hr |
| Fish | 45 min |
| Vegetable | 30 min |

\* approximate

 **16-5** *Required cooking time for a stock varies depending upon the size of bones used.*

**Types of Stocks.** The way a stock is prepared will determine the outcome of the stock. Some stocks have ingredients that are roasted prior to use. A roasted ingredient will produce a stock with a rich, brown color and a roasted flavor. Other stocks are made simply by simmering uncooked ingredients together. Uncooked ingredients yield a light or semiclear stock. The taste of a stock should be strong enough to be identifiable but not so strong as to overpower the taste of other items it is ultimately used in, such as a soup. Also, a stock made with larger and thicker bones, such as a beef stock, requires more cooking time than stock made with small, thin bones. A longer cooking time is needed to extract the marrow and nutrients of dense bones than is needed for small bones.

Beef stocks usually simmer for about 8 hr. A chicken stock made with many small, less dense bones usually cooks for about 4 hr. A fish stock made from thin, fragile fish bones usually requires a cooking time of only 30 min to 45 min. ***See Figure 16-5.*** No matter the type of stock prepared, following some simple steps helps ensure a tasty finished product.

### General Procedure for Making Stock

There are many different types of stock produced in the professional kitchen. Although there are some differences between types of stock, there is a basic set of guidelines that can be applied to any stock. For example, stock is made in a pot that is taller than it is wide to allow for slower evaporation.

*Practicing the following guidelines yields a consistent and great-tasting stock that is easy to produce:*

- Always begin a stock with cold water.
- Bones should be completely covered with water throughout preparation.
- A stock should always be simmered, never boiled.
- Never stir the stock while simmering.
- Always skim impurities and fat as they rise to the top.
- Always strain stock thoroughly and gently.
- Always cool a hot stock rapidly.
- Always follow proper storage procedures.

Cold water added to bones at the beginning stages of stock preparation helps dissolve any impurities and blood proteins that could later cloud a stock. As the stock is heated, these impurities coagulate or thicken and begin to float to the surface. The impurities can be skimmed from the surface and discarded.

Beginning a stock with hot water causes impurities to coagulate, or thicken into a mass, on the surface of the bones and inside the bones. The coagulated impurities are commonly referred to as "scum." As the stock cooks, the coagulated impurities begin to break up into smaller pieces and eventually cloud the stock. In this case, the impurities do not rise to the top, so they cannot be skimmed off.

Bones should be completely submerged in water during preparation. A bone that is above the surface of the water does not add any flavor or nutrients to the stock and is wasted. Bones that are not submerged darken from exposure to the air.

Simmering the stock at a temperature of between 180°F to 185°F helps to draw flavor and nutrients from the bones. Boiling a stock (even if only for a short time) breaks up coagulated impurities as they release from bones and forces the impurities to break up and cloud the stock.

The solid components of a stock should not be disturbed. Stirring the stock during cooking causes impurities in and around the solid ingredients to be broken up throughout the stock and results in cloudiness.

As impurities coagulate and rise to the surface, they need to be skimmed and discarded. Fat, as it is rendered (extracted) from meat scraps and bones, also rises to the surface and needs to be removed. A skimmer is used to remove impurities and a solid spoon or ladle is used to remove fat. If fat or impurities are left in the stock during the cooking process, they cloud the stock.

The solid material in a stock must remain intact during cooking and throughout the process of draining the stock. Many stockpots have a lower spigot drain that makes it easy to drain all liquid out through the bottom of the stockpot without needing to tip the pot. This gentle draining of unstrained stock helps ensure that impurities remain in the pot and are not disturbed, which would cause them to break up and cloud the stock.

If a stockpot with a drain spigot is not available, a large ladle is used to remove the stock. It is then strained through a china cap lined with multiple layers of cheesecloth. Cheesecloth acts like a fine strainer and removes any small particles that can cloud the stock.

**Cooling Stocks.** Food should be brought out of the temperature danger zone as soon as possible. Stocks, like all potentially hazardous foods, should be cooled prior to storing. The two most common methods for cooling stocks are in an ice bath and with a cooling wand.

An ice bath surrounds the hot food with ice to rapidly cool the food. The stock is placed in a clean, nonreactive stainless steel container (heavy-gauge plastic is also acceptable but does not cool as quickly as metal) in a clean and sanitized sink. The sink is then filled with ice cubes and cold water to surround the stock container. Precaution is needed to not tip the container or allow ice water to enter it. The stock is stirred to

*Chef's Tip:*

Never place a hot stock stored in a large container in a refrigerator. The large container not only takes a very long time to cool completely, but it also raises the temperature inside the refrigerator. Raising the temperature inside the refrigerator creates a potentially hazardous environment for other stored foods.

## Cooling Wands

San Jamar

A cooling wand may be used to cool stocks quickly for storage.

speed the cooling process. Stirring uses the process of convection to mix cooled stock near the outer sides of the container with warmer stock in the center of the container.

Another common method for cooling a stock is to use a cooling wand. A *cooling wand* is a heavy-gauge, hollow plastic paddle with a screw-on cap at the top of the handle that is filled with water and frozen prior to use. *See Figure 16-6.* The clean, frozen wand can then be inserted into the hot stock and used to stir the stock until cooled. The wand is removed, washed, and sanitized after each use to prevent cross-contamination and foodborne illness. Always check that the screw-on cap is securely fastened to prevent the water in the wand from contaminating the stock.

When stock has cooled, it should be wrapped or covered, labeled, dated, and stored in a refrigerator. As the stock becomes colder, any remaining fat from bones or meat scraps solidifies on the surface of the stock. Solid fat can be removed with a kitchen spoon or skimmer. *Degreasing* is the process of removing surface fat from a food. Many professional kitchens use solidified fat in place of butter or oil when sautéing meats as the fat will add additional flavor to the dish.

### White Stocks

A *white stock* is stock produced by gently simmering raw poultry, beef, veal, pork, or fish bones in water with vegetables and herbs. Because the bones and vegetables have not been roasted or browned, the stock will stay almost colorless. A white mirepoix is often used to avoid a possible color change in the stock. Chefs have historically disagreed about whether or not bones should be blanched prior to being used in a white stock. Some chefs think that blanching removes many impurities from the bones. Other chefs believe that blanching the bones wastes nutrients and flavor. The most common practice is to simply rinse the bones in clean, cold water to remove any loose impurities that are present on the surface of the bones before use.

CULINARY PROCEDURES

To make a white stock, apply the following procedure:

1. Rinse the bones in cool, clean water and place in a stockpot.
2. Completely cover the bones with cold water.
3. Bring water to a simmer. Continually skim off impurities as they rise to the surface.
4. Add the white mirepoix (celery, onions, parsnips, and leeks) and sachet d'épices to the pot.
5. Maintain a simmer and continually skim rising impurities. A stock with smaller bones such as poultry will only need to simmer 1 hr to 1½ hr, while large veal or beef bones will require 4 hr to 6 hr of simmering.

6. Carefully drain the stock to minimize disturbing the cooked ingredients.

7. Strain the stock with a china cap lined with cheesecloth.

8. Quickly cool the strained stock with either an ice bath or a cooling wand to bring it out of the temperature danger zone.

9. Label the stock with the correct product name, time, and date, and store in the refrigerator.

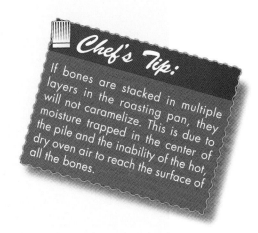

**Chef's Tip:**
Never let the stock boil, as boiling can disperse impurities into the stock, causing cloudiness and loss of quality.

## Brown Stocks

A brown stock is produced by roasting and browning poultry, beef, veal, or game bones; vegetable mirepoix; and a tomato product (if desired). The roasted items are placed in a stockpot covered with cool, clean water and brought to a simmer. The caramelization of sugar naturally present in the bones and vegetables adds the rich, brown color, roasted flavor, and high gelatin content that are characteristic of a well-made brown stock.

## Use the following procedure for making a brown stock:

1. Caramelize the bones. Place a single layer of 2″ to 3″ lengths of bones in a roasting pan. Place the pan in a 400°F oven for 45 min to 1 hr, remembering to stir occasionally to produce even browning. Reserve rendered fat for use in step 3. *See Figure 16-7.*

2. Transfer the roasted bones to a stockpot.

3. Place the roasting pan on burners and add the mirepoix to the pan. Using fat rendered in step 1, sauté mirepoix until caramelized while continually stirring. Use caution to not burn the mirepoix, as it will contribute a bitter taste in the stock.

4. Pour off all excess rendered fat from the roasting pan and reserve it for later use.

5. Deglaze the roasting pan. Place the roasting pan with all of the caramelized food particles (known as the fond) on open burners over medium heat until hot. Carefully add about ½″ of water to the pan and gently scrape the particles loose from the bottom. After the pan has been deglazed, pour the rich, brown deglazed liquid (known as the deglazing liquor) into the stockpot with the roasted bones.

6. After the mirepoix has been caramelized, add a small amount of tomato product (tomato sauce or paste, for example) if desired. Cook slowly until the tomato product caramelizes. *Note:* Caramelizing a tomato product in this manner is known as pincé and adds additional richness and flavor to the brown stock. Another benefit of caramelizing a tomato product is that the caramelizing process reduces some of the acidic characteristics of the tomato.

**Chef's Tip:**
If bones are stacked in multiple layers in the roasting pan, they will not caramelize. This is due to moisture trapped in the center of the pile and the inability of the hot, dry oven air to reach the surface of all the bones.

1. Caramelize bones.

2–4. Transfer roasted bones to stock-pot. Caramelize mirepoix. Pour off excess fat.

5–7. Deglaze roasting pan. Add tomato product. Deglaze pan again.

8. Add mirepoix to stockpot.

9–10. Bring contents to simmer; skim impurities. Carefully drain stock.

11–13. Strain with a china cap lined with cheesecloth. Quickly cool stock. Label and store in refrigerator.

**16-7**   *A brown stock is produced by gently simmering roasted and browned poultry, beef, veal, or game bones, vegetable mirepoix, and a tomato product (if desired).*

7. Once again, deglaze the hot roasting pan with water, scraping gently.

8. Add the caramelized mirepoix, fond, and deglazing liquor to the stockpot.

9. Bring contents to a simmer. Continually skim the surface of the stock to prevent any floating impurities from clouding the final product. Do not let the stock boil as boiling can disperse small particles of impurities into the stock, causing cloudiness and loss of quality. A stock with smaller bones, such as from poultry, needs only 1 hr to 1½ hr simmering time, while larger veal or beef bones require 4 hr to 6 hr of simmering.

10. Carefully drain the stock to minimize disturbing the cooked ingredients.

11. Strain the stock with a china cap lined with cheesecloth.

12. Quickly cool the strained stock with either an ice bath or a cooling wand to bring it out of the temperature danger zone.

13. Label the stock with the correct product name, time, and date, and store in the refrigerator.

A *remouillage* (French for "rewetting") is a stock made from using bones that have already been used once to make a stock. Used bones are placed in water with vegetables and aromatics and simmered to extract any additional flavor they may contain. The resulting stock is less flavorful than the original stock made from unused bones. A remouillage is often used in place of water when starting a new stock or when making a soup or sauce as it already has flavor compared to plain water.

A *bouillon* is the liquid that remains after simmering meats. Bouillon is the French word for broth. Bouillon results when stock is used to poach meats. The flavor of the stock is strengthened by the addition of the meat juices released from the meat as it cooks. The most common use of a bouillon is when making a type of soup known as a consommé.

## Glaces

A *glace* (pronounced glahss) is a highly reduced stock that results in an intense flavor. Glace comes from the French word "glaçage," which means glazing or icing. As much as 1 gal. of a stock is slowly reduced to one-eighth or one-tenth of its original volume to yield just 1 or 2 cups.

The most common glace is a *glace de viande*, or beef glaze, made from caramelized beef bones and mirepoix. *Viande* is the French term for meat. Other classic and popular glaces are *glace de volaille* (chicken glaze), made from a concentrated reduction of chicken stock, *glace de veau* (veal glaze), made from a veal stock reduction, and *glace de poisson* (fish glaze), made from an incredibly reduced fish stock. These classic glaces are added to soups or sauces to intensify the flavor. When cold, they have a rubberlike texture that allows them to be cut into cubes, which makes them easy to store and add to another product. ***See Figure 16-8.***

Once completely cooled, glace has a rubbery, firm texture that makes it easy to handle. Glace can be stored using either of the following techniques:

- Pour approximately 1″ deep into a hotel pan or sheet pan and cut into cubes using a pastry wheel.

- Pour into ice cube trays and allow to set up into cubes.

*Glace*

**16-8** *Cold glace has a rubberlike texture that allows it to be cut into cubes for easy storage and use.*

## To make a glace, apply the following procedure:

1. Choose the appropriate-size pan. Use of the appropriate-size pan is necessary to make a perfect glace. The saucepan should be just large enough so that not too much surface area is exposed. The greater the size of the pan, the quicker the evaporation of the stock. Reducing a stock too quickly through evaporation can ruin a glace.

2. Simmer the stock slowly to allow a slow reduction. Continually check the bottom of the saucepan to ensure that the reduction does not begin to scorch or burn.

3. Continually skim any surface impurities while the glace is reducing.

4. As the glace reduces, strain and transfer it into progressively smaller pans. Straining the glace each time it is transferred into a smaller pan helps to ensure the product is as free from impurities as possible. Not straining a glace during the reduction process will result in a gritty, cloudy sauce with an almost curdled appearance.

5. When reduction is finished, the glace should be strained carefully to remove any impurities or solids.

6. Quickly cool the strained glace with either an ice bath or a cooling wand to bring it out of the temperature danger zone.

7. Label the glace with the correct product name, date, and time and store properly in the refrigerator.

It is important to store cubes of glace in sanitized containers and to avoid handling the glace with bare hands. Improper storage and handling can greatly shorten the amount of time that glace can be stored. When properly prepared, stored, and handled, a glace will last for several months in refrigeration.

## Fish Stocks and Fumets

Fish stocks and fumets are made by simmering fish bones or shellfish shells in water. A *fumet* is a concentrated stock made from fish bones or shellfish shells and vegetables that is used to make soups and sauces. The main difference from making a white stock is that the cooking time for fish stocks and fumets is much shorter because the fish bones and shellfish shells are thinner and more delicate than the bones used in a white stock. Bones used for a fish stock should be those from a lean fish, such as sole or snapper, as opposed to an oily fish, such as salmon. Fish heads may be used; however, eyes and gills should be removed. The fish must also be properly eviscerated. To *eviscerate* is to remove all entrails and organs from a carcass. ***See Figure 16-9.***

Fish bones must be completely washed prior to use in a stock in order to keep the stock very clear. Vegetables used in a mirepoix for a fish stock or fumet should be cut very small due to the quick cooking time (usually no more than 45 min total). Often mushroom stems or trimmings are added to fish stocks and fumets to increase the flavor and aromatic characteristics.

*Eviscerating Fish*

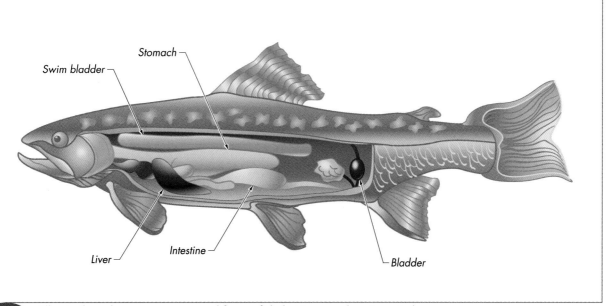

Swim bladder
Stomach
Liver
Intestine
Bladder

16-9    *All entrails and organs are removed from a fish that is properly eviscerated.*

A fish stock is a basic stock prepared by adding bones, vegetables, a sachet, and cold water to a stockpot and bringing to a simmer over medium heat. The resulting stock is clear and mild in flavor.

To make a fish stock, apply the following procedure:

1. Rinse all bones and shells in cold water to remove any surface debris or impurities.
2. Place the bones, small diced mirepoix (including mushroom pieces if desired), sachet, and cold water in a stockpot over medium heat and bring to a simmer.
3. Simmer for 30 min to 45 min, using caution to never let the stock reach a boil. Continually skim the surface of the stock to prevent any floating impurities from clouding the final product.
4. Strain the stock carefully to remove any impurities or solids.
5. Quickly cool the strained stock with either an ice bath or a cooling wand to bring it out of the temperature danger zone.
6. Label the stock with the correct product name, time, and date, and store in the refrigerator.

A fumet differs from a fish stock in that the small diced vegetables, bones, and shells are sweated in a stockpot over medium-low heat prior to use. Sweating the ingredients results in a strongly flavored yet still relatively colorless stock. Care is taken to not discolor the vegetables, bones, or shells in the sweating process. Fish stock or water and a sachet are added to the sweated ingredients. A fumet will also have wine and/or lemon juice added for flavor.

To make a fumet, apply the following procedure:

1. Rinse all bones and shells in cold water to remove any surface debris or impurities. *See Figure 16-10.*
2. Sweat vegetables, bones, and shells. Melt a small amount of butter in a stockpot over low-to-medium heat. Place onions, parsley stems, thyme, mushroom pieces, garlic, trimmings, shells, and bones in the pot and begin to slowly sweat them, being careful not to allow them to gain color.
3. Gently deglaze the pot. When onions are translucent, deglaze pot with a small amount of lemon juice and white wine.

4. Add a well-made fish stock (or water if stock is not available) and a few lemon slices and bring to a simmer. Using a previously prepared fish stock gives the fumet a much more intense flavor.

5. Simmer for about 30 min. Continually skim the surface of the stock to prevent any floating impurities from clouding the final product.

6. Quickly cool the strained stock with either an ice bath or a cooling wand to bring it out of the temperature danger zone.

7. Label the stock with the correct product name, time, and date, and store in the refrigerator.

*Preparing a Fumet*

1. Rinse bones and shells in cold water.

2. Sweat vegetables, bones, and shells in a small amount of butter.

3. Deglaze pot with lemon juice and white wine.

4–7. Add fish stock and lemon slices. Simmer for 30 min; skim off impurities. Quickly cool stock. Label and store in refrigerator.

**16-10** When preparing a fumet, the small-diced vegetables, bones, and shells are sweated in a stockpot over medium-low heat prior to use.

An *essence* is a fish stock similar to a fumet but it uses a greater amount of aromatic ingredients such as celery, morels, a bouquet garni, a sachet, and even fennel root. The flavor of an essence is highly concentrated and extremely aromatic. It can often be used as a sauce itself.

A *court bouillon* (French for "short broth") is a highly flavored and aromatic vegetable broth made from simmering vegetables with herbs and a small amount of an acidic liquid (usually vinegar or wine). Court bouillon is used to gently cook fish, seafood, or vegetables using the poaching method. Court bouillon is also used to produce a contemporary sauce called a nage (pronounced nahj).

## THICKENING METHODS

Stocks are thickened to create sauces. Sauces can add flavor or texture to a dish or can be the binding ingredient that pulls a dish together. Sauces are thickened through reduction or through the use of a thickening agent. Simmering a sauce causes evaporation to occur and can reduce a sauce up to one-fourth of its original volume. This reduction results in concentrated flavors and lessens or eliminates the need for a thickening agent. Thickening agents are used to give body and texture to a sauce. A thickening agent is what turns chicken stock into chicken gravy. Proper use of a thickening agent or reduction yields a sauce that has a smooth, lump-free texture, does not taste starchy, and thinly coats the back of a spoon or spatula.

The most common thickening agents are flour, cornstarch, arrowroot, and a liaison. The first three of these are starches. Starches function as thickening agents because the starch granules absorb moisture when placed in a liquid and heated. *Gelatinization* is the process of a starch absorbing moisture. ***See Figure 16-11.*** Flour is one of the most popular thickeners used in the professional kitchen. Two main thickening products are made from flour: roux and beurre manié.

### Gelatinization

┌ Thickening agent ┌ Thickened liquid

16-11   *A starch can be added to a liquid to act as a thickening agent. The resulting thickened liquid is thick enough to coat the back of a spoon or spatula.*

Whenever using any thickening method, the final sauce must be strained prior to serving. It is possible that some of the thickening agent did not dissolve completely in the sauce. Straining removes any unappetizing lumps from a finished sauce.

## Roux

The most common thickening agent is a roux. A *roux* (pronounced roo) is a cooked mixture of equal amounts by weight of flour and fat that is used to thicken sauces and soups. Pastry flour and cake flour are the most effective types of flour to use in a roux as they both contain higher starch content than bread flour. All-purpose flour (a combination of pastry and bread flour) can also be used.

Cooking flour in fat coats the starch granules in fat. This coating of fat helps prevent the starches from lumping when added to a hot liquid. If flour is added by itself to a liquid such as a hot stock, clumps of flour stick together as the outside surface of the starch goes though the gelatinization process. The result is miniature balls of gooey flour. Once the flour begins to clump, the improperly gelatinized flour pieces will never be distributed or broken down, no matter how much the mixture is whisked. When the roux is made properly, by cooking the flour and fat together first, the fat-coated starch granules dissolve smoothly in the hot stock.

There are three types of roux made in the professional kitchen: white, blonde, and brown. ***See Figure 16-12.*** White roux is made by cooking equal amounts (by weight) of fat and flour until a bubbly, slightly foamy appearance is achieved. It is then immediately removed from the heat and allowed to rest for 5 min to 10 min before it is added to a liquid. White roux is used in white sauces when no color is desired. There should be a faint smell of freshly baked bread.

A blonde roux is made by cooking the same mixture of fat and flour for a somewhat longer time until the roux exhibits a slightly golden or tan color. The roux also develops an aroma similar to freshly popped popcorn. Blonde roux is used in sauces that have a slight or tan color such as those made with fresh chicken or light veal stock. Blonde roux not only adds some of its color to a sauce but also adds a slightly toasted flavor from starches in the flour beginning to caramelize.

A brown roux is made by cooking the same mixture of fat and flour for an even longer period of time until the roux changes to a deep brown color and develops a nutty aroma. Due to its intense flavor and color, a brown roux is used with sauces that are made with a rich veal or beef stock. It is important to use caution when making a brown roux to avoid overcooking it. A roux that is too dark has an unpleasant, bitter taste and will leave black flecks in a prepared dish.

Chef's Tip:

As a roux is cooked more (turning from white to brown), it will lose some of its gelatinizing strength. Therefore, more brown roux is needed than white roux to thicken sauces and soups.

## Roux

| Type | Cooking Time | Characteristics | Uses |
|------|-------------|-----------------|------|
| **White roux** | 2 min<br>5 min to 10 min rest | White, bubbly, slightly foamy<br>Fresh bread smell | White sauces |
| **Blonde roux** | 4 min to 7 min | Golden tan color<br>Slightly nutty popcorn smell | Light-colored sauces, such as chicken-based or veal-based |
| **Brown roux** | 10 min to 12 min | Deep brown color<br>Very nutty smell | Dark sauces, such as beef-based |

16-12 *There are three types of roux used in the professional kitchen—white, blonde, and brown.*

### To make a roux, apply the following procedure:

1. Using a heavy-bottomed saucepan to prevent scorching, heat clarified butter or other desired fat until hot, using caution to not burn the fat. *See Figure 16-13.*

2. Add equal amounts (by weight) of sifted pastry or cake flour to the hot fat.

3. Stir to form a paste-like consistency while cooking over medium heat.

4. Continue stirring while cooking until desired color of roux is achieved.

## Preparing a Roux

1. Heat clarified butter until hot.

2. Add sifted flour.

3. Stir over medium heat.

4. Cook until desired color is achieved.

### Proportions of Roux Ingredients to Liquid

| Sauce Consistency | = | (Flour* | + | Fat | = | Roux) | + | Liquid |
|---|---|---|---|---|---|---|---|---|
| **Light sauce** | | 6 oz | | 6 oz | | 12 oz | | 1 gal. |
| **Medium sauce** | | 8 oz | | 8 oz | | 16 oz | | 1 gal. |
| **Heavy sauce** | | 10 oz | | 10 oz | | 20 oz | | 1 gal. |

**Light sauce** refers to the consistency of most fine and delicate sauces. A light sauce should be nappe, or just thick enough to coat the back of a spoon.

**Medium sauce** Medium sauce refers to the consistency of a cream soup or standard smooth gravy, such as would be served over mashed potatoes.

**Heavy sauce** refers to the consistency of a binding sauce, such as may be used in a turkey pot pie.

\* Remember that the type of flour used will also affect the thickening power of a roux. The above ratios are for use with pastry or cake flour. If using all-purpose flour, more flour may be needed to achieve the same results.

16-13    A roux is made by cooking equal parts of clarified butter and sifted flour over medium heat. A sauce can be made light or heavy, depending on the amount of roux added.

**Chef's Tip:**
Anything thickened with a roux will form a skin on the surface if left to sit in a steam table. To prevent a skin from forming, place a few pats of butter on the surface of the sauce or drizzle a small amount of melted butter over it.

*The following tips are helpful when making a roux:*

- Use heavy-bottomed stainless steel cookware to make a roux, as stainless steel is nonreactive and therefore does not affect taste or color. Also, a heavy-bottomed pot helps prevent the roux from scorching.

- Do not use excessively high heat when making a roux. Roux cooked over medium heat cooks more evenly and has a more consistent color, texture, and flavor than a roux cooked over high heat.

- Do not overthicken a sauce by adding more roux than needed. Use 1 lb of roux to thicken 1 gal. of liquid to achieve a medium consistency.

- After adding a roux, always wait for the liquid to reach a full boil. Roux cannot thicken to its fullest capacity until the liquid reaches a boil and is then allowed to simmer for a few minutes.

*Chef's Tip:*

After adding a roux to a liquid, the sauce will need to be heated for a short time in order to fully incorporate the roux and cook out the flour taste. Most chefs typically recommend simmering a sauce for 15 min to 20 min after adding a roux. Take care to not allow the sauce to scorch.

Adding a roux to a liquid can be difficult. Adding the roux correctly results in a smooth and silky sauce that has a nappe consistency. *Nappe* is the consistency of a liquid that is thick enough to coat the back of a spoon. Adding roux incorrectly results in lumps and a pasty or floury-tasting sauce. When adding roux to a liquid, the roux and liquid must be at different temperatures. If they are both extremely hot or both very cold, the roux may create lumps and not work properly. There are two accepted methods for adding roux, and the appropriate method depends upon the temperature of the roux.

With a hot roux, use cold or room-temperature stock. The roux is made in a pot large enough to incorporate the cold stock. The cold stock is carefully added to the hot roux while whisking vigorously. This eliminates the need to use numerous pots as well.

Many restaurants make large batches of roux and store it for future use as needed. If using a previously made roux that is at room temperature or colder, a hot stock should be used. Add room-temperature roux carefully to the hot stock while whisking vigorously. Whisking vigorously dissolves the roux completely and incorporates it into the stock.

### Beurre Manié

Beurre manié is French for kneaded butter. A *beurre manié* is made with equal amounts by weight of pastry or cake flour and softened butter and is whisked into a sauce just before service. Unlike a roux, the beurre manié is not cooked. Instead, the flour and soft butter are kneaded together until a smooth texture is achieved. The mixture is separated into small balls and whisked into a sauce to finish it. The butter component adds additional sheen and richness to the sauce. The process of kneading butter and flour together acts to coat each granule of starch with fat, which allows the starch to absorb moisture without lumping. ***See Figure 16-14.*** A beurre manié is often used to thicken a small amount of sauce, as when making a pan sauce, while a roux is used more often for larger quantities.

*Preparing a Buerre Manié* _____

1. Add equal parts butter and flour.     2. Mix well.

16-14   *A beurre manié is prepared by kneading together equal amounts of sifted flour and soft whole butter.*

## Cornstarch

Cornstarch is the finely ground, powdery, pure starch derived from corn. Unlike a roux, which is only used to thicken hot foods, cornstarch can be used to thicken hot or cold foods. It has a white color, but turns translucent as it is incorporated into a dish and cooked. It is very popular as a thickening agent in Asian stir-fry dishes where liquid ingredients such as sesame or soy sauce are used as the base ingredient for a sauce or glaze.

Cornstarch has double the thickening power of flour, so less cornstarch is needed to thicken a liquid than when using flour. A sauce thickened with cornstarch is not nearly as stable as a sauce thickened with a well-made roux. Cornstarch-thickened sauces have a tendency to break down if held over heat for long periods of time. Therefore, it is not recommended to reheat a sauce thickened with cornstarch.

Cornstarch must be mixed with a cool liquid prior to being added to a hot liquid. Unlike a roux that is made from hot fat and flour cooked together slowly, cornstarch is made into a slurry to thicken a liquid. A *slurry* is a mixture of equal parts of cool water and cornstarch. Some restaurants use a cornstarch slurry to thicken sauces and gravies because a slurry does not require fat like a roux and is therefore less expensive.

If cornstarch alone is sprinkled into a hot liquid, the starch granules clump together and form lumps. These lumps cannot be whisked or completely dissolved in the sauce and have to be removed by straining. Making a slurry by mixing cornstarch into a cool liquid separates each starch granule and allows the starch to absorb liquid most efficiently. The milky white slurry is added to either a hot or cold liquid with a whisk and heated thoroughly. Once the mixture reaches a boil, the slurry thickens to its full potential. It is not accurate to determine a sauce's consistency until the sauce has reached a boil, as the sauce will continue to thicken until it reaches a boil. After reaching a boil, the sauce needs to be simmered for about 5 min to completely cook out the taste of cornstarch.

## Arrowroot

Arrowroot has a taste and effect similar to cornstarch. It comes from the root structure of tropical plants. Arrowroot is added to a sauce in a slurry, like cornstarch, but a sauce made from arrowroot is clearer than a sauce thickened with cornstarch. Arrowroot is flavorless and more expensive than cornstarch.

## Liaison

A *liaison* is a mixture of egg yolks and heavy cream that is used to thicken sauces. Because a liaison does not contain a starch ingredient, it does not use the process of gelatinization. A liaison is typically used for cream- or milk-based dishes such as custards, puddings, or rich cream sauces. A liaison is composed of three parts heavy cream added to one part egg yolk. The cream and yolk are whisked together. Egg yolks begin to coagulate at temperatures above 145°F, but the addition of cream raises the coagulation point of the liaison to approximately 180°F. The difference in temperature between the liaison and a hot liquid could cause the egg yolks to coagulate. Therefore, a small amount of the hot liquid to be thickened is added to the liaison and whisked in to temper the yolks from an extreme temperature change.

After the liaison has been whisked well, it is added to the hot liquid while whisking vigorously. If the liaison-thickened sauce needs to be heated, the temperature should be kept between 140°F and 185°F. The sauce must be heated to a minimum of 140°F to cook the egg yolks to a temperature where they are safe for consumption. The liaison should not be heated above 185°F as the yolks will cook and coagulate to the point of appearing curdled. If the yolks cook and coagulate, the liaison must not be used or the sauce will also appear curdled.

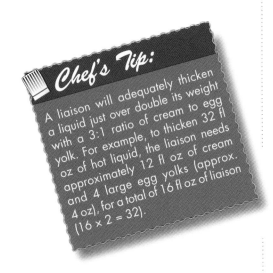

*Chef's Tip:*
A liaison will adequately thicken a liquid just over double its weight with a 3:1 ratio of cream to egg yolk. For example, to thicken 32 fl oz of hot liquid, the liaison needs approximately 12 fl oz of cream and 4 large egg yolks (approx. 4 oz), for a total of 16 fl oz of liaison (16 × 2 = 32).

To add a liaison to a liquid, apply the following procedure:

1. Heat liquid to be thickened to a simmer. Use caution to not overheat the liquid. *See Figure 16-15.*
2. Remove the liquid from the heat.
3. Whisk together one part egg yolk and three parts heavy cream.
4. Add a small amount of the heated liquid to the liaison while whisking rapidly. Continue adding the hot liquid while whisking until the liaison is fairly warm.
5. Add the warmed liaison into the remainder of the heated liquid while whisking vigorously.
6. Carefully heat the thickened sauce to a temperature above 140°F but not over 185°F.

CULINARY PROCEDURES

## Adding a Liaison

1–2. Heat liquid to a simmer. Remove from heat.

3. Whisk egg yolk and heavy cream together.

4. Add a small amount of heated liquid to liaison.

5. Add warmed liaison to remainder of liquid, whisking vigorously.

6. Carefully heat thickened sauce to a temperature between 140°F and 185°F.

Note: Thickened sauce should coat the back of a spoon.

16-15   A liaison does not contain a starch.

## SAUCES

Many chefs consider sauces to be the most important component of a dish. A great sauce can change an ordinary dish into something fantastic. A sauce is meant to complement a dish and should never overpower it. It is similar to icing on a cake. A cake can taste very good by itself, but when icing is added, the cake can be delectable. The same can be said about a great sauce. Any dish can be very good, but add an excellent sauce to a dish and something incredible is created.

With a few exceptions, most sauces are composed of just a few key components: a flavorful liquid, some sort of thickening agent, and additional flavorings and seasonings. Classical sauces are usually separated into two different categories: mother (leading) sauces and small (compound) sauces. In addition to the classical sauces, contemporary sauces, such as salsas, relishes, chutneys, coulis, nages, flavored oils, and foams, continue to emerge.

### The Mother Sauces

A *mother sauce* is one of five sauces from which all classical sauces described by Escoffier are produced. The five mother sauces are béchamel, velouté, espagnole, tomato, and hollandaise. Mother sauces are rarely used by themselves. Instead, they commonly serve as the base for many different small (compound) sauces.

**Béchamel.** *Béchamel* (also known as a cream sauce or a white sauce) is a mother sauce that is made by thickening milk with a white roux and seasonings. The principal ingredient in a béchamel is scalded milk and the thickener is a white roux. An onion piquet (half of an onion studded with cloves and a bay leaf) is added to the milk. ***See Figure 16-16.*** The milk is scalded, the onion piquet is removed, and a white roux is whisked in. The sauce is finished with a little nutmeg and seasoned. As with any sauce, béchamel is strained prior to use in a dish.

*Onion Piquet*

16-16  *An onion piquet is a half of an onion studded with cloves and a bay leaf.*

A well-made béchamel should be silky smooth and have a nappe consistency. It has a very mild flavor with only a subtle taste of onion. A béchamel was traditionally made by adding hot, heavy cream to a thick, reduced white veal stock. Today, the simpler version of béchamel prepared with scalded milk and roux is used in many vegetable and gratin dishes.

With the addition of different seasonings, ingredients, and garnishes, a simple but well-made béchamel can be transformed into countless smaller sauces. Each of the following small (compound) sauces are based on a foundation of béchamel:

- Mornay sauce: made by adding grated cheese to béchamel

- Soubise sauce: made by adding puréed cooked onions to béchamel

**Velouté.** *Velouté* (pronounced veh-loo-TAY) is a mother sauce made from a flavorful light stock (either veal, chicken, or mild fish stock) and a blonde roux. Traditionally, only veal stock was used. Today, chicken stock is almost always used instead of veal stock because of the versatility of chicken stock, and cost. Fish stock is not often used as the flavor may overpower a dish. No matter what type of stock is used, a velouté should have a very smooth and silky texture similar to a béchamel.

A velouté is typically made into a middle (intermediate) sauce before being used for a small sauce. There are two middle sauces, suprême and allemande. A *suprême sauce* is a chicken velouté with the addition of cream. A suprême sauce can be wonderful by itself. It also is used alone as a binding sauce in many casseroles. An *allemande* (pronounced ah-leh-MAHND) *sauce* is made by adding fresh lemon juice and a yolk-and-cream liaison to velouté. The liaison makes allemande a rich, wonderful sauce.

Suprême and allemande sauces can be used to make an almost countless number of small sauces. Because both middle sauces are derived from velouté, almost any small sauce made from one middle sauce can also be made with the other. The following small (compound) sauces are made from velouté and allemande sauces:

- Aurore sauce: made from a velouté with tomato purée

- Bercy sauce: made from a fish velouté

- Poulette sauce: made from an allemande flavored with mushroom essence

**Espagnole.** *Espagnole* (pronounced ess-spah-nyol) is a mother sauce made from a full-bodied brown stock, brown roux, tomato purée, and a hearty caramelized mirepoix. Although espagnole can be used alone, it is typically made into one of two middle sauces, a demi-glace or a jus lié. These two sauces are then broken down into the small (compound) sauces of the espagnole family.

*Chef's Tip:*
A béchamel, like any other thickened sauce, can be made into a lighter or heavier version by adding slightly less or slightly more roux.

A demi-glace is the most widely used middle sauce. Escoffier described demi-glace as an espagnole sauce "taken to the extreme limit of perfection." A *demi-glace* is made by adding equal parts of espagnole and brown stock together and reducing the mixture by half. This richly flavored reduced sauce is often finished with a small amount of Madeira or sherry to further boost its flavor. A properly made demi-glace should be a nappe consistency when hot and somewhat rubbery when cold.

A *jus lié* (also called a fond lié) is made from equal parts of espagnole and brown stock. The main difference between jus lié and demi-glace is that roux is omitted from the espagnole. The jus lié is thickened either by adding a cornstarch or arrowroot slurry or simply by a slow reduction. Some chefs prefer a jus lié to a demi-glace as it does not have the taste of roux or flour and is easier to make. A disadvantage of a jus lié is that it will only be as rich and flavorful as the brown stock from which it was produced because it is thickened with only a cornstarch or arrowroot slurry. In comparison, an espagnole or demi-glace has a richer taste because of the butter used in the roux of the espagnole. When completed, a jus lié is not quite as thick as a demi-glace but should still be able to coat foods nicely.

An espagnole sauce is converted into either a demi-glace or a jus lié before being made into a small (compound) sauce. The following is a list of some popular espagnole (compound) sauces.

- Bordelaise sauce: made from a demi-glace flavored with red wine, herbs, and shallots
- Chateaubriand sauce: made from a demi-glace flavored with white wine, shallots, tarragon, cayenne, and lemon juice
- Madeira sauce: made from a demi-glace flavored with shallots, Madeira wine, and butter
- Périgueux sauce: made from a demi-glace flavored with Madeira wine and truffles
- Robert sauce: made from a demi-glace flavored with onions, white wine, vinegar, and Dijon mustard

**Tomato Sauce.** A *tomato sauce* is a mother sauce made by sautéing mirepoix and tomatoes, adding white stock, and thickening with a roux. A well-made tomato sauce is thick, flavorful, and tastes like ripe tomatoes. It is coarser than the other classical sauces, but should have similarly rich characteristics that allow it to cling to the foods it is served with. Typically, plum (Roma) tomatoes are used to make tomato sauce, as they have less water, fewer seeds, and much more meat than other varieties. Tomatoes are most often peeled and seeded before use in the sauce.

Some chefs do not use a roux to thicken a tomato sauce. Instead, tomato sauces are typically either puréed or ground in a food mill, resulting in a hearty sauce. A frequent problem with modern sauces, however, is that water can separate from the puréed pulp and form an unattractive pool beneath pasta on a prepared plate. This can be avoided by using a thickening agent such as a roux or a slurry, which helps by causing any water that separates from the tomato pulp to bind with the starch and thicken. A properly thickened sauce makes an attractive presentation on a pasta dish, such as spaghetti. ***See Figure 16-17.***

There may be as many tomato sauce recipes as there are chefs who make tomato sauce. Some chefs rely heavily on the beautiful flavor of seasonal, ripe tomatoes cooked for a relatively short period of time to produce an amazingly rich and fresh sauce. Others use canned tomato products sautéed with mirepoix, roasted pork or veal bones, and rendered bacon, cooked for a long time over low heat. Some tomato sauces are chunky, while others are puréed smooth.

Many chefs also add a small amount of sugar or gastrique to a tomato sauce to mask the acidic qualities of the sauce. *Gastrique* is a sugar syrup made by caramelizing a small amount of granulated sugar in a saucepan and deglazing the pan with a small amount of vinegar. The sugar or gastrique does not reduce the amount of acid in tomato sauce, but merely covers up or softens the harshness of the acidic qualities. To reduce acid in the sauce, baking soda can be added while the sauce simmers. Tomato sauce immediately foams when baking soda is added, because of chemical reactions taking place as the soda neutralizes some of the acid. The chemical process is similar to the process of an antacid reducing the amount of acid in a person's stomach. When the foaminess from the baking soda disappears, the acid in the sauce has been reduced.

Tomato sauce can be made into many small (compound) sauces, the majority of which are called tomato sauce. Many tomato sauce variations simply have a regional, cultural, or ethnic description attached to identify the sauce as a different sauce in the tomato family. Tomato sauce variations include the following:

- Creole
- Marinara
- Mexican
- Milanaise
- Portuguese
- Spanish
- Vegetarian

**Hollandaise.** *Hollandaise* is a mother sauce that is thickened with egg yolks. Hollandaise is different from other sauces thickened with yolks because hollandaise relies on emulsification rather than coagulation. Egg yolks contain a natural emulsifier called *lecithin*. The emulsification occurs when warm, clarified butter is rapidly whisked into heat-tempered egg yolks. The lecithin in the yolk emulsifies (coats) each droplet of fat (butter), suspending the fat in the mixture. When the sauce is made

*Tomato Sauce*

**16-17** *Using a roux results in a hearty tomato sauce.*

properly, each yolk should completely emulsify approximately 4 oz of clarified butter. The hollandaise sauce is finished with a small amount of water and lemon juice or vinegar. The resulting hollandaise should be light and silky, a very pale lemon color, and should have a well-rounded flavor, more complex than that of just butter.

A classical hollandaise is started with a flavorful reduction of dry white wine, white wine vinegar, minced shallots, and freshly cracked peppercorns, reduced over medium heat. A quicker version made in many restaurants omits this step and simply seasons the sauce at the end to taste.

**To make a hollandaise sauce using the classical method, apply the following procedure:**

1. Add dry white wine, white wine vinegar, minced shallots, and freshly cracked peppercorns to a saucepan. Reduce over medium until almost all of the liquid has evaporated. ***See Figure 16-18.***

2. Strain and slightly cool the reduction. Transfer the reduction to a stainless steel bowl.

3. Add egg yolks to the reduction, whisking vigorously until light and foamy.

4. Add a small amount of warm water (2 oz per every 6 yolks) and whisk to incorporate.

5. Place the mixture over a gently simmering water bath and whisk rapidly until the mixture reaches approximately 145°F (at this temperature, the mixture usually begins to emit a small amount of steam) and has tripled in volume. *Note:* The bowl may be momentarily removed from the water bath to allow a few seconds of more thorough whisking without overcooking the yolks.

6. Remove the bowl from the water bath, place bowl on a rolled up, moistened towel, and begin gradually whisking in warm clarified butter very slowly and in a fine, steady stream. As the clarified butter is incorporated, the mixture begins to thicken. *Note:* If the sauce begins to get too thick, add a small amount (1 tsp to 2 tsp) of warm water to thin the sauce slightly. Continue whisking in the clarified butter until about 4 oz of clarified butter per yolk has been added.

7. Add a small amount of lemon juice, salt, white pepper, cayenne pepper, and Worcestershire sauce to finish.

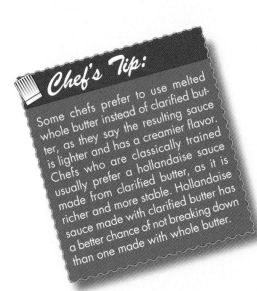

*Chef's Tip:* Some chefs prefer to use melted whole butter instead of clarified butter, as they say the resulting sauce is lighter and has a creamier flavor. Chefs who are classically trained usually prefer a hollandaise sauce made from clarified butter, as it is richer and more stable. Hollandaise sauce made with clarified butter has a better chance of not breaking down than one made with whole butter.

Hollandaise should be held for no more than 2 hr at 145°F. If the sauce is held above 150°F, the emulsification breaks down, as egg yolks curdle and solidify at higher temperatures. If the sauce is held at 100°F or below, the sauce thickens and could promote bacterial growth. The following are small (compound) sauces that can be made from hollandaise:

- Béarnaise sauce: made with a reduction of vinegar, wine, tarragon, and shallots and finished with egg yolks and butter
- Choron sauce: hollandaise blended with tomato purée
- Mousseline sauce: hollandaise blended with unsweetened heavy creme
- Maltaise sauce: hollandaise blended with Maltese orange juice and grated orange rind

## Preparing Hollandaise

*1–2. Reduce white wine, vinegar, shallots, and pepper over medium heat; strain and cool the reduction.*

*3. In a stainless steel bowl, combine reduction and egg yolks, whisking vigorously.*

*4–5. Add a small amount of water; whisk rapidly over a hot water bath.*

*6. Remove from heat and whisk in warm clarified butter.*

*7. Add lemon juice, salt, pepper, cayenne pepper, and Worchestershire to finish.*

*Serve immediately.*

**16-18** *Hollandaise is a mother sauce made from a white wine reduction, egg yolks, clarified butter, lemon juice, and seasonings.*

When making a hollandaise for the first time, the sauce may break and appear thin, curdled, or separated. This most often happens either because the yolks are heated at too high of a temperature, resulting in overcooked yolks, or because butter is added too quickly and does not emulsify properly. Sometimes if a hollandaise is kept too cold it becomes thick but does not break until it is served over a warm menu item, such as broiled asparagus. To repair a broken hollandaise, the first step is to determine whether it is broken because of overcooked yolks or from holding a completed sauce at too low a temperature.

If the hollandaise is broken because of overcooked yolks, add 1 yolk to a clean mixing bowl and whisk with 1 tsp of warm water until thick. Slowly add the broken hollandaise to the fresh egg yolk while whisking rapidly to incorporate. Continue adding broken sauce to the mixture until all of the broken hollandaise has been incorporated.

If the hollandaise is broken from being held at too low a temperature, reheat the sauce over a double boiler while whisking continuously until the sauce is soft and warm. In a separate stainless steel bowl, add 2 tsp of warm water and a few tablespoons of the broken sauce. Whisk rapidly to emulsify and then slowly whisk in the remainder of the broken hollandaise sauce until all of the sauce has been added.

## Butter Sauces

The butter sauce family includes three major types of sauces: beurre blanc, compound butter, and broken butter. In a beurre blanc sauce, cold softened butter is whisked into a flavorful wine reduction and finished with a slight bit of cream. Compound butter sauces are made by mixing cold, softened butter with flavorings, such as fresh herbs and garlic. These flavored butters are chilled and then sliced to order. A broken butter sauce is prepared by slowly whisking whole butter until the fat, milk solids, and water separate. Broken butters are finished with a small amount of an acid, such as lemon juice, and seasoning.

**Beurre Blanc.** *Beurre* is the French word for butter. Sauces in the beurre blanc sauce family are classical sauces that originated in the Brittany region of France from a sauce produced as part of the fish-poaching process. The three main components of a beurre blanc sauce are shallots, wine, and whole butter. The butter becomes the sauce, while the shallots and wine are the flavoring ingredients. Beurre blanc sauces are lighter and thinner than hollandaise but should be slightly thicker than heavy cream.

In the traditional preparation, after a fish is poached, a small amount of the court bouillon (poaching liquid) is reduced with shallots, white wine, and a small amount of white wine vinegar. The reduction is removed from the stove and small pieces of whole butter are whisked into the reduction. Trace amounts of lecithin and other emulsifiers naturally found in butter help to create an emulsification. The key to adding butter to the reduction is constant whisking.

**To make a beurre blanc sauce, apply the following procedure:**

1. In a small saucepan, combine dry white wine, white wine vinegar, shallots, and a few whole peppercorns. ***See Figure 16-19.***

2. Over medium heat, reduce the mixture until only a few tablespoons of liquid remain. *Note:* If too much of the reduction remains, the resulting sauce will be too thin.

3. Add tablespoon-size pieces of cold whole butter over a very low heat while whisking constantly. *Note:* Cold butter is whisked into the mixture to keep the sauce between 100°F and 120°F. At temperatures above 120°F, the emulsification can break down. At temperatures below 100°F, the sauce becomes too thick.

4. Strain the sauce through cheesecloth or a chinois.

5. If desired, finish with a small amount of heated heavy cream and season to taste. *Note:* Heated cream is an optional addition that adds some stability and richness to the sauce.

CULINARY PROCEDURES

*Preparing Beurre Blanc*

1–2. Combine white wine, vinegar, shallots, and pepper over medium heat; reduce to a few tablespoons.

3. Whisk in cold butter.

4–5. Strain through cheesecloth or a chinois; finish with a small amount of warm heavy cream if desired.

**16-19** *The three main components of a beurre blanc sauce are shallots, wine, and whole butter.*

Other sauces in the beurre blanc family include the following:

- Beurre rouge: Dry red wine and red wine vinegar are substituted for the white wine and white wine vinegar.

- Beurre nantais: Muscadet wine is substituted for the dry white wine.

- Beurre citron: The juice of one lemon per pound of butter is substituted for the wine and vinegar. A small amount of lemon zest is added to finish the sauce.

- Dill beurre blanc: Fresh dill is added to the beurre blanc just prior to service. This sauce is often finished with a touch of lemon juice or zest.

- Peppercorn beurre blanc: The strained beurre blanc is finished with cracked pink, green, or black peppercorns.

**Compound Butters.** A *compound butter* is a flavorful butter made by working flavoring ingredients such as fresh herbs, vegetable purées, dried fruits, preserves, or wine reductions into whole butter. Compound butters can be served alone or as a sauce over grilled meat, fish, or vegetable dishes. They can also be added to other classic sauces for additional flavor. Some herbs, such as tarragon, turn black quickly when chopped; however, these delicate herbs can be incorporated into whole butter without a change in color.

Typically, compound butters are mixed well, rolled into a cylinder shape on a sheet of plastic wrap or parchment paper, and refrigerated until needed. When needed, the cylinder is sliced and the butter is placed on top of a freshly cooked item. The dish is served immediately so that the butter finishes melting in front of the customer. ***See Figure 16-20.*** These sauces are used often in restaurants because they can be prepared in advance and held in the refrigerator for a few days, or in the freezer for longer periods, before the flavors begin to deteriorate.

## Compound Butters

16-20 *A compound butter is chilled, sliced, and served immediately after being placed.*

## To make a compound butter, apply the following procedure:

1. In a food processor, combine and process the flavoring ingredients.
2. Add whole butter, salt, and pepper, and purée until all ingredients are well incorporated.
3. Remove butter and place on a sheet of parchment paper or plastic wrap. Roll the mixture into a cylinder shape, approximately 1″ to 1½″ thick.
4. Place in refrigerator until needed, or freeze for later use.
5. When needed, slice off a disk of compound butter about ¼″ to ½″ thick and place on an item just prior to serving.

Although the variations of compound butters are virtually endless, the following are some common preparations:

- Almond butter: Finely grind 4 oz almonds in a food processor until very smooth. Add 8 oz whole butter and mix well.

- Anchovy butter: Soak 1 oz anchovies (approximately 8 to 10 anchovies) in cold water for 15 min to remove excess oil and salt. Pat anchovies on paper towels to dry, and place in processor with 8 oz whole butter. Purée until smooth. *Note:* A small amount of anchovies is usually sufficient for flavor.

- Bercy butter: Reduce 1 cup of white wine and 2 minced shallots until only about 2 tbsp of liquid remain. Cool and add 8 oz whole butter and 2 tbsp freshly diced parsley. Purée until smooth.

- Colbert butter: Place 1 tbsp chopped fresh tarragon, 3 tbsp glace de viande, 3 tbsp chopped fresh parsley, and 1 tsp lemon juice in a food processor and purée until smooth.

- Herb butter: Add 2 tbsp to 4 tbsp of desired chopped fresh herb to 8 oz of whole butter and purée until smooth.

- Lobster, shrimp, or crayfish butter: In a meat grinder, grind 6 oz of lobster, shrimp, or crayfish shells with 8 oz of whole butter. Place in small saucepan and melt butter over low heat. Keep on very low heat for 30 min and strain. Chill the melted butter until solid. Season as desired, and purée until smooth.

- Maître d'hôtel butter: Place 3 tbsp of freshly chopped parsley, 1 tsp of fresh lemon juice, and 8 oz of whole butter in a food processor and purée until smooth. This is the most well-known and most commonly used compound butter.

- Red wine (Marchand de Vin) butter: Combine 1 cup of dry red wine, 1 tbsp glace de viande, 1 minced shallot, 1 clove garlic, and 6 cracked black peppercorns. Reduce until to about 2 tbsp of liquid remain and chill. Blend in a food processor with 8 oz whole butter. Red wine butter is wonderful to serve on red meat.

**Broken Butters.** Broken butter sauces are easy to prepare and are intended to be made à la minute (French for "in the minute") as a pan sauce. This means that the sauce is made reusing the pan that an item was cooked in. By reusing the pan, some of the flavors from the dish are incorporated into the sauce. To make a broken butter sauce, whole butter is heated until the fat, milk solids, and water separate or "break." The broken butter mixture is then flavored with an acid, such as lemon juice or wine vinegar. The two most common broken butter sauces are beurre noisette and beurre noir.

Beurre noisette is French for "hazelnut butter." This butter sauce is commonly referred to as "nut-brown butter" because of its color and wonderful, nutty aroma. To prepare a beurre noisette, whole butter is cooked over medium heat until the white milk solids begin to turn brown. As the solids change color, they add a toasted nut flavor to the butter. The butter is then allowed to cool slightly and fresh lemon juice is added. The juice from one lemon is ample for a cup of butter.

*Chef's Tip:*

A classic dish served with a broken butter is Dover sole à la Munière. In this dish, sole fillets are dredged in seasoned flour and lightly sautéed. The fish is removed from the pan (any brown specks of flour should also be removed from the pan at this time) and whole butter is added. The butter is cooked until the milk solids begin to brown and lemon is squeezed into the butter. The fillets are then returned to the sauté pan and coated with the sauce.

Beurre noir is French for "black butter." Beurre noir is prepared in a manner similar to preparation of beurre noisette. The main differences between the two preparations are the cooking time and the acid used. In a beurre noir, the butter is cooked a bit longer, until the milk solids turn a rich, dark brown, and the lemon juice is replaced with wine vinegar. Chefs use many different kinds of vinegar; however, sherry vinegar, champagne vinegar, and balsamic vinegar seem to provide the best overall flavor.

Almost any flavoring ingredient can be added to a broken butter sauce to change the flavor profile. For example, chopped herbs, julienne of leeks, crushed garlic, or sliced truffles can each be added to give a broken butter sauce an exciting flavor.

## Contemporary Sauces

Sauces from around the world are gaining popularity due to increasing emphasis on the benefits of healthy cuisine and a growing influence from diverse cultures. Although the mother sauces are still popular, many contemporary sauces are lighter, have less fat, and are gaining wide acceptance. Today's chefs do not have to stay within the bounds of traditional sauces such as béchamel or espagnole. Sauces such as salsas, relishes, chutneys, coulis, nages, flavored oils, and foams are appearing on menus around the globe.

**Salsas, Relishes, and Chutneys.** Some of the best-known sauces are more often thought of as condiments. Whether eating salsa at a Mexican restaurant or bright green pickle relish on a Chicago-style hot dog, chances are most people have had some experience with at least two of these contemporary sauces. Salsa (the Spanish word for "sauce") has become one of the most popular modern sauces. Salsas and relishes are cold, diced vegetables and/or fruits, herbs, and spices, mixed together. Salsas are considered a healthier sauce due to the low fat content and high nutrient values of the fresh ingredients. Chutneys are similar to salsas and relishes except that the ingredients in chutney are cooked with sugar and spices to yield a sweet-and-sour flavor. The texture of chutney is often similar to that of a pie filling. Salsas, relishes, and chutneys can have textures that range from almost puréed to very coarse.

**Coulis.** Coulis sauces are among the oldest contemporary sauces. A *coulis* is a sauce typically made from either raw or cooked puréed fruits or vegetables. A coulis can be served either warm or cold. Vegetable coulis is typically served as a sauce over grilled or sautéed items, while fruit coulis is commonly served with desserts.

A properly made coulis should have the consistency of a fresh tomato sauce and a texture that can range from fairly smooth to

slightly coarse. A fruit coulis is usually made from frozen fruit that is puréed with a bit of simple syrup and a hint of lemon or vanilla.

**Nages.** A *nage* is a very aromatic, concentrated, and slightly reduced broth, often court bouillon, used as a finished sauce. A broth used as a sauce is also commonly called an au jus or an essence on a menu. Nage is becoming popular in contemporary cuisine for its light yet flavorful nature. A nage is prepared by reserving and straining the liquid in which the main item was cooked. Vegetables and aromatics are added to the broth and it is reduced by up to half of the original volume. The resulting liquid is a very aromatic broth that is often served to accompany fish. Some chefs choose to clarify the broth, like when making a consommé. A chef may even whisk in whole softened butter or a bit of heavy cream to add richness to a nage. To add a bit of moisture to a dish, only a small amount of a nage is needed. A small amount of broth can be considered a delicate sauce, while a greater amount would be thought of as a soup.

**Flavored Oils.** Flavored oils have become one of the most popular contemporary sauces. Although more commonly used as a flavorful garnish on a plate to add a hint of contrasting flavor and a bold color, oils flavored with herbs, spices, and vegetables are used everywhere. Typically a neutral-flavored oil such as canola or grape seed oil is used, although olive oil can be used if the flavor pairs well with the flavoring ingredients.

Blanching the flavoring items in salt water gives them a more vibrant color, resulting in a more colorful oil. Flavored oils can be used to add flavor, moisture, color, and aroma to a dish. Since flavored oil is generally used only in small amounts, only a small amount of fat and calories are added to the finished dishes.

## To make a flavored oil, apply the following procedure:

1. In salted water, quickly blanch any herbs or vegetables to be used to retain their color. Spices are not blanched.

2. Remove flavoring ingredients from the hot water and immediately shock in ice water to stop the cooking process.

3. After the items are cool, remove them from the ice water and squeeze out any excess water.

4. Place all herbs, spices, and vegetables in a blender with the desired oil and purée completely, until no portions of flavoring ingredients remain as visible solids.

5. Strain the oil through a chinois and reserve until needed.

**Foams.** Foams are the newest addition to the sauce family. Foams are produced by making a reduction of a flavoring ingredient, shallots, garlic, and wine, and puréeing the result. The reduced sauce has a very light texture. A touch of cream can be added if desired, and the sauce is then whipped with a stick mixer until light and foamy. The resulting foam is spooned over an item or drizzled on a plate.

## SUMMARY

Sauces and the stocks from which they are made are the foundation of cuisine. Flavorful and intense stocks are produced from bones, water, and a small amount of vegetables. Some of the most famous classical chefs, such as Escoffier and Carême, spent a great deal of time writing about and constructing the many different flavors and styles of sauces that are still used in modern cuisine. Whether it is a classical mother sauce or a contemporary salsa, the major purpose of a sauce is to complement the food it is served with. A sauce should never be used to hide a poor or lesser-quality food. With practice, the right quality ingredients, and patience, a chef can produce amazing stocks and sauces to complement prepared foods.

Refer to the CD-ROM for Quick Quiz® questions related to chapter content.

## Review Questions

1. Identify the specific components of a stock and their respective proportions.

2. Describe the purpose of adding cold water to bones at the beginning stages of stock preparation.

3. Contrast the two most common methods for cooling stocks.

4. Describe the eight-step procedure for making a white stock.

5. List the main ingredients of brown stock.

6. Describe two different techniques for storing glace.

7. What are the main differences between a fumet and an essence?

8. Identify the four most common thickening agents.

9. Contrast the different uses for white, blonde, and brown roux.

10. Describe the four-step procedure for making a roux.

11. Distinguish a beurre manié from a roux.

12. Describe the six-step procedure for adding a liaison to a liquid.

13. List the five main types of mother sauces.

14. Describe the main ingredients of an espagnole sauce and the two middle sauces in which it is used.

15. Describe the seven-step procedure for making a hollandaise sauce.

16. List the three main components of a beurre blanc sauce.

17. Demonstrate the five-step procedure for making a compound butter.

18. Describe common ways to serve coulis.

19. Contrast the main features of salsas, relishes, and chutneys.

20. Describe how foams are prepared.

# Recipes

## Béchamel

*yield:* 1 qt (eight 4-oz servings)

| | |
|---|---|
| 1 | onion piquet |
| 1 qt | milk |
| 4 oz | pastry or cake flour |
| 4 oz | clarified butter |
| pinch | nutmeg |
| to taste | salt and white pepper |

1. Add milk and onion piquet to a heavy-bottomed stainless steel saucepan; scald milk.
2. In a separate pan, heat clarified butter and flour together to make a white roux, being careful not overcook or color the roux. Allow the roux to cool slightly.
3. Remove onion piquet from the scalded milk. Add white roux to the milk with a whisk.
4. Bring sauce to a boil; lower heat and simmer the sauce for at least 30 min, stirring occasionally to prevent scorching.
5. Season with salt, white pepper, and a pinch of nutmeg.
6. Strain sauce through fine cheesecloth or a chinois to remove any small lumps.

### Nutrition Facts

**Per serving:** 229.3 calories; 59% calories from fat; 15.6 g total fat; 42.7 mg cholesterol; 87.2 mg sodium; 202.8 mg potassium; 17.3 g carbohydrates; 0.3 g fiber; 6.8 g sugar; 17 g net carbs; 5.3 g protein

## Recipe Variations:

**86** **Cream Sauce:** Add 8 oz of scalded heavy cream and a few drops of fresh lemon juice to béchamel. Season to taste.

**87** **Caper Sauce:** Sauté 2 oz to 3 oz of diced shallots with 2 tbsp of capers. Deglaze the pan with 3 oz white wine and reduce by half. Add 4 oz heavy cream, bring to a boil, and add béchamel. Season to taste.

**88** **Mornay:** Add 4 oz grated Gruyère cheese and 2 oz grated Parmesan cheese to béchamel. Thin with a little scalded cream if necessary.

**89** **Cheddar Cheese Sauce:** Add 8 oz grated cheddar or American cheese, 1 tsp Worcestershire sauce, and 1 tsp dry mustard to béchamel.

**90** **Creamy Dijon Sauce:** Sauté 2 oz to 3 oz diced shallots. Deglaze pan with 3 oz white wine, and reduce wine by half. Add 2 oz Dijon mustard and béchamel. Season to taste.

**91** **Soubise:** Sweat 1 lb of finely diced onions in 2 tbsp butter until translucent. Add béchamel and simmer until onions are tender. Strain through a chinois. Season with salt and white pepper to taste.

**92** **Tomato Soubise:** Sweat 1 lb of finely diced onions in 2 tbsp butter until translucent. Add 4 oz tomato paste to sweated onions, cook for a few minutes, and then add béchamel. Simmer until onions are tender. Strain through a chinois. Season with salt and white pepper to taste.

## 93

## Velouté

*yield: 1 qt (eight 4-oz servings)*

*Daniel NYC*

| 4 oz | clarified butter |
| 4 oz | pastry or cake flour |
| 1 qt | light stock |
| to taste | salt and white pepper |

1. In a heavy-bottomed stainless steel saucepan, add clarified butter and flour and cook to make a blonde roux. Remove from heat and allow roux to rest for 10 min.
2. Return roux to medium heat and slowly add light stock to roux while whisking.
3. Bring sauce to a boil; lower heat and simmer the sauce for at least 30 min, stirring occasionally to prevent scorching.
4. Season with salt and white pepper.
5. Strain velouté through fine cheesecloth or a chinois to remove any small lumps.

### Nutrition Facts

**Per serving:** 196.2 calories; 58% calories from fat; 13.1 g total fat; 34.1 mg cholesterol; 209.8 mg sodium; 144.3 mg potassium; 15.3 g carbohydrates; 0.2 g fiber; 1.9 g sugar; 15.1 g net carbs; 4.3 g protein

## Suprême Sauce

*yield: 1 qt (eight 4-oz servings)*

| 1 qt | chicken, veal, or fish velouté |
| 1 c | heavy cream |
| to taste | salt and white pepper |

1. In a heavy-bottomed stainless steel saucepan, add velouté and simmer until slightly reduced.
2. Slowly whisk in heavy cream until well incorporated.
3. Season with salt and white pepper.
4. Strain sauce through fine cheesecloth or a chinois to remove any small lumps.

### Nutrition Facts

**Per serving:** 76.1 calories; 82% calories from fat; 7.2 g total fat; 24.7 mg cholesterol; 68.2 mg sodium; 29.2 mg potassium; 2.3 g carbohydrates; 0 g fiber; 0.3 g sugar; 2.3 g net carbs; 0.8 g protein

# Recipes

## Allemande

*yield: 1 qt (eight 4-oz servings)*

| | |
|---|---|
| 1 qt | veal or chicken velouté |
| 2 | egg yolks |
| 6 oz | heavy cream |
| 1½ tsp | lemon juice |
| to taste | salt and white pepper |

1. In a heavy-bottomed saucepan, add velouté and simmer.
2. In a separate stainless steel mixing bowl, add yolks and heavy cream. Whisk to blend thoroughly.
3. Temper the yolks by adding a small amount of the hot velouté to the liaison mixture while whisking vigorously to prevent yolks from coagulating.
4. Pour the tempered liaison into the hot velouté slowly while whisking continuously to prevent coagulation of the yolks.
5. Bring the sauce to a simmer and remove from the heat (do not heat over 180°F).
6. Add lemon juice and season with salt and white pepper to taste.
7. Strain the sauce through a fine cheesecloth or chinois.

### Nutrition Facts

**Per serving:** 115.4 calories; 84% calories from fat; 11 g total fat; 87.3 mg cholesterol; 73.1 mg sodium; 40.6 mg potassium; 2.8 g carbohydrates; 0 g fiber; 0.3 g sugar; 2.7 g net carbs; 1.7 g protein

### Recipe Variations:

**96** **Aurore:** Gently sweat 6 oz tomato paste in a small amount of clarified butter and add 1 qt suprême or allemande sauce.

**97** **Bercy:** Sweat 2 oz to 3 oz minced shallots in clarified butter. Deglaze with ½ cup white wine and reduce by half. Add 1 qt velouté, suprême, or allemande sauce. Finish with 3 tbsp whole butter, 1 tbsp chopped fresh parsley, and 1 tbsp lemon juice. *Note:* In classical preparations, bercy is made with a fish velouté.

**98** **Poulette:** Dice 1 cup of wild or button mushrooms and simmer in 1 cup of water. Reduce mushrooms and cooking liquid to ¼ cup. Add mushrooms and cooking liquid to prepared suprême or allemande sauce and simmer. Finish sauce by whisking in 1 tbsp lemon juice, 2 tbsp chopped parsley, and 2 tbsp whole butter.

**99** **Dugléré:** Sauté ¾ cup diced wild mushrooms and ¼ cup diced tomatoes in clarified butter and deglaze with 3 oz white wine. Reduce wine by half and add suprême sauce. Whisk together, finish with fresh chopped parsley, and season to taste.

**100** **Curry:** Sauté 2 oz to 3 oz brunoise diced onions in clarified butter with 1 tbsp minced garlic. Add 2 tbsp curry powder and ½ tsp fresh minced thyme, and sauté 1 min more. Deglaze with 3 oz white wine and reduce by half. Add this reduction to a velouté and continue with the basic preparation of a suprême or allemande sauce. Finish with 1 tbsp chopped parsley and season to taste.

## Espagnole (Brown Sauce)

*yield: 1 qt (eight 4-oz servings)*

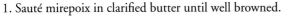

| ½ lb | mirepoix |
| 2 oz | clarified butter |
| 2 oz | flour |
| 1½ qt | brown stock |
| 2 oz | tomato purée |

*Sachet d'épices*

| ½ leaf | bay leaf |
| ⅛ tsp | thyme |
| 2 sprigs | parsley |
| 2 | peppercorns, cracked |

1. Sauté mirepoix in clarified butter until well browned.
2. Add flour and heat to make a brown roux.
3. Carefully add the hot brown stock and the tomato purée, using a whisk to incorporate and dissolve the roux.
4. Add the sachet, bring to a boil, and lower temperature to a simmer.
5. Move saucepot over to one side of the burner, forcing the sauce to simmer only on one side of the pot. *Note:* This causes any impurities floating to the surface to be forced to one side, making them much easier to skim.
6. Simmer for 1½ hr until reduced to approximately 1 qt.
7. Strain through a chinois or china cap lined with layers of cheesecloth.
8. Season to taste.

### Nutrition Facts
**Per serving:** 114.1 calories; 46% calories from fat; 6.1 g total fat; 15.4 mg cholesterol; 378.8 mg sodium; 462.6 mg potassium; 10.8 g carbohydrates; 1 g fiber; 2.6 g sugar; 9.8 g net carbs; 4.8 g protein

## Jus Lié

*yield: 1 qt (eight 4-oz servings)*

| 1 qt | brown veal stock |
| ¼ oz | arrowroot or cornstarch |

1. Bring brown veal stock to a simmer.
2. Dilute arrowroot or cornstarch with cool, clean water or light stock to make a starch slurry.
3. Whisk starch slurry into stock and return stock to a simmer.
4. Simmer for 15 min, until sauce is completely thickened and starch flavor has disappeared.
5. Strain with a chinois or cheesecloth-lined china cap and cool and store properly.

### Nutrition Facts
**Per serving:** 16.2 calories; 5% calories from fat; 0.1 g total fat; 0 mg cholesterol; 237.8 mg sodium; 226 mg potassium; 1.6 g carbohydrates; 0 g fiber; 0.6 g sugar; 1.5 g net carbs; 2.4 g protein

## Demi-Glace

*yield: 1 qt (eight 4-oz servings)*

103

1 qt      brown stock
1 qt      espagnole sauce

1. Combine brown stock and espagnole sauce in saucepot over medium heat, remembering to offset the pot to allow simmering on one side of the pot only.
2. Continue to simmer and skim as needed until sauce is reduced by half.
3. Strain with a chinois or cheesecloth-lined china cap and use or cool and store properly.

**Nutrition Facts**

**Per serving:** 128.2 calories; 42% calories from fat; 6.1 g total fat; 15.4 mg cholesterol; 607.9 mg sodium; 656.9 mg potassium; 12 g carbohydrates; 0.8 g fiber; 3 g sugar; 11.1 g net carbs; 7.1 g protein

**Recipe Variations:**

**104  Bordelaise:** In a medium saucepan over medium heat, combine 12 oz dry red wine, 2 oz minced shallots, 1 sprig of fresh thyme, 1 bay leaf, and ¼ tsp ground peppercorns. Reduce by three-quarters and add 1 qt demi-glace. Simmer for 15 min to 20 min, strain through a chinois, and finish by whisking in 2 oz whole butter. *Note:* A classical bordelaise is served with a slice of poached beef marrow.

**105  Madeira:** Prepare a bordelaise, substituting 6 oz Madeira for the red wine.

**106  Périgourdine:** Prepare a Madeira sauce and add sliced truffles.

**107  Périgueux:** Prepare a Madeira sauce and add finely diced truffles.

**108  Chasseur (Hunter's Sauce):** In whole butter, sauté 6 oz sliced wild or button mushrooms and 2 oz finely diced shallots. Add 1 cup white wine and reduce by three-quarters. Add 1 qt demi-glace and 6 oz diced tomatoes and simmer for 15 min more. Add 1 tbsp chopped fresh parsley. Do not strain the sauce.

**109  Champignon (Mushroom Sauce):** Sauté 8 oz mushrooms in a small amount of butter until browned. Add 2 oz finely diced shallots and cook 1 min. Deglaze the pan with 6 oz white wine and reduce by three-quarters. Add 1 qt demi-glace and simmer for 15 min. Finish by whisking in ½ tsp lemon juice, 1 tbsp sherry, and 2 oz whole butter.

**110  Chateaubriand:** In a medium saucepan over medium heat, combine 12 oz dry white wine and 2 oz minced shallots and reduce by three-quarters. Add 1 qt demi-glace. Simmer and reduce slightly for 30 min. Add 2 tsp fresh lemon juice, ¼ tsp cayenne pepper, and 1 tbsp fresh chopped tarragon. Finish by whisking in 2 oz whole butter.

**111  Marchand de Vin (Wine Merchant Sauce):** Reduce 8 oz dry red wine and 2 oz diced shallots by two-thirds. Add 1 qt demi-glace and strain.

**112  Piquant:** Combine 2 oz diced shallots, 4 oz white wine, and 4 oz white wine vinegar and reduce by two-thirds. Add 1 qt demi-glace and simmer for 15 min. Add 2 oz diced cornichons, 1 tbsp fresh parsley, and 2 tsp fresh tarragon.

**113  Robert:** Sauté 6 oz diced onion in whole butter, taking care not to brown the onions. Add 1 cup of dry white wine and reduce by two-thirds. Add 1 qt demi-glace and simmer for 15 min. Strain and add ½ tbsp sugar dissolved in 1 tsp lemon juice and 2 tsp dry mustard.

**114  Roquefort:** Sauté 2 oz minced shallots in whole butter until slightly golden brown. Add 1 tsp minced garlic and cook 1 min more. Deglaze the pan with 1 cup dry white wine and reduce by three-quarters. Next add 3 oz to 4 oz crumbled Roquefort cheese and whisk in 1 qt demi-glace.

**115  Stroganoff:** Sauté 3 oz minced shallots and 6 oz sliced mushrooms in whole butter until slightly golden brown. Deglaze the pan with 8 oz dry white wine and reduce by three-quarters. Add 1 qt demi-glace and simmer for 15 min. Finish by whisking in 4 oz sour cream, 1 tbsp fresh parsley, and ½ tsp dry mustard.

## Tomato Sauce

**116**

*yield: 1 qt (eight-4 oz servings)*

| | |
|---|---|
| 1 oz | salt pork (or bacon), diced |
| 8 oz | mirepoix, diced small |
| 1 clove | garlic, crushed |
| 1 qt | tomatoes (canned or fresh) |
| 1 c | tomato purée, canned |
| ½ tbsp | salt |
| 1 tsp | granulated sugar |
| 1 c | white (light) stock |
| 4 oz to 6 oz | roasted veal or pork bones (optional) |
| 1 tsp | baking soda |
| 2 tsp | cornstarch slurry |

*Sachet d'épices*

| | |
|---|---|
| 1 leaf | bay leaf |
| ¼ tsp | thyme, dried |
| 3 sprigs | parsley |
| ¼ tsp | peppercorns, cracked |

1. Render salt pork (or bacon) in heavy-bottomed saucepot until it is slightly crisp, but not browned.
2. Add mirepoix and sauté until tender but not browned.
3. Add garlic and cook for 1 min to 2 min without browning.
4. Add tomatoes, tomato purée, salt, sugar, white stock, roasted bones (if desired), and sachet d'épices.
5. Bring sauce to a boil and immediately lower heat. Simmer sauce for 1 hr.
6. Add baking soda and stir to incorporate. Simmer until no longer foaming.
7. Whisk in cornstarch slurry and simmer for 10 min.
8. Remove sachet (and bones, if used) and either pass sauce through a food mill or purée to reach desired consistency.
9. Adjust seasonings as necessary.

---

**Nutrition Facts**

**Per serving:** 70.2 calories; 26% calories from fat; 2.1 g total fat; 5.4 mg cholesterol; 741.7 mg sodium; 428.4 mg potassium; 10.4 g carbohydrates; 2 g fiber; 5.5 g sugar; 8.4 g net carbs; 3.5 g protein

---

**Recipe Variations:**

**117  Creole:** Sauté 4 oz small-diced onion, 4 oz small-diced bell peppers, 2 oz thinly sliced celery, and 1 tsp crushed garlic until tender. Add ½ tsp chili powder, ¼ tsp cayenne pepper, 1 bay leaf, ¼ tsp thyme, and a dash of hot pepper sauce. Add tomato sauce and simmer for 15 min. Adjust seasonings to taste.

**118  Marinara:** Sauté 4 oz small-diced onion and 3 cloves garlic in 3 oz of whole butter until softened. Add ½ cup of fresh chopped parsley and 1 cup diced tomatoes and sauté 1 min. Add 1 qt vegetarian tomato sauce (see vegetarian variation) and simmer for 15 min. Adjust seasonings to taste.

**119  Mexican:** Sauté 4 oz small-diced onion, 4 oz small-diced bell peppers, ½ small-diced jalapeno, and 1 tsp crushed garlic until tender. Add ½ tsp chili powder, ½ tsp cumin, ¼ tsp cayenne pepper, ¼ tsp coriander, and 1 bay leaf. Add tomato sauce and simmer for 15 min. Adjust seasonings to taste.

**120  Milanaise:** Sauté 1 cup sliced mushrooms, 3 oz diced mushrooms, 2 cloves crushed garlic, and 5 oz cooked smoked ham in 2 tbsp butter until tender. Add tomato sauce and simmer for 15 min. Add 2 tbsp chopped parsley and adjust seasonings to taste.

**121  Portuguese:** Sauté 4 oz small-diced onion in 1 tbsp olive oil. Add 2 cloves of crushed garlic and ½ lb crushed tomatoes (concassé). Simmer for 20 min to 30 min. Add tomato sauce and simmer for 15 min. Add ¼ cup chopped fresh parsley and seasonings to taste.

**122  Spanish (Vera Cruz):** Sauté 4 oz small-diced onion, 4 oz small-diced bell peppers, 2 oz thinly sliced celery, 4 oz sliced mushrooms, and 1 tsp crushed garlic until tender. Add ½ tsp thyme, ¼ tsp cayenne pepper, 1 bay leaf, and a dash of hot pepper sauce. Add tomato sauce and simmer for 15 min. Add 3 oz sliced green olives and adjust seasonings to taste.

**123  Vegetarian:** Omit the salt pork and roasted bones. Sweat the vegetables in 1 tbsp olive oil.

# Recipes

## Hollandaise

*yield: 1 qt (eight-4 oz servings)*

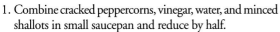

| | |
|---|---|
| ½ tsp | peppercorns, cracked |
| 2 oz | white wine vinegar |
| 2 oz | water |
| 1 | shallot, minced |
| 4 | egg yolks |
| 2 tsp | lemon juice |
| 16 oz | clarified butter, warm |
| 2 shakes | Tabasco® sauce |
| ½ tsp | Worcestershire sauce |
| to taste | salt and white pepper |

1. Combine cracked peppercorns, vinegar, water, and minced shallots in small saucepan and reduce by half.
2. Place yolks in a stainless steel mixing bowl and whisk until light and creamy in color and volume has increased by almost double.
3. Strain the vinegar reduction into the yolks and whisk rapidly to incorporate.
4. Place the mixture in the mixing bowl over a double boiler to begin heating the sauce while whisking rapidly. As the mixture cooks, it will begin to thicken. When the mixture has thickened so that the whisk leaves a thick trail behind it, the yolks have cooked enough. *Note:* Use caution to not allow the yolks to cook too rapidly and curdle. It is acceptable to remove the bowl from the double boiler from time to time to allow thorough mixing and to prevent overcooking.
5. Remove the mixing bowl from the double boiler and place on top of a saucepot lined with a kitchen towel to stabilize the bowl.
6. Quickly whisk 1 tsp of the lemon juice and 1 tbsp lukewarm water into the cooked yolk mixture. The water and lemon juice will both help stop the cooking process of the yolks as well as thin the mixture slightly to aid in incorporating the clarified butter.
7. While whisking rapidly, slowly begin to drizzle in the warm clarified butter in a narrow, steady stream until all of the clarified butter has been added. Do not stop whisking until all the butter has been incorporated, or the hollandaise could separate and break. *Note:* If the mixture gets too thick at any point when adding the clarified butter, add a small amount of lukewarm water to thin the mixture slightly.
8. Whisk in the remainder of the lemon juice and add the Tabasco®, Worcestershire, and seasonings to taste.
9. Hold for service over a warm bain marie.

### Nutrition Facts

**Per serving:** 471.5 calories; 90% calories from fat; 48.3 g total fat; 226.9 mg cholesterol; 57.7 mg sodium; 199.5 mg potassium; 9.2 g carbohydrates; 0 g fiber; 0.1 g sugar; 9.2 g net carbs; 3.1 g protein

## Recipe Variations:

**125** **Béarnaise:** Reduce 2 oz minced shallots, 5 tbsp chopped tarragon, 1 tsp ground peppercorns, and 8 oz white wine vinegar by three-quarters. Add the reduction to the egg yolks and prepare hollandaise as directed. Strain through cheesecloth or a chinois and season to taste.

**126** **Choron:** Add 2 oz freshly stewed tomato purée to béarnaise preparation.

**127** **Dugléré:** Add sautéed mushrooms, tomatoes, and chopped parsley to taste.

**128** **Grimrod:** Add ¼ tsp saffron.

**129** **Maltaise:** Add 2 oz concentrated orange juice and 2 tsp orange zest.

**130** **Mousseline:** Fold in 8 oz whipped heavy cream just prior to serving. Items dressed with mousseline sauce can be placed under the broiler to brown the surface, resulting in a very light golden glaze known as a glaçage.

**131** **Noisette:** Substitute browned butter for clarified butter.

**132** **Valois:** Add 8 oz glace de viande to prepared béarnaise.

## Cooked Asparagus and Rosemary Coulis

*yield: 1 qt (eight-4 oz servings)*

| | |
|---|---|
| 2 tbsp | vegetable oil |
| 1 | shallot, minced |
| 1 clove | garlic, crushed |
| 2 lb | asparagus |
| ¼ c | white wine |
| 2 sprigs | fresh rosemary |

1. In a saucepan, sweat crushed garlic and minced shallot in a small amount of vegetable oil.
2. Add tender portion of asparagus (do not use the woody, fibrous ends as they are too stringy) and sweat until bright green in color.
3. Deglaze pan with a small amount of white wine.
4. Place ingredients in a blender with freshly chopped rosemary and purée. *Note:* Rosemary is not added until this point so that it retains color and freshness.
5. Strain through a china cap and season to taste. *Note:* If coulis is too thick, it can be thinned with a small amount of light stock.

---

**Nutrition Facts**

**Per serving:** 60.3 calories; 53% calories from fat; 3.6 g total fat; 0 mg cholesterol; 2.9 mg sodium; 241.8 mg potassium; 4.8 g carbohydrates; 2.4 g fiber; 2.1 g sugar; 2.4 g net carbs; 2.6 g protein

---

## Eggplant Foam

*yield: 10 servings*

| | |
|---|---|
| 1 | shallot, minced |
| ½ tsp | fresh garlic, crushed |
| 2 tbsp | whole butter |
| 2 oz | eggplant, small diced |
| 1 tbsp | white wine |
| 2 oz | light stock |

1. In a saucepan, sweat shallots and garlic in butter.
2. When shallots are tender, add eggplant and sweat until tender.
3. Deglaze the pan with white wine and light stock. Reduce liquid by half.
4. Add 1 tbsp heavy cream and cook gently for 2 min.
5. Using a stick mixer, purée all ingredients until very smooth and foamy.
6. Season to taste and serve immediately. *Note:* Foam can be remixed with a stick mixer if more volume is needed.

---

**Nutrition Facts**

**Per serving:** 53.5 calories; 39% calories from fat; 2.4 g total fat; 6.3 mg cholesterol; 13.4 mg sodium; 153.7 mg potassium; 7.2 g carbohydrates; 0.2 g fiber; 0.2 g sugar; 7 g net carbs; 1.2 g protein.

---

# Soups

National Onion Association

*Soups* are versatile foods that can range from flavorful broths to hearty chowders. Soup can be a nutritious meal in itself, served as an appetizer prior to a main course, or as an accompaniment to a sandwich or salad. Soup is considered a comfort food because it is nourishing and can be easily digested when we are ill. Soups are important menu items that include clear, thick, and specialty varieties.

## Chapter Objectives:

1. Describe common types of clear and thick soups.
2. Demonstrate the follow procedures: preparing a broth, clarifying a consommé, preparing a purée soup, and preparing a chowder.
3. Demonstrate two different procedures for preparing a cream soup.
4. Contrast the three varieties of specialty soups.
5. Demonstrate two different methods for preparing a bisque.

## Key Terms
- **clear soups**
- **broth**
- **consommé**
- **clarify**
- **clearmeat**
- **oignon brûlé**
- **thick soups**
- **bisque**
- **chowder**

## SOUP VARIETIES

Soup is often the first course of a customer's dining experience. Just as a dessert can be an amazing finale or a ho-hum conclusion to a great meal, a wonderfully prepared soup can set the stage for an unforgettable meal, while a poorly made soup may suggest to the customer that the courses to follow will be only mediocre at best. Soup, just like a main course entrée, should reflect the staff's level of skill, attention to detail, and commitment to providing high-quality food and customer satisfaction. *See Figure 17-1.*

Soup is one of the most versatile categories of food and allows chefs to be creative while demonstrating classical stock and sauce preparation techniques. Many of the most popular soups are produced by using a high-quality stock as a base and adding other ingredients to make the dish unique. Soups can be made using fresh and seasonal ingredients but can also be made using high-quality leftovers. Items such as leftover baked chicken can be diced and used to create a wonderful chicken noodle soup. Using leftover ingredients helps to produce soups at an economical cost. The key to using leftovers, however, is quality. Items that are not of the highest quality should be discarded and should never be used in a soup.

Most soups can be classified into one of three varieties: clear (unthickened), thick, or specialty. The different varieties of soups are characterized by the overall finished appearance and consistency of the soup, and the preparation and cooking methods used to make the soup. Specialty soups include bisques, chowders, and cold soups, as well as unique regional varieties. Although many specialty soups are thick in consistency, they are produced using different methods than are used for traditional thick soups.

National Chicken Council

### Soup Varieties

$\mathscr{Soups}$

**chilled English cucumber soup**
with charentais melons, yuzu pearls, salmon ceviche and roe

**light lobster bisque**
with bomba rice and summer vegetables

**creamy yellow spring pea soup**
with house-smoked trout

**leek and potato potage**
with Wellfleet clams and fines herbes

**cream of Alsace cabbage soup**
with home-smoked sturgeon and caviar

**17-1** *The varieties of soup an establishment offers should reflect the staff's skill level, attention to detail, and commitment to high-quality food and customer satisfaction.*

## CLEAR SOUPS

A *clear (unthickened) soup* is a stock-based soup with a thin, watery consistency. Clear soups include broths and consommés. Broths, which are produced from well-made stocks, can be made from meat, poultry, seafood, or vegetables. When meat is used, a broth will have a much more concentrated flavor and color than a stock made from bones. Consommés are made from high-quality broths that have been further clarified to remove all impurities and surface fat. *See Figure 17-2.*

| Components of Clear Soups | | |
|---|---|---|
| Preparation | Liquid | Ingredients |
| **Broth** | Prepared stock | Meat, mirepoix, sachet d'épices or bouquet garni |
| **Consommé** | High-quality prepared broth of roasted meat or poultry | Clearmeat, mirepoix, sachet d'épices or bouquet garni, and oignon brûlé (for clarity) |

**17-2** *Stocks are used to prepare broths, which are subsequently used to prepare consommés.*

## Broths

A *broth* is a flavorful liquid made by simmering stock along with meat or vegetables and seasonings. A stock is used for the liquid ingredient in a broth, as opposed to water. Meat or vegetables are added to the stock and simmered to create a more intense flavor, color, and overall body. A broth can be served as a finished product or made into a broth-based soup or consommé.

Typically, tougher cuts of meat or poultry are used to produce a full-flavored, rich broth. In addition, not allowing the broth to boil vigorously, continually skimming impurities from the surface, and careful straining of the cooked broth produce a broth of excellent quality.

## To prepare a broth, apply the following procedure:

1. Prepare the main ingredient (meat, poultry, vegetables) to desired size and truss if necessary to keep all parts (such as wings and legs on poultry) intact and connected. Rinse in cold water to remove any impurities or loose particles. *See Figure 17-3.*

2. If a dark color and roasted flavor are desired, carefully brown the meat and mirepoix. If no color is desired, simply sweat the mirepoix in a stockpot.

3. Completely cover the main ingredient and mirepoix with cold stock. Add a sachet d'épices.

4. Bring stock to a simmer. Continually skim off impurities as they rise to the surface.

5. Reduce heat to a low simmer and cook until the meat is tender and a good flavor has developed. Occasionally taste the broth to monitor progress.

6. Carefully drain the broth to minimize disturbing the cooked ingredients. The used flavoring ingredients can be stored for future use.

7. Strain the broth with a china cap lined with cheesecloth.

8. Quickly cool the strained broth using an ice bath or a cooling wand to bring it out of the temperature danger zone. When the broth is cold, excess fat and grease will harden on the surface and can be easily removed.

## Preparing Broths

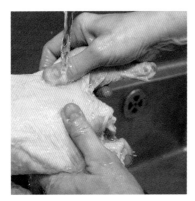
1. Rinse prepared flavoring ingredient in cold water.

2. If desired, brown the meat and mirepoix, or if no color is desired, sweat the mirepoix.

3–4. Add cold broth to meat and mirepoix; bring to a full simmer.

5. Reduce to a low simmer; skim surface often; simmer until desired flavor is achieved.

6–7. Carefully drain broth; strain broth through damp cheesecloth.

8. Cool broth and remove excess fat and grease.

17-3   A broth is prepared by simmering a flavoring ingredient, such as poultry, and mirepoix in prepared stock.

If a broth is needed for service immediately, the completed broth is brought to a boil. Seasonings are adjusted to taste. Fat droplets present on the surface of the broth can be removed with a clean paper towel. ***See Figure 17-4.*** The completed broth is appropriately garnished and served. *Note:* It is important to heat any garnish prior to adding it to the broth to avoid cooling the broth down.

Mirepoix may be cut larger or smaller depending on the main ingredient. If the broth contains a large cut of meat that requires a long cooking time, a large-cut mirepoix can be added halfway through the cooking process. Beef broth, for example, typically simmers for 2 hr or longer. If the main flavoring ingredient is something that cooks more quickly, such as chicken, a smaller mirepoix can be added at the beginning of the cooking process. Chicken broth typically is finished after 30 min to 40 min of simmering or as soon as the chicken is cooked through. The size of the mirepoix should be such that it can be cooked thoroughly in the allotted time without being overcooked. ***See Figure 17-5.***

## Removing Fat from Broth

**17-4** *A clean paper towel can be used to absorb fat droplets from the surface of broth.*

## Using Mirepoix

**17-5** *The size of mirepoix added can vary depending on cooking time.*

### Consommés

A *consommé* (pronounced con-so-may) is a very rich and flavorful broth that has been further clarified to remove any impurities or particles that could cloud the finished product. The flavor of the main ingredient should be pronounced and the color of the consommé should be very rich. Consommé made from poultry should have a deep golden or amber color, while consommé made from red meat should be dark with a roasted meat color. It is very important that a consommé be made with the highest-quality ingredients and be properly prepared, as a finished consommé is only as good as the broth and ingredients it was made from. Although the clarification process adds a small bit of flavor, the bulk of a consommé's flavor is derived from the quality of the broth.

**Clarifying a Consommé.** To *clarify* is to remove impurities, sediment, cloudiness, and particles to leave a very clear and pure liquid. For this to occur, very cold stock or broth is combined with a mixture known as a clearmeat.

A *clearmeat* is very cold, lean ground meat, fish, or poultry that is combined with an acid such as wine, lemon juice, or tomato product and cold ground mirepoix, egg whites, and an oignon brûlé (pronounced onion brew-lay). An *oignon brûlé* is half an onion that is charred on the cut side in a heavy-bottomed pan. The term oignon brûlé is French for "burnt onion." An oignon brûlé is used to give an intense roasted flavor and deeper color to a darker-colored consommé, broth, or sauce.

To clarify a consommé, the clearmeat is mixed well and then whisked into the cold broth it will eventually clarify, in an appropriately sized stockpot, and brought to a low simmer. As the temperature increases, the mixture is occasionally stirred until the clearmeat begins to coagulate. Once the clearmeat begins to form a solid mass, the mixture cannot be disturbed or stirred as the clearmeat could break apart and cloud the consommé. As the clearmeat mixture gently simmers, it rises to the surface and, in doing so, will strain the impurities from the consommé. When the clearmeat rises, it is known as a raft. ***See Figure 17-6.*** In addition to clarifying the consommé, the raft enhances the flavor, color, and richness of the final consommé.

Once the raft forms, a hole is gently poked into the top of it. This hole allows the progress of the consommé to be monitored as it cooks. After gently simmering for 1 hr to 1½ hr, the consommé is removed through the hole in the raft using a ladle. The consommé is then strained through multiple layers of dampened cheesecloth to remove any small impurities that may not have been trapped by the raft. The consommé should be cooled immediately and stored properly. Once completely cooled, any remaining fat will solidify on the surface and can be easily removed. The resulting liquid should be a crystal clear and intensely flavored consommé. The process of clarification takes some practice to master, but a clear and flavorful consommé is worth the effort.

If the consommé is not going to be used immediately, it can be cooled quickly and completely, allowing the fat to solidify on the surface for easy removal. If any fat remains on the surface when the consommé is reheated for service, it can be removed by blotting the surface with a clean paper towel.

## Rafts

**17-6** When clearmeat rises to the top of a simmering consommé, it is known as a raft.

## To clarify a consommé, apply the following procedure:

1. Prepare the clearmeat by combining freshly ground lean meat, mirepoix, and egg whites with any acid (such as a tomato product or lemon juice) and herbs or seasonings, if desired. ***See Figure 17-7.***

2. Blend the cold clearmeat well with a small amount of cold broth to loosen the mixture. *Note:* Both the clearmeat and broth should be between 32°F and 35°F. A small amount of crushed ice can be added to bring down the temperature, if needed.

3. In a heavy-bottomed stockpot (preferably with a drain spigot), combine the clarification mixture and cold broth. Stir well.

4. While occasionally stirring, bring the mixture to a simmer over medium heat. The clearmeat will begin to form a solid mass (raft).

5. Once a solid mass forms, stop stirring and gently poke a hole into the top of the newly formed raft. This hole allows the progress of the consommé to be monitored as it cooks.

6. Adjust the heat to allow the consommé to gently simmer. Continue to simmer for 1 hr to 2 hr until a desired flavor and color are achieved. Occasionally baste the raft with consommé. Basting helps prevent the raft from drying out or breaking apart.

7. Carefully strain the consommé through multiple layers of clean cheesecloth and degrease completely.

CULINARY PROCEDURES

**Chef's Tip:**
When clarifying a consommé, the acid added to the clearmeat depends on the main ingredient of the broth. Tomato purée is used as the acid ingredient for a dark consommé such as beef, veal, or lamb. Lemon juice is used as the acid ingredient for a light-colored consommé such as poultry or fish.

**Rescuing an Unsuccessfully Clarified Consommé.** Occasionally, various factors can cause a consommé to not be as clear as it should be. Factors such as the consommé boiling instead of simmering, stirring the consommé after the raft has formed, and starting the consommé with clearmeat and broth that were not cold enough can cause the final consommé to be cloudy. When this happens, it is necessary to conduct a secondary clarification.

To perform a secondary clarification on a gallon of consommé, another clarification mixture consisting of 4 lightly beaten egg whites, ¼ cup of finely ground mirepoix, and either 1 tbsp tomato purée or 2 tsp lemon juice is added to the consommé. The secondary clarification is added to the consommé, and then the clarification procedure is repeated. The mixture is brought to a simmer and the secondary clarification forms a raft. After the raft forms, the consommé is gently strained and degreased.

**Classical Consommé Variations.** Consommés have been around for a very long time and were once used as a way to show off a chef's skill to people of nobility or wealth. A perfectly clear and greaseless consommé announces to the customer that the chef has refined soup-making skills. Although there are countless garnishes appropriate for a consommé, it is important that any garnish is precisely prepared, as it can be clearly seen through the crystal clear liquid of the consommé.

**Nutrition Note**

A secondary clarification is not used unless absolutely necessary, because the second clarification also removes some of the flavor and liquid of the consommé along with the impurities and sediment left behind from the first clarification. This is why it is important to get it right the first time!

## Preparing Consommés

1. Prepare the clearmeat.

2. Blend the clearmeat and cold broth.

3. Add the clearmeat and remaining broth to stockpot.

4. Bring the mixture to a simmer.

5-6. When clearmeat forms a raft, poke a hole in the raft. Simmer for 1 hr to 2 hr; occasionally baste the raft.

7. Carefully strain through multiple layers of clean cheesecloth.

**17-7** Consommé is clarified by preparing a high-quality broth with clearmeat.

A consommé can be made with any meat, poultry, or seafood and is often named after the item it was produced from, such as pheasant consommé or quail consommé. The following are some examples of classical consommés:

- Consommé bouquetière: garnished with a petite bouquet of assorted fresh vegetables tied with a string of leek
- Consommé brunoise: garnished with a brunoise cut of carrot, celery, leek, and turnip and finished with minced chervil
- Consommé célestine: thickened slightly with tapioca and garnished with a julienne of crêpe, chopped truffles, and savory herbs
- Consommé chasseur: garnished game consommé with quenelles (poached puréed meat dumplings) of game meat or petite profiteroles (cream puffs) studded with game forcemeat

- Consommé diplomate: chicken consommé thickened slightly with tapioca and garnished with julienne of truffles and quenelles (poached puréed meat dumplings) of chicken and crawfish butter
- Consommé Grimaldi: clarified consommé with a greater quantity of tomato purée; garnished with cooked custard (royale) cut into various small shapes and a julienne of blanched celery
- Consommé julienne: garnished with a blanched julienne of leeks, carrots, turnips, celery, and cabbage and finished with a chiffonade of chervil or sorrel
- Consommé mikado: garnished chicken consommé that has a slight tomato flavor, with peeled, seeded diced tomato and cooked chicken meat
- Consommé madrilène: served as a tomato-based consommé in cold jelly (aspic) form
- Consommé printanier: garnished with petite blanched balls of turnip, carrots, peas, and chiffonade of chervil
- Consommé royale: garnished with large cubes or batonnet of cooked custard (royale)

## THICK SOUPS

As the name suggests, a *thick soup* is a soup having a thick texture and consistency. The two basic types of thick soups are cream soups and purée soups. The main difference between a cream soup and a purée soup is that most cream soups use an added starch, such as the flour in a roux, to thicken them. Purée soups, on the other hand, are thickened by puréeing the main ingredients, which have starchy properties. Additionally, purée soups are generally thicker and coarser than cream soups and are seldom strained. Due to the slightly lower starch content of vegetable purée soups, rice or potatoes are sometimes added to increase the texture.

When making a cream soup or purée soup, always use a heavy-bottomed saucepot, preferably constructed of nonreactive stainless steel, to avoid burning or scorching. A blender or a food mill can be used to purée the ingredients and finish the soup.

### Cream Soups

Cream soups are traditionally made using one of two methods. The first method uses velouté, and the second uses a roux preparation. These two variations work in much the same way, and the choice of which to use is a matter of preference. Regardless of the method chosen, the main ingredients need to be simmered until tender in order to be able to be puréed smooth.

A velouté preparation consists of simmering the main ingredient, such as asparagus for a cream of asparagus soup in a light velouté (stock thickened with a roux) until tender. Then the main ingredient and velouté are puréed, finished with cream, and seasoned to taste.

**Action Lab**

Working in teams, prepare cream of broccoli soup two different ways and compare the final products. Have one team prepare a cream of broccoli soup, using the roux preparation method without puréeing the soup. Have another team prepare a cream of broccoli soup using the velouté method, and have them purée the soup after it is prepared. Taste and evaluate each soup for flavor, appearance, and texture. Discuss the results.

To prepare a cream soup using the velouté preparation, apply the following procedure:

1. In heavy-bottomed saucepot, sweat the white matignon with the chopped main ingredient in butter until slightly tender. ***See Figure 17-8.***

2. Add prepared hot velouté sauce and simmer until vegetables are completely tender (approximately 30 min), stirring often.

3. Place soup in blender and carefully purée until smooth, using caution as the hot soup can splatter from the blender when the blender is turned on, causing burns.

4. Strain through a china cap or chinois.

5. Return strained soup to stove, thin with stock if needed, and return to a simmer.

6. Finish with hot cream and season to taste.

## Preparing Cream Soups—Velouté Preparation

1. Sweat matignon and flavoring ingredient in butter.

2. Add prepared velouté; simmer 30 min.

3. Purée in blender until smooth.

4. Strain through china cap or chinois.

5–6. Return to simmer; finish with hot cream and season to taste.

Garnish and serve warm.

**17-8** In the velouté preparation, the main ingredient is simmered in velouté and then puréed.

A roux preparation consists of sweating the main ingredient, such as asparagus, with an aromatic white matignon in butter. Next, flour is added to the mixture to make a roux. Then hot stock is added to the roux, and the mixture is allowed to simmer and thicken. Finally, the soup is puréed until smooth and finished with cream.

## To prepare a cream soup using the roux preparation, apply the following procedure:

1. In heavy-bottomed saucepot, sweat the white matignon and the main ingredient in enough fat to also make the roux. *See Figure 17-9.*

2. When the aromatics and main ingredient are slightly tender, add flour to make a roux.

3. Cook the roux just long enough to achieve a pale golden color in the flour and then add warm light stock while whisking vigorously. Bring to a slight simmer and cook until the taste of flour is gone and the vegetable ingredients are completely tender, about 30 min.

4. Place soup in blender and carefully purée until smooth, using caution as the hot soup can splatter from the blender when the blender is turned on, causing burns.

5. Strain through a china cap or chinois.

6. Return strained soup to stove, thin slightly with additional hot stock if needed, and return to a simmer.

7. Finish with hot cream and season to taste.

CULINARY PROCEDURES

### Purée Soups

Purée soups are made by cooking starchy vegetables such as potatoes, squash, turnips, or carrots, or dried legumes such as lentils, beans, or split peas in broth until tender. When the starchy ingredient is tender, the soup is puréed using all or a portion of the starchy ingredients to thicken the soup. Often a portion of the starchy ingredient is not puréed, but is reserved and added into the puréed mixture to add texture to the soup. Neatness and accuracy of vegetable cuts is not a priority in purée soups as the vegetables are puréed to make the final soup. However, preparing vegetables in a relatively consistent size ensures uniform cooking.

Purée soups can contain many different ingredients, from chiles to roasted vegetables. Pork products such as bacon, salt pork, and smoked ham are also often found in purée soups.

1. Sweat matignon and flavoring ingredient in butter.

2. Add flour to make a roux and cook a few minutes more.

3. Add warm stock, whisking vigorously; simmer 30 min.

4. Purée in blender until smooth.

5. Strain through china cap or chinois.

6–7. Return to simmer; finish with hot cream and season to taste.

**17-9** *In the roux preparation, a roux is made of the main ingredient, matignon, and flour; stock is added, and the soup is puréed.*

## To prepare a purée soup, apply the following procedure:

1. Depending on the recipe, either sweat or lightly brown the mirepoix in butter. ***See Figure 17-10.***

2. Add the cooking liquid and bring to a simmer.

3. Add the main ingredients and the sachet and simmer until the main ingredients are tender enough to purée. Use caution to not allow the soup to scorch or burn on the bottom.

4. Remove the sachet and reserve about 20% of the liquid from the pot.

5. Purée the remainder of the ingredients by processing them in a food mill, food processor, or blender until the desired texture is achieved, using the reserved liquid to thin as necessary.

6. Return the soup to a simmer and adjust seasonings as necessary.

## Preparing Purée Soups

1-2. Sweat or brown the mirepoix. Add desired liquid and bring to simmer.

3. Add flavoring ingredients and sachet d'epices; simmer until tender.

4. Remove sachet and reserve about 20% of the liquid.

5. Purée remainder of ingredients; use reserved liquid to thin to desired consistency.

6. Return to simmer; season to taste.

Garnish and serve.

**17-10** A purée soup is thicker and coarser than a cream soup.

**Controlling the Consistency of Soups.** Thick soups tend to further thicken if they are made in advance and stored. When it is time to reheat, a small amount of hot water, stock, broth, hot milk, or cream may be added to thin the soup to the desired consistency. If a thick soup is too thin, a small amount of roux, beurre manié, or a cornstarch slurry may be whisked in and the soup brought to a simmer. If a starch is added to thicken the soup, it should be simmered for 20 min to 30 min to cook out the starch flavor and thicken the soup to the fullest potential.

Florida Tomato Committee

## SPECIALTY SOUPS

Many soups do not fall into the traditional categories of thin or thick soups. These other soups may be considered specialty soups and can fall into three categories: bisques, chowders, or cold soups. A bisque is another form of cream soup that is typically made from shellfish. A chowder may be a cream- or broth-based soup, but is usually served as a very hardy bowl of soup with chunks of both potato and other main ingredients. Cold soups include traditional cold cream soups but can also be very different, such as soups made from a purée of fruit and yogurt.

### Bisques

A *bisque* is a form of cream soup that is typically made from shellfish. Although the name bisque traditionally applies to soup made from shellfish, in modern cuisine many cream soups are referred to as bisques. Bisques are typically prepared using one of two methods, using either a roux or a heavy cream reduction.

**To prepare a bisque using the roux preparation, apply the following procedure:**

1. In a heavy-bottomed saucepot, sweat the aromatic matignon in a small amount of fat. **See Figure 17-11.**
2. Add a small amount of tomato paste to the pan and sauté lightly. Sautéing tomato paste in this way is known as *pincé*.
3. Add flavoring ingredients and cook until caramelized.
4. Deglaze pan with a small amount of brandy, cognac, or white wine.
5. Add a sachet d'épices and a light stock. Simmer for 20 min to 25 min or until a strong flavor and color is achieved. *Note:* A mild seafood velouté can be added in place of the light stock. If velouté is used, omit step 6.
6. Add roux to thicken while whisking vigorously.
7. Bring to a slight simmer and cook 15 min to 20 min, or until the taste of flour is gone.
8. Purée and strain through a china cap, chinois, or cheesecloth.

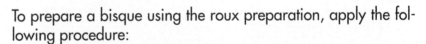

9. Return strained soup to stove, thin slightly with additional hot stock if needed, and bring to a simmer once again.

10. Finish with hot cream and season to taste.

## Preparing Bisques—Roux Preparation

1-2. Sweat matignon in fat; add tomato paste and sauté lightly.

3. Add flavoring ingredients and caramelize.

4. Deglaze pan with brandy, cognac, or white wine.

5. Add sachet d'epices and stock; simmer 20 min to 25 min.

6-7. Add roux while whisking vigorously; simmer 15 min to 20 min.

8. Purée and strain.

9. Return to a simmer.

10. Finish with hot cream and season to taste.

Garnish and serve hot.

**17-11**    A bisque prepared with roux should be simmered 15 min to 20 min or until the taste of flour is gone.

To prepare a bisque using the heavy cream reduction preparation, apply the following procedure:

1. In a heavy-bottomed saucepot, sweat the aromatic matignon in a small amount of fat. ***See Figure 17-12.***

2. Add a small amount of tomato paste to the pan and sauté lightly.

3. Add flavoring ingredients and cook until all are caramelized.

4. Deglaze pan with a small amount of brandy, cognac, or white wine.

5. Add a sachet d'épices and a small amount of light stock. Simmer for 15 min to 20 min, or until reduced by half.

6. Add heavy cream; simmer gently until reduced to a nappe consistency.

7. Remove sachet d'épices, place soup in blender, and purée slightly, using caution as the hot soup can splatter from the blender when the blender is turned on, causing burns.

8. Strain through a china cap, chinois, or cheesecloth.

9. Return strained soup to stove, thin slightly with additional hot stock if needed, and return to a simmer. Garnish and serve hot.

## Chowders

A *chowder* is a very hearty soup with large chunks of potatoes and other main ingredients. Although most chowders are cream-based and very similar to a cream soup, a few chowders are broth-based and served thin. The majority of cream-based chowders are thickened with a roux. A major difference between cream soups and chowders is that chowders are not puréed. Chowder is typically so hardy that it can almost be thought of as a variety of stew.

To prepare a chowder, apply the following procedure:

1. Render diced bacon or salt pork until it begins to caramelize. ***See Figure 17-13.***

2. Add mirepoix and sweat until onions are translucent.

3. Add flour to make a roux.

4. Add stock or liquid.

5. Add the main flavoring ingredients, seasonings, and sachet d'épices.

6. Simmer until potatoes and main ingredients are fully cooked.

7. Finish with hot cream.

## Preparing Bisque—Heavy Cream Reduction Preparation

1–2. Sweat matignon in fat; add tomato paste and sauté lightly.

3. Add flavoring ingredients and caramelize.

4. Deglaze pan with brandy, cognac, or white wine.

5. Add sachet dépices and stock; simmer 20 min to 25 min.

6. Add heavy cream; simmer to nappe consistency.

7. Purée in blender.

8–9. Strain through china cap, chinois, or cheesecloth; return to a simmer and season to taste.

Garnish and serve hot.

**17-12** A bisque prepared with a cream reduction is simmered for 15 min to 20 min or until the liquid has reduced by half.

## Preparing Chowder

1. Render bacon or salt pork until it begins to caramelize.

2. Add mirepoix and cook until onions turn translucent.

3. Add flour to make a roux.

4–5. Add stock; add flavoring ingredients, and sachet d'epices.

6. Simmer until potatoes and main ingredients are fully cooked.

7. Finish with hot cream.

 **17-13** A chowder is a hearty soup that is similar to a stew.

### Historical Note

In 1917, Executive Chef Louis Diat decided to prepare and serve his mother's recipe for vichyssoise, named after the town in France where he was born. The guests at the event had never been served cold soup before and many of them returned the soup to the kitchen, complaining that the soup was not hot.

### Cold Soups

Cold soups come in a variety of styles. Regardless of which variety of cold soup is being prepared, they are typically classified into two categories: those that require cooking the main ingredients and those that use fresh raw ingredients that are puréed. A refreshing cold fruit soup such as cold cantaloupe soup can be simply made by puréeing fresh cantaloupe with mint and yogurt. Many cold soups are made using fruit or vegetable juice. One popular cold soup made using juice is gazpacho. To prepare gazpacho, fresh raw vegetables are puréed with tomato juice and spices and served cold. Another variety of cold soup is simply a classic cream soup served cold. The most famous cold cream soup is vichyssoise, or cold potato and leek soup.

## SUMMARY

Soups are often the first course of a dining experience and can set the stage for a wonderful meal. Soups can be a good way to make use of high-quality leftovers such as freshly cooked chicken or properly cooked vegetables. However, a soup will only be as good as the ingredients used to make it, so poor-quality ingredients or ingredients that are not fresh should never be used. When preparing soup in a restaurant, the chef will use professional stock- and sauce-making techniques. Soups are generally grouped into clear soups, thick soups, and specialty soups. Clear soups include broths and consommés. Cream soups and purée soups are known as thick soups. Speciality soups include bisques, chowders, and cold soups. No matter what type of soup is being served, it is important to serve hot soup very hot, and cold soup very cold.

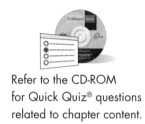

Refer to the CD-ROM for Quick Quiz® questions related to chapter content.

## Review Questions

1. Contrast the basic textures and ingredients of soups classified as clear versus those classified as thick.

2. Describe the procedure for preparing a broth.

3. Describe the function of an oignon brûlé in a consommé, sauce, or broth.

4. Identify factors that can contribute to producing an unsuccessfully clarified consommé.

5. Describe how to prepare a cream soup using the velouté method.

6. Describe various methods for controlling the consistency of cream and purée soups.

7. Identify the three categories of specialty soups.

8. Describe the difference in initial simmering times between a bisque prepared using the roux preparation and a bisque prepared using the heavy cream reduction preparation, before adding the roux or heavy cream.

9. Describe the procedure for preparing a chowder.

10. Identify two common types of cold soup.

# Recipes

## 135

## Chicken Broth

*yield: 1 gal. (sixteen 8-oz servings)*

| | |
|---|---|
| 1 lb | mirepoix, medium dice |
| 2 tbsp | vegetable oil |
| 3 lb | stewing hen |
| 1 gal. | cold chicken stock (or water) |

*Sachet d'épices*

| | |
|---|---|
| 2 | bay leaves |
| 1 sprig | fresh thyme |
| 3 stems | fresh parsley |
| 1 small clove | garlic, crushed |
| ¼ tsp | peppercorns, cracked |

1. Sweat mirepoix in vegetable oil over medium heat without browning.
2. Rinse chicken in cool water and place in stockpot with mirepoix.
3. Cover chicken and mirepoix with cold chicken stock (or water if no stock is available).
4. Add sachet d'épices and bring pot to a full simmer.
5. Reduce to a low simmer and skim occasionally.
6. Continue cooking until a full flavor and color have developed and chicken is cooked completely, about 35 min to 45 min.
7. When fully cooked, remove chicken and cool appropriately for later use.
8. Carefully strain the chicken broth through dampened cheesecloth and refrigerate until needed, taking caution not to disturb the remainder of the solid ingredients during straining.
9. Carefully remove all meat from the chicken carcass and dice or prepare as needed to garnish the broth for service.
10. Garnish prior to serving.

---

**Nutrition Facts**

**Per serving:** 244.3 calories; 55% calories from fat; 15 g total fat; 43.3 mg cholesterol; 398.8 mg sodium; 444.8 mg potassium; 11.1 g carbohydrates; 0.7 g fiber; 5 g sugar; 10.4 g net carbs; 15.3 g protein

---

## 136

## Beef Consommé

*yield: 1 gal. (sixteen 8-oz servings)*

| | |
|---|---|
| 8 | egg whites |
| 3 lb | ground beef (lean shank, shoulder, or neck meat) |
| 1 lb | mirepoix, ground fine with meat if possible |
| 2 | tomatoes, seeded and diced |
| 5 qt | beef stock or broth, very cold |
| ½ onion | oignon brûlé |
| to taste | salt |

*Sachet d'épices*

| | |
|---|---|
| 3 | bay leaves |
| 4 stems | fresh parsley |
| 2 sprigs | fresh thyme |
| ½ tsp | peppercorns, cracked |

1. Slightly whip egg whites until frothy.
2. Prepare the clearmeat by combining very cold ground meat, mirepoix, tomato, and egg whites in mixing bowl.
3. Mix in a small amount of very cold stock or broth to loosen the clearmeat.
4. Place clearmeat in appropriate sized stockpot (with bottom spigot if possible) with the rest of the very cold stock or broth and mix well.
5. Add the oignon brûlé and sachet d'épices.
6. Over medium heat, bring mixture to a gentle simmer while stirring occasionally.
7. Simmer and continue to occasionally stir until clearmeat mixture begins to form a solid mass (raft) and then discontinue stirring. *Note:* Do not stir the broth after the raft forms to avoid clouding the consommé.
8. Carefully poke a 2″ hole in the top of the raft to allow steam and pressure to escape.
9. Simmer gently for 1 hr to 2 hr until a full-bodied and rich-colored consommé develops.
10. Carefully strain the consommé through multiple layers of moistened cheesecloth, degrease surface, and adjust seasonings as needed. *Note:* If consommé is not needed immediately, completely chill to allow any additional surface fat to solidify and be removed.

---

**Nutrition Facts**

**Per serving:** 186.3 calories; 49% calories from fat; 10 g total fat; 44.1 mg cholesterol; 539.1 mg sodium; 712.9 mg potassium; 5.3 g carbohydrates; 0.6 g fiber; 2.5 g sugar; 4.8 g net carbs; 18.2 g protein

---

**Variation:**

**137** **Consommé Bouquetière:** Garnish with a petite bouquet of assorted fresh vegetables tied with a string of leek.

**138** **Consommé Brunoise:** Garnish with a brunoise cut of carrot, celery, leek, and turnip and finish with minced chervil.

**139** **Consommé Célestine:** Thicken slightly with tapioca and garnish with a julienne of crêpe, chopped truffles, and savory herbs.

**140** **Consommé Chasseur:** Garnish game consommé with quenelles (poached puréed meat dumplings) of game meat or petite profiteroles (cream puffs) studded with game forcemeat.

**141** **Consommé Diplomate:** Thicken chicken consommé slightly with tapioca and garnish with a julienne of truffles and quenelles (poached puréed meat dumplings) of chicken and crawfish butter.

**142** **Consommé Grimaldi:** Clarify consommé with a greater quantity of tomato purée. Garnish with cooked custard (royale) cut into various small shapes and a julienne of blanched celery.

**143** **Consommé Julienne:** Garnish with a blanched julienne of leeks, carrots, turnips, celery, and cabbage and finish with a chiffonade of chervil or sorrel.

**144** **Consommé Mikado:** Garnish chicken consommé that has a slight tomato flavor with peeled, seeded, diced tomato and cooked chicken meat.

**145** **Consommé Madrilène:** Serve a tomato-based consommé in cold jelly (aspic) form.

**146** **Consommé Printanier (Parisienne):** Garnish with petite blanched balls of turnip, carrots, peas, and a chiffonade of chervil.

**147** **Consommé Royale:** Garnish with large cubes or batonnet of cooked custard (royale).

## Cream of Asparagus Soup (Velouté)

*yield: 1 gal. (sixteen 8-oz servings)*

| | |
|---|---|
| 3 oz | whole butter |
| 1 gal. | chicken velouté |
| 8 oz | heavy cream, hot |
| to taste | salt and white pepper |
| 6 oz to 8 oz | asparagus tips, blanched |

*White matignon*

| | |
|---|---|
| 8 oz | onion, small dice |
| 4 oz | celery, small dice |
| 2¼ lb | asparagus, diced |

1. Sweat the white matignon in the butter until slightly tender.
2. Add hot velouté and whisk to incorporate all ingredients.
3. Simmer for approximately 30 min or until all vegetables are tender, making sure to stir often to avoid scorching on the bottom of the pot.
4. Place hot soup in blender and, using caution, purée until smooth.
5. Place back in pot over medium heat, bring to a simmer, and add cream. (It may be necessary to add a small amount of stock if soup is too thick after puréeing.)
6. Adjust seasonings to taste.
7. Garnish with warm blanched asparagus tips on the surface of each cup.

---

**Nutrition Facts**

**Per serving:** 393.5 calories; 61% calories from fat; 27.5 g total fat; 76.2 mg cholesterol; 345 mg sodium; 471.6 mg potassium; 29 g carbohydrates; 2.8 g fiber; 5.5 g sugar; 26.2 g net carbs; 9.3 g protein

---

**Variation:**

**149** **Cream of Asparagus Soup (Roux):** Instead of velouté, add 8 oz of flour to matignon and cook until golden brown. Turn off flame and add 1 gal. warm chicken stock while whisking continuously to avoid lumping.

# Recipes

## Purée of Navy Bean Soup with Bacon

*yield: 1 gal. (sixteen 8-oz servings)*

| | |
|---|---|
| 8 oz | bacon, medium dice |
| 1 lb | mirepoix, medium dice |
| 2 oz | leeks, diced |
| 1 tbsp | garlic, chopped |
| 1 gal. | chicken or ham stock |
| 1 lb | ham hock or meaty ham bone |
| 1 lb | navy beans, sorted, washed, and soaked overnight |
| 8 oz | diced canned tomatoes, drained |
| to taste | salt and pepper |

*Sachet d'épices*

| | |
|---|---|
| 2 | bay leaves |
| 1 tsp | peppercorns, cracked |
| 2 sprigs | fresh thyme (or ½ tsp, dried) |

1. In a heavy-bottomed saucepot, cook bacon slowly over medium heat to render out some of the fat and brown slightly. *Note:* Do not cook until crisp and well done.
2. Add mirepoix, leeks, and garlic, and cook over low heat until slightly tender.
3. Add the stock and the ham hock or meaty ham bone and bring to a simmer.
4. Add drained beans and sachet d'épices.
5. Cover the pot and simmer until the beans are tender, about 1½ hr to 2 hr.
6. Add tomatoes and cook an additional 15 min.
7. Remove ham hock and the sachet d'épices. Reserve about 20% of the soup.
8. In a food mill, blender, or food processor, purée the remainder of the soup until the desired texture is achieved. Add back the reserved soup as needed to reach the desired finished consistency.
9. Place soup back in pot and bring to a simmer over medium heat.
10. Remove meat from ham hock or bone. Dice the meat and add to soup.
11. Garnish with diced crisp bacon and croutons.

### Nutrition Facts

**Per serving:** 134.1 calories; 48% calories from fat; 7.1 g total fat; 10.4 mg cholesterol; 764.2 mg sodium; 359.8 mg potassium; 9.1 g carbohydrates; 2.8 g fiber; 1.3 g sugar; 6.3 g net carbs; 8.1 g protein

## Shrimp Bisque

*yield: 1 gal. (sixteen 8-oz servings)*

| | |
|---|---|
| 1 lb | mirepoix, small dice |
| 3 oz | clarified butter |
| 2 lb | shrimp shells (no eyes or veins) |
| 3 cloves | garlic, minced |
| 3 oz | tomato paste or purée |
| ½ c | flour |
| 1 c | dry white wine |
| 1 gal. | seafood velouté, mild |
| 1 pt | heavy cream, hot |
| 1 lb | shrimp, diced and sautéed (for garnish) |
| 2 oz | chives, finely diced |

*Sachet d'épices*

| | |
|---|---|
| 4 sprigs | thyme, fresh |
| 2 | bay leaves |
| 5 | peppercorns, cracked |

1. In a heavy-bottomed saucepot over medium heat, caramelize mirepoix in clarified butter.
2. Add shells and continue to cook until pink.
3. Add garlic and cook for 1 min.
4. Add tomato paste or tomato purée and cook until slightly caramelized, stirring frequently to prevent scorching. Reduce by half.
5. Deglaze the pan with white wine. Reduce liquid by half.
6. Add velouté and sachet d'épices. Gently simmer for 30 min to 45 min.
7. Remove the sachet. Carefully strain the stock, reserving the solids and liquids separately.
8. Place the solids in a blender with just enough liquid to purée. Purée until desired texture is achieved.
9. Place all ingredients back in pot and gently simmer for 15 min.
10. Whisk in heated heavy cream.
11. Strain soup through a fine chinois. Season to taste.
12. Place shrimp garnish and a small amount of chives in a bowl. Fill with bisque.

### Nutrition Facts

**Per serving:** 239.7 calories; 49% calories from fat; 13.5 g total fat; 142.4 mg cholesterol; 235.1 mg sodium; 291.1 mg potassium; 11.7 g carbohydrates; 0.8 g fiber; 2.1 g sugar; 10.9 g net carbs; 15.7 g protein

## New England-Style Clam Chowder

**152**

*yield: 1 gal. (sixteen 8-oz servings)*

*National Fisheries Institute*

| | |
|---|---|
| 6 oz | salt pork or bacon, minced fine |
| 4 oz | flour |
| 2 qt | clams, canned, with juice |
| 1 qt | clam stock or fish stock, mild |
| 1 lb | potatoes, small dice |
| 1 qt | heavy cream, hot |
| ½ tsp | Tabasco® sauce |
| ½ tsp | Worcestershire sauce |
| to taste | salt and white pepper |

*White matignon*

| | |
|---|---|
| 4 oz | celery, small dice |
| 4 oz | leeks, small dice |
| 8 oz | onions, small dice |

*Sachet d'épices*

| | |
|---|---|
| 2 | bay leaves |
| 2 sprigs | thyme, fresh |
| 4 stems | parsley, fresh |
| 4 | peppercorns, crushed |

1. In a heavy-bottomed saucepot, render fat from pork until pork begins to caramelize.
2. Add white matignon and sweat until onions are translucent.
3. Add flour and stir well while cooking slowly to make a roux; cook roux for 5 min to 7 min.
4. Strain clams from juice and reserve clams until needed.
5. Add juice from the canned clams and clam or fish stock while whisking continuously to avoid lumping. Bring to a simmer.
6. Add potatoes and sachet d'épices. Simmer until potatoes are slightly tender.
7. Remove sachet and add reserved clams, hot cream, Tabasco®, and Worcestershire. Season to taste.
8. Garnish with a dollop of chive sour cream or savory herb petite croutons.

---

### Nutrition Facts

**Per serving:** 176.7 calories; 61% calories from fat; 12.3 g total fat; 43.3 mg cholesterol; 484.4 mg sodium; 400.1 mg potassium; 10.8 g carbohydrates; 1 g fiber; 1 g sugar; 9.8 g net carbs; 6.1 g protein

---

## Manhattan-Style Clam Chowder

**153**

*yield: 1 gal. (sixteen 8-oz servings)*

| | |
|---|---|
| 4 oz | salt pork or bacon, small dice |
| 8 oz | celery, medium dice |
| 4 oz | carrots, medium dice |
| 4 oz | leeks, medium dice |
| 8 oz | onions, medium dice |
| 4 oz | green bell peppers, medium dice |
| 3 cloves | garlic, minced fine |
| 1 qt | tomatoes, canned with juice, medium dice |
| 2 qt | clams, canned, with juice |
| 1 qt | clam stock or fish stock, mild |
| 1 lb | potatoes, small dice |
| ½ tsp | Tabasco® sauce |
| ½ tsp | Worcestershire sauce |
| ½ tsp | Old Bay® seasoning |
| to taste | salt and white pepper |

*Sachet d'épices*

| | |
|---|---|
| 2 | bay leaves |
| 2 sprigs | thyme, fresh |
| 4 stems | parsley, fresh |
| 4 | peppercorns, cracked |

1. In a heavy-bottomed saucepot, render fat from pork until pork begins to caramelize.
2. Add celery, carrots, leeks, and onions. Sweat until onions are translucent.
3. Add bell peppers and garlic. Sauté for 1 min to 2 min.
4. Add tomatoes with juice and deglaze pan.
5. Strain clams from juice and reserve clams until needed.
6. Add clam juice and clam or fish stock to pot.
7. Add potatoes and sachet d'épices. Simmer until potatoes are slightly tender.
8. Remove sachet and add clams, Tabasco®, Worcestershire, and Old Bay®. Season to taste.
9. Carefully skim the surface of the liquid to remove any fat or impurities.

---

### Nutrition Facts

**Per serving:** 93.3 calories; 29% calories from fat; 3.1 g total fat; 9.1 mg cholesterol; 551.7 mg sodium; 532.4 mg potassium; 12.2 g carbohydrates; 1.9 g fiber; 3.4 g sugar; 10.3 g net carbs; 4.8 g protein

---

# Recipes

## 154

## Roasted Corn and Poblano Chowder

*yield: 1 gal. (sixteen 8-oz servings)*

| | |
|---|---|
| 2 lb | kernel corn, thawed |
| 6 oz | salt pork or bacon, minced fine |
| 3 cloves | garlic, minced |
| 3 | poblano peppers, small dice |
| 2 oz | flour |
| 2 qt | chicken stock |
| 1 lb | potatoes, small dice |
| 1 qt | heavy cream, hot |
| ½ tsp | Tabasco® sauce |
| ¼ tsp | Worcestershire sauce |
| 1 tbsp | cilantro, chopped |
| to taste | salt and white pepper |

*White matignon*

| | |
|---|---|
| 6 oz | celery, small dice |
| 6 oz | leeks, small dice |
| 8 oz | onions, small dice |

*Sachet d'épices*

| | |
|---|---|
| 2 | bay leaves |
| 2 sprigs | thyme, fresh |
| 4 stems | parsley, fresh |
| 4 | peppercorns, cracked |

1. Spread thawed kernel corn on a sheet pan and place under broiler to caramelize. Stir to ensure even caramelization without burning. Reserve 4 oz of caramelized corn for garnish and set remainder aside until needed.
2. In a heavy-bottomed saucepot, render fat from pork until pork begins to caramelize.
3. Add white matignon; sweat until onions are translucent.
4. Add garlic and diced poblano peppers. Sauté 1 min.
5. Add flour and stir well while cooking slowly to make a roux; cook roux for 5 min to 7 min.
6. Add stock while whisking continuously to avoid lumping. Bring to a simmer.
7. Add potatoes, corn, and sachet d'épices. Simmer until potatoes are slightly tender.
8. Remove sachet d'épices. Remove half of soup mixture and purée in blender.
9. Return puréed soup to pot and bring to a simmer.
10. Carefully skim the surface to remove any fat or impurities.
11. Add hot cream, Tabasco®, and Worcestershire. Season to taste.
12. Garnish with reserved corn and chopped cilantro.

### Nutrition Facts
**Per serving:** 231.5 calories; 51% calories from fat; 13.3 g total fat; 42.8 mg cholesterol; 458 mg sodium; 437.8 mg potassium; 21.6 g carbohydrates; 2.3 g fiber; 4.3 g sugar; 19.3 g net carbs; 8 g protein

## Vichyssoise

*yield: 1 gal. (sixteen 8-oz servings)*

| | |
|---|---|
| 6 oz | whole butter |
| 3 qt | chicken stock, mild |
| 2 lb | potatoes, medium dice |
| 1 qt | heavy cream |
| 4 oz | sour cream (for garnish) |
| 2 tbsp | chives, fresh, chopped |
| to taste | salt and white pepper |

*White matignon*

| | |
|---|---|
| 4 oz | celery, small dice |
| 8 oz | onions, small dice |
| 2 lb | leeks, white part only, julienned |

*Sachet d'épices*

| | |
|---|---|
| 2 | bay leaves |
| 2 sprigs | thyme, fresh |
| 4 stems | parsley, fresh |
| 4 | peppercorns, cracked |

1. Melt whole butter in a heavy-bottomed saucepot over medium heat.
2. Add white matignon. Sweat until onions are translucent and leeks have wilted.
3. Add stock, potatoes, and sachet d'épices. Simmer 30 min to 40 min or until potatoes are almost tender.
4. Remove sachet d'épices. Purée soup in blender or food processor until smooth.
5. Strain mixture through cheesecloth or a chinois to remove any solids.
6. Chill soup in an ice bath and reserve until needed.
7. When ready for service, add cold heavy cream and chives, and season to taste.

### Nutrition Facts
**Per serving:** 243.5 calories; 53% calories from fat; 14.8 g total fat; 40.8 mg cholesterol; 248.3 mg sodium; 503.2 mg potassium; 22 g carbohydrates; 2 g fiber; 4.8 g sugar; 20 g net carbs; 6.7 g protein

## Cold Curried Carrot and Coconut Milk Soup

*yield: 1 gal. (sixteen 8-oz servings)*

| | |
|---|---|
| 4 oz | whole butter |
| 8 oz | onions, small dice |
| 3 lb | carrots, small dice |
| 6 oz | celery, small dice |
| 3 tbsp | ginger, small dice |
| 8 oz | green onions, small dice |
| 3 tbsp | curry powder |
| 2 qt | chicken stock |
| 3 cup | coconut milk, unsweetened |
| 2 to 3 tbsp | lime juice, fresh |

1. In a heavy-bottomed saucepot over medium heat, sweat onions, carrots, celery, and ginger in whole butter until onions are translucent.
2. Add green onions and curry powder. Sauté 1 min.
3. Add stock. Cover and simmer 20 min to 30 min or until carrots are soft.
4. Add coconut milk and bring to a boil.
5. Purée mixture in a blender in small batches until very smooth. Transfer puréed soup to a clean bowl.
6. Stir in lime juice. Chill until very cold.
7. Garnish with crystallized ginger or diced green onions.

**Nutrition Facts**

**Per serving:** 174.9 calories; 62% calories from fat; 12.7 g total fat; 14.4 mg cholesterol; 189.4 mg sodium; 461.3 mg potassium; 13.3 g carbohydrates; 2.7 g fiber; 5.3 g sugar; 10.6 g net carbs; 4.1 g protein

## 157

## Gazpacho

*yield: 1 gal. (sixteen 8-oz servings)*

| | |
|---|---|
| 2 lb | tomatoes, peeled and seeded, concassé |
| 4 oz | onions, medium dice |
| 3 oz | green onions, medium dice |
| 1 tbsp | jalapenos, minced fine |
| 1 | green bell pepper, medium dice |
| 1 | red bell pepper, medium dice |
| 1 lb | cucumbers, peeled and seeded, medium dice |
| ½ lb | celery, medium dice |
| 1 tbsp | garlic, minced |
| 2 tbsp | basil, chiffonade |
| 2 tsp | tarragon, chopped |
| 3 oz | balsamic vinegar |
| 3 oz | olive oil |
| 2 oz | lemon juice |
| 1 tbsp | Worcestershire sauce |
| 2 tsp | Tabasco® sauce |
| 2 qt | tomato juice |
| as needed | vegetable stock |
| to taste | cayenne pepper |
| to taste | salt and pepper |

1. Combine all ingredients except tomato juice and vegetable stock in food processor and purée, leaving a bit of texture to ingredients.
2. Add tomato juice and pulse to incorporate.
3. Thin with vegetable stock if necessary and chill until needed.
4. Garnish with croutons and cilantro or avocado.

**Nutrition Facts**

**Per serving:** 72.5 calories; 51% calories from fat; 4.2 g total fat; 0 mg cholesterol; 45.7 mg sodium; 435.8 mg potassium; 8.8 g carbohydrates; 1.7 g fiber; 5.7 g sugar; 7.1 g net carbs; 1.6 g protein

# Poultry

is the classification of all edible domestic birds. Poultry meats have always been popular food items because they are low in cost and can be prepared using most cooking methods. Poultry meats are quite tender and easy to digest when cooked properly. Poultry is low in fat, calories, and cholesterol if the skin is removed before cooking. The two most common poultry meats are chicken and turkey. Other varieties of poultry include duck, goose, guinea fowl, squab, ostrich, pheasant, and quail.

## Chapter 18

## Chapter Objectives:

1. Contrast the eight varieties of poultry.
2. Describe common forms of market poultry and list factors to consider before deciding which form to purchase.
3. Explain the meaning of the USDA inspection stamp and the USDA grades for poultry.
4. Identify the main precautions to take when handling and storing poultry.
5. Demonstrate techniques used for fabricating poultry.
6. Identify factors to consider when determining the appropriate cooking method for poultry.
7. Give examples of cuts of poultry that are commonly deep-fried or pan-fried.
8. Identify considerations for sautéing poultry.
9. List the main factors to consider when grilling or broiling poultry.
10. Explain the role of temperature in roasting small, medium, and large birds.
11. Demonstrate techniques used to carve poultry.
12. Compare the simmering method of cooking poultry to the poaching method.
13. Describe two different ways poultry can be braised or stewed.

## Key Terms

- **poultry**
- **ratite**
- **game bird**
- **giblets**
- **foie gras**
- **confit**
- **cassoulet**
- **squab**
- **tenderloin**
- **tip**
- **flat wing tip**
- **wing drumette**
- **drumstick**
- **thigh**
- **leg**
- **breast quarter**
- **leg quarter**
- **caul fat**
- **mark**
- **marinade**
- **barding**

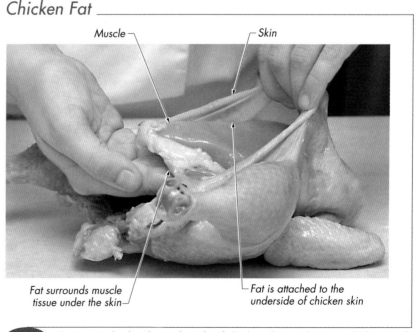

National Chicken Council

## POULTRY COMPOSITION

*Poultry* is the collective term for the various kinds of birds raised for human consumption. Poultry is an extremely popular menu choice because of its versatility, availability, and low cost. Some birds have both light and dark meat, while others have only dark meat. Leg and thigh meat are always dark meat, but breast and wing meat can be either light or dark. Whether the bird can fly determines the color of the meat. If a bird flies, the breast and wing muscles get more exercise, which causes more blood to flow through them. Blood contains the protein myoglobin, which causes darkening of muscle tissue. The more a particular muscle group is worked, the tougher and darker the muscle becomes.

If a particular bird, such as a chicken, does not fly, the breast and wing muscles get less exercise and do not receive as much blood flow. The result is that those areas have lighter-colored "white" meat. However, chickens spend their entire day walking around. Thus, chickens have dark meat in their legs and thighs. Birds that fly, such as ducks, have dark meat throughout their entire bodies, because all the muscle groups get exercise when a bird flies.

In all meats, there are two different kinds of fat. The first kind of fat is highly visible and is located outside of each muscle portion. For example, the skin that surrounds the breast meat of the chicken has a layer of fat attached to the underside. ***See Figure 18-1.*** This layer of fat serves to protect the muscle from damage and to insulate it from the outside environment. The second kind of fat is intramuscular fat, or marbling. This type of fat makes a piece of meat juicy and moist.

## Chicken Fat

Muscle — | — Skin

Fat surrounds muscle tissue under the skin —

└ Fat is attached to the underside of chicken skin

**18-1** *Fat is attached to the underside of chicken skin and surrounds the muscle tissue beneath the skin.*

While most animals have both kinds of fat, poultry has almost no marbling. The fat in poultry can be found just beneath the skin and around the tail and the abdomen. Because poultry has little intramuscular fat, poultry can become very dry if it is even slightly overcooked.

As poultry ages, the breastbone changes from being flexible to being thick and hard. The meat and skin also toughen, and the overall flavor of the bird intensifies. Therefore, younger birds are often desired for their tenderness and mild flavor.

## KINDS OF POULTRY

The six different kinds of poultry recognized by the United States Department of Agriculture (USDA) are chicken, turkey, duck, goose, pigeon, and guinea fowl. Each kind of poultry is subdivided into classes based on the age and tenderness of a bird. Although the USDA identifies only these six kinds of birds as poultry, there are two additional groups of edible birds: ratites and game birds are also served in food service establishments. ***See Figure 18-2.***

A *ratite* is any of a large variety of flightless birds that have small wings in relation to body size and flat breastbones. Ratites include ostriches, emus, and rheas. Although ratite is considered poultry, the flavor of the meat is similar to beef, but sweeter. Ratite meat is dark red and looks like beef after it is cooked.

A *game bird* is a wild bird, such as quail, partridge, pheasant, or grouse, that is hunted for human consumption. While it is legal to hunt these animals for personal use, it is illegal to sell game that has been hunted in the wild. Farm-raised game birds are game birds that are raised for sale and can be sold legally under state regulations. Game birds are considered poultry. Their meat is white, but darker than chicken or turkey breast. Poultry, ratites, and game birds can be cooked using many of the same cooking methods used to prepare other meats and meat dishes.

## Chicken

Chicken is the most popular kind of poultry. Chickens can be cooked whole or in parts, which allows great flexibility in preparation. Younger chickens are very tender and can be cooked using almost any cooking method. Older chickens have developed more connective tissue and tougher muscular tissue, and are therefore better either stewed or braised.

*National Chicken Council*

| | Classifications of Poultry | | |
|---|---|---|---|
| Bird | Class | Weight | Common Cooking Methods |
| **Chicken** | Cornish game hen | less than 2 lb | Broil, grill, or roast |
| | Broiler | 1½ lb to 2 lb | Any |
| | Fryer | 2 lb to 3½ lb | Any |
| | Roaster | 3½ lb to 5 lb | Any |
| | Capon | 5 lb to 8 lb | Roast |
| | Hen/stewing chicken | 3 lb to 8 lb | Stew or braise |
| **Turkey** | Fryer or roaster | 4 lb to 8 lb | Roast, sauté, or pan-fry |
| | Young turkey | 12 lb to 16 lb | Roast or stew |
| | Yearling | 10 lb to 30 lb | Roast or stew |
| | Mature turkey | 10 lb to 30 lb | Stew |
| **Duck** | Fryer or broiler | 3 lb to 4 lb | Roast or sauté |
| | Roaster | 4 lb to 6 lb | Roast |
| | Mature | 4 lb to 6 lb | Braise |
| **Goose** | Young | 4 lb to 10 lb | Roast |
| | Mature | 10 lb to 18 lb | Braise or stew |
| **Guinea fowl** | Young | 1 lb to 1½ lb | Roast |
| | Mature | 1 lb to 2 lb | Braise or stew |
| **Pigeon** | Squab | ¾ lb to 1½ lb | Broil, sauté, or roast |
| | Pigeon | 1 lb to 2 lb | Not commonly prepared |
| **Ratites** | Ostrich | up to 300 lb | Grill, broil, sauté, stew, or braise |
| | Emu | up to 100 lb | Grill, broil, sauté, stew, or braise |
| | Rhea | up to 55 lb | Grill, broil, sauté, stew, or braise |
| **Game Birds** | Grouse | 1 lb to 10 lb | Sauté or roast |
| | Pheasant | 1½ lb to 2¼ lb | Roast or braise |
| | Partridge | 1 lb | Broil, sauté, or roast |
| | Quail | ¼ lb | Grill, broil, sauté, or roast |

**18-2** *Birds raised for human consumption include poultry, ratites, and game birds.*

Chickens are classified by age and weight. Chicken classifications include Cornish game hens, fryers and broilers, roasters, capons, and hens or stewing chickens. *See Figure 18-3.*

A Cornish game hen is a male or female chicken between 5 and 6 weeks old. The meat of a Cornish game hen is very tender and flavorful, with minimal fat. They can be prepared in a wide variety of ways including broiling, grilling, or roasting.

## Chickens

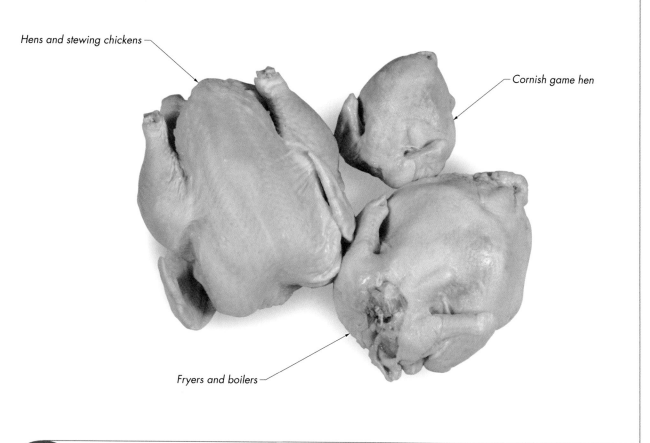

Hens and stewing chickens

Cornish game hen

Fryers and boilers

**18-3** *Chickens are classified by age and weight.*

Fryers and broilers are young male or female chickens under 13 weeks old. They have very tender flesh and smooth skin, and contain a slightly higher percentage of fat than a Cornish game hen. Although either term is used to represent this class, the weight of a broiler usually is between 1½ lb and 2 lb, while a fryer typically weighs between 2 lb and 3½ lb. Any cooking method can be used to prepare fryers and broilers.

Roasters are young male or female chickens ranging in age from 3 months to 5 months old. They have tender meat and smooth skin, but have a higher ratio of meat to bone than do birds in the previously mentioned categories. This means that if a broiler and a roaster had the same amount of bone, the roaster would have more meat than the broiler. Roasters range in weight from 3½ lb to 5 lb. Roasters can be prepared using any cooking method.

A *capon* is a surgically castrated male chicken. Capons are castrated to produce a large, well-formed breast with flesh that is more tender and better flavored than that of an average chicken. The result is thicker portions of meat, more moisture (due to a higher fat content), and a higher proportion of light meat to dark meat than in other chicken varieties. Once a bird is castrated, the tenderness of the flesh is affected very little as the bird ages. Capons are usually between 5 months and 8 months old. The average weight of a capon is 5 lb to 8 lb. They are most often prepared by roasting.

A *hen,* or stewing chicken, is a female chicken that has laid eggs for one or more seasons and is usually more than 10 months old. The flesh and skin of these birds are tough due to the age of the birds. The tougher flesh and skin require slow moist-heat cooking methods, such as stewing or braising, to make the meat tender enough to consume. Hens average in weight from 3 lb to 8 lb. Many people do not care for the flavor of hens, as they are stronger in flavor than any other type of chicken.

## Turkey

Turkeys are native to North America and are the second-most-popular poultry consumed in the United States. Turkeys are economical, regularly available, and have low fat content. They are most often roasted, sometimes stuffed. Other preparations include sautéed turkey cutlets, grilled turkey tenderloins, and ground turkey meat. Virtually any method can be used for preparing ground turkey meat.

Turkeys are bred as lightweights and heavyweights. Lightweights are bred for fast growth and a more marketable size. Heavyweights are bred to be larger-sized turkeys and are used mostly in food service establishments. Heavyweight turkeys have more meat in proportion to bone and sell at a lower cost per pound than lightweights. ***See Figure 18-4.***

Like chicken, turkeys are also classified by age and weight. Common classes of turkeys include fryers or roasters, young turkeys, yearlings, and mature turkeys.

Fryers and roasters are lightweight male or female turkeys less than 16 weeks old. They are very tender with soft, flexible skin. They average in weight from 4 lb to 8 lb. Male fryers and roasters are commonly called "toms." Female turkeys are commonly called "hens." Common preparation methods include roasting, frying, or broiling.

A young turkey is a male or female turkey less than 8 months old. These turkeys have tender, lean meat, with smooth skin and a flexible breastbone. Young lightweights average in weight from 6 lb to 10 lb. Young heavyweight turkeys average 12 lb to 16 lb. Young turkeys are best when roasted, boiled, or stewed.

## Nutrition Note

Ground turkey has become a popular substitute for ground beef or ground pork. By using ground turkey in place of ground beef or pork, you can prepare dishes that are lower in cholesterol and saturated fat.

## Turkeys

Turkeys are bred as lightweights or heavyweights

Ground turkey

USDA

Breast

Thighs

Wings

Legs

**18-4** *Turkeys are available whole or fabricated into breasts, legs, thighs, wings, or ground meat.*

### Historical Note

Where do you think the first Thanksgiving turkey came from? One might assume that Pilgrim hunters shot the turkey in the wild and brought it home for their first Thanksgiving dinner. However, that was probably not the case. While wild turkeys are native to North America, the Pilgrims brought domesticated turkeys with them from Europe. Spanish explorers are believed to have brought the first turkeys to Europe in the 1500s, just as they brought spices and other foods from the countries they visited. Around that time, Spanish ships traveling to England introduced the turkey to English farmers. The English farmers began to breed and raise their own turkeys for food, and in 1620, the Pilgrims brought these turkeys with them on the Mayflower. The Pilgrims selectively bred their domesticated turkeys to yield tastier, tenderer, and meatier birds.

A yearling is a mature bird between 8 months and 15 months old. The meat of a yearling is still tender, but it is very lean. Yearlings range in weight from 10 lb to 30 lb and are commonly sold in supermarkets at a cheaper price per pound than young turkeys or roasters.

Mature turkeys are turkeys more than 15 months old, with toughened meat and a hardened breastbone. Like yearlings, mature turkeys can range in weight from 10 lb to 30 lb. However, mature turkeys are usually used only in processed foods or to produce lean ground turkey meat. Mature turkeys have a strong flavor that does not lend itself to dry-heat cooking methods.

When purchasing whole poultry, it is common to find a small bag inside the cavity of the bird containing the giblets. *Giblets* is the collective name for a bird's neck, heart, liver, and gizzard (the bird's second stomach). ***See Figure 18-5****. The neck, heart, and gizzard are often used to make giblet gravy. The neck contains a good amount of

## Giblets

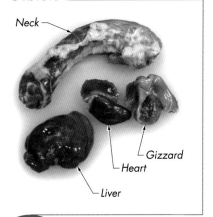

Neck

Gizzard

Heart

Liver

18-5 *Giblets include a bird's neck, heart, liver, and gizzard.*

## Duck

18-6 *Whole duck can be roasted, or individual cuts of duck can be prepared using a variety of cooking methods.*

gelatin, which when used in stock makes the stock rich and flavorful. The gizzard, heart, and liver are not used to make stock because their strong flavors would overpower the flavor of the stock. The heart of a turkey can be served with a béchamel sauce in a classic preparation called creamed hearts. Livers are classically used in pâtés, can be breaded and fried, or can be sautéed and served with caramelized onions. Gizzards can be cooked and served on their own by trimming off the tough connective tissue, and then breading or battering and frying them.

### Duck

Ducks are the third-most-popular type of poultry consumed worldwide. Consumption of duck outranks that of turkey throughout much of Europe. Because they fly, ducks do not have white breast meat. Ducks also have less meat in proportion to bone and fat than do most other birds. A duck yields half as much meat as a chicken of the same size. This means that a 3 lb to 4 lb duck provides only two servings of meat, while a 3 lb to 4 lb chicken provides four servings of meat. Ducks are also very high in fat. The majority of fat in a duck is located in and just beneath the skin.

When cooking duck, the fatty skin must be rendered slowly or it becomes nearly inedible. Duck breasts are commonly seared in a sauté pan then finished in the oven and served medium rare. Duck legs are most often braised or roasted until well done and tender. Whole ducks are usually roasted on a rack to allow the fat to render, such as when making Duck à l'orange or Peking duck. **See Figure 18-6.**

Like other forms of poultry, ducks are classified by age and weight. Classes of duck include broilers or fryers, roaster ducklings, and mature ducks. Broilers or fryers are ducks less than 8 weeks old that weigh between 3 lb and 4 lb. Broilers and fryers have very tender meat, a soft bill, and a soft windpipe. Roaster ducklings are ducks that are between 8 weeks and 16 weeks old that weigh between 4 lb and 6 lb. The meat of a roaster duckling is tender, and the bill and windpipe are just starting to harden. A mature duck is a duck more than 6 months of age that, on average, weighs between 4 lb and 6 lb. The flesh of a mature duck is fairly tough and the bill and windpipe are hardened.

Ducks are not raised in as tight a living space as chickens; therefore, they are not as prone to infection or contamination by biological hazards such as salmonella. Even though it is poultry, duck breast is most commonly served medium rare. When cooked, duck breast meat has the same appearance as a lean cut of beef. Duck breast is very dry and tough if overcooked.

*Foie gras* (pronounced fwah grah) is the fattened liver of a duck or goose. Foie gras is a delicacy that is considered a staple of classic fine dining. ***See Figure 18-7.*** To make foie gras, ducks or geese are force-fed a rich diet of corn and other feed through a funnel and are prevented from exercising. Under these conditions, the liver swells to at least four times its normal size and becomes almost solid fat. The fattened liver is removed and sold as the most expensive component. Because the foie gras is almost all fat, it is typically seared quickly in a very hot sauté pan and served immediately. Foie gras can also be poached, cooled, and puréed to make a pâté de foie gras.

Duck legs and thighs are most commonly prepared confit. *Confit* (pronounced cone-FEE) is a French term for meat that has been cooked and preserved in its own fat. The duck leg and thigh are salted, seared until well browned, and then simmered in rendered duck fat until tender. The confit can be served whole or the meat can be pulled from the bone and used in salads and other dishes as a flavorful ingredient. A traditional cassoulet is a dish that consists of white beans stewed with duck fat, fresh sausage, and whole duck confit.

### Goose

Geese, like ducks, contain entirely dark meat and a large amount of fat in both the skin and flesh. Geese weighing less than 11 lb are considered light. Geese over 12 lb are considered heavy. They are classed into two groups: young and mature.

Young geese are usually less than 6 months of age. They have tender, dark flesh and a windpipe that is easily dented. The meat has a rich taste, due to a high fat content. Young geese weigh from 4 lb to 10 lb. There are large flaps of thick, fatty skin around the neck and tail portions of geese that should be removed with a knife or kitchen shears prior to cooking. Geese are commonly roasted at very high temperatures to aid in rendering some of the fat from the skin and the flesh. Young roasted goose is frequently served with an acidic fruit sauce to cut some of the fatty richness.

Mature geese are over 6 months of age. The flesh is tough and can be prepared only by braising or stewing. As in older ducks, the windpipe of a mature goose is hardened. Mature geese average from 10 lb to 18 lb and are not often used by food service establishments.

### Guinea Fowl

Guinea fowl (or simply "guineas") have been domesticated in most parts of the world and are related to one of the most well-known game birds, the pheasant. Guinea fowl are agile, colorful birds with flesh that is darker than that of chickens. Guineas taste much like wild game. Guinea fowl have both light and dark meat that is lean and tender.

### Foie Gras

*Daniel NYC*

**18-7** *Foie gras is fattened duck or goose liver.*

More popular in Europe than in the United States, guinea fowl are available as either young or mature guineas. Young guineas have tender, lean flesh with a flexible breastbone and weigh, on average, between 1 lb and 1½ lb. Young guineas are most commonly roasted whole. Mature guineas are old guinea fowl that have tough flesh and a hardened breastbone and weigh, on average, between 1 lb and 2 lb. Mature guineas must be braised or stewed to make the meat tender enough to eat.

## Pigeon

Pigeons are classified into two groups based on age: squab and pigeon. A *squab* is a young pigeon bred and raised for human consumption. These young birds have never flown, but they consist of exclusively dark meat because of a special diet used to produce meat that is extra tender. Squabs are marketed when they are 3 weeks to 4 weeks old and weigh between 6 oz and 14 oz. Squabs are expensive and are found on only upscale menus. Squabs are most commonly prepared using the roasting method, but can also be prepared by sautéing or broiling. ***See Figure 18-8.***

A pigeon is a mature squab over 4 weeks old. The flesh of a pigeon is tough and strong in flavor, and can be prepared using only the stewing or braising methods. Pigeons weigh between 1 lb and 2 lb and are almost never served in food service establishments.

## Ratites

Ratites are a group of flightless birds that have recently gained popularity as a food. Types of ratite birds include the ostrich, the rhea, and the emu. Since 2002, the USDA has required ratite meat to be inspected before being sold for human consumption. Ratites have dark red meat that is almost identical in color to beef. Ratites have almost no breast meat. Rather, meat is taken from the back, thigh, and forequarter. The tenderest cut is called the fan and is taken from a muscle in the thigh. Ratite meats are sold as steaks, fillets, medallions, and whole roasts, or as ground meat. ***See Figure 18-9.***

Ratite meats are prepared similarly to lean cuts of beef. Most cuts of ratite meat are best when prepared medium rare or medium using a quick dry-heat cooking method such as grilling, broiling, or sautéing. If ratite meat is desired well done, it is best to prepare it using a slow moist-heat method, or a combination cooking method such as braising or stewing. Because ratite meat is very lean, preparing it well done using a dry-heat cooking method results in a dry, tough piece of meat. Cooked ratite meat is complemented by savory flavors such as mushrooms, garlic, red wine, and herbs. The rich, sweet flavors of sauces made with dried fruit, such as a cherry compote, also complement the flavor of ratites. Ratite meat is available year-round from some local and online sources.

## Squab

*Daniel NYC*

18-8    *Squabs are young pigeons bred and sold for human consumption.*

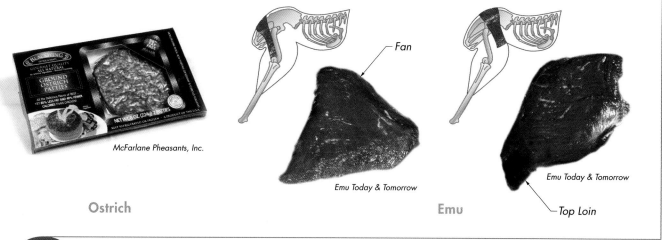

McFarlane Pheasants, Inc.

*Fan*

Emu Today & Tomorrow

*Ostrich*

*Emu*

Emu Today & Tomorrow

*Top Loin*

**18-9** *Ostriches and emus are two types of ratites.*

**Ostrich.** Ostrich is the most popular of the ratites and the most readily available in the United States. Ostriches are also the largest of the ratites, yielding the most meat. A male ostrich can tip the scales at over 300 lb and can reach 9′ in height. The tenderest cuts of meat are found in the loin. These cuts are best prepared with fast, dry-heat cooking methods. Other, tougher cuts are found in the leg, and these cuts are best prepared by first marinating or tenderizing the meat and cooking it with a dry-heat method until well done and tender, or cooking slowly using a combination cooking method.

**Emu.** The emu is a native to Australia and is second in size to only the ostrich. Commercial farming of emus began in 1987, and emu meat became available commercially in 1990. Today a number of countries produce emus, with over 1 million head of emu being raised each year in the United States alone. Emu chicks gain about 2½ lb per week and are fully grown in about a year. Emu meat is prepared in the same manner as ostrich meat, with tender cuts cooked quickly using dry-heat methods, and tougher cuts slow-cooked using a combination cooking method and moist heat. The flavor of emu is slightly different from that of ostrich, although the two are very similar.

**Rhea.** The rhea is smaller than the emu, but is still one of the largest birds in existence. One variety of rhea, the American rhea, is the largest bird native to South America. Like all ratites, rhea chicks grow quite quickly and are considered fully grown at 1 year of age. Meat from a rhea is prepared using the same cooking methods as for ostrich and emu. Because they are the smallest of the ratites, they yield less meat than ostriches or emus. The meat is very lean, is similar in flavor to ostrich and emu, and is available through online suppliers.

U.S. Department of Agriculture

## Game Birds

Game birds are birds that are found naturally in the wild and that have been hunted for sport and food. In the United States, birds caught or hunted in the wild cannot be sold, so it is illegal to serve such birds in restaurants. However, some farmers have started to domesticate and raise game birds on farms. These farm-raised birds can be purchased from licensed food service distributors and sold in restaurants. Farm-raised birds are inspected for wholesomeness by the USDA. It is important to note that, although they are inspected for wholesomeness, farm-raised game birds are not graded by the USDA and are not classified as poultry.

Because of their scarcity, farm-raised game birds are high in price. They are prepared similarly to domestic poultry, but require slightly different preparations to help preserve their game flavor. Older male game birds are tougher and have a stronger flavor than older female game birds. Because game birds are generally leaner than common poultry, overcooking these birds results in tough, stringy, dry meat. Game birds are usually aged or ripened for a short time in open air. The term "high" is used to refer to a bird that has been ripened for one to two weeks. Game birds lack a sufficient amount of fat covering to be aged longer. When a bird is high, the tail feathers pull out easily. Game birds include waterfowl, such as wild ducks and geese, and field fowl, such as wild turkeys, pheasants, quails, grouses, and partridges. ***See Figure 18-10.***

Game Birds

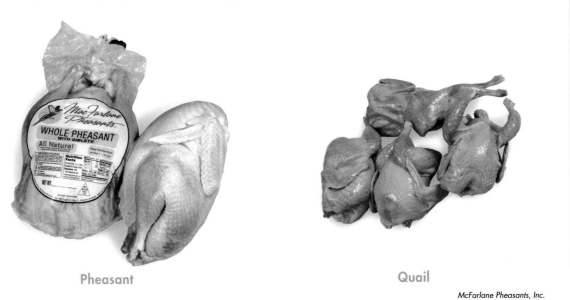

Pheasant

Quail

*McFarlane Pheasants, Inc.*

**18-10** *Game birds such as pheasant and quail that have been farm-raised can be sold for consumption.*

**Pheasant.** Pheasant has long been the most popular game bird. The pheasant is a fairly large, long-tailed bird with dark, rich, gamey-tasting meat. Pheasants range in size from 1½ lb to 2¼ lb and are prepared by roasting or braising. The female bird is preferred over the male because the female is more tender and moist. Pheasant meat is considered possibly the best game bird meat. A consommé made from a pheasant carcass is also extremely flavorful.

**Quail.** Quail is the smallest of the common game birds. With breast meat weighing only about 1 oz to 2 oz, they are close in size to chicks. Because the amount of meat on a quail is so small, they are commonly boned out and cooked in one of the following ways: filled with stuffing, rice, or a forcemeat, then roasted and served whole; broiled; skewered and grilled; or sautéed. One of the most common quail preparations is to season the boneless quail, sauté it over high heat, and deglaze the pan with balsamic vinegar and sherry.

**Grouse.** A grouse resembles small domestic fowl in appearance, but has thicker and stronger legs. There are more than 40 species of grouse found in North America. The most common types of grouse are the ruffled grouse, the sage grouse, and the blue or dusty grouse. All species have a fairly long, feathered tail, a medium-size wingspan, and a short, thick bill.

The grouse has dark meat that is recognized by gourmets and connoisseurs as one of the finest meats from any type of bird. Grouse can be prepared using different methods; however, large grouse are best roasted, and small grouse are best sautéed. The flesh of the female bird is usually superior in flavor to that of the male.

**Partridge.** Partridges are smaller than pheasants and usually provide only enough meat to serve two people. A typical partridge weighs about 1 lb. The meat is white and must be cooked slightly on the done side to develop the desired succulent, gamey flavor. Roasting is the most popular method of cooking partridges, but they can also be broiled or sautéed. The flavor of partridge is gamier than that of pheasant and the flesh is quite a bit tougher and chewier.

*U.S. Department of Agriculture*

## MARKET FORMS

Birds are sold after they have been inspected and the head, neck, feet, and feathers have been removed. They are eviscerated (the entrails and viscera are removed), washed, and sanitized. Birds can be either packed fresh or frozen, whole or cut up, boneless or bone in, whole, ground, or processed into another form such as chicken nuggets or deli meat. Many food service operations purchase fabricated (cut up) poultry or prepared forms to save on labor costs and for the convenience of having the processing work already done.

National Chicken Council

Common prepared forms of poultry include breasts, tenders, tenderloins, wings, drumsticks, thighs, legs, halves, and quarters:

- A *breast* is the top front portion of the meat above the rib cage, consisting of white meat in flightless birds, or dark meat in birds that fly.

- A *tender* is a small strip of breast meat.

- A *tenderloin* is the inner pectoral muscle that runs alongside the breastbone. Each chicken has two wings, which are divided into tips, flat wing tips, and wing drumettes.

- A *tip* is the outermost section of wing. It is often reserved for making stocks as it contains little or no meat.

- A *flat wing tip (paddle)* is the second section of wing located between the two wing joints.

- A *wing drumette* is the innermost section of wing located between the first wing joint and the shoulder.

- A *drumstick* is the lower section of leg located below the knee joint.

- A *thigh* is the upper section of leg located below the hip and above the knee joint.

- A *leg* consists of a drumstick and thigh.

- A *poultry half* consists of a full half-length of bird split down the breast and spine.

- A *breast quarter* is a half of a breast, a wing, and a portion of the back.

- A *leg quarter* is a thigh, a drumstick, and a portion of the back.

Purchasing whole birds and fabricating them in-house can save a great deal of money, as the purchase price per pound is substantially reduced. Furthermore, poultry fabrication is relatively easy. With some training, kitchen staff can learn the many different ways to fabricate poultry. **See Figure 18-11.**

Before choosing to purchase either fabricated (cut up) chicken or whole chicken, it is important to consider factors that can affect the outcome of the product, such as employee skill level, storage space, menu offerings, and labor costs. For example, it may be less expensive per pound to purchase whole chickens and debone them in-house. A manager, food service director, or food service buyer should ask the following questions:

- Do the kitchen personnel have sufficient knife skills to process the chickens correctly and efficiently?

- Does the establishment have storage space for the bones, trimmings, and scraps that can be made into stock or stored for later use?

- Does the establishment have storage space for the fabricated pieces? (*Note:* Fabricated poultry products come packaged in boxes, allowing a chef to remove only as much product as needed and leave the rest in frozen or refrigerated storage.)

- When purchasing fabricated poultry products, will the increase in cost per pound be greater than the savings on labor cost?

- Can the establishment use all parts of the chicken, including the fabrication by-products, in-house? Does the establishment have a use for leg and thigh meat? Does the establishment make fresh stock in-house?

## Poultry Cuts

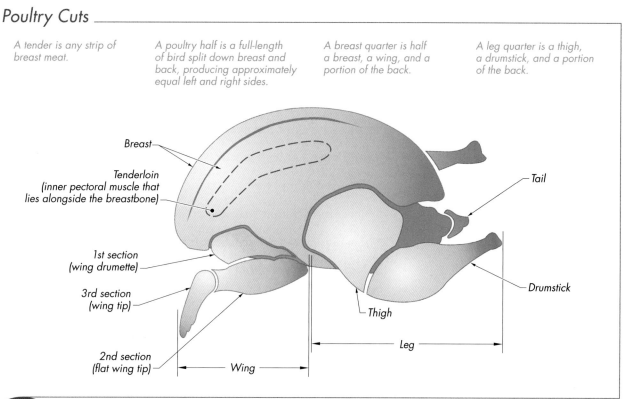

A tender is any strip of breast meat.

A poultry half is a full-length of bird split down breast and back, producing approximately equal left and right sides.

A breast quarter is half a breast, a wing, and a portion of the back.

A leg quarter is a thigh, a drumstick, and a portion of the back.

Breast
Tenderloin (inner pectoral muscle that lies alongside the breastbone)
1st section (wing drumette)
3rd section (wing tip)
2nd section (flat wing tip)
Wing
Thigh
Leg
Tail
Drumstick

**18-11** *Poultry is available as prefabricated cuts.*

## Poultry Inspection

All poultry produced for human consumption must be inspected for wholesomeness by the USDA. The USDA inspection stamp is a guarantee by the USDA that the meat was processed under proper conditions, that it was inspected by trained personnel who verified its fitness for human consumption, and that the meat was wholesome at the time of inspection. The USDA inspection stamp is a round stamp. *See Figure 18-12.* The round USDA inspection stamp is found either on a tag attached to the wing of the bird or on a label on the packaging.

While this inspection is mandatory, it does not indicate the tenderness or overall quality of the bird. It guarantees only that the product was fit for human consumption at the time of processing. It is imperative to monitor the handling and processing of poultry when it enters a food service establishment to ensure that it continues to be safe to prepare and eat.

## USDA Inspection Stamp

INSPECTED FOR WHOLESOMENESS BY U.S. DEPARTMENT OF AGRICULTURE P-42

**18-12** *The USDA inspection stamp guarantees wholesomeness at the time of inspection.*

## USDA Grade Stamp

**18-13** *The USDA grading stamp indicates the quality of the poultry.*

*Chef's Tip:*

Quality grades address the overall quality and appearance of poultry, but they have nothing to do with tenderness or flavor. For indicators of these, a chef must look at a bird's class. For example, a Cornish hen is more tender and has a better flavor than a stewing hen, because a Cornish hen is a younger bird.

## Poultry Grading

In addition to the round inspection stamp, a shield-shaped USDA grading stamp may also be on a tag clipped to the poultry or stamped on the packaging material, indicating the quality of the bird. *See Figure 18-13.* Although grading inspection is not mandatory, most poultry sold in restaurants and the marketplace has been graded. Poultry is graded based on overall quality with respect to shape, distribution of fat, condition of the skin, and general appearance of the bird. USDA grades for poultry are Grade A, Grade B, and Grade C. For the most part, poultry sold to consumers is Grade A. To be classified as Grade A, the following standards must be met:

- the bird must be free from deformities
- the bird must not contain any unplucked pin feathers
- the bird must not have any discoloration or bruising
- the flesh should not have any cuts or tears
- the carcass should not have any broken bones
- if the poultry is frozen, it should be free from freezer burn and ice crystals
- the bird should have plump, meaty flesh and a thin layer of fat under the skin.

Poultry not meeting these criteria receives a grade of either B or C, depending on the extent of the flaws. Grade B and Grade C birds are typically used for processed poultry items such as soups, chicken pot pies, or chicken nuggets.

## Poultry Handling and Storage

Poultry spoils rapidly, and spoiled poultry is extremely dangerous. Although poultry develops an odor as it spoils, poultry may be unsafe for consumption prior to developing any odor. Fresh poultry should always be delivered packed in crushed ice. After it is received, it should be washed, packed with fresh crushed ice, and refrigerated as soon as possible. Poultry should be stored in the coldest part of the refrigerator between 32°F and 34°F and should always be stored beneath other foods to prevent poultry juices from dripping onto and contaminating other foods. It is generally best to store poultry in drip pans to catch any juices that may overflow or drip from it.

Fresh poultry should be purchased the day before use to eliminate a long holding period and decrease the chance of spoilage. If fresh poultry is not going to be used for two or three days, it should be frozen immediately after it is received to prevent loss of quality and possible spoilage. Frozen poultry, such as turkey or chicken breasts, should be kept frozen in the original packaging, at or below 0°F. Frozen poultry products can safely be stored frozen for up to six months before their quality deteriorates.

When frozen poultry is needed for use, it should be moved into a refrigerator a day before use to thaw overnight to ensure proper food safety. Whole turkeys may need an additional day or two to thaw because of their size and density, but they should still be thawed in a refrigerator. Thawing frozen poultry in a refrigerator reduces the risk of spoilage. Poultry should never be refrozen once thawed.

Chickens are often raised in crowded living conditions and are often carriers of viruses and bacteria such as salmonella. To help reduce the potential for foodborne illness, all poultry should be washed prior to preparation. In addition, poultry must be cooked thoroughly to an internal temperature of 165°F (or to between 150°F and 155°F to allow for carryover cooking).

*Chef's Tip:*

Salmonella and campylobacter jejuni are serious bacterial concerns when handling, processing, and cooking raw poultry. It is extremely important to wash and sanitize all work areas and tools (including hands) when handling, processing, or cooking raw poultry to prevent cross-contamination. Never use the same cutting boards or tools that were used with raw poultry for other products until they have been washed and sanitized.

## POULTRY FABRICATION

Fabricating poultry is a relatively easy skill to master with some practice. All birds have similar overall body and bone structure. Chicken is typically the easiest, most economical, and most readily available bird for fabrication training. Once the skill of poultry fabrication is learned, it can be applied to any bird raised for human consumption.

### Trussing a Whole Bird

When roasting a whole bird, the bird is usually tied, or trussed, to keep it compact in shape. Trussing the bird pulls the legs and wings tightly to the body, which helps the bird to roast evenly and retain moisture as well as gives it a pleasing finished appearance. Prior to trussing the bird, a few steps are taken to produce the best results:

Refer to the CD-ROM for the **Trussing Poultry** Media Clip.

• Remove the giblet bag from the cavity of the bird and refrigerate for later use.

• Trim any excess or thick pockets of fat around the neck area and the tail portion of the bird.

• Pull the skin tightly and evenly across the breast to fully cover any exposed breast meat. This helps prevent the breast meat from drying out during cooking.

• If desired, remove the wing tips, as they have a tendency to burn during roasting. If the wings are left intact, tuck the first joint behind the second joint for a neater appearance.

After these steps have been performed, the bird is ready to be trussed and roasted. Butcher's twine is the most commonly used trussing string.

**To truss a whole bird, apply the following procedure:**

1. Cut a length of butcher's twine approximately three times the length of the bird to be trussed. ***See Figure 18-14.***
2. With the breast up and the neck facing you, place the center of the twine beneath the bird, about 1"under the tail.
3. Bring the twine up around the legs and cross the ends, creating an "X" between the legs.
4. Pass the ends of the twine under the legs and pull tight.
5. Flip the bird over and tuck the wings, if they are still attached. Pull the twine across the wings and cross at the neck.
6. Tie a square knot around the neck to secure the truss.

*Trussing a Whole Bird* _____

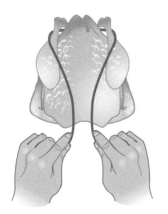

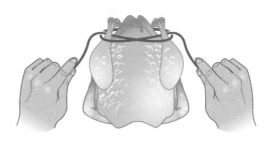

*1–2. Measure twine three times the length of the bird; place the twine under tail of the bird.*

*3–4. Bring the twine up around the legs and cross the ends, creating an "X" between the legs. Pass the ends of the twine under the legs and pull tight.*

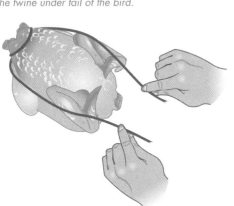

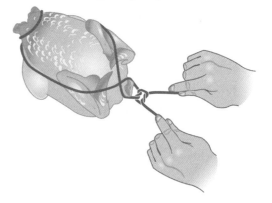

*5. Flip the bird over and tuck the wings, if they are still attached. Pull the twine across the wings and cross it at the neck.*

*6. Tie a square knot around the neck to secure the truss.*

**18-14**  *Trussing a bird allows it to roast evenly, retain moisture, and have a beautiful finished appearance.*

5. Bring the string up and around the top of the legs. Tie a knot to secure the legs together.

6. Flip the bird over. Bring the string around to the back of the bird and tie a square knot to secure the truss.

## Cutting Techniques

There are many ways that poultry and other birds can be cut into controlled portion sizes or forms for use with various cooking methods. Typical cutting methods for birds include cutting them into halves, quarters, or eighths. They can also be cut to produce boneless breasts, boneless skinless breasts, and airline breasts or suprêmes. These cutting techniques can be applied to almost any type of bird because all birds have similar body structure.

Birds are commonly cut into halves for roasting or broiling. The bird is split from top to bottom between the breasts and along the backbone to the tail. This results in two equal portions.

## To cut a bird in half, apply the following procedure:

1. Square the bird with hands, firmly squeezing the legs and wings in toward the body. *See Figure 18-15.*

2. Place the bird back-side up and use a stiff boning knife to split the bird along both sides of the backbone from the neck to the tail.

3. Remove the backbone.

4. Open both sides of the bird to reveal the keel bone (breastbone). Cut through the keel bone and wishbone lengthwise from neck to tail. Use force by hitting the handle of the knife with the heel of the hand if necessary.

5. Using a knife, cut through the meat and skin behind the breastbone to separate the bird into two halves.

6. Make a small slit in the skin just beneath the leg and insert the end of the drumstick into the slit. This will hold the leg firmly to the body when cooking.

Birds are commonly cut into quarters with bones in for roasting, broiling, or grilling. The bird is divided and separated into leg and thigh sections and wing and breast sections. There are two of each section, yielding four quarters.

## Cutting a Bird into Halves

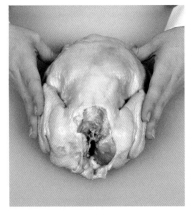

*1. Square up bird.*

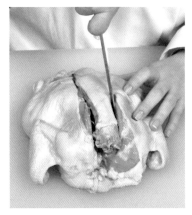

*2. Split bird along both sides of the backbone from neck to tail.*

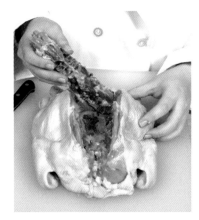

*3. Remove backbone.*

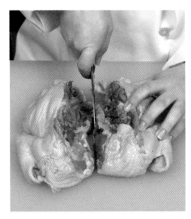

*4. Cut through the keel bone and wishbone.*

*5. Cut through meat and skin behind keelbone.*

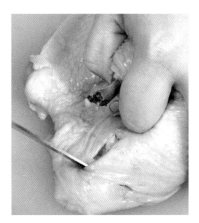

*6. Cut a slit beneath leg bone and insert bone in slit.*

**18-15** *Birds are commonly cut into halves for roasting or broiling.*

### To cut a bird into quarters, apply the following procedure:

1. Stand the bird on end and, with a stiff boning knife, split the bird along either side of the backbone from the neck to the tail. *See Figure 18-16.*

2. Lay the bird breast-side down and remove the backbone.

3. Open both sides of the bird to reveal the keel bone. Cut through the keel bone and wishbone lengthwise from the neck to tail. Use force by hitting the handle of the knife with the heel of the hand if necessary.

4. Using a knife, cut through meat and skin behind the keel bone to separate the bird into two halves.

5. Using a stiff boning knife, cut inside the joint between the breast and the thigh to separate the thigh from the breast. *Note:* Use care to not cut the small, tender portion of meat, known as the oyster, located just along the backbone in the hollow of the hip joint and at the top of each thigh. This tender, usable meat should be removed and reserved for later use.

## Cutting a Bird into Quarters

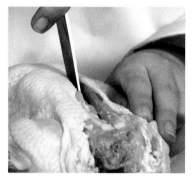
1. Split bird along either side of backbone from neck to tail.

2. Remove backbone.

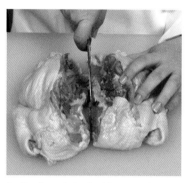
3. Cut through the keel bone and wishbone.

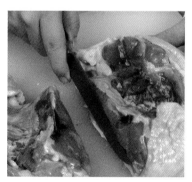
4. Cut through meat and skin behind keel bone.

5. Cut inside joint between breast and thigh.

**18-16** *Birds are cut into quarters for roasting, broiling, or grilling.*

Birds are commonly cut into eighths (bone in) for roasting, broiling, grilling, deep-frying, and pan-frying. To cut a bird into eighths, quarters are cut into two individual pieces (breast and wing, thigh and leg), yielding eight pieces from each bird. It is important to realize that the cooking time for individual pieces is less than for larger sections.

Refer to the CD-ROM for the **Cutting Poultry into Eighths** Media Clip.

## To cut a bird into eighths, apply the following procedure:

1. Start with the bird cut into quarters. ***See Figure 18-17.***
2. Using one hand to hold the wing section, cut the wing from the breast at the wing joint with a stiff boning knife. Repeat for other breast and wing.
3. Holding the leg so that the inside thigh bone is visible, locate the thin line of fat that separates the leg and thigh muscles. With a stiff boning knife, cut along this line and separate the leg and thigh joint. Repeat for other leg and thigh.

### Cutting a Bird into Eighths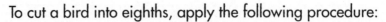

*1. Start with a bird already cut into quarters.*

*2. Cut wing from breast at wing joint; repeat with other breast and wing.*

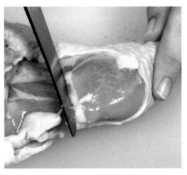

*3. Cut along the line of fat separating the leg and thigh; repeat with other leg and thigh.*

**18-17** *Birds are cut into eighths for roasting, broiling, grilling, deep-frying, and pan-frying.*

Boneless breasts are the most popular poultry cut. They are also the most versatile poultry cut. They can be grilled, sautéed, broiled, baked, pan-fried, poached, stuffed, or skewered, and can be used in an endless number of recipes and presentations. Unlike other cuts of poultry, they are almost as popular served cold on salads as they are when served hot. Larger breasts found on bigger birds such as turkey can be roasted whole or sliced into thin medallions (scallops) and sautéed.

## To cut boneless breasts, apply the following procedure:

1. Cut alongside the keel bone, following the rib cage, to remove the wing and breast meat completely. ***See Figure 18-18.***
2. Separate the leg and thigh sections from the bird by bending the legs and thighs backward until they touch, dislocating the hips.
3. Place the bird breast-side down. Using the tip of a stiff boning knife, cut along the backbone from neck to tail.

4. Cut horizontally at the pelvis, slightly above the oysters. Oysters are two edible bits of muscle located in the hollow of the pelvic bone. Pull the thighs and legs away from the carcass.

5. With a boning knife, cut along the edge of the wishbone to remove completely.

6. Cut alongside the keel bone, following the rib cage, to remove the wing and breast meat completely. *Note:* When removing the breast meat, ensure that the small tenderloin is included and does not tear free of the breast meat. Repeat procedure on the other side to remove the second breast.

7. The first section of the wing (drumette) can be removed or left on if desired for presentation. If the drumette is left on, chop off the end knuckle and push back the meat for a finished presentation.

## Preparing Boneless Breasts

*1. Remove middle wing section and wing tips.*

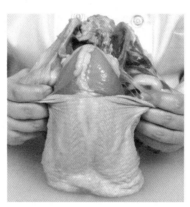

*2. Separate leg and thigh sections from bird by bending backwards until they dislocate from hips.*

*3–4. Cut down along the backbone from the neck to tail; cut horizontally at the pelvis slightly above the oysters.*

*5. Cut alongside the wishbone.*

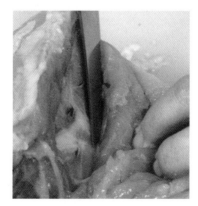

*6. Cut alongside the keel bone to remove wing and breast meat.*

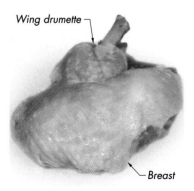

Wing drumette

Breast

*7. If the wing drumette is left on, chop off the end knuckle and push the meat back prior to cooking.*

**18-18** *Boneless breasts are the most popular cut of poultry.*

Refer to the CD-ROM for the **Fabricating Boneless Poultry Legs and Thighs** Media Clip.

If skinless breasts are desired, the skin can be removed by pulling it firmly from the breast meat. A flattened breast may also be desired, such as to be stuffed or sautéed. To flatten a breast, a heavy piece of plastic wrap is wrapped around the breast meat and it is pounded with a mallet until the desired thinness or diameter is achieved. Plastic wrap makes this task easier because the mallet does not stick to or tear the flesh.

Boneless legs and thighs are a great way to use these less-utilized cuts of poultry. Leg and thigh meat are actually moister and more flavorful than breast meat, especially compared to white-meat breasts such as chicken breasts or turkey breasts. Legs and thighs have more flavor and moisture because they are dark meat and contain additional fat. Boneless legs and thighs are often stuffed and roasted. When stuffing and roasting legs and thighs, the bones are removed, taking care to keep the meat intact.

To prepare partially boneless legs and thighs to be stuffed, apply the following procedure:

1. Separate the leg and thigh sections from the bird, using the same procedure as for cutting a bird into quarters. *See Figure 18-19.*
2. With a stiff boning knife, cut down the length of the thigh bone.
3. Scrape the thigh meat down off the thigh bone to the knee joint with the blade of the knife.
4. Cut between the leg and thigh bones at the knee joint to remove the thigh bone.
5. Chop off the knee joint. Push the meat back away from the joint for a finished presentation. The thigh portion can be stuffed as desired.

Leg and thigh meat can also be cut up into smaller pieces for use in dishes such as stir-fries, soups, and casseroles. If the meat is to be cut up, care does not need to be taken to keep the meat intact while removing the bone.

To prepare boneless legs and thighs, apply the following procedure:

1. Separate the leg and thigh sections from the bird, using the same procedure as for cutting a bird into quarters. *See Figure 18-20.*
2. Place the leg and thigh on the cutting board with the inside of the thigh facing upward. With a stiff boning knife, cut down the length of the thigh and leg bones on each side and around the cartilage of the knee to free it from the thigh meat.

3. Pull the meat away from the leg bone and cut the meat off where it connects to the end joint (shin).

4. With smooth, even strokes, cut around the knee joint until the L-shaped leg and thigh bones and the knee cartilage are free from the meat. Pull bones to remove and reserve for stock. Repeat with other leg.

## Preparing Partially Boneless Legs and Thighs for Stuffing

1–2. Separate leg and thigh sections from bird; cut down the length of the thigh bone.

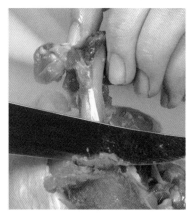

3. Scrape the meat down the thigh bone to the knee joint with the blade of the knife.

4. Cut between the leg and thigh bones at the knee joint to remove thigh bone.

5. Chop off the knee joint; push the meat back prior to cooking.

**18-19** Boneless legs and thighs can be stuffed and roasted or used in other dishes.

If the skin is to be removed from a boneless leg or thigh, firmly grab the leg or thigh meat in one hand and the skin in the other and pull the skin from the meat. Flattened leg or thigh meat may also be desired if it is to be stuffed or sautéed. The same procedure as for flattening breast meat is used. A heavy piece of plastic wrap is wrapped around the meat, and it is pounded with a mallet.

## Preparing Boneless Legs and Thighs

1–2. Separate leg and thigh sections from bird; cut down length of thigh bones and around cartilage of knee.

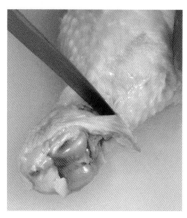

3. Cut off leg meat where it connects to the end of the shin.

4. Cut with smooth, even strokes around the knee joint until bones can be pulled free from meat.

**18-20** Leg and thigh meat can be diced for use in stir-fry dishes, soups, and casseroles.

Leg and thigh meat can be stuffed with forcemeat, grain, stuffing, or another combination of ingredients. The stuffed meat can be seared in a sauté pan and finished in the oven, or it can be braised or roasted. Before roasting, brush each stuffed leg with oil to provide moisture and to aid in achieving a golden brown color.

Boneless stuffed legs and thighs are commonly trussed or wrapped in caul fat before cooking to help to maintain the shape of the meat. *Caul fat* is a meshlike fatty membrane that surrounds sheep or pig intestines. It can be wrapped around meat to be roasted or seared to add additional moisture and maintain a consistent shape. Caul fat melts almost completely away during the cooking process, but not before it has helped set the shape of the item being cooked.

## COOKING POULTRY

Poultry can be cooked using nearly any cooking method. The trick is to determine which cooking method is best for a particular part of the bird. As when cooking any cut of meat, it is important to understand the characteristics of a poultry cut to determine which cooking method would be best to prepare it. Smaller cuts of meat cook more quickly than larger cuts of meat. Leg and thigh meat have more connective tissue than breast or wing meat, lending them to moist-heat cooking methods and longer combination cooking methods. Breast meat must be cooked quickly but thoroughly, so as not to overcook. Because white meat has little fat or connective tissue, it cooks quickly and dries out if even slightly overcooked. Some birds, such as duck and pheasant, have breast meat that is considered red meat.

Cooking procedures for poultry are similar to those used for other types of meat. Tougher meats are cooked using lengthier cooking methods, and more tender meats are cooked using quicker cooking methods. Regardless of the cooking method used, poultry should always be cooked well done, except in the case of duck, squab, and the breast meat of a few other game birds. This red meat should be cooked no more than medium to avoid drying it out.

Larger poultry should be cooked slowly to reduce shrinkage and retain moisture. Smaller poultry should be cooked at temperatures between 375°F and 400°F to avoid drying out the meat while cooking. In the professional kitchen, it is a common practice to stuff small birds, such as Cornish hens and squabs, but not larger birds. If stuffing is to be served with large birds, it is strongly recommended that it be prepared and baked separately.

## Determining Doneness

Although most meat is cooked to a customer's requested temperature, such as medium rare or medium, poultry is always cooked well done. The common exceptions to this rule are for duck and squab breasts, which are considered red meat and are most often cooked medium rare or medium. Preparing perfectly well done poultry is tricky. Perfectly well done meat should still be moist and juicy. Cooking even 10 min past perfectly well done results in dry, overcooked meat. There are four common ways of checking the doneness of poultry: temperature, touch, joints, and juices (TTJJ). *See Figure 18-21.*

**Temperature.** Temperature is taken using a standard instant-read thermometer inserted into the thickest part of the meat. The thermometer should be inserted close to any large bones in the thickest portion of the flesh. Poultry should be cooked to an internal temperature of 165°F. The difficulty with using temperature to indicate doneness is that small items, such as a boneless chicken breast or a butterflied quail, may be too thin for the thermometer to provide an accurate reading.

**Touch.** Touch, although not as accurate as temperature, is a relatively good method for an experienced chef to determine doneness. Poultry, like any other meat, is soft and pliable when raw, but firmer and more solid when cooked. The firmness of the poultry increases in proportion to doneness. When raw or undercooked meat is touched, the finger sinks slightly into the meat, and the meat feels soft. When poultry is thoroughly cooked, the flesh springs back when touched and is much firmer. The difference is similar to the difference between touching the bicep of a relaxed arm and touching the bicep of a flexed arm.

### Checking Poultry for Doneness

**T**emperature—Should be 165°F internally

**T**ouch—Should be firm and solid

**J**oints—Should be soft and tender

**J**uices—Should be clear with no signs of blood

**18-21** *The four common indicators of doneness of poultry are temperature, touch, joints, and juices (TTJJ).*

**Joints.** Joints between bones become soft and tender as the cartilage that holds them together cooks. The dissolving of the tough connective tissue between joints is an indication that the meat is fully cooked. When raw or undercooked, joints are firmly connected. If the ends of a connected leg and thigh portion are taken, one in each hand, twisting the joint typically does not cause the knee joint to come apart. However, if the same motion is applied to a perfectly cooked leg and thigh portion, the bones twist and separate at the knee joint.

**Juices.** Juices from raw poultry are red, just as from any raw meat. When poultry is fully cooked, the juices turn clear as the protein myoglobin coagulates. The change in color as poultry cooks can be used to explain why chicken stock, made from juices extracted from chicken meat and bones, is relatively clear. Tipping a roasted bird to see the juices run out or inserting a thermometer and watching the juices that run from the hole left behind are good methods of checking whether the item is fully cooked or not. When poultry is cooked well done, the juices run clear and have no signs of pink color.

*Perdue Foodservice, Perdue Farms Incorporated*

## Deep-Frying and Pan-Frying Poultry

Of all the cooking methods, pan-frying or deep-frying result in moist poultry. This is because portioned cuts are either breaded or battered to seal in the moisture when fried. Poultry cooked using either of these methods should have a crisp, golden brown exterior. The most popular cut used with both of these cooking methods is chicken that has been cut into eighths. ***See Figure 18-22.*** Boneless cuts of poultry such as breaded chicken breasts or turkey cutlets can also be deep-fried or pan-fried.

The pan-frying method is most commonly chosen to help fried foods retain a flat shape. For example, a boneless chicken breast that has been deep-fried curls and does not sit flat on a plate, while a boneless breast that has been pan-fried sits perfectly flat.

**Seasoning.** The most common seasoning method for fried poultry is simply to add salt and white pepper to the flour, bread crumbs, or batter. The following are ways to add flavor to deep-fried or pan-fried fowl:

- Prior to breading, the meat can be soaked in a marinade containing salt and various herbs, seasonings, or flavorings. Salt and white pepper are most commonly used.

- The flour, bread crumbs, or batter can be, and almost always are, seasoned to add flavor to the crispy coating.

*Frying Poultry*

*National Chicken Council*

**18-22** *Pan-frying or deep-frying results in moist poultry preparations.*

- The fried poultry can be served with a flavorful sauce such as barbeque sauce, honey, or a Buffalo-style hot sauce. The fried items can be tossed and coated in the sauce, or the sauce can be served on the side for dipping.

**Time and Temperature Requirements.** When either deep-frying or pan-frying, the key to achieving proper cooking is having the fat hot enough to crisp the exterior, yet not so hot as to overcook or burn the exterior of the meat in the time required for the interior meat to fully cook. Fat used to fry poultry should be between 300°F and 375°F. Pan-frying usually requires a lower temperature and a slightly longer cooking time than deep-frying. The food is turned over when the side being fried becomes golden brown. Pan-frying is typically done closer to 300°F, while deep-frying is usually done closer to 375°F. When determining the appropriate temperature for the oil, it is important to remember that larger items need lower heat, while smaller items, especially those without bones, do best with higher heat.

The best method for determining doneness is to insert an instant-read thermometer into the thickest part of the flesh. The touch method does not work with fried poultry because the crunchy exterior prevents an accurate measure of firmness. The joint twisting method is also not recommended for testing doneness of fried poultry, because it would break the golden coating. Lastly, the juice method is not recommended, because the crunchy coating prevents most of the juice from dripping out, and the clear frying oil could easily be misinterpreted as clear juices.

*National Chicken Council*

## Sautéing Poultry

Because sautéing is a fast dry-heat cooking process, it is important that meat to be sautéed is tender and not too thick. Breast meat can be dredged in seasoned flour and sautéed in dishes such as chicken Marsala and chicken saltimbocca. Turkey breast medallions are sautéed for turkey picatta, and boneless leg and thigh meat from any bird can be sautéed or stir-fried for a quick Asian meal. ***See Figure 18-23.*** Although it is most common to sauté boneless cuts of meat, smaller birds such as quail and squab are often sautéed bone in.

Sautéed items are often served with a pan sauce made in the same pan that was used to cook the food. Once the desired exterior color is achieved and the food is cooked through, it is removed from the

*Sautéing Poultry*

*National Chicken Council*

18-23 *Sautéed chicken is used in many Asian dishes.*

## Pan Sauces

National Chicken Council

**18-24** *Pan-fried poultry is often served with a pan sauce.*

pan. While the pan is still hot, flavoring ingredients such as garlic, shallots, tomato product, or spices are added to the pan and sautéed in the remaining fat. The pan is then deglazed with stock, wine, or another flavorful liquid and reduced to make a flavorful pan sauce. Other ingredients such as cream or fresh herbs can be added to the pan and reduced to make a richer and more flavorful sauce. The meat is usually added back to the pan to be glazed in the sauce and reheated for service. *See Figure 18-24.*

**Seasoning.** Poultry meats are typically neutral in flavor and are lower in fat than most other meats. They greatly benefit from seasoning, basting, or marinating as a means of imparting moisture and flavor. Poultry seasonings can be as simple as salt and pepper or as complex as a Cajun spice blend. Many prepared spice blends are available and make creative seasoning much easier and more consistent from one dish to the next.

Prior to being sautéed, poultry cuts are commonly dredged in seasoned flour. The seasoned flour helps seal in moisture and promotes even browning of the food while allowing seasonings to stick to the food. Fresh herbs, seasonings, and other flavorings can also be added to the pan while an item is being sautéed.

**Time and Temperature Requirements.** Items to be sautéed should be cooked over medium-high heat in a sauté pan. If a pan is too hot, the food or particles of flour can burn. If the sauté pan is not hot enough, the food absorbs some of the cooking fat and sweats in its own juices, preventing the appropriate color from being achieved. Larger and thicker items are not recommended for sautéing, because a thermometer may be required to determine their doneness. The best cuts of poultry for sautéing are boneless breasts and boneless dark meat, scallops, and medallions. An experienced chef can tell the doneness of thinner products such as these simply by touch.

## Grilling or Broiling Poultry

Many types of poultry can be prepared by broiling or grilling. Very small game birds, such as quail, are typically skewered prior to cooking on a grill or broiler to make them lie flat across the grates. Any small bird, such as a Cornish hen, should be split in half (butterflied) and then broiled or grilled. Birds larger than a Cornish hen should be cut into quarters or eighths for broiling or grilling. Larger birds, such as whole turkeys, are too thick to be grilled or broiled because the exterior would burn and dry out long before the interior would be fully cooked. The heat from a broiler or a grill is intense, so care must be taken to not overcook or dry out poultry. In banquet and large-scale catering operations, grilled items are marked. To *mark* means to cook an item on a broiler or grill just long enough to produce crosshatch marks, and then finish the item in an oven later, just prior to service.

When grilling or broiling poultry, it is important to remember that most bird meat contains very little fat and is prone to drying out when cooked using dry-heat cooking methods such as grilling and broiling. Both of these cooking methods use metal grates that leave caramelized markings on the meat being cooked. These markings, called crosshatch marks, are a signature of broiling or grilling in a professional kitchen. ***See Figure 18-25.*** Broiled or grilled items should have a well-browned exterior and should be moist, juicy, and cooked well throughout (with the exception of duck and squab breasts, which can be cooked to medium rare or medium).

The skin of most poultry burns quite easily on the grill or broiler due to the intense heat, so care should be taken to watch the skin closely. Some chefs prefer to remove the skin prior to cooking to avoid burning the skin as well as cut fat from the dish. For example, boneless and skinless grilled chicken breasts are used in many restaurants on salads and in many dishes. However, the chef must be aware of the potential for skinless cuts to become dry or overcooked due to the low amount of fat in the meat.

**Seasoning.** Grilled or broiled cuts of poultry are typically seasoned prior to cooking so that the flavors of the seasoning can penetrate the surface of the meat during cooking. A marinade is an effective way to add flavor to poultry and other birds, and works especially well with grilling and broiling methods. A *marinade* is a liquid commonly used to soak meat to add moisture and flavor to the meat. Poultry may be marinated by completely immersing the poultry in the marinade. Because poultry cuts are relatively thin and low in fat, they do not need to be left in a marinade for an extended period. Poultry cuts are marinated for approximately 30 min to 60 min prior to cooking. Typical marinades for poultry contain an acidic component such as wine or citrus juice. They also usually include an oil to add fat, and some combination of herbs and spices in addition to salt and pepper. Popular marinades for poultry include white wine and herb marinades, citrus marinades, and ethnically inspired marinades such as Asian (teriyaki) or Hispanic (mojo, pronounced mo-ho) marinades.

Marinade must be discarded after it is used. Care should be taken to make only enough marinade to flavor the amount of meat being prepared. Any remaining marinade used on raw poultry must be discarded immediately to avoid cross-contamination. Used marinade should never be used as a sauce. If stuffing poultry, marinate the poultry first.

## Roasting Poultry

Roasting is one of the most common methods of cooking whole poultry. Larger birds such as turkeys or geese need longer times and lower temperatures to allow them to cook thoroughly without the outer meat becoming dry and overcooked. Smaller birds such as Cornish hens and chickens can be roasted at higher temperatures for shorter cooking times. The trick to roasting poultry perfectly is to determine how long the bird should be roasted and at what temperature. ***See Figure 18-26.***

*Grilling Poultry*

*National Chicken Council*

18-25 *Crosshatch marks are a signature of broiled or grilled items.*

*Roasting Poultry*

*National Chicken Council*

18-26 *Roasting is one of the most common methods for preparing whole birds.*

Some poultry and small game birds benefit greatly from barding during roasting, particularly small birds cooked at high temperatures. *Barding* is a preparation method where a food item is wrapped in a thin layer of fat before cooking. This additional layer of fat helps to keep the thin skin of the bird from burning while providing fat to very lean meat.

**Seasoning.** All poultry should be seasoned prior to cooking, but the size of the bird determines where the seasoning should be placed. For large birds such as turkey or capon, whole herbs or mirepoix are added to the internal cavity. These aromatics release their aromas and flavors to infuse the meat as the bird slowly roasts. Smaller birds can be seasoned directly on the skin with salt and pepper, as they are not in the oven long enough to release flavors of aromatics stuffed inside. Herbs should never be placed on the skin of any bird to be roasted, as they would simply burn in the oven.

**Time and Temperature Requirements.** The appropriate cooking temperature is determined by the product being roasted. Smaller birds can be roasted for shorter periods at higher temperatures, while larger birds need lower temperatures and longer times to be perfectly cooked.

Low-temperature roasting is used for larger birds such as turkeys and capons. In this case, the birds have a larger amount of meat on the carcass and therefore need a longer time to cook completely. If these large birds were roasted at higher temperatures, the skin would burn and the outer surface of the flesh would dry out before the minimum internal temperature of the bird was achieved. Large birds are roasted between 275°F and 325°F. Some chefs prefer to start larger birds at a temperature greater than 400°F for a few minutes to allow the skin to begin to brown, but this is not necessary as the extended cooking time should be ample to provide a beautiful golden brown skin, even at lower temperatures.

Medium-high-temperature roasting is used mainly for chickens, as they commonly fall into the 2½ lb to 3 lb weight range. Chickens are smaller than the largest birds, so they require a shorter cooking time and a higher temperature. Chickens should be roasted at 350°F to 375°F to ensure a golden color and crisp skin by the time the meat is cooked perfectly.

High-temperature roasting is used for birds that are 2 lb or under, which includes small birds such as squab, pheasant, and Cornish hens. The birds are roasted between 375°F and 400°F to produce a crisp, golden brown skin quickly without drying out the meat. Because the birds are small, if they are roasted at lower temperatures for longer periods, they may not develop good color and crisp skin before they become overcooked.

Very-high-temperature roasting is used for very thick- or fatty-skinned birds such as duck or goose. The goal of very-high-temperature roasting is to render as much fat as possible from the skin, making the skin crispy and less greasy. If a duck or goose were roasted at a low temperature, the skin would remain thick and fatty and would not be pleasant to eat. Roasting at a temperature between 400°F and 425°F allows the fat to melt out and away from the skin. As the hot fat seeps out of the skin, it fries the skin slightly, making the skin crispy and brown. It is important to prick the skin of fatty- or thick- skinned birds prior to roasting and again during the roasting process. Pricking the skin ensures that the skin is not seared in the high heat, which would prevent the rendered fat from escaping.

**Basting.** Basting poultry while roasting helps add additional moisture to the bird as it cooks. Birds should be basted only with fat, either with fat that escapes from the bird itself during cooking, or with butter or other fat that was added to the roasting pan prior to cooking. A common mistake is to baste a bird with stock or wine. It is important to remember that poultry and game birds differ from other meats, in that they do not have marbling or much internal fat. Basting with stock or wine dries out the bird by washing away any fat that would otherwise be absorbed by the meat.

*National Chicken Council*

The purpose of basting is to add fat to a lean bird as it roasts. Fatty birds such as duck or goose should not be basted. Basting can make fatty birds greasy and unpleasant to eat.

**Carving.** Poultry that is roasted or cooked whole using another cooking method are typically carved for service. Carving is cutting a large portion of cooked meat into smaller pieces. Roasted chickens and other small birds are commonly carved into quarters.

## To carve a bird into quarters, apply the following procedure:

1. Cut on the inside of the thigh just below the breast meat to remove the leg and thigh portion. This is similar to the cut made when carving raw poultry into quarters. **See Figure 18-27.**

2. With a carving fork and knife, bend the leg and thigh portion back to dislocate it at the hip joint, then cut it free from the carcass by cutting through the hip joint with the tip of the knife. Repeat on other leg and thigh portion.

3. While holding the bird with the fork against the backbone, begin to slice down along one side of the breastbone, alongside one of the breasts, following the rib cage. Continue to cut along the breast until reaching the wing joint.

4. With the tip of the knife, cut through the wing joint to completely separate the breast from the carcass. Repeat on other side of breast.

CULINARY PROCEDURES

## Carving into Quarters

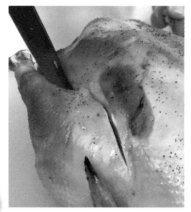

1. Cut inside of thigh just below breast.

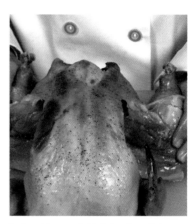

2. Bend leg and thigh back to dislocate at hip; cut through joint with knife.

3. Slice along breastbone following rib cage to wing joint.

4. Cut through wing joint to remove wing and breast; repeat on other side.

**18-27** Whole roasted birds can be carved into quarters for individual servings.

**To carve a bird into eighths, apply the following procedure:**

1. Separate the leg and thigh sections from the bird, using the same procedure as for carving a bird into quarters. *See Figure 18-28.*

2. Remove each breast section from the bird, using the same procedure as for carving a bird into quarters.

3. Cut through the knee socket at the top of each drumstick to separate the legs and thighs.

4. Cut through the wing joint at the base of each wing to separate the wings and breasts.

## Carving into Eighths

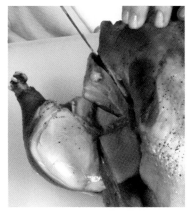

1. Separate leg and thigh sections.

2. Remove breast and wing sections.

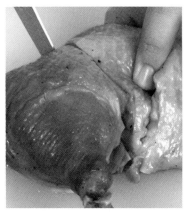

3. Cut through knee socket to separate legs and thighs.

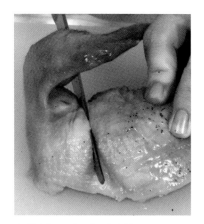

4. Cut through wing joint to separate wing and breasts.

**18-28** *Whole roasted birds can be carved into eighths for smaller individual servings.*

To carve a large bird, apply the following procedure:

1. Remove both leg and thigh portions, as when cutting raw poultry into quarters. ***See Figure 18-29.***

2. To slice and portion thigh and leg meat, hold the leg with one hand for stability and slice the thigh in thin slices parallel to the bone. When the thigh meat is completely sliced, continue to the leg meat by slicing parallel to the bone. Repeat on other leg and thigh portion.

3. With the tip of the knife, trim along the wishbone around the neck and completely remove it.

CULINARY PROCEDURES

4. Begin to slice along the breastbone all the way down along the breast meat. Follow the natural curve of the rib bones to completely remove the breast meat from the carcass. Note: The breast can also be carved directly on the bird if desired, by making a horizontal slice just above the wing joint where it is joined to the breast. Starting at the neck end of the bird, begin to slice the breast on the bias.

5. At the thickest end of the breast, begin to slice on a bias. Slicing on a bias makes the slices appear larger. Repeat on other breast half.

## Carving Large Birds

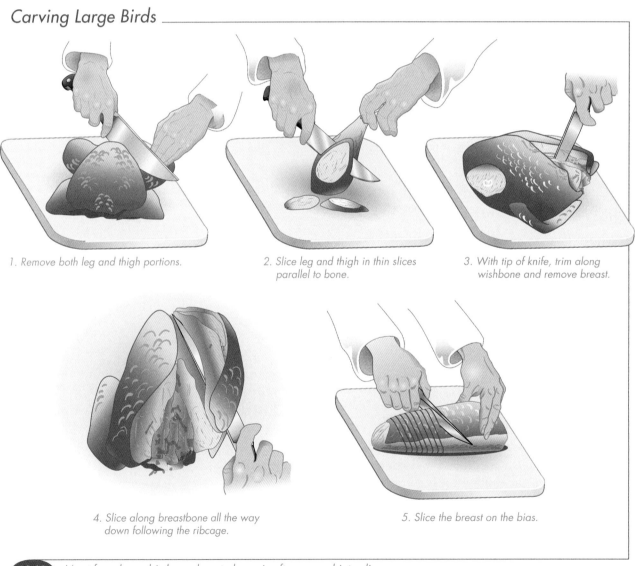

1. Remove both leg and thigh portions.

2. Slice leg and thigh in thin slices parallel to bone.

3. With tip of knife, trim along wishbone and remove breast.

4. Slice along breastbone all the way down following the ribcage.

5. Slice the breast on the bias.

18-29  Meat from large birds, such as turkeys, is often carved into slices.

## Simmering and Poaching Poultry

Simmering and poaching are similar moist-heat cooking methods. The main differences between simmering and poaching are in the temperature and the time required to fully cook an item. Simmering uses higher heat over a longer period, with moderate water movement. Poaching uses a lower heat over a shorter period, with almost no water movement. The appropriate cooking method is determined by the tenderness of the meat being cooked.

Birds that are a little older are tougher and are cooked using the simmering method. The slightly higher heat combined with a longer cooking time can help a tough piece of meat become more tender. Simmering is not recommended for more tender cuts, such as a chicken breast. Cooking a chicken breast using the simmering method can result in tough, stringy, overcooked meat. The flesh of the tender breast can shrink, curl up, and become dry. Simmering removes all of the internal moisture from the meat, because the moisture is forced out as the meat becomes pressurized in the hot water.

Poaching is used for tender cuts or parts of a bird that can be cooked slowly to keep them tender. Poaching a tender cut such as a chicken breast allows it to cook slowly while remaining moist and tender. Poaching generally takes less time than simmering. No matter which cooking method is chosen, it is important to ensure that the meat is completely submerged in the cooking liquid. If a portion of the meat is not submerged, it will not cook properly and may become tough and dry.

**Seasoning.** Seasoning the meat with salt and pepper has little or no effect on simmered or poached meat, because the seasonings would simply be washed off during the cooking process. Although the cooking liquid absorbs some flavor from the meat, the liquid also needs some additional flavor or seasoning added to it at the beginning of the cooking process. The flavor of the cooking liquid becomes the main flavor of the cooked meat. Typically, stock is used for the cooking liquid, because it has a great deal more flavor than plain water. There are many different ingredients that can be added to the poaching or simmering liquid to help flavor the meat such as a mirepoix, a bouquet garni, fresh herbs, wine, or lemon juice.

**Time and Temperature Requirements.** Poaching is a gentle cooking method used to produce moist and tender meat. A cooking temperature between 160°F and 170°F, with almost no steam or bubbles being formed in the pot, typically produces the most desirable results. Simmering is used to prepare tougher and older birds that can benefit from extended cooking times at higher temperatures. Simmering liquid is usually heated between 185°F and a maximum of 200°F. It is essential that simmering liquid never be allowed to come to a boil while an item is being simmered. Boiling would result in tough, stringy, and undesirable meat.

Whichever method is used, the most accurate method of determining doneness is to insert an instant-read thermometer into the thickest portion of the meat, near a large bone if possible. Most poultry should be cooked to an internal temperature of 165°F.

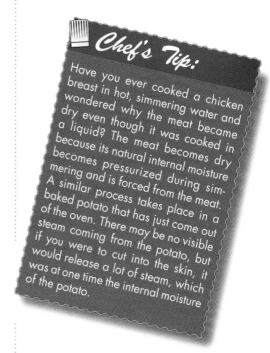

**Chef's Tip:**
Have you ever cooked a chicken breast in hot, simmering water and wondered why the meat became dry even though it was cooked in a liquid? The meat becomes dry because its natural internal moisture becomes pressurized during simmering and is forced from the meat. A similar process takes place in a baked potato that has just come out of the oven. There may be no visible steam coming from the potato, but if you were to cut into the skin, it would release a lot of steam, which was at one time the internal moisture of the potato.

## Braising and Stewing Poultry

As with other meats, braising and stewing are used to prepare tough or less flavorful poultry that will benefit from cooking in a flavorful liquid for a long period. *Braising* is a cooking method that requires the food to be sautéed before it is simmered in a small amount of liquid. *Stewing* is a cooking method that requires the food to be boiled or simmered slowly in a slightly thickened cooking liquid. Braised and stewed poultry should be tender but should not fall off the bone when cooked properly. If the meat is falling apart, it has been overcooked. Almost all poultry that is prepared by braising or stewing is cooked either whole or as bone-in cuts. For example, whole bone-in duck leg and thigh portions are often braised, resulting in a moist and flavorful dish.

The majority of the flavor of the dish comes from the liquid the item is cooked in. Braised and stewed dishes are served in the cooking medium. ***See Figure 18-30.*** After the meat has been properly cooked, it is usually removed from the cooking liquid. The liquid is then checked for the proper consistency and is seasoned. Cooked poultry is always served in or with the braising or stewing liquid as its sauce.

**Time and Temperature Requirements.** Poultry can be braised or stewed in two different ways. Either the meat can be added raw to the cooking liquid and cooked at the proper temperature until done, or it can be seared or browned prior to adding it to the liquid. Whether or not to sear the meat depends on whether the sauce is to remain light colored or if it would benefit from the color and caramelized flavor of seared meat being cooked in it. For example, if the resulting sauce is to be a white sauce, raw meat is added to the cooking liquid. If the sauce is to be a rich, red wine sauce, such as is served with a coq au vin, the meat should be seared to caramelize it prior to adding it to the cooking liquid.

Any of the four indicators of doneness (temperature, touch, joints, and juices) can be used to determine the doneness of braised or stewed meat. Juices can be checked to see if they run clear, an instant-read thermometer can be inserted into the thickest portion to make sure it has reached 165°F, or the joints can be checked for tenderness. The touch method may also work, but the meat may be very hot while covered in the thickened sauce. Although the finished product should be tender and moist, the meat should not be falling apart or unable to stay on the bone for service. The temperature of each of these cooking methods should be between 205°F and 212°F.

## SUMMARY

Poultry is popular due to ease of preparation, versatility, availability, and low cost. Poultry is separated into six different classifications recognized by the USDA. They are chicken, turkey, duck, goose, pigeon, and guinea. Each of these classifications is subdivided into classes based on the age and tenderness of the bird. In addition, ratites and game birds are also bred for human

---

*Braising and Stewing Poultry*

National Turkey Federation

**18-30** *Mature poultry is tougher than young poultry and is often braised or stewed.*

consumption and referred to as poultry. Ratites are large flightless birds that have very small wings relative to their body size. This group includes the ostrich, emu, and rhea. Game birds include four varieties of birds that are sold for human consumption: quail, partridge, pheasant, and grouse.

Some birds have both light and dark meat, while others have only dark meat. Leg and thigh meat are always dark meat. The breast and wing meat can be either light or dark meat, depending on whether the bird flies. Fabricating poultry is a relatively easy skill to master with practice. Fabrication methods include techniques for trussing and cutting.

Poultry, ratites, and game birds are prepared using many of the same cooking methods used to prepare other meats. Cooking methods used for poultry are determined by the cut or part of the bird. Whole poultry requires more cooking time to produce evenly cooked meat. Poultry that has been cut into parts cooks more quickly. In addition, the age of the bird and the tenderness of the meat also determine the cooking technique used.

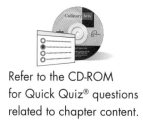

Refer to the CD-ROM for Quick Quiz® questions related to chapter content.

## Review Questions

1. List the six different classifications of poultry recognized by the United States Department of Agriculture (USDA).

2. List two other categories of birds that can be cooked using many of the same cooking methods as poultry.

3. What does the round USDA inspection stamp signify?

4. Identify the standards that must be met in order for poultry to be marked Grade A.

5. Between what temperatures should poultry be stored in the refrigerator?

6. Explain why poultry must be thawed in the refrigerator rather than at room temperature.

7. Explain the procedure for trussing a whole bird.

8. Explain the procedure for cutting a bird in half.

9. What are some of the reasons why boneless breasts are the most popular poultry cut?

10. Describe the four methods of determining the doneness of poultry.

11. Describe common seasoning methods used when sautéing poultry.

12. Explain what it means to mark a piece of poultry.

13. Explain the procedure for carving a bird into eighths.

14. Highlight the differences between simmering and poaching poultry.

15. Describe the two states in which poultry or other birds can be started for braising or stewing.

#  Recipes

## Herbed Pan-Fried Chicken

158

*yield: 2 servings*

| | |
|---|---|
| 2 c | flour |
| 1 tbsp | dried oregano |
| 1 tbsp | dried basil |
| 1 tbsp | dried thyme |
| 1 tbsp | dried sage |
| ½ tbsp | dried mustard |
| 2 tsp | garlic powder |
| 2 tsp | onion powder |
| 2 quarters | chicken, leg and thigh portion |
| to taste | salt and pepper |
| 1 c | buttermilk |
| 1–2 c | vegetable oil |

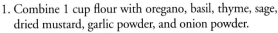

*National Chicken Council*

1. Combine 1 cup flour with oregano, basil, thyme, sage, dried mustard, garlic powder, and onion powder.
2. Season chicken with salt and pepper.
3. Dredge chicken in plain flour, dip in buttermilk, and dredge in seasoned flour, pressing firmly to adhere the flour to the chicken. Place in refrigerator for 15 min.
4. Heat ½″ of vegetable oil in sauté pan over low to medium heat.
5. Place chicken in pan, skin-side down. Pan-fry until light golden brown.
6. Turn chicken over and finish cooking in 350°F oven. Check frequently to see if bottom is browning too quickly and turn chicken over if necessary.
7. Check thigh with a fork or knife. When juices run clear, remove from oven. Drain on paper towels to remove excess fat.

### Nutrition Facts

**Per serving:** 2952.5 calories; 65% calories from fat; 218 g total fat; 492.5 mg cholesterol; 782.1 mg sodium; 1829.4 mg potassium; 110.3 g carbohydrates; 7 g fiber; 7.9 g sugar; 103.3 g net carbs; 133.2 g protein

## Chicken Kiev

159

*yield: 4 servings*

| | |
|---|---|
| 4 | boneless, skinless chicken breasts |
| 8 tbsp | butter |
| ½ clove | garlic, minced |
| 1 tsp | fresh chives, minced |
| to taste | salt and pepper |
| pinch | fresh marjoram |
| 1½ c | flour |
| 2 eggs | beaten |
| ½ c | milk |
| 1½ c | bread crumbs |
| 4–6 tbsp | cooking oil |

1. Wrap each chicken breast in plastic wrap. Flatten using a mallet. Remove plastic wrap.
2. Mix chilled butter in mixer at slow speed until a smooth consistency is achieved.
3. Add minced garlic, chives, salt, and pepper to butter. Crush the marjoram in the palm of the hand and add to mixture. Blend at low speed until evenly seasoned. If needed, chill for a short time in refrigerator.
4. Roll chilled butter into four pieces, each about the size and shape of a finger. Place a roll of butter in the center of each breast.
5. Roll the chicken around the butter and fold in the ends to completely enclose the butter. Wrap tightly in plastic wrap and place in freezer until butter is very firm.
6. With one hand, coat a piece of chicken with flour.
7. With opposite hand, dip the floured chicken in egg wash. *Note:* Keep one hand reserved for dry mixtures and one hand reserved for wet mixtures. This helps to prevent breading sticking to the hands while working.
8. With the dry hand, coat egged chicken in bread crumbs. Repeat for remaining pieces of chicken.
9. Heat oil in a large skillet. Fry chicken in hot oil until golden brown and cooked through. Turn as needed to cook evenly.
10. Transfer chicken to paper towels or a wire rack to drain excess oil.

### Nutrition Facts

**Per serving:** 968.9 calories; 41% calories from fat; 45.5 g total fat; 306.7 mg cholesterol; 574.3 mg sodium; 817.9 mg potassium; 66.7 g carbohydrates; 3.1 g fiber; 4.5 g sugar; 63.5 g net carbs; 69.1 g protein

## Sautéed Chicken Provençale

*yield: 1 serving*

| | |
|---|---|
| 1 tbsp | olive oil |
| 1 | partially boneless chicken breast, skin on |
| to taste | salt and pepper |
| ½ c | onion, diced |
| 1 clove | garlic, minced or sliced thin |
| ½ c | fresh tomatoes, concassé |
| 2 oz | white wine |
| 2 oz | chicken stock |
| ½ tsp | capers, optional |
| 1 tbsp | kalamata olives, sliced |
| 1 tbsp | fresh basil |
| 1 tbsp | parsley |

1. Heat a sauté pan over medium heat. Add ½ tbsp olive oil.
2. Season chicken with salt and pepper to taste.
3. Place skin-side down in oil and cook until skin is crisp, golden brown in color, and fat is completely rendered out.
4. Turn breast over and place in oven until slightly underdone.
5. Remove chicken from pan, and discard oil.
6. In ½ tbsp fresh olive oil, sweat onions until translucent. Add garlic and cook until fragrant.
7. Add tomatoes and deglaze the pan with white wine. Reduce by half.
8. Add stock, capers, olives, basil, and parsley. Bring to a boil.
9. Season with salt and pepper to taste. Reduce heat and cook chicken until completely cooked through.

---

**Nutrition Facts**

**Per serving:** 443.9 calories; 45% calories from fat; 22.6 g total fat; 94.5 mg cholesterol; 743.5 mg sodium; 800.2 mg potassium; 16.4 g carbohydrates; 2.5 g fiber; 6.8 g sugar; 13.9 g net carbs; 33.7 g protein

---

## Chicken Marsala

*yield: 4 servings*

| | |
|---|---|
| 4 | boneless, skinless chicken breasts |
| 2 tbsp | extra virgin olive oil |
| ½ c | all-purpose flour, seasoned to taste with salt and pepper |
| ½ lb | porcini or button mushrooms, sliced thick |
| ¼ lb | prosciutto, sliced thin and cut into ½" squares |
| to taste | salt and pepper |
| 4 oz | Marsala wine |
| 4 oz | fresh chicken stock |
| 2 tbsp | butter |
| 1 tbsp | fresh parsley, chopped |

1. Spread a 2′ long sheet of plastic wrap on a clean cutting board. Arrange the chicken breasts on half of the plastic wrap and fold the other half over the breasts to cover.
2. With a mallet, pound the breasts to about half their original thickness or about ¼" thick.
3. In large sauté pan, heat olive oil over medium-high heat.
4. Dredge the breasts in the seasoned flour. Shake off any excess flour.
5. Place breasts carefully into the hot oil and sauté until golden brown. Turn breasts and sauté on other side. When golden brown on both sides, remove chicken from pan and set aside.
6. Add mushrooms to the pan and sauté in remaining oil until slightly browned. Remove mushrooms.
7. Lower the heat to medium and add the prosciutto to the sauté pan. Sauté for 1 min to render some of the fat.
8. Return the mushrooms to the pan and sauté until the moisture has evaporated, about 5 min. Season with salt and pepper.
9. Add Marsala and boil down for a few seconds to cook out the alcohol.
10. Add chicken stock and simmer for 1 min to reduce the sauce slightly.
11. Stir in butter and return the chicken to the pan; simmer gently for 1 min to heat the chicken through.
12. Season with salt and pepper to taste and garnish with chopped parsley before serving.

---

**Nutrition Facts**

**Per serving:** 680.7 calories; 24% calories from fat; 18.8 g total fat; 172.8 mg cholesterol; 1041.3 mg sodium; 1696.7 mg potassium; 56 g carbohydrates; 6.9 g fiber; 0.5 g sugar; 49.1 g net carbs; 70.2 g protein

---

# Recipes

## Chicken Cacciatore

*yield: 4 servings*

| | |
|---|---|
| 1 c | flour |
| to taste | salt and pepper |
| 1 | frying or roasting chicken, cut in eighths |
| 2 tbsp | cooking oil |
| 1 c | mushrooms, medium dice |
| 1 c | onions, minced |
| 1 clove | garlic, minced |
| ½ c | Marsala wine |
| 16 oz | crushed tomatoes in juice |
| 1 c | tomato purée |
| 1 tsp | fresh basil, crushed |
| 1 tsp | fresh oregano, crushed |
| ½ tsp | fresh chives, minced |

*National Chicken Council*

1. In a medium bowl, mix flour, salt, and pepper.
2. Coat each piece of chicken in the seasoned flour.
3. Heat oil in a large skillet. Sauté chicken in oil until golden brown. Transfer chicken to paper towels or a wire rack to drain excess oil.
4. Using the same oil, sauté mushrooms and garlic until tender, but do not brown.
5. Add Marsala and simmer for 5 min.
6. Add the tomatoes with juice, tomato purée, basil, oregano, and chives. Season with additional salt and pepper to taste.
7. Cover the bottom of a roasting pan with the sautéed chicken. Pour the sauce over the chicken. Bake at 325°F for 45 min or until chicken is tender.
8. To serve, place two pieces of chicken on a plate and cover with sauce. Garnish as desired.

### Nutrition Facts
**Per serving:** 1000 calories; 54% calories from fat; 60.6 g total fat; 245.1 mg cholesterol; 471.5 mg sodium; 1462 mg potassium; 43.5 g carbohydrates; 5.3 g fiber; 5.2 g sugar; 38.3 g net carbs; 64.5 g protein

## Chinese Five-Spice Duck Breast with Cherry and Ginger Lacquer

*yield: 4 servings*

| | |
|---|---|
| 4 | boneless duck breasts, skin intact |
| 2 tsp | Chinese five-spice powder |
| ¼ tsp | ginger powder |

*Cherry and Ginger Lacquer*

| | |
|---|---|
| 2 oz | red wine |
| 1 tsp | fresh ginger, minced |
| 1 oz | soy sauce |
| 1 oz | teriyaki sauce |
| 3 oz | cherry jelly or preserves |
| 1 oz | water |

1. Score fatty skin on breast, using caution to not cut into meat.
2. Rub duck with five-spice powder and ginger to coat well.
3. In sauté pan over medium heat, begin to render fat from duck breast, skin-side down.
4. When rendered and slightly cooked on the fatty side, turn over and sear the bottom of the breast.
5. Cook until internal temperature of the breast reaches 135°F.
6. Remove from pan and set aside to rest.
7. Place red wine in small saucepan with ginger and reduce by half.
8. Add soy sauce, teriyaki sauce, cherry jelly or preserves, and water and cook until mixture is slightly thickened and smooth.
9. Slice duck breast on a bias. Top with cherry and ginger lacquer and serve.

### Nutrition Facts
**Per serving:** 339.6 calories; 12% calories from fat; 4.8 g total fat; 271.7 mg cholesterol; 852.6 mg sodium; 71.2 mg potassium; 17.7 g carbohydrates; 0.3 g fiber; 11.2 g sugar; 17.4 g net carbs; 53.5 g protein

## Grilled Chicken Breast with Sweet Teriyaki Marinade

*yield: 4 servings*

| | |
|---|---|
| 4 oz | soy sauce |
| ½ c | mirin (sweet rice wine) or sweet sherry |
| 2 tbsp | rice wine vinegar or cider vinegar |
| 2 tbsp | vegetable oil |
| 2 oz | granulated sugar |
| 2 tbsp | grated or minced fresh ginger |
| 2 oz | lager-style beer (optional) |
| 4 | boneless chicken breasts |

*National Chicken Council*

1. Mix soy sauce, mirin or sherry, vinegar, vegetable oil, sugar, ginger, and beer in a saucepan and bring to boil to dissolve the sugar.
2. Remove from heat and cool over an ice bath.
3. Place chicken in cold marinade for 1 hr.
4. Remove breasts from marinade. Grill or broil until cooked through.

### Nutrition Facts

**Per serving:** 446 calories; 19% calories from fat; 10 g total fat; 136.9 mg cholesterol; 1356.6 mg sodium; 763.2 mg potassium; 23.7 g carbohydrates; 0.3 g fiber; 14.8 g sugar; 23.4 g net carbs; 56.5 g protein

## Grilled Chicken with Ginger-Lemongrass Marinade

*yield: 4 servings*

| | |
|---|---|
| 2 tbsp | fresh ginger, grated |
| 1 tbsp | fresh lemongrass (white part only), minced |
| 1 tbsp | garlic, minced |
| 2 oz | rice wine vinegar |
| 4 tbsp | brown sugar |
| ½ tsp | hot sauce (such as Tabasco®) |
| 2 oz | lime juice (unsweetened) |
| 1 tbsp | cilantro, minced |
| 2 oz | canola oil |
| 1 tsp | kosher salt |
| ¼ tsp | pepper |
| 4 | boneless chicken breasts |

1. Mix ginger, lemongrass, garlic, vinegar, brown sugar, and hot sauce in a saucepan and bring to boil to dissolve the sugar.
2. Remove from heat and cool over an ice bath.
3. Add lime juice, cilantro, oil, salt, and pepper to mixture and whisk.
4. Place chicken in cold marinade for 1 hr.
5. Remove breasts from marinade. Grill or broil until cooked through.

### Nutrition Facts

**Per serving:** 450.3 calories; 32% calories from fat; 17.2 g total fat; 136.9 mg cholesterol; 637.6 mg sodium; 808.2 mg potassium; 22.6 g carbohydrates; 0.2 g fiber; 13.5 g sugar; 22.3 g net carbs; 54.8 g protein

# Recipes

## 166

## Grilled Chicken with Garlic-Herb Marinade

*yield: 4 servings*

| | |
|---|---|
| 2 tbsp | fresh garlic, minced |
| 1 tbsp | fresh thyme leaves, stems removed |
| 1½ tbsp | fresh rosemary leaves, stems removed |
| 1 tbsp | fresh parsley |
| 1 tbsp | kosher salt |
| 6 oz | fresh lemon juice |
| 1 c | extra virgin olive oil |
| ½ tsp | pepper |
| 4 | boneless chicken breasts |

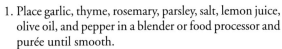

*National Chicken Council*

1. Place garlic, thyme, rosemary, parsley, salt, lemon juice, olive oil, and pepper in a blender or food processor and purée until smooth.
2. Place chicken in cold marinade for 30 min to 1 hr.
3. Remove breasts from marinade. Grill or broil until cooked through.

### Nutrition Facts
**Per serving:** 761 calories; 66% calories from fat; 57.1 g total fat; 136.9 mg cholesterol; 1567.8 mg sodium; 701.7 mg potassium; 6.7 g carbohydrates; 0.9 g fiber; 1.2 g sugar; 5.9 g net carbs; 55.2 g protein

## Roast Chicken

*yield: 4 servings*

| | |
|---|---|
| 1 | roaster chicken, trussed |
| 1 tbsp | butter, melted |
| to taste | salt and pepper |

*Mirepoix*

| | |
|---|---|
| ½ c | onions, small dice |
| ¼ c | celery, small dice |
| ¼ c | carrots, small dice |

*National Chicken Council*

1. Place the mirepoix in the bottom of the roasting pan.
2. Rub or brush the surface of the chicken with butter. Place in roasting pan, breast-side up.
3. Season with salt and pepper to taste.
4. Roast at 375°F for approximately 1 hr or until thoroughly cooked.
5. Strain drippings through a china cap. Deglaze the roasting pan and save both liquids for preparing gravy.
6. Carve chicken into quarters. Serve with bread dressing, mashed potatoes, and gravy.

### Nutrition Facts
**Per serving:** 233.9 calories; 30% calories from fat; 8 g total fat; 122.8 mg cholesterol; 211.8 mg sodium; 451.5 mg potassium; 3 g carbohydrates; 0.6 g fiber; 1.4 g sugar; 2.4 g net carbs; 35.5 g protein

## Roasted Chicken Roulades with Prosciutto

*yield: 2 servings*

| | |
|---|---|
| 2 | boneless, skinless chicken breasts |
| 2 slices | prosciutto |
| 2 tbsp | olive oil |
| to taste | salt and black pepper, cracked |
| ½ tsp | thyme, chopped |
| ½ tsp | sage, chopped |
| ½ tsp | rosemary, chopped |
| 2 cloves | garlic, creamed |

1. With chicken breasts between plastic wrap, pound chicken to ¼″ thick.
2. Lay a slice of prosciutto on each breast and drizzle with 1 tbsp olive oil. Season with salt and pepper.
3. Combine herbs and garlic and rub on oiled breast.
4. Using the plastic wrap, roll up chicken breasts tightly into an even log shape.
5. Discard plastic and tie the roulade with butcher's string to hold the shape.
6. Heat the remaining oil in a sauté pan and gently sear chicken on all sides.
7. Roast in a 400°F oven until an internal temperature of 165°F is achieved.
8. Remove from oven and allow to rest, covered, for 5 min to 10 min.
9. Slice crosswise to desired thickness.

**Nutrition Facts**

**Per serving:** 220 calories; 38% calories from fat; 9.4 g total fat; 78.4 mg cholesterol; 532 mg sodium; 381.5 mg potassium; 0.7 g carbohydrates; 0.1 g fiber; 0 g sugar; 0.6 g net carbs; 31.3 g protein

## Chicken à la King

*yield: 4 servings*

| | |
|---|---|
| ¾ lb | boneless chicken |
| ¼ c | green peppers, diced |
| 1 tbsp | pimientos, diced |
| ⅓ c | mushrooms, diced |
| 2 tbsp | butter |
| ¼ c | butter |
| 2 oz | flour |
| 1 c | chicken stock |
| 1 c | milk |
| ⅓ c | light cream |
| 2½ tbsp | dry sherry |
| to taste | salt |
| 3-4 drops | egg shade food color |

1. Dice chicken into 1″ cubes and sauté until cooked through. Set aside.
2. Sauté the peppers, pimientos, and mushrooms in 2 tbsp butter and deglaze with sherry. Set aside.
3. In a saucepot, melt ¼ cup of butter. Add flour to make a roux. Cook for 5 min, stirring constantly.
4. Add hot chicken stock and whip with a wire whisk until thickened and smooth.
5. In a separate saucepot, heat the milk and cream, being careful not to scald it.
6. Add the hot milk and cream to the thickened chicken stock. Whip with a wire whisk until the sauce is smooth.
7. Add salt to taste and the egg shade to achieve a desired color. Mix well.
8. Add the chicken and sautéed vegetables. Mix well.
9. Serve hot over a slice of toast or in a prepared pastry shell.

**Nutrition Facts**

**Per serving:** 411.2 calories; 53% calories from fat; 25.1 g total fat; 116.2 mg cholesterol; 249.9 mg sodium; 455.7 mg potassium; 17.8 g carbohydrates; 0.6 g fiber; 4.6 g sugar; 17.2 g net carbs; 25.6 g protein

**Recipe Variation:**

**170 Turkey à la King:** Substitute turkey for chicken and prepare as directed.

# Recipes

## 171

## Chicken Tetrazzini

*yield: 4 servings*

| | |
|---|---|
| 1 lb | boneless chicken |
| 16 oz | thin spaghetti |
| ⅓ lb | mushrooms, sliced |
| 3 tbsp | butter |
| ¼ c | flour |
| 2½ c | chicken stock |
| ⅓ c | cream |
| 1 tbsp | dry sherry |
| ⅓ c | Parmesan cheese |
| to taste | salt and pepper |

1. Simmer the chicken until thoroughly cooked. Dice into 1″ cubes. Set aside.
2. Boil the spaghetti. Drain in a colander and rinse in cold water. Reheat in warm water. Season with salt and pepper. Set aside.
3. Sauté the mushrooms in 1 tbsp of butter until tender. Set aside.
4. In a saucepot, melt 2 tbsp of butter. Add flour to make a roux. Cook for 5 min, stirring constantly.
5. Add hot chicken stock and whip with a wire whisk until thickened and smooth.
6. In a separate saucepot, heat the cream, being careful not to scald it.
7. Add the heated cream and the sherry to the thickened chicken stock. Whip with a wire whisk until the sauce is smooth.
8. Add the chicken and sautéed mushrooms. Season with salt and pepper to taste. Mix well.
9. Arrange spaghetti on the bottom of each serving dish. Place chicken mixture over spaghetti. Sprinkle with Parmesan cheese and brown slightly in the broiler. Serve immediately.

### Nutrition Facts

**Per serving:** 788 calories; 22% calories from fat; 19.9 g total fat; 114.2 mg cholesterol; 503.1 mg sodium; 781.2 mg potassium; 98.1 g carbohydrates; 3.4 g fiber; 5.2 g sugar; 94.7 g net carbs; 50 g protein

## Stuffed Breast of Chicken Provençale with a Tomato-Pesto Cream Reduction

*yield: 2 servings*

| | |
|---|---|
| 2 | boneless, skinless chicken breasts |
| 4 oz | ground turkey meat |
| 3 oz | frozen chopped spinach, thawed and drained |
| 1 tsp | garlic |
| to taste | salt and white pepper |
| 1 tsp | cream |
| 2 tbsp | seasoned flour |

*Tomato-Pesto Cream Sauce*

| | |
|---|---|
| 1 tbsp | tomato paste (or tomato purée) |
| 1 tsp | olive oil |
| 1 clove | garlic, minced |
| 2 oz | white wine |
| 2 oz | light cream |
| 1 tbsp | pesto sauce (or fresh basil purée) |
| to taste | salt and pepper |

1. Pound breasts thin between two sheets of plastic wrap and refrigerate until needed.
2. In a food processor, combine cold turkey meat and spinach. Purée until smooth.
3. Add garlic, salt, and white pepper and mix well.
4. Add cream and pulse to incorporate.
5. Divide the forcemeat evenly between the two chicken breasts. Place forcemeat in the center of each breast and roll up. Secure with toothpicks or butcher's twine.
6. Roll stuffed chicken in seasoned flour.
7. Sauté in a small amount of olive oil to slightly brown the exterior.
8. Place chicken in oven to finish cooking at 325°F for 10 min to 15 min, until internal temperature reaches 165°F.
9. Meanwhile, for the sauce, sauté tomato paste in olive oil until slightly browned.
10. Add garlic and sauté for 1 min.
11. Deglaze the pan with white wine.
12. Add cream and reduce slightly.
13. Finish sauce with pesto, salt, and pepper, whisking to incorporate thoroughly.

### Nutrition Facts

**Per serving:** 570.2 calories; 31% calories from fat; 20.5 g total fat; 205.6 mg cholesterol; 574.3 mg sodium; 1352.5 mg potassium; 17.7 g carbohydrates; 6 g fiber; 1.6 g sugar; 11.7 g net carbs; 73.6 g protein

## 173

## Lemongrass Poached Chicken Breasts

*yield: 2 servings*

| | |
|---|---|
| 2 | boneless, skinless chicken breasts |
| to taste | salt and white pepper |
| 1 tbsp | butter |
| ½ | shallot, minced |
| 3 oz | white wine |
| 1 oz | white wine vinegar |
| 12 oz | chicken stock |
| 1" piece | ginger, chopped |
| 5 pods | star anise |
| 8 | peppercorns |
| 1 tbsp | lemongrass, chopped |
| 4 oz | heavy cream |
| to taste | fresh lemon |
| 2 tsp | basil |
| 2 tsp | cilantro |

1. Season chicken with salt and white pepper.
2. Melt butter in a medium sauté pan.
3. Add shallots, white wine, vinegar, stock, ginger, star anise, peppercorns, and lemongrass. Slowly heat to 160°F.
4. Add chicken breasts and poach until no longer pink but still moist.
5. Remove chicken breasts. Reduce liquid to 4 oz, add heavy cream and reduce to a nappe consistency (until it coats the back of a spoon).
6. Strain the sauce. Add lemon, basil, cilantro, and salt and pepper to taste.
7. Return chicken breast to pan with sauce and reheat.
8. Serve warm with sauce poured over breast.

---

**Nutrition Facts**

**Per serving:** 729.7 calories; 41% calories from fat; 34.4 g total fat; 238.8 mg cholesterol; 584.2 mg sodium; 1474.8 mg potassium; 35.5 g carbohydrates; 2.8 g fiber; 3.4 g sugar; 32.7 g net carbs; 64.8 g protein

---

# Meat

*Meat* generally refers to the edible muscles of beef, veal, lamb, and pork. A skilled chef must possess thorough knowledge of muscle composition, cuts of meat, grading standards, storage requirements, and preparation techniques in order to prepare savory meat dishes. Some cuts of meat are tender, while others are tough. A chef must know which cooking methods produce the best results for each cut.

## Chapter Objectives:

1. Describe the structure and composition of muscle.
2. Describe the major categories of domesticated cattle.
3. Describe how to determine the quality of pork, veal, and lamb.
4. Describe basic cuts of meat.
5. Contrast the advantages and disadvantages of purchasing various cuts of meat.
6. Explain the USDA inspection and grading system for meat.
7. Describe the main factors to consider before purchasing meat.
8. Identify the major storage and preparation requirements for meat.
9. Demonstrate the following procedures: boning a tenderloin; trimming a tenderloin; cutting a tenderloin into various cuts; cutting lamb tenderloin into noisettes, trimming and cutting a boneless strip loin into steaks; grinding fresh meat; separating a primal rack into two individual racks; boning a leg of veal or lamb, carving a roast leg of lamb or steamship round roast; carving a bone-in prime rib; and slicing a prime rib roast.
10. Differentiate among rubs, marinades, and meat tenderizers.
11. Describe trussing, barding, larding, and aging.
12. Determine the best cooking method for various cuts of meat and explain how to determine doneness.

## Key Terms

- collagen
- elastin
- shrinkage
- marbling
- fat cap
- Institutional Meat Purchase Specifications
- whole carcass
- dressed
- side
- primal cut
- fabricated cut
- specification
- irradiation
- barding
- larding
- wet aging
- dry aging

## MUSCLE COMPOSITION

The carcasses of animals such as cattle, veal, sheep, and hogs are made up of bones, muscle tissue, connective tissues, and fat. Understanding structure and composition of animal muscles helps a chef to better understand cooking methods and fabrication techniques appropriate for use with each part of an animal. For example, a leg is supported by bones that allow an animal to stand, while groupings of muscle tissue held together by connective tissues contract and relax to make the leg bend and straighten at the joint. Bones, muscle tissue, and connective tissues are surrounded by a layer of protective fat and a thin covering of skin.

### Muscle Fibers

Muscle tissue of all animals is made up of bundles of long, stringlike fibers held together by connective tissues. Fine-grained meats such as beef tenderloin have small muscle fibers that are bound in small bundles. Coarse-grained meats such as a corned beef brisket contain larger muscle fibers bound in larger bundles. ***See Figure 19-1.*** These bundles of fibers are similar to the many pieces of string that make up a rope. The direction of grain is parallel to the length of muscle fibers.

## Muscle Fibers

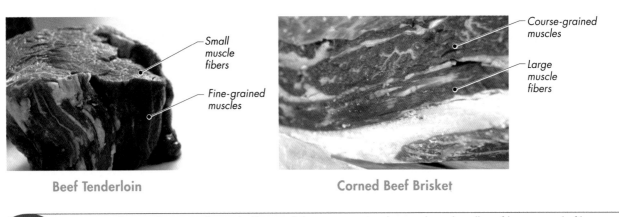

Beef Tenderloin
- Small muscle fibers
- Fine-grained muscles

Corned Beef Brisket
- Course-grained muscles
- Large muscle fibers

**19-1** *Fine-grained meats have small muscle fibers, while coarse-grained meats have bundles of large muscle fibers.*

### Connective Tissues

Connective tissues are strong tissues made from protein that bind bundles of muscle fibers together. They form a thin, sleevelike material that covers each individual string of muscle fiber. A second thin membrane of connective tissue forms around each muscle group and is continuous with the tendons that hold each muscle in place.

Connective tissues are generally found around heavily worked muscle groups. Generally, the more a muscle works, the tougher it gets and the more connective tissue that forms. Therefore, the shoulder and leg areas that work the hardest in animals have the most connective tissue and generally yield tougher meat than other areas of the body. Muscle groups that receive less exercise are more tender and have less connective tissue. A powerful and muscular bull, for example, yields meat that is very tough and has lots of connective tissue. In comparison, a young calf (veal) yields very tender meat with little connective tissue. The amount of connective tissue also increases and becomes harder as an animal gets older.

The two types of connective tissues found in meat are called collagen and elastin. *Collagen* is a soft, white, connective tissue that breaks down into gelatin when heated. Collagen and fat give meat a rich taste and help to thicken and flavor sauces made from meat juices. *Elastin* is a tough, rubbery, yellowish connective tissue that does not break down with heat. The yellowish color makes it easy to distinguish elastin from collagen. Ligaments and tendons are examples of the connective tissue elastin. Elastin should be completely trimmed from meat prior to cooking, as it is relatively inedible.

## Nutritional Composition

Each muscle fiber is made up of many cells that contain mostly water, protein, and fat. ***See Figure 19-2.*** Approximately 75% of the cell is water, 20% is protein, and 5% is fat. Muscle also contains trace amounts of carbohydrates.

Because the majority of muscle is made up of water, overcooking meat can cause the water to evaporate, resulting in dry meat. As meat loses water, it undergoes shrinkage. *Shrinkage* is the loss of volume and weight of a piece of food as the food cooks. Shrinkage explains why a 20 lb roast may be only 18.5 lb when fully cooked. Meats roasted at low temperatures for longer periods lose less water during the cooking process than meats cooked at high temperatures for shorter periods.

Protein shrinks (coagulates) and becomes firmer when subjected to heat. Cooking a piece of meat at too high a temperature toughens the protein. Cooking methods such as grilling or frying use very high temperatures for short periods and result in only the exterior of an item receiving high amounts of heat. High heat quickly cooks the exterior of meat to a crispy texture while slowly cooking the interior of the meat. This is the reason that a grilled steak is crispy and somewhat dry on the outside yet remains very tender and juicy on the inside.

Fat is found in two forms in meat, as marbling and as fat cap. As an animal eats more and gets fatter, fat cells take the place of some of the water and protein cells in the muscle tissue. *Marbling* is trace amounts of fat speckled throughout the muscle. Marbling accounts for close to 5% of the makeup of muscle tissue. ***See Figure 19-3.*** Marbling affects the taste, tenderness, and quality of meat. Marbled fat within meat tissue is the main source of flavor and moisture in meat.

## Nutritional Composition

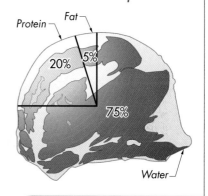

**19-2** *The nutritional composition of muscle fibers are made up of water, protein, fat, and trace amounts of carbohydrates.*

## Marbling

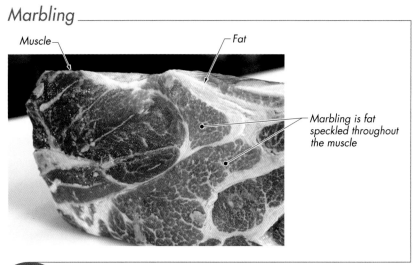

Muscle

Fat

Marbling is fat speckled throughout the muscle

**19-3** *Marbling adds flavor, moisture, and desirable texture to meat.*

Fat cap is the second form of fat, and is not found within the tissues of the muscle. *Fat cap* is a thick layer of fat that surrounds a muscle. This layer of fat provides muscles with a source of energy and serves as an insulation layer to keep animals warm. Animals that live in cold climates, such as polar bears, have a very thick layer of fat cap under their skin, while animals that come from hot climates, such as zebras, have a very thin layer of fat cap. As an animal gets heavier and older, layers of fat develop around and on top of groups of muscle tissues. A typical carcass of beef may contain more than 30%. Fat cap is left on meats while cooking as a means of making them juicier and preventing them from drying out.

Carbohydrates are found in only very small amounts in muscle tissue. From a nutritional standpoint, they are relatively insignificant. However, these small amounts of carbohydrates in meat are what allow the meat to caramelize when subjected to high heat. Without carbohydrates, meat would not gain any color during a dry-heat cooking process such as grilling or broiling.

### BEEF

Beef is the flesh of domesticated cattle, which includes steers, heifers, cows, bulls, and stags. The age and sex of these animals has a great effect on the taste and quality of the meat. Most consumable beef comes from steers. ***See Figure 19-4.***

A *steer* is a male calf that is castrated prior to reaching sexual maturity. Steers produce the best-quality beef and a high yield (amount of salable meat). Steers are about 2 to 3 years old when they are marketed. They typically weigh between 650 lb and 1250 lb, and produce 400 lb to 800 lb of meat. Most steers are grain-fed, which increases the marbling and improves the taste of the meat.

| Types of Beef | | |
|---|---|---|
| **Type** | **Description** | |
| Steers | Immature, castrated male calf | |
| | Largest source of consumable beef | |
| | Marketed at 2 to 3 years of age | |
| Heifers | Young female cow that has not borne a calf | |
| | Mature faster than steers | |
| | Marketed at 2 to 3 years of age | |
| Cows | Mature female that has borne one or more calves | |
| | Primarily raised for calf and milk production | |
| | Marketed at an older age, producing poor-quality meat | |
| Bulls | Sexually mature, uncastrated male | |
| | Commonly used to make sausage and dried beef | |
| Stags | Sexually mature, castrated male | |
| | Poor-quality meat | |
| | Marketed at an older age | |

**19-4** *Domesticated cattle include steers, heifers, cows, bulls, and stags.*

A *heifer* is a young female cattle that has not borne a calf. Heifers produce high-quality meat and are generally marketed when 2 to 3 years old. Heifers mature faster than steers, but steers are preferred because they usually produce a higher yield.

A *cow* is a mature female cattle that has borne one or more calves. Cows are primarily raised for calf and milk production. Cows are marketed at an older age because they are kept as long as the calves they bear or they are used for milk production. Cows are not commonly used for meat because the meat contains an uneven distribution of yellow fat.

A *bull* is a sexually mature, uncastrated male cattle. Bull meat is dark red and commonly used in the making of sausage and dried beef.

A *stag* is a male cattle that is castrated after it has become sexually mature. The quality of meat from a stag generally lacks the characteristics necessary to achieve even a mediocre grade. Stag meat is not used in professional kitchens.

In addition to the major categories of domesticated cattle, other factors such as an animal's diet may be designated. *Grain-fed beef* is meat obtained from cattle that were grain-fed in dry lots (feeding pens) for a period of 90 days to a year. Grain-fed beef are marketed in April and May. *Grass-fed beef* is meat obtained from cattle that were raised on grass with little or no special feed. Most grass-fed beef cattle are marketed during the fall months of the year.

*Baby beef* is a term applied to beef from cattle less than 18 months of age. These young cattle produce approximately 400 lb to 550 lb of meat. This beef is tender, but it lacks the pronounced flavor of mature beef. Calf carcasses or beef-dressed veal are terms for meat from animals too large to be sold as veal, but too small to be sold as beef. These animals produce between 150 lb and 375 lb of meat.

*Branded beef* is beef with a trademark or trade name that is used by some packers to indicate their own grades. These brands are sometimes placed on a product even though the meat is also graded by the United States Department of Agriculture (USDA).

## PORK

*Pork* is the meat of hogs usually slaughtered less than a year old. The best pork on the market comes from hogs between 6 months and 8 months old. Pork is the second-most-consumed meat in the United States. In addition, pork is the only meat of which all wholesale cuts can be cured. *Curing* is the processing of an item with salt or chemicals to retard the action of bacteria and to preserve the meat. More than two-thirds of all pork is marketed in a cured form as bacon or ham. *See Figure 19-5.* At one time, pork had the highest fat content of any meat. In recent years, selective breeding has resulted in much leaner meat with a much milder flavor.

## Cured Pork

Ham is cured, smoked hog thigh and buttock

High percentage of lean meat

**Ham**

High-quality bacon contains 40% to 45% fat

Bacon is cured, smoked hog belly

**Bacon**

19-5 *More than two-thirds of all pork is marketed in a cured form as bacon or ham.*

Pork is tender because the animal is very young when it is slaughtered. It can be cooked using almost any cooking method. A female pig, or sow, produces its first litter when it is just a year old, so sows can be marketed at a young age that is not possible with animals such as cows or sheep. Most hogs are marketed during the fall and winter months. A *barrow* is a castrated male hog. A *gilt* is an immature female hog.

Pork is a light pinkish color. When cooked, most cuts of pork are very close in color to white-meat poultry. Pork is always cooked well done, except for the tenderloin, which is most commonly cooked medium.

## VEAL

*Veal* is the meat of young, usually male, milk-fed calves. Veal has very little fat covering and a high moisture content, and is extremely tender. It has a very light pink color with no visible marbling. ***See Figure 19-6.*** Most calves destined to be veal result from the demand for milk. A cow does not produce any milk until bearing a calf. In addition, a cow must bear a calf annually in order to continue producing milk. Male dairy cows are of little value to dairy farmers except as meat. Most veal is slaughtered between 8 and 16 months of age. Veal must be slaughtered prior to the animal eating solid foods, because the iron content in solid foods turns the meat more reddish and causes marbling. Once this happens the animal can no longer be sold as veal, but instead must be marketed as calf.

Veal _____

Light pink color

No visual marbling

National Cattlemen's Beef Association

**19-6** *Milk-fed veal is pale pink in color with no visible marbling.*

## LAMB

*Lamb* is the meat of immature sheep slaughtered before one year of age. Lamb is lower in fat and cholesterol than other red meats and higher in iron than poultry and fish, which makes it a healthy alternative protein source. Because the animal is so young at the time of processing, lamb is very tender and delicate in flavor. It is typically served with savory herbs or rich flavorful sauces. The tender meat works well with most cooking methods. *Mutton* is meat of sheep slaughtered over one year of age. Mutton from sheep marketed between the age of 12 months and 20 months is sometimes called yearling mutton. Mutton has a strong flavor and odor that many people find undesirable, and is rarely used in the professional kitchen.

The age and diet of a lamb affect the quality of its meat. When a lamb is born, it feeds on its mother's milk until it is about 5 months old. A lamb under 5 months old produces the highest-quality meat. The flesh is red like beef and has a smooth grain. The bones are soft and porous, and the fat is firm. ***See Figure 19-7.***

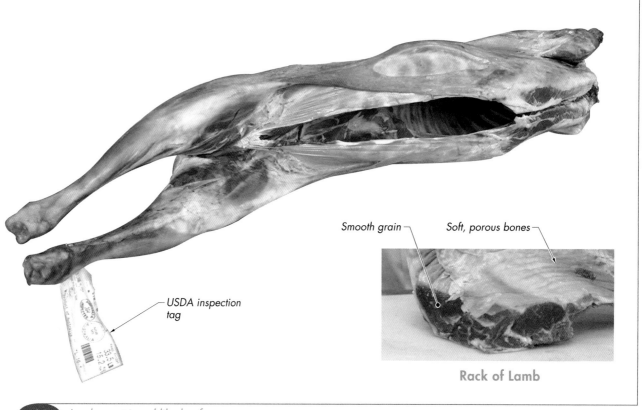

Smooth grain —     Soft, porous bones —

USDA inspection tag

Rack of Lamb

**19-7**   *Lamb meat is red like beef.*

After 5 months, a lamb is sent out to pasture, where it feeds on grass and grain and starts to mature. As a lamb matures from lamb to mutton, changes occur in the character, color, and consistency of the flesh, bones, and fat of the animal. The flesh becomes darker, and the bones become harder and whiter. The fat becomes soft and slightly greasy. When an animal reaches the mutton stage, the meat is dry and strong in flavor. Three types of lamb are available on the market: genuine spring lamb, spring lamb, and yearling lamb.

A *genuine spring lamb* is a lamb marketed between the age of 3 and 5 months that has been fattened primarily on its mother's milk. Genuine spring lambs are also classified as milk lambs and Easter lambs. Genuine spring lamb is available on the market from April to July and is considered the best lamb. A *spring lamb* is a lamb marketed between the age of 5 months to 10 months. Spring lambs are marketed during fall and winter and are fattened on some milk but then weaned off to eat grass and grain. The meat is darker and slightly tougher than genuine spring lamb.

A *yearling lamb* is a lamb marketed between the age of 10 and 12 months. Yearling lambs are too young to be sold as mutton and are the most economical lamb to use. However, special preparation may be required to achieve the desired taste.

## BASIC MEAT CUTS

Beef, pork, veal, and lamb are available in a variety of forms, from whole carcasses to portion-controlled cuts. Thorough knowledge of each of these market forms is necessary for accurate product ordering. Using standard terminology and industry standards for product identification when ordering helps ensure that a restaurant receives the correct products from a supplier.

The *Institutional Meat Purchase Specifications (IMPS)* are specifications published by the United States Department of Agriculture (USDA) for commonly purchased meats and meat products. All large cuts of meat are numbered by category, with beef in the 100 series, lamb in the 200 series, veal in the 300 series, and pork in the 400 series. Portioned cuts of each type of meat are listed in series numbered by 1000s (i.e., beef, 1000 series; lamb, 2000 series; veal, 3000 series; and pork, 4000 series). *See Figure 19-8.* Most food service establishments purchase primarily fabricated cuts and some primal cuts. The National Association of Meat Purveyors (NAMP) publishes a reference book titled *The Meat Buyer's Guide,* which lists all IMPS with color photos and descriptions. *See Appendix.*

Beef, pork, veal, and lamb share many of the same market forms, although some forms are specific to particular types of meat. Common market forms include whole carcasses, partial carcasses, primal cuts, fabricated cuts, and variety meats. Beef, veal, and lamb are available in all five of these market forms. Pork is commonly marketed as fabricated cuts rather than by the whole carcass or primal cut as it has many extra cuts that are not desirable for use in the professional kitchen. In addition, pork spoils more quickly than other types of meat. Having too much pork on hand could be very costly for a restaurant operator if the meat spoiled before it could be used.

### Whole Carcasses

A *whole carcass* is a whole animal with the hide, head, entrails, and feet removed. However, a whole carcass of pork has only the head and entrails removed, and a lamb carcass has a fell. *Fell* is the thin, paperlike covering on the outside of a lamb carcass. A carcass can be purchased at a cheaper cost per pound than other market forms because less labor and handling have been performed. A whole carcass is not a common purchase for a food service operation because of the skill, labor, time, and space needed to store, process, and fabricate the meat. Also, a whole carcass may yield many cuts of meat that are not needed by the operation.

## USDA Institutional Meat Purchase Specifications (IMPS)

| Series | Item No. | Product | Item No. | Product |
|---|---|---|---|---|
| **100**<br>**(fresh beef)** | 100 | Carcass | 132 | Triangle |
| | 101 | Side | 134 | Beef bones |
| | 102 | Forequarter | 135 | Diced beef |
| | 103–112 | Rib | 136 | Ground beef |
| | 113–114 | Chuck | 138–139 | Beef trimmings |
| | 117 | Foreshank | 155 | Hindquarter |
| | 118–120 | Brisket | 157 | Hindshank |
| | 121 | Plate, short plate | 158–170 | Round |
| | 122 | Plate, full | 172 | Loin, full loin, trimmed |
| | 123 | Short ribs | 173 | Loin, short loin |
| **200**<br>**(lamb)** | 204 | Rack | 238–239 | Trimmings |
| | 206 | Shoulders | 242–243 | Loins, full |
| | 207–208 | Shoulders, square-cut | 244 | Loin, boneless, 3-way |
| | 209 | Breast | 245 | Sirloin |
| | 210 | Foreshank | 246 | Tenderloin |
| **300**<br>**(veal and calf)** | 304 | Foresaddle, 11 ribs | 341 | Back, 9 ribs, trimmed |
| | 304A | Forequarter, 11 ribs | 342 | Back, strip, boneless |
| | 306 | Hotel rack, 7 ribs | 344 | Loin, strip loin, boneless |
| | 307 | Rack, ribeye, 7 ribs | 346 | Leg, butt tenderloin, defatted |
| | 308–311 | Chuck | 347 | Loin, short tenderloin |
| | 312 | Foreshank | 349–363 | Leg, top round, cap on |
| | 313–314 | Breast | 389 | Mixed bones |
| **700 (variety**<br>**meats)** | 710 | Lamb liver | 725 | Beef, lamb, or pork brains |
| | 713 | Veal sweetbreads | 726 | Beef tripe, scalded, bleached |
| | 715 | Beef tongue, short cut | | (denuded) |
| | 716 | Tongue, Swiss cut | 727 | Beef tripe, honeycomb, bleached |
| **1100 (fresh beef**<br>**portion cuts)** | 1112 | Rib, ribeye roll steak | 1170 | Round, bottom (gooseneck) round steak |
| | 1114 | Chuck, shoulder-clod, top blade steak | 1173 | Loin, porterhouse steak |
| | 1116 | Chuck, chuck eye roll steak | 1174 | Loin, T-bone steak |
| | 1121 | Plate, skirt steak | 1179–1180 | Loin, strip loin steak |
| | 1123 | Short ribs, flanken style | 1184 | Loin, top sirloin butt steak, boneless |
| **1200 (lamb**<br>**portion cuts)** | 1202 | Braising steak, Swiss | 1234 | Leg chops, boneless |
| | 1204 | Rib chops | 1296 | Ground lamb patties |
| | 1207 | Shoulder chops | 1297 | Lamb steaks, flaked and formed, frozen |
| **1300 (veal and**<br>**calf portion cuts)** | 1302 | Veal slices | 1337 | Osso buco, hindshank |
| | 1306 | Rack, rib chops | 1338 | Veal steak, flaked and formed, frozen |
| | 1309 | Chuck, shoulder arm chops | | |
| **1500 (cured,**<br>**smoked, and fully**<br>**cooked pork**<br>**portion cuts)** | 1513 | Ham patties (cured), fully cooked | | |
| | 1531 | Ham steaks (cured and smoked), boneless | | |
| | 1545 | Pork loin chops (cured and smoked) | | |
| | 1548 | Pork loin chops, boneless, center cut (cured and smoked) | | |
| | 1596 | Pork patty, precooked | | |

**19-8**  *The USDA Institutional Meat Purchase Specifications (IMPS) number all cuts of meat by animal category.*

Typically, the only time when an establishment would purchase a whole carcass would be to roast a whole pig or lamb on a spit. A *suckling pig* is a pig between 4 weeks and 6 weeks old. A suckling pig weighs from 20 lb to 35 lb dressed. *Dressed* is a term used to describe an animal that has been slaughtered and cleaned for consumption by removing all blood and inedible parts. Suckling pigs are purchased with the head on and are priced per pig rather than by the pound. They are roasted whole and are used for ornamental purposes, such as for a centerpiece on a banquet table.

## Partial Carcass Cuts

A *partial carcass cut* is a cut of meat that is a primary division of a whole carcass. Partial carcass cuts are named differently for beef, veal, and lamb. Common partial carcass cuts include sides, forequarters, hindquarters, foresaddles, and hindsaddles. Beef is cut into sides, forequarters, and hindquarters. Veal and lamb are cut into foresaddles and hindsaddles. Lamb is also cut into backs and bracelets. Pork is not processed into any of these larger cuts. **See Figure 19-9.**

## Partial Carcass Cuts

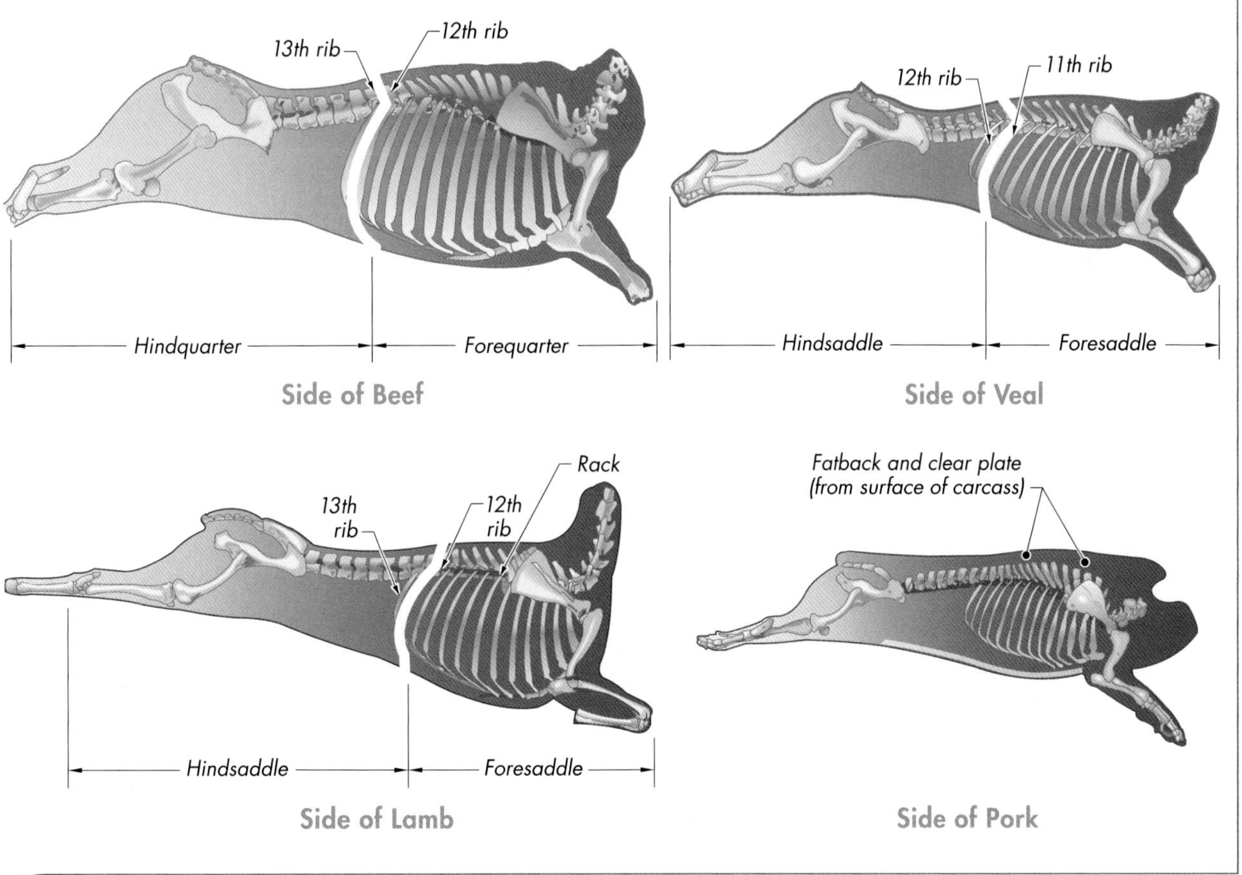

Side of Beef

Side of Veal

Side of Lamb

Side of Pork

19-9　Partial carcass cuts include sides, forequarters, hindquarters, foresaddles, and hindsaddles.

Like whole carcasses, partial carcass cuts are rarely purchased by a food service establishment because of the skill, labor, time, and space needed to process the meat, as well as because partial carcasses may yield cuts of meat that are not needed by the operation. Veal is occasionally purchased in partial carcass cuts if the establishment has the capability of cutting, storing, and utilizing all the meat.

**Sides.** A *side* is a half of a complete carcass split along the backbone. There are two sides to each carcass, a right side and a left side. A side can be purchased at a cheaper cost per pound than any smaller market form. However, many factors must first be considered. The proper facilities and equipment for storage and fabrication must be available, and staff capable of cutting a side of meat into individual cuts is required. Because of the amount of work, time, labor, and space needed to break down a side of meat, sides are almost never purchased by food service operations.

**Forequarters and Hindquarters.** A *quarter* is one side of a beef carcass divided into two parts between the 12th and 13th rib bones. The *forequarter* is the front quarter of beef consisting of the rib, chuck, shank, brisket, and short plate. The *hindquarter* is the back quarter of beef consisting of the short loin, flank, sirloin, rump, and round. The meat of the hindquarter is more desirable, as it contains more underutilized muscles than the forequarter. Consequently, the hindquarter sells at a higher price. The forequarter does not contain as much tender meat and typically costs less per pound than the hindquarter.

**Foresaddles and Hindsaddles.** Lamb and veal are typically not split into sides, but rather are split into head and tail sections known as a foresaddle and a hindsaddle. The foresaddle and hindsaddle have both sides of each cut joined together. Veal is split between the 11th and 12th ribs, while lamb is split between the 12th and 13th ribs. The *foresaddle* is the front half of the carcass consisting of the shoulder, rack, breast, and shank. The *hindsaddle* is the tail or rear half of the carcass consisting of the loin and leg.

**Backs and Bracelets.** In addition to foresaddles and hindsaddles, lamb is also commonly divided into backs and bracelets. A *back* is a rack and loin still joined together and well trimmed. A back is especially desired by operations that sell a lot of lamb chops, as chops can be cut from the end of the rack to the end of the loin.

When a rack is not divided between the two sides and remains joined along the backbone, it is called a double rack or hotel rack. A *bracelet* is a hotel rack (double rack) with the breast still attached. Leaving the breast attached to the rack yields a greater amount of meat but also a greater amount of fat. A bracelet is not typically prepared whole but is further fabricated into a hotel rack and a breast.

## Primal Cuts

A *primal (wholesale) cut* is a division of a partial carcass cut. A single carcass has two sets of primal cuts (one set on the left side and one set on the right side). Primal cuts are ordered by food service establishments that have at least one person on staff who is skilled in cutting meat. Primal cuts are large, but are still very manageable. The size of a primal cut is ideal for mastering proper meat-cutting methods. *See Figure 19-10.*

Before beef is divided into primal cuts, it is first split into sides. A forequarter of beef contains four primal cuts: the chuck (shoulder), rib (rack), short plate (navel), and brisket and shank. The hindquarter consists of five primal cuts: the short loin, sirloin, flank, round, and rump. Of these hindquarter primal cuts, the short loin and sirloin (together called the loin) are the most prized cuts for their tenderness and flavor. The loin contains all the best steak meats and is always in demand.

Veal and lamb are not split into sides before being divided into primal cuts. Rather, the left and right primal cuts remain joined together and are purchased as a single cut. For example, the leg primal cut is delivered as two joined legs. Veal is fabricated into one of six primal cuts: shoulder, rack (rib), breast, shank, loin, and leg. Lamb is fabricated into the same cuts; however, the shank is not separated from the breast. Lamb is most commonly purchased as primal cuts unless specifically requested otherwise from the customer.

**Chuck or Shoulder.** A *chuck* (shoulder clod chuck) is a shoulder of an animal. In beef, it is called chuck, and in veal and lamb, it is called shoulder. In beef and veal, a primal chuck or shoulder contains the first five rib bones, some of the backbone, and a small amount of the arm and blade bones. In lamb, the components of the cut are the same except that the cut contains only the first four rib bones. Even though the chuck contains some rib bones, this meat is not nearly as tender as meat fabricated from the rib or loin. The shoulder is the largest primal cut of the carcass, and accounts for 26% of total carcass weight in beef, 21% in veal, and 36% in lamb.

The shoulder is one of the most-exercised muscles on the animal body, so it is a fairly tough cut of meat with many connective tissues. However, it is also quite lean and possesses excellent flavor. When trimmed for cooking, the chuck or shoulder produces about 75% to 80% usable meat. This cut is never cooked whole, as it contains many smaller bones and connective tissues that would be hard to carve and serve. Instead, it is trimmed well and separated into several smaller sections.

Larger pieces of chuck or shoulder lend themselves to combination cooking methods such as braising or stewing. Smaller tough pieces and trimmings of chuck or shoulder produce very flavorful ground meat. A small, thin muscle section on top of the beef chuck provides a cut known as flat iron steak. This cut is somewhat tender and lends itself to marinating and then grilling or broiling. In lamb, most shoulder meat is cubed for stews or ground for forcemeats. *See Figure 19-11.*

*Chef's Tip:*
Primal cuts of beef forequarters constitute approximately 59% of the total weight of a carcass.

# Primal Cuts

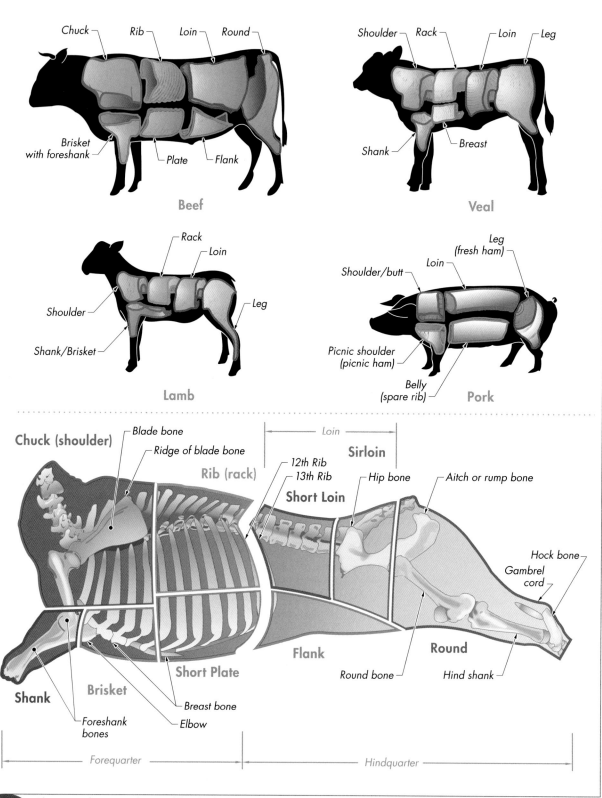

**Beef**

- Chuck
- Rib
- Loin
- Round
- Brisket with foreshank
- Plate
- Flank

**Veal**

- Shoulder
- Rack
- Loin
- Leg
- Shank
- Breast

**Lamb**

- Rack
- Loin
- Shoulder
- Leg
- Shank/Brisket

**Pork**

- Shoulder/butt
- Loin
- Leg (fresh ham)
- Picnic shoulder (picnic ham)
- Belly (spare rib)

**Chuck (shoulder)**
- Blade bone
- Ridge of blade bone

Loin

Sirloin

**Rib (rack)**
- 12th Rib
- 13th Rib

**Short Loin**
- Hip bone
- Aitch or rump bone

- Hock bone
- Gambrel cord

**Flank**

**Round**
- Round bone
- Hind shank

**Shank**
- Foreshank bones

**Brisket**
- Breast bone
- Elbow

**Short Plate**

Forequarter

Hindquarter

**19-10** *There are five primal cuts of beef which are often further divided into nine sub-primal cuts.*

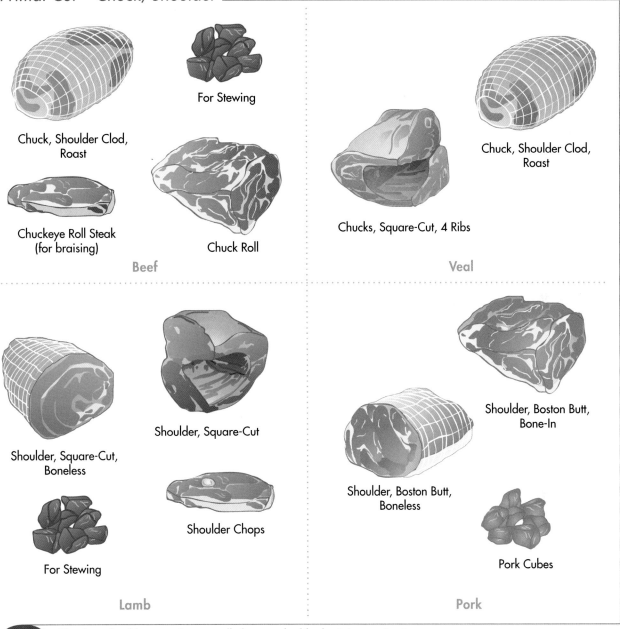

For Stewing

Chuck, Shoulder Clod, Roast

Chuckeye Roll Steak (for braising)

Chuck Roll

**Beef**

Chucks, Square-Cut, 4 Ribs

Chuck, Shoulder Clod, Roast

**Veal**

Shoulder, Square-Cut, Boneless

Shoulder, Square-Cut

Shoulder Chops

For Stewing

**Lamb**

Shoulder, Boston Butt, Bone-In

Shoulder, Boston Butt, Boneless

Pork Cubes

**Pork**

**19-11** *Chuck or shoulder meat is generally best cooked by braising or stewing.*

**Rib or Rack.** The primal rib (rack) is located between the shoulder and loin. In beef, it is called a rib, and in veal and lamb, it is called a rack. The rib or rack is the best cut from the forequarter or foresaddle of the carcass, because it comes from an area of the back where the muscles are not worked much. As a result, the meat is tender and well marbled. The primal rack is the most popular of all the lamb cuts. In beef and veal, the rib or rack contains seven rib bones, while in lamb the rack has eight rib bones. The rib bones (finger bones) make the rib cut the best cut for

roasting, because they form a natural rack for the meat to cook on. A veal rib is different from a beef rib in that veal is not split into two halves along the backbone (chine bones). A double rib rack, commonly called a hotel rack, consists of two very tender veal rib loins. Lamb rack is also available as a double rack.

The average weight of a beef rib is 20 lb to 25 lb. The majority of a beef rib (75% to 80%) is usable meat. A beef rib contains the well-known prime rib roast, which is the only cut from the forequarter that can be improved by aging. The rib bones can be left on for roasting, as doing so helps to produce a moister roast. Alternatively, the bones can be removed to produce a boneless rib eye roast, which, if desired, can be further cut into boneless rib eye steaks. A *rib eye* is a large, eye-shaped muscle within the rib that is a continuation of the sirloin muscle. Meat from eye-shaped muscles such as the rib eye or tenderloin is often referred to as "eye meat." If the meaty bones are removed, they can be prepared as smoked or barbequed beef ribs.

Beef short ribs are popular in the southern United States and are produced from the small rib bone ends that were sawed off and left on the primal rib as the rib roast was removed. These short ribs make excellent cuts since there is generally a sizable portion of lean meat left on them. Another preparation is to remove the seven rib bones, roll up the meat, truss it with butcher twine, and roast it. This preparation is known as a rolled rib roast. ***See Figure 19-12.***

*Beef Rib*

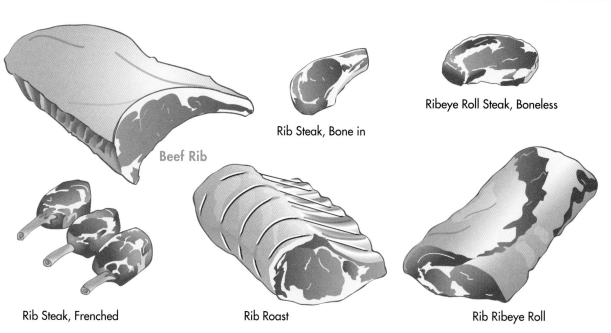

Ribeye Roll Steak, Boneless

Rib Steak, Bone in

Beef Rib

Rib Steak, Frenched

Rib Roast

Rib Ribeye Roll

19-12    *Beef rib can be fabricated into various steaks and roasts.*

The veal rack can be split into two halves and trussed into a circle to form a crown rib roast. In another preparation, the loins are trimmed, Frenched (meat and fat are removed from the bones), and cut into veal chops. Veal ribs can also be boned out to yield a small portion of the tenderloin, known as the short tenderloin, and the very tender boneless veal rib eye roast.

Rack of lamb is popular for the lamb chops it provides. The primal rack can be split in two and the two sides tied in a circle to form a crown roast of lamb. Rack of lamb is sometimes coated with herbs, roasted whole, and then sliced to order, but in most cases, it is cut into single or double chops and grilled or broiled. ***See Figure 19-13.***

*Rack of Lamb*

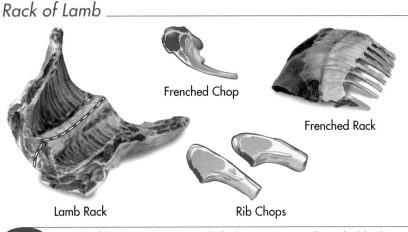

Frenched Chop

Frenched Rack

Lamb Rack

Rib Chops

**19-13** *A rack of lamb can be prepared whole or cut into single or double chops.*

**Short Plate.** A *short plate* is a thin portion of the beef forequarter that lies just beneath the rib cut. It is sometimes referred to as the navel, and accounts for only 9% of the carcass weight. The bones attached to this cut are the remaining sections of the rib bones. In the professional kitchen, the small bones from the short plate are also used in the preparation of short ribs, but these short ribs are not quite as meaty as those from the rib cut. In addition to beef short ribs, the short plate yields the skirt steak and small, miscellaneous portions of meat best suited for grinding.

**Breast.** A *breast* is a thin, flat cut of meat located under the shoulder and ribs in veal and lamb. It contains the breastbone, tips of the rib bones, and cartilage. The breastbone is typically still cartilage because the animal is so young. In lamb, the breast also includes the shank. The breast accounts for approximately 16% of total carcass weight in veal and 17% of the total carcass weight in lamb.

The breast is not a very popular cut, but is used in the professional kitchen for such preparations as stuffed breast of veal or lamb. The thin, flat shape and ample connective tissue make the breast portion a great piece to stuff, roll, tie, and braise, which breaks down the connective tissue and yields a very tender rolled roast. In lamb, the small section of 12 rib tips can also be braised until tender for Denver ribs or lamb riblets.

**Brisket and Shank.** In beef, the *brisket* (breast) is a thin section of meat that contains the ribs, the breastbone, and layers of lean muscle, fat, and connective tissue. A *shank* is a cut of meat from the lower front or rear leg. It contains a large amount of bone that is surrounded by a smaller amount of very tough but flavorful meat. The front arm shank is referred to as the foreshank and the rear leg shank is commonly called the hindshank.

In beef, the brisket and shank are two separate muscle groups that make up one primal cut. Located just below the primal chuck, the brisket-and-shank cut adds up to just about 8% of the total weight of the beef carcass. The ribs and the breastbone are always removed prior to cooking. The brisket is a tough cut of beef but has excellent flavor when cooked properly. It has long muscle fibers that run in several directions, making it difficult to slice. The brisket is best prepared using moist-heat cooking methods such as braising or simmering. In the professional kitchen, brisket is prepared in many ways, such as curing and peppering to make pastrami, braising in the preparation of pot roast and sauerbraten, simmering for New England-style boiled beef brisket, and corning or pickling to prepare classic Irish corned beef. ***See Figure 19-14.***

Shanks are prized for making stocks and rich reduction sauces. Shank meat is most desired ground, to flavor and clarify a consommé, because the meat has a high concentration of collagen, which converts to gelatin when cooked. To be utilized in the professional kitchen, shanks are generally cut perpendicularly across the bone, braised in a rich liquid, and served with jardinière-cut vegetables. Veal shank can be sliced in cross-sections across the bone. ***See Figure 19-15.***

**Loin.** A *loin* (saddle cut) is a primal cut located between the primal rack and leg, and includes the 12th and 13th rib in beef and veal (the 13th rib in lamb), the loin eye muscle, the center section of the tenderloin, the strip loin, and some flank meat. A *tenderloin* is an eye-shaped muscle running from the primal rib cut into the primal leg. In beef, the loin is the combination of a short loin primal cut and a sirloin primal cut that have not been split apart. A complete, unsplit primal loin from a veal or lamb carcass is commonly known as a saddle of veal or a saddle of lamb. ***See Figure 19-16.***

In beef, the tenderloin can be carefully removed prior to cutting the short loin and sirloin apart to produce club steaks, bone-in strip steaks, and T-bone steaks. The result is that the whole tenderloin is available to be cut into steaks or roasted whole. Alternatively, the bones of the beef loin can be completely removed and both the tenderloin and strip loin can be carefully removed separately. Beef tenderloin can be cut into filets or tournedos, or roasted whole as chateaubriand. Beef strip loin can be cut into boneless New York strip steaks or roasted whole and carved or sliced. Cuts from the short loin are the most popular for aging, as they have ample fat covering and marbling and are very tender. Aging any beef loin cut intensifies the flavor and tenderness of the meat, but also raises the price of the meat.

*Beef Brisket*

Corned beef

*Perdue Foodservice, Perdue Farms Incorporated*

**19-14** Brisket is often corned and pickled to make corned beef.

*Veal Shank*

*The Beef Checkoff*

**19-15** Veal shank is sliced in cross-sections.

## Loin

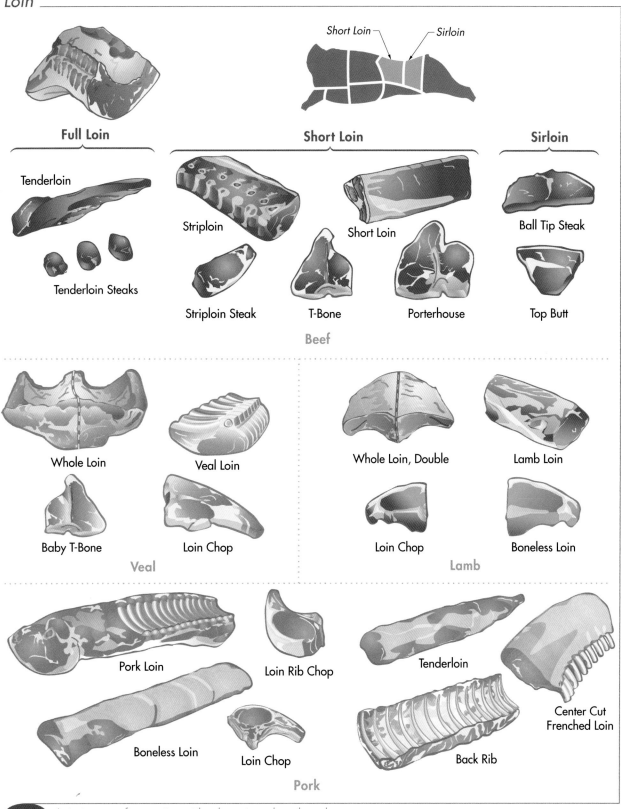

**Full Loin**

Tenderloin

Tenderloin Steaks

**Short Loin**

Short Loin ─ ─ Sirloin

Striploin

Striploin Steak

Short Loin

T-Bone

Porterhouse

**Sirloin**

Ball Tip Steak

Top Butt

Beef

Whole Loin

Baby T-Bone

Veal Loin

Loin Chop

Veal

Whole Loin, Double

Loin Chop

Lamb Loin

Boneless Loin

Lamb

Pork Loin

Boneless Loin

Loin Rib Chop

Loin Chop

Tenderloin

Back Rib

Center Cut
Frenched Loin

Pork

 19-16  *Loin is most often cut into either bone-in or boneless chops.*

Veal loins are sometimes roasted in the professional kitchen, but in most cases they are converted into either bone-in or boneless veal chops. The cuts from the loin are best prepared using dry-heat cooking methods such as grilling, broiling or roasting.

In lamb, an *English lamb chop* is a 2″-thick cut taken along the entire length of the unsplit loin. The tenderloin can be boned out and cut into medallions or noisettes or can be roasted whole, but it is usually cut into boneless or bone-in chops and cooked using a dry-heat cooking method such as grilling or broiling. A *noisette* (pronounced nwa-ZET) is a small, round, boneless medallion of meat cut from the rib.

**Short Loin.** In beef, a *short loin* is the first primal cut in the hindquarter and is located just to the rear of the primal rib. It accounts for approximately 8% of the carcass weight and includes the 13th rib and a small section of the backbone. When the rib eye muscle is included in this primal cut, its name changes to the strip loin. The tenderloin is located just beneath the strip loin and is the most tender piece of meat on the entire carcass. When the short loin and sirloin are split apart, the largest end of the tenderloin (butt tenderloin) remains in the sirloin primal cut, while the smaller part of tenderloin is in the short loin. However, sometimes the entire tenderloin is removed from the loin prior to dividing the short loin and sirloin.

The short loin can be carefully cut in cross-sections across the muscles and bone to produce some of the best (and most expensive) steaks from the beef carcass. Starting from the end nearest the primal rib, cross-section cuts produce club steaks (which do not include any tenderloin), T-bone steaks (which include a small portion of tenderloin and strip steak), and porterhouse steaks (which include large sections of tenderloin and strip steak). **See Figure 19-17.**

*Beef Short Loin*

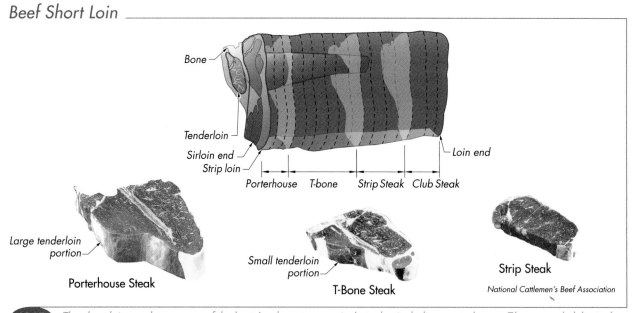

**19-17** *The short loin produces some of the best (and most expensive) steaks, including porterhouse, T-bone, and club steaks.*

**Sirloin.** A *sirloin* is a primal cut of beef situated just behind the short loin and accounts for approximately 7% of the carcass weight. It contains the top sirloin butt, the bottom sirloin butt, bottom sirloin butt tri-tip, and the butt tenderloin. The sirloin also contains some of the backbone and hipbone.

With the exception of the butt tenderloin muscle, meat from the sirloin is not quite as tender as meat from the short loin. However, this meat still works well for roasting, can be cut into butt steaks that work well with dry-heat methods such as grilling or broiling, or can be marinated and then skewered.

**Flank.** A *flank* is a thin, flat section of the beef hindquarters located beneath the loin. It has long, coarse fibers and a large percentage of fat to lean meat. A flank contains one thin, oval-shaped, boneless flank steak that has long muscle fibers running through it. A flank steak weighs about 1 lb. It should be scored or cubed before it is cooked, and the fat covering should be removed. It can also be marinated and grilled to produce a more tender and flavorful piece of meat. Except for the flank steak, flank meat is rarely used in professional kitchens. ***See Figure 19-18.***

**Round.** A *round* is a very large grouping of muscles that represent the hind leg (hip and thigh) of the beef carcass. The round weighs approximately 200 lb and accounts for about 24% of the carcass weight. Due to its size, the round is commonly broken down at the meat packer and sold as separate subprimal cuts. It contains some large bones including the round bone (leg), aitch bone (pelvis), shank, and tailbone. It can be broken down into six major cuts: the top (inside) round, bottom round, knuckle (tip), rump, heel, and shank. It can be broken down into six major cuts: the top (inside) round, bottom round, knuckle (tip), rump, heel, and shank.

The round, like the chuck, is a versatile piece of meat in the professional kitchen. When the round is trimmed for cooking, about 70% to 75% of the cut is usable. The top (inside) round and knuckle portions can be roasted nicely, but the bottom round, outside round, and eye of round are better braised, and the shanks are best used ground to clarify consommés or whole to flavor stocks and rich sauces. The bottom round contains two smaller cuts, the outside round and the eye of round. The round with the shank and rump removed is referred to as a Chicago round or steamship round roast. When the shank meat is cleaned from the shank bone end to create a handle, the primal round is referred to as a baron of beef. A round of beef can be slow-roasted whole and looks amazing being carved at a banquet or action station in the dining room.

**Rump.** A *rump* is the tail section of the beef carcass, located behind the sirloin and above the round. The rump is actually part of the bottom round. It is called a rump roast if the bone is removed and it is rolled and trussed. ***See Figure 19-19.*** If the bone is left in the rump, it is called a standing rump roast.

*Flank Steak*

Should be scored or cubed before cooking

Long, coarse fibers

**19-18** *Flank steak has long, coarse fibers and a large percentage of fat to lean.*

*Boneless Rump Roast*

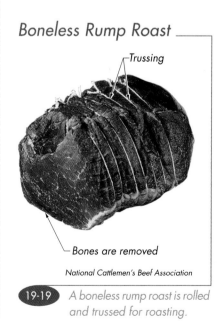

Trussing

Bones are removed

National Cattlemen's Beef Association

**19-19** *A boneless rump roast is rolled and trussed for roasting.*

**Leg.** A *primal leg* is a hind leg of veal or lamb. In lamb, the primal leg is not split into two separate legs; rather, they remain joined at the hip. Lamb legs can also be ordered as single legs. The primal leg is different from a beef round because it contains the leg, rump, and sirloin. It accounts for approximately 42% of the total carcass weight in veal and 34% of the total carcass weight in lamb. The primal leg contains the last portion of the backbone, the pelvis (hip bone and aitch bone), round bone, hind shank and tailbone. The *aitch bone* (pronounced like the letter "h") is the buttock or rump bone and is located at the top of the leg. The highest-quality veal legs weigh approximately 20 lb and come from animals weighing between 100 lb and 150 lb. The highest-quality lamb legs weigh between 4 lb and 9 lb.

The leg is the most versatile of all cuts of veal or lamb because it contains solid, lean, fine-textured meat. Meat that is more tender is located near the sirloin end and tougher meat is located toward the shank. ***See Figure 19-20.*** In veal, the entire leg is typically boned and cut into scallops or cutlets rather than being roasted whole. The leg is boned by following the muscle structure of the meat so that pieces of equal tenderness and less muscle fiber are removed. The legs can be fabricated into cuts of each major muscle, including top round, bottom round, knuckle, eye round, sirloin, and butt tenderloin. These cuts are commonly sliced against the grain (perpendicular to the length of the muscle fibers) and pounded thin to tenderize.

In lamb, the primal leg is commonly split in two and then either partially boned, stuffed, and roasted, or completely boned, tied, and roasted. Meat from the sirloin end can also be cut into lamb steaks. Meat from the shank end is most commonly used for stews or ground for patties. Shank meat is also commonly cut in cross-sections known as the foreshank, braised, and served in a rich, flavorful sauce.

## Fabricated Cuts

A *fabricated (portion-controlled) cut* is a small, ready-to-cook cut that is packaged to certain specifications for quality, size, and weight. Fabricated cuts include items such as roasts, chops, steaks, cutlets, stew meat, and ground meat. They can be cut to any specification to meet the needs of the customer. Fabricated cuts are the most convenient and popular way of purchasing meat because they eliminate trimming waste, provide uniform portions, reduce labor costs, eliminate the need for expensive cutting equipment, and control overall cost. However, the price per pound is much greater for fabricated cuts than for primal cuts.

Hogs are generally not sold as whole carcasses, partial carcasses, or primal cuts. Instead, pork is sold in 11 different wholesale cuts, including loin, ham, bacon, spareribs, Boston butt, picnic ham, jowls, feet, hocks, fat back, and clear plate. ***See Figure 19-21.*** As with other meats, pork cuts range in quality and expense.

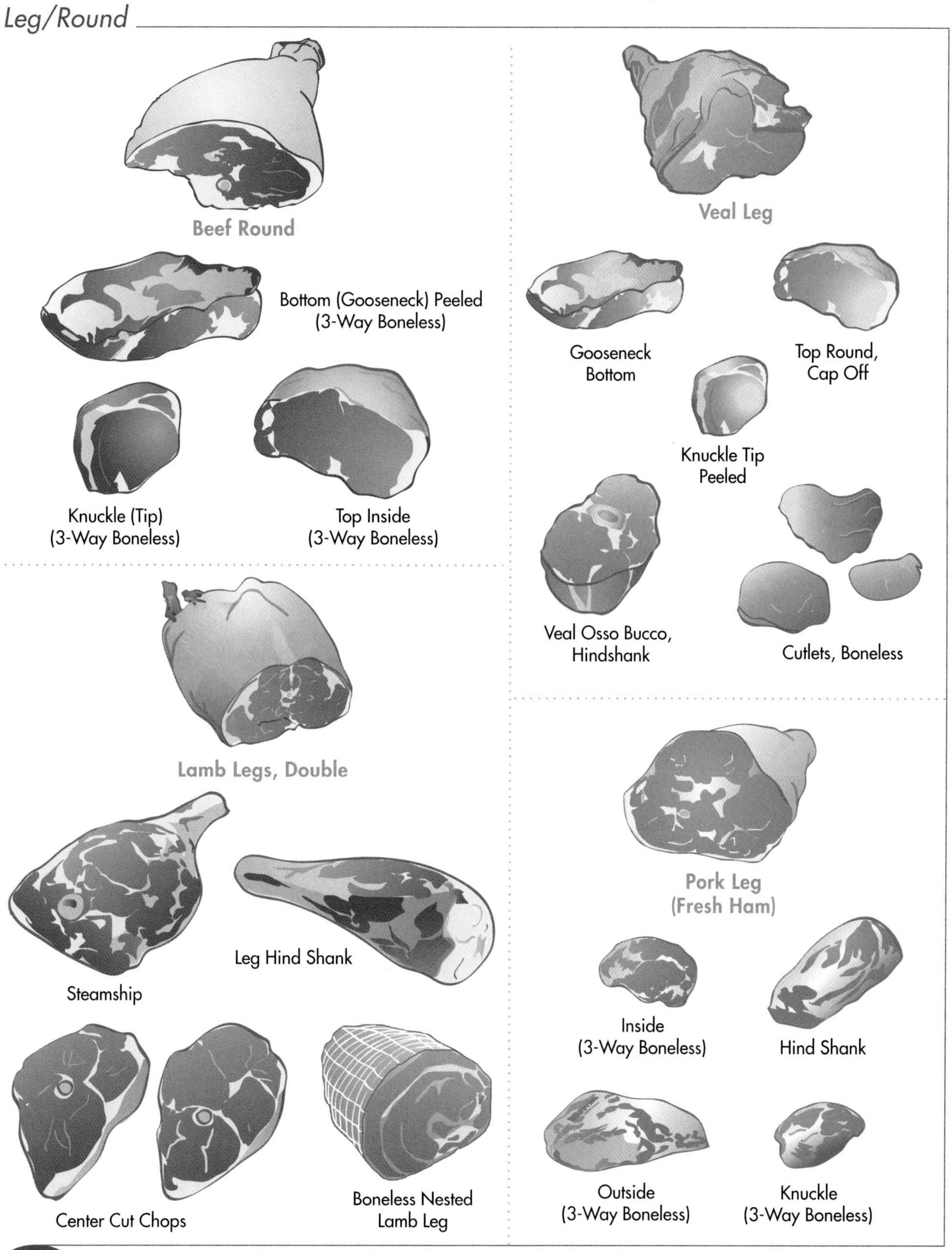

Leg/Round

**Beef Round**

Bottom (Gooseneck) Peeled
(3-Way Boneless)

Knuckle (Tip)
(3-Way Boneless)

Top Inside
(3-Way Boneless)

**Veal Leg**

Gooseneck
Bottom

Top Round,
Cap Off

Knuckle Tip
Peeled

Veal Osso Bucco,
Hindshank

Cutlets, Boneless

**Lamb Legs, Double**

Steamship

Leg Hind Shank

Center Cut Chops

Boneless Nested
Lamb Leg

**Pork Leg
(Fresh Ham)**

Inside
(3-Way Boneless)

Hind Shank

Outside
(3-Way Boneless)

Knuckle
(3-Way Boneless)

**19-20** *Many cuts can be made from a leg of veal, but it is most often boned out and cut into scallops or cutlets.*

## Wholesale Cuts of Pork

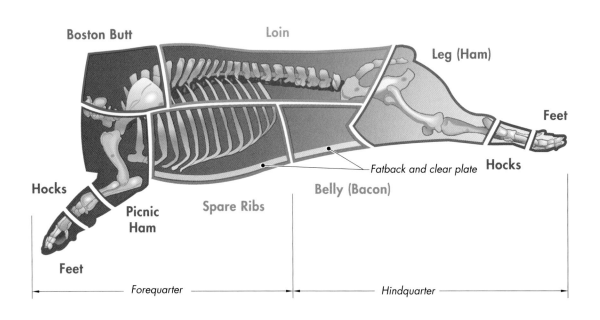

Boston Butt

Loin

Leg (Ham)

Feet

Hocks

Fatback and clear plate

Hocks

Belly (Bacon)

Picnic
Ham

Spare Ribs

Feet

Forequarter

Hindquarter

**19-21** *Hogs are blocked out for five different primal cuts which are often further divided into eleven sub-primal cuts.*

Refer to the CD-ROM for the **Fabricating Boneless Pork Loin** Media Clip.

Refer to the CD-ROM for the **Preparing Boneless Pork Chops** Media Clip.

**Loins.** A pork loin consists of the rib end, loin end, and tenderloin. The loin cut extends along the greater part of the backbone, from about the third rib, through the rib and loin area. It is one of the leanest and most popular pork cuts. Pork loin under 15 lb is considered to be the highest in quality because it is tender and has the most flavor.

The tenderloin, which is the tenderest of all pork cuts, is taken from the underside of the loin. It is a fairly long, tapered, narrow strip of lean meat and weighs about 8 oz to 12 oz. Tenderloins are typically cut from the loin and stored until enough meat is available to place a dish on the menu. Tenderloin is a versatile cut that can be broiled, sautéed, braised, roasted, or fried.

The loin is typically separated into a rib end and a loin end by cutting through the last two ribs. This leaves one rib on the loin end. All bones except the rib bones are then removed from each cut. The rib bones are left on the rib end, mainly for appearance, when the cut is served. Roast loin of pork, pork chops, or pork cutlets can be fabricated from the loin cuts. *Canadian bacon* is the trimmed, pressed, and smoked boneless loin of pork. The average weight of Canadian bacon is 4 lb to 6 lb. It can be cooked by baking, sautéing, or broiling. *Loin backs* are the rib bones removed from the loin. In some cases, loin backs are used in place of spareribs because they contain more meat.

**Hams.** A *ham* is the thigh and buttock of the hog. It contains a high proportion of lean meat. Ham is primarily marketed in two forms: smoked and cured, and fresh (uncooked). In addition, hams are available cooked, partly cooked, raw, smoked, canned, boneless, tenderized, and shankless. The most popular form of ham is cured in a solution of salt, sugar, and sodium nitrate and then smoked. The skin may be left on or may be removed.

Some hams, such as Virginia ham and country ham, are cured but not smoked. *Virginia ham* is a ham that is cured in salt for a period of about 7 weeks. It is rubbed with a mixture of molasses, brown sugar, sodium nitrate, and pepper, and cured for an additional 2 weeks. The ham is then hung, hock down, for a period of 30 days to a year. In the professional kitchen, hams are baked, cut into steaks and broiled or panbroiled, boiled, sliced, or used for sandwiches. *Prosciutto ham* is a type of dry and hard cured Italian ham. It is packaged ready to eat, usually sliced very thin, and is often used in the making of hors d'oeuvres and appetizers.

**Bacon.** *Bacon* is the cured and smoked belly of a hog. The size of the belly determines the quality of the bacon. Small bellies from very young hogs yield the most desirable bacon. High-quality bacon should contain about 40% to 45% fat and 5% rind. Bacon is a versatile item in the professional kitchen. It is used for sandwiches, garnishes, appetizers, and entrée dishes, and is, of course, a featured item in many breakfast entrées. Grease rendered from cooked bacon can be used to season or sauté other foods.

**Spareribs.** *Spareribs* are the whole rib section removed from the belly (side) of the hog carcass. They are located above the bacon section. Spareribs have little meat, but the meat is tender and has an excellent flavor. The highest-quality spareribs weigh 3 lb or less. Spareribs may be purchased fresh or smoked, although fresh spareribs are the more popular variety. Spareribs can be served as an appetizer or as an entrée and are commonly prepared boiled, barbecued, broiled, or roasted.

**Boston Butts.** A *Boston butt* is a square, compact cut of hog shoulder located just above the lower half of the shoulder (picnic ham). It weighs between 6 lb and 10 lb and contains the blade bone and a large portion of lean meat with a long fiber and a coarse grain. Boston butt is usually sold fresh when the bone is left in. Boston butt can be roasted, boiled, or cut into cutlets and sautéed or fried.

When boneless, the blade section (located above the blade bone) is usually sold as cottage ham. *Cottage ham* is smoked, boneless meat extracted from the blade section of the Boston butt. Cottage hams typically weigh 1 lb to 4 lb. For best results, cottage ham should be boiled or steamed.

**Picnic Hams.** A *picnic (callie) ham* is the lower half of the shoulder of the hog carcass. The picnic ham resembles the full ham in shape, but is smaller and contains more bone and less lean meat. It is marketed in two forms: fresh and smoked. The average weight of a picnic ham is 3 lb to 6 lb. Picnic hams can be purchased at a very low cost per pound. Fresh picnic hams can be used for preparing chop suey, pork patties, or country sausage. Smoked picnic hams can be prepared as creamed ham, deviled ham, and ham salad.

**Jowls.** A *jowl* is the cured and smoked cheek meat of a large hog. It is sometimes referred to as jowl bacon. Jowls can be cured and smoked in the same manner as bacon or braised. The jowl weighs about 2 lb to 4 lb and is used for bacon crackling, seasoning, or for flavoring items such as baked beans and string beans. Jowls can be purchased at a very reasonable price.

**Feet.** The feet are the cut below the shank. Both front and hind feet are marketed. Feet average from ¾ lb to 1½ lb and are usually sold in a pickled form, although they may also be purchased fresh or smoked. Pig feet are commonly broiled or served boneless on a cold plate.

**Hocks.** A *ham hock* is the knee joint of a hog. Hocks are removed from both the front and hind legs. They have little meat, but good flavor. Hocks are purchased fresh or smoked and often cooked with sauerkraut.

**Fat Back and Clear Plate.** Fat back and clear plate are both fairly solid, rectangular slabs of fat extracted from the surface of a hog carcass. The two cuts are very similar, except that in some cases the clear plate may have a few more strips of lean meat running through it. *Salt pork* is fat back or clear plate that contains lean meat and has been cured in salt. In the professional kitchen, salt pork is used in preparations as a flavoring ingredient for dishes such as beans, or it is used to add juice to a preparation. Lard is usually rendered from these cuts.

### Unique Cuts

Fabricated cuts unique to veal include racks, saddles, cutlets, scallopines, and baby T-bone steaks. A *rack* is a complete rib from a whole veal carcass containing two unsplit primal ribs. A *saddle* is a complete loin from a whole veal carcass containing two unsplit loins. A *cutlet* is a thin, boneless slice of veal. Veal cutlets can be prepared as Wiener schnitzel (Viennese veal steak) by pounding them very thin and then breading and frying them. Wiener schnitzel is commonly served with lemon or can be topped with a variety of traditional vegetables or sauces, such as mushrooms and woodland cream sauce. A *scallopine* is a small, ¼″-thick slice of veal (generally leg meat) about 2″ to 3″ in diameter. A *baby T-bone steak* is a 6 oz to 8 oz steak cut from the loin of veal. It contains loin meat on one side of the small T-bone and tenderloin on the other side. Baby T-bone steaks are similar to beef T-bone steaks or porterhouse steaks, but are much smaller.

Fabricated cuts unique to lamb include English lamb chops, double lamb chops, crown roast, riblets, and hotel racks. An *English lamb chop* is a 2″-thick cut of meat taken along the entire length of the unsplit lamb loin. A *double lamb chop* is a rib chop cut to a thickness equal to that of two standard rib chops.

A *crown roast* is an unsplit rack of lamb with the rib ends Frenched and the ribs formed into a circle to resemble a crown. *Frenching* is a method of removing the meat and fat from the end of a rib bone, and is generally applied to chops. In addition to crown roast, sometimes the leg bone of a roast leg of lamb is Frenched. A *riblet* is a rectangular strip of meat containing part of a rib bone, cut from the breast of lamb. A *hotel rack* is an unsplit rib section of a lamb carcass.

### Variety Meats

A *variety meat* (offal) is an edible part of an animal that is not part of a wholesale or primal cut. Variety meats are sometimes called meat sundries or glandular meats and are regarded as delicacies within various cultures.

Variety meats include tongue, liver, tripe, oxtail, heart, brains, sweetbreads, and kidneys. Variety meats are typically prepared using moist-heat cooking methods, with the exception of liver. Veal variety meats are more tender than those from beef; with the exception of tongue, they are always more desirable and slightly more expensive. Lamb variety meats are not as popular.

Beef tongue can be purchased smoked, pickled, fresh, and corned, but smoked is the most popular form. ***See Figure 19-22.*** Tongue is always prepared using moist-heat cooking methods. To determine doneness, the tip of the tongue is felt to test tenderness. When it is soft, the tongue is done. After it is cooked, the tongue is cooled in cold water, skinned, covered with water, and stored in the refrigerator. Tongue can be served cold, or can be reheated and served hot. Lamb tongue, if not utilized in sausage, can be pickled and marketed as pickled lamb's tongue.

Liver is covered with a thin membrane that should be removed before slicing. For best results, liver should be somewhat frozen when it is sliced. For larger slices, it is cut at a 45° angle. Liver is best broiled, sautéed, or pan-fried. Veal liver is best when broiled, pan-fried, or sautéed, and should always be cooked medium unless otherwise specified by the customer. ***See Figure 19-23.***

*Tripe* is the muscular inner lining of a stomach of an animal. Tripe comes from edible animals such as cattle or sheep. The most desirable tripe is known as honeycomb tripe. *Honeycomb tripe* is the lining of the second stomach in cattle. Tripe may be purchased pickled, fresh, or canned. Tripe may be fried, creamed, used in soup, or served cold with vinaigrette dressing.

An *oxtail* is the tail from a beef carcass. It is sometimes referred to as oxjoint or beef joint. Oxtail has considerable bone but also possesses a good portion of meat with a rich flavor. When cutting the tail into sections for cooking, a French knife is used to cut between the natural joints. The bone should not be splintered by using a cleaver. Oxtail is most popular when used in stew or when braised in a rich cooking liquid. The thin end of the tail can be used in oxtail soup. ***See Figure 19-24.***

### Beef Tongue

U.S. Wellness Meats

19-22    *Beef tongue is the most popular beef variety meat.*

### Veal Liver

U.S. Wellness Meats

19-23    *Veal (calf) liver is more tender than beef liver and has a milder flavor.*

## Oxtail Soup

**19-24** *Oxtail is traditionally used in oxtail soup.*

Heart is the toughest of all beef variety meats. The heart should be washed thoroughly in warm water and some of the arteries and veins should be cut away before cooking. Soaking the heart in vinegar improves tenderness. The heart is prepared by slow, moist-cooking methods, such as simmering and braising. A cooking time of 3 hr produces proper tenderness. The heart is rarely used in food service establishments, but is popular in European cuisine.

Brains have a soft texture and are covered by a tough, thin membrane enclosing small individual segments of flesh. Brains do not keep well, so they should be used as soon as possible after purchasing. When brains are delivered they are first placed in a solution of cold water, salt, and lemon juice or vinegar and left to soak for approximately 1 hr. This is done so it is easier to remove the outer membrane. After the outer membrane has been removed, the brains are parboiled for about 15 min in another solution of water, vinegar or lemon juice, and salt. The presence of acid in the solution keeps the brains white and firm. The brains are then broken apart and breaded and fried or floured and sautéed for a more attractive appearance. Cooked brains have the texture of a cooked potato and a mild taste.

*Sweetbreads* are the thymus glands of an animal, located in the throat of the animal. Sweetbreads come primarily from young calves, veal, and lamb, because as an animal matures the thymus gland disappears. High-quality sweetbreads should be plump and somewhat firm with a thin protective membrane. The usual procedure used to prepare sweetbreads is to soak them in cool water overnight and then blanch them in a poaching liquid of water, salt, and lemon juice or vinegar for about 10 min. The presence of lemon juice or vinegar keeps the sweetbreads white and firm. After blanching, all membranes should be removed and the sweetbreads should be pressed to condense their texture. After 24 hr, they can be cut apart and used with any cooking method. Sweetbreads are most popular sautéed in brown butter or flash-fried.

Kidneys have many irregular lobes divided by deep cracks and average in weight between ¾ lb to 1 lb. Before cooking, the kidney is split lengthwise and all suet (fat) and urinary canals must be carefully removed. Veal kidneys can be broiled with excellent results, compared to tougher beef kidneys, which must be cooked by moist heat. Veal kidneys are highly desirable and are used to prepare entrées such as kidney stew, kidney pie, and kidney steak.

Pork variety meats are often used in the preparation of various types of sausage. *Chitterlings* are the large and small intestines of the hog. They are emptied and thoroughly rinsed before boiling or frying. *Head cheese* is the jellied, spiced, pressed meat from the head of a hog. It is covered with a natural casing and sold as a luncheon meat. Pork variety meats are not commonly used in the professional kitchen.

## INSPECTION AND GRADING

The USDA requires inspection of all meats sold commercially. Inspection of meat is used to determine the wholesomeness of the product. Grading is an optional process that results in a designation of the overall quality of the meat. Having a thorough understanding of inspection and grading helps a consumer make smart choices when selecting meat products.

### Inspecting for Wholesomeness

All meat used in food service operations must be inspected by the USDA. At the time of slaughter and after inspection, the carcass is stamped with the round USDA inspection stamp. The round USDA stamp indicates only that the animal was not diseased and that the meat was clean and wholesome for human consumption at the time of the inspection. It is important to note that this stamp does not indicate anything about the tenderness or quality of the meat.

The round inspection stamp is used for whole carcasses or fabricated and processed meats, and is found either on the meat itself or on the case it is packed in. A purple vegetable dye is used to stamp the meat. The stamp indicates that the animal passed the inspection and includes a number identifying the plant where the animal was processed. ***See Figure 19-25.*** Slightly different stamps are used depending on whether the meat is in the form of a whole carcass or fabricated and/or processed meats. The stamp for fabricated or processed meats includes the same information as the stamp for whole carcasses, except that all information is spelled out, rather than abbreviated, on the stamp for processed meats.

### Grading Meat

Government grading of meat provides consumers with a means of comparison for quality and price. Grading involves a voluntary inspection that is at the sole discretion of the meat processor. The grading stamp has a shield shape and designates the overall quality of the meat. ***See Figure 19-26.*** Quality and yield grading stamps are stamped onto carcasses with a purple vegetable dye in the same manner as wholesomeness inspection stamps.

There are two different inspections included in a grading process, a quality inspection and a yield inspection, and both are represented by the grade inspection stamp. If a processor chooses to have meat graded, it is graded for quality, yield, or both, to ensure a standard level of quality. Yield is the amount of salable meat that can be obtained from a carcass. Quality is the tenderness and palatability of the meat. Beef and lamb can be graded for quality, yield, or both. Veal and pork can only be graded for quality.

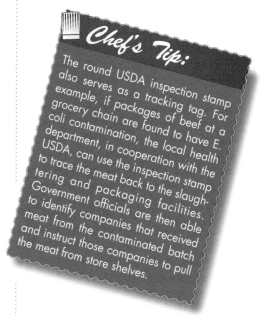

**Chef's Tip:**

The round USDA inspection stamp also serves as a tracking tag. For example, if packages of beef at a grocery chain are found to have E. coli contamination, the local health department, in cooperation with the USDA, can use the inspection stamp to trace the meat back to the slaughtering and packaging facilities. Government officials are then able to identify companies that received meat from the contaminated batch and instruct those companies to pull the meat from store shelves.

*USDA Inspection Stamps*

Mark for raw whole carcass meat

Mark for fabricated or processed meats

38
U.S.
INSP'D&P'S'D

U.S.
INSPECTED
AND PASSED BY
DEPARTMENT OF
AGRICULTURE
EST. 38

**19-25** *A round USDA inspection stamp indicates the animal was inspected for wholesomeness.*

*USDA Quality Grade Stamps*

USDA CHOICE

**19-26** *USDA quality grades are shown in a shield.*

**Historical Note**

Prior to government involvement in meat grading and inspection, all meat was slaughtered, butchered, and fabricated at local meat-cutting and butcher shops. Consumers had no way of knowing if the meat they were buying was fresh or even safe to eat. In 1906, the federal government passed the *Meat Inspection Act.* This law required all meat sold and/or transported across state lines to be inspected and certified as wholesome and not diseased at the time of slaughter.

## Quality Grading

Quality grading was initiated by the USDA in 1927. Quality grading is based on the overall tenderness, juiciness, and flavor of the meat. The criteria used in USDA quality grading are based on the following characteristics:

- texture
- color of the lean meat
- firmness of the flesh
- age (maturity) of the animal at the time of slaughter
- amount of marbling (percentage of fat) in the lean meat
- overall shape of the carcass

All six of these characteristics must be considered together in order for the meat to be graded. For example, an old steer carcass may have good marbling in the meat, but its age and other factors would cause it to earn a lower grade. Conversely, a young steer may have age on its side, but the marbling or firmness of the meat may not be desirable. Although the grading for each animal is based on specific details, the overall terms have relatively the same criteria.

The grades of meat most commonly used in professional food service operations are USDA Prime, USDA Choice, and USDA Select. The grade of USDA Good is used less often. Pork is graded differently from beef, veal, and lamb in that it is graded with numbers. Pork is graded as U.S. 1, U.S. 2, or U.S. 3, with U.S. 1 being the most desirable. The grade of pork is less important because all pork comes from young animals. ***See Figure 19-27.***

| USDA Quality Meat Grades | | | | |
|---|---|---|---|---|
| Beef | Veal | Lamb | Pork | Description |
| USDA Prime | USDA Prime | USDA Prime | U.S. 1 | Juicy, flavorful, well-marbled, thick fat covering, very tender |
| USDA Choice | USDA Choice | USDA Choice | | Juicy, good marbling, tender |
| USDA Select | USDA Good | USDA Good | U.S. 2 | Minimal marbling, soft fat covering |
| USDA Standard | | | | Little fat covering, thin flesh, large bones in proportion to flesh |
| USDA Commercial | USDA Standard | Utility | U.S. 3 | Tough meat from an old animal |
| Utility | Utility | | Utility | Tough meat from an old animal; large bones in proportion to flesh |
| Cutter and Canner | | Cull | | Inferior tough meat, watery |

**19-27** *Pork is graded differently than beef, veal, or lamb.*

**USDA Prime.** USDA Prime meats are known for being extremely well marbled and having a thick, firm fat covering. Prime meats are the juiciest and most flavorful of all meats but they are expensive and produced in limited quantity. The best beef available comes from prize steers and heifers. Of all beef marketed in the United States, only 4% is graded Prime. Prime beef has a very high fat content, which makes it costly when it is trimmed for cooking. For veal to be graded Prime, the animal must be rated as superior for yield and quality. The fat surrounding the areas of the shoulders, rump, and kidneys must be thick and white. The meat must be firm and very light pink in color, and must possess a smooth surface when cut. It should not appear grainy or stringy as does beef. USDA Prime veal is not available in great quantities.

For a lamb to be stamped Prime, the animal must be 3 months to 5 months old and only milk fed. A USDA Prime lamb carcass is compact and has plump legs. The back is wide and thick, and the neck is short and thick. The interior displays pale pink flesh with a smooth grain, soft and porous bones, and firm, white fat around the kidneys.

**USDA Choice.** USDA Choice is the most popular grade of beef. USDA Choice meat is slightly less marbled than meat graded Prime and has less of a fat covering. Its good marbling ensures it is still a juicy and tender cut. Choice is the most common grade in upscale and quality-focused establishments and better retail market places. It is preferred by most professional establishments because there is less waste than with Prime beef, making it more cost effective. Veal graded Choice is derived from compact, thick-fleshed, plump animals. The bones are small in proportion to the size of the animal. Prime and Choice are the only grades of veal that are typically graded and used in food service establishments.

USDA Choice is the most popular grade of lamb used in food service establishments. It is high-quality lamb and has excellent eating qualities. To be graded Choice, the lamb carcass must have a slightly compact body, the legs must be short and plump, and the back slightly wide and thick. The neck must be slightly short and thick. The interior of the animal has a pink flesh just slightly darker than the prime grade. The flesh grain is smooth, the bones soft and porous, and there is a generous amount of white fat around the kidney.

**USDA Select.** USDA Select is a grade used only for beef. It is a relatively inexpensive grade compared to Prime and Choice and can produce a good product if cooked with care. Most USDA Select beef comes from grass-fed steers and heifers. In some cases, corn-fed cows are graded Select. Select-grade beef has a soft fat covering that is generally slightly yellowed, with minimal marbling in the lean meat. This grade of beef is used commonly in inexpensive food service establishments.

*U.S. Department of Agriculture*

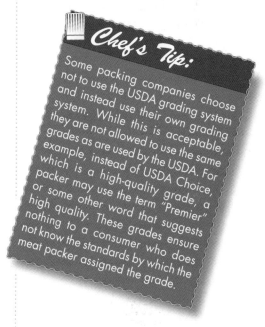

**Chef's Tip:**

Some packing companies choose not to use the USDA grading system and instead use their own grading system. While this is acceptable, they are not allowed to use the same grades as are used by the USDA. For example, instead of USDA Choice, which is a high-quality grade, a packer may use the term "Premier" or some other word that suggests high quality. These grades ensure nothing to a consumer who does not know the standards by which the meat packer assigned the grade.

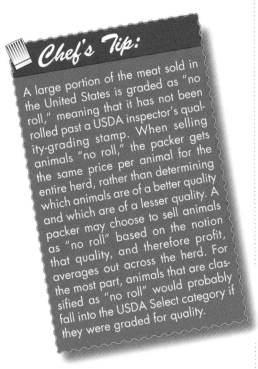
**USDA Good.** USDA Good meat is rarely used in the professional kitchen. A veal carcass graded Good is thin-fleshed and somewhat slender in appearance. The flesh is slightly soft to the touch, and the cut surface displays some roughness. The bones are large in proportion to the size of the animal. USDA Good lamb is slightly rangy and bony, with somewhat thin legs, a fairly narrow, thin back, and a slightly long, narrow neck. The interior of the carcass has dark pink flesh with a slightly rough grain. The bones are a little hard and are lighter in color, with little or no pink tinge visible. There is less fat than in other grades and the kidney may be slightly exposed, because of the lack of fat.

**USDA Standard.** USDA Standard is a grade that was created by the USDA in 1959 and falls between Good and Commercial grades. Most Standard beef is from young steers, heifers, and cows. It has very poor conformation (overall shape) and little fat covering. USDA Standard is not handled by many meat purveyors because it is not suitable for use in commercial food service establishments. It is used mostly in canned meat products. In veal, USDA Standard is a commercial grade that is rarely used in the professional kitchen except when the preparation is being stewed or braised. Veal carcasses of this grade are thin-fleshed, rough, and sunken in appearance. The flesh is soft and the surface of the flesh is rough when cut.

**Lesser Quality Grades.** Meat graded Commercial, Utility, Cutter, Canner, or Cull is rarely or never used in professional food service establishments. Meat of these grades comes from thin, rangy animals. The meat is watery, dark in color, and rough in grain, with little or no fat. Commercial, Utility, Cutter, and Canner grades are primarily used in processed or canned goods. Meat graded Cull is commonly used in dog food.

In pork, barrows (castrated male) and gilts (immature female) are graded U.S. 1. Young sows are graded U.S. 2. Old sows, which have soft fat and oily carcasses, are generally graded U.S. 3. Pork grading, like beef grading, is based on quality and yield. Young hogs usually have a very high proportion of usable meat, so the yield grade is high.

### Yield Grading

Yield grading was initiated in 1965 for beef and lamb, to measure the percentage of edible meat to fat and bone. In other words, yield grading measures how much meat a carcass yields compared to the amount of inedible fat and bone it contains. The yield grade is shown in the form of a shield with the yield number in the center of the shield. Yield grade shields are numbered 1 to 5 and indicate how much usable meat can be obtained from a carcass. A grade of 1 indicates the highest yield of meat, and a grade of 5 indicates the lowest yield of meat. In other words, a yield grade of 1 indicates the leanest meat, while a yield grade of 5 indicates the fattiest meat. ***See Figure 19-28.*** It is important to remember that fat is desirable in meat for juiciness and flavor. Beef of the best quality grades usually has a yield grade of 4 or 5, while USDA Select beef may have a yield grade of 1.

## PURCHASING

Meats are typically the most expensive ingredients used in the kitchen. Knowing how much meat needs to be purchased for any given day is only a small part of the information that a food service professional must master. Meats are highly perishable and can be purchased in many forms and in many stages, from whole carcasses to fabricated cuts. When purchasing meat, it is important to know the factors that determine the best choice of product and how to safely handle and store meat.

Typically, the more the carcass has been processed, the higher the cost per pound of meat. For example, a whole carcass of beef may cost less than $1.00 per pound. A high-quality center-cut filet mignon that is already completely cut and trimmed might cost nearly $1.00 per ounce.

### Determining Needs

Before purchasing meat for a food service establishment, a menu must be developed. It is important to know which foods will be served before buying the ingredients. Once the menu has been developed, but before the first meat order is placed, the following questions should be considered:

- What is the skill level of the employees who will be working with the meat? Are they able to fabricate meat from large cuts into food service cuts or portion-controlled units?

- Is there ample freezer and refrigeration space for the temporary storage of all of the raw meat trimmings, fat, and bone that are the result of meat fabrication?

- Is there enough time, labor, and space to utilize all trimmings, fat, and bones to make items such as stocks and ground meats before spoiling?

- Considering the additional labor cost, will buying larger cuts of meat and fabricating them in-house save money? Is it more cost effective to purchase meat in the fabricated units needed for service?

Purchasing larger cuts and fabricating them in-house does not necessarily result in cost savings. Less-skilled employees may require a great deal of time to fabricate meats, substantially increasing labor costs. Conversely, if an operation has a person who is skilled in cutting meats, has storage space for the byproducts of fabrication, and has the time and staff to utilize these byproducts, it might be beneficial to purchase large cuts and fabricate them in-house. It is important to perform a cost analysis to determine the most profitable way to purchase meat for a particular food service operation.

*USDA Yield Grades*

**19-28** *The number in a yield grade stamp indicates, on a scale of 1 to 5, how much salable meat can be obtained from the carcass.*

Chapter 19 — Meat **679**

## Making the Purchase

Purchasing meat can be a relatively easy task if the meat specifications have been completely determined. A *specification* is a particular characteristic of a product that is required to meet a specific production or service need. For example, the IMPS/NAMP name includes such descriptions as 110 rib, roast-ready, or boneless. When it is time to purchase meat for the operation, the product IMPS/NAMP name, meat grade, and size need to be presented. Size specifications include weight range, portion range, thickness, chilled state, and trim fat. Weight range is used to specify a range of weight for a large roast or other cut of meat. Portion weight is used to specify the weight of an individual fabricated cut, such as a 12 oz steak. Thickness of portion-controlled cuts can be indicated, such as 1″-thick pork chops. Chilled state refers to whether the meat is refrigerated or frozen. Trim fat is typically ¼″ unless otherwise specified.

The cooking method used to prepare each meat dish determines what sort of cut is needed. It is important to understand which cooking methods are best for each cut of meat in order to produce great dishes. If meats are to be grilled, broiled, sautéed, fried, roasted, or poached, a more tender cut of meat must be used that works well with short cooking times. If meats are to be stewed or braised, then tougher, more flavorful cuts with ample connective tissue should be used as they can withstand longer cooking times without drying out. For example, if the menu lists Irish lamb stew, meat suitable for stewing, such as lamb shoulder, should be purchased. If the menu lists a grilled steak sandwich, meat such as rib eye, sirloin, or butt steak should be purchased because it is suitable for quick, dry-heat cooking but remains tender.

Pairing the cut of meat with the appropriate cooking method creates great cuisine. Similarly, choosing either the wrong cut or the wrong cooking method can result in a poor-quality meal that is tough or undesirable to eat. Using a less expensive, tougher cut of meat to save money when making a grilled steak sandwich can be like eating a slice of a gym shoe on a sesame seed bun. Using an expensive filet mignon to make beef stew can result in dry, shredded meat with a mushy texture.

## MEAT STORAGE AND PACKAGING

Meats are composed of proteins, fats, and water, which means they are susceptible to spoilage and contamination. Maintaining a safe storage temperature and environment is very important. When receiving meat, the temperature and condition of the meat should be inspected at the point of delivery. Meat that is in the temperature danger zone when received should be rejected.

Refrigerated meats should be kept at temperatures between 30°F and 35°F. Frozen meats should be kept at temperatures between –25°F and –50°F. Vacuum-packed meats should never be opened until needed for service or preparation. Cryovac® is a type of vacuum packaging that is

often used in food service. It is a durable, sealed, airtight plastic that helps to preserve meat for 3 to 4 weeks. Once the vacuum seal is broken, meat has a shelf life of only 2 or 3 days. Cut meats that are not vacuum-sealed should be wrapped tightly in butcher's paper. Cut meats should never be wrapped in plastic wrap because it is not airtight and the presence of moisture allows bacteria to grow rapidly.

In addition to packaging, meat can also be irradiated to reduce the risk of potentially harmful microorganisms. *Irradiation* is a process where food is subjected to ionizing radiation to kill insects, bacteria, and parasites. Irradiation is sometimes called cold pasteurization. The process does not affect the flavor or appearance of the food and can help to prolong shelf life. The USDA requires all irradiated foods be labeled to indicate that they have been treated with irradiation.

When frozen meats need to be thawed, they should be placed in the refrigerator overnight. Larger cuts of meat may take more than one day to thaw under refrigeration. Freezer burn on meat is the result of a loss of moisture on the surface of the meat. Freezer burn can be prevented by freezing meats in vapor-proof packaging designed for use in freezing temperatures.

## FABRICATING MEAT

A food service establishment can save money by purchasing primal cuts if staff have the skills to fabricate them in-house without ruining or wasting meat. Understanding basic procedures and techniques involved in fabricating portion cuts of meat allows an establishment to more closely control food costs and product quality.

A popular preparation for veal, lamb, or pork loin is a boneless, stuffed roast. The bone must be removed from the loin before it is trussed and stuffed.

## To bone a loin for trussing and roasting, apply the following procedure:

1. Trim the surface fat evenly, to approximately ¼″ thick. Turn the loin over (top-side down) and again trim fat and connective tissue from tenderloin. ***See Figure 19-29.***

2. Slide the knife under the exposed rib and slice through to the backbone to separate the tenderloin eye muscle from the vertebrae.

3. Cut carefully down the center of the backbone between the muscle and the exposed vertebrae. Continue cutting to the end of the vertebrae.

4. Continue until tenderloin is completely removed from the primal loin.

CULINARY PROCEDURES

## Boning a Loin

*1–2. Trim the surface fat to ¼ thick. Slide the knife under the exposed rib and slice through to backbone to separate the tenderloin eye muscle from the vertebrae.*

*3. Cut carefully down the center of the backbone between the muscle and the exposed vertebrae. Continue cutting to the end of the vertebrae.*

*4. Continue until tenderloin is completely removed from the primal loin.*

**19-29** *The tenderloin that results from boning a loin is one of the most prized cuts of meat.*

The roast can be stuffed or spread with a forcemeat if desired. With the top side still down, the flank sides are rolled tightly towards the center, ending at the location where the backbone was removed. The rolled roast is tied with butcher's twine to secure.

The tenderloin is often trimmed for roasting and in preparation for fabrication into smaller portion-controlled cuts.

**To trim tenderloin for roasting, apply the following procedure:**

1. With a rigid boning knife, carefully remove the chain muscle from the side of the tenderloin, and reserve. The *chain muscle* is a thin strip of tender meat surrounded with fat and is located next to the tenderloin. Once trimmed, the chain muscle can be used in many preparations such as soup, ground meat, and stock. Some chefs prefer to leave the chain muscle intact when preparing pork tenderloin. ***See Figure 19-30.***

2. Trim and pull the thick fat covering away from tenderloin to expose the loin.

3. Position the tenderloin so that the head (wide end) of the tenderloin is nearest the cutting hand, and the tail (narrow end) is nearest the free hand. Insert the tip of the boning knife just beneath the silverskin against the meat at the tail end of the tenderloin. *Silverskin* is very tough, silver-colored elastin that does not break down while cooking.

4. Draw the knife along the length of the tenderloin, just beneath the silverskin, toward the head of the tenderloin with the blade angled slightly upward. With the free hand, firmly hold the freed silverskin, continuing to cut toward the head. Repeat until all silverskin is removed, using caution to not cut away any meat. *Note:* Because silverskin can taper into the tenderloin muscle, it may be necessary to scrape muscle from the top of the silverskin in order to completely remove the silverskin.

## Trimming Tenderloin

Tenderloin —
Chain muscle

1. Remove chain muscle if desired.

2. Trim and pull away thick fat.

Silverskin —

3. Insert knife beneath silverskin against meat of tenderloin.

4. Draw knife beneath silverskin to remove.

 **19-30** Tenderloin is often trimmed into smaller cuts in preparation for roasting or fabrication.

Refer to the CD-ROM for the **Preparing Beef Tenderloin** Media Clip.

Once beef tenderloin has been trimmed, it can be fabricated into smaller portion-controlled cuts, such as chateaubriands, filets mignons, tournedos, and tenderloin tips.

To cut a trimmed beef tenderloin into various cuts, apply the following procedure:

1. Starting at the largest end, cut off the uneven tip of the tenderloin, and reserve for tenderloin tips. ***See Figure 19-31.***

2. The largest section of the tenderloin, which looks like a double-muscled section, is used for chateaubriand. Make a cut crosswise just after the large portion ends to remove the chateaubriand.

3. The center of the tenderloin is cut crosswise to desired thickness to produce filets mignons. *Note:* The entire tenderloin can be cut into filet mignons if the chateaubriand and tournedos are not desired.

4. The smallest third of the tenderloin is cut crosswise to desired thickness to produce tournedos. Tournedos are 2½" in diameter and ½" to ¾" thick. *Note:* If desired, tournedos can be cut from the entire length of tenderloin.

5. The last 3" to 4" of the tail end is cut crosswise into tenderloin tips for items such as kebobs or stroganoff.

## Fabricating Beef Tenderloin

1–2. Cut off uneven tip and reserve; make a crosswise cut to remove chateuabriand.

3. Cut center section crosswise into filet mignon steaks.

4–5. Cut next section crosswise into tournedos steaks; cut smallest end crosswise into tenderloin tips.

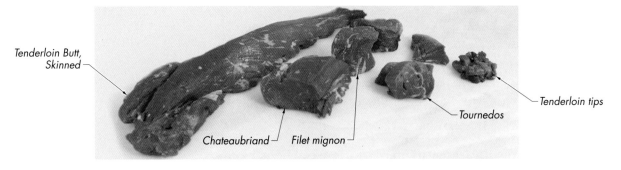

**19-31** *Tenderloin can be fabricated into chateaubriand, filet mignon, tournedos, and tenderloin tips.*

Lamb tenderloin can be cut into noisettes if desired.

## To cut lamb tenderloin into noisettes, apply the following procedure:

1. Slide the knife along the backbone of the loin in long, smooth motions to cut away the connective tissue that holds the tenderloin in place. *See Figure 19-32.*
2. Continue cutting until the connective tissues are cut completely through and the tenderloin is completely separated from the bone.
3. Trim the fat off the loin, leaving only a ¼" trim of exterior fat. Remove any silverskin.
4. Cut the loin crosswise to produce noisettes of desired thickness.

*Cutting Lamb Noisettes*

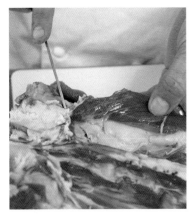

*1–2. Slide knife along backbone to remove tenderloin.*

*3. Trim exterior fat to ¼" and remove any silverskin.*

*4. Cut crosswise into noisettes of desired thickness.*

**19-32** *Lamb tenderloin is often cut into noisettes.*

A boneless strip loin is often trimmed and fabricated into smaller portion-controlled cuts. In beef, the last few inches of the tenderloin in the sirloin end of the strip loin contain a line of connective tissue that does not break down during cooking. Steaks cut from this end of the beef loin are referred to as vein steaks and, although the meat is as tender as the rest of the loin, the connective tissue is very tough, making these steaks harder to eat and not as desirable as steaks from the rest of the strip loin.

To trim and cut a boneless strip loin into steaks, apply the following procedure:

1. Trim the surface fat to a uniform thickness, approximately ¼″ thick. **See Figure 19-33.**

2. Square up the strip loin by trimming away some of the tail (lip) by making a cut down along the side of the strip loin, leaving about 1″ to 1½″ from the tenderloin intact.

3. Turn the loin over and trim off any additional fat or connective tissue that could make the steaks tough.

4. Cut the loin crosswise into steaks of desired weight or thickness.

*Fabricating Boneless Strip Loin*

*1–2. Trim surface fat to ½″; square up sides by trimming tail to within 1″ to 1½″ of eye muscle.*

*3. Turn over and trim any additional fat.*

*4. Cut loin crosswise into steaks of desired weight or thickness.*

**19-33** *Boneless strip loin is trimmed and cut into steaks of desired thickness.*

Refer to the CD-ROM for the **Fabricating Boneless Strip Loin** Media Clip.

Ground meat is a very versatile ingredient and can be used in the preparation of items such as meatballs, sausages, meatloaf, meat fillings, etc. Meat to be ground should be very cold before grinding, as should all parts of the grinder that are in contact with the meat during the process. Some chefs like to place these items in the freezer for a short time before grinding, because the colder the meat, the better the result. Room temperature meat or even meat that is just somewhat cool has a tendency to be smashed while grinding, losing its texture and appearing puréed instead of ground.

during cooking. Even poultry is often trussed when roasting, pushing the wings against the breast and helping them to not overcook by the time the dark meat is done.

## To truss a roast, apply the following procedure:

1. Tie a loop of string around the roast, with the knot on the top center of the meat at the nearest end. ***See Figure 19-39.***
2. Loop the string around an outstretched opened hand, pulling the string from the spool and then back behind the hand.
3. Twist the loop around backward with the hand so that the loop twists around itself twice.
4. Spread the fingers wider to open the loop wider.
5. Once the loop is wide enough, place the loop around the meat and position the loop 1″ or 1½″ from the previous loop.
6. Pull the strings to tighten the loop around the meat. Repeat down the entire roast.

CULINARY PROCEDURES

## Trussing a Roast

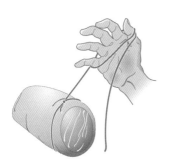

1. Tie loop of string around roast with knot on top center of meat at nearest end.

2. Loop string around outstretched hand and back behind hand.

3. Twist loop around backwards with hand so loop twists around itself twice.

4. Spread fingers to make loop wider as needed.

5. Place loop around meat and position 1″ to 1½″ from previous loop.

6. Pull strings to tighten; repeat down entire roast.

19-39   *Trussing meat helps to maintain its shape while it is cooking.*

## To truss a loin, apply the following procedure:

1. Starting with a piece of string long enough to tie completely around the loin, pass the string around the loin and cross one end over the other. ***See Figure 19-40.***

2. Make a loop by passing the loose end of the string around a finger on the left hand (or the right hand, if left-handed).

3. Loop the string back underneath itself and pass the tail end of the string back through the hole where the finger was.

4. Pull both ends of the string to tighten until the string sits firmly against the meat. Tie a basic knot to lock the butcher's knot in place.

5. Loop the string around an outstretched open hand, pulling the string from the spool and then back behind the hand. Twist the loop around backward with the hand, so that the loop twists around itself twice.

6. Once the loop is wide enough, place the loop around the meat and position the loop 1″ or 1½″ from the previous loop. Pull the strings to tighten the loop around the meat. Repeat down the entire roast.

### Trussing a Loin

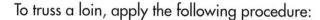

1–2. Pass string around loin and cross end over end. Make a loop by passing loose end of string around finger.

3. Loop string underneath itself and pass tail of string through loop.

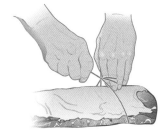

4. Pull both ends to tighten firmly.

5. Twist loop around backwards with hand so loop twists around itself.

6. Place loop around meat and position 1″ to 1″ from previous loop; pull both strings to tighten; repeat down loin.

**19-40** Loin meat is commonly trussed for preparation.

**Barding.** *Barding* is the process of laying a piece of pork fat back across the surface of a lean cut of meat, such as a roast, to add moisture and flavor. The fat back is trussed to the roast to prevent it from falling off the roast in the oven. As the roast cooks, the fat is rendered from the fat back and absorbed into the meat. This works similarly to basting a piece of meat or poultry while roasting. However, while the fat back both prevents the meat from drying out, it also prevents the meat from caramelizing. Occasionally, the fat back is removed for the last 10 min of roasting to allow the item to gain some color. Strips of raw bacon can also be used, but bacon adds a strong flavor to the dish.

**Larding.** *Larding* is the process of inserting thin strips of pork fat into lean meat with a larding needle. For example, beef tenderloin is often larded with salt pork. The salt pork is cut into thin strips and drawn through the meat to increase the juiciness when the meat is roasted. Larded meats are most commonly cooked by braising, as the process allows ample time for the fat to be rendered and absorbed by the meat. The pork fat adds moisture and flavor to the lean meat.

**Aging.** The period of rest that occurs for a few days after the slaughter is known as the aging period. It is the minimum amount of time that meat must be allowed to rest in order to be edible. Immediately after an animal is slaughtered, the carcass is very limp. However, within the first 5 hr to 24 hr, the carcass goes through a period known as rigor mortis where natural enzymes in the meat cause the tissues to seize and become stiff. The limpness of the carcass disappears, and both the joints and the muscle become stiff. This stiffening stage lasts for 2 to 3 days under normal refrigeration. Meat that is experiencing rigor mortis is commonly called green meat ("green" because it is so freshly slaughtered). Eating green meat is not pleasant, as the meat is extremely tough and almost flavorless. If the meat is frozen immediately after slaughter, the process is put on pause and rigor mortis sets in once the meat is thawed. Natural enzymes in the tissue of the meat begin to relax and tenderize the muscles 2 to 3 days after slaughter. The meat is now ready to be cooked, frozen, or processed.

Pork and veal typically are not aged any longer than the initial 2 to 3 days, because the fat on pork spoils relatively quickly, and veal lacks a protective covering of fat that would allow it to endure a longer aging process. However, beef and lamb can benefit from longer aging processes, as the enzymes in both of these meats continue to break down and tenderize the tissue, making it even more tender and flavorful. Beef and lamb can be aged for an extended period using either wet aging or dry aging.

*Wet aging* is the process of aging meat in vacuum-sealed plastic. Most carcasses are broken down into smaller cuts that are more manageable to handle in a typical food service operation. After these meats are broken down, they are commonly placed into vacuum-sealed plastic packaging so that they are cleaner, easier to handle, and protected from bacteria and mold. Beef and lamb are allowed to age from 1 week to 6 weeks in vacuum packaging under normal refrigeration. During this period, natural enzymes further break down and tenderize the meat, intensifying its natural flavor.

*Dry aging* is the process of aging larger cuts of meat that are hung in a very well-controlled environment where temperature, humidity, and constant air flow are monitored around the clock for up to 6 weeks. The natural enzymes break down the tissue of the meat, making it extremely flavorful and very tender. It is important that the meats are not wrapped and do not touch each other during the dry aging process, because moisture and harmful bacteria could form. Dry aging commonly results in 5% to 20% weight loss through the evaporation of moisture and the eventual trimming of inedible surface material. The surface of the meat dries out like leather in the process and usually develops a layer of harmless mold, both of which must be trimmed off and disposed off prior to cooking. Because of the time, labor, and waste that goes in to producing dry-aged meats, they are very expensive and typically are found at only premium butchers, high-end distributors, and the most elite of steak houses.

## COOKING MEAT

After the cut of meat to be prepared is determined, the best cooking method for that cut needs to be identified. With so many different cuts of meat available, it can be challenging to know which cooking methods produce the best results. The key to determining the appropriate cooking method is knowing which part of the animal the cut is from.

As previously discussed, the more a muscle was worked and the older the animal it came from, the tougher the meat is and the more connective tissue it has. Cattle, hogs, calves (for veal), and lambs work certain muscle groups of their bodies more than others. For example, the shoulders and legs generate most of power for walking, running, and standing. Muscles in the ribcage and chest are worked through breathing. Cuts of meat that come from these strong muscle areas usually are tougher and have a greater amount of connective tissue. Tougher meat benefits from being slowly cooked using moist-heat cooking methods. Cuts taken from areas that are not heavily worked, such as the back of an animal (animals do not use their backs as people do), are more tender and can be cooked by dry-heat cooking methods.

### Determining Doneness

One of the more difficult tasks a cook faces is determining the doneness of meats. Customers have a choice as to how much or how little they prefer items such as steak or roasted meats to be cooked. On the other hand, meat cooked using a combination cooking method should always be cooked until tender. Cooking meats to the desired degree of doneness

*Carlisle FoodService Products*

is a skill that takes practice to master, as it is determined by multiple factors: the type of meat, the intensity of the heat, the thickness of the meat, and the temperature of the meat when it begins to cook. Because of the many factors involved in cooking meat properly, timing alone is not an accurate gauge of doneness.

To determine a degree of doneness for larger pieces of meat, a thermometer should be inserted into the thickest part of the meat with the exception of braised or stewed meats, which are generally cooked until thoroughly tender. The temperature reading on the thermometer determines how much or how little the meat is cooked. *See Figure 19-41.* Cuts of meat that are grilled, broiled, or sautéed to a certain degree of doneness could also be tested with the touch method. The touch method, as the name implies, uses the sense of touch to check the texture and firmness of the cooked meat. For example, a steak cooked rare is soft and slightly mushy to the touch. The texture should feel almost the same as when the meat was raw. A steak that is cooked well done is firm and springs back immediately when gently pressed with the fingertip.

### Action Lab

One way to practice the touch method of determining doneness is on your own hands. Gently touch the ball of the right thumb (located beneath the thumb at the base of the palm) with the left index finger. This is similar to what a rare steak should feel like. Next, gently touch the right index finger and right thumb together and again touch the ball of the right thumb with the left index finger. This feels similar to what a medium rare steak should feel like. Next, gently touch the right middle finger and right thumb together. This feels similar to what a steak cooked medium should feel like.

| Determining Doneness | | |
| --- | --- | --- |
| Degree of Doneness | Internal Temperature* | Cooking Time Per Pound† |
| Very rare | 120 to 125 | 12 to 15 |
| Rare | 125 to 130 | 15 to 18 |
| Medium | 135 to 140 | 18 to 20 |
| Medium well | 145 to 150 | 20 to 23 |
| Well done | 150 to 160 | 20 to 25 |
| Meat should always be allowed to come to room temperature before roasting | | |

\* in °F
† in min.

 *Checking the temperature is the preferred method of testing the doneness of large pieces of meat.*

## Dry-Heat Cooking Methods

Dry-heat methods should only be used for more tender cuts of meat that have little connective tissue, such as any cuts from the primal rib or loin, including the tenderloin. Cuts from these areas include strip steaks, butt steaks, porterhouse steaks, filets mignons, chops, and rib eye steaks. This is because the high heat and short cooking time will not assist in tenderizing meats or breaking down connective tissues. Foods are subjected to high direct heat while being cooked using dry-heat cooking methods.

**Pan-Frying and Deep-Frying.** Pan-frying and deep-frying use a larger amount of hot oil to cook a product than is used in other dry-heat methods. Meats to be deep-fried should be tender and are commonly breaded. The breading protects the meat from the hot oil and helps the meat to retain moisture. Meats to be fried using either method should not be so thick as to prevent them from fully cooking. Meats are typically cut into cutlets, scallops, steaks, or chops for frying or pan-frying.

**Sautéing.** Sautéing uses a small amount of hot fat to sear and cook the meat. Only very tender and uniformly sized cuts should be chosen for sautéing to ensure even cooking. Sautéed meat items are often accompanied by a pan sauce made in the same pan after the cooked meat has been removed. When sautéing, care should be taken to not burn the oil or the surface of the item being sautéed by using too high of heat.

**Grilling and Broiling.** Grilling and broiling use a hot flame to quickly sear and cook foods. Only very tender cuts of meat should be used with these methods. Meats cooked by either of these methods should also be well marbled so they will turn out moist and flavorful, rather than drying out from the intense heat. Even so, well-done meats cooked with these methods will probably be somewhat dry, because most of the moisture cooks out as the meat cooks to that stage. Meats to be grilled or broiled should be properly trimmed of fat, while too little fat may cause the meat to dry out. Meats to be grilled or broiled are commonly taken from the rib and loin cuts.

**Roasting.** Roasting uses hot air to cook meat. Meats to be roasted should be tender and well marbled. Very lean meats will become dry when roasted unless some form of fat is added to them. If the meat to be roasted is very lean, additional fat can be added by either barding or larding.

Meats to be roasted are commonly taken from the rib, loin, and leg cuts. Smaller roasts should be cooked at higher temperatures, between 400°F and 450°F, to allow them to caramelize nicely on the exterior without overcooking the interior. Larger roasts require a longer cooking time and should be roasted at lower temperatures, between 275°F and 325°F, to prevent excessive shrinkage and dryness. It is important to allow for carryover cooking when roasting meats to a desired temperature.

## Moist-Heat Cooking Methods

Moist-heat methods are used for meats that can benefit from extended cooking times by making them more tender or breaking down connective tissue. Foods cooked using moist-heat cooking methods are subjected to longer cooking times, lower temperatures, and some form of moisture. Meat should never be boiled, as the intensity of the high temperature would cause the meat to shrink excessively and become tough. Simmering is the main moist-heat cooking method used to cook meat.

Simmering uses long cooking times and lower heat to cook tougher cuts of meat and make them tender. The simmering method is used only for certain cuts of meat, including briskets of beef, fresh or cured ham, pork butts, ham hocks, and tongue. Simmered meats should be cooked in liquid that is between 180°F and 195°F. Meat should be simmered until it is tender or until a fork inserted into the meat can be easily removed. If the fork inserted into the meat cannot be easily pulled out, the meat is still too tough and needs to be simmered longer.

## Combination Cooking Methods

Braising and stewing are combination cooking methods often used to cook tougher cuts of meat with ample connective tissues. The meat is first seared using a dry-heat method and cooked for a long cooking time with liquid help tenderize the cut. Searing the meat helps to add a caramelized flavor to the meat and the resulting sauce.

Braising is typically used for larger (roast-size) cuts of meat that are well marbled and have a relatively good amount of fat. Both tender and tougher cuts can be braised. The difference between preparing tender cuts or tougher cuts is the amount of time the meat is cooked. These larger cuts are first browned in a small amount of hot fat and then a cooking liquid is added, to about halfway up the surface of the meat. The meat is slow cooked in the liquid until tender. The resulting liquid is then slightly thickened if necessary and served as a sauce to accompany the meat. Tender cuts that are commonly braised are veal or pork chops, and tougher cuts are taken from the shank or chuck.

Stewing involves the same methods and style of cooking as does braising, with two primary differences. First, stewing involves bite-size pieces of meat, rather than roast-size pieces as used in braising, but the meat for both comes from the same cuts. Second, when using the stewing method, the meat is completely covered with the cooking liquid. In braising the liquid only comes halfway up the product. In both procedures, the meat should be trimmed of most surface fat prior to cooking.

Different from this brown stew are two additional versions of white stew called a fricassee and a blanquette. In a fricassee, the meat is seared in a small amount of hot fat but not allowed to brown. The cooking liquid is then added and the meat is cooked until tender. In a blanquette, the meat is blanched in simmering water, rinsed to remove any impurities, and then added to the cooking liquid. White stews should have an ivory color when finished.

## CARVING AND SLICING ROASTED MEATS

Many cuts of meat are roasted whole and sliced to order. Some cuts, such as a beef inside round, are relatively easy to slice as long as a carver slices against the grain of the meat. Slicing against the grain produces a cut of meat that is more tender and less stringy than if it was sliced in the direction of the grain. Additionally, some cuts of meat contain bones that either must be removed prior to slicing or else the meat must be sliced off the bone by hand. The most common types of roasted meats that are sliced for serving include roast leg of lamb, steamship round roast, and prime rib.

### Carving a Roast Leg of Lamb or Steamship Round Roast

A bone-in roast leg of lamb can be difficult to carve because large bones are present in the center of the roasted meat. A steamship round roast is the largest type of whole roast. It is typically prepared for large buffets and special events and can be impressive to carve at a carving station in front of guests. Attractive cuts of meat can be produced if a chef understands the structure of the leg and the proper way to remove the cooked meat.

**To carve a roast leg of lamb or steamship round roast, apply the following procedure:**

1. After meat is roasted to desired degree of doneness, place the meat on a cutting board or carving station, with the large hip-joint end facing downward and the long exposed arm bone facing upward. *Note:* Use a ball of aluminum foil to stabilize the bottom of the roast if necessary.

2. With the free hand, hold the shank bone firmly in place.

3. Trim excess fat from exterior of roast to expose the meat.

4. With a slicer, make a vertical cut along the bone, about 1″ from the end of the shank meat. (This wedge of meat acts as a bumper in case the knife slips toward the hand holding the bone.) *See Figure 19-42.*

5. Use long smooth stokes to slice horizontally toward the bone. As slices become larger near the sirloin end of the leg, begin to slightly angle the slicer to create a smaller surface.

6. Rotate the leg to remove slices from the other sides of the roast.

## Carving a Roast Leg of Lamb

*1–4. Hold shank firmly in place; trim any execss fat; make a vertical cut along bone.*

*5–6. Use long, smooth strokes and slice on an angle towards bone; rotate the leg to remove slices from other sides of the roast.*

**19-42** *A roast leg of lamb is carved into slices cut at a slight angle to the leg bone.*

### Carving Bone-in Prime Rib

A bone-in prime rib is a tender, juicy cut of meat. Although the bone is inedible, it can be left in the cut of meat for presentation.

### To carve a bone-in prime rib, apply the following procedure:

1. Remove netting from the roast (if present) and trim away any vertebrae from the bottom of the roast. ***See Figure 19-43.***

2. Trim off the majority of the fat cap, allowing only about ¼" to ½" to remain.

3. Turn the roast over and trim away excess fat from the top of the roast.

4. Using a long slicer and making long, smooth cuts, begin to slice the rib roast starting at one end to remove the end cut without a bone. The next slice should include a bone. The third slice should not include a bone, and so on. A bone-in slice is often called a "king cut," while a boneless slice may be referred to as a "queen cut." Some establishments may carve each slice with a bone in if a very large portion is to be served to each guest.

CULINARY PROCEDURES

## Carving Bone-in Prime Rib

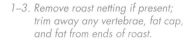

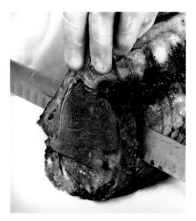

1–3. Remove roast netting if present; trim away any vertebrae, fat cap, and fat from ends of roast.

4. Make long smooth cuts with slicer to remove an end without a bone; the next slice includes a bone and the following does not, and so on.

**19-43** When carving bone-in prime rib, alternating slices contain a bone.

## Slicing Prime Rib for Large Functions

If many portions of prime rib are needed—for a wedding with 400 guests, for example—slicing by hand is not feasible, as it would take too long and might yield inconsistent cuts. The best method of slicing prime rib roasts in quantity is to use an electric slicing machine.

**Chef's Tip:**

Prime rib is often roasted with the bones intact because the bones allow the meat to roast more slowly, resulting in a juicier piece of meat. This principle is true for all types of roasted meats, as well as grilled meats such as bone-in strip steaks. The bones keep the meat moist and flavorful. The bones can be removed after roasting if desired for presentation.

To slice a prime rib roast using an electric slicing machine, apply the following procedure:

1. Remove netting, vertebrae, and fat (if present) as described in previous procedure. ***See Figure 19-44.***

2. With a long slicer, cut evenly and as close as possible along the rib bones to remove them from the rib eye roast, being careful to cut as close as possible to not remove too much meat with the bones.

3. Place the rib eye on the slicing machine, adjust blade to desired thickness, and begin slicing. *Note:* The blade depth must be checked often and adjusted as necessary to ensure that the appropriate-size slice is being produced. Always follow manufacturer safety precautions, use all provided blade guards, and keep fingers away from the cutting blade when operating slicing machines.

## Slicing Prime Rib for Large Functions

**Safety Note:** Always follow manufacturer safety precautions. Use all provided blade guards and keep fingers away from the cutting blade when operating slicing machines.

1–2. Remove netting, chine bones, and fat if present; cut close to the rib bones to remove completely.

3. Place ribeye on slicer and slice to desired thickness.

**19-44** A slicing machine is often used when cutting prime rib for large functions.

## SUMMARY

Meats are probably the most expensive ingredients commonly used in the professional kitchen. A skilled food service employee must possess a good understanding of the characteristics of meats, the grading standards, specific cuts, and cooking methods in order to prepare wonderful dishes for customers. Some establishments have kitchen staff with knowledge of meat cutting while other establishments choose to purchase meats in portion-controlled units.

The most common types of meat used in food service include beef, veal, lamb, and pork. Beef, veal, and lamb are often purchased as primal cuts. These primal cuts include the chuck or shoulder, rib or rack, short plate, breast, brisket, shank, loin, short loin, sirloin, flank, leg, round, and rump. Pork is more commonly purchased as fabricated cuts. Variety meats such as tongue, liver, tripe, oxtail, heart, brain, sweetbreads, and kidneys are often regarded as delicacies within various cultures.

Meat is inspected by the USDA for wholesomeness, quality grade, and yield grade. The round USDA wholesomeness inspection stamp on a carcass indicates that the animal was not diseased and that the meat was clean and wholesome for human consumption at the time of the inspection. Quality grading is based on the overall tenderness, juiciness, and flavor of the meat. Yield grading measures the percentage of edible meat to fat and bone.

When purchasing meats, the order must specify the product IMPS/NAMP name, meat grade, and size. Because meats are a perishable and a potentially hazardous food, proper handling and storage are vital. Meats may be packaged in Cryovac® packaging or preserved with irradiation.

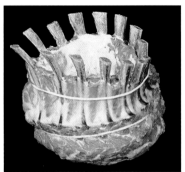

*Beef Checkoff*

Refer to the CD-ROM for Quick Quiz® questions related to chapter content.

Prior to cooking, primal cuts of meat are often fabricated into smaller portions. Cuts that are often fabricated in a food service establishment include the primal loin, leg, and rack. Tenderloin and strip loin are often cut into smaller portion cuts before cooking. Special preparation techniques include using rubs, marinades, tenderizers, barding, larding, and aging. These techniques are used to improve flavor, tenderness, and texture.

Choosing the correct cooking method to prepare a cut of meat is critical to the quality of the finished product. Knowing how to determine the amount of doneness is also essential. Meat dishes are often prepared to customer order on a scale of rare to well-done. Dry-heat cooking methods are preferred for tender cuts of meat that have little or no connective tissue. Moist-heat cooking methods are used for tougher cuts of meat that can benefit from extended cooking times that help break down connective tissue. Combination cooking methods are often used for the toughest cuts of meat with the most connective tissues.

Meat is often served in individual portions that are plated in the kitchen. However, certain cuts such as bone-in legs or prime rib are carved tableside for an impressive presentation. Procedures for carving larger cuts of meat help ensure uniform portion size and shape.

## Review Questions

1. Contrast the two types of connective tissues found in meat.

2. Explain the causes and effects of shrinkage.

3. List the five major categories of domesticated cattle.

4. At what age are hogs slaughtered to yield the best quality of pork?

5. Describe the main features of veal.

6. Describe the differences between lamb and mutton.

7. List common market forms of beef, veal, and lamb.

8. Explain why pork is commonly marketed as fabricated cuts.

9. Contrast the advantages and disadvantages of purchasing a whole carcass.

10. Identify the partial carcass labels for beef, veal, and lamb.

11. Among the hindquarter primal cuts, identify the two that are the most prized for their tenderness and flavor.

12. Describe the difference between a beef rib and a beef short plate.

13. Describe the advantages and disadvantages of purchasing fabricated cuts.

14. Identify the distinguishing features of pork tenderloin.

15. Describe the differences among Virginia hams, prosciutto hams, and picnic hams.

16. List the criteria used by the USDA to grade the quality of meat.

17. Contrast the main characteristics of USDA Prime beef, USDA Choice beef, and USDA Select beef.

18. List five meat grades that are rarely or never used in professional food service establishments.

19. How does the planned cooking method of meat affect purchasing decisions?

20. List the required storage temperatures of both refrigerated meats and frozen meats.

21. Describe how irradiation affects meat.

22. Describe the main differences between a wet rub and a marinade.

23. Describe five methods commonly used to physically tenderize meat.

24. Contrast the main features of wet aging and dry aging.

25. Explain how to determine doneness for larger pieces of meat.

26. Identify common dry-heat cooking methods for meat.

27. Describe the criteria to consider when braising meat.

# Recipes

## 174

## Veal Picatta with a Lemon Caper Sauce

*yield: 4 servings*

| | |
|---|---|
| 4 | veal cutlets, 3 oz each |
| 2 tbsp | all-purpose flour |
| to taste | salt and white pepper |
| 1 tbsp | olive oil |

*Sauce*

| | |
|---|---|
| ⅔ c | dry white wine |
| 2 tbsp | fresh lemon juice |
| 2 tsp | capers, drained |
| 3 oz | cream |

1. Pound veal cutlets to ⅛" thickness.
2. In a small bowl, combine flour, white pepper, and salt.
3. Lightly coat cutlets with flour mixture.
4. In a large skillet, heat oil over medium heat. Add cutlets. Cook 3 min to 4 min for medium doneness, turning once.
5. Remove cutlets and keep warm. Drain any excess oil from pan.
6. Add wine and lemon juice to the skillet to deglaze the pan. Cook until sauce begins to thicken.
7. Remove sauce from heat. Stir in capers and cream. Reduce until nappe is achieved.
8. Spoon sauce over cutlets and serve hot.

**Nutrition Facts**

**Per serving:** 249.5 calories; 50% calories from fat; 14.3 g total fat; 96.9 mg cholesterol; 145.5 mg sodium; 374.6 mg potassium; 4.7 g carbohydrates; 0.2 g fiber; 0.2 g sugar; 4.5 g net carbs; 18.8 g protein

## 175

## Sautéed Beef Tenderloin Tips in Mushroom Sauce

*yield: 4 servings*

| | |
|---|---|
| 2 tbsp | vegetable oil |
| 5 oz | mushrooms, sliced thick |
| 2 oz | onion, diced |
| 2½ c | Espagnole sauce (see chapter 16) |
| 2½ tbsp | Burgundy wine |
| 3 to 4 tbsp | salad oil |
| 1½ lb | beef tenderloin tips, sliced ¼" thick on bias |
| 4 sprigs | parsley |

1. Heat 2 tbsp oil in a saucepan over medium heat and then add the beef tips. Cook until browned. Remove meat from heat and set aside.
2. Add mushrooms and onions and sauté until translucent and slightly browned.
3. Add 2 tbsp oil to pan
4. Add meat back into the saucepan and deglaze with wine. Reduce wine by half.
5. Add Espagnole sauce and bring to a simmer until heated through (about 15 min).
6. Serve over hot buttered noodles and garnish with parsley.

**Nutrition Facts**

**Per serving:** 731.2 calories; 67% calories from fat; 55.2 g total fat; 129.8 mg cholesterol; 551.1 mg sodium; 1206.9 mg potassium; 16 g carbohydrates; 1.7 g fiber; 4.3 g sugar; 14.3 g net carbs; 40.9 g protein

## Broiled Lamb Chops

*yield: 4 servings*

| | |
|---|---|
| 4 | lamb chops, 5 oz each |
| 2½ tbsp | salad oil |
| 1 sprig | fresh rosemary (leaves only), diced |
| 1 clove | garlic, minced |
| to taste | salt and pepper |

1. Mix oil, rosemary, garlic, salt, and pepper to create marinade.
2. Coat lamb chops with marinade mixture and let sit for 30 min.
3. Place chops in the broiler. Brown on each side, turning once, to desired degree of doneness.

**Nutrition Facts**

**Per serving:** 461.8 calories; 77% calories from fat; 39.8 g total fat; 102.1 mg cholesterol; 152.5 mg sodium; 332 mg potassium; 0.3 g carbohydrates; 0.1 g fiber; 0 g sugar; 0.3 g net carbs; 24.1 g protein

## Shish Kebabs

*yield: 4 servings*

| | |
|---|---|
| 20 oz | lamb leg meat, cut in 1" cubes |
| 4 | small tomatoes, cut in quarters |
| 8 | medium mushroom caps |
| 4 | small onions, cut in quarters |
| | metal skewers |

*Marinade*

| | |
|---|---|
| ⅔ c | salad oil |
| ⅓ c | olive oil |
| 2½ tbsp | red wine vinegar |
| 1 tbsp | fresh lemon juice |
| ½ tsp | garlic, minced |
| to taste | thyme |
| to taste | marjoram |
| to taste | basil |
| to taste | oregano |
| to taste | salt and pepper |

1. Mix marinade ingredients well and set aside.
2. Place meat, tomatoes, mushrooms, and onions in marinade and let sit for 2 hr.
3. Remove ingredients from the marinade.
4. Place lamb, tomato slices, mushroom caps, and onion wedges on metal skewers, alternating meat and vegetables. Each skewer should contain five cubes of lamb, two wedges of tomato, two mushroom caps, and two wedges of onion.
5. Drain the shish kebabs, reserving marinade. Place on a broiler for approximately 15 min on low heat. Brush frequently with reserved marinade. Turn as needed.
6. Serve immediately on a bed of cooked rice or pilaf.

**Nutrition Facts**

**Per serving:** 709.6 calories; 76% calories from fat; 61.4 g total fat; 93.6 mg cholesterol; 170.7 mg sodium; 696.5 mg potassium; 10 g carbohydrates; 1.9 g fiber; 4.5 g sugar; 8.1 g net carbs; 30.7 g protein

# *Recipes*

## 178

## Roast Standing Rib of Beef

*yield: 8 servings*

| | |
|---|---|
| 6 lb | boneless standing rib of beef, trussed for roasting |
| to taste | salt and pepper |
| 4 oz | onions, rough cut |
| 2 oz | celery, rough cut |
| 2 oz | carrots, rough cut |

*Sauce*

| | |
|---|---|
| 2 tsp | Worcestershire sauce |
| to taste | salt |

1. Place beef in a roasting pan, rib-side up. Season with salt and pepper to taste. *Note:* This recipe could also be used for bone-in rib roasts of lamb, pork, or veal.
2. Roast in 350°F oven for 1 hr until the roast is evenly browned on the surface. Remove from oven.
3. Turn roast over so rib side is down. Add onions, celery, and carrots to roasting pan.
4. Roast at 350°F until garnish becomes light brown.
5. Reduce temperature to 325°F and continue to roast to desired doneness.
6. Remove the roast and allow to rest for 20 min. before slicing.
7. Meanwhile, pour the drippings and vegetables into a large saucepot. Deglaze the roasting pan by adding 1 cup of fresh water and bring to a boil while scraping free the "fond" from the bottom of the pan. Add deglazed juices to reserved drippings. Add Worcestershire sauce and salt. Simmer for 15 min. Skim off fat and strain sauce through a china cap.
8. Using a knife, remove butcher twine and feather bones from the roast.
9. Stand the roast up by placing the large end down. Carve the roast as desired.
10. Serve each portion with 1½ oz of sauce.

> **Nutrition Facts**
> **Per serving:** 730.8 calories; 64% calories from fat; 51.2 g total fat; 196.8 mg cholesterol; 215.6 mg sodium; 909.1 mg potassium; 2.5 g carbohydrates; 0.5 g fiber; 1.1 g sugar; 2 g net carbs; 60.7 g protein

## Herbed Veal Roast with Apricot-Thyme Chutney

*yield: 4 servings*

| | |
|---|---|
| ½ tbsp | sage, fresh, chopped |
| 1 clove | garlic, crushed |
| ¼ tsp | black pepper |
| 2 lb | veal roast |

*Chutney*

| | |
|---|---|
| ½ tbsp | vegetable oil |
| 1 | medium onion, sliced |
| 3 oz | dried apricots, coarsely chopped |
| ½ c | chicken broth |
| ½ tbsp | sugar |
| ½ tsp | cider vinegar |
| ½ tsp | thyme, dried, crushed |

1. Combine sage, garlic, and pepper. Press evenly into the surface of the veal roast.
2. Place roast, rib-ends down, in roasting pan. Insert ovenproof meat thermometer in the thickest part of the meat, not touching bones or resting in fat.
3. Roast for 45 min at 325°F or until thermometer registers 155°F.
4. Meanwhile, heat oil in a large skillet over medium heat. Add onions and cook slowly for 15 min to 20 min, stirring occasionally.
5. Add remaining chutney ingredients to onions. Cover and simmer 20 min to 25 min.
6. Remove roast and allow to rest 15 min.
7. Carve into slices and serve with hot chutney.

> **Nutrition Facts**
> **Per serving:** 158.8 calories; 22% calories from fat; 4 g total fat; 45.2 mg cholesterol; 150.2 mg sodium; 509.4 mg potassium; 18.7 g carbohydrates; 2.2 g fiber; 14.3 g sugar; 16.5 g net carbs; 13.1 g protein

## Roasted Rack of Lamb

180

*yield: 8 servings*

| | |
|---|---|
| 1 | rack of lamb, Frenched |
| to taste | marjoram |
| to taste | salt and pepper |
| 1 clove | garlic, minced |
| 1 pint | veal stock |

1. Place scrap lamb bones or a roasting rack in a roasting pan. Place trimmed rack of lamb on top of bones or roasting rack, fat-side up. Season with marjoram, salt, and pepper to taste.
2. Roast in 400°F oven for 30 min to 45 min or until desired doneness is achieved.
3. Transfer lamb to a bake pan and hold in a warm place.
4. To make an au jus, deglaze the roasting pan with veal stock. Pour drippings and stock into a saucepan. Add garlic. Simmer over low heat.
5. Strain sauce through a china cap. Season to taste.
6. Cut the rack of lamb between the ribs. Serve two ribs per serving with au jus.

### Nutrition Facts

**Per serving:** 104.2 calories; 45% calories from fat; 5.3 g total fat; 37.4 mg cholesterol; 196 mg sodium; 262.9 mg potassium; 0.8 g carbohydrates; 0 g fiber; 0.3 g sugar; 0.8 g net carbs; 12.5 g protein

## Meat Loaf

181

*yield: 4 servings*

Beef Checkoff

| | |
|---|---|
| 1 tbsp | vegetable oil |
| 1 stalk | celery, small diced |
| 1 | medium onion, small diced |
| ½ c | breadcrumbs |
| 2½ tbsp | milk |
| 1 | egg, beaten |
| to taste | salt and pepper |
| ½ tsp | thyme |
| 1 lb | ground beef |

1. Heat oil in a small pan. Sauté celery and onions until tender. Allow to cool.
2. Add egg, celery, onions, breadcrumbs, salt, thyme, and pepper.
3. Add ground beef and mix thoroughly by hand. *Note:* If the mixture is too moist, add additional bread crumbs or cracker crumbs as necessary.
4. Form mixture into a loaf and place in a greased roasting pan.
5. Bake at 350°F for about 1 hr. Bake for an additional 30 min.
6. Remove from the oven.
7. Slice loaf in ½" slices. Top with espagnole if desired.

### Nutrition Facts

**Per serving:** 366.8 calories; 56% calories from fat; 22.8 g total fat; 130.9 mg cholesterol; 276.6 mg sodium; 462.6 mg potassium; 13.7 g carbohydrates; 1.3 g fiber; 2.9 g sugar; 12.5 g net carbs; 25.1 g protein

# *Recipes*

## 182

## Corned Beef Brisket

*yield: 9 servings*

| | |
|---|---|
| 3 lb | corned beef brisket |
| 1 tsp | pickling spices |

1. Place the brisket in a stockpot and cover with cold water.
2. Add pickling spices and simmer until meat is tender.
3. Remove any scum that rises to the surface.
4. When tender, turn off heat and allow brisket to remain in cooking liquid to cool slightly.
5. Slice the brisket against the grain. Serve while hot.

### Nutrition Facts

**Per serving:** 300.5 calories; 67% calories from fat; 22.6 g total fat; 81.6 mg cholesterol; 1840.2 mg sodium; 449.1 mg potassium; 0.3 g carbohydrates; 0 g fiber; 0 g sugar; 0.3 g net carbs; 22.2 g protein

## Italian Meatballs

*yield: 4 servings*

| | |
|---|---|
| 1 tbsp | olive oil |
| 4 oz | onions, small diced |
| ½ tsp | garlic, minced |
| 2 slices | bread, cubed |
| 2½ tbsp | milk |
| 1 lb | ground beef |
| 1 tbsp | Parmesan cheese, grated |
| ¼ tsp | oregano |
| to taste | basil |
| to taste | salt and pepper |
| 1 | egg, beaten |
| 2 tsp | fresh parsley, chopped |

1. In a skillet, heat olive oil over medium heat. Sauté onions and garlic until tender. *Note:* Do not allow the onions to brown.
2. In a mixing bowl, combine bread and milk. Mix well with a kitchen spoon.
3. Add sautéed onions and garlic, ground beef, Parmesan cheese, oregano, basil, salt, and pepper to the bread and milk. Mix well by hand until evenly combined.
4. Pass the mixture through the food grinder twice while using the medium-hole chopper plate. *Note:* If the mixture is too wet or loose, additional bread crumbs can be added as needed.
5. Add egg and parsley. Blend thoroughly. Adjust seasoning as necessary.
6. Form into balls approximately 2″ in diameter and place on greased sheet pans. *Note:* Rubbing hands with olive oil will prevent the meat from sticking to the hands.
7. Bake at 350°F until cooked through. Serve with pasta and an Italian-style sauce.

### Nutrition Facts

**Per serving:** 349.3 calories; 59% calories from fat; 22.7 g total fat; 132 mg cholesterol; 274.4 mg sodium; 426.6 mg potassium; 10 g carbohydrates; 0.8 g fiber; 2.4 g sugar; 9.2 g net carbs; 24.7 g protein

# Beef Stroganoff

*yield: 4 servings*

| | |
|---|---|
| 5 oz | wide egg noodles |
| 2 tbsp | vegetable oil |
| 1 lb | beef round steaks, cut into thin strips |
| 1 clove | garlic, crushed |
| ¼ tsp | salt |
| ¼ tsp | pepper |
| ⅔ c | Espagnole sauce (see chapter 16) |
| ¼ c | sour cream |
| 1 tsp | Worcestershire sauce |

1. Cook noodles according to directions on package. Keep warm.
2. In a large skillet, heat 1 tbsp vegetable oil over medium heat. Stir-fry beef and garlic 1 min.
3. Remove beef from skillet and season with salt and pepper.
4. In the same skillet, heat 1 tsp vegetable oil over medium heat. Sauté mushrooms for 2 min.
5. Stir in Espagnole sauce and bring to a simmer. Add sour cream and Worcestershire sauce. Mix thoroughly.
6. Return beef to skillet and heat thoroughly.
7. Serve over egg noodles.

**Nutrition Facts**

**Per serving:** 299 calories; 42% calories from fat; 14.3 g total fat; 83.8 mg cholesterol; 445.4 mg sodium; 511.2 mg potassium; 11.9 g carbohydrates; 0.6 g fiber; 0.1 g sugar; 11.3 g net carbs; 29.5 g protein

## 185

# Oxtail Stew

*yield: 4 servings*

| | |
|---|---|
| 2 oz | shortening |
| 2 lb 6 oz | oxtails |
| to taste | salt and pepper |
| 1 tbsp | all-purpose flour |
| 3 oz | onions, small diced |
| ½ tsp | garlic, minced |
| 2½ c | beef stock |
| ⅓ c | tomato purée |
| ½ tsp | thyme |
| 1 | bay leaf |
| 4 oz | carrots, julienned |
| 4 oz | celery, julienned |
| 4 oz | turnips, julienned |
| 4 oz | peas |
| 4 | small onions |
| 2 tbsp | butter |

1. Put the shortening in the braiser and heat in 400°F oven. Heat thoroughly.
2. Add oxtails and season with salt and pepper. Turn as needed until evenly browned on all sides.
3. Sprinkle flour over meat and blend thoroughly with a kitchen spoon. Cook 5 min.
4. Add diced onions and garlic. Cook until tender.
5. Add brown stock, tomato purée, thyme, and bay leaf. Cover and cook in braiser until the meat is tender and the sauce is slightly thick. Remove bay leaf.
6. Meanwhile, in a separate saucepan, simmer carrots, celery, turnips, and peas in water until tender. Drain and add to the stew, reserving some peas for garnish.
7. Sauté the whole onions in butter. Add to the stew.
8. Serve in large soup bowls and garnish with peas.

**Nutrition Facts**

**Per serving:** 508.7 calories; 44% calories from fat; 25.3 g total fat; 294.9 mg cholesterol; 597.1 mg sodium; 1272.2 mg potassium; 26.5 g carbohydrates; 5.9 g fiber; 11.7 g sugar; 20.6 g net carbs; 44 g protein

# Fish and Shellfish

## Chapter 20

*Fish and shellfish* are classified as seafood. Seafood includes any edible animal that lives in freshwater or saltwater. There are more than 25,000 species of fish, and hundreds of species of shellfish. Seafood can be purchased whole or processed. Because seafood spoils quickly, it can be kept fresh for a maximum of one or two days, or can be frozen for a few months. The cooking methods for seafood vary depending on the species and the fattiness of the fish or shellfish.

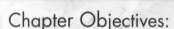

## Chapter Objectives:

1. Describe the major classifications of fish, based on external shape and bone structure.
2. Contrast the qualities of major types of freshwater fish.
3. Describe common types of saltwater fish.
4. Contrast crustaceans and mollusks, giving examples of each.
5. List other items categorized as seafood on restaurant menus.
6. Describe the various market forms in which fish and shellfish may be purchased.
7. Contrast the three types of optional inspections that a seafood producer can elect.
8. Identify factors to consider when purchasing and storing fish and shellfish.
9. Demonstrate common techniques for preparing fish and shellfish before cooking.
10. List factors to consider when cooking fish and shellfish.
11. Demonstrate how to poach a whole fish.
12. Demonstrate how to poach a whole lobster and how to remove cooked lobster meat from the shell.
13. Contrast the methods for cooking mollusks and cephalopods.

## Key Terms
- **fish**
- **round fish**
- **flatfish**
- **boneless fish**
- **aquafarming**
- **anadromous fish**
- **shellfish**
- **crustacean**
- **mollusk**
- **molting**
- **univalve**
- **bivalve**
- **adductor muscle**
- **cephalopod**
- **tranche**
- **shucking**
- **en papillote**

## SEAFOOD

At one time, seafood was available only to people who lived near a coast, lake, pond, or river. Today seafood is widely available due to advances in catching, processing, and transportation, and is increasing in popularity because it is low in fat and high in essential fatty acids, making it a great choice for those who are heath and weight conscious. Seafood can be divided into two categories, fish and shellfish.

Both fish and shellfish have flesh that is naturally tender. Compared to meats such as beef or pork, fish and shellfish require very little cooking time. A fish or shellfish provides only one or at most a few edible cuts, while animals such as cattle, lambs, or hogs yield a large number of different cuts, with different cuts from the same animal often requiring different cooking methods.

Fish and shellfish are highly perishable and do not store well. It is important for a chef to know the body structure as well as proper handling, storage, fabrication, preparation, and cooking methods associated with different varieties of fish and shellfish.

## FISH

*Fish* is a classification for aquatic animals that have fins for moving through the water, gills for breathing, and an internal bone structure with a backbone (or vertebrae). Fish are classified by their external shape and bone structure as well as by habitat (freshwater or saltwater).

### Composition and Bone Structure

Fish have an edible flesh consisting of protein, carbohydrates, fat, water, and trace amounts of vitamins and minerals. A significant difference between fish and meat is that fish flesh does not have connective tissues. Because fish flesh lacks connective tissues, it is naturally tender and can be cooked quickly using low heat. However, because there are no connective tissues holding it together, cooked fish is fragile and comes apart easily.

All fish have fins for swimming, gills for breathing, and an internal bone structure. However, the bodies of fish can be very different from one variety to another, internally and externally. Fish are classified as round fish, flatfish, or boneless fish, based on external shape and internal bone structure. ***See Figure 20-1.***

*Round fish* are fish with a long, cylindrical body and an eye located on each side of the head. A backbone runs from head to tail along the upper length of the body, just beneath the dorsal fin. The round fish body shape is the most common shape for fish. Round fish are found in freshwater lakes and streams as well as in saltwater. Large round fish have large, inedible scales that are typically removed prior to preparation. Trout and salmon are examples of round fish.

## Fish Shape and Bone Structure

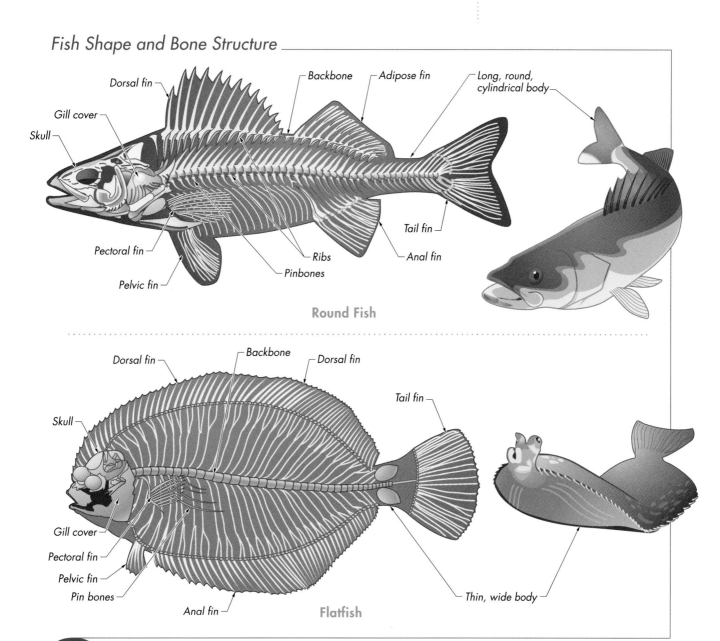

Dorsal fin

Backbone

Adipose fin

Long, round, cylindrical body

Gill cover

Skull

Pectoral fin

Ribs

Pinbones

Anal fin

Tail fin

Pelvic fin

**Round Fish**

Dorsal fin

Backbone

Dorsal fin

Tail fin

Skull

Gill cover

Pectoral fin

Pelvic fin

Pin bones

Anal fin

Thin, wide body

**Flatfish**

**20-1** *The shape and bone structure of round fish is different from that of flatfish.*

*Flatfish* are thin, wide fish with both eyes located on one side of the head. A backbone runs from head to tail down the center of the body, rather than at the top edge as in round fish. Flatfish swim with the body horizontal, parallel to the surface of the water, with one side facing down and the other side having both eyes facing toward the surface. The skin on the top side is typically dark greenish-brown and may change colors to blend in with its environment, while the bottom side is light in color. Many flatfish prefer to lay hidden on the ocean floor, with the colorless blind side in the sand, to catch prey that swim over them. Flatfish are called left-handed fish if the eyes are on the left side of the body or right-handed fish if the eyes are on the right side. Flounder are an example of flatfish.

*Boneless fish* are fish shaped like round fish; however, they have thick cartilage skeletons instead of bones. Boneless fish often have smooth, tough outer skin without scales. Sharks are an example of boneless fish.

## Lean Fish and Fat Fish

Fish are categorized as lean fish or fat (oily) fish, based on the amount of fat contained in the flesh. ***See Figure 20-2.*** Lean fish contain as little as 0.5% fat. Walleye are an example of lean fish. Fat fish contain up to 20% fat and are rich in omega-3 fatty acids and vitamins A and D. Examples of fat fish include orange roughy, salmon, and tuna. The fat content of a particular type of fish can vary slightly by season. Fat content in fish can affect both flavor and the cooking method required. Lean fish are often best prepared using moist-heat methods, while fat fish can be prepared using dry-heat cooking methods.

| Fat Content of Common Fish | | | |
|---|---|---|---|
| Lean Fish | | Fat Fish | |
| Species | Fat Content* | Species | Fat Content* |
| Bluefish | 4g | Brook trout | 6g |
| Codfish | 1g | Butterfish | 7g |
| Croaker | 3g | Catfish | 6g |
| English and Dover sole | 1g | Lake trout | 6g |
| Flounder | 1g | Pompano | 8g |
| Haddock | 1g | Rainbow trout | 5g |
| Hake | 0 | Salmon | 9g |
| Halibut | 2g | Shad | 12g |
| Mullet | 3g | Spanish mackerel | 5g |
| Northern pike | 1g | Whitefish | 5g |
| Red snapper | 1g | | |
| Sea bass | 2g | | |
| Smelt | 2g | | |
| Striped bass | 2g | | |
| Walleye pike | 1g | | |
| Yellow perch | 1g | | |

* per 3 oz serving

 Lean fish contain lower amounts of fat in their flesh than is found in fat fish.

## FRESHWATER FISH

There are a wide variety of freshwater fish on restaurant menus. Freshwater fish come from freshwater lakes and rivers. Some varieties of freshwater fish have become so popular that they are being farm raised, or aquafarmed. *Aquafarming* is the raising of fish or other seafood in a controlled environment. Most freshwater fish have a round body shape and are classified as round fish. ***See Figure 20-3.***

## Common Freshwater Fish

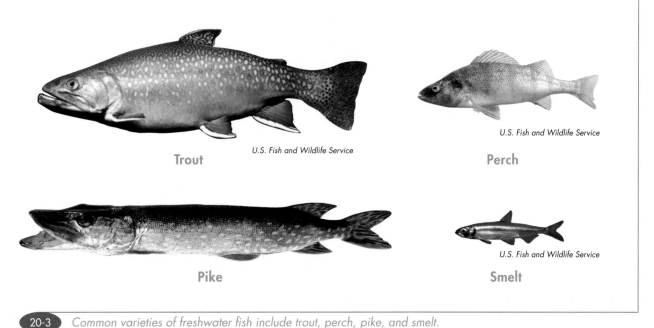

Trout

*U.S. Fish and Wildlife Service*

Perch

*U.S. Fish and Wildlife Service*

Pike

Smelt

*U.S. Fish and Wildlife Service*

**20-3** *Common varieties of freshwater fish include trout, perch, pike, and smelt.*

### Trout

*Trout* are a family of round, fat fish with tender flesh that is rich yet delicate tasting. Like salmon, trout spend the majority of their life at sea but come back to freshwater rivers and streams to spawn. A good portion of trout sold commercially have been farmed. Commonly available types of trout include brook trout, lake trout, and rainbow trout. ***See Figure 20-4.*** Trout is best when broiled or sautéed.

Brook trout is a medium-fat fish and one of the finest eating fish available, provided it is taken from ice-cold water. Brook trout have silver-gray, slightly speckled skin, and a square, slightly forked tail. Brook trout average in weight from 8 oz to 10 oz, which is the perfect size to provide a generous portion.

Lake trout is the largest trout, having dark to pale gray skin that is covered with white spots, and a fairly large head. The flesh may be red, pink, or white depending on the lake from which the fish was taken. Lake trout weighing 4 lb to 10 lb with pink flesh are considered the best.

Rainbow trout are named for the characteristic purplish red band that extends along their sides from the head to the tail. The average weight of a rainbow trout is about 2 lb, but the size of the fish depends to a great degree on the size of the body of water from which the fish is taken. The larger the body of water, the larger the fish.

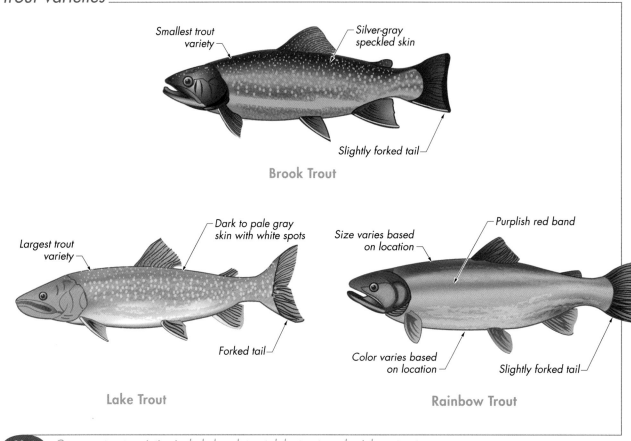

Smallest trout variety

Silver-gray speckled skin

Slightly forked tail

**Brook Trout**

Largest trout variety

Dark to pale gray skin with white spots

Forked tail

**Lake Trout**

Size varies based on location

Purplish red band

Color varies based on location

Slightly forked tail

**Rainbow Trout**

**20-4** *Common trout varieties include brook trout, lake trout, and rainbow trout.*

Perch Varieties

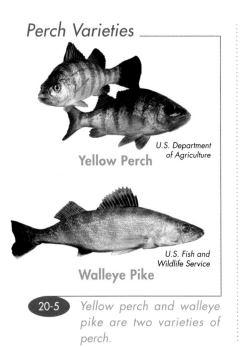

*U.S. Department of Agriculture*

**Yellow Perch**

*U.S. Fish and Wildlife Service*

**Walleye Pike**

**20-5** *Yellow perch and walleye pike are two varieties of perch.*

## Perch

*Perch* are a family of lean freshwater fish native to the Great Lakes and northern Canada. The most common variety is the yellow perch. Yellow perch have skin that is dark olive green on the back, blending into golden yellow on the sides, and becoming lighter near the belly. Their green skin is marked with six to eight dark, broad vertical bands that run from the back to just above the belly. Yellow perch average about 12″ in length and weigh about 1 lb. They have a mild-tasting white meat. Yellow perch can be sautéed, fried, or broiled.

Another popular perch variety is walleye pike. A *walleye pike* (jack salmon) is a lean fish that resembles a pike but is a member of the perch family. The color of the walleye pike varies, but it is usually a dark olive green on the back, shading into a light yellow on the sides and belly. Walleye pike have exceptionally large, shiny eyes and vary considerably in size, with an average weight between 2 lb and 5 lb. The flesh is relatively fine-grained and has an excellent flavor, but many small bones are present. Walleye pike is popular fried or sautéed. *See Figure 20-5.*

## Pike

A *pike* is a lean fish with a long, lean body, a long, broad, flat snout, and bands of sharp teeth. It has a firm, flaky flesh that contains many bones. The most common variety of pike is the Northern pike. A Northern pike consumes one-fifth of its own weight in food daily. It averages from 2 lb to 4 lb, but specimens as large as 10 lb to 15 lb are not uncommon. The Northern pike inhabits cold freshwater and is found mainly in Canadian lakes. Northern pike is available all year and is most plentiful in June, but its flesh is firmer and sweeter when caught during the colder months of the year. Northern pike is a popular menu item whether fried, sautéed, broiled, or baked.

## Smelt

A *smelt* is a small, lean member of the salmon family having a slender body, a long, pointed head, a large mouth, and a deeply forked tail. It is olive to dark green along the top, blending into a lighter shade with a silver cast along the sides. Its belly is silver and its fins are speckled with tiny spots. Smelt are classified as both freshwater and saltwater fish. The largest catch of freshwater smelt comes from Lake Michigan. Saltwater smelt are found off the Atlantic coast, from New York to Canada. They can reach a size of about 10″ long and weigh as much as 1 lb, although most are much smaller. Smelt used in the professional kitchen averages about 6 to 8 fish per pound. Smelt are marketed whole and, once eviscerated, are prepared whole by frying or sautéing.

## Tilapia

A *tilapia* is a freshwater fish with lean, firm white meat. Tilapia are primarily aquafarmed, as wild tilapia tends to have a muddy taste. An average tilapia weighs between 2 lb and 3 lb. Tilapia are sometimes marketed as cherry snapper, although they are not members of the snapper family. Tilapia are prepared by sautéing, grilling, and frying.

## Catfish

A *catfish* is a freshwater fish named for the whiskerlike barbels that protrude off the sides of its face. Catfish are fat fish that provide a firm, white flaky meat with excellent eating qualities. They do not have scales, but have a tough skin that adheres tightly to the flesh, which can make cleaning difficult. The skin is always removed prior to cooking.

Most commercially available catfish are farm raised and have a milder taste than wild-caught catfish. In the commercial kitchen, the most common size of catfish used is between ¾ lb and 1 lb. It is usually either fried or sautéed whole. Catfish is also available as breaded fillets, breaded strips, and breaded nuggets.

## SALTWATER FISH

There are multitudes of saltwater fish on restaurant menus. However, some are more common than others. Saltwater fish may be round, flat, or boneless. Some varieties of saltwater round fish begin their lives in freshwater but spend most of their lives in saltwater, returning to freshwater to spawn. *See Figure 20-6.* Saltwater flatfish do not leave their natural habitat.

### Common Saltwater Fish

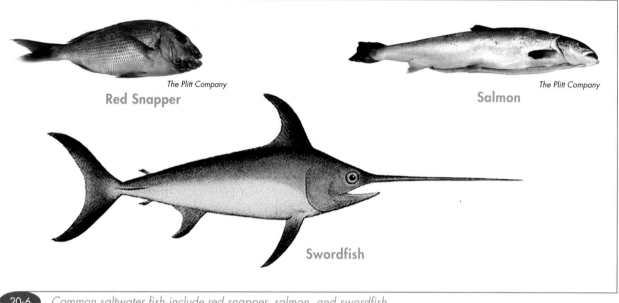

*The Plitt Company*

**Red Snapper**

*The Plitt Company*

**Salmon**

**Swordfish**

**20-6** *Common saltwater fish include red snapper, salmon, and swordfish.*

### Flatfish

**Flounder**

**Sole**

*The Plitt Company*

**20-7** *Flounder and sole are both flatfish.*

### Sole

A *sole* is lean flatfish regarded as the most flavorful and smoothest textured of all flatfish. It has a narrower body than that of a flounder. True Dover sole is the most expensive variety of flatfish. It averages 10″ in length, with brown to pale brown skin on the colored side. Dover sole has a delicate, pearly white flesh that pairs wonderfully with many classical sauces. The fish is often dressed, sautéed whole, and then filleted tableside in classic preparations such as sole a la Munière (sole with a lemon-butter sauce) or sole amandine (sole with almonds).

### Flounder

A *flounder* is a lean saltwater flatfish. There are hundreds of species of flounder. In the United States, some varieties of flounder may be marketed as "sole" or "dab." *See Figure 20-7.* Flounder is best when sautéed, broiled, or fried. The largest flounder comes from the waters off the New England coast. Species of Atlantic flounder include summer flounder, winter flounder, yellowtail flounder, gray sole, and sand dab.

Summer flounder can reach sizes up to 15 lb; the average fish caught ranges from 2 lb to 5 lb. They are found in waters as far north as Maine but are more plentiful south of Cape Cod down to the shores of Florida. Winter flounder is noted for its thick, meaty fillets. These flounder average about 2 lb each. Winter flounder is also called George's Bank flounder. Yellowtail flounder is named for its yellow tail. The average yellowtail weighs between 1 lb and 2 lb. It has a fairly thin body and is considered a less desirable food fish than other varieties.

Gray sole (fluke) is a dark brown or russet gray right-handed flounder with fins that are almost black. Gray sole average 2 lb to 3 lb, although specimens weighing up to 4 lb are not uncommon. Gray sole is typically a very lean flatfish with excellent flavor. Sand dab (windowpane flounder) is considered an excellent pan fish, weighing between 1 lb and 2 lb. They have bone-free fillets and sweet-tasting, oil-free meat.

In addition to the Atlantic varieties, several varieties of flounder are found in the Pacific. These include petrale sole, domestic Dover sole, English sole, and California halibut.

Petrale sole has thick, meaty fillets and excellent flavor. It is regarded as the best American flounder. Domestic Dover sole is a flounder that closely resembles the true Dover sole. It is one of the least-popular varieties of flounder, as it lacks the desirable flavor and texture found in other varieties. English sole is also considered to be of mediocre quality. It is commonly marketed in the United States as "fillet of sole" and in Canada as "lemon sole." California halibut is a large flounder, weighing 12 lb on average, and has a thick, lean flesh that is typically cut into steaks. California halibut are similar to halibut in taste and texture.

## Halibut

A *halibut* is a very large flatfish that resembles a giant flounder; only swordfish, tuna, and some sharks reach a larger size. Mature halibut range in weight between 50 lb and 150 lb, with specimens weighing less than 100 lb generally being richer in flavor than those weighing more. Halibut purchased for use in the professional kitchen typically weigh between 25 lb and 70 lb. Chicken halibut are very young halibut weighing between 4 lb and 12 lb. Chicken halibut is desirable for its fine eating qualities.

Halibut is found in both the Atlantic and Pacific Oceans. Although other fish may be referred to as halibut, only two varieties, Atlantic halibut and Pacific halibut, are recognized by the FDA as true halibut. Both varieties provide thick, lean flesh and white, sweet-tasting meat.

Halibut can be fried, broiled, steamed, poached, sautéed, grilled, or baked with excellent results. Since halibut is fairly dry, it should always be served with a sauce.

## Common Salmon Varieties

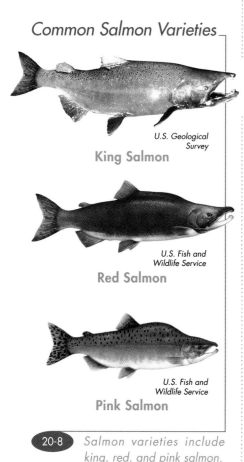

*U.S. Geological Survey*

**King Salmon**

*U.S. Fish and Wildlife Service*

**Red Salmon**

*U.S. Fish and Wildlife Service*

**Pink Salmon**

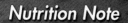 **20-8** *Salmon varieties include king, red, and pink salmon.*

### Nutrition Note

Salmon contains omega-3 fatty acids, which are poly-unsaturated and may help prevent heart disease, inflammatory processes, and certain cancers.

## Salmon

A *salmon* is an anadromous fish found in both the northern Atlantic Ocean and the Pacific Ocean. ***See Figure 20-8.*** An *anadromous fish* is a saltwater fish that migrates into freshwater to spawn. Salmon instinctively swim many miles, sometimes hundreds of miles, against currents and up rivers to return to the freshwater location where it was originally spawned. Pacific salmon die after the fish lays its eggs. This is not true for Atlantic salmon.

Atlantic salmon is found mainly in the north Atlantic and in rivers and streams along the coasts of Maine, Nova Scotia, Quebec, Labrador, and New Brunswick. Atlantic salmon differ from the Pacific varieties in that Atlantic salmon return to sea after spawning and often live to spawn a second time. The flesh of an Atlantic salmon is medium pink with excellent eating qualities. An average Atlantic salmon weighs between 10 lb and 15 lb, but can easily weigh as much as 20 lb. Atlantic salmon and Pacific (Alaska) salmon varieties include king, sockeye, coho, pink, and chum.

King (Chinook) salmon is the largest of the Pacific salmon and has a reddish-orange flesh and a superb, rich flavor. King salmon range in size from 5 lb to 25 lb, with an average king salmon weighing 20 lb. King salmon is considered the finest Pacific salmon. It can be baked, grilled, sautéed, or poached, and can be served with little additional flavoring due to its high oil content.

Sockeye (red) salmon, like the king salmon, has a deep red flesh and relatively high fat content. An average sockeye salmon weighs between 3 lb and 8 lb. It is excellent for broiling, grilling, baking, steaming, poaching, or sautéing.

Coho (silver) salmon has deep pink flesh and an above-average flavor. Aquafarmed coho are much smaller than those that are wild, and weigh between 1 lb and 2 lb on average. Coho salmon is desirable for smoking. This variety of salmon has low oil content compared to some of the other varieties, so care should be taken when cooking to not dry it out.

Pink salmon is sometimes called humpback salmon because of the noticeable hump that develops in front of the dorsal fin of males during the spawning season. Pink salmon is the smallest of the Pacific salmon, weighing only 3 lb per fish on average. It has soft, pink flesh and a lower fat content than higher-quality salmon. Care should be taken not to overcook pink salmon because it will dry out more quickly than other varieties.

Chum (keta) salmon has pale yellow, soft flesh containing a moderate amount of oil. A typical chum salmon weighs 8 lb. It is used in dishes where low cost is important.

A large percentage of salmon is marketed either fresh or frozen. Frozen salmon is held at 32°F until it is flash-frozen. It is protected from dehydration by glazing. *Glazing* is the process of covering an item with water to form a protective coating of ice before the item is frozen. Proper glazing of frozen salmon results in a product that tastes as fresh as the day it was caught.

Alaska produces more than 90% of the salmon consumed in North America. Approximately 30% to 35% of Alaska salmon is canned. Nothing is added to the product except perhaps a trace amount of salt. Other popular salmon products include the gravlax, which is a traditional Scandinavian salmon dish that is cured for 24 hr to 48 hr with salt, sugar, and dill; lox, which is salmon that is brine cured for 24 hr to 48 hr and then cold-smoked; nova, which is brine-cured, cold-smoked salmon that is less salty than lox; and salmon roe, which are bright orange-colored salmon eggs.

### Tuna

*Tuna* are very large fish that are members of mackerel family and that have a low to moderate fat content and bright red flesh. There are many different varieties of tuna, some weighing more than 300 lb. The most common varieties of tuna include albacore, bigeye, yellowfin, skipjack, and bluefin.

Albacore (longfin) tuna is the variety most often used for canned white tuna. Because it contains higher levels of mercury than other types of tuna, pregnant women are warned to refrain from consuming albacore tuna. Bigeye and yellowfin tuna are both marketed as ahi tuna, which is most often seared as a steak and served rare or medium rare. Yellowfin and skipjack varieties are used most often in canned tuna products that are not specified as white or albacore tuna. Bluefin tuna is prized for its use in sushi and sashimi because of its high fat content. ***See Figure 20-9.***

*Bluefin Tuna*

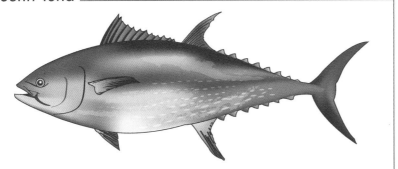

20-9   *Bluefin tuna is often used in sushi and sashimi.*

Fabrication of tuna results in four separate loins, which are commonly cut into steaks, with the smaller end portions being cut into smaller cubes. The belly loins are commonly fattier than the back loins and are therefore more desirable, especially for sashimi. A dark red, almost black, muscle runs along the lateral line of the fish and should be removed prior to preparation of the fish. Running along the loin, there may also be a blood vein that should be removed, as it has a strong flavor and is not desirable.

Cooked tuna should have a pink center similar to that of a medium to medium-rare filet mignon. It is common for tuna to be seared and served rare or medium rare; however, to decrease the risk of foodborne illness, only tuna that is graded Grade A and has been processed and handled under superior conditions should be served in this manner. Tuna steaks are best prepared by being brushed with oil and seasoned, or marinated just before grilling or broiling.

### Cod

The cod family includes Atlantic cod, haddock, Pacific cod, pollock, hake, and whiting. ***See Figure 20-10.*** Although cod can reach over 200 lb, they are commonly available weighing about 10 lb. Scrod is a term for young cod between 1 lb and 2½ lb in size. Scrod is sweeter and moister than mature cod and is delicious when broiled or sautéed.

In the professional kitchen, cod is usually purchased as skinless, boneless fillets. The meat is lean, white, and dry, with a flaky grain, and is best cooked by poaching, steaming, or baking. Because of its dryness, cod should always be served with a sauce.

Atlantic cod is used most in prepared frozen fish products and sold in fillets, steaks, or drawn. Pollock is a cod variety with pinkish flesh when raw that turns white when cooked. It is commonly referred to as blue cod or Boston bluefish. Pollock is the main ingredient in processed fish items such as surimi, crab sticks, and some fish sticks.

Haddock is a cod variety with meat that is slightly darker (though still considered white) and more fibrous than that of Atlantic cod, yet remains firm and lean with an excellent flavor. The average weight of a haddock is 4 lb. Haddock is available year-round, but is in peak season in the spring. Steaming, baking, and broiling are the best cooking methods for haddock, and a sauce is typically served.

Pacific cod, also known as grey cod, is most often marketed as "true cod." Hake is a type of cod with lean flesh that is darker and more fibrous than that of Atlantic cod or Pacific cod. Hake have a slender body with two sets of dorsal fins, a forked tail, and a pointed snout. Hake can be purchased whole, drawn, dressed, or as fresh or smoked fillets. Although hake is considered inferior to other types of cod and is less popular in the professional kitchen, it can be improved if baked or poached and served with an appropriate sauce, such as Creole or Dugléré. Hake is

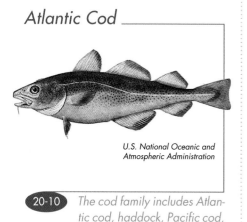

*Atlantic Cod*

U.S. National Oceanic and Atmospheric Administration

**20-10** *The cod family includes Atlantic cod, haddock, Pacific cod, hake, and whiting.*

sometimes substituted for haddock because the flesh and eating qualities are somewhat similar.

Whiting is a type of cod that produces lean, delicate fillets. It is one of the most delicate fish, with fillets that break apart very easily if not handled gently. Whiting suffers in the professional kitchen because the flesh is so fragile. It is more often used in processed fish items.

## Herring

A *herring* is a long, thin fish that is somewhat fatty with shiny, silvery-blue skin. Herring are generally found in the northern Atlantic and in parts of the Pacific. A typical herring weighs only 8 oz. Herring is most commonly smoked or brined because it spoils rapidly. It is also great when broiled or grilled. **See Figure 20-11.** A sardine is a very small young herring. Sardines are fatty and oily and are typically sold packed in oil and used on salads and sandwiches. However, sardines can be prepared fresh by grilling, broiling, or even smoking.

## Bass

*Bass* are a common variety of spiny-finned fish with white-colored lean flesh that produces a sweet-tasting, delicate fillet. There are many species of bass in both saltwater and freshwater. The most common varieties of bass available commercially are black sea bass and striped bass. **See Figure 20-12.**

Sardines _____

Washington Department of Fish and Wildlife

**20-11** *Sardines are small, young herring that are often packed in oil.*

Common Bass Varieties _____

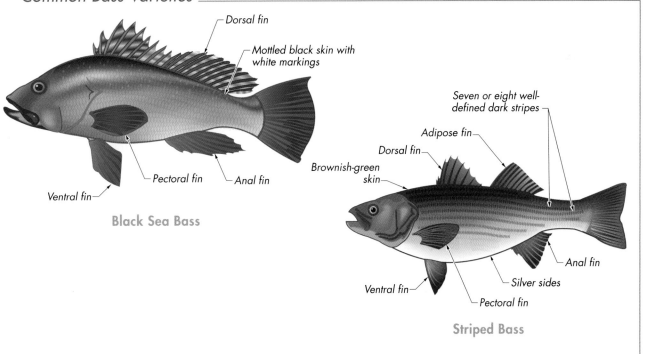

Black Sea Bass

Dorsal fin

Mottled black skin with white markings

Pectoral fin

Anal fin

Ventral fin

Striped Bass

Seven or eight well-defined dark stripes

Adipose fin

Dorsal fin

Brownish-green skin

Ventral fin

Pectoral fin

Silver sides

Anal fin

**20-12** *The most common varieties of bass include black sea bass and striped bass.*

A black sea bass is a saltwater fish with mottled black skin. It has lean, white, flavorful meat and averages 1 lb to 3 lb in weight. Although it is caught in both the Atlantic and Pacific Oceans, black sea bass taken from the Atlantic is leaner and is considered to be of better quality. Bass weighing between ¾ lb and 1 lb are considered best and are usually sautéed, broiled, or baked whole. These fish are often cooked whole in Asian restaurants.

Striped bass is a variety that begins life as a freshwater fish and migrates to saltwater at maturity. It is native to the Atlantic coast and marked with seven to eight well-defined dark stripes running from the head to the tail. It is a lean, white-fleshed fish that is sweet and rich tasting, and is best when sautéed, baked, or broiled. A hybrid striped bass was developed in response to restrictions on commercial fishing of striped bass. Hybrid striped bass has a somewhat oilier and darker flesh than that of a striped bass.

## Grouper

A *grouper* is a lean, white-fleshed fish that has excellent eating qualities, similar to those of a red snapper. The skin is tough, has an unpleasant strong flavor, and is always removed. Although some varieties can weigh as much as 700 lb, grouper is typically available weighing between 5 lb and 15 lb. The most common varieties of grouper are black, red, and yellow grouper. Grouper is best prepared using dry-heat cooking methods such as grilling or sautéing and is commonly prepared blackened.

## Red Snapper

A *red snapper* is a lean fish with a juicy, fine-flavored pink flesh that becomes pearly white when cooked and flakes easily. Red snapper is named for its deep red fins and red skin, which fades slightly on the belly and around the throat. The red snapper averages between 4 lb and 7 lb, but fish weighing up to 25 lb are not uncommon. The hard, tough bones of the red snapper make it slightly difficult to fillet. Red snapper is best prepared by baking, broiling, steaming, or poaching. The head and bones of the red snapper are prized for their wonderful flavor when used in stocks and soups.

## Mahi-mahi

A *mahi-mahi* is a high-quality lean fish that has colorful skin and firm, pink flesh with a sweet flavor. Mahi-mahi is also marketed as dorado or dolphin fish but is not related to the dolphin, which is a mammal. Mahi-mahi weigh between 12 lb and 15 lb on average, but can grow as large as 50 lb. Mahi-mahi becomes unpleasantly dry when overcooked, so it is often served with a flavorful sauce. The preferred cooking methods for mahi-mahi are broiling, grilling, baking, or poaching.

## Orange Roughy

An *orange roughy* is a saltwater fish with orange to light gray skin that has a rough appearance. ***See Figure 20-13.*** The firm, lean flesh is pearly white, possesses a sweet, delicate flavor, and flakes nicely. In North America, most orange roughy is imported from New Zealand. It is available year-round and is most often sold as skinless, boneless fillets. Orange roughy is most commonly baked, or stuffed and baked. It can also be broiled, steamed, sautéed, and fried with excellent results.

## Mackerel

A *mackerel* is a long, streamlined fish that tapers toward the rear with firm, grayish pink flesh that is rich in flavor. The skin is a dark bluish brown on black with golden spots. Boston and Spanish mackerel are fatty fish that measure between 14″ and 18″ long and weigh between ½ lb and 5 lb on average. King mackerel is a larger variety commonly sold cut into steaks. For best results, mackerel should be cooked by broiling, baking, or grilling. It is also wonderful smoked.

## Swordfish

A *swordfish* is a very large, fatty-fleshed fish that has a long, swordlike bill extending from its head. The flesh is quite dense, not flaky, and has a sweet, mild flavor. It is slightly pinkish and turns white when cooked. Swordfish is most commonly sold as steaks and prepared by grilling or broiling. ***See Figure 20-14.***

## *Swordfish Steaks*

20-14  *Swordfish is commonly sold as steaks.*

## Orange Roughy

20-13  Orange roughy is commonly served as fillets.

### Historical Note

In the early 1990s, orange roughy consumption reached an all-time high. Restaurants across the United States had a high demand, and orange roughy supplies dwindled as the adult fish were over-harvested without enough time for spawning and repopulation of the species. A ban was temporarily placed on the sale and distribution of orange roughy until supplies increased and the species was no longer in danger of depletion.

## Sturgeon

A *sturgeon* is a saltwater fish with a wedge-shaped snout and a body style similar to that of a shark. However, sturgeons have no teeth and are bottom feeders. Although prized for its flesh as well, sturgeon is primarily caught to obtain its roe, better known as caviar. Russian or Ossetra sturgeon is a prized variety found in rivers throughout Russia and provides one of the highest-quality caviars available, as well as a firm, flavorful fillet. Caviar from this variety of sturgeon bears the name Ossetra.

Starry sturgeon, also referred to as Sevruga sturgeon, is found primarily in the Black and Caspian Seas. Its caviar and flesh are of high quality and demand a high price. The Beluga sturgeon is the most prized sturgeon of all. Its caviar and firm, oily flesh are the highest priced from any of the sturgeon varieties. Some beluga sturgeons can weigh up to 2000 lb, although most do not weigh over 1000 lb. Lake sturgeon can be found throughout North America and are commonly prepared smoked or brined.

## Monkfish

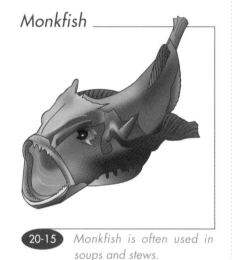

*Monkfish*

A *monkfish* is a lean saltwater fish with a sweet, firm-textured flesh that when cooked has qualities similar to those of lobster meat. Monkfish is also known as goosefish, bull mouth, devilfish, rape, *lotte* (in French), "poor man's lobster," anglerfish, and belly fish. **See Figure 20-15.** It is a large fish, weighing up to 50 lb, but only the tail section is edible. It is primarily sold as whole tails and fillets. Monkfish can be cooked using any cooking method and is often used in soups and stews. For best results, monkfish should be served with an appropriate sauce, as the lean flesh dries out if overcooked.

**20-15** *Monkfish is often used in soups and stews.*

## Shark

A *shark* is a boneless fish, with a skeleton composed of cartilage rather than bone. **See Figure 20-16.** There are approximately 250 species of shark, but only a few provide quality meat. Mako and blue sharks from the Atlantic coast supply most of the quality shark meat to the U.S. market. The flesh of mako and blue sharks is pinkish white when raw, turning white when cooked, and is relatively firm textured. The blacktip shark is the premier species found in the Gulf of Mexico. Blacktip shark has a snowy white meat but is somewhat drier than mako. Thresher sharks are found off the West Coast and have a coarser-textured meat. Sharks used for human consumption usually range in weight between 30 lb and 200 lb. Other shark varieties can weigh well over 1000 lb. Shark is most commonly grilled or broiled but can also be baked, poached, or fried.

**Chef's Tip:**
Shark meat has a slight ammonia smell. Chefs often soak shark meat in milk or cream before cooking it because lactic acid neutralizes the enzyme that creates this odor. If the meat has a strong ammonia odor it should not be used, as this indicates the fish was not handled properly after it was caught.

## Shark

**20-16** *Sharks are boneless fish that are commonly grilled or broiled.*

### Skate

A *skate (ray)* is a lean, boneless fish with two winglike sides. ***See Figure 20-17.*** The edible meat of a skate is divided by a layer of cartilage that runs through the center of the wing. Each wing has two separate fillets, separated by the layer of cartilage. Skate are colored similarly to flatfish in that they are grayish brown on top and white on bottom. Typically only the wings are removed and marketed. The majority of skate sold for human consumption weigh between 2 lb and 4 lb; however, skate of some varieties can weigh up to a ton. Skate wing is commonly sautéed or grilled.

### Eel

An *eel* is a long, slender fish with a body that resembles that of a snake. The skin of an eel is similar to that of a catfish, in that it is smooth, tough, tight to the flesh, and does not have scales. Eel is a high-fat fish with a mild, slightly sweet flavor, and can be prepared using almost any cooking method.

### SHELLFISH

*Shellfish* is a classification for aquatic invertebrates that may have a hard external skeleton or shell. Shellfish are not fish. They are invertebrates, meaning that they do not have an internal skeleton. Shellfish are commonly categorized as either crustaceans or mollusks. A *crustacean* is a shellfish that has a segmented, hard external shell. ***See Figure 20-18.*** The external shell functions as a skeleton and is called an exoskeleton. Lobsters, shrimp, crabs, and crayfish are examples of crustaceans. A *mollusk* is a shellfish with a soft unsegmented body. Some mollusks, such as clams or oysters, have a hard external shell to protect the soft inner body. Other mollusks have a thin internal shell called a cuttlebone. Mollusks can be divided into three classifications: univalves, bivalves, and cephalopods. ***See Figure 20-19.***

## Skate

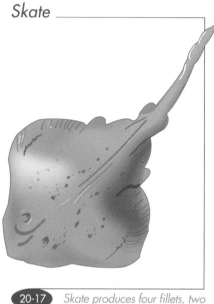

**20-17** *Skate produces four fillets, two from each wing.*

## Crustaceans

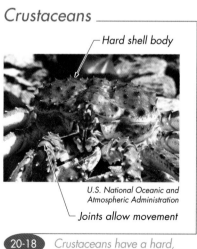

Hard shell body

*U.S. National Oceanic and Atmospheric Administration*

Joints allow movement

**20-18** *Crustaceans have a hard, external shell.*

## Mollusks

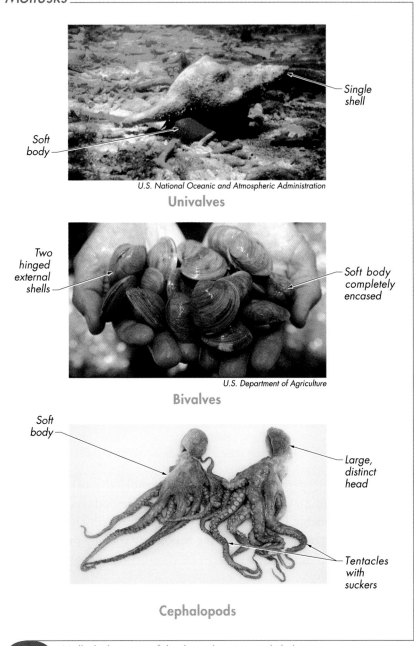

Univalves

*Single shell*

*Soft body*

*U.S. National Oceanic and Atmospheric Administration*

Bivalves

*Two hinged external shells*

*Soft body completely encased*

*U.S. Department of Agriculture*

Cephalopods

*Soft body*

*Large, distinct head*

*Tentacles with suckers*

20-19 *Mollusks have a soft body and no internal skeleton.*

## CRUSTACEANS

Crustaceans have a hard exterior shell to provide shape and form to a soft body that has no internal bone structure. They are found in both fresh and saltwater and can breathe underwater. Unlike fish, crustaceans can live out of water for a few days but need to be kept moist by being covered with a wet towel, wet newspaper, or seaweed. The most common forms of crustaceans are shrimp, prawns, lobster, crab, and crayfish.

## Shrimp and Prawns

Shrimp and prawns are crustaceans with a tender white meat and a distinctive flavor. Both shrimp and prawns are found in saltwater and freshwater, but prawns are more commonly identified with freshwater sources, while shrimp are commonly considered saltwater crustaceans. In culinary terms, the words shrimp and prawns are often used interchangeably. In North America and Asia, larger shrimp are commonly marketed as prawns. In the United Kingdom and Australia, the term prawns is used almost exclusively for both. ***See Figure 20-20.***

The four most common types of shrimp used in the professional kitchen are white, brown, pink, and black tiger. White (common) shrimp is a greenish gray color. Brown (Brazilian) shrimp is a brownish-red color. Pink (coral) shrimp is a medium or deep pink color. Black tiger shrimp are gray-and-black striped and are some of the largest shrimp available on the market. Although these four types of shrimp vary in color when caught, they differ very little in appearance when cooked.

All shrimp have a similar flavor and nutritional value. Poaching is the most common method of cooking shrimp. Grilling, breading and frying, sautéing, and broiling are other popular methods for preparing shrimp. Cooked shrimp can be chilled and served cold, as in shrimp cocktail. Shrimp is frequently used in appetizers, entrées, and salads. Only the tail section of shrimp is edible.

### Shrimp and Prawns

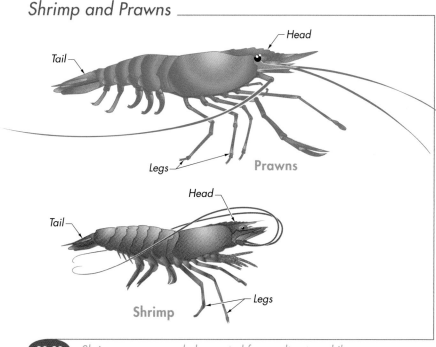

**20-20** *Shrimp are commonly harvested from saltwater while prawns are commonly harvested from freshwater, but the terms are often used interchangeably.*

## Lobsters

Lobsters are one of the most prized varieties of edible crustaceans. The tail of a lobster contains a sweet-tasting white meat. The external shell ranges in color from brown to bluish-black, depending on the variety of lobster, but all lobster shells turn red when cooked. The whole lobster is edible except for a small section of membranes (the stomach) located around the eye area inside the shell. Lobster is most often prepared by broiling, poaching, or steaming. *See Figure 20-21.*

### Lobster

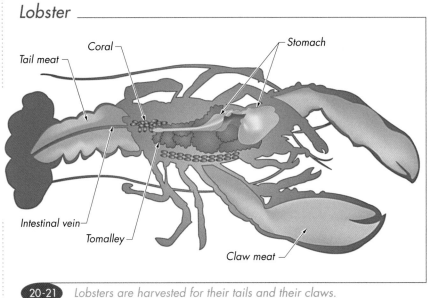

**20-21** *Lobsters are harvested for their tails and their claws.*

Lobsters are sold alive and should be kept alive up to the time of cooking. A *sleeper* is a lobster that is dying. Sleepers are sold at a reduced price and must be cooked immediately to retain the firm flesh. Common varieties of lobster include the Maine lobster, spiny lobster, slipper lobster, and langoustine (Norway lobster).

**Maine Lobsters.** A *Maine lobster* is a cold-water lobster that has a dark bluish green shell, two large heavy claws, medium-size antennae, and four slender legs on each side of the body. Maine lobsters are harvested from cold North Atlantic waters along the coast of the northeastern United States. They are most plentiful during the summer months when they migrate closer to shore.

**Spiny Lobsters.** A *spiny (rock) lobster* is a lobster with many prominent spines on the body and legs, long slender antennae, very small claws, and five very slender legs on each side of the body. Only the tail section of a spiny lobster is edible, and the flesh is coarser in texture and not as delicate in flavor as that of a Maine lobster. The tail section weighs between 4 oz and 1 lb and is shipped to market frozen. Spiny lobsters from waters around Florida, Brazil, and Cuba have slightly smooth shells with large yellowish spots on the brownish green tail section and are sold as warm-water tails. Spiny lobsters from South Africa, Australia, and New Zealand have rough shells with no spots on the brownish maroon tail section and are sold as cold-water tails. Of these two varieties, cold-water tails are superior in taste and texture to warm-water tails.

**Slipper Lobsters.** A *slipper lobster* is a lobster with large disk-shaped antennae that protrude from the front of the head. Slipper lobsters do not have claws. The meat of a slipper lobster is not nearly as tender and moist as that of a Maine lobster or a spiny lobster and has a flavor that is stronger and slightly fishier.

**Langoustines.** A *langoustine* is a lobster that is similar in appearance to a shrimp, except it has very long front arms with long, thin claws. Only the tail section of a langoustine is eaten. Langoustines are found in the northeastern Atlantic Ocean and the North and Mediterranean Seas. Although they are less expensive than other lobsters, langoustines are often considered a fancier meal.

## Crabs

Crabs are a popular shellfish with tender, sweet-tasting meat that can be used in many menu items. Edible crabs can be distinguished from inedible varieties by counting the pairs of legs. Edible crabs have five pairs of legs; four for walking and one that serves as arms. Inedible crabs have four pairs of legs; three for walking and one used as arms.

Crabs are available fresh (live), frozen, and in canned products. When using fresh crabs, only live crabs should be used. Fresh crabs that have died should be rejected or discarded. Hard-shell crab is commonly prepared by simmering in a flavorful poaching liquid until the shell turns bright red and the meat is cooked through. Soft-shell crab is best either fried or sautéed. The five common varieties of crab taken from the North American coastal waters are blue crab, Dungeness crab, king crab, snow (spider) crab, and stone crab. *See Figure 20-22.*

**Blue Crabs.** Blue crabs are found along the Atlantic coast. A typical blue crab measures approximately 5″ across its blue-grey shell and weighs approximately 5 oz. Hard-shell blue crab is marketed live, precooked and then frozen, or as cooked and picked canned crab meat.

Blue crab is typically the variety of crab used for soft-shell crab in North America. A soft-shell crab is a crab harvested within 6 hr of molting. *Molting* is the process of a crab shedding its hard shell to grow a larger shell. Molting season is from mid-May until the beginning of September each year. Soft-shell crab is available fresh (live) only during the molting period. Soft-shell crabs are handled and packed with special care to ensure that they arrive at their destination alive. They must be kept alive up to the time of cooking, unless they are cleaned and quick-frozen. The frozen variety of soft-shell crab is available year-round.

**Historical Note**

Blue crab is a popular regional food in the Chesapeake Bay area of Maryland and Virginia. Because of high demand, the blue crab population of Chesapeake Bay has been threatened by overfishing.

## Common Crab Varieties

U.S. National Oceanic and Atmospheric Administration
**King**

U.S. National Oceanic and Atmospheric Administration
**Snow**

U.S. Fish and Wildlife Service
**Dungeness**

**20-22** *Common varieties of crab include king crab, snow crab, and Dungeness crab.*

**Dungeness Crabs.** A *Dungeness crab* is a Pacific crab that has desirable, sweet-tasting meat. Approximately 25% of the body weight of a large Dungeness crab is meat, which is the highest percentage found in any variety of crab. Most of the edible meat in a Dungeness crab is body meat (rather than leg meat). Dungeness crabs are larger than blue crabs, weighing approximately 1¾ lb to 4 lb each. Dungeness crab is most commonly available as either frozen or canned cooked meat, but is also available live.

**King Crabs.** *King crabs* are the largest-size variety of crabs, typically weighing between 6 lb and 20 lb and measuring as much as 6′ from the tip of one leg to the tip of the opposite leg. The meat has a pinkish tinge when raw but becomes snowy white when cooked. King crab is available in the shell as cooked frozen clusters (legs), legs and claws, leg and body meat, or shredded meat. Canned or frozen king crab meat is also available shucked or pulled (removed from the shell). King crab is not typically available live or fresh, as the legs are removed at sea and flash-frozen to preserve quality.

**Snow Crabs.** A *snow (spider) crab* is similar to the king crab, but is smaller, lower in cost, and available in greater supply. It is sold as cooked, frozen leg clusters. A snow crab has only a fraction of the meat of a king crab. Frozen leg clusters are commonly steamed or poached just enough to warm the meat since they are already fully cooked. Snow crab is a versatile product that can be used as a main ingredient, steamed and served whole, or served cold in a crab cocktail similarly to shrimp cocktail.

**Stone Crabs.** A *stone crab* is an Atlantic crab with a brownish red shell and large claws of unequal size. Unlike other varieties of crab, when a stone crab is caught, one claw is removed, and the crab is placed back into the water. In approximately 12 to 16 months, the crab generates a new claw in place of the one that was removed. Each time a claw is removed and allowed to grow back, it comes back slightly larger than before. The claws are the only part of the stone crab that is harvested. Claws range from 2 oz to 5 oz, depending on the length of time that the claw has been growing. Stone crab claws are typically served cold with most of the shell removed, and garnished similar to a shrimp cocktail. However, they can also be served hot and have a similar taste and texture to lobster.

## Crayfish

A *crayfish (crawfish, crawdad)* is a freshwater crustacean that resembles a small lobster. Crayfish can range from 3″ to 7″ long, but are typically available between 3″ and 4″. About 98% of the crayfish harvest comes from aquafarming in Louisiana and the Pacific Northwest. They are most commonly used in Creole, Cajun, and French cuisine. Crayfish has a flavor similar to that of shrimp, with a slightly tougher texture. Only the tail portion contains enough meat to eat. *See Figure 20-23.*

## MOLLUSKS

Like crustaceans, mollusks do not have an internal skeleton. A mollusk's soft body is often protected by an external shell or supported by an internal shell. Mollusks should always be alive prior to cooking unless frozen or canned. If the mollusk is dead prior to cooking, it should be discarded and should not be eaten under any circumstance, as illness could result. Edible mollusks are divided into three classifications: univalves, bivalves, and cephalopods.

## Univalves

A *univalve* is a mollusk that has a single solid shell and a single foot. A univalve uses its foot to move along the surface of underwater structures. Univalves include abalone and conch. *See Figure 20-24.*

**Abalone.** An *abalone* is a univalve contained in a bowl-shaped shell with a brown exterior and an iridescent multicolored interior. California abalone is considered highest in quality but all varieties have a sweet, slightly salty flesh, similar to that of a clam. Aquafarmed abalone are raised in saltwater pens for 3 to 4 years and reach an average size of 3″ to 4″ when mature. Abalone is also imported from the coastal waters of Japan, New Zealand, and Mexico and sold in canned and frozen form. The texture of abalone is tender like lobster but has a creamier taste. It is very tough when overcooked, so care should be taken to cook it delicately.

Crayfish _____

**20-23** *Crayfish are small freshwater crustaceans that resemble lobsters.*

U.S. National Oceanic and
Atmospheric Administration
**Pink Abalone**

Soft unsegmented body from
inside of shell is used in cuisine

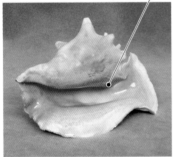

U.S. Fish and Wildlife Service
**Conch**

**20-24**  Univalves live inside a one-piece shell.

**Littleneck Clams** _____

U.S. Department of Agriculture

**20-25**  Littlenecks are small hard-shell clams.

**Conch.** A *conch* is a univalve that has a pinkish-orange shell and resembles a large snail. Conchs are found in the warm waters off the Florida Keys and throughout the Caribbean. The meat of the conch is rubbery and benefits from being sliced thin and tenderized. Conch has a sweet flavor similar to that of a clam. Very fresh conch is commonly eaten raw with lime juice and hot sauce.

### Bivalves

A *bivalve* is a mollusk that has a top shell and a bottom shell connected by a central hinge, which the animal can close for protection. Live bivalves should close tightly when gently tapped on the shell. Bivalves that do not close when tapped are dead and should be discarded. A bivalve that is noticeably heavier than others it is shipped with is also probably dead, its shell filled with mud, and should be discarded. Common types of bivalves include clams, cockles, mussels, oysters, and scallops.

**Clams.** A *clam* is a bivalve found in both freshwater and saltwater. Clams can be fried or steamed with excellent results, and are the key ingredient in clam chowder. Different varieties are found in the Atlantic than in the Pacific. The Atlantic coast produces many species, including soft-shell, hard-shell, and surf clams.

A *soft-shell (long-neck, steamer) clam* is a clam with a thin, brittle shell that breaks easily. Soft-shell clams have a protruding siphon that prevents the shell from closing completely, causing them to dry out more quickly and to not live as long as hard-shell clams once removed from the water. They also have a tendency to be gritty due to excess sand settling inside the shell. It is important to purge soft-shell clams of their sand by soaking them in a solution of salted water (⅓ c of salt per gallon of clean water) and cornmeal (1 c per gallon). They have tender, sweet-tasting meat and are commonly steamed or fried. Soft-shell clams should be steamed until they open.

A *hard-shell clam (quahog)* is an Atlantic clam with a blue-grey shell that contains a chewy flesh. These clams are rarely sold by the name hard-shell clam or quahog, but rather by names given to differentiate between the different sizes. The four common sizes of Atlantic hard-shell clams are littlenecks, cherrystones, topnecks, and chowders. ***See Figure 20-25.***

A *littleneck* is a hard-shell clam 1½″ to 2″ in size. Littlenecks are the smallest, tenderest, and most expensive variety of hard-shell clam. These bite-size clams are commonly served raw on the half shell (similar to oysters) or steamed. A *cherrystone* is a hard-shell clam 2″ to 3″ in size. Cherrystones can be served on the half shell but are most commonly steamed. A *topneck* is a hard-shell clam approximately 3″ in size. Topneck is the most popular size of clam used for stuffing and baking. A *chowder* is a hard-shell clam of the largest size classification. Chowder clams are typically cut into strips or minced for soups and chowders.

A *surf clam* is a species of Atlantic clam known for its large size and desirable flavor. Surf clams are 8″ at full size and have a sweet, mild flavor, but are fairly tough at full size. Half of the shucked clam is the tongue, which is often cut into clam strips to be breaded and fried. The other half of the clam is the adductor muscle. The *adductor muscle* is a muscle that opens and closes the shell of a bivalve. The adductor muscle is most commonly chopped or ground and used in chowders.

Clams found on the Pacific coast are tougher in texture than the Atlantic varieties. Types of Pacific clams include the butter clam, geoduck, manila clam, and razor clam.

A *butter clam* is a Pacific hard-shell clam with a mild, desirable flavor. About 75% of all butter clams are harvested by hand in the wild, while about 25% are aquafarmed in shallow tide pools. A *geoduck* is a Pacific clam with a large siphon that protrudes from inside the shell and provides sweet, rich-tasting meat that is relatively tender. Geoducks are the largest Pacific clams and typically weigh between 2 lb and 2½ lb. The meaty siphon is the edible portion of the clam. The siphon is cut away from the body of the clam and split in half lengthwise. Then it is cut into very thin slices and served sashimi- or Carpaccio-style, or it can be steamed, poached, or sautéed quickly. Geoduck becomes very tough if overcooked.

The *manila clam* is the most common variety of Pacific clam. ***See Figure 20-26.*** The shell of a manila clam is similar in appearance to that of a hard-shell Atlantic clam, with slight ridges from the lip to the hinge. The manila clam can be steamed or served raw and has a sweet and salty flavor. A *razor clam* is a Pacific clam named for the sharp edge of its narrow, oval-shaped shell. Razor clams are expert diggers that anchor themselves deep in the sand or gravel for protection. They have a sweet, desirable flavor.

**Cockles.** A *cockle* is a bivalve with a 1″-wide shell with deep, straight ridges that surround a small, fleshy meat. Cockles are often served freshly shucked with just a squeeze of fresh lemon juice. They are also often used in paella. Cockles are less common in North America than in Europe.

**Mussels.** A *mussel* is a freshwater or saltwater bivalve with whiskerlike threads (byssal threads) that extend outside of the shell that allow the animal to attach to items underwater for protection. The mussel's threads are referred to as a beard. In the wild, mussels are commonly found attached by their beards to rocks. Aquafarmed mussels attach their beards to ropes and hang until they are large enough to be harvested. Common types of mussels include blue mussels and greenlip mussels.

A blue mussel is the most common variety of edible mussel. ***See Figure 20-27.*** Aquafarmed blue mussels have a thinner shell with a blue-black color, while wild-caught blue mussels have a thicker shell with a silver-blue color. Blue mussels have tender, sweet meat with a bright orange color. They vary in size, but are commonly available between 10 and 20 mussels per pound. Blue mussels can be steamed and served either hot or cold, or shucked and sautéed or simmered in a sauce.

## Manila Clams

Discard any open shells

Shells should be tightly closed

The Plitt Company

**20-26** *Manila clams are found in the Pacific ocean.*

**Historical Note**

A mussel's beard must be removed as close to preparation time as possible, because removing it too early can cause the mussel to die.

## Blue Mussels

New Zealand Greenshell™

**20-27** Blue mussels are raised on ropes in aquafarms.

## Atlantic Oysters

Florida Department of Agriculture and Consumer Services, Bureau of Seafood and Aquaculture Marketing

**20-28** Atlantic oysters are plump and tender.

A greenlip mussel is a mussel with a distinctive green-edged shell and is larger than a blue mussel, with 8 to 12 mussels per pound on average. They have a sweet, plump, and tender flesh. Greenlip mussels are best prepared steamed with white wine, lemon, and herbs, or they can be fully cooked and served cold with cocktail sauce and lemon.

**Oysters.** An *oyster* is a saltwater bivalve with a very rough shell that is coated with calcium deposits. Although oysters are not found in freshwater, they can be found in brackish water (a mixture of freshwater and saltwater). Most oysters are cultivated in beds that require much care and attention if they are to continue to produce.

Many different varieties of oysters are available. The most common varieties include Atlantic oysters, Pacific oysters, Olympia oysters, and European oysters. *See Figure 20-28.* The two most significant varieties are the Atlantic oyster and the Pacific oyster. Dozens of additional varieties are derived from these two varieties. The variety of oyster and where it comes from can make significant differences in the flavor profile of an oyster. The different amounts of minerals, ocean salt, and nutrients present in the environment determine the taste of a particular oyster.

An *Atlantic (Eastern, American) oyster* is an oyster with a distinctive salty flavor and plump, tender meat contained in a flatter shell than other oyster varieties. Atlantic oysters account for roughly 90% of all oyster production and are between 3″ and 4″ on average. Common varieties of Atlantic oysters include blue point, Chesapeake Bay, and Long Island oysters.

A *Pacific (Japanese) oyster* is a large oyster with a thick, curvy shell that yields a briny, sweet, and mild-tasting meat. The flesh is plump and moist, with a silver, gold, or white color. Pacific oysters are aquafarmed in large quantities and rarely harvested from the wild. Common varieties of Pacific oyster include Hamma-Hamma, Penn Cove, and Kumamoto oysters.

An Olympia oyster is a very small oyster, about 1½″ wide. Olympia oysters contain a small, plump flesh that is mild and delicate. A European oyster is an oyster with a relatively flat, cup-shaped shell and meat with a creamy texture and a salty-sweet taste. European oysters are most commonly served on the half shell, as they are prized as having a finer flavor than oysters of any other variety.

Oysters are available year-round, but most are at peak quality from September to April. Oysters of good quality are plump, well shaped, and surrounded by a clear, gelatinous fluid. Oysters must have a tightly closed shell to be of good quality. If the shell is open, or does not close fully when lightly tapped, the oyster is dead and is not edible. Shells that are uncommonly heavy probably contain mud, and should be discarded.

Oysters may be eaten raw or poached, breaded and fried, baked, or roasted. One of the most common preparations is to top an oyster with a savory filling, such as a creamed spinach, and either bake or broil it. *See Figure 20-29.* The secret to proper oyster preparation is to apply just enough heat to heat them through, leaving them plump and tender and avoiding overcooking.

## Oysters Rockefeller

Florida Department of Agriculture and Consumer Services, Bureau of Seafood and Aquaculture Marketing

**20-29** *Oysters Rockefeller was named after John D. Rockefeller, as the richness of the dish was likened to him being the richest man of the day.*

**Scallops.** A *scallop* is a bivalve with a highly regular fan-shaped shell and a well-developed adductor muscle. The cream-colored adductor muscle is lean and juicy and possesses a sweet, delicate flavor, and is the portion of the scallop most commonly eaten. The rest of the scallop body is made up of either white or red roe, referred to as coral. The coral is considered a delicacy and is sometimes served with scallops in upscale restaurants, but is usually removed prior to sale of scallops in the shell. Many chefs use the fan-shaped scallop shells as serving dishes when featuring seafood appetizers and entrées. There are two primary varieties of scallops, bay scallops and sea scallops. *See Figure 20-30.*

A bay scallop is a fairly small, cold-water scallop taken from shallow saltwater. Bay scallops average 70 to 100 scallops per pound. They are rarely served in a shell, but are typically completely cleaned prior to sale, leaving only the adductor muscle. They are usually packed wet (soaked in a preservative that whitens the scallop and helps prevent spoiling), dry (untreated, without any preservatives), or individually quick-frozen. *Individually quick-frozen (IQF)* is a designation for products preserved using a method whereby each item is glazed with a thin layer of water and frozen individually. Once completely frozen, IQF portions can be packaged together without sticking to other portions. IQF packaging is convenient because it allows as much of an item to be used as is needed without having to thaw an entire box or bag. Bay scallops are considered to have better flavor than sea scallops and usually are higher in price. They are commonly prepared by sautéing or frying in butter, or are served seviche-style, where they are marinated in lemon juice and seasonings until cured.

## Scallops

Bay scallops

Sea scallops

**20-30** *The two primary varieties of scallops are bay scallops and sea scallops.*

A *sea scallop* is a scallop found in deep saltwater and is large with a coarse texture. Sea scallops are typically two to five times larger than bay scallops, averaging 20 to 30 scallops per pound. Sea scallops have a sweet, somewhat briny taste.

Color is the best way to judge the quality of scallops. The best-quality scallops have a creamy, almost translucent color. If scallops are white, they have been packed in ice water, and the flavor and texture have been impaired. If they are a brownish color, they are too old and should not be used. Scallops are best when sautéed, poached, or broiled.

## Cephalopods

A *cephalopod* is a mollusk with arms that extend from the base of a distinct head. Cephalopods do not have an outer shell; however, squid and cuttlefish have an internal flat shell referred to as a cuttlebone. ***See Figure 20-31.*** Cephalopod is derived from a Greek word meaning literally head-foot. Cephalopods have a distinct head, well-developed eyes, and a mouth that is surrounded by a sharp, birdlike beak used for cracking the shells of shrimp, crab, lobster, and other prey. The most common varieties of cephalopods are squid, octopus, and cuttlefish.

## Cephalopods

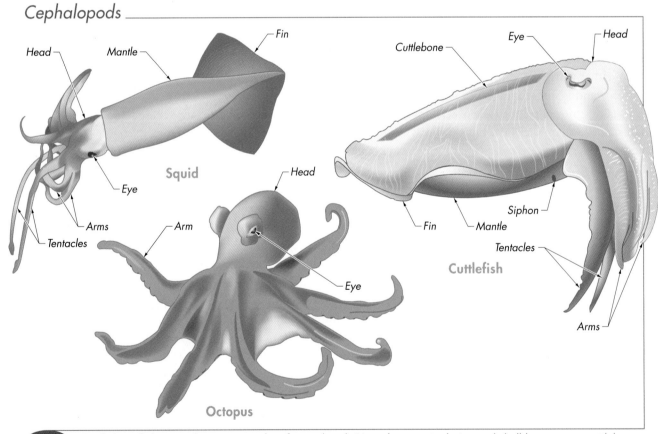

**20-31** Cephalopods have eight arms extending from a head. Many have a single internal shell known as a cuttlebone.

A *squid* is a cephalopod having an internal cuttlebone, a distinct head, a beak, well-developed eyes, fins, eight arms, and two tentacles. Squid is the most popular cephalopod in Western cuisine. Most whole squid at market range between 8 and 10 squid per pound. Larger squid varieties are commonly cut up into steaks and sold frozen. The arms, tentacles, mantle (body tube), and fins are edible, but as with the other cephalopods, the flesh can become tough if overcooked. Squid is often called by its Italian name, *calamari*. Squid is commonly sautéed, breaded and fried, or used in rich stews and pasta dishes.

An *octopus* is a cephalopod with a defined head, well-developed eyes, eight arms, and a beak that is used to crack open the shells of its prey. An octopus does not have an internal or external shell. Octopus is usually sold whole by the pound and is available fresh or frozen. When cooked, the grayish skin turns deep purple or reddish. The flesh beneath the skin is white, firm, and sweet. Octopus becomes tough if overcooked so care should be taken to cook it gently. Cold cooked octopus is commonly used to make a *pulpo* (Spanish for "octopus") salad.

A *cuttlefish* is a cephalopod with well-developed eyes, a cuttlebone, and eight arms, and two tentacles. Cuttlefish are the least common cephalopods used in the professional kitchen, but are often used in Asian dishes. Cuttlefish is also commonly used to make cuttlefish risotto, which is often black in color due to ink from the cuttlefish. Other than the ink, the arms and tentacles are the only parts of the cuttlefish that are eaten. They are typically cut into rings and fried, sautéed, or cooked into stews, soups, or rice dishes.

## Historical Note

The cuttlebone from a cuttlefish is often used as a calcium-rich treat hung in bird cages.

## OTHER SEAFOOD

Escargot and frog legs are two examples of food items that appear on menus under "seafood," even though they are not fish or shellfish. These items are prepared like fish and shellfish items.

### Escargot

*Escargot* is the French term for "snail" and refers to any variety of land snail fit for human consumption. A snail is a type of mollusk that has a coiled or spiral-shaped shell that grows as the snail grows. Although there are many different varieties of snails, only a few are commonly eaten. The *petit-gris* snail is the most popular variety, originating in France. Because snails consume decayed matter on the ground, it is important to purge the stomach of the snail prior to preparing, as the stomach contents may be toxic to humans. This is done by feeding the snails cornmeal and allowing them to purge. After purging, snails are commonly removed from the shell, eviscerated, and sautéed quickly with garlic, white wine, and butter. They may also be stuffed back into the shell, topped with garlic butter and wine, and baked.

## Frog Legs

*Frog legs are classified as seafood.*

### Frog Legs

Although frogs are not fish or shellfish, frog legs are often listed on menus as seafood. Only the hind legs of the frog are commonly eaten. The best frog legs come from bullfrogs raised on frog farms. Bullfrogs produce hind legs with a large amount of white meat. Frog legs are on the market all year, but are most plentiful from April to October. A large percentage of the frog legs used in the professional kitchen come from India or Japan. Frog legs are sold by the pair, with the most desirable legs averaging 2 or 3 pairs per pound. Frog legs are typically sautéed with garlic, parsley, wine, and butter or can be breaded and deep-fried. *See Figure 20-32.*

### MARKET FORMS

Fish and shellfish may be purchased in various forms. Each form has certain advantages and disadvantages in terms of cost, convenience, and labor. The more preparation that is done before delivery to the establishment, the higher the cost per pound. The most suitable market form to order depends on the type of seafood required, preparation, storage facilities, employee skill, and equipment.

### Market Forms of Fish

The quantity of fish required is determined by the number of people being served, the portion size, and the market form. Fresh fish are commonly available in several market forms including whole, drawn, dressed, steaks, and fillets. *See Figure 20-33.* Frozen, canned, and specialty fish items are also served in food service establishments.

A *whole (round) fish* is a fish marketed the way it was taken from the water. Nothing has been done to process it. Whole fish have the shortest shelf life of any market form because all of the internal organs (viscera) are still present. Fish purchased whole cost less per pound than other forms but require more preparation and yield more waste. When purchasing whole fish, allow 1 lb per serving.

A *drawn fish* is a fish that has had only the viscera removed, with the scales still on. Drawn fish can be prepared whole. Most fish are sold in this form as it results in the longest shelf life. When purchasing drawn fish, allow ¾ lb per serving.

A *dressed fish* is a fish that has been scaled and has had the viscera, gills, and fins removed. Dressed fish are a popular market form in the professional kitchen. When purchasing dressed fish, allow ½ lb per serving. Pan-dressed fish are scaled, with the viscera and gills removed and the fins and tail trimmed. The head may be removed. Pan-dressed fish are prepared by pan-frying. Smaller fish are often pan-dressed with the head attached.

## Market Forms of Fish

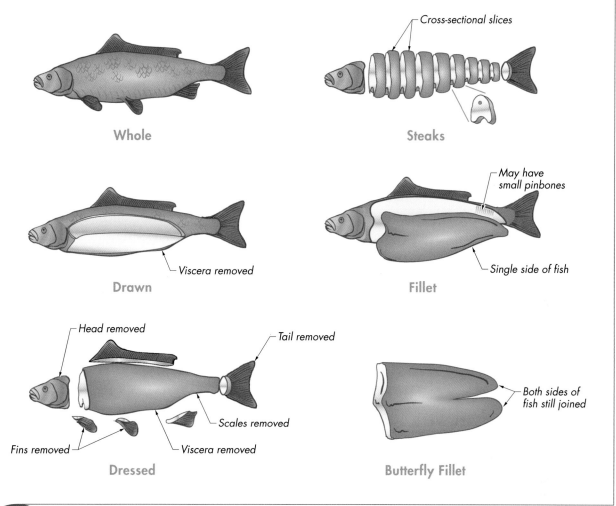

Whole

Steaks
- Cross-sectional slices

Drawn
- Viscera removed

Fillet
- May have small pinbones
- Single side of fish

Dressed
- Head removed
- Fins removed
- Tail removed
- Scales removed
- Viscera removed

Butterfly Fillet
- Both sides of fish still joined

**20-33** *The market form of fish best suited for a food service establishment depends on several factors including storage facilities, staff skill, and equipment.*

Fish are available in a variety of convenience forms. Fish may be purchased stuffed, breaded, or topped with assorted foods. However, convenience forms are always more expensive, and this must be considered when determining cost. Steaks and fillets are processed cuts of fish that require minimal preparation. When purchasing steaks or fillets, allow ⅓ lb per serving.

A *fish steak* is a cross-sectional slice of a larger-size dressed fish. Steaks are ready to cook when purchased. Generally, the only bone present in a steak is a small section of the backbone. Steaks from very large fish such as swordfish are boneless. Tuna, salmon, swordfish, and shark are among the types of fish that are commonly available as steaks.

A *fish fillet* is the fleshy side of a fish, cut lengthwise away from the backbone. There are two fillets to each round fish, one on each side. Flatfish have four fillets, two on each side. Fillets can be purchased with or without bones (boneless fillets being more expensive) and with or without skin.

Fillets with the skin left on are sold scaled. A *butterflied fillet* is two single fillets from a dressed fish that are held together by the uncut back or belly of the fish. The shape of a butterflied fillet resembles an open book.

Frozen fish is served in restaurants more often than fresh fish. Using frozen fish allows a food service establishment to serve a greater variety of fish year-round. If handled and prepared properly, frozen fish is very comparable to fresh. Frozen fish cannot be purchased whole. The portion allowance is the same for frozen fish as it is for fresh fish. A ⅓ lb to ½ lb portion of the edible part should be allowed per person.

Canned fish should be checked for signs of damage or bulging before they are used, and discarded if such signs are observed. Canned salmon, tuna, anchovies, and sardines are the most common varieties of canned fish.

Specialty fish items include smoked, cured, salted, and pickled fish. Popular smoked and cured fish are cod, haddock (finnan haddie), salmon, sturgeon, and herring. Popular salted fish are cod and mackerel. Popular pickled fish are salmon and herring. Specialty fish items are commonly used in hors d'oeuvres and canapés and on seafood buffets.

## Market Forms of Shellfish

Shellfish are available in several market forms including live and fresh in the shell, shucked fresh, frozen, or canned. Processed items have been fabricated into an easy-to-use form. Cooked, peeled, and deveined shrimp is an example of processed shellfish. Shucked shellfish are marketed with the shell removed. Shucked oysters, scallops, and clams are available both fresh and frozen. Some shucked shellfish, such as mussels and oysters, can be purchased canned.

Whole shrimp and prawns may be sold fresh and live, but the majority of the catch is processed by removing the head and thorax (body) and freezing the tail while still at sea to ensure maximum freshness. Typically, only very upscale restaurants purchase live shrimp or prawns as they are quite expensive, although other establishments may use fresh shrimp if the establishment is near a supply source. Shrimp and prawns are most commonly sold frozen in 5-lb blocks, IQF, or in smaller (consumer-size) packages. Shrimp may also be purchased cooked and canned.

Frozen shrimp can be purchased green (uncooked), peeled or unpeeled; cooked and peeled; or peeled, cleaned, and breaded. Frozen shrimp are sold by the pound. The cost of frozen shrimp is based on how much preparation has been done before purchase. *Peeled and deveined* is a term used to describe shrimp that has been peeled and has had the sand vein (intestinal tract) removed prior to freezing. These shrimp are ready to cook and serve. Cooked shrimp may be purchased peeled or in the shell. Cooked shrimp is sold by the pound. Canned shrimp is seldom used in the professional kitchen.

Crustaceans are commonly packed by count (number of items) per pound. For example, shrimp can be packed at 21 to 25 shrimp per pound.

## Nutrition Note

Colossal or jumbo shrimp are the most expensive shrimp but take less time to peel and clean. Smaller shrimp are less expensive but take longer to peel and clean because more pieces are needed to fill an order. A standard rule of thumb is that 1 lb of raw, shell-on shrimp yields ½ lb of cooked and shelled shrimp.

This package of shrimp may be labeled as 21/25 ct. This indicates that there are an average of 21 to 25 shrimp per pound in the package. *See Figure 20-34.*

Lobsters are most often purchased live. Prepared lobster tails can also be purchased frozen. Cooked lobster meat is sometimes available canned. Lobsters are available in four sizes:

- Chicken—between ¾ lb and 1 lb

- Quarter—between 1 lb and 1½ lb

- Select—between 1½ lb and 2½ lb

- Jumbo—over 2½ lb

Crabs are purchased in three forms: live, cooked meat (fresh or frozen), and canned. On the coasts, both hard-shell and soft-shell crabs are sold live. Away from the coasts, only soft-shell crabs are available live.

Cooked crabmeat may be purchased in the shell, fresh or frozen, or as fresh, cooked meat. The fresh, cooked meat is available as lump, flake, lump and flake, and claw meat, categorized as follows:

- Lump meat—large pieces of white meat from the body of the crab

- Flake meat—small pieces of white meat from the remaining parts of the body

- Lump and flake meat—combination of lump and flake meat

- Claw meat—meat from the claws, which has a brownish tint

Fresh, cooked crabmeat should be kept packed in ice and refrigerated until it is used. Crabmeat is purchased frozen if it is not going to be used immediately. Canned cooked crabmeat is available in various-size cans and is commonly used in commercial kitchens. The biggest advantage of canned cooked crabmeat is a long shelf life. It can be used interchangeably with cooked frozen crabmeat.

A synthetic crabmeat known as surimi is also available. *Surimi* is a fish product that looks, cooks, and tastes like crabmeat. Surimi is made from a mixture of pollock, snow crab, turbot, wheat starch, egg whites, vegetable protein, and other ingredients. It is low in calories, sodium, fat, and cholesterol and high in protein. Surimi is marketed precooked and frozen to protect its flavor and can be purchased as legs, chunk meat, or flake meat. *See Figure 20-35.*

Oysters and clams are purchased in four forms: live in the shell, shucked, frozen, or canned. Oysters and clams live in the shell are usually packed according to count (number of items) per a specific unit of volume, such as gallon, bushel, or pint. The higher the count, the smaller the size of each piece. For example, if one can of oysters is marked 350 per gallon and another can of equal size is marked 225 per gallon, the 225-count can contains fewer, larger oysters than the 350-count can. Shucked oysters and

| Sizing Shrimp | |
| --- | --- |
| Count | Size |
| 25 or fewer per lb | Jumbo |
| 25 to 30 per lb | Large |
| 30 to 42 per lb | Medium |
| 42 or more per lb | Small |

**20-34** *Shrimp is sold according to size or grade.*

## Surimi

Harbor Seafood

**20-35** *Surimi is a fish product that looks, cooks, and tastes like crabmeat.*

clams are usually packaged in gallon containers. Frozen oysters and clams are used if fresh ones are unavailable. Canned oysters and clams are rarely used in the professional kitchen.

Scallops are most often purchased shucked and are available either fresh or frozen. Fresh scallops can be purchased by the gallon or pound. Frozen scallops are usually purchased in 5-lb blocks.

## INSPECTION AND GRADING

Fresh fish and shellfish are not subject to a mandatory federal inspection as is required for meat and poultry. Instead, all seafood inspections are optional and, when desired, are paid for by the producer. These optional inspections are carried out by the National Marine Fisheries Service, which is part of the United States Department of Commerce (USDC), not the USDA. There are three types of optional inspections that a seafood producer can elect: Type 1, Type 2, and Type 3.

Type 1 inspection guarantees that the fish or shellfish product is safe and wholesome for human consumption, is accurately labeled, has a good odor, and was processed in a sanitary, inspected facility. After being processed under a Type 1 inspection, the product is marked with a "processed under federal inspection" (PUFI) stamp on a tag or slip affixed to the carton it is packed in. Type 1 inspection involves continual inspection of the fresh product, the processing plant, and the packaging methods, from the time the product arrives to the moment it is packaged for sale. ***See Figure 20-36.***

Type 2 inspection takes place in a warehouse or cold storage facility where product is randomly inspected to ensure it meets the product specifications listed on a specification sheet. ***See Figure 20-37.*** Type 3 inspection involves examination of the fishing boats and processing plants to ensure that they are adhering to strict sanitation guidelines when handling and processing fish and seafood. The purpose of a Type 3 inspection is solely to ensure proper sanitation.

*PUFI Stamps*

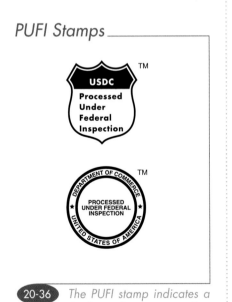

20-36 *The PUFI stamp indicates a fresh product processed at a regulated plant according to approved packaging methods.*

*Inspection Stamps*

20-37 *The USDC inspection stamp ensures that a product meets specifications.*

Seafood grading is also optional. Only fish and shellfish that were processed under a Type 1 inspection are eligible for grading. Because there are so many varieties of fish, the USDC sets grade standards for the most common varieties only. Products may be graded A, B, or C. Fish and shellfish graded Grade A are of the best quality, with no visible bruising, damage, or defects. Fresh and frozen seafood used in food service establishments is typically Grade A, while Grades B and C typically are used for canned or processed products only. *Figure 20-38.*

## PURCHASING FISH

Fresh fish spoils even more rapidly than poultry and meat, so it is essential to ensure that fish is received in the best condition and then stored and handled properly. Doing so helps ensure the best-quality fish and seafood are being presented to customers. When purchasing fresh fish, all parts of the fish should be considered to determine freshness. *See Figure 20-39.* Because most fish is not inspected, it is important to inspect fresh fish and shellfish deliveries for quality and freshness upon receiving a shipment. Even if the fish being delivered is from an inspected facility, it may not have been held at proper temperatures since it left the plant, and there is no way of knowing how long it has been in process or storage before sale.

*Grading Stamp*

20-38 A grading stamp indicates the quality grade of a product.

| Considerations for Purchasing Fish | | | |
|---|---|---|---|
| | Look | Touch | Smell |
| Fresh fish | • Round, clear, bulging eyes<br>• Bright red or reddish-pink gills<br>• Properly covered flesh<br>• Moist and solid fillets | • Wet, slightly slippery exterior surface<br>• Smooth scales laying flat against body<br>• Moist, intact fins<br>• Firm flesh | • Slight smell of seaweed or the ocean<br>• Slight smell of fish |
| Signs of age or deterioration | • Cloudy or sunken eyes<br>• Brown gills or missing gills<br>• Bruised, discolored, or damaged flesh<br>• Flesh of fillets separates when slightly bent | • Slimy internal cavity<br>• Slimy fillets<br>• Dried-out fins<br>• Mushy flesh<br>• Rough scales | • Strong fishy smell<br>• Ammonia odor |

20-39 Fresh fish should look, feel, and smell fresh.

The smell of a fish is considered one of the easiest ways to determine freshness, but it can also be misleading. Fresh fish have a light fishy smell. Strong fishy smells or ammonia odors are signs of deterioration, and fish that give off such smells should be discarded.

The external surface of a fish should be wet and a little slippery. Scales should be smooth and should lay flat against the body of the fish. Scales that are sticking out or feel rough are signs of age, and fish showing these signs should be rejected. Fins should be moist and intact. If the fins are dried out or removed, the fish may be older and the flesh may be starting to dry out.

The eyes and gills also give indications of age. The eyes of the fish should be round, slightly bulging, and clear. Cloudy or sunken eyes are signs of age. The gills of a fresh fish should be bright red or reddish pink. If the gills are brown or missing altogether, the fish is probably not fresh.

If fish has been cut into fillets, they should be moist and solid. The flesh should be firm and should spring back when touched. Gently pressing on the fillet should not leave an indentation. A fillet should not be slimy, and the flakes of flesh should not separate when the fillet is bent slightly. The flesh should be of the proper color for the particular variety of fish. There should not be any bruises or other signs of visible damage.

## PURCHASING SHELLFISH

Shellfish are typically purchased live, fresh, frozen, canned, or smoked. Shellfish such as lobsters, crabs, clams, oysters, and mussels should be alive if purchased in the shell. Shellfish that are purchased live should show movement when received and must be kept alive until they are cooked.

Lobsters must be kept alive until they are ready to be cooked. Lobsters are purchased through a local dealer or shipped express by rail or air in containers filled with seaweed. When received, the lobsters must be carefully inspected. Live lobsters have a tightly curled tail. Lobsters that are dying must be cooked immediately to save as much meat as possible. When a lobster dies, the moisture evaporates from the meat quickly. Lobsters that have just died but that are very fresh can be purchased at a lower price as stills.

Oysters and clams are purchased in four forms: live in the shell, shucked, frozen, and canned. Oysters and clams in the shell must be alive when purchased, which is indicated by a tightly closed shell. If the shell is open and does not close when handled, the oyster or clam is dead and no longer fit for human consumption. Fresh shucked oysters and clams are packed in metal containers and must be kept refrigerated and packed in ice at all times to prevent spoilage. If handled in the proper manner, oysters and clams remain fresh for about a week.

## STORAGE

Fish and shellfish are such perishable ingredients that it is of the utmost importance to inspect them when they are received in the operation. Once accepted, it is imperative to store them between 30°F and 34°F and to use them as quickly as possible to maintain their quality and freshness.

### Fresh Fish and Shellfish

Fresh fish should always have the viscera removed before storing as it helps the fish to remain fresh much longer. Fresh fish can be stored a maximum of 1 to 2 days. If the fish is not cooked immediately upon receipt, it should be removed from the original ice it was packaged in and rinsed under cool, clean water. The rinsed fish should then be stored in a perforated insert in a double hotel pan. Plastic wrap should be placed over the fish, and the ice poured on top of the plastic wrap. *See Figure 20-40.* The pan should be placed in the coldest part of the refrigerator as soon as possible. The ice helps hold the proper temperature and reduces deterioration. The plastic wrap helps prevent the flesh of the fish from freezer burn and from absorbing water as the ice melts. All ice should be changed daily and the water from the melted ice discarded. Fish must be stored away from other foods in the refrigerator to prevent odors from the fish from affecting the taste or smell of other foods.

Live lobsters, live crabs, and other live crustaceans should be covered and stored with wet seaweed or damp newspaper to keep them from drying out. Storing crustaceans in a saltwater tank is the best method of keeping them alive but is also the most costly and most difficult. If stored under ideal conditions in the absence of a tank, live crustaceans can live for two to four days.

Scallops, oysters, and clams that are purchased live in the shell should be kept under refrigeration at 41°F. They should be kept in the original box or netted bag and placed in a pan to prevent drips from contaminating other foods. They should never be covered with or placed on top of ice, because the intense cold from the ice would cause them to die. Live mollusks also should not be stored in a sealed container or plastic bag, because they would likely die from lack of air. Under ideal conditions, fresh mollusks can live in refrigerated storage for up to a week. *See Figure 20-41.*

### Frozen Fish and Shellfish

When receiving frozen fish, it is important to check the quality of the frozen product. Frozen fish should be received frozen solid. If the fish is even slightly thawed on the edges, it should be rejected, as refreezing will cause a loss in quality. If there are dry-looking spots on the fish, it means it has thawed slightly and been refrozen and should also be rejected. There should be no signs of moisture on the outside of the box. Signs of

## Storing Fish

Cover with crushed ice and store

Remove fish from delivery cartons and rinse

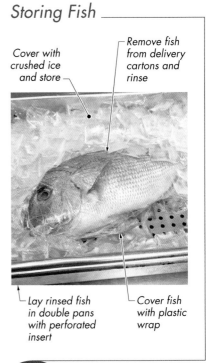

Lay rinsed fish in double pans with perforated insert

Cover fish with plastic wrap

**20-40** *Following proper storage procedures helps ensure freshness and reduce contamination and spoilage.*

## Storing Fresh Mollusks

Fresh clams

Placed in drip pan

Netted bag

20-41 *Fresh scallops, oysters, and clams can be refrigerated for up to a week.*

water or moisture may mean that the product has thawed and refrozen. Fish that has been refrozen will have a poor texture and stronger flavor and will be undesirable. Ice crystals on the fish are another sign that the fish thawed or was not stored and held at 0°F or below. Fish quickly deteriorate at temperatures above 0°F, so fish displaying ice crystals should be rejected.

Fish stored in the freezer must be wrapped properly with moisture-proof wrapping to help prevent freezer burn. Fat fish should be stored in the freezer for no more than 2 months as it quickly starts to deteriorate, while lean fish can be stored frozen for up to 6 months. Stock should be rotated so that the oldest fish is used first. Frozen fish should be thawed in the refrigerator at a temperature of 38°F to 40°F as close as possible to the time the fish is needed. Once frozen fish is thawed, it must be used or discarded. Frozen fish should never be refrozen after it has thawed.

Frozen fish can be packaged block-frozen, layerpacked, shatterpacked, cellopacked, or individually quick-frozen (IQF). When block-frozen, seafood is placed in a block-shaped form, which is placed between two hollow stainless steel plates that have refrigerant flowing through them. The plates freeze the package of seafood into a solid block within two to four hours. Shrimp, scallops, crabmeat, and ungraded fish are often packaged block-frozen. Block-frozen items are prepared breaded or battered and fried.

Layerpacks, also known as shatterpacks, consist of high-quality graded fish fillets layered on polyethylene sheets with the edges slightly overlapping so that entire layers of frozen seafood can be removed when desired.

Cellopacks contain ungraded fish fillets frozen as packets, typically one to three fillets per packet, wrapped in cellophane, frozen, and packaged six packets per box. Fillets packaged in this manner may be inconsistent in size and are relatively inexpensive.

Individually quick-frozen (IQF) seafood is loosely packaged fillets or shellfish that have each been glazed with a thin film of water and blast-frozen. IQF packaging makes it easy to remove the exact number of items needed. IQF products are graded and packaged according to an average size, such as 2-oz to 3-oz ocean perch fillets.

## PREPARATION TECHNIQUES

Seafood may be purchased whole and fabricated in-house or may be purchased in a processed and portion-controlled form. Anytime an operation purchases items processed, it will pay a higher price per pound. Knowing and using proper processing and preparation techniques for fish and shellfish can save a good deal of money in the professional kitchen.

When preparing fish, the scales must always be removed prior to preparation.

## To scale a fish, apply the following procedure:

1. Place the fish in a large sanitized sink.
2. Firmly hold the tail with one hand. Use a fish scaler or the back of a fillet knife to scrape against the scales toward the head. The scales will pop off. *Note:* Use caution to not use too much pressure, which could damage the flesh of the fish.
3. When finished with one side, turn the fish over and repeat on the other side.
4. Rinse the fish thoroughly under cold running water to remove any loose scales that remain.

A fish steak is a popular cut for many round fish varieties.

## To cut a round fish into steaks, apply the following procedure:

1. Remove the viscera, scales, and fins from the fish. **See Figure 20-42.**
2. Make a crosscut slice of desired thickness all the way through the fish.
3. Continue to cut additional steaks, ensuring consistency in the thickness and size of the steaks.

*Cutting a Round Fish into Steaks*

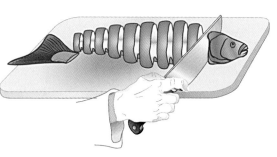

*1. Remove viscera, scales, and fins.*

*2–3. Make a crosscut slice of desired thickness; continue to cut, ensuring consistency in size and thickness.*

**20-42** *Round fish may be cut into steaks for portioning.*

The most common serving portion from fish is the fillet.

## To fillet a round fish, apply the following procedure:

1. Make a cut about ½" deep along the backbone from just behind the head all the way to the tail. **See Figure 20-43.**

2. Make a second cut just behind the gills to the backbone. Do not cut through the backbone.

3. When the blade of the knife hits the backbone, turn the blade horizontally and begin to make smooth slices along the backbone toward the tail. Continue to cut away the flesh until the fillet is completely removed.

4. Turn the fish over and repeat the procedure on the other side.

5. Trim any rib bones away from the fillet.

6. Run the fingers gently along the surface of the flesh to raise the ends of any pin bones that may remain. Remove any pin bones with a clean, sanitized pair of tweezers or needle-nose pliers.

## Filleting a Round Fish

1–2. Cut ½" deep along backbone; make a second cut just behind gills.

3. Slice toward tail along backbone until flesh is completely removed.

4. Turn over and repeat on other side.

5. Trim any rib bones away from fillet.

6. Run fingers along surface to raise any pin bones. Remove any remaining pin bones.

 **20-43** Filleting a round fish removes the meat from the backbone to produce a boneless cut of fish.

A fillet can be prepared with or without the skin intact. For a particular presentation or dish, it may be desirable to remove the skin.

## To skin a fillet, apply the following procedure:

1. Place the fillet on a clean work surface, skin-side down. ***See Figure 20-44.***

2. Using a knife that is at least 50% wider than the fillet itself and starting at the tail, carefully cut down through the flesh just to the skin.

3. Turn the blade toward the head end of the fillet. With the other hand, hold the tail skin with a towel for better grip.

4. With the edge of the blade angled slightly downward, begin to make smooth slices between the skin and the flesh toward the head area. Slice very carefully to ensure that not too much flesh is left on the skin and that the knife does not cut through the skin.

5. Continue all the way to the head end until the flesh is completely freed from the skin. Store fillet properly until needed.

CULINARY PROCEDURES

## Removing Skin from a Round-Fish Fillet

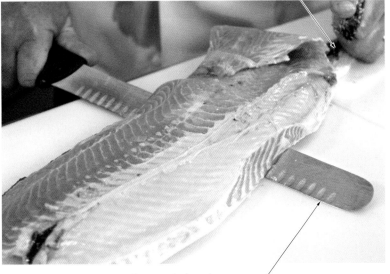

*Place fillet skin down; cut through flesh just to skin*

*Make smooth slices between skin and flesh; continue until skin is completely removed*

**20-44** *The skin from a round-fish fillet may be removed prior to cooking.*

Refer to the CD-ROM for the **Filleting Flat Fish** Media Clip.

Flatfish yield four fillets, two on each side of the fish. The backbone of a flatfish runs down the center of the fish rather than the top as in a round fish.

## To fillet a flatfish, apply the following procedure:

1. Place the fish on a clean cutting board with the darker side facing upward. With the tip of a flexible boning knife, make a cut along the backbone from the head to the tail. ***See Figure 20-45.***

2. Starting at the gill area, with the blade of the knife in the slice above the backbone, turn the blade toward the tail and use consistent and smooth strokes to slice the flesh away from the bones. *Note:* The flexible blade of the knife should glide along the surface of the bones without cutting through them.

3. Remove the second fillet on the dark side using the same process as was used on the first fillet.

4. Turn the fish over and repeat the same process on the other side.

*Filleting a Flatfish* _____

1. Place fish dark-side up on cutting board; cut along backbone from head to tail.

2. Turn blade toward tail and slice flesh away from bones.

3–4. Remove second fillet using the same process; turn over and repeat on other side.

**20-45** Flatfish yield four fillets, two on each side of the fish.

A true Dover sole is somewhat unusual, in that the skin can be removed prior to filleting simply by pulling it off.

**To remove the skin from a Dover sole, apply the following procedure:**

1. Make a very shallow slice at the end of the fillets, just in front of the tail. ***See Figure 20-46.***
2. Use the tip of a boning knife or fingers to slightly scrape the edge of the skin where the slice was made to begin to free the skin from the flesh beneath.
3. Using a towel, grab the freed edge of the skin and pull firmly to completely free the skin from the flesh.
4. Turn the fish over and repeat on the other side.

*Removing Skin from a Dover Sole* _____

  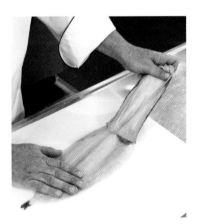

1. Make a shallow cut at end of fillet just before tail.

2. Scrape edge of skin up with tip of knife or fingers.

3–4. With a towel, grab edge of skin and pull firmly to remove; repeat on other side.

**20-46** *The skin of a true Dover sole can be removed prior to filleting the fish.*

A *tranche* is a boneless, bias-cut slice from the fillet of a large fish. Tranches can be cut from both round fish and flatfish. The diagonal cut of a tranche makes the portion appear larger.

**To portion-cut tranches, apply the following procedure:**

1. Place the boneless whole fillet (skin on or skinless) on a clean cutting board, skin-side down. ***See Figure 20-47.***

2. Using a long slicer, slice using a long stroke at a steep angle to cut a tranche to the desired thickness and weight.

3. Continue slicing along the fillet, adjusting the thickness of the tranches as necessary to maintain a consistent portion weight from the wide end to the narrow end.

## Cutting Tranches

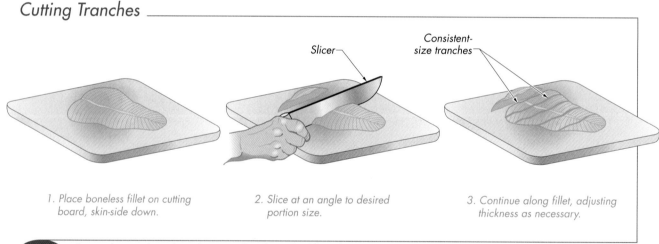

*Slicer*

*Consistent-size tranches*

1. Place boneless fillet on cutting board, skin-side down.

2. Slice at an angle to desired portion size.

3. Continue along fillet, adjusting thickness as necessary.

**20-47** *A tranche is a boneless, bias-cut slice from the fillet of a large fish.*

Shrimp and prawns contain a sand vein, or intestinal tract, that must be removed prior to preparation. The shells are also removed prior to preparation to make the meat easier to eat in a prepared dish.

Refer to the CD-ROM for the **Deveining Shrimp** Media Clip.

## To clean and devein shrimp or prawns, apply the following procedure:

1. Hold the tail fins in one hand with the underside of the shrimp facing upward. Pinch the small finlike legs and the edge of the shell and pull to remove. ***See Figure 20-48.***

2. Remove the shell from the shrimp or prawn, leaving the tail fins and first section of shell intact if desired. *Note:* The tail fins can be removed or can be left on to serve as a handle when eating.

3. With a paring knife, make a shallow slice along the back to expose the sand vein. *Note:* If the shrimp or prawn is to be butterflied, instead of making a shallow slice, make a deeper one to completely open the meat so that it lays flat.

4. Place the shrimp or prawn under cold running water and rinse while pulling the vein to remove. Discard the vein.

CULINARY PROCEDURES

## Cleaning and Deveining Shrimp

1–2. Pinching legs and edge of shell; pull to remove legs; remove shell, leaving tail fins and first section intact if desired.

3. Make a shallow slice along the back to expose the vein.

4. Rinse shrimp under cold running water while pulling vein to remove.

**20-48** *The tail may be left on or removed when cleaning and deveining shrimp.*

A common preparation for lobster tail is to broil the tail. When broiling the tail, the shell is split open to reveal the meat inside. The finished dish is served with the tail meat intact in the shell.

To prepare a lobster tail for broiling, apply the following procedure:

1. Place the tail underside down on a clean work surface. ***See Figure 20-49.***
2. With a French knife, stick the tip of the knife into the center of the tail just above the bottom fins.
3. With the opposite hand on the front end of the knife to hold it in place, push the handle of the knife down to slice just through the shell lengthwise.
4. Spread the cut shell all the way open to reveal the tail meat.
5. Pull the tail meat through the cut shell.
6. Make a few shallow slices along the underside of the tail just deep enough to pierce the tough lower membrane.
7. Place the tail on top of the shell, underside down, and brush with clarified butter before broiling.

*Preparing Lobster Tail for Broiling*

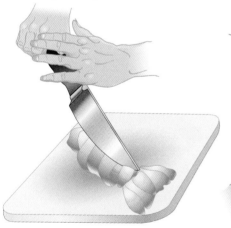

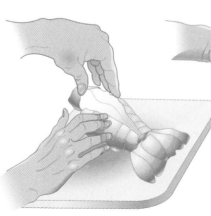

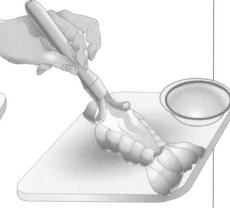

*1–3. Place tail underside down; stick tip of French knife into tail just above fins; push handle of knife down to slice through shell lengthwise.*

*4–6. Spread cut shell open; pull tail meat through cut shell; make shallow slices on underside of tail.*

*7. Place tail underside down and brush with clarified butter.*

 *A lobster tail is cut open to reveal the tail meat prior to broiling.*

Cleaning soft-shell crab is much easier than cleaning hard-shell varieties because there are no hard shells to crack and pick apart. Cleaning a soft-shell crab simply involves removing the inedible portions of the crab, such as the eyes, stomach, and abdomen or apron, which is tucked under the body to carry and conceal eggs.

## To clean a soft-shell crab, apply the following procedure:

1. Peel back the pointed top shell to reveal the stringlike gills beneath. Carefully scrape away the gills completely. Repeat on other side.

2. Locate the eyes on the top of the head. With a French knife or kitchen shears, cut to remove the head and mouth just behind the eyes and discard.

3. Gently squeeze the body just behind where the head was removed to squeeze out a green bubble of fluid. Rinse. *Note:* This fluid must be squeezed out and washed away, as it has a bitter, unpleasant taste.

4. Turn the crab over and locate the tail (apron). Firmly twist the tail and pull to remove. *Note:* The intestinal tract will be attached and should be discarded with the tail.

*Shucking* is the process of opening a bivalve such as an oyster or clam. Oysters and clams must be shucked to access the edible meat inside the shell if it is desired raw. Oysters are shucked using the a similar method to that for clams. ***See Figure 20-50.*** An oyster is shucked using an oyster knife and a clam is shucked using a clam knife. Clams may be steamed to open and access the meat, while oysters must be shucked.

## To shuck an oyster or a clam, apply the following general procedure:

1. Hold the shell firmly with a clean kitchen towel in the palm of the hand, with the hinge resting against the base of the thumb and the lip opening toward the fingertips.

2. Insert the blade of the knife between the edges of the opening between the top shell and bottom shell.

3. Holding the oyster or clam in one hand, squeeze the blade edge between the shells until at least halfway inserted. *Note:* Do not force the tip of the knife in between the shells in a stabbing motion as the knife may slip and cut the hand, as well as damage the meat inside the shell.

4. Carefully pop the shells open slightly with the knife.

5. Use the tip of the knife to separate the meat (abductor muscle) from the top shell.

6. Completely open the two shells and discard the top shell.

7. The meat can either be completely removed from the bottom shell or left in the bottom shell to serve or bake.

## Shucking Oysters and Clams

1–3. Hold oyster firmly with towel; insert tip of oyster knife with gentle pressure near the back of the shell; twist knife to pop hinge open.

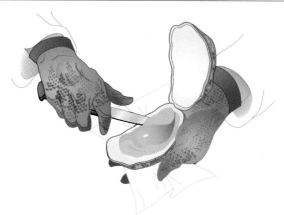

4–7. Insert tip of knife to cut the meat (adductor muscle) free from shell; remove top shell and discard; remove oyster or leave in bottom shell if desired.

**Shucking Oysters**

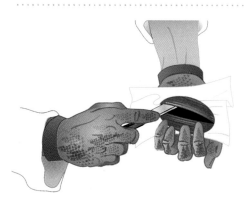

1–3. Hold clam firmly with towel; insert blade of clam knife; squeeze blade edge between shells.

4–7. Pop shell open; cut the meat (adductor muscle) free from shell; discard top shell; completely remove clam or leave on shell to bake.

**Shucking Clams**

**20-50** A similar procedure is used to shuck oysters and clams.

A mussel has hairlike threads, called a beard, that are not eaten and must be removed prior to preparation.

**To clean and debeard a mussel, apply the following procedure:**

1. Under clean running water, brush the mussel and rinse to remove excess mud, sand, or debris. ***See Figure 20-51.***

2. With a clean, sanitized pair of needle-nose pliers, grab the beard and gently pull to tear the beard free from the shell. *Note:* The beard can also be removed by pinching it with the fingers and pulling, but it may be difficult.

## Cleaning Mussels

Beard

**20-51** *Mussels are cleaned and de-bearded before cooking.*

## Cleaning Squid

Squid is often shipped cleaned and ready for cooking. If the squid is white and the eyes have been removed, it has already been cleaned. If it is darker in color and the eyes are intact, it must be cleaned before being prepared.

**To clean a squid, apply the following procedure:**

1. Gently pull the arms and tentacles away from the mantle (body). When they are removed, the intestines, eyes, and ink sac will still be attached.

2. Cut the arms and tentacles just below the eyes to remove them from the head.

3. Remove the ink sac from the head, reserve if desired, and discard the remainder of the head.

4. Pull back the tentacles to expose the beak. With the fingers, carefully pull out the beak and discard. The tentacles may be left whole or cut into smaller sections as desired.

5. Hold the mantle in one hand and carefully pull out the transparent quill with the other hand and discard.

6. With the fingers, pull as much skin as possible from the mantle to reveal the lighter-colored flesh below. Discard the dark-colored skin. The mantle can be cut into rings or sections or left whole. Wash the cleaned squid before use.

### Cleaning Octopus

Octopus typically is shipped cleaned. However, occasionally it is shipped as caught and additional cleaning is necessary.

**To clean an octopus, apply the following procedure:**

1. Locate the beak of the octopus and use a paring knife to carefully remove it from the body. Discard the beak.
2. Locate the eyes of the octopus and cut just beneath the eyes to remove the head. Discard the head.
3. Carefully and firmly pull the dark-colored skin from the body. Discard the dark skin.
4. Carefully and firmly pull the suction cups and skin from the arms. Discard the cups and skin.

## COOKING FISH AND SHELLFISH

Fish and shellfish are nutritious foods that can be prepared with exciting results. However, fish and shellfish are more fragile than many other protein-rich foods. The flesh of the fish flakes apart easily and overcooks quickly. Because fish lacks the internal marbling found in red meat, even slight overcooking can result in a tough, dry portion. Shellfish also require special care when cooking. Understanding the proper cooking methods for each variety helps the chef to serve tender, flavorful seafood.

### Fish

The fat content of a fish produces a difference in flavor and affects the cooking method required. Fat fish, such as mackerel, tuna, trout, and salmon, produce superior eating qualities if baked, broiled, or grilled because their natural fat prevents them from drying out during dry-heat cooking. Lean fish, such as red snapper, haddock, halibut, and perch, are best prepared with moist-heat methods such as poaching. Both fat and lean fish can be sautéed or fried with excellent results. Exceptions can be made to cooking procedures for fat and lean fish if allowances are made for the fat content. For example, lean fish can be baked or broiled if basted during the cooking process. Fat fish may also be steamed or poached with special care in cooking and handling the fish.

Fish steaks work well for grilling or broiling as the thickness of the flesh makes them more durable than fillets and allows them to handle the intense heat of the open flame. Fish fillets are best cooked using other cooking methods as the tenderness of a fillet may cause it to break or stick on the grill or broiler and fall apart. When cooking a fish fillet with the skin on, the skin should be scored with a sharp knife first so that the fillet does not curl during cooking. ***See Figure 20-52.***

*Chef's Tip:*
Fish should always be seasoned lightly so as to not overpower the delicate taste of the fish. It is best to season fish immediately before cooking.

## Considerations for Cooking Fish

| Cooking Method | Effect on Lean Fish | Effect on Fat Fish |
|---|---|---|
| **Dry Heat with Fat—sautéing and frying** | Lean fish work well when cooked with dry-heat methods such as sautéing or frying, where fat is used during the cooking process; the addition of fat (even if it is only on the exterior breading or batter) adds moisture and flavor to a lean fish. | Smaller species of fat fish can be sautéed or pan-fried but larger fat fish are usually not cooked this way due to excess fat absorption; exceptions occur—for example, some chefs may sauté a salmon fillet as long as the sauce is not rich in fat as well. |
| **Dry Heat—grilling, broiling, and roasting** | Lean fish can be baked or broiled, but will benefit from basting with fat during cooking to prevent drying out; grilling and broiling are not recommended, as the high heat will end to dry out the flesh of lean fish; use caution to not overcook as the fish will become dry and tough. | Fat fish work very well when prepared using dry heat methods such as broiling or baking; the extra fat content of the flesh helps to keep the fish moist while the high heat helps to increase flavor. |
| **Moist Heat—poaching and steaming** | Poaching is a preferred cooking method for lean fish as it helps the flesh of fish to retain moisture; however, if the fish is overcooked it will become dry. | Special care must be taken when using moist-heat cooking methods such as poaching or steaming. |

**20-52** *Recommended cooking methods for fish vary based on fat content.*

*En papillote* a French term that refers to food cooked in greased parchment paper. When baking fish en papillote, the fish is often topped with julienne vegetables, butter, white wine, and lemon, the paper folded over, and the edges twisted together to seal in the fish. The paper package is then placed in a 350°F oven and baked for approximately 10 min per inch of thickness, until the paper is browned slightly and puffed. The package is removed from the oven, plated, and opened tableside to allow the aromas to escape in front of the customer. Fish is commonly poached in a court bouillon or fumet. These flavorful poaching liquids add a delicate flavor and additional moisture to the fish. When poaching, it is essential to ensure that the poaching liquid never reaches a simmer, as the intense heat of a simmer would cause the fish to toughen and dry out.

Steaming is the healthiest way to prepare fish, because it adds no additional calories or fat. Steaming also helps the fish to retain the most nutritional value, as the steam gently cooks the fish without immersing the fish

**Chef's Tip:**

The problem with cooking fat fish using moist heat is that the hot water can remove some of the rich fish oils from the flesh. If fat fish is to be cooked using moist-heat methods, care should be taken to cook fat fish perfectly and over a gentle heat so that it retains as much of its natural oil as possible.

in a liquid, which could wash away valuable nutrients. Items can be steamed on tea leaves or fresh herbs so that the steam that penetrates and cooks the fish while the leaves or herbs impart a delicate aroma to the fish.

Wine and lemon juice are common ingredients in poaching liquid. The small amount of acid they add to the liquid makes items taste fresher and retain more of their natural color. Fish such as salmon and trout are sometimes poached whole and then chilled for a cold foods presentation or cold seafood buffet.

## To poach a whole fish, apply the following procedure:

1. Lightly oil a rack or glazing screen and place a whole fish on it. Secure the fish to the rack with butcher's twine to prevent the fish from tipping over and to ensure the fish cooks in the desired shape.

2. Lower the rack into a well-seasoned court bouillon and gently poach the fish at a temperature between 170°F and 180°F degrees for 30 min.

3. Turn off the heat and let the fish rest in the warm court bouillon for an additional 20–30 min until completely cooked through. Allowing the fish to rest without additional heat results in the fish absorbing more moisture and flavor from the court bouillon and yields a superior product.

4. Carefully remove the rack or glazing screen and allow the fish to drain completely. If the fish is to be served warm, serve immediately. If it is to be served cold, immediately chill the fish. *Note:* Whole poached fish is much less fragile when chilled, so it is important to leave it on the rack or glazing screen until cold. Removing the fish while hot may result in the fish breaking or falling apart.

### Crustaceans

Crustaceans can be prepared using a variety of cooking methods. Lobsters are commonly either poached whole and then cracked tableside, or the tail is removed from the shell, placed on top of it, and broiled. Lobster meat is not as delicate as the flesh of a fish and can handle the high heat of a broiler. Hard-shell crabs are most commonly steamed or poached (commonly referred to as boiled, but boiling uses too high a heat and results in tough crab meat). Individual crab legs or clusters of legs are sold precooked and need only to be heated through prior to serving. Soft-shell crab is best dusted lightly in seasoned flour and sautéed or lightly breaded and fried. Because the shell is soft, the entire crab (excluding the eyes and lungs, which are removed in advance) can be eaten.

Live lobster is typically poached in court bouillon.

## To poach a whole lobster, apply the following procedure:

1. Place the live lobster on its belly and, using the tip of a French knife, pierce the center of its head. ***See Figure 20-53.***
2. Drop lobster into a stockpot filled with 180°F court bouillon and poach for 5 min per pound.
3. When time is up, turn off heat and allow lobster to rest in the liquid for an additional 10 min to 15 min.
4. Remove the lobster and drain excess liquid. Serve with drawn butter and fresh lemon.

## Poaching Whole Lobsters

1. Pierce head with French knife.

2–3. Poach in 180°F court bouillon 5 min per lb; turn off heat and allow to rest in liquid 10 min to 15 min.

4. Remove lobster and drain excess liquid.

**20-53** *Live lobsters are poached whole in court bouillon.*

Many preparations use lobster meat as an ingredient. Cooked lobster meat can be removed from the shell and reserved for later use.

## To remove cooked lobster meat from the shell, apply the following procedure.

1. Place drained, cooked lobster in an ice bath to end the cooking process.
2. Remove the claws by firmly pulling and twisting them away from the body. Twist the claw off from each arm. ***See Figure 20-54.***

3. With the back of a French knife, carefully crack the claw just beneath the pinchers, cracking all the way around the claw.

4. Carefully remove the claw meat from inside the shell. *Note:* The claw meat can be removed in one piece and used for a nice presentation.

5. Twist the tail section away from the rest of the body.

6. Remove the thin membrane from the underside of the shell.

7. Split the tail shell and remove the tail meat.

## Removing Lobster Meat from the Shell

1–3. Remove claws from body; crack each claw with the back of a French knife.

4. Remove meat from claw portion.

5–7. Twist the tail section off of the body; remove the membrane from the underside; split tail shell and remove tail meat.

**20-54** *The claws and tail of a lobster are removed and split open to access the meat.*

### Mollusks

Mollusks are prepared using a variety of cooking methods. Scallops are most commonly removed from the shell and sautéed to produce a golden exterior. They are tender and moist when not overcooked. Oysters are most commonly eaten raw on the half shell. It is especially important when serving oysters raw that they are from safe, uncontaminated water. If oysters are to be eaten cooked, they are most commonly baked with a topping or filling or lightly breaded and fried. Clams are most commonly shucked, stuffed with a filling, or topped with butter and seasoned breadcrumbs and baked. Clams can also be chopped and canned for use in soups and chowders. Mussels are commonly steamed or simmered in the shell until the shell opens. Mussels can also be sautéed for use in pasta and rice dishes.

## Cephalopods

Cephalopods are prepared using various methods. Squid is commonly either grilled as the whole tube, or cut into rings and either quickly sautéed or lightly breaded and fried. The tube or mantle can also be stuffed whole and baked. Care needs to be taken to avoid overcooking squid, as it becomes rubbery and almost inedible when overcooked. The ink sac of the squid is often reserved and used as a black food coloring for items such as risotto or pasta.

Cooking octopus quickly, for five minutes or so, yields a chewy but not unpleasant texture, which is the style used for sushi. Like squid, if octopus is overcooked it becomes tough and rubbery and is inedible. Poached octopus is often chilled, tossed with vinaigrette, and served as a cold salad. Tenderized octopus can also be sautéed in butter and garlic and served with lemon as a side dish or appetizer.

Cuttlefish is used most often in risotto and in Asian cuisine. As with squid, it is commonly cut into rings and lightly sautéed. It can also be stir-fried, or gently simmered in cooking liquid when making risotto.

## SUMMARY

Fish and shellfish are increasingly popular menu choices because they are healthy, nutritious foods and because improvements in transportation and storage have increased their availability. A wide variety of seafood is readily available including many types freshwater fish, saltwater fish, crustaceans, and mollusks.

Available market forms of fish range from whole or drawn, to frozen or canned. Similarly, shellfish are available live and fresh in the shell, shucked fresh, frozen, or canned. The most suitable market form to order for any given commercial kitchen depends on the type of seafood required, preparation requirements, storage facilities, and equipment available.

Inspection and grading of seafood is optional and is carried out by the National Marine Fisheries Service. Only fish and shellfish that were processed under a Type 1 inspection are eligible for grading. Regardless of whether they have undergone inspection, fish and shellfish are highly perishable, so following the proper purchasing and storage requirements is crucial.

Being able to identify wholesome, high-quality seafood and the best storage and cooking methods for each type of seafood ensures that an establishment serves the freshest, most flavorful fish and shellfish. Both fish and shellfish have flesh that is naturally tender and requires very little cooking time compared to the flesh of land

Refer to the CD-ROM for Quick Quiz® questions related to chapter content.

animals. Preparation and cooking methods vary with the type of fish or shellfish being cooked, and many types of seafood can be cooked in several ways, from steaming or poaching to sautéing or frying. The cooking method chosen depends partly on the skill set of the kitchen staff and the needs of the customer.

## Review Questions

1. Contrast the differences between round fish, flatfish, and boneless fish.

2. Identify characteristics and common examples of fat fish.

3. What is aquafarming?

4. Contrast different types of trout.

5. Describe the most common variety of perch.

6. List the main features of catfish.

7. Identify common species of Atlantic and Pacific flounder.

8. Compare Atlantic halibut and Pacific halibut.

9. Contrast king (Chinook) salmon and pink salmon.

10. Identify types of tuna that are often seared as a steak and served rare or medium rare.

11. Describe the main qualities of hake.

12. Contrast the definitions of crustacean and mollusk, giving examples of each.

13. List four different types of lobster.

14. What are the differences between king crabs and snow crabs?

15. Describe three categories of mollusks.

16. Describe how escargot is often served.

17. List various market forms in which fish are available.

18. Identify different market forms of shrimp.

19. Contrast the three types of optional inspections that a seafood producer can elect.

20. How can inspection of the eyes and gills of fish give indications of quality and freshness?

21. List general considerations for purchasing shellfish.

22. At what temperature range should nonfrozen fish and shellfish be stored?

23. How should live lobsters, live crabs, and other live crustaceans be stored?

24. Describe the use of the tranche cut in fish preparation.

25. Describe how to clean a soft-shell crab.

26. Describe the process of cooking fish en papillote.

27. Describe common ways to cook cephalopods.

# *Recipes*

## 186

## Battered Fish and Chips

*yield: 4 servings*

| | |
|---|---|
| 3 | russet potatoes, large |
| 1½ c | all-purpose flour |
| 2 tsp | baking powder |
| 1 tsp | salt |
| 1 | egg, beaten just enough to emulsify |
| 12 oz | soda water |
| ½ tsp | white vinegar |
| 8 | 3-oz cod portions, cut 1"–2" wide and 3" long |
| ½ c | all-purpose flour |
| to taste | kosher salt |
| | canola oil (for deep-frying) |

1. Wash potatoes and slice on a mandolin to about ¹⁄₁₆" in thickness. Store in cold water to prevent enzymatic browning.
2. Mix 1½ c flour, baking powder, salt, and beaten egg in a mixing bowl. Add soda water and vinegar and mix just enough to moisten. Let mixture rest for 15 min.
3. Fry potatoes in 225°F oil for 5 min to blanch. After 5 min, remove from oil with strainer or spider and transfer onto sheet pans to cool.
4. Raise oil temperature to 375°F.
5. Dip cod portions into ½ c flour and then into batter mixture.
6. Remove from batter, allowing excess batter to run off. Use swimming method to place cod into the hot frying oil. When a portion floats it is done cooking.
7. After all cod portions have been fried, place blanched potato chips in the fryer and fry until golden brown and crisp.
8. When potatoes are removed from the fryer, season with kosher salt. Serve immediately.

---
**Nutrition Facts**

**Per serving:** 1567.6 calories; 63% calories from fat; 112.2 g total fat; 115.8 mg cholesterol; 1006.8 mg sodium; 2004.6 mg potassium; 96.5 g carbohydrates; 6.8 g fiber; 0.3 g sugar; 89.7 g net carbs; 44.4 g protein

---

## 187

## Fried Calamari with Lemon-Garlic Butter

*yield: 6 servings*

| | |
|---|---|
| 1 lb | calamari rings and tentacles |
| 2 c | buttermilk |
| 2 c | all-purpose flour |
| 2 tsp | chopped parsley |
| ⅛ tsp | cayenne pepper |
| ½ tsp | garlic powder |
| to taste | salt and pepper |
| 2 tbsp | butter |
| 1 clove | garlic, minced |
| 1 tbsp | fresh lemon juice |
| | canola oil (for deep-frying) |

1. Soak calamari in buttermilk for 20 min to 30 min.
2. While calamari is soaking, mix flour, parsley, cayenne, garlic powder, salt, and pepper together in bowl.
3. Remove calamari from marinade and shake off excess buttermilk.
4. Dip calamari in flour mixture and shake off excess flour.
5. Fry in 350°F to 375°F oil until slightly golden, approximately 1 min to 2 min. *Note:* Use caution, as calamari overcooks quickly and becomes tough and rubbery when overcooked.
6. While calamari is frying, mix butter, garlic, salt, and pepper in a small saucepan and cook for 2 min, using caution to not brown the garlic.
7. Add lemon juice, salt, and pepper to butter and remove from stove.
8. Remove calamari from the fryer and drain well. Season with salt and pepper as needed and toss with lemon-garlic butter. Serve immediately.

---
**Nutrition Facts**

**Per serving:** 932.6 calories; 74% calories from fat; 78.7 g total fat; 189.6 mg cholesterol; 169.2 mg sodium; 365.8 mg potassium; 38.6 g carbohydrates; 1.2 g fiber; 4.2 g sugar; 37.4 g net carbs; 18.9 g protein

---

## Sautéed Tilapia with Watercress Cream Sauce

188

*yield: 4 servings*

| | |
|---|---|
| 4 | 6-oz tilapia fillets |
| to taste | salt and pepper |
| 2 oz | all-purpose flour |
| 1 tbsp | vegetable oil |

*Watercress cream sauce*

| | |
|---|---|
| 1 tbsp | butter (or 2 tsp vegetable oil) |
| ½ bunch | fresh watercress |
| 1 | shallot, minced |
| 1 clove | garlic, minced |
| 2 oz | white wine (optional) |
| 4 oz | heavy cream |
| to taste | salt and white pepper |

1. Season tilapia on both sides with salt and pepper.
2. Dredge tilapia in flour and shake off excess.
3. Heat oil in sauté pan to medium-high heat and carefully lay fish in pan. Cook fish until golden brown, turning to brown both sides.
4. While fish is cooking, begin to sauté the shallots in the butter (or oil) until translucent. Add garlic and cook 1 min more.
5. Add white wine to deglaze. Reduce by half.
6. When wine is reduced, add cream and simmer until reduced by about one-quarter. *Note:* Cream can be omitted and replaced with whole milk or chicken stock and then thickened with a cornstarch slurry for a lower-calorie sauce.
7. Transfer cream mixture to a blender with watercress. Purée until very green and smooth. Season to taste and strain if desired.
8. Plate fish on top of cream sauce and garnish with watercress.

### Nutrition Facts

**Per serving:** 393.1 calories; 49% calories from fat; 22 g total fat; 176.8 mg cholesterol; 275.5 mg sodium; 496.9 mg potassium; 12.4 g carbohydrates; 0.4 g fiber; 0.1 g sugar; 12 g net carbs; 32.3 g protein

## Crab Cakes

189

*yield: 12 (4-oz servings)*

| | |
|---|---|
| 4 oz | mayonnaise |
| 2 oz | onion, minced |
| 1 oz | green onion, minced |
| 1 tsp | fresh dill, minced |
| 4 | eggs, lightly beaten |
| 1 tsp | Worcestershire sauce |
| ½ tsp | dry mustard |
| ½ tsp | kosher salt |
| ¼ tsp | cayenne pepper |
| ½ tsp | Old Bay® seasoning |
| 2 c | panko bread crumbs (or soda crackers, finely crushed) |
| 2 lb | lump crabmeat, picked over to remove any shell pieces |
| 2 tbsp | unsalted butter |
| 4 oz | canola oil (for pan-frying) |
| | fresh lemon, cut into wedges (for garnish) |

1. Add mayonnaise, onions, dill, egg, Worcestershire, mustard, salt, cayenne, and Old Bay® to mixing bowl and mix well to incorporate.
2. Gently fold in ½ c of the panko (or crackers) and the crabmeat into the mayonnaise mixture, using caution to not break apart the chunks of crab.
3. Gently form the crab mixture into desired size cake, approximately 1" thick.
4. Dip the crab cakes into the remaining crumbs and coat well on all sides.
5. Place cakes in refrigerator for at least 30 min to firm. *Note:* This is an important step, as the cakes will fall apart easily if not allowed to rest. Resting will allow the crumbs, both inside and outside, to absorb moisture, which aids in binding the ingredients.
6. In a large skillet over medium heat, melt the butter into the oil and wait for the foam from the butter to disappear. Once gone, the oil should be hot enough to begin pan-frying the crab cakes.
7. Add crab cakes to pan and pan-fry until golden brown in color, turning to cook both sides.
8. Place cooked cakes on paper towels to drain excess oil.
9. Serve immediately with wedges of fresh lemon and sauce remoulade, if desired.

### Nutrition Facts

**Per serving:** 290.3 calories; 56% calories from fat; 18.5 g total fat; 145.3 mg cholesterol; 550.7 mg sodium; 345.2 mg potassium; 11.5 g carbohydrates; 0.5 g fiber; 1 g sugar; 11 g net carbs; 18.9 g protein

# Recipes

## 190

## Deep-Fried Coconut-Breaded Shrimp with Minted Orange Dipping Sauce

*yield: 4 servings*

| | |
|---|---|
| 1 c | all-purpose flour |
| ¾ tsp | Madras curry powder |
| ¼ tsp | cayenne pepper |
| 1 tsp | baking powder |
| ½ tsp | salt |
| 2 | eggs, slightly beaten |
| 12 oz | club soda |
| 2 c | shredded unsweetened coconut |
| 20 | 21–25 count (or larger) shrimp, peeled and deveined with tail on |
| | canola oil (for deep-frying) |

*Minted orange dipping sauce*

| | |
|---|---|
| 4 oz | orange marmalade |
| 1 oz | lime juice, fresh |
| 1 tsp | rice wine vinegar |
| ¼ tsp | Madras curry powder |
| ⅛ tsp | kosher salt |
| 2 tsp | fresh mint leaves, cut chiffonade |

1. In a large bowl, combine flour, curry, cayenne, baking powder, and salt and mix to distribute.
2. Add beaten eggs and club soda. Whisk just to incorporate while still leaving batter somewhat lumpy. Let batter rest for 10 min to 15 min.
3. Heat frying oil to 375°F while batter is resting.
4. Spread coconut evenly in hotel pan and set aside.
5. One at a time, dip shrimp into the batter while holding the tail up so as not to coat the tail as heavily.
6. Remove shrimp from batter, shake off excess and roll in coconut to coat evenly.
7. Fry shrimp in hot oil for 2 min to 3 min, or until completely cooked.
8. Remove from hot oil and drain on paper towels.
9. To prepare sauce, place orange marmalade, lime juice, vinegar, curry, and salt in small saucepan and heat until completely melted.
10. Add fresh mint and stir to incorporate.
11. Serve shrimp with sauce.

### Nutrition Facts

**Per serving:** 1976.9 calories; 81% calories from fat; 185.7 g total fat; 161.6 mg cholesterol; 639 mg sodium; 806.1 mg potassium; 72 g carbohydrates; 20 g fiber; 25.6 g sugar; 52 g net carbs; 21.9 g protein

## Oysters Rockefeller

*yield: 20 servings*

*Florida Department of Agriculture and Consumer Services, Bureau of Seafood and Aquaculture Marketing*

| | |
|---|---|
| 20 | fresh oysters |
| 4 oz | bacon, cooked and cut into small dice |
| 1 tbsp | unsalted butter |
| 2 | shallots, minced |
| 2 oz | fennel, minced |
| 2 oz | celery, minced |
| 2 cloves | garlic, minced |
| 1 tsp | Pernod (or ⅓ tsp anise extract) |
| 2 oz | white wine |
| to taste | cayenne |
| 1 tbsp | flat-leaf parsley, finely chopped |
| 2 oz | baby spinach, stems removed and finely chopped |
| 2 oz | watercress leaves, finely chopped |
| 4 oz | béchamel sauce, heated |
| 4 tbsp | bread crumbs, coarsely chopped |
| to taste | salt and pepper |
| 1½ c | rock salt (as needed for serving) |

1. Shuck fresh oysters and scrub shells to clean thoroughly. Reserve shells and oysters under refrigeration until needed.
2. In a large sauté pan, render bacon until crisp and golden brown, then remove from hot pan and drain on paper towels.
3. To same sauté pan, add butter, shallots, fennel, and celery. Sweat until translucent.
4. Add garlic and cook for 1 min more.
5. Add Pernod, white wine, and cayenne and reduce by half.
6. Add parsley, spinach, and watercress and cook until wilted and bright green, about 1 min.
7. Add hot béchamel, bread crumbs, and diced bacon. Season and stir well to incorporate.
8. Remove from heat and place on a sheet pan to cool in the refrigerator until needed.
9. Place each oyster back into shell and top with 2 tsp of the spinach mixture, using care to spread mixture over entire surface of oyster.
10. Bake at 450°F until bubbly, about 5 min to 7 min.
11. Serve hot oysters on a bed of rock salt.

### Nutrition Facts

**Per serving:** 131.3 calories; 38% calories from fat; 5.6 g total fat; 32.5 mg cholesterol; 8268.8 mg sodium; 337.9 mg potassium; 13 g carbohydrates; 1.4 g fiber; 0.6 g sugar; 11.6 g net carbs; 7.5 g protein

## Steamed Mussels in Tomato Kalamata Olive Sauce

192

*yield: 4 servings*

| | |
|---|---|
| 2 tbsp | olive oil |
| 1 tbsp | butter |
| 2 | shallots, minced |
| 3 oz | celery, minced |
| 3 tbsp | garlic, minced |
| 1 qt | canned diced tomatoes, in juice |
| 4 oz | kalamata olives, pitted and cut in half |
| ¼ tsp | crushed red pepper |
| 2 tsp | thyme, fresh |
| 3 tbsp | basil, cut chiffonade |
| 4–5 lb | mussels, scrubbed and debearded |
| 3 oz | dry white wine |
| 4 oz | fish fumet |
| 2 tbsp | fresh parsley, chopped fine |

1. Heat olive oil and butter in saucepot. Add shallots and celery and sweat for 1 min to 2 min.
2. Add garlic and cook for 1 min more.
3. Add tomatoes, olives, crushed red pepper, thyme, basil, and mussels.
4. Deglaze with white wine and fish fumet, mix well, and cover.
5. Reduce heat to medium, and steam until mussels have opened, about 5 min to 7 min. Discard any mussels that do not open.
6. Move mussels to serving bowls and spoon sauce over mussels.
7. Serve sprinkled with chopped fresh parsley.

---

**Nutrition Facts**

**Per serving:** 803.2 calories; 31% calories from fat; 28.3 g total fat; 134.9 mg cholesterol; 2429.1 mg sodium; 3024.9 mg potassium; 73.4 g carbohydrates; 5.1 g fiber; 10.8 g sugar; 68.3 g net carbs; 62.7 g protein

---

## Baked Clams with Pesto and Pine Nut Crust

193

*yield: 6 servings*

| | |
|---|---|
| ½ c | cornmeal |
| 2 tbsp | salt |
| 3 lb | littleneck clams |
| 2 tbsp | butter |
| 2 | shallots, minced |
| 3 tbsp | garlic, crushed |
| 4 oz | dry white wine |
| 2 oz | lemon juice, fresh |
| 2 tbsp | fresh parsley, chopped fine |
| 4 tbsp | pesto |
| 2 tbsp | whole pine nuts |
| 3½ c | coarse bread crumbs |
| 4 oz | Parmesan, freshly grated |
| 4 oz | olive oil |
| 4 oz | melted butter |
| 1 tsp | kosher salt |
| ½ tsp | black pepper |

1. Mix cornmeal and salt in a large bowl and mix with ½ gal. of cold water.
2. Add clams to bowl and adjust water as necessary to completely submerge clams. *Note:* This step helps purge the clams of sand and grit, as the clams will open just enough to eat the cornmeal, rinsing sand and grit from inside the shells.
3. Add 2 tbsp butter to a sauté pan.
4. Add shallots and garlic and sweat for 2 min.
5. Deglaze with wine and lemon juice and reduce by half.
6. Remove from heat and add parsley, pesto, and pine nuts and stir to incorporate.
7. In a separate bowl, combine bread crumbs, Parmesan, olive oil, melted butter, salt, and pepper and mix well.
8. Add pesto mixture and mix well. Reserve until needed.
9. Shuck clams, discard top shells, and carefully free clams from bottom shells.
10. Top each clam with the pesto and pine nut mixture and bake in a 350°F oven until golden brown, about 12 min to 15 min.

---

**Nutrition Facts**

**Per serving:** 916.4 calories; 51% calories from fat; 54.1 g total fat; 81.9 mg cholesterol; 3502.1 mg sodium; 839.6 mg potassium; 81.1 g carbohydrates; 4.1 g fiber; 4.6 g sugar; 77 g net carbs; 26.8 g protein

---

# Recipes

## 194

## Sautéed Sole Meunière

*yield: 4 servings*

| | |
|---|---|
| 4 | 10-oz pan-dressed lemon sole or Dover sole |
| 2 oz | all-purpose flour, seasoned with kosher salt and white pepper |
| 2 tbsp | canola oil (for sautéing) |
| 2 oz | clarified butter |
| 1 oz | yellow squash, cut petit Parisian |
| 1 oz | zucchini, cut petit Parisian |
| 1 oz | carrot, cut petit Parisian and blanched |
| 2 oz | fresh lemon juice |
| 1 tbsp | fresh parsley, chopped |

1. Pat sole dry with paper towel and dredge in seasoned flour.
2. In a hot sauté pan over medium heat, add oil and sauté sole until golden brown on both sides and cooked through, about 4 min to 5 min.
3. Remove fish from pan and keep warm.
4. Wipe remaining oil out of pan.
5. Lower heat and add the butter and petit Parisian vegetables to pan. Sauté lightly. Remove from pan and set aside.
6. Continue cooking remaining butter until it reaches a light brown color and has a nutty aroma.
7. Add lemon juice to deglaze the pan.
8. Remove from heat and add chopped parsley.
9. Plate fish and top with Parisian vegetables.
10. Spoon sauce over the top and serve.

### Nutrition Facts

**Per serving:** 289 calories; 59% calories from fat; 19.5 g total fat; 64.5 mg cholesterol; 66.2 mg sodium; 358.4 mg potassium; 13.4 g carbohydrates; 0.8 g fiber; 1 g sugar; 12.5 g net carbs; 15.3 g protein

## 195

## Poached Pistachio-Crusted Salmon

*yield: 4 servings*

| | |
|---|---|
| 2 lb | salmon fillet |
| 1 | egg |
| 2 oz | heavy cream |
| to taste | salt and white pepper |
| ½ lb | shelled pistachios, coarsely chopped |
| 1 | lemon, sliced |

*Court bouillon*

| | |
|---|---|
| 1 | bay leaf |
| ½ tsp | thyme |
| 4 | peppercorns |
| 1 qt | water |
| 2 oz | white wine |
| 1 | shallot, minced |

1. To begin the court bouillon, make a sachet of bay leaf, thyme, and peppercorns.
2. Place water, wine, shallot, and sachet in a large saucepan and bring to a boil. Lower temperature and hold at a simmer (between 175°F and 180°F).
3. From salmon fillet, cut four 6-oz portions.
4. Place remaining salmon in a food processor with 2 tsp ice-cold water and purée.
5. Add egg, cream, salt, and white pepper to food processor and pulse to incorporate ingredients evenly.
6. Remove puréed salmon from processor and spread on top of salmon fillets, coating evenly.
7. Dip purée-topped side into chopped pistachios to coat well.
8. Slowly lower fillets into court bouillon and poach for approximately 12 min to 14 min until cooked through.
9. Remove the fillets and plate immediately. Garnish with lemon slices.

### Nutrition Facts

**Per serving:** 289 calories; 59% calories from fat; 19.5 g total fat; 64.5 mg cholesterol; 66.2 mg sodium; 358.4 mg potassium; 13.4 g carbohydrates; 0.8 g fiber; 1 g sugar; 12.5 g net carbs; 15.3 g protein

## Shrimp Cocktail

*yield: 4 servings*

| | |
|---|---|
| 20 | large shrimp, peeled and deveined with tail on |
| 1 gal | water |
| 1 tbsp | pickling spice (equal parts of black peppercorns, red pepper flakes, mustard seed, coriander seed, dill seed, clove and bayleaf) |
| 1 | lemon, cut into quarter wedges |
| 2 oz | white wine |
| 4 oz | white mirepoix |

*Harbor Seafood*

1. Add water, pickling spice, lemon, wine, and mirepoix to a stockpot and bring to 180°F.
2. Drop shrimp in water and poach for 5 min at no more than 180°F. Turn off heat.
3. Add a scoop of ice to poaching liquid to cool it slightly. Let shrimp sit in water for an additional 3 min to 5 min. *Note:* Two tricks to a perfect shrimp cocktail are never letting the shrimp boil and letting them cool slightly in the poaching liquid.
4. Remove shrimp and place in a bowl covered with ice until completely cool.
5. For service, hang shrimp off the side of a glass dish that also contains a serving of cocktail sauce and a wedge of fresh lemon.

**Nutrition Facts**

**Per serving:** 71.4 calories; 13% calories from fat; 1.2 g total fat; 53.2 mg cholesterol; 92.5 mg sodium; 204.9 mg potassium; 6.9 g carbohydrates; 2.2 g fiber; 1.3 g sugar; 4.7 g net carbs; 8 g protein

## Seafood Ceviche Tacos

*yield: 4 servings*

| | |
|---|---|
| 2 oz | red bell peppers, small diced |
| 2 oz | green bell pepper, small diced |
| 2 oz | red onion, small diced and rinsed |
| 1 | jalapeño, roasted and small diced |
| 1 tbsp | garlic, minced |
| ¼ c | cilantro, minced |
| ¼ tsp | coriander, ground |
| 6 tbsp | fresh lime juice |
| 1½ tbsp | olive oil |
| to taste | salt and pepper |
| 4 oz | bay scallops, raw |
| 4 oz | shrimp, peeled, deveined, and cut into medium dice |
| 4 oz | fish fillet, firm with white flesh |
| 4 oz | crabmeat |
| 8 | taco shells |

1. Combine all ingredients except seafood in a large bowl and mix well.
2. Add seafood and gently toss in the mixture to coat well.
3. Refrigerate for at least 1 hr.
4. Check seasoning once more and adjust as needed.
5. Serve in taco shells with a fresh cilantro garnish and a mild white Mexican grating cheese such as cotija.

**Nutrition Facts**

**Per serving:** 294.8 calories; 37% calories from fat; 12.4 g total fat; 79.6 mg cholesterol; 365 mg sodium; 520.9 mg potassium; 23.5 g carbohydrates; 3.1 g fiber; 1.6 g sugar; 20.4 g net carbs; 23.2 g protein

# Recipes

## 198

## Pan-Fried Buttermilk Walleye

*yield: 4 servings*

| | |
|---|---|
| 2 oz | buttermilk |
| 1 tsp | Dijon mustard |
| 4 | 6-oz to 8-oz walleye fillets |
| 4 oz | all-purpose flour |
| ¼ tsp | cayenne |
| ¼ tsp | paprika |
| ½ tsp | Old Bay® seasoning |
| ½ tsp | kosher salt |
| ¼ tsp | white pepper |
| | canola oil (for pan-frying) |

1. Combine buttermilk and Dijon mustard in a large bowl and mix well.
2. Add fish fillets to mixture and coat. Let fish marinate for 30 min.
3. In another bowl combine flour, cayenne, paprika, Old Bay®, salt, and white pepper and mix well.
4. Drain the fish from the buttermilk mixture and dredge in seasoned flour. Let rest for a few minutes.
5. Heat oil in a saucepan to 375°F.
6. Dredge the fish in the seasoned flour once again and then place fish directly into the hot oil.
7. Fry for 5 min to 7 min or until cooked through.
8. Plate and garnish as desired. Serve immediately.

### Nutrition Facts

**Per serving:** 655.6 calories; 74% calories from fat; 55.6 g total fat; 55.3 mg cholesterol; 343.6 mg sodium; 311.5 mg potassium; 22.7 g carbohydrates; 0.9 g fiber; 0.8 g sugar; 21.8 g net carbs; 16.8 g protein

## Grilled Indian-Spiced Mahi-Mahi with Mango and Mint Chutney

*yield: 4 servings*

| | |
|---|---|
| ½ tsp | cumin, ground |
| ½ tsp | nutmeg, ground |
| ½ tsp | cardamom, ground |
| ½ tsp | green curry paste |
| ¼ tsp | ginger, ground |
| 1 clove | garlic, crushed |
| 1 tbsp | water |
| 1 tbsp | olive oil |
| to taste | salt and pepper |
| 4 | 6-oz to 8-oz mahi-mahi fillets |

*Florida Department of Agriculture and Consumer Services, Bureau of Seafood and Aquaculture Marketing*

*Mango and mint chutney*

| | |
|---|---|
| 4 oz | mangoes, small diced |
| 1 tbsp | dried currants |
| ½ tsp | fresh ginger root, grated |
| 1 | jalapeño pepper, roasted, peeled, and seeded |
| 2 tsp | fresh mint, cut chiffonade |
| 2 tsp | lime juice |
| ¼ tsp | yellow curry powder |
| to taste | salt and pepper |

1. Make a paste by mixing the cumin, nutmeg, cardamom, curry paste, ginger, garlic, water, olive oil, salt, and pepper.
2. Rub the paste over the fish fillets.
3. Let fish marinate for 30 min.
4. Combine the chutney ingredients in a bowl and mix well.
5. Place fish on a well-oiled grill and grill for 4 min to 5 min per side or until cooked through.
6. Serve topped with mango and mint chutney.

### Nutrition Facts

**Per serving:** 471.2 calories; 14% calories from fat; 7.7 g total fat; 186.1 mg cholesterol; 294.6 mg sodium; 1720 mg potassium; 8.2 g carbohydrates; 1.3 g fiber; 5.9 g sugar; 6.9 g net carbs; 87.7 g protein

## Wasabi and Peanut-Crusted Ahi Tuna with Soy-Peanut Reduction

*yield: 4 servings*

| | |
|---|---|
| 4 | 4-oz ahi tuna steaks |
| 1 | egg white |
| to taste | salt and pepper |
| 3 tbsp | wasabi peas, coarsely chopped |
| 1 tbsp | roasted peanuts, chopped fine |

*Soy-peanut reduction*

| | |
|---|---|
| 1 oz | white wine |
| ½ oz | soy sauce |
| 1 oz | teriyaki sauce |
| 1 oz | water |
| 2 oz | creamy peanut butter |

1. Beat egg white in small bowl.
2. Season fish on both sides with salt and pepper.
3. Mix wasabi peas and peanuts.
4. Dip fish in egg white to coat.
5. Roll fish in wasabi peas and peanut mixture to coat completely.
6. To prepare soy-peanut reduction, add white wine to small saucepan and reduce by half.
7. Add all other ingredients and bring to a simmer.
8. Heat a sauté pan over medium heat and add oil. Sauté tuna until golden brown on both sides and cooked medium rare or medium.
9. Serve tuna drizzled with soy-peanut reduction.

### Nutrition Facts
**Per serving:** 176.5 calories; 40% calories from fat; 8.5 g total fat; 12.7 mg cholesterol; 659. 9 mg sodium; 369.1 mg potassium; 10.7 g carbohydrates; 3.4 g fiber; 3.1 g sugar; 7.3 g net carbs; 14.6 g protein

## Smoked Fish Croquettes with Watercress-Almond Pesto

*yield: 20 2-oz croquettes*

| | |
|---|---|
| 1 lb | smoked salmon (or any smoked fish) |
| 1 lb | cooked, peeled potatoes, riced |
| 3 | egg yolks |
| 1 oz | shredded horseradish |
| 2 tsp | dry mustard |
| to taste | salt and white pepper |

*Breading*

| | |
|---|---|
| 4 oz | seasoned all-purpose flour |
| 4 | whole eggs, slightly beaten |
| 1 c | seasoned panko bread crumbs |

*Watercress-almond pesto*

| | |
|---|---|
| 2 bunches | fresh watercress, blanched, shocked, and drained well |
| 4 cloves | garlic, minced |
| 1 c | extra virgin olive oil |
| 3 oz | almonds, sliced and toasted |
| 1 c | shredded Parmesan |
| to taste | salt and pepper |

1. Coarsely flake salmon and then mix gently with potatoes, egg yolks, horseradish, and mustard. Season with salt and white pepper.
2. Form salmon into 2-oz patties.
3. Mix the breading and follow standard breading procedure to bread the patties.
4. To prepare pesto, combine watercress, garlic, and half the oil in a blender or food processor, and process as quickly as possible to retain the bright green color.
5. Add almonds, Parmesan, and remaining oil and purée to achieve proper consistency. Set aside.
6. Fry in 350°F oil. Serve the fish patties immediately with pesto.

### Nutrition Facts
**Per croquette:** 257.9 calories; 60% calories from fat; 17.7 g total fat; 90.2 mg cholesterol; 683.1 mg sodium; 201.9 mg potassium; 14.4 g carbohydrates; 1.4 g fiber; 1 g sugar; 12.9 g net carbs; 10.7 g protein

# Recipes

## Crab and Corn Pancakes

*yield: 25 pancakes*

| | |
|---|---|
| ¾ c | all-purpose flour |
| 2 tbsp | baking powder |
| 1 c | cornmeal |
| 1 c | boiling water |
| 2 tbsp | honey |
| 2 | whole eggs, slightly beaten |
| 1 c | whole milk |
| 2 tbsp | melted butter |
| ½ lb | lump crabmeat |
| ¼ c | leeks (white only), minced |
| ¼ c | puréed cooked sweet corn |
| 2 tsp | parsley, chopped fine |
| to taste | salt and pepper |

1. Sift flour and baking powder together and set aside.
2. Combine cornmeal, boiling water, and honey and mix well, then remove from heat and let cool.
3. In a separate bowl, mix eggs, milk, and melted butter. Combine with the cornmeal mixture.
4. Add the flour mixture and stir just enough to combine, using caution to not overmix.
5. Fold in crabmeat, leeks, corn, parsley, salt, and pepper.
6. Preheat griddle to 400°F and scoop out batter into 3-oz portions.
7. When bubbles appear on the top of each pancake, flip the pancake over and then cook until done.
8. Serve immediately.

### Nutrition Facts

**Per pancake:** 68 calories; 26% calories from fat; 2 g total fat; 28.4 mg cholesterol; 168.5 mg sodium; 77.9 mg potassium; 9.4 g carbohydrates; 0.5 g fiber; 2.1 g sugar; 8.9 g net carbs; 3.3 g protein

---

## Cornmeal-Crusted Oysters with Spicy Fruit Salsa

*yield: 4 servings*

| | |
|---|---|
| 4 oz | cornmeal (or polenta) |
| 2 oz | flour |
| to taste | salt and white pepper |
| 12 | shucked oysters, shells scrubbed and reserved |
| ½ c | flour |
| 3 | eggs |
| | canola oil (for deep-frying) |
| | rock salt (for plating) |

*Salsa*

| | |
|---|---|
| 2 oz | mango, diced |
| 2 oz | papaya, diced |
| 2 oz | banana, diced |
| 2 oz | honeydew melon, diced |
| ½ tsp | lime juice |
| 2 tsp | cilantro, minced |
| ½ tsp | sugar |
| ½ tsp | cider vinegar |
| 2 shakes | Tabasco® sauce |
| to taste | salt and white pepper |

1. Mix the cornmeal with 2 oz flour and season with salt and pepper.
2. Bread the oysters using standard breading procedure.
3. In a separate pan, mix ½ c flour and eggs.
4. Prepare the salsa by tossing the fruit with rest of the salsa ingredients.
5. Deep-fry the oysters until crispy and place back into cleaned half oyster shells (3 per plate).
6. Plate oysters on a bed of rock salt.
7. Serve salsa in a bowl in the center of the plated oysters, or place a spoonful of salsa on top of each oyster and serve.

### Nutrition Facts

**Per serving:** 904.5 calories; 61% calories from fat; 63.1 g total fat; 233.6 mg cholesterol; 372.7 mg sodium; 562.9 mg potassium; 61.2 g carbohydrates; 3.9 g fiber; 6.9 g sugar; 57.3 g net carbs; 24.7 g protein

## Seafood Cioppino with Grilled Fennel Root

*yield: 4 servings*

| 2 tbsp | olive oil |
| 2 oz | onions, small diced |
| 2 oz | celery, small diced |
| 2 oz | red peppers, diced |
| 2 oz | green peppers, diced |
| 2 oz | fennel root, diced |
| 1 tbsp | minced garlic |
| 2 | bay leaves |
| 2 tsp | dried oregano |
| 1 tsp | dried red pepper flakes |
| ½ tsp | thyme |
| 1 tbsp | tomato paste |
| 1 lb | canned diced tomatoes, with juice |
| ¾ c | white wine |
| 1 qt | mild fish stock |
| 8 | mussels |
| 4 | slipper lobsters, cut in half |
| 12 oz | thick-fleshed fish (such as snapper or salmon), cut into 1-oz cubes |
| 12 | medium shrimp, peeled and deveined |
| 8 wedges | fresh fennel root, cut into 1½"-wide wedges, blanched, oiled, and grilled (for garnish) |
| 1 tbsp | parsley, chopped fine |

1. In a saucepan, sweat all the vegetables in the olive oil until slightly cooked.
2. Add the herbs and tomato products and gently sauté.
3. Deglaze with the white wine and reduce by half.
4. Add fish stock. Season well and let simmer.
5. Add seafood, stir well, cover, and simmer for 15 min or until cooked through.
6. Serve in a deep plate garnished with wedges of grilled fennel and crusty bread. Sprinkle with fresh chopped parsley.

### Nutrition Facts

**Per serving:** 480.1 calories; 23% calories from fat; 12.9 g total fat; 217.2 mg cholesterol; 1333 mg sodium; 2105.1 mg potassium; 21.4 g carbohydrates; 5.9 g fiber; 7 g sugar; 15.5 g net carbs; 62 g protein

## 205

## Poached Dover Sole and Salmon Roulade

*yield: 4 servings*

| 4 | Dover sole fillets |
| 20 | fresh spinach leaves |
| 6 oz | boneless skinless salmon, cut into 1" chunks (keep on ice) |
| 2 oz | cream |
| 1 tsp | salt |
| ½ tsp | white pepper |

1. Place the four Dover fillets side by side on plastic wrap and lay another sheet of plastic over the top.
2. Gently pound the wrapped fillets to make about half the thickness and to make them wider. Set aside when finished.
3. With an ice bath standing by, submerge spinach leaves for 3 sec to 5 sec in a saucepan of salted boiling water until just limp. Remove immediately and shock in ice bath.
4. When spinach is cold, remove from bath and lay on paper towels to drain. Set aside until needed.
5. Place ice-cold salmon in bowl of food processor and purée. When fairly smooth, add cream, salt, and white pepper and purée until very smooth. *Note:* Use caution to not overpurée, as cream will curdle.
6. Unwrap top sheet of plastic from sole fillets and spread fillets evenly with salmon mixture.
7. Lay spinach leaves over sole to completely cover salmon.
8. Using bottom sheet as an aid, begin to tightly roll each fillet into a log with the salmon and spinach inside.
9. When completely rolled, wrap the roulades tightly in silicone paper and poach for 15 min to 20 min until cooked through, using caution to not allow the water to reach a boil.
10. Remove the roulades from the water and let rest for a few minutes.
11. Remove silicone paper and slice into ¼"-thick slices. Serve immediately.

### Nutrition Facts

**Per serving:** 261.8 calories; 32% calories from fat; 9.5 g total fat; 108.4 mg cholesterol; 1609.1 mg sodium; 953.3 mg potassium; 2.4 g carbohydrates; 1.2 g fiber; 0.2 g sugar; 1.3 g net carbs; 40.2 g protein

# Potatoes, Pasta, and Grains

_Potatoes, pasta, and grains_ are considered staples of the human diet. These foods, commonly called starches, are low in fat and rich in carbohydrates. Because of their mild taste, starches are easily flavored by other ingredients. Potatoes are root vegetables and are rich in certain vitamins and minerals. Pastas are products made from a smooth, flour-based dough. Grains including rice, corn, and wheat are harvested from various wild grasses. Although many diet trends limit consumption of carbohydrate-rich foods, most nutritionists maintain that carbohydrates are an essential component of a well-balanced meal and a healthy diet.

## Chapter 21

## Chapter Objectives:

1. Describe the major classifications of potatoes.
2. Identify the guidelines for purchasing and storing potatoes.
3. Demonstrate the procedures for preparing deep-fried potatoes, baked potatoes, potato casseroles, simmered potatoes, and mashed potatoes.
4. Describe common types and subclassifications of pasta.
5. Demonstrate the procedures for making pasta dough with a mixer and by hand.
6. Describe how to cook and store pasta.
7. Contrast the procedures for preparing ravioli and tortellini.
8. Identify the four parts of a grain.
9. Contrast the various ways that grains are processed.
10. Identify the main characteristics of major types of rice, corn, wheat, and other common grains.
11. Explain how to store grain.
12. Demonstrate common grain preparation methods.

## Key Terms

- **potato**
- **fingerling**
- **solanine**
- **pasta**
- **gluten**
- **extrusion**
- **grain**
- **parcooked**
- **milling**
- **refined grain**
- **converted rice**
- **hominy**
- **grits**
- **buckwheat**
- **farro**
- **kamut**
- **millet**
- **quinoa**

## Russet Potatoes

Idaho Potato Commission

**21-1** Russet potatoes are used frequently for French fries, baked potatoes, and mashed potatoes.

## POTATOES

Potatoes are the world's second-largest food crop (behind rice). A *potato* is a round, oval, or elongated tuber vegetable that grows underground. Potato preparations are perhaps the most common side dishes. Potatoes are commonly categorized as starchy or mature potatoes and waxy or new potatoes. Sweet potatoes and yams are considered a separate category of potatoes due to their distinct color and difference in taste.

### Starchy or Mature Potatoes

The starchy potato, commonly called a "mealy potato," is higher in starch and lower in moisture than potatoes of other varieties. The most well known of the starchy potatoes is the russet potato. After cooking, starchy potatoes become light and fluffy inside. Because of this, they are the preferred choice for baking, mashing, puréeing, casseroles, and deep-frying. The low moisture in these potatoes causes less spattering in hot fat, making them the best choice for frying. Starchy potatoes are the type of potatoes most commonly served at restaurants as baked potatoes.

A *russet potato* is a potato with a reddish brown, thin skin, a long shape, and shallow eyes. Russets have excellent white flesh that is high in starch. Because a large percentage of these potatoes are grown in the state of Idaho, they are sometimes called Idaho potatoes. Russets are the best potatoes for French-frying, but can also baked or used to make mashed potatoes. Russet potatoes are also sometimes referred to as "bakers." ***See Figure 21-1.***

A *white potato* is an oblong starchy potato with a thin, white or light brown skin and tender white flesh. White potatoes have less starch than russet potatoes. They are excellent for baking, roasting, or making scalloped and au gratin potatoes. Popular varieties include the White Rose and the Cascade. ***See Figure 21-2.***

### Waxy or New Potatoes

Waxy potatoes include red potatoes, yellow potatoes, some varieties of fingerling potatoes, and purple potatoes. Potatoes of this type have thin skin and a slightly waxy interior. Compared to starchy potatoes, waxy potatoes stay much firmer in the center when fully cooked and retain their shape much better. Waxy potatoes can be prepared by boiling, steaming, roasting, braising, stewing, or sautéing. They are used more often than starchy potatoes for almost everything except deep-frying or baking. Because they keep their shape so well after being cooked, they are a great choice for potato salad. ***See Figure 21-3.***

A *red potato* is a round, red-skinned, waxy potato excellent for roasting, broiling, or grilling. Popular varieties of red potato include Chieftain, Norland, Red La Soda, and Klondike Rose.

A *yellow potato* is an oval waxy potato with yellowish skin and flesh and pink eyes. The flesh has a buttery, nutty flavor and remains a yellowish color after cooking. Yellow potatoes are excellent for roasting, boiling, or mashing. Popular varieties of yellow potato include Yukon Gold, Yellow Finn, and Provento.

A *fingerling* is a small, tapered, waxy potato. Three of the most popular types of fingerlings are the Russian Banana, Ruby Crescent, and French fingerling. The Russian Banana has a tapered shape and smooth tan or yellow skin with a butter-colored flesh and buttery flavor. Russian Bananas are excellent for boiling, baking, roasting, or steaming and are a great addition to salads. Ruby Crescents have rose-colored skin and yellow flesh and are excellent for boiling, roasting, or steaming. French fingerlings are small, long potatoes with purple-pink skin and yellow waxy flesh with a pink streak through the center. They are versatile potatoes that can be used to add visual interest to a meal.

## White Potatoes

21-2   *White potatoes are best for baking or roasting, or making scalloped or au gratin potatoes.*

## Waxy Potatoes

Red potatoes

Fingerling potatoes

Purple potatoes

21-3   *Waxy potatoes include red, yellow, and purple varieties.*

A *purple potato* (blue potato) is a potato with smooth, thin, purple skin and deep purple flesh. They have a fine waxy texture and a nutty flavor. Purple potatoes can be prepared by baking, mashing, frying, or steaming.

The term *new potato* (early crop potato) refers to any variety of potato that is harvested before its starches fully develop. New potatoes are potatoes typically harvested before they grow larger than 1½″ in diameter. Because they are harvested so small, they are relatively uniform in size. They have delicate skin and tender, waxy flesh that works well for roasting, steaming, and boiling. They keep their shape beautifully after cooking. New potatoes cannot be stored as long as other potatoes because of their high moisture content. ***See Figure 21-4.***

## New Potatoes

21-4   *New potatoes are harvested before they reach 1½″ in diameter.*

## Sweet Potatoes and Yams

**21-5** Sweet potatoes and yams, although often confused for one another, are actually different species of plants.

## Potato Eyes

**21-6** Any part of the potato that is green or has eyes should be removed prior to preparation.

### Sweet Potatoes and Yams

Sweet potatoes and yams share a similar appearance but are actually different species of plants. ***See Figure 21-5.*** A *sweet potato* is a low-starch/high-moisture tuber that has sweet, orange-colored flesh and a thin skin. The ends of the sweet potato are tapered. Sweet potatoes are commonly used in the professional kitchen because they can be prepared in many different ways. They bake, mash, fry, and sauté with excellent results. They complement pork and poultry, providing an exciting taste experience when served with these items. Sweet potatoes are usually roasted or boiled. Though they taste sweet when boiled, their sweetness can be greatly intensified by roasting or baking, as the sugars begin to caramelize and concentrate. Their insides are then mashed, with the addition of brown sugar and butter.

Yams are often confused with sweet potatoes, although the yam actually belongs to an entirely different family of tubers. A *yam* is a low-starch/high-moisture tuber that is not nearly as sweet as a sweet potato. Yams have much thicker skin that is harder to peel, with very hard, almost woody, flesh. Like sweet potatoes, yams can be round or oblong in shape. They are much starchier and have an earthier flavor and yellower flesh than that of a sweet potato. They can be prepared using the same methods used for other tubers, such as boiling and mashing.

### PURCHASING AND STORING POTATOES

Potatoes are purchased in many market forms including fresh, canned, frozen, and dehydrated. Fresh potatoes are purchased in large quantities to save on cost. Fresh potatoes should be stored in a cool, dry place.

Good-quality potatoes are firm when pressed in the hand. They should be clean and have shallow eyes. The eyes are the areas where the potato is beginning to sprout. Eyes should be removed from potatoes prior to cooking. When cut, they should display a yellow-white color, and moisture should appear on the cut side. Areas of a potato that are starting to sprout or that have a greenish color to the flesh or skin need to be removed and should not be eaten because they contain a toxin called solanine. *Solanine* is a natural toxin that is not harmful when consumed in small amounts but can be harmful if eaten in large quantities. Any part of the potato that has sprouts, eyes, or a green-colored surface should be removed. The remainder of the potato, however, is completely safe to consume. ***See Figure 21-6.***

### Market Forms of Potatoes

Potatoes can be purchased in five market forms: fresh, fresh processed, canned, frozen, and dehydrated. The market form to purchase depends on the need, the time element of preparation, equipment, storage, and cost. Many chefs believe there is no substitute for the fresh potato. However, various forms of potato available on the market may reduce time and labor and can sometimes result in cost savings. ***See Figure 21-7.*** Frozen precooked French fries, dehydrated mashed potatoes, and canned cooked whole potatoes are used extensively with good results. Establishments can even purchase real (not dehydrated) mashed potatoes if they choose to not make them from scratch.

## Purchasing Potatoes

Potatoes are generally sold to food service establishments in 50-lb cases. Starchy potatoes are packed according to the size of the potato or by how many potatoes make up a 50-lb case. ***See Figure 21-8.*** They may be packaged as 40-count potatoes, 60-count potatoes, and 90-count potatoes. In a 50-lb box of 90-count potatoes, there are approximately 90 potatoes weighing around ½ lb each. When receiving a shipment of potatoes, the potatoes must be inspected to ensure they are firm and undamaged and show no signs of sprouting or green patches. Potatoes that are purchased cleaned have shorter storage life because cleaning removes a protective outer coating that deters bacteria. In addition, it is impractical for a food service establishment to purchase cleaned potatoes because they are expensive and would need to be washed again regardless.

## Grading Potatoes

Potatoes, like other vegetables, can be voluntarily graded by the USDA. US Fancy is the highest grade and is found only in high-end restaurants. Most wholesale potatoes sold have a grade of US No. 1. Potatoes that are sold in retail grocery stores are usually US Grade A, but can be Grade B.

## Potato Storage

Under the proper conditions, potatoes can be stored for a period of up to four months. The conditions under which potatoes are stored determine the length of time they keep. Potatoes should be stored in a cool, dark, dry place with temperatures between 39°F and 50°F. The higher the temperature under which potatoes are stored, the shorter the lifespan of the potato, because potatoes begin to sprout and dehydrate under higher temperatures. Refrigeration is not recommended, as excessively cold temperatures increase the concentration of starch in potatoes and would result in a gummy texture. Potatoes are best stored in cardboard or brown paper bags. Plastic bags are not recommended because they do not allow adequate ventilation and support the growth of mold. ***See Figure 21-9.*** Containers of potatoes should be kept out of sunlight to prevent the formation of chlorophyll, which turns potatoes green.

## Market Forms of Potatoes

Hash brown potatoes

Tater tots

Instant mashed potatoes

*U.S. Department of Agriculture*

**21-7** *Frozen precooked potato products and instant mashed potatoes may save time and reduce labor cost in the professional kitchen.*

## Purchasing Potatoes

*Idaho Potato Commission*

**21-8** *Potatoes such as russets are packaged by count per 50 lb case.*

## Potato Storage

Idaho Potato Commission

**21-9** Potatoes should be stored between 39°F to 50°F in cardboard boxes or paper bags to allow adequate ventilation.

Chef's Tip:
Rinse excess starch from cut potatoes by again discarding the water they were stored in and covering the potatoes with clean cold water. Rinsing the potatoes in this way removes additional starch and helps the potatoes to not stick together when frying.

## POTATO PREPARATION

Potatoes are one of the most popular food items prepared in the professional kitchen. They are served for breakfast, lunch, and dinner and can be prepared in a variety of ways. Baked potatoes are popular to serve with dinner entrées such as fish or beef. French-fried potatoes are popular to serve with hot sandwiches. Potatoes can be deep-fried, sautéed, grilled, roasted, steamed, simmered, or used in a braised or stewed dish. Many potato preparations use more than one cooking method. The cooking method used is determined by the form and type of equipment available, the recipe used, and the desired result.

Uncooked potato begins to discolor immediately after it is cut or peeled. A cut raw potato first turns a pinkish shade and then quickly turns tan or grey, eventually discoloring to black. To prevent discoloration, potatoes should be placed in cold water as soon as they are cut or peeled and kept in cold water until they are ready to be cooked.

### Deep-Frying Potatoes

Deep-fried potatoes such as French fries and potato chips are a popular accompaniment to many menu items. French fries can be prepared in a variety of shapes including waffle-cut fries, shoestring fries, steak fries, and cottage fries. Low-moisture potatoes such as russet potatoes produce the best results when deep-frying, becoming crispy on the outside and light and fluffy on the inside. Compared to other types of potatoes, low-moisture potatoes spatter the least in hot oil and stay crisp for the longest period after being cooked.

Very thin-cut potatoes, such as potato chips or shoestring fries, can be deep-fried in a single step, usually in hot oil between 350°F and 375°F. Potatoes that are cut slightly thicker, such as French fries, first need to be blanched in a slightly lower-temperature oil around 275°F. Blanching removes some of the natural internal moisture from a potato, causing the potato to cook more evenly, obtain a more uniform golden brown color, and stay crisper for much longer. If the potato were cooked in 375°F oil without blanching, the exterior would instantly sear, preventing the interior moisture from escaping. The trapped internal moisture would cause the potato to become limp and soggy soon after frying.

### To deep-fry potatoes, apply the following procedure:

1. Wash, peel (if desired), and cut potato to desired size for frying. Store peeled potatoes in cold water. When ready to fry, completely drain the precut potatoes and carefully pat dry. ***See Figure 21-10.***
2. Blanche the cut, rinsed, and dried potatoes in 275°F oil until just slightly undercooked.

3. Remove potatoes from hot oil and drain well.

4. Place the drained fried potatoes on sheet pans and store in refrigerator until cool. *Note:* Potatoes can be held in this state in the refrigerator for a few hours or can be frozen for approximately one month.

5. When fried potatoes are needed, place the partially cooked potatoes in 375°F oil and fry until golden brown and crispy. If using a basket in a deep-fat fryer, use caution to not overfill the basket.

## Deep-Frying Potatoes

1. Wash, peel, and cut to desired size; store in cold water; drain well before use.

2. Blanch in 275°F oil until slighty undercooked.

3–4. Remove from oil and drain well; refrigerate until cool.

5. When needed, fry in 375°F oil until golden brown.

**21-10** *Potatoes are blanched, cooled, and stored before being deep-fried.*

### Sautéing or Pan-Frying Potatoes

Waxy potatoes such as red potatoes work best for sautéing. Sautéed potatoes are usually parboiled or steamed before being finished in the sauté pan to give them a golden brown, crunchy exterior. If they are boiled or steamed first, they should be removed from the water and allowed to steam dry before they are sautéed. They should not be rinsed under cold water to cool, because they would absorb too much water and would not brown properly. ***See Figure 21-11.***

### Grilling Potatoes

Potatoes can be grilled to produce a golden color and smoky flavor. When grilling potatoes, they can be sliced into ½"-thick wedges or ¼"-thick slices, coated with oil and seasonings, and placed on a preheated grill surface. The potatoes should be grilled until al dente. Overcooking potatoes makes them soft and difficult to remove without breaking.

## Potato Preparations

*National Potato Promotion Board*

**Pan-Fried Potatoes**

*Idaho Potato Commission*

**Roasted Potatoes**

**21-11**   *Sautéed or pan-fried potatoes are usually parboiled or steamed first.*

### Roasting and Baking Potatoes

Potatoes can be roasted or baked. "Roasted potatoes" usually refer to potatoes that are cut up, tossed with oil or other seasonings, and cooked with dry heat in an oven. "Baked potatoes" refers to potatoes that are cooked whole in an oven. In addition, various potato dishes such as casseroles are prepared by baking in an oven.

Any kind of potato can be baked, but some varieties work better for different preparations. The true "baked potato" that is served with butter and sour cream is produced from a russet potato. These low-moisture potatoes make the best baked potatoes, which are fluffy and light and, since they are low-moisture potatoes, absorb butter and sour cream, making them taste even better.

The common baked potato is baked with the skin intact. Potatoes can be baked directly on the oven rack or can be placed on sheet pans for easy removal from the oven (which is especially convenient when preparing many potatoes for a large function). The skin is washed thoroughly and pricked with a fork prior to baking. The tiny holes left by the fork allow moisture to escape during the baking process. An average-size 90-count potato takes about 1 hr to bake.

Waxy potatoes can also be roasted in the oven with wonderful results. Potatoes such as red, fingerling, or yellow potatoes can be tossed lightly in oil, seasoned, placed in a roasting pan, and cooked until golden brown on the outside and soft and creamy on the inside. Starchy potatoes could also be roasted in this way but would have a slightly tougher exterior. The potatoes are cooked properly when a fork inserted into a potato has little resistance.

Baked potatoes should be fluffy on the inside and are traditionally served with sour cream, butter, salt, pepper, and chives. Baked or roasted potatoes should be served as quickly as possible because their quality deteriorates if they are held for too long after being cooked.

**To prepare baked potatoes, apply the following procedure:**

1. Select uniform-size starchy potatoes such as russet potatoes.

2. Wash potatoes thoroughly.

3. Pierce the potatoes with a fork. ***See Figure 21-12.*** *Note:* Piercing the potatoes with a fork allows moisture to escape during the baking process, which is important to the texture of the finished product.

4. Place on an oven rack or sheet pan and bake at a temperature of 400°F for approximately 45 min to 1 hr, until the potatoes become slightly soft when squeezed gently.

CULINARY PROCEDURES

HELPFUL TIPS

*The following tips are helpful in preparing baked potatoes:*

- Prior to baking, some chefs lightly rub the potatoes with salad oil and sprinkle them with salt to crisp the skin and add flavor.

- Wrapping potatoes that are to be baked is not recommended. Wrapping potatoes in foil before they are baked causes them to cook by steaming rather than by baking and results in an undesirably dense, mealy texture. Baking potatoes on simply a sheet pan or the oven rack results in crisp skin and a fluffy interior.

A *casserole* is a baked dish of mixed foods containing a starch (such as potatoes, pasta, or rice), other ingredients (such as meat or vegetables), and a sauce. Starchy potatoes are the best choice for use in a baked casserole recipe. Because these potatoes are low in moisture, they can absorb the most flavor from a sauce. Potatoes used in a casserole can be sliced raw or can be partially cooked prior to slicing. Partially cooking the potatoes prior to baking decreases required baking time. The type of sauce varies and can be anything from a cream sauce to a stock.

A casserole can be baked with or without a topping of other ingredients such as bread crumbs or cheese. In French, *gratinée* (pronounced grahteen-YAY) refers to the process of topping a dish with a thick sauce, cheese, or bread crumbs and then browning it in a broiler or high-temperature oven. The French term *gratin* refers to any dish prepared using the gratinée method. Common casserole dishes using potatoes are scalloped potatoes and potatoes au gratin.

*Baking Potatoes*

*1–3. Select potatoes, wash thoroughly, and pierce with fork.*

*4. Place on sheet pan or rack and bake at 400°F until slightly soft.*

**21-12** *Baked potatoes are pierced with a fork to allow steam to escape while cooking.*

To prepare a potato casserole, apply the following procedure:

1. Wash and peel the potatoes. Store peeled potatoes in cold water to avoid browning. ***See Figure 21-13.***

2. If precooked potatoes are desired, steam or simmer them until half cooked. Drain any liquid from the potatoes and allow them to cool slightly.

3. Slice potatoes to a uniform thickness of ¼″ and layer uniformly in a casserole dish.

4. Cover the potatoes with desired sauce and any other ingredients.

5. If desired, top with cheese or bread crumbs. Bake according to recipe instructions.

*Preparing Potato Casserole*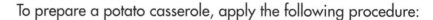

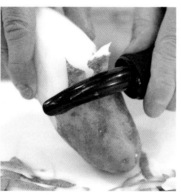

1–2. Wash, peel, and store potatoes in cold water; steam or simmer until half cooked.

3. Slice to uniform thickness and layer in casserole dish.

4. Cover with desired sauce and other ingredients.

5. Top with cheese or bread crumbs.

**21-13** Scalloped potatoes and au gratin potatoes are examples of potato casseroles.

*The following tips are helpful when preparing potato casseroles:*

- Adding the sauce ingredient already warm decreases cooking time.
- The sauce ingredient should be well seasoned as it is the dominant flavor of the dish. Typically salt and white pepper are added.
- A mandolin can be used to cut the potatoes for a casserole very consistently and quickly.

## Moist-Heat Cooking Methods

Any kind of potato can be simmered with good results. However, simmered potatoes of certain types work best for different uses. If the simmered potatoes are to be mashed potatoes, starchy potatoes work best. If they will not be mashed, such as whole or quartered potatoes to be served with a corned beef and cabbage dinner, or a potato to be used in potato salad, then waxy potatoes or new potatoes work well, because they retain their shape and do not fall apart as easily as starchy potatoes.

If simmered potatoes are to be served whole (not mashed, whipped, or puréed), place them on a sheet pan in a single layer to steam dry. Potatoes are sufficiently dried and are ready to serve or garnish when steam no longer rises from the potatoes. The most common preparation for simmered potatoes is to season them with butter, chopped fresh parsley, salt, and pepper.

## To simmer potatoes, apply the following procedure:

1. Either scrub the surface of the potatoes or peel to remove the skin. All eyes, sprouts, or green surfaces must also be removed. *See Figure 21-14.*

2. If the potatoes are large, cut it into consistent-size pieces so that they cook evenly. If the potatoes are to be cooked whole, make sure that all the potatoes are fairly similar in size.

3. Always start potatoes in cold liquid with salt added (usually water, although other liquids such as a light stock can be used). Using cold liquid when beginning to simmer the potatoes helps ensure the potatoes cook evenly.

4. Bring the liquid containing the potatoes to a boil, and then lower to a simmer.

5. Potatoes should be simmered until very tender. Remember that waxy or new potatoes retain their shape when fully cooked, while starchy potatoes begin to break up.

6. To test for doneness, pierce the potatoes with a fork. If the fork penetrates a potato with little or no resistance, that potato is cooked fully.

7. Drain the potatoes in a colander to remove all liquid.

## Simmering Potatoes

1–2. Scrub or peel potato; cut to uniform size.

3–4. Start potatoes in cold liquid with salt; bring to a boil and reduce to a simmer.

5–6. Test for doneness with a fork; drain to remove all liquid.

**21-14** Simmering is one of the most common methods of potato preparation.

CULINARY PROCEDURES

**To mash or whip simmered potatoes, apply the following procedure:**

1. Return drained, cooked potatoes to the pot they were cooked in and warm them uncovered over low heat until all visible water evaporates and steam no longer rises. *See Figure 21-15.*

2. In a separate pot, heat any additional ingredients such as milk, cream, stock, or butter to a temperature equal to that of the potatoes.

3. Place potatoes in a mixer and mash with a paddle attachment until fairly smooth.

4. Add hot ingredients. For mashed potatoes, mix well with paddle attachment. If whipped potatoes are desired, change the paddle attachment to a whip attachment and whip until light and fluffy.

5. Adjust seasonings to taste.

To purée potatoes, simmered potatoes are drained and then placed in a ricer or food mill and forced through until a fine, smooth texture is achieved. Any hot ingredients and seasonings such as milk, butter, and salt are added and mixed into the puréed potatoes.

## Mashing or Whipping Potatoes

1–2. Heat cooked potatoes until water is evaporated or steam no longer rises; heat additional ingredients in a separate pan to same temperature as hot potatoes.

3. Place potatoes in mixer and mash with paddle attachment.

4–5. Add hot ingredients using whip attachment; adjust seasonings.

21-15 A paddle is used to mash potatoes, while a whip is used to whip the potatoes to a finer texture.

*The following tips are helpful when whipping, mashing, or puréeing potatoes:*

- Use caution to not overwhip potatoes or overwork a potato purée. Overworking potatoes causes excess starch to be released and results in a gummy, sticky texture.
- Additional flavors can be added to the potatoes to make a dish more interesting or complex. Ingredients such as puréed roasted garlic, chopped basil or other fresh herbs, sun-dried tomato purée, various cheeses, or pesto can be added.

## Combination Cooking Methods

Potatoes are commonly added to braised or stewed dishes. Not only do they add additional texture, flavor, and nutrients to a stewed or braised dish, but potatoes also help to thicken sauces by releasing starch during the cooking process. It is important to ensure potato size is appropriate to the cooking time. For example, a dish that cooks for a long period requires larger pieces of potato than a dish that cooks for a short period. If a dish that has a long required cooking time calls for small diced potatoes, the potatoes should be added late in the cooking process to avoid overcooking.

## PASTA

*Pasta* is a term for rolled or extruded products made from a dough of flour, water, salt, oil, and sometimes eggs. Pasta is one of the most versatile food products used in the professional kitchen because it is inexpensive, stores well, and has a mild flavor that does not compete with the flavor of sauces, herbs, or other ingredients.

Most pasta products are made from wheat flour such as semolina flour or farina flour. These flours are made from hard wheat, which contains a very high percentage of gluten. *Gluten* is a combination of proteins in flour that provides the strength to hold the shape, form, and texture of bakery products when cooked. Pasta products are formed into many different shapes and sizes for variety and function. Ingredients are sometimes added to pasta dough to produce colored pasta. For example, tomato paste produces red pasta, spinach purée produces green pasta, or squid ink produces black pasta.

### Types of Pasta

Pasta can be made in countless shapes and sizes. ***See Figure 21-16.*** When pasta dough is soft, it is shaped by rolling it flat and cutting it to desired size, or by extrusion. *Extrusion* is the process of shaping pasta by pushing the dough through dies that create various shapes. The dough is then allowed to dry, resulting in a hard, mold-resistant product that can withstand shipping and storage. The shape of pasta used in a given preparation is often determined by the sauce and how it clings to the particular pasta shape. In addition, the pasta shape should complement the appearance of the finished preparation. Pasta products are classified by size and shape into four general categories: ribbon, tube, extruded shape, and hand formed.

**Tube Pasta.** Tube pastas are pushed through a tube-shaped pasta extruder. A *pasta extruder* is a machine that forces dough through a shaped die to create pasta of various shapes. The extruded pasta dough is then fed through a tube cutter that cuts the tubes to desired length.

- Elbow macaroni—Straight, round, hollow tubes of pasta that are slightly curved. Elbow macaroni is one of the most popular types of pasta and can be topped with any sauce, baked, or put in soups and salads.

- Penne—Hollow, diagonally cut tubes, 1½″ to 2″ in length with a smooth surface. Penne is usually baked in pastas with meat sauces.

- Mostaccioli—Hollow, diagonally cut tubes 1½″ to 2″ in length with a ribbed surface.

## Types of Pasta

| Type, size, and shape | | Type, size, and shape | |
|---|---|---|---|
| **Spaghetti** Long, round, solid rod; ³⁄₃₂" diameter |  | **Manicotti** Large, round tubes of pasta, straight or diagonally cut; 1" to 1½" in diameter *Dakota Pasta Growers Co.* |  |
| **Vermicelli** Long, round, very thin solid rod; ¹⁄₃₂" diameter |  | **Cannelloni** Large, round tubes of pasta, straight cut; 1" in diameter; ribbed exterior *National Pasta Association* |  |
| **Capellini** Very fine; round, strandlike rod |  | **Conchiglie** Miniature shells *Dakota Pasta Growers Co.* |  |
| **Linguine** Long, thin, flat strip; ⅛" wide |  | **Jumbo Shells** Large shells *Dakota Pasta Growers Co.* |  |
| **Fettuccini** Long, thin, flat strip; ¼" wide |  | **Fusilli** Corkscrew shape *Dakota Pasta Growers Co.* |  |
| **Lasagna** Flat, extra wide, rippled-edge strip; 2" wide *Dakota Pasta Growers Co.* |  | **Orzo** Small, oval shape, similar to a grain *Dakota Pasta Growers Co.* |  |
| **Egg Noodles** Flat strip of varying length; classified as thin, medium, or wide according to width *Dakota Pasta Growers Co.* |  | **Farfalle** Bow tie shape *Dakota Pasta Growers Co.* |  |
| **Elbow Macaroni** Straight, round, hollow tubes that are slightly curved *National Pasta Association* |  | **Ravioli** Filled envelopes of pasta |  |
| **Penne** Hollow, diagonally cut tubes 1½" to 2" in length; smooth exterior *Dakota Pasta Growers Co.* |  | **Tortellini** Ring-shaped filled pasta |  |
| **Mostaccioli** Hollow, diagonally cut tubes 1½" to 2" in length; ribbed exterior |  | **Rotini** Twisted-shaped pasta *Dakota Pasta Growers Co.* |  |

 **21-16** *Pasta is made in a wide variety of shapes and sizes.*

- Manicotti—Large, round tubes of pasta. The diameter of the tube is approximately 1″ to 1½″. Manicotti can be either straight or diagonal cut. They are stuffed with a meat, poultry, or cheese filling and baked in a rich sauce.

- Cannelloni—Similar to manicotti, with the differences being a ridged outer surface and that the ends are cut straight.

**Ribbon Pasta.** Ribbon pastas are thin, round strands or flat, ribbon-like strands of pasta. To form ribbon pasta, the pasta dough is rolled out by hand or with a pasta machine, and then cut to desired width. A *pasta machine* is a motor-driven or hand-cranked machine with a set of rollers through which pasta dough is passed to flatten the dough to a desired thickness. ***See Figure 21-17.*** Round strands of pasta are formed by pressing the dough through round dies. The following are common types of ribbon pastas:

- Spaghetti—Long, round, solid rods of pasta. The diameter of the rod is approximately 3/32″. Spaghetti is one of the most popular pastas, and can be used with any sauce. Very thin strands of spaghetti are referred to as spaghettini.

- Vermicelli—Long, round, thin solid rods of pasta. The diameter of the rod is approximately 1/32″. Vermicelli is sold in both straight and coiled forms, though the straight variety is more popular. Vermicelli is used most often in soup.

- Capellini (angel hair)—Very fine, round, strand pasta, thinner than spaghetti or vermicelli. It works well with lighter sauces or olive oil.

- Linguine—Long, thin, flat strips of pasta, about ⅛″ wide. Linguine is a good shape for all sauces.

- Fettuccine—Long, thin, flat strips of pasta, about ¼″ wide. It is perfect for heavier sauces, such as a white sauce.

- Lasagna—Flat, ripple-edge pasta, about 2″ wide. Lasagna is most often layered with meat, sauces, or ingredients.

- Egg noodles—Flat, ribbon-shaped pastas that can be cut long or short and are classified as thin, medium, or wide according to width. Egg noodles must contain at least 5.5% egg solids, so a pound of noodles must contain the equivalent of approximately two or more eggs.

**Extruded Shape Pasta.** Extruded shape pastas are also pushed through a pasta extruder, but are made into various different shapes such as corkscrews, shells, or wheels. Conchiglie is a miniature shell-shaped pasta commonly used in soups or cold salads. Jumbo shells are large shell-shaped pasta that can be served plain or stuffed with meat, poultry,

## Pasta Machine

Browne-Halco (NJ)

**21-17** A pasta machine is used to roll pasta dough to a consistent thickness.

seafood, vegetables, or cheeses. Fusilli is shaped like a corkscrew. The twisted shape holds bits of meat, vegetables, and cheeses very well. Rotelle is a wheel-shaped pasta most often used in soups. Orzo is small oval pasta with an appearance similar to that of a grain. Orzo is available in a variety of colors and can be used in soups, salads, and side dishes.

**Hand-Formed Pasta.** Some types of pasta are prepared from soft dough that is formed by hand or a machine. These shapes are often filled with meats, cheeses, or vegetables. Farfalle, ravioli, and tortellini are examples of hand-formed pastas.

Farfalle (bow tie pasta) is a flat square of pasta that is pinched in the center to make the shape of a bow tie. Farfalle is commonly served with a medium-density sauce, such as a marinara. Ravioli is also formed from flat pasta, in equal-size squares or other shapes. To make ravioli, one sheet of pasta is topped with a filling, and a second sheet of pasta is laid on top. The top sheet is pressed down around the fillings, and individual ravioli are formed by cutting around the fillings with a pastry wheel, ravioli cutter, or knife. Ravioli is commonly served with a rich red or white sauce and Parmesan cheese.

Tortellini, like ravioli, is a filled pasta. When making tortellini, very thin pasta dough is cut into circles and a filling is placed in the center of each circle. The circle is folded in half, and the edges are pressed together to hold the filling in place. The half circle is then formed into a ring by wrapping it around a finger and pressing the two ends together.

## Asian Noodles

Asian noodles are unique both in the types of flour used and in preparation methods. Because of the unique ingredients or methods used in making many types of Asian noodles, they are most often purchased rather than made in the kitchen. Asian noodles are commonly made from buckwheat flour, rice flour, or bean starch. Many Asian noodles are prepared for use by soaking the noodles in warm water rather than boiling them. Types of Asian noodles include soba, mung bean, rice, ramen, somen, udon, hokkien, and mian noodles.

**Soba Noodles.** Soba noodles are brownish-grey noodles made with buckwheat flour and are traditional in Japanese cuisine. They can be used hot or cold, in soups, main dishes, and side dishes. Soba noodles can be substituted for spaghetti if desired.

**Cellophane Noodles.** Cellophane (Mung bean) noodles are made from bean starch. They are somewhat transparent in appearance and are commonly rehydrated in hot water before use. Mung bean noodles can be deep-fried to produce a crispy and crunchy delicate noodle. If deep-frying, they do not need to be presoaked. Mung bean noodles are used hot or cold, in soups or stir-fries or as fillings, especially in vegetarian dishes. Similar noodles are available made from sweet potato starch.

*Chef's Tip:*
Mung bean noodles are extremely hard before they are soaked, and break into sharp fragments if broken apart by hand. To avoid injury, use kitchen shears to cut the noodles into pieces, rather than pulling them apart by hand.

**Rice noodles.** Rice noodles are made from ground rice flour and are available dried in various shapes, including sheets, flat sticks of various widths, or vermicelli. They should be soaked in hot water before they are cooked and thoroughly rinsed after cooking. If they are not rinsed after cooking, excess starch remains on the noodles, causing them to turn into a sticky clump. Thin rice noodles can be deep-fried to produce a crispy, delicate noodle that can be used as a base for an entrée, in salads, or even in some desserts.

**Ramen.** Ramen noodles are long, thin wheat noodles, made with or without egg. ***See Figure 21-18.*** While in North America ramen noodles are often associated with instant noodles, ramen can take many forms. Ramen noodles are available dehydrated, fresh, or frozen, and may be packaged in a formed brick shape or as straight sticks. Instant ramen noodles are deep-fried and dehydrated, and are most often sold with a packet of seasoning or vegetables. Ramen noodles are used with different combinations of ingredients and flavors in various Asian cuisines. They are most often used to make soups and can also be used in salads.

**Somen.** Somen noodles are long, thin wheat noodles that are white in color and are sold as rods, similar to vermicelli. Varieties of somen are available that include egg yolk or green tea. Somen is often sold bundled in a dried form. It is used in stir-fries and soups or served cold in noodle dishes.

**Udon.** Udon noodles are thick, white wheat noodles. They are available in fresh, dried, or instant form, as flat, round, or square strands. Udon noodles have a chewy, slippery texture that works well in soups, stews, or stir-fries, with meat dishes, or served cold.

**Hokkien.** Hokkien noodles are thick, yellow egg noodles resembling spaghetti. They are sold fresh, dried, or vacuum-sealed. Hokkien noodles work well in soups or stir-fries or served with a sauce.

**Mian.** Mian noodles are Chinese wheat noodles sold in round strands or ribbons of various widths. Mian are available fresh or dried. Some varieties include egg yolk. Mian are popular in a wide range of dishes including soups, stir-fries, and chow mein.

## PASTA PREPARATION

Some pasta preparations are often named after the pasta ingredient used in the dish. For example, macaroni and cheese, chicken fettuccine, and lasagna are named after macaroni, fettuccine, and lasagna pasta, respectively. Pasta is often purchased ready to use because it is convenient, inexpensive, available in many different varieties, and much less labor intensive when compared to pasta made from scratch. Some chefs prefer to make pasta from scratch because fresh pasta is more tender and allows the chef to flavor, color, and fill it as desired. If pasta is to be made from scratch, the dough can either be processed in a mixer or kneaded by hand.

*Ramen Noodles*

The Beef Checkoff

**21-18** Ramen noodles are used with different combinations of ingredients in various Asian cuisines.

To prepare pasta dough with a mixer, apply the following procedure:

1. In a mixer with a paddle attachment, combine eggs, salt, and oil, and mix well. *See Figure 21-19.*

2. Begin by adding approximately ⅓ of the flour and mix until the mixture forms a smooth, soft dough.

3. Replace the paddle attachment with the dough hook and add the rest of the flour to knead the mixture thoroughly.

4. Remove the dough from the mixer and place in a bowl. Cover the bowl with plastic wrap and allow the dough to rest for 20 min.

5. Once the dough has rested, it can be rolled and cut into shapes.

CULINARY PROCEDURES

## Preparing Pasta Dough with a Mixer

*1. Mix eggs, salt, and oil in mixer with paddle attachment.*

*2. Add ⅓ of flour and mix until smooth; replace paddle with dough hook.*

*3–4. Add remaining flour and knead with dough hook until dry; place dough in bowl; cover and let rest.*

*5. After allowing gluten to rest; roll and cut into desired shapes.*

**21-19** *A mixer can be used to prepare pasta dough from scratch.*

To prepare pasta dough by hand, apply the following procedure:

1. On a clean work surface, place flour in a mound and form a well in the center. *See Figure 21-20.*

2. Place eggs, oil, and salt in the well.

3. Slowly begin working a small amount of the flour into the well, incorporating with the egg mixture. Continue mixing in the flour until all wet and dry ingredients are mixed.

4. Knead the mixture until a smooth, dry ball of dough has formed.

5. Cover the dough and allow it to rest for 20 min on the work surface, or place the dough in a bowl, cover, and allow to rest for 20 min.

6. Roll dough to desired thickness and cut into desired shapes.

## Preparing Pasta Dough by Hand

1. Place flour in a mound and form a well in center.

2. Place eggs, oils, and salt in the well.

3. Work flour into well to incorporate eggs; continue mixing until all wet and dry ingredients are mixed.

4–5. Knead mixture into a smooth, dry ball of dough; cover and let rest for 20 min.

6. Roll dough to desired thickness and cut into shapes.

21-20 If small amounts of pasta dough are to be prepared from scratch, the standard procedure can be modified for hand mixing.

Most pasta is cooked in boiling water. Pasta should be checked for doneness frequently as the cooking process nears completion. Pasta should be cooked al dente, which means "to the tooth" and refers to a texture that requires a bit of a bite to chew, rather than being mushy. Pasta can be tested by removing a piece from the pot and pressing it between the thumb and forefinger. It should be soft but firm, not mushy.

A pound of raw pasta yields approximately 2 lb to 2½ lb when cooked. A rule of thumb is that pasta at least doubles in volume when cooked. Cooking time varies, depending on the shape, size, and quality of the pasta. Fresh pasta requires less cooking time than dried pasta. Pasta should always be cooked uncovered, being careful not to allow a large head of froth to boil over from the pot. When the minimum cooking time is reached, the pasta should be tested frequently to ensure the proper doneness. All pasta products are cooked using the same basic procedure. ***See Figure 21-21.***

| Approximate Cooking Times For Dried Pasta | | | | | | | |
|---|---|---|---|---|---|---|---|
| Pasta | Minutes* | Pasta | Minutes* | Pasta | Minutes* | Pasta | Minutes* |
| Spaghetti | 10 to 12 | Fettuccini | 10 to 12 | Penne and Mostaccioli | 9 to 11 | Fusilli | 12 to 14 |
| Vermicelli | 5 to 7 | Lasagna | 11 to 13 | Manicotti and Cannelloni | 10 to 12 | Orzo | 5 to 7 |
| Capellini | 3 to 5 | Egg Noodles | 8 to 14* | Conchiglie | 9 to 12 | Farfalle | 9 to 12 |
| Linguine | 9 to 12 | Elbow Macaroni | 9 to 12 | Jumbo shells | 20 to 25 | Tortellini | 10 to 12 |

* Depending on width

 Cooking time varies for different types of pasta.

## To cook pasta, apply the following procedure:

1. Determine the approximate amount of water needed (1 gal. of water per pound of pasta). Place the water in a stockpot or steam-jacketed kettle.

2. Add salt and bring the water to a rolling boil.

3. Add the pasta and stir gently with a paddle. *Note:* If the pasta is long, such as spaghetti or vermicelli, it is best not to break it, but to spread it out around the inner wall of the pot and then stir and lift it gently with the paddle or a kitchen fork until the pasta becomes submerged.

4. Return the water to a boil, stirring and lifting the pasta occasionally during the cooking period.

5. When the cooking period ends, drain the pasta using a colander and rinse the pasta in cold water. *Note:* Rinsing the pasta in cold water ends the cooking process and helps to remove any starch residue from the exterior of the cooked pasta. If the starch residue is not rinsed away, the pasta may become sticky.

To prepare ravioli, apply the following procedure:

1. Roll out a sheet of pasta dough as thin as possible on a floured bench or a piece of heavy canvas, until the dough is about 12″ square. Alternatively, use a pasta machine to roll out a 1′ to 2′ sheet of pasta dough. *See Figure 21-22.*

2. Using a pastry bag, deposit ¼ oz portions of filling on the sheet of dough, spaced evenly approximately 2″ apart. One-quarter pound of filling should produce 16 equal portions.

3. With a pastry brush, brush straight lines of egg wash (6 eggs to 1 qt milk) around each portion of filling.

4. Roll out a second sheet of dough as thin as the first, and place the second sheet directly on top of the first sheet. Starting in the center, press down around each portion of filling and all edges to seal the two sheets of dough together.

5. Use a pastry wheel or knife to cut between the mounds.

6. Separate the squares and place on a sheet pan covered with wax paper, wax-side up. Cover, and refrigerate or freeze for later use.

*Preparing Ravioli* _____

1. Roll out a sheet of dough as thin as possible.

2. Use pastry bag to space filling evenly 2″ apart.

3. Brush egg wash between filling and around edges of dough.

4. Place a second sheet of dough directly on top of first sheet; press down around filling.

5. Use pastry wheel to cut between mounds of filling.

6. Place squares on sheet pan; cover and refrigerate for future use.

21-22 *Ravioli consists of envelopes of pasta wrapped around a filling of meats, cheeses, or vegetables.*

To prepare tortellini, apply the following procedure:

1. Roll out pasta dough as thin as possible on a floured bench or a piece of heavy canvas. ***See Figure 21-23.***
2. Using a 2″ biscuit cutter or round cutter, cut the dough into rounds.
3. Moisten the edge of each round with water or egg wash.
4. Place a small portion of filling in the center of each round.
5. Fold the circles in half and press the two edges together until they are tightly sealed.
6. Shape the half-circles into small rings by stretching the tips of each half circle slightly and forming the dough around a finger. Press the tips together until they are tightly sealed.
7. Cover, and refrigerate or freeze for later use.

CULINARY PROCEDURES

*Preparing Tortellini* _____

*1. Roll out dough as thin as possible.*

*2. Cut dough into rounds.*

*3. Place small portion of filling in center of each round.*

*4. Moisten edge of round with egg wash.*

*5. Fold circles in half; press edges together until tightly sealed.*

*6–7. Shape into rings; cover and refrigerate for future use.*

**21-23** *Tortellini is made by folding a circle of pasta around a filling and forming a ring shape.*

## Storing Prepared Pasta

Pasta is best when cooked for immediate use. In the professional kitchen, pasta is often cooked in advance, rinsed in cold water to stop the cooking process, tossed with a small amount of oil to prevent sticking, and stored covered until needed. Stored cooked pasta can be rinsed in warm water or dropped in simmering water to bring it up to temperature for serving as needed. If pasta is to be stored and then reheated in simmering water, it may be slightly undercooked prior to storage to avoid overcooking during reheating.

## GRAINS

Grains have been a staple food source since the time of the earliest civilizations. A *grain* is the edible fruit (kernel or seed) of a grass. The name of the type of grass is often the same name given to the grain from that plant. For instance, the wheat plant produces a grain that is called wheat. Common grains include rice, wheat, corn, and barley, but there are many other edible grains from around the world.

## Grain Composition

Rice, wheat, and corn are the most popular edible grains. Also called cereals, grains are composed of four parts: husk, bran, endosperm, and germ. **See Figure 21-24.** When grain is harvested, some of these parts are removed to make the grain easier to eat.

## Grain Composition

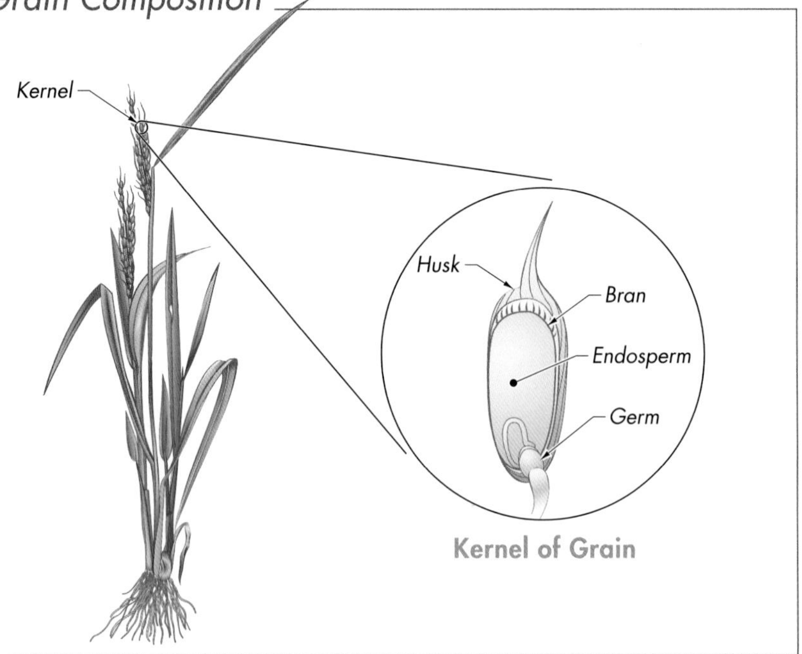

Kernel

Husk
Bran
Endosperm
Germ

**Kernel of Grain**

21-24 *Kernels of grain are composed of husk, bran, endosperm, and germ.*

The *husk (hull)* is the inedible, protective outer covering of grain. The *bran* is the tough outer layer of grain that covers the endosperm. While it is often removed in grain processing, bran provides necessary fiber, complex carbohydrates, and B-complex vitamins. Studies indicate that bran can help lower cholesterol. The *endosperm* is the largest component of a grain kernel and is milled to produce flours and other products. Endosperm consists of carbohydrates (in the form of starches) and a small amount of protein. The *germ* is the smallest part of a grain kernel and contains a small amount of natural oils (fat) as well as vitamins and minerals.

## Processing Grains

Grains come in many shapes and sizes, from large kernels to small seeds. Most grains are hard, and all have a husk that is indigestible in its natural form. It is necessary to process grains to some extent to make them easier to digest. However, the amount of processing can affect the nutritional value of a grain. Processing often goes beyond removing the husk and can include removing other parts such as the nutrient-rich bran and germ. Whole grains are the most minimally processed form of grain. Processing may involve various types of grinding, tumbling, or rolling to produce pearled grains, cracked grains, flaked grains, meal, or flour. ***See Figure 21-25.***

*Processing Grains*

Husk removed

Scrubbed exterior

Broken into coarse pieces

**Whole Grains**    **Pearled Grains**    **Cracked Grains**

*Indian Harvest Specialty Foods, Inc.*

**21-25** *Grinding, tumbling, or rolling grains often removes parts such as the husk, bran, or germ.*

**Whole Grains.** Whole grains are grains where only the husk is removed. Because the rest of the grain is intact, whole grains require more time to cook and soften than do grains that are further processed. Some whole grains are available parcooked. *Parcooked* is a term for slightly or partially cooked. Grains are parcooked to reduce the required amount of cooking time at the final preparation.

**Processed Grains.** Processed grains are grains that are processed to some extent in order to make them easier to digest, prolong their shelf life, change their color, or make them easier to use in the preparation of other foods. For example, some grains are processed by *milling,* which is the successive grinding of grain into a fine meal and finally into a fine powder. Flour and meal (such as cornmeal) are examples of milled grains.

A *refined grain* is a grain that has undergone processing to remove components such as the germ or bran. Removing the germ can help preserve the product so that it does not turn rancid or spoil quickly. The bran is removed from common all-purpose flour to make the product white in color instead of brown. A problem with refined grains is that in processing they lose nutritional value, as vitamins, minerals, and fiber are removed.

**Pearled Grains.** Pearled grains actually have received this name because they are scrubbed during processing to remove the bran, which results in the grain having a pearl-like appearance.

**Cracked Grains.** Cracked grain refers to a whole kernel that is cracked or broken into smaller, coarser pieces during processing. Bulgur wheat is an example of a cracked grain.

**Flaked (Rolled) Grains.** Some par-cooked grains can be rolled instead of ground to produce flakes of grain. Oatmeal flakes are a common example of flaked grains.

## TYPES OF GRAINS

The three most abundant grains are rice, wheat, and corn. Although rice is the most important grain in the world as a part of everyday diets, corn and wheat are close behind. Globally, wheat and corn are the largest cultivated grains, with rice being third.

### Rice

Rice is a popular grain in many cuisines and is the staple food source for more than half the population of the world. Rice is different from other grains in that the seed is produced from grasses grown in paddies (flooded fields) instead of in typical soil. It can be used as a side dish, main dish, or dessert. The high starch content of rice allows it to absorb flavor from whatever ingredients it is cooked with. A rule of thumb is that rice will triple when cooked.

All types of rice are milled whole and are naturally brown in color. After the rice is harvested from the fields, it is rolled between two rollers to remove the husk. After the husk is removed, the bran is still attached to the grain, leaving the rice with a brown appearance. Rice that is milled and left with the bran attached is commonly called brown rice. To produce white rice from brown rice, the bran is removed. There are three primary categories of rice: short grain, medium grain, and long grain. *See Figure 21-26.*

## Rice

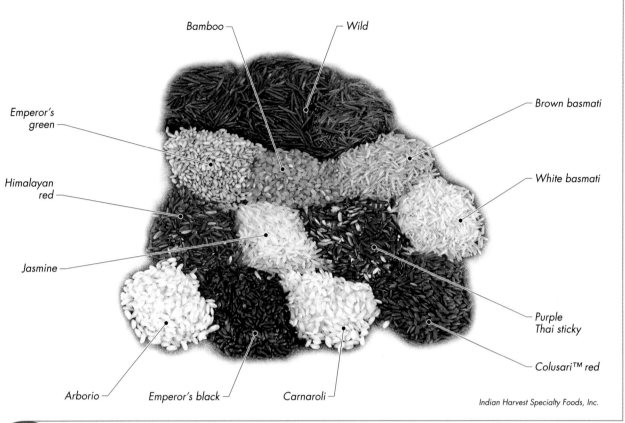

Bamboo — Wild

Emperor's green

Himalayan red

Jasmine

Arborio — Emperor's black — Carnaroli —

Brown basmati

White basmati

Purple Thai sticky

Colusari™ red

Indian Harvest Specialty Foods, Inc.

**21-26** *Rice is categorized as short grain, medium grain, or long grain.*

**Short-Grain Rice.** Short-grain rice is high in starch content and becomes soft and tender when cooked properly. A high starch content also makes this rice sticky when cooked. Short-grain varieties of rice are commonly used to make risotto and sushi. Arborio rice and carnaroli rice are types of short-grain rice that are widely used for risotto. Types of exotic short-grain rice include bamboo rice and Colusari™ red rice.

Sticky rice (also called sweet rice) is a short-grain rice used in Asian cuisines in desserts. Sticky rice is not used in sushi dishes. The rice is soaked for several hours then steamed rather than boiled. Sticky rice is sometimes ground into flour to make Asian dumplings or pastries.

**Medium-Grain Rice.** Medium-grain rice contains slightly less starch than short-grain rice but is still sticky when cooked. It has a similar appearance to short-grain varieties, with the grains being only slightly longer. Granza rice and Valencia rice are examples of medium-grain rice and are often used in the making of paella.

Types of exotic medium-grain rice include emperor's rice and purple Thai sticky rice. Emperor's rice is available with either black or green granules. It is a versatile rice that can be used when making a pilaf or served as part of an

Asian meal. Purple Thai sticky rice is a multicolored variety with granules in shades of dark purple and brown with flecks of white, and turns a deep purple when cooked. Colorful exotic rice can add interest to a special dish.

**Long-Grain Rice.** Long-grain rice has long, slender granules that remain light and fluffy if cooked properly. Basmati rice is a long-grain rice that is widely used in Indian cuisine. It has a creamy, yellow color with a sweet, nutty flavor. Jasmine rice is a long-grain rice widely used in Thai cuisine. It is fragrant and has a floral taste. Jasmine rice is used in popular dishes such as chicken curry.

**Wild Rice.** Wild rice, a long black grain, is not true rice, but rather is a grass that grows from a reedlike plant. However, it is prepared in the same manner as rice and is referred to as rice. Wild rice is considerably more expensive than white rice due to its limited availability. It has a rich, nutty flavor and is high in nutrients. Wild pecan rice, unrelated to pecans, is a long-grain rice with a rich nutty flavor. It is very aromatic and pairs well as a side dish. Himalayan red rice and Wehani® rice are varieties of long-grain red rice that are aromatic and nutty in flavor.

**Converted Rice.** Converted rice is specially processed long-grain rice that is parboiled to remove the surface starch and hull, dried, and further milled to produce either brown or white rice. This process results in rice with a high nutrient content because nutrients from bran are forced into the endosperm. Converted rice will hold up well for a long period in a steam table or warmer without becoming sticky and starchy. It is widely used in food service establishments.

**Instant Rice.** Instant rice is formulated by precooking rice and then drying it, so that it can be prepared quickly. During this process, some of the nutrients may be removed. Instant rice is quick and easy to prepare and may be useful in a pinch, although texture and taste may suffer.

## Corn

Corn is one of the only grains that is eaten both dried and as a fresh vegetable. Corn has a sweet flavor that can stand alone without adding sugars or seasonings. Corn can be ground and made into white or yellow cornmeal, depending upon the origin of the corn kernel. *See Figure 21-27.* Cornmeal is a coarsely ground grain that is inexpensive, can be used to give products a gritty texture, and is popular in breads and fried-food coatings. Corn can also be processed into grits and hominy.

*Hominy* is a corn product treated with lye, which causes the kernels to swell to twice their normal size before the hull is stripped. The stripped kernels are dried, leaving a semisoft swollen kernel that resembles a popcorn kernel. Hominy is popular in Mexican cuisine and is used in a soup called *posole*. Hominy can be ground into flour and made into tortillas or used to make tamales.

## Corn

White Cornmeal    Yellow Cornmeal    Grits

**21-27** *Corn can be ground into cornmeal or grits.*

*Grits* are a type of meal made from ground corn or hominy. Instant or quick-cooking varieties are available in addition to traditional slow-cooking grits. As a breakfast side dish, grits are popular in the southern United States and are traditionally served with butter, salt, and pepper. Because grits have a mild flavor, additional ingredients such as cheese, bacon, or hot sauce can be added. Some people prefer to add butter and sugar or honey to plain grits to bring out the sweetness of the corn.

### Wheat

The most common form of wheat used in cooking is milled flour. However, there are many other wheat products that can be used as side dishes, including wheat berries, bulgur wheat, and couscous. *See Figure 21-28.*

**Wheat Berries.** A wheat berry is a chewy wheat kernel with only the husk removed. It contains both the bran and germ. Wheat berries are cooked by simmering, using a procedure similar to that for rice. After simmering, they can be finished by sautéing in a pan with various herbs and spices to create a delicious side dish. Wheat berries can be found through specialty food distributors.

**Bulgur Wheat.** Bulgur wheat is golden brown, nutty-tasting wheat kernels with the husk and bran removed. It can be purchased in fine, medium, and coarse granulations. Bulgur wheat is commonly simmered and seasoned with herbs, spices, and vegetables, and cooks in about half the time required for wheat berries. Tabouli is a Middle Eastern side dish made from cooked, chilled bulgur wheat that is mixed with mint, lemon, olive oil, and parsley.

**Durum Wheat.** Durum wheat is a hard variety of wheat that is high in protein and gluten. It is ground into semolina, which is often used to make pasta. Durum wheat is also available pearled.

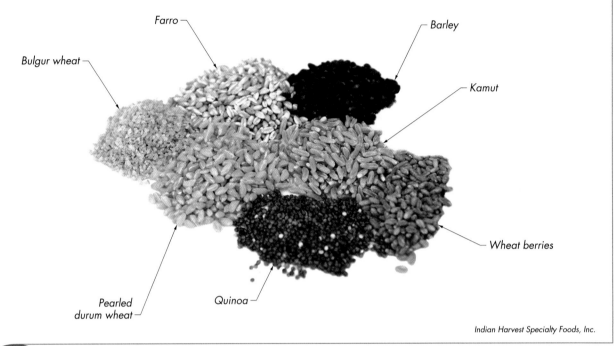

Bulgur wheat

Farro

Barley

Kamut

Wheat berries

Pearled durum wheat

Quinoa

Indian Harvest Specialty Foods, Inc.

**21-28** *There are many wheat products other than flour, including wheat berries and bulgur wheat. Other types of grains include barley, farro, kamut, and quinoa.*

**Couscous.** Couscous is made from durum wheat with both the bran and germ removed. The endosperm is formed into tiny round pellets and steamed. Common couscous is somewhat fine in texture, similar to cornmeal. Israeli couscous is a larger-size variety of couscous. Couscous is prepared by steaming or simmering. It can then be finished by sautéing and adding various herbs and spices. Couscous can also be added as a stuffing to chicken and turkey or served as a side dish.

## Other Grains

In addition to rice, corn, and wheat, there are many other types of edible grains. Some of these grains are widely available in North America, such as barley, buckwheat, or oats. Other grains from countries around the world are becoming more readily available due to improvements in transportation. These include farro, kamut, millet, and quinoa.

**Barley.** Barley is a very hardy grain that can grow in extremely cold or warm climates. Hulled barley takes twice as long as rice to cook and has a chewy texture. Pearled barley is polished barley with the bran removed. Barley is often added to soups and stews for its hardy, earthy flavor and is a natural thickener.

**Buckwheat.** Although commonly regarded as a grain, *buckwheat* is an herb with edible seeds. Whole buckwheat is commonly ground into flour. Buckwheat groats are whole-grain buckwheat crushed into coarse pieces and prepared similarly to rice. *Kasha* is toasted buckwheat. Buckwheat flour cannot be substituted for wheat flour, as buckwheat lacks gluten-forming properties.

**Oats.** Oats are derived from the berry of oat grass. Oats can be purchased in many different forms and have a variety of uses in items including hot cereal, muffins, breads, and cookies. The whole oat grain is referred to as an oat groat and includes both the bran and germ with only the husk removed. Steel-cut oats are groats that have been toasted and cut into small pieces. Steel-cut oats are cooked as a hot cereal, requiring at least a 3:1 ratio of water to oats. Rolled oats (old-fashioned oats) are a form of oats that have been steamed and flattened between rollers and formed into small, flat flakes. Rolled oats require less cooking time than steel-cut oats. Quick-cooking oats have been rolled, cut into small pieces, and partially cooked to reduce cooking time even more than for rolled oats.

**Farro.** *Farro* (spelt) is an ancient Italian grain that is high in protein and similar in taste to barley. Farro has been grown and consumed in Italy since Roman times and is common in the Mediterranean diet.

**Kamut.** *Kamut* is an ancient variety of wheat that has a natural sweetness and a buttery quality. It remains firm and chewy when cooked.

**Millet.** *Millet* is a small, round, slightly nutty grain that is very high in protein. It is eaten primarily in Africa and Asia. Millet looks similar to couscous and is prepared like rice.

**Quinoa.** *Quinoa* is a gluten-free grain that is high in vitamins and minerals and forms a complete protein. Quinoa is one of the oldest known grains. It is native to the South American Andes and was a staple food of the Incas. Quinoa can be eaten as a hot breakfast cereal, added to a salad, or eaten as a side dish.

## GRAIN STORAGE

Grains should be stored in a cool, dry place in an airtight container to keep moisture out and prevent insects from getting into the product. Grains that contain germ should be kept in a refrigerator or freezer to prevent the germ from becoming rancid, necessitating disposal of the product. Milled grains such as white rice will keep for many months in a cool, dry place.

## GRAIN PREPARATION

Grains are most commonly prepared by simmering. Pilaf and risotto are two common preparations using a combination of sautéing and simmering. A risotto preparation begins with a light sautéing of grain. A liquid is then incorporated gradually in small additions. In a pilaf preparation, a sautéing technique is also used prior to adding a simmering liquid. A pilaf can be finished on the stove or in an oven.

### Simmering Grains

Simmering is the most common method for preparing any type of grain. The grain is simply added to a measured amount of boiling water. The appropriate amount of water depends on the type and amount of grain being cooked. When the water returns to a boil, it is lowered to a simmer and the pot is covered until the grain is fully cooked. In most cases, the grain is cooked until all the liquid has been absorbed.

### Risotto Method

Risotto is a classic Italian dish traditionally made with Arborio (short-grain) rice. Rice cooked using the risotto method has a puddinglike texture. Risotto is cooked slowly to release the starches from the rice, resulting in a rich creaminess in the finished product.

### To prepare a risotto, apply the following procedure:

1. Melt fat in saucepan and sweat onions and garlic. *See Figure 21-29.*
2. Add grain and stir to coat with fat. Sauté until the grains appear translucent.
3. Add white wine to coat and cook until wine is almost completely reduced.
4. Pour in a small quantity of liquid (usually chicken stock) into the saucepan and continue to stir rice until all the liquid has been absorbed.
5. Repeat step number 4 until all of the liquid has been absorbed and the grain is cooked al dente. This slow cooking process releases starches in the grain, resulting in a creamy finished product.

**Pilaf Method.** When using the pilaf method, the flavoring ingredients and grains are sautéed in fat. Hot liquid or stock is then added along with any seasonings, and the grain is covered and left to simmer until the liquid has been absorbed. Coating the grains in fat before adding liquid prevents them from clumping together. A classic pilaf is finished on the stove, although it can also be finished in the oven.

## Preparing Risotto

1. Melt fat; sweat onions and garlic.

2. Add grain and stir to coat with fat.

3. Add white wine to coat; cook until almost completely reduced.

4. Add small quantity of chicken stock; stir until all liquid is absorbed.

5. Repeat step 4 until all liquid is absorbed and the grain is al dente.

Garnish as desired.

 21-29 Risotto is cooked slowly, adding small amounts of liquid at a time.

To prepare a pilaf, apply the following procedure:

1. Melt fat in saucepan and sweat onions and garlic. **See Figure 21-30.**
2. Add the desired grain, and stir to coat until slightly toasted.
3. Add hot liquid or stock to saucepan.
4. Bring liquid to boil and then reduce heat to a simmer.
5. Cover and allow to simmer or place in a hot oven until all the liquid has been absorbed. Remove from heat and fluff with a fork when finished.

CULINARY PROCEDURES

## Preparing Pilaf

1. Melt fat; sweat onions and garlic.

2. Add grain; stir until slighty toasted.

3–4. Add hot liquid; bring to a boil and reduce heat to simmer.

5. Cover and allow to simmer or place in hot oven until all liquid is absorbed; remove from heat and fluff with fork.

**21-30** Pilaf is made by sautéing grain in fat before simmering with seasonings.

## SUMMARY

Potatoes, pastas, and grains are considered some of the basic foods of the human diet. People and cultures around the world rely on these starches to meet nutritional needs. Use of potatoes, pasta, and grains can result in increased profits for a food service establishment, as they are relatively inexpensive to purchase and easy to prepare.

Potatoes are categorized as starchy or mature, and waxy or new. Starchy potatoes such as russet potatoes are best for frying, baking, or use in casseroles. Waxy potatoes such as red potatoes, yellow potatoes, fingerlings, or purple potatoes are versatile in use. Low-starch/high-moisture potatoes such as new potatoes are excellent for roasting, steaming, or boiling.

Pasta is a product made from a flour-based dough and can be purchased in dried form or can be made fresh from scratch. It is available in many shapes that are categorized as ribbon, tube, extruded shape, and hand-formed pastas. Filled pastas include ravioli and tortellini. Pasta is usually prepared by boiling, while many Asian noodles are prepared for use by soaking in hot water.

Grains include staples such as rice, wheat, and corn. A grain is composed of a husk, bran, endosperm, and germ. Grains are often processed to remove one or more parts of the natural grain. The most nutritious parts of grain are the bran and the germ. Heavily processed grains include milled flour and meal. Less-processed forms include pearled, cracked, and flaked grains. Grains are most commonly prepared by simmering or by combination methods such as the pilaf or risotto methods.

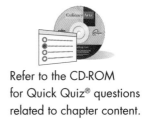

Refer to the CD-ROM for Quick Quiz® questions related to chapter content.

## Review Questions

1. Identify common uses for starchy potatoes.

2. Describe the common characteristics of waxy or new potatoes.

3. Contrast the characteristics of a sweet potato with those of a yam.

4. Describe the criteria to consider when purchasing fresh potatoes.

5. How should potatoes be stored?

6. Explain the five-step procedure for deep-frying potatoes.

7. Describe how to prepare a potato before baking it.

8. Describe the five-step procedure for preparing a potato casserole.

9. Why should potatoes be started in a cold liquid when being simmered?

10. Identify common types of ribbon pasta.

11. What is the difference between extruded shape pastas and hand-formed pastas?

12. List common types of Asian noodles.

13. Describe the common characteristics and uses for soba noodles.

14. Contrast ramen noodles with somen.

15. Describe the five-step procedure for preparing pasta dough by hand.

16. Describe the four parts of a grain.

17. What are the advantages and disadvantages of refined grains?

18. Identify dishes that commonly include short-grain rice as a main ingredient.

19. Describe common types of exotic medium-grain rice.

20. List different varieties of long-grain rice.

21. Describe the main characteristics of grits.

22. Contrast the qualities and preparations of wheat berries with those of couscous.

23. How is bulgur wheat usually prepared?

24. Describe common types of oats.

25. How should grains that contain germ be stored?

26. Describe the five-step procedure for preparing risotto.

## Candied Sweet Potatoes

*yield: 8 servings*

206

| | |
|---|---|
| 3 lb | sweet potatoes |
| 6 tbsp | butter |
| ½ lb | brown sugar |
| 2 oz | water |
| 2 oz | orange juice |
| 1 tsp | salt |
| ⅛ tsp | cinnamon |
| 2 tsp | vanilla extract |

1. Wash sweet potatoes and roast in a 350°F oven until almost cooked but still firm in the center, approximately 30 min to 40 min.
2. Remove potatoes from oven and let stand to cool slightly, making them easier to handle.
3. While potatoes are cooling, place butter, brown sugar, water, orange juice, salt, and cinnamon in a saucepot and bring to a boil to dissolve the sugar.
4. Add the vanilla and remove from heat.
5. Peel the potatoes and slice into ¼"-thick slices. Shingle-layer potato slices in a baking dish or hotel pan and top with hot sugar syrup mixture.
6. Bake at 350°F, basting with the sugar syrup until bubbly and potatoes are completely tender, about 20 min to 30 min.

### Nutrition Facts

**Per serving:** 390.6 calories; 20% calories from fat; 8.9 g total fat; 22.9 mg cholesterol; 319.6 mg sodium; 1506 mg potassium; 76 g carbohydrates; 7 g fiber; 28.9 g sugar; 69 g net carbs; 2.7 g protein

## American Fries (Home Fries)

*yield: 8 servings*

207

*Idaho Potato Commission*

| | |
|---|---|
| 2 lb | red potatoes or new potatoes, peeled |
| 2 tbsp | canola oil |
| 1 tbsp | parsley, chopped |
| to taste | salt and pepper |

1. Boil potatoes until they become tender, or steam until they are al dente. Allow to cool overnight in the refrigerator.
2. Slice cooled potatoes to a thickness of about ¼" thick.
3. Heat 2 tbsp canola oil in a saucepan and sauté potatoes in skillet until golden brown on both sides.
4. Season and serve garnished with chopped parsley.

### Nutrition Facts

**Per serving:** 145.4 calories; 21% calories from fat; 3.5 g total fat; 0 mg cholesterol; 27.1 mg sodium; 786.2 mg potassium; 27 g carbohydrates; 1.9 g fiber; 0 g sugar; 25.1 g net carbs; 2.8 g protein

### Recipes Variations

**208** **Lyonnaise Potatoes:** Boil, cool, and slice potatoes as for American fries. Heat skillet and add 6 oz to 8 oz julienned onion and cook until softened but not browned. Remove onions and add 2 oz clarified butter to skillet, add potatoes, and cook until browned on both sides. Return onion to skillet and continue to cook until slightly brown. When the potatoes are golden brown, add the sautéed julienned onions and continue to sauté to marry the flavors. Season with salt and pepper to taste.

**209** **Hashed Brown Potatoes:** Peel and boil whole potatoes, boiling until cooked through. Grate cooled potatoes and place into hot oil and sauté. When brown, remove and season with salt, pepper, and parsley if desired.

**210** **Hash Lyonnaise:** Prepare hashed brown potatoes and, when the potatoes are golden brown, add sautéed julienned onions. Continue to sauté until the onions have blended with the potatoes.

# Recipes

## Potato Pancakes

*yield: 16 (2-oz pancakes)*

| | |
|---|---|
| 2 lb | russet potatoes, peeled and grated finely |
| 1 | medium onion, grated |
| 2 | large eggs, beaten |
| 1 c | flour |
| 1 tsp | baking powder |
| 1 tsp | kosher salt |
| | canola oil (for frying) |
| to taste | kosher salt (if desired) |

1. Place potatoes, onions, and beaten eggs into large mixing bowl.
2. Sift baking powder and salt into flour.
3. Add flour mixture to potato mixture and mix until incorporated.
4. Place canola oil into skillet, filling ⅛″ deep, and heat canola oil. *Note:* Test a small drop of potato mixture in the oil to check temperature before pan-frying. Potato mixture should bubble when added to oil.
5. Using a #16 portion control scoop, drop mixture into medium-hot oil. Flatten each pancake with the back of an oiled spatula.
6. When browned, turn over and brown on other side.
7. Transfer pancakes to paper towels to drain excess oil.
8. Sprinkle with salt (if desired) and serve immediately.

**Nutrition Facts**

**Per serving:** 459.5 calories; 80% calories from fat; 41.6 g total fat; 26.4 mg cholesterol; 513.3 mg sodium; 419.6 mg potassium; 20.3 g carbohydrates; 1.2 g fiber; 0.4 g sugar; 19.1 g net carbs; 3.1 g protein

## Scalloped Potatoes

*yield: 8 servings*

| | |
|---|---|
| 2 tbsp | unsalted butter |
| 1 pt | thin or light béchamel sauce |
| to taste | kosher salt |
| to taste | white pepper |
| pinch | nutmeg |
| 2¼ lb | russet potatoes |

1. Oil the inside and bottom of a 2″-deep half pan and set aside.
2. Heat béchamel in a separate pan until hot. Add nutmeg, salt, and pepper to taste.
3. Peel potatoes and then slice to about ⅛″ thickness. Store slices in water to prevent browning.
4. Remove sliced potatoes from water, drain well, and place a layer of the potatoes on the bottom of the oiled pan.
5. Pour hot béchamel over potatoes, and then add another layer of potatoes. Repeat adding layers of sauce and potatoes until finished.
6. Shake the pan so the potatoes are evenly distributed.
7. Cover and bake in a 350°F oven for 30 min.
8. Uncover and bake an additional 30 min or until golden brown and cooked through.
9. Remove from the oven and let cool for 10 min to 15 min before serving.

**Nutrition Facts**

**Per serving:** 155.3 calories; 27% calories from fat; 4.9 g total fat; 13 mg cholesterol; 20.5 mg sodium; 558.4 mg potassium; 25.3 g carbohydrates; 1.7 g fiber; 1.6 g sugar; 23.6 g net carbs; 3.4 g protein

### Recipe Variation

**213** **Potatoes au Gratin:** Add 3 cloves of mashed or creamed garlic and ¼ cup Gruyère or Parmesan cheese to the hot béchamel before pouring over sliced potatoes.

**214** **Hash-in-Cream Potatoes:** Peel red potatoes and boil until they become tender. Allow to cool. Shred the potatoes. Add the béchamel and season with salt and a touch of nutmeg. Garnish with paprika.

**215** **New Potatoes in Cream:** Peel and cook in heavily salted water until they are just tender. Drain thoroughly. Add béchamel and season with salt and white pepper. Garnish with chopped parsley.

**216** **Au Gratin New Potatoes:** Peel red potatoes and cook until tender. Allow to cool overnight. Chop or hash the potatoes into medium-size pieces. Add béchamel and season with salt and a touch of nutmeg. Place mixture in a hotel pan and sprinkle with grated Cheddar cheese and paprika. Bake in a 350°F oven until the cheese melts and is slightly brown.

# Recipes

## 217

## French-Fried Potatoes

*yield: 8 servings*

| 2 lb | potatoes |
| | canola oil (for frying) |
| to taste | salt |

1. Peel potatoes and cut into desired shape and size. Store in cold water to prevent browning.
2. Strain potatoes and place in deep-fry baskets.
3. Blanch potatoes for 5 min in 250°F fat that is deep enough to completely cover potatoes. *Note:* Do not allow potatoes to brown.
4. Drain potatoes and place on sheet pans to cool completely.
5. Before serving, fry again in deep fat at a temperature of 350°F to 375°F until golden brown and crisp.
6. Season with salt to taste.

### Nutrition Facts
**Per serving:** 810 calories; 89% calories from fat; 81.9 g total fat; 0 mg cholesterol; 26.2 mg sodium; 477.4 mg potassium; 19.8 g carbohydrates; 2.5 g fiber; 0.9 g sugar; 17.3 g net carbs; 2.3 g protein

## 218

## Rissole (Oven-Browned) Potatoes

*yield: 8 servings*

| 2 lb | whole new potatoes or red potatoes, cut into quarters |
| 1 tbsp | vegetable or canola oil |
| to taste | salt |
| to taste | pepper |

1. Place oil in mixing bowl with potatoes, paprika, salt, and pepper and toss to coat.
2. Place potatoes in roasting pan and roast in a 375°F oven, turning frequently until potatoes become golden brown and tender.

### Nutrition Facts
**Per serving:** 105.1 calories; 15% calories from fat; 1.9 g total fat; 0 mg cholesterol; 26.2 mg sodium; 615.8 mg potassium; 20.4 g carbohydrates; 1.8 g fiber; 0 g sugar; 18.6 g net carbs; 2.3 g protein

### Recipe Variations:

**219  Minute (Cabaret) Potatoes:** After potatoes are removed, sprinkle with a small amount of garlic powder and 1 tbsp melted butter and toss to coat evenly.

**220  O'Brien Potatoes:** Prepare in the same manner as minute potatoes, but omit the garlic and add sautéed fine-diced onion, green peppers, and red peppers.

**221  Vesuvio Potatoes:** Peel a raw potato and cut it into 6 wedges (in half lengthwise and then each half into 3 equal wedges). Roast potato as described above. When almost cooked, toss potatoes with a mixture of 2 tsp lemon juice and 2 tsp olive oil, season with oregano, salt, and black pepper, and roast until fully cooked.

**222  Château Potatoes:** In a sauté pan, sauté a blanched, tournéed potato in clarified butter and season it with salt and white pepper, cooking until it is golden on all sides. Alternatively, to prepare in the oven, peel an entire russet potato, cutting the entire potato into a tourné shape and then cutting it in half lengthwise from end to end. Brush it with clarified butter, season it, and roast it in an oven at 350°F until tender. Sprinkle with finely chopped parsley (if desired) for service.

# Recipes

## Duchess Potatoes

*yield: 14 (3-oz servings)*

| | |
|---|---|
| 3 lb | baking potatoes |
| 4 tbsp | unsalted butter, softened |
| ⅛ tsp | nutmeg, grated |
| ¼ tsp | white pepper |
| to taste | salt |
| 3 | egg yolks |
| 1 | large egg |
| 2 oz | half-and-half or milk, heated |
| 2 tsp | unsalted butter, melted |

1. Peel potatoes and cut into quarters.
2. Boil the potatoes in heavily salted water until very tender.
3. Drain potatoes from water and allow to steam dry on a sheet pan. *Note:* This will allow most of the water remaining in the potatoes to evaporate, which will result in a lighter potato mixture.
4. After potatoes have stopped emitting steam and while still warm, put them through a ricer or mash them very well.
5. Add potatoes to an electric mixer and beat in the butter, nutmeg, white pepper, and salt just until incorporated.
6. Add the egg yolks and the whole egg, one at a time, and the half-and-half and beat the mixture until it is combined well and smooth.
7. Transfer the potato mixture to a pastry bag fitted with a large star tip. Pipe pyramid-shaped mounds, about 3 oz in weight, onto a sheet pan.
8. Brush with the 2 tsp melted butter and bake in 375°F oven until ridges are golden brown, approximately 10 min.

### Nutrition Facts
**Per serving:** 131.6 calories; 38% calories from fat; 5.7 g total fat; 71.8 mg cholesterol; 25.9 mg sodium; 424.8 mg potassium; 17.4 g carbohydrates; 2.2 g fiber; 0.8 g sugar; 15.2 g net carbs; 3.2 g protein

### Recipe Variations

**224** **Croquette Potatoes:** Pipe duchess mixture from a pastry bag, without a tip, onto a sheet pan. Score with a knife about every 1½″ to 2″ and place in the freezer to firm up. Once firm, remove from freezer, snap apart at the score, and bread the resulting cork-shaped croquette potatoes using the standard breading procedure. *Note:* Partially freezing or firming the potatoes makes them easier to handle without destroying the shape when breading.

## Twice-Baked Potatoes

*yield: 10 potatoes*

*Idaho Potato Commission*

| | |
|---|---|
| 10 | russet potatoes |
| 4 oz | unsalted butter, softened |
| ¼ tsp | fresh nutmeg, grated |
| ½ tsp | white pepper |
| 2 tsp | salt |
| 6 | egg yolks |
| 1 | large egg |
| 2 oz | half-and-half or milk, heated |
| 3 oz | Parmesan cheese, finely grated |
| 1 tbsp | parsley, minced and rinsed |
| 1 tbsp | scallions, minced |
| 2 tsp | unsalted butter, melted |

1. Scrub potatoes well and rinse.
2. Pierce potatoes with a fork, rub with vegetable oil, and season with kosher salt.
3. Bake potatoes at 400°F until tender and cooked through, about 1 hr.
4. Remove potatoes from the oven, slice the top of the potatoes, and scoop out the potato into the bowl of an electric mixer.
5. Beat in the butter, salt, white pepper, and nutmeg into the potatoes just until incorporated.
6. Add the egg yolks (1 yolk per pound) and the whole egg, one at a time, the hot half-and-half, Parmesan, and parsley and beat the mixture until it is very smooth.
7. Transfer the potato mixture to a pastry bag fitted with a large star tip. Pipe the potato mixture back into the potato skin shells in a decorative manner until filled to the top.
8. Brush or drizzle with the 2 tsp melted butter and bake in 375°F oven until ridges are golden brown, approximately 10 min.

### Nutrition Facts
**Per serving:** 465.3 calories; 31% calories from fat; 16.6 g total fat; 183.2 mg cholesterol; 629.6 mg sodium; 1582.1 mg potassium; 67.9 g carbohydrates; 4.9 g fiber; 2.5 g sugar; 63 g net carbs; 13.7 g protein

# Recipes

## Pasta

**226**

*yield: 1 lb*

*Frieda*

| 4 | eggs |
| 2 tbsp | olive oil |
| 1 tsp | salt |
| 1 tbsp | water |
| ½ lb | bread flour |
| ½ lb | semolina flour |

*Mixer method*

1. In a mixer with paddle attachment, combine eggs, salt, and oil and mix well.
2. Begin by adding approximately ⅓ of each flour and mix until the mixture forms a smooth, soft dough.
3. Replace the paddle attachment with the dough hook and begin adding the rest of the flour while kneading with the dough hook. The dough mixture should be very dry (not tacky) when all of the flour has been added.
4. Remove the dough from the mixer and place in a bowl covered with plastic wrap for at least 15 min to allow the gluten to rest.
5. Once rested, the dough can be rolled and cut into desired shapes and sizes.

*By hand method*

1. Mix salt and flour on work surface and form a well in the center.
2. Add cracked eggs and oil to well.
3. Begin mixing in a small portion of the flour into the wet ingredients and continue mixing until all of the wet ingredients have been mixed into the flour.
4. Begin kneading the pasta dough and continue until the dough becomes very smooth and soft, not tacky. The dough mixture should be very dry (not tacky) when all of the flour has been added.
5. Place the dough in a bowl covered with plastic wrap to rest for at least 15 min to relax the gluten.
6. Once rested, the dough can be rolled and cut into desired shapes and sizes.

---

**Nutrition Facts**

**Entire recipe:** 2214.9 calories; 22% calories from fat; 56.2 g total fat; 981.4 mg cholesterol; 2658.2 mg sodium; 960.3 mg potassium; 331.5 g carbohydrates; 14.3 g fiber; 2.5 g sugar; 317.2 g net carbs; 85.1 g protein

---

**Recipe Variations:**

**227** **Simple Pasta Dough:** Use 1 lb bread flour, instead of the bread flour and semolina flour combination. The mixture is easier to knead but the resulting pasta is not as firm. Dough made from exclusively bread flour will turn soft and mushy if even slightly overcooked.

**228** **Green (Spinach) Pasta Dough:** Blanch 2 oz of fresh spinach (per 1-lb dough recipe) in heavily salted water just until limp (about 5 seconds) and refresh immediately in cold water to retain color. Squeeze out all water from spinach and purée completely in blender. Add to pasta dough in place of water.

**229** **Red (Tomato) Pasta Dough:** Add 3 oz of tomato paste to dough in place of water and knead well to incorporate.

**230** **Black (Squid Ink) Pasta Dough:** Add ½ oz of squid ink to pasta dough and knead well to incorporate.

# Recipes

## Fettuccine Alfredo

*yield: 6 servings*

| | |
|---|---|
| 1 lb | fettuccine, dry |
| 3 oz | whole butter |
| 2 | garlic cloves, creamed |
| 4 oz | white wine (optional) |
| 16 oz | heavy cream |
| 3 oz | Parmesan cheese |
| to taste | salt and white pepper |
| 2 tsp | fresh parsley, minced and rinsed |

*Wisconsin Milk Marketing Board*

1. Boil the pasta in 1 gal. heavily salted water until almost al dente, then drain, refresh, and drain again.
2. In a large sauté pan, add the whole butter over medium heat. Add the garlic and cook for 1 min to 2 min without browning.
3. Deglaze pan with white wine and reduce by half.
4. Add cream and Parmesan and reduce by a third.
5. Add the pasta and heat through until sauce is thick enough to coat the pasta.
6. Adjust seasonings and fresh parsley and serve immediately.

### Nutrition Facts
**Per serving:** 732 calories; 55% calories from fat; 46.1 g total fat; 151.7 mg cholesterol; 281 mg sodium; 225.2 mg potassium; 59.8 g carbohydrates; 1.8 g fiber; 0.2 g sugar; 57.9 g net carbs; 17 g protein

### Recipe Variations:

**232  Chicken Tetrazzini:** Cook penne pasta instead of fettuccine to al dente in salted water, then drain and reserve. Sauté 1 lb chicken breast meat, cut into 1″ pieces, with ½ lb quartered button mushrooms. Follow Fettuccine Alfredo recipe beginning at step 2, adding all of the ingredients to the mushrooms and chicken. Finish the dish by topping with mozzarella cheese and browning it under a broiler.

## Penne Arrabiata

*yield: 6 entrée servings*

| | |
|---|---|
| 1 lb | penne pasta, cooked and drained |
| | Arrabiata sauce |
| 2 tbsp | olive oil |
| 1 tbsp | butter |
| 4 oz | onion, diced small |
| 3 | garlic cloves, minced |
| 2 tsp | crushed red pepper flakes |
| 4 oz | white wine (or light stock) |
| 2 lb | canned diced peeled tomatoes |
| 1 tbsp | basil, chiffonade |
| ½ tbsp | parsley, chopped fine and rinsed |
| 2 | anchovy fillets, crushed |
| 2 tsp | kosher salt |
| ½ tsp | black pepper |
| 2 oz | Parmesan cheese, grated |

1. Heat oil and butter in medium saucepot over medium heat.
2. Add onion and sweat without browning until translucent.
3. Add garlic and pepper flakes and cook for 1 min longer.
4. Add white wine and reduce by half, then add rest of the ingredients and simmer for 15 min to 20 min. Adjust seasonings as needed.
5. Toss sauce over cooked penne pasta and garnish each serving with 1 tsp grated Parmesan cheese.

### Nutrition Facts
**Per serving:** 423.6 calories; 20% calories from fat; 9.7 g total fat; 14.5 mg cholesterol; 830.3 mg sodium; 445.1 mg potassium; 64.7 g carbohydrates; 5 g fiber; 5 g sugar; 59.7 g net carbs; 15.2 g protein

# Recipes

## Macaroni and Cheese

*yield: 24 (6-oz servings)*

| | |
|---|---|
| 2 lb | elbow macaroni |
| 2 qt | milk |
| 4 oz | butter |
| 4 oz | all-purpose flour |
| 4 shakes | Tabasco® sauce |
| ½ tsp | Worcestershire sauce |
| ½ tsp | dry mustard |
| pinch | nutmeg |
| 2 lb | cheddar cheese, grated |
| 5 oz | panko bread crumbs (or other coarse bread crumbs) |
| 2 oz | whole butter, melted |

*Wisconsin Milk Marketing Board*

1. Boil pasta in heavily salted water as per package directions. When almost fully cooked, strain, refresh and drain again.
2. While pasta is cooking, heat milk in heavy-bottomed saucepan.
3. In another pot, melt butter and add flour. Cook to make a blond roux until cooked but not allowed to brown. When finished, allow roux to rest for 10 min without heat.
4. When milk reaches a boil, add roux while whisking continuously. After all roux is added, add Tabasco®, Worcestershire, mustard, and nutmeg and simmer for 3 min to 5 min to cook roux completely.
5. When thick and smooth, add the cheddar cheese slowly with a whisk, stirring continuously to incorporate.
6. After all cheese has been added, remove from heat and pour over cooked pasta and stir well.
7. Pour macaroni and cheese into a buttered 2″ full hotel pan.
8. In another bowl, mix melted butter with panko bread crumbs and sprinkle over top of macaroni and cheese.
9. Bake at 350°F uncovered for approximately 30 min until golden and bubbly.
10. Allow to rest 10 min to 15 min prior to serving.

### Nutrition Facts

**Per serving:** 362.9 calories; 52% calories from fat; 21.6 g total fat; 63.1 mg cholesterol; 314.2 mg sodium; 203.6 mg potassium; 26.2 g carbohydrates; 0.9 g fiber; 5.2 g sugar; 25.3 g net carbs; 15.8 g protein

## Pasta Carbonara

*yield: 6 servings*

| | |
|---|---|
| 1 lb | fettuccine, dry |
| 1½ oz | whole butter |
| 6 oz | bacon, diced small |
| 3 cloves | garlic, minced |
| 4 oz | white wine |
| 18 oz | heavy cream |
| 6 oz | Parmesan cheese, grated |
| to taste | salt and pepper |
| 3 | egg yolks |

1. Cook fettuccine in 1 gal. salted boiling water until cooked al dente. When done, drain completely and reserve.
2. While pasta is cooking, render diced bacon in a large sauté pan until crisp and browned.
3. Add the garlic to the pan and sauté briefly without browning.
4. Deglaze pan with white wine and reduce by half.
5. Add cream, bring to a boil, and add pasta, Parmesan, salt, and pepper.
6. Add the yolks and remove from heat. Stir well to incorporate and thicken the sauce. Serve immediately.

### Nutrition Facts

**Per serving:** 934.3 calories; 59% calories from fat; 63.1 g total fat; 286.6 mg cholesterol; 740.8 mg sodium; 316.5 mg potassium; 61.3 g carbohydrates; 1.8 g fiber; 0.4 g sugar; 59.4 g net carbs; 27.2 g protein

# *Recipes*

## Linguine with Scallops, Capers, and Sun-dried Tomatoes

*yield: 4 servings*

| | |
|---|---|
| 2 tbsp | olive oil |
| 12 | sea scallops |
| 2 tbsp | garlic, minced |
| 2 tbsp | capers |
| 4 tbsp | sun-dried tomatoes, oil packed, small diced |
| 4 oz | white wine |
| 2 tbsp | fresh lemon juice |
| 2 tsp | lemon zest |
| ¾ tsp | fresh thyme, minced |
| 2 tbsp | fresh basil, chiffonade |
| 2 tbsp | unsalted butter |
| 8 oz | linguine pasta, cooked, rinsed in cold water, and drained. |
| to taste | kosher salt and pepper |

1. Heat large sauté pan and add 1 tbsp olive oil.
2. Season scallops with kosher salt and pepper and sear on both sides until golden brown. Remove immediately from pan.
3. Add garlic and capers back to pan with another 1 tbsp olive oil and cook for 1 min.
4. Add sun-dried tomatoes and cook for 1 min more.
5. Add white wine and lemon juice and reduce by half.
6. Drop pasta back into boiling water to reheat.
7. Add lemon zest, thyme, basil, and remaining butter and oil, whisk well, and season.
8. Add scallops back to sauce and toss to coat well.
9. Strain pasta from water and add to sauce and scallops. Toss and serve.

### Nutrition Facts
**Per serving:** 375.9 calories; 34% calories from fat; 14.7 g total fat; 30.1 mg cholesterol; 230.9 mg sodium; 398 mg potassium; 41.9 g carbohydrates; 2.1 g fiber; 0.3 g sugar; 39.8 g net carbs; 14.7 g protein

## Lasagna Bolognaise

*yield: 24 servings (one 2" hotel pan)*

*Wisconsin Milk Marketing Board*

| | |
|---|---|
| 3 qt | tomato sauce |
| 1 lb | ground beef |
| 1 lb | ground Italian sausage |
| 3 lb | lasagna noodles, uncooked |
| 4 lb | ricotta cheese |
| 2 lb | grated mozzarella cheese |
| 8 oz | grated Parmesan |
| 6 | whole eggs, slightly beaten |
| 2 tbsp | garlic powder |
| 1 tbsp | salt |
| 2 tbsp | fresh parsley, coarsely chopped |
| 1 lb | mozzarella cheese, grated |

1. Brown ground beef and sausage in a sauté pan until cooked through and drain off fat.
2. Add tomato sauce to ground beef and stir well.
3. In a large bowl, add the ricotta, mozzarella, Parmesan, eggs, garlic, salt, and parsley and mix well.
4. Spray a 2" full hotel pan with nonstick cooking spray and place a small amount of sauce on the bottom of the pan.
5. Place the uncooked pasta sheets on the bottom of the pan with the edges overlapping. *Note:* The noodles are not cooked in this recipe so that they will absorb moisture from the sauce and cheese.
6. Place about a 1"-thick layer of cheese filling all across the noodles and then top with another layer of meat sauce.
7. Place another layer of uncooked noodles on top of the meat sauce and repeat with another cheese layer and sauce layer and another layer of pasta. Top the last layer of pasta with meat sauce and sprinkle the remaining mozzarella and Parmesan cheese on top.
8. Bake lasagna at 350°F until internal temperature reaches 165°F, approximately 1 hr 10 min.
9. Remove lasagna from the oven and let rest for 20 min before serving. This lasagna should stand firm when sliced.

### Nutrition Facts
**Per serving:** 654.6 calories; 37% calories from fat; 27.7 g total fat; 150.2 mg cholesterol; 1684.9 mg sodium; 726.9 mg potassium; 58.2 g carbohydrates; 3.3 g fiber; 6.4 g sugar; 54.9 g net carbs; 42.8 g protein

### Recipe Variations:

**238 Cheese Lasagna:** Omit the meat from this recipe.

**239 Stuffed Manicotti:** Use same cheese filling in flat pasta dough and roll to form small rolls.

**240 Stuffed Shells:** Use the same filling and stuff into precooked shell pasta, top with sauce, and bake until hot.

# Recipes

## 241

## German Spaetzle

*yield: 8 (4-oz servings)*

| | |
|---|---|
| 3 | eggs, beaten |
| 3 oz | whole milk |
| 2 oz | water |
| ½ tsp | salt |
| ⅛ tsp | white pepper |
| pinch | nutmeg |
| 1 tbsp | fresh parsley, chopped fine and rinsed |
| ½ lb | all-purpose flour |

1. Combine the eggs, milk, water, salt, pepper, nutmeg, and parsley.
2. In a mixer, begin adding the flour while mixing until all flour has been added. Beat slightly until smooth.
3. Allow the batter to rest for at least 1 hr.
4. Bring a pot of salted water to a boil and, using a spaetzle maker or a perforated hotel pan, push the batter through the holes into the boiling water.
5. When a spaetzle floats to the surface it is ready to be removed. Using a spider or slotted spoon, remove spaetzle and shock in ice water to stop the cooking process.
6. Heat whole butter in a sauté pan and add cooked spaetzle. Sauté until spaetzle is heated through and the butter begins to turn slightly brown.
7. Season and serve immediately.

### Nutrition Facts
**Per serving:** 138.1 calories; 16% calories from fat; 2.5 g total fat; 80.5 mg cholesterol; 178.2 mg sodium; 74.6 mg potassium; 22.4 g carbohydrates; 0.8 g fiber; 0.8 g sugar; 21.6 g net carbs; 5.7 g protein

## Gnocchi with Pancetta and Asparagus

*yield: 6 servings*

| | |
|---|---|
| 2 lb | russet potatoes |
| 2 | eggs, slightly beaten |
| 1½ oz | whole butter |
| 1½ tsp | kosher salt |
| ¼ tsp | black pepper, ground coarse |
| pinch | nutmeg, grated |
| 10 oz | all-purpose flour |
| 3 oz | pancetta, diced (or bacon) |
| ½ tbsp | olive oil |
| 1 tbsp | butter |
| 2 | garlic cloves, crushed |
| 4 oz | asparagus, cut into 1" sections |
| 1 tbsp | parsley, rough chopped and rinsed |
| 1 tbsp | fresh Parmesan cheese, grated |
| to taste | salt and pepper |

1. Peel potatoes, cut into quarters, and boil in salted water until very tender.
2. Strain potatoes in colander, then place potatoes on a sheet pan and put into a 250°F oven for 10 min to dry them out completely.
3. While potatoes are still hot, purée them using a ricer or food mill.
4. Add the eggs, butter, salt, pepper, and nutmeg and mix to incorporate.
5. Add the flour and mix well to form a firm dough.
6. Using both hands flattened, begin to roll the dough into long ropes about ½" to ¾" thick.
7. Cut the rope into 1½"-long sections and use the tines of a fork or a gnocchi board to press grooves in each piece.
8. Cook the gnocchi in a pot of salted boiling water for about 3 min until they float to the surface. Strain the gnocchi from the water and drain in a colander to remove excess water.
9. While gnocchi is draining, brown the pancetta in a large sauté pan. After pancetta is browned, add the butter, garlic, gnocchi, and asparagus and sauté until gnocchi begins to turn golden brown.
10. Deglaze pan with white wine and reduce until almost completely evaporated.
11. Season with salt, pepper, parsley, and Parmesan cheese and serve.

### Nutrition Facts
**Per serving:** 468.2 calories; 33% calories from fat; 17.7 g total fat; 101.2 mg cholesterol; 660.5 mg sodium; 783.2 mg potassium; 64.8 g carbohydrates; 3.7 g fiber; 1.6 g sugar; 61.1 g net carbs; 12.8 g protein

# Recipes

## Shrimp Fried Rice

*yield: 4 servings*

| | |
|---|---|
| 1½ tbsp | vegetable oil |
| 1 tsp | sesame oil |
| 2 oz | onions, small diced |
| 2 oz | carrots, blanched and cut medium dice |
| 2 oz | napa cabbage, medium diced |
| 4 oz | shrimp, medium diced |
| 1 tsp | ginger, grated fine |
| 1 tsp | garlic, minced |
| 1 lb | cooked long grain rice, cooled |
| to taste | salt and black pepper |
| 2 oz | snow peas, medium diced |
| 1 | egg, beaten |
| 1 tbsp | tamari soy sauce |

1. Heat 1 tbsp oil and the sesame oil in wok or very hot sauté pan, add onions, and stir-fry until onions begin to turn brown.
2. Add carrots and brown slightly, then add cabbage and stir-fry 1 min.
3. Add shrimp and cook until almost cooked through, then add the ginger and garlic and cook for 10 sec.
4. Add the cooked rice and cook until rice begins to brown, then quickly add the snow peas and cook for 1 min more.
5. Push rice to sides of wok and add remaining ½ tbsp of oil to the bottom of the wok. Add beaten egg directly on top of the oil and let stand for 10 sec for the egg to begin cooking. Stir egg into the rice to incorporate.
6. Add the soy sauce, adjust seasonings as needed, and serve immediately.

### Nutrition Facts
**Per serving:** 275.8 calories; 27% calories from fat; 8.5 g total fat; 96 mg cholesterol; 363.2 mg sodium; 229.8 mg potassium; 37 g carbohydrates; 1.5 g fiber; 2.1 g sugar; 35.6 g net carbs; 11.7 g protein

## Sesame and Ginger Wild Rice with Almonds

*yield: 6 servings*

| | |
|---|---|
| ½ c | wild rice |
| 12 oz | water |
| 1 tsp | kosher salt |
| ½ tsp | sesame oil |
| ½ tsp | vegetable oil |
| 1 tsp | fresh garlic, minced |
| 1 tsp | fresh ginger, grated |
| 2 tsp | sliced almonds |
| 2 | scallion, diced small |
| 1 tbsp | fresh cilantro, chiffonade |
| to taste | salt and black pepper |

1. In a medium saucepan, combine rice, water, and salt and bring to a boil. Cover, lower to a simmer, and cook for 1 hr.
2. In medium sauté pan, add vegetable oil with the sesame oil, and sauté ginger and garlic for 1 min.
3. Add cooked wild rice and sauté for a few seconds.
4. Add almonds and scallions and toss with cilantro.
5. Season to taste.

### Nutrition Facts
**Per serving:** 63.8 calories; 19% calories from fat; 1.4 g total fat; 0 mg cholesterol; 351.5 mg sodium; 94.4 mg potassium; 11 g carbohydrates; 1.2 g fiber; 0.6 g sugar; 9.8 g net carbs; 2.4 g protein

# Recipes

## Wheat Berries with Walnuts and Dried Cherries

*yield: 6 servings*

| | |
|---|---|
| 1 c | wheat berries |
| 16 oz | chicken stock (or water) |
| 2 tsp | shallot, minced |
| 1 tsp | fresh garlic, minced |
| 1 tbsp | walnuts |
| ½ tsp | thyme |
| 2 tsp | vegetable oil |
| to taste | kosher salt and black pepper |

1. In a small saucepot, bring wheat berries and chicken stock to a boil, cover, lower heat to a simmer, and cook for 15 min to 18 min until al dente.
2. In another pan, add vegetable oil and shallots and sauté until shallots are translucent.
3. Add garlic and walnuts and sauté for 1 min more.
4. Add cooked wheat berries and sauté until heated through.
5. Toss with thyme and season to taste.

### Nutrition Facts

**Per serving:** 159.9 calories; 20% calories from fat; 3.8 g total fat; 2.3 mg cholesterol; 112.8 mg sodium; 229.7 mg potassium; 27.5 g carbohydrates; 4 g fiber; 1.4 g sugar; 23.5 g net carbs; 5.8 g protein

## Spicy Quinoa and Cilantro Salad

*yield: 6 servings*

*Salad*

| | |
|---|---|
| 1 c | quinoa, rinsed well |
| 12 oz | stock or water |
| 2 | scallions, bias cut |
| 4 oz | tomato, peeled, seeded, and medium diced |
| 1 | red bell pepper, roasted, peeled, seeded, and medium diced |
| 2 tbsp | pumpkin seeds, toasted |

*Dressing*

| | |
|---|---|
| 1 oz | cilantro leaves |
| 1 oz | parsley leaves |
| 1 clove | garlic, minced |
| 1 | jalapeño, seeded |
| 2 oz | white vinegar |
| 4 oz | olive oil |
| 1 tsp | water |
| to taste | salt and black pepper |

1. In a saucepan, bring rinsed quinoa and stock to a boil, lower heat, cover, and simmer until quinoa is tender and liquid is absorbed. Let cool.
2. Combine all dressing ingredients in a blender and blend until smooth.
3. Toss remaining salad ingredients with cooled quinoa and toss lightly with dressing.

### Nutrition Facts

**Per serving:** 488.5 calories; 51% calories from fat; 28.5 g total fat; 9.9 mg cholesterol; 2135 mg sodium; 846.2 mg potassium; 46.5 g carbohydrates; 2.7 g fiber; 0.8 g sugar; 43.8 g net carbs; 14.4 g protein

# Recipes

## 247

## Grilled Sun-dried Tomato Polenta

*yield: 8 side servings*

| | |
|---|---|
| 16 oz | whole milk |
| 16 oz | water |
| 1 tsp | kosher salt |
| ¼ tsp | white pepper |
| 8 oz | fine cornmeal |
| 2 tbsp | sun-dried tomato, small diced |
| 2 tbsp | Parmesan cheese, grated |
| 2 tbsp | olive oil |

1. Bring milk and water to a boil and add salt and pepper.
2. Begin adding cornmeal in a fine stream while stirring constantly. *Note:* Adding the cornmeal too quickly or not stirring constantly while adding will result in lumpy polenta.
3. After all cornmeal has been added, add sun-dried tomatoes, reduce heat, and simmer for 35 min to 40 min or until polenta is smooth and grains are tender.
4. Add Parmesan cheese and stir well to incorporate.
5. Pour polenta onto a parchment-lined sheet pan and spread until about ½″ thick. Spray top surface with a nonstick cooking spray and lay another piece of parchment on top of the polenta and against the surface to prevent a skin from forming. Refrigerate until needed.
6. For service, cut the cooled polenta into desired shape and size, brush with olive oil, and grill on each side until golden and heated through.

### Nutrition Facts

**Per serving:** 176.7 calories; 33% calories from fat; 6.8 g total fat; 7.2 mg cholesterol; 315.6 mg sodium; 198.9 mg potassium; 25.1 g carbohydrates; 2.2 g fiber; 3.7 g sugar; 22.9 g net carbs; 4.9 g protein

### Recipe Variations:

**248** **Creamy Polenta:** Follow grilled polenta recipe, increasing the milk to 18 oz. Serve immediately.

**249** **Creamy Gorgonzola Polenta:** Add 2 oz of heavy cream and 2 oz of Gorgonzola at the end of the cooking process and stir well to incorporate.

**250** **Creamy Mushroom Polenta:** 3 oz of mushrooms are sliced and sautéed in the pot prior to adding the milk and water.

## Basmati Rice with Pineapple and Coconut

*yield: 4 side servings*

| | |
|---|---|
| 1 oz | celery, minced |
| 2 oz | onion, minced |
| ½ tbsp | hot chile pepper (such as jalapeño, Serrano, or Thai chile), seeded and minced |
| 4 oz | basmati rice |
| 7 oz | water |
| ½ oz | lime juice |
| ½ oz | tamari soy sauce |
| to taste | salt and pepper |
| 1 tbsp | grated coconut |
| 1 tsp | sesame oil |
| 2 tsp | vegetable oil (or peanut oil) |
| 2 oz | pineapple, small diced |
| 1 | scallion, small diced |
| ½ tsp | cilantro, chopped coarse |

1. Sauté celery and onion in sesame and vegetable oils in a medium saucepot until translucent.
2. Add garlic and chile and cook for 1 min.
3. Add rice and sauté until translucent and grains are coated with oil.
4. Add water, soy, lime juice, and coconut and season to taste.
5. Bring mixture to a boil, cover, and lower to a simmer. Cook for 20 min.
6. Turn off heat and allow mixture to rest for 5 min covered. After 5 min, remove lid and fold in pineapple, scallion, and cilantro and stir well. Serve immediately.

### Nutrition Facts

**Per serving:** 94.8 calories; 36% calories from fat; 3.9 g total fat; 0 mg cholesterol; 264.7 mg sodium; 144.8 mg potassium; 13.8 g carbohydrates; 1.1 g fiber; 3.3 g sugar; 12.6 g net carbs; 1.7 g protein

# *Recipes*

## 252

## Curry and Dried Apricot Couscous

*yield: 6 servings*

| | |
|---|---|
| 10 oz | couscous |
| 12 oz | chicken stock, hot |
| 2 tsp | curry powder |
| 5 oz | dried apricots, small diced |
| 1.5 oz | scallions, small diced |
| 2 tbsp | mint, rough chopped |
| 1 tsp | lemon juice |
| 2 tbsp | olive oil |
| to taste | kosher salt and black pepper |

1. Add curry powder to chicken stock and bring to a simmer.
2. Place the couscous in a medium-size bowl, pour hot chicken stock over it, and stir to mix well.
3. Cover bowl and let stand for 10 min until all stock has been absorbed.
4. Fluff couscous with a fork and stir in remaining ingredients gently to combine.
5. Season to taste.

### Nutrition Facts
**Per serving:** 282.7 calories; 17% calories from fat; 5.6 g total fat; 1.7 mg cholesterol; 96.2 mg sodium; 451.2 mg potassium; 51.1 g carbohydrates; 4.2 g fiber; 13.9 g sugar; 47 g net carbs; 7.9 g protein

**Recipe Variation:**

**253 Curry and Dried Apricot Couscous (Chilled):** To serve as a cold side salad, add 2 tsp sherry vinegar with other ingredients and toss to combine. Cool under refrigeration and serve.

## Risotto

*yield: 6 servings*

| | |
|---|---|
| 2 oz | onions, minced |
| 1 tsp | garlic, minced |
| 1 oz | butter |
| 8 oz | Arborio rice |
| 2 oz | white wine |
| 20 oz | chicken stock, hot |
| 1 oz | cream |

1. Sweat onions and garlic in butter until onions are translucent without browning.
2. Add Arborio rice and stir to coat all grains with butter. Cook until a light toasted aroma can be detected, without browning.
3. Add white wine and cook, stirring continuously with a wooden spoon or a high-temperature spatula until almost dry.
4. Begin adding about ⅓ of the hot stock and again stir continuously until all of the moisture has been absorbed. Continue adding the stock in thirds while stirring and allowing it to absorb until all has been added.
5. Add cream and stir well to incorporate.
6. Season with salt and white pepper and serve immediately.

### Nutrition Facts
**Per serving:** 229.8 calories; 26% calories from fat; 6.8 g total fat; 19.8 mg cholesterol; 145.1 mg sodium; 127.4 mg potassium; 34 g carbohydrates; 0.1 g fiber; 1.9 g sugar; 33.9 g net carbs; 5.3 g protein

**Recipe Variations:**

**255 Risotto Milanaise:** Add ¼ tsp of saffron to the hot stock that will be used and finish with 2 oz grated Parmesan cheese and 1 tbsp butter.

**256 Mushroom Risotto:** Sweat 2 oz diced mushrooms with the onions and garlic and continue with recipe.

**257 Asparagus Risotto:** Blanch 3 oz of asparagus in salted water until al dente. Remove and shock in cold water and drain. Purée asparagus in processor or blender. Prepare basic risotto recipe and, when finished, add puréed cooked asparagus to risotto, stir well, and serve immediately. *Note:* Preparing the risotto in this manner will help to maintain the bright green color of the asparagus. Tips of blanched asparagus can also be added at this point if desired.

# Yeast Breads and Quick Breads

## Chapter 22

## Yeast breads and quick breads

*Yeast breads and quick breads* can be some of the most challenging foods to prepare, as they require the highest level of accuracy both in measuring ingredients and preparation techniques. As much as cooking relies on recipes, baking relies on formulas. The science of baking is based on how ingredients chemically react with one another. The proper combinations of flour, fat, liquid, leavening agents, eggs, and other ingredients are fundamental to the successful preparation and baking of yeast breads and quick breads as well as cakes, pastries, and desserts.

## Chapter Objectives:

1. Identify common types of bakeshop equipment, tools, and bakeware.
2. Identify the basic ingredients used to create baked products.
3. Use the formula for calculating the baker's percentage of each ingredient in a yeast bread or quick bread recipe.
4. Contrast the three categories of yeast doughs.
5. Describe the 12 steps used to produce yeast dough.
6. Demonstrate how to use a baker's scale.
7. Describe popular types of quick breads.
8. Define common terms used to describe different aspects of mixing.
9. Demonstrate how to prepare quick breads using the biscuit method, the muffin method, and the creaming method.
10. Identify guidelines to follow for baking quick breads.

## Key Terms

- **gluten**
- **hydrogenation**
- **leavening agent**
- **formula**
- **baker's percentage**
- **lean doughs**
- **rich doughs**
- **rolled-in doughs**
- **kneading**
- **fermentation**
- **punching**
- **scaling**
- **rounding**
- **panning**
- **proofing**
- **scoring**
- **docking**
- **quick bread**

## BAKESHOP EQUIPMENT

A bakeshop is equipped with the various types of equipment, hand tools, and smallwares necessary for the preparation of yeast breads, quick breads, cakes, pastries, and desserts. Some of the most frequently used pieces of equipment include rolling pins, pastry brushes, rings, sheet pans, proofing cabinets, mixers, and ovens. ***See Figure 22-1.***

### Bakeshop Equipment

Matfer Bourgeat USA
**French Rolling Pin**

Matfer Bourgeat USA
**Pastry Brushes**

Matfer Bourgeat USA
**Ring**

Matfer Bourgeat USA
**Sheet Pan**

Cres Cor
**Proofing Cabinet**

Hobart
**Floor Mixer**

Blodgett Oven Company
**Convection Oven**

**22-1** *Common bakeshop equipment includes rolling pins, pastry brushes, rings, sheet pans, proofing cabinets, mixers, and ovens.*

**Mixers.** Most bakeshops have both tabletop mixers for mixing small volumes and floor mixers for mixing large quantities. A *mixer* is an electric appliance with attachments that can be used to mix, beat, knead, whip, or cream ingredients. Mixers come in a variety of sizes that range from 4-qt tabletop mixers to 100-qt floor mixers.

**Convection Ovens.** A convection oven is one of the most basic pieces of equipment used in the bakeshop. It is preferred over conventional ovens because the air circulation within the oven produces a more evenly baked product.

**Reel Ovens.** A *reel oven* is an oven that has shelves that rotate in a manner similar to a Ferris wheel. Reel ovens are convenient when baking large quantities of the same item. For example, when baking 20 apple pies, the pies could be baked all at once in a reel oven where temperature and humidity controls help produce consistent baked goods.

**Sheeters.** A *sheeter* is a piece of equipment used to roll out large pieces of dough. Sheeters are useful when making large quantities of breads or pastry. Using a sheeter reduces manual labor costs compared to the cost of rolling dough by hand.

**Proofing Cabinets.** A *proofing cabinet* is a tall, narrow stainless steel box in which temperature and humidity can be controlled. It creates the perfect environment for yeast doughs to proof, or rise. Proofing allows a yeast dough to increase in volume before baking. Yeast growth is controlled mostly by temperature. A proofing cabinet helps to ensure a consistent product because the temperature and humidity can be controlled.

**Bakeware.** Bakeware is available in a variety of shapes and sizes, including round, square, and rectangular pans. The size and shape of pans used depends on the items being prepared.

**Sheet Pans.** A *sheet pan* is an aluminum pan with very low sides that is used to bake large amount of products at once. Sheet pans are available in full, half, or quarter sizes. They are commonly used for baking biscuits, rolls, sheet cakes, cookies, and pastries.

**Springform Pans.** A *springform pan* is a pan with a metal clamp on the side of the pan that allow the sides of the pan to be removed. Springform pans are most often used to bake cheesecakes.

**Tart Pans.** *Tart pans* are shallow pans with short sides that are used to bake tarts. The sides of a tart pan are often fluted and the bottom may be removable.

**Molds.** *Molds* are pans that have distinctive shapes. For example, ramekins are small round dishes that are used to make a variety of desserts, such as crème brûlée. Soufflé molds are similar to ramekins but are a little larger.

**Rings.** *Rings* are pans without a bottom that are used to give products a round shape. Rings come in a variety of sizes and are used mostly when preparing cakes or specialty desserts.

**Rolling Pins.** *Rolling pins* are long cylinders used to roll out many types of bread, pie, cookie, and pastry dough. French rolling pins do not have handles and are tapered on each end.

**Pastry Bags.** Pastry bags and tips are used when decorating cakes and pastries. *Pastry bags* are cone-shaped paper, canvas, or plastic bags with two open ends, the small of which is fitted with a pastry tip. Disposable pastry bags can be used to eliminate some of the cleanup.

**Pastry Tips.** *Pastry tips* are cone-shaped tips that are fitted into the narrow end of pastry bags and used to create decorative shapes and patterns with icing. Hundreds of different pastry tips are available.

**Pastry Brushes.** A *pastry brush* is a small, narrow brush used to apply liquids such as egg wash or butter onto baked products. Pastry brushes may be made out of natural, nylon, or silicone bristles. Nylon brushes cannot be used with hot items because the nylon could melt.

**Pastry Cloth.** A *pastry cloth* is a large piece of canvas that is used as a surface for rolling out dough, as the dough does not stick to it easily.

## BAKESHOP INGREDIENTS

Baking is one part art and two parts science. In order to successfully prepare attractive and delicious breads, cakes, pastries, and desserts there are some basic concepts and terms that must be learned.

Different types of flours, sugars and sweeteners, eggs, fats, and milk produce different results when used in baked goods. Thickeners and leavening agents are used to control texture and lightness of breads. Flavoring ingredients allow a chef to control the flavor of a product. Ingredients cannot be substituted in baked goods as easily as they can be in other dishes. For instance, substituting cake flour with bread flour results in an end product with an entirely different look, texture, and taste. Baking is based upon chemical reactions, and bakeshop recipes are known as formulas. When preparing a formula, ingredients must be measured precisely and substitutions should be limited to ensure a successful end product.

### Flours

*Flour* is the fine powder that is left after grinding grain. Flour supplies strength to a dough, and structure and nutritional value to a baked product. It also acts as an absorbing agent. Flour is characterized as being hard or soft, depending on the protein content of the grain. The hardest wheat generally contains more protein and fewer starch granules. A higher protein content results in a higher amount of gluten in a finished product. *Gluten* is the protein in flour that, when combined with water, gives structure, strength, and elasticity to a baked product. ***See Figure 22-2.*** Elasticity is needed so that the dough can hold gases produced through the fermentation of the yeast.

**Bread Flour.** *Bread flour* (hard wheat flour) is a type of flour that has a high protein content and that produces the most gluten when water is added. The longer a product is kneaded or mixed, the more gluten forms, resulting in a more structurally sound finished product. *Cake flour* (soft wheat flour) is a type of flour characterized by low amounts of protein and high amounts of starch. Cake flour does not form gluten as readily as bread flour, resulting in a more delicate finished product. Cake flour is used in the preparation of cakes, cookies, pies, and pastries.

**All-Purpose Flour.** *All-purpose flour* is a type of flour that is a mixture of both hard and soft wheat flours. Most large bakeshops do not use all-purpose flour, as they prefer using flour that is specific to their products. All-purpose flour is used more frequently in small bakeshops. *Self-rising flour* is a type of flour that that contains measured amounts of baking powder and salt, so additional baking powder and salt do not need to be added.

**Whole-Wheat Flour.** *Whole-wheat flour* is unbleached flour that contains the bran, endosperm, and germ of the wheat kernel. It has a nuttier flavor than other flours. Whole-wheat flour is used more readily in breads than pastries and has more nutritional value.

**Rye Flour.** *Rye flour* is a dark-colored flour milled from rye seeds. Whole rye flour lacks protein to form gluten so it is almost always combined with wheat flour when making bread. White rye flour comes from the center of the endosperm. Cream rye flour comes from the second layer of the endosperm. Dark rye flour comes from the outermost portion of the endosperm. Rye bread traditionally has a strong flavor and usually contains caraway seed, which adds to its pungency. *Pumpernickel flour* is a meal form of rye made by grinding the whole grain.

**Other Flours.** Flours can also be made from corn, soybeans, rice, barley, oats, millet, potatoes, buckwheat, spelt, quinoa, and amaranth. However, these flours lack the gluten-forming properties of wheat, so they must always be combined with wheat flour when making bread.

## Sugars and Sweeteners

Sugars and sweeteners add color, texture, and taste to baked goods. Sugar, when used in yeast dough, stimulates the growth of yeast, supplies moisture, and helps prolong freshness. Sugar comes from a variety of plant sources and is available in many different forms. Granulated sugar, confectioners' sugar, molasses, corn syrup, honey, and maple syrup are just a few of the different forms of sugar and sweeteners used in the bakeshop. ***See Figure 22-3.***

**Gluten Formation**

22-2 Gluten is formed when the protein in wheat flour combines with water.

**Nutrition Note**

Celiac disease is a disease where the body cannot absorb nutrients from foods that contain gluten. A person with this disease cannot eat anything made from wheat flour such as yeast breads, quick breads, cakes, cookies, pies, pastries, or pasta. Food service establishments should have ingredient information available so that customers can determine whether a dish contains gluten. A celiac can become quite ill after eating anything that contains gluten.

**22-3** *Many types of sugars and sweeteners are used in a bakeshop.*

**Granulated Sugar.** *Granulated sugar* is an all-purpose, white sugar composed of small, uniform-size crystals. *Turbinado sugar* is a pale brown sugar that is purified and cleaned through a steaming process before being dried into coarse-grain crystals. Because of its large crystals and refining process, it is advertised as "sugar in the raw" and is more expensive than granulated sugar. In this case, the term "raw" simply refers to minimal processing that results in a more natural product. The USDA has not approved actual "raw sugar" for human consumption.

**Brown Sugar.** Brown sugar can be substituted for granulated sugar if more moisture and a stronger flavor is desired. *Dark brown sugar* is a moist sugar product that contains approximately 7% molasses. *Light brown sugar* is a moist sugar product that contains approximately 3.5% molasses. Brown sugar should be kept in an airtight container, as it can dry out quickly and turn very hard.

**Confectioners' Sugar.** *Confectioners' (powdered) sugar* is granulated sugar that has been ground into a very fine powder. A small amount of cornstarch is added to confectioners' sugar to prevent lumping. Confectioners' sugar works well in icings and meringues, as the added cornstarch absorbs excess moisture and helps prevent the product from weeping.

**Superfine Sugar.** *Superfine sugar* is granulated sugar with very small crystals. Superfine sugar dissolves quickly because of its fineness and is preferred for use in recipes where the sugar needs to dissolve completely, such as for an angel food cake or a meringue.

**Molasses.** *Molasses* is the dark residual syrup that is left after sugar cane has been processed. Cane syrup is processed three to four times, each time yielding another grade of molasses. Each time the syrup is boiled and processed the flavor of the molasses becomes more concentrated and caramelized. Blackstrap molasses is the strongest, most processed form of molasses. Molasses is often added to bread, cake, and cookie recipes, not to add sweetness, but to add a strong, distinct flavor.

**Corn Syrup.** *Corn syrup* is a sweet syrup made by adding an enzyme to starch extracted from corn kernels. Corn syrup is used in a variety of recipes and helps keep food moist due to its hygroscopic abilities. *Hygroscopic* is the ability to attract water. When corn syrup is used in recipes, the product does not lose as much water as it would if granulated sugar were used. *Glucose syrup* is a type of corn syrup that contains a high amount of glucose (sugar). *Dark corn syrup* is a corn syrup with added caramel color and flavor to give it an aroma similar to that of molasses.

**Honey.** *Honey* is a sweet, thick fluid made by honey bees from flower nectar and is 1½ times as sweet as granulated sugar. The flavor of honey depends on the flowers from which the nectar was derived. Clover, buckwheat, and lavender each yield a different flavor of honey.

**Maple Syrup.** *Maple syrup* is a sweet syrup derived from the sap of maple trees. Maple syrup is graded as Grade AA, Grade A (Grade A Light or Grade A Fancy), or Grade B. The lightest grade is Grade AA, which can be used as a sweet topping for breakfast foods such as waffles, French toast, or pancakes. Grade B is the darkest, most robust maple syrup and is used primarily in baking. Maple-flavored syrups made from corn syrup and maple flavoring are also available.

## Eggs

Eggs serve a variety of important roles in the formation of a final baked product. They contribute to the structure, aeration, flavor, and color of an item. The protein in eggs can be used to give structure to a product, and eggs can be used to thicken mixtures such as custard sauces. Eggs that are beaten or whipped trap pockets of air inside a dough or batter that expand when heated, helping baked products to rise.

Egg yolks are natural emulsifiers, meaning that they hold or bind ingredients together that normally would not combine, such as water and oil. Finally, egg yolks add a rich flavor and a golden color to a final baked product and help in the browning process. An egg wash, consisting of a combination of egg yolk and either milk or water, is often brushed on top of pie crusts or bread doughs before baking to add a glossy sheen and a golden brown color to the baked good. *See Figure 22-4.*

## Egg Washes

An egg wash adds a glossy sheen to baked goods.

### Fats

Fat is one of the most highly scrutinized ingredients due its potential negative effects on health, but it is an essential ingredient in the bakeshop. Fat adds richness and tenderness, improves grain and texture, develops flaky layers in puff and Danish pastry, and improves the shelf life of baked products. Butter, lard, shortening, margarine, and oil all contain fat and are used to make tender, moist, baked products. ***See Figure 22-5.***

**Butter.** *Butter* is a fat product made from cream that has a butterfat content of at least 80%. Bakeries prefer unsalted butter because it allows the chef to control of the amount of salt that is added to a recipe. Salted butter, however, has a longer shelf life. *European-style butter* is a type of butter that contains 82% to 86% butterfat. It is considered premium butter and is more expensive than regular butter.

## Fats

| Butter | Lard | Shortening | Margarine | Oil |

**22-5** Fats used in a bakeshop include butter, lard, shortening, margarine, and oils.

**Lard.** *Lard* is a 100% fat product made from rendered pork fat. A very high-quality lard can be used primarily to make light and flaky pastry crusts.

**Shortening.** Most yeast dough recipes call for hydrogenated shortening because it produces the best results. *Hydrogenation* is a process that changes the molecular structure of vegetable oil into a creamy or solid state. When hydrogenation occurs, the melting point of fat is raised, which makes it pliable when it is blended with flour. Another effect of hydrogenation is that the fat becomes less likely to turn rancid. Hydrogenated shortening does not contain any water but about 10% air by weight. Because of its air content, shortening is a much more consistent fat for aerating batters.

Some recipes call for an emulsified shortening. Emulsified shortening contains emulsifiers that contribute to better aeration of batters and a more tender texture of the final product. Emulsified shortenings assist with moisture absorption and are often used in icings and high-ratio cakes.

**Margarine.** *Margarine* is a fat product made from hydrogenated vegetable oils. Margarine is available in solid and spreadable forms, depending on the extent of the hydrogenation process. Although it is flavored and colored to imitate butter, many pastry chefs do not use margarine, as they prefer the taste and performance of real butter in their products.

**Oil.** Oils maintain moisture in a product. An *oil* is a fat that remains in a liquid state at room temperature. Oils are most often taken from plant sources rather than animal fats. A wide variety of oils are available. For baking, a mild-flavored oil, such as corn, vegetable, or canola oil, is preferred. Oils with a strong flavor, such as olive oil, can overpower the flavor of a baked product.

*Chef's Tip:* A negative effect of hydrogenation is that a portion of the fat is chemically changed from unsaturated fats to trans fats. Trans fats are considered unhealthy fats.

## Milk

Milk and other dairy products have many different effects on bakery and pastry items. First, milk provides a smooth, rich flavor. Milk improves the texture of dough, supplies moisture, improves flavor, and adds nutrients. The emulsifiers and proteins in milk assist in producing a finer crumb (texture) with smaller air bubbles in yeast breads. When breads or pastry items containing milk are baked, the proteins and sugars in the milk absorb more moisture, thereby producing a softer crust.

These same proteins and lactose (milk sugar) break down during baking and begin to caramelize, which adds a rich color to the crust of the baked good. Milk also adds important nutrients such as calcium, vitamin D, and protein.

Milk used in the preparation of rolls, breads, and sweet doughs may be purchased in liquid or dry form. Dry milk must be reconstituted in water before or during the mixing period. Milk-like products such as soy milk, rice milk, or almond milk can also be used in place of milk. However, unlike fats, thickeners, eggs, and sweeteners, the effects of milk are not as critical to the finished product. Water can be substituted for milk in most bread recipes.

## Thickening Agents

Thickening agents provide structure and stability to baked goods. Common thickening agents found in the bakeshop include cornstarch, arrowroot, flour, tapioca, and gelatin. ***See Figure 22-6.*** The various thickening agents are derived from different food sources and each has properties that make it favorable for use in some foods and not others.

### Thickeners

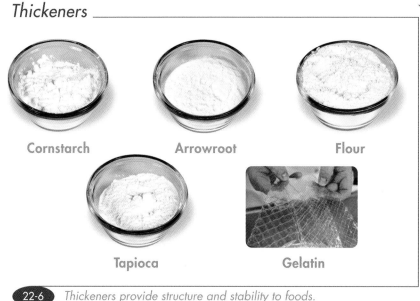

Cornstarch     Arrowroot     Flour

Tapioca     Gelatin

**22-6**   *Thickeners provide structure and stability to foods.*

**Cornstarch.** Cornstarch is one of the most popular thickening agents because it will not cloud a clear sauce and is therefore preferred for use in pie fillings. Cornstarch should be mixed with a small amount of cold water to create a slurry before it is added to a hot mixture. A slurry prevents any lumps from forming in the finished product. Cornstarch should be thoroughly cooked or it can leave an unpleasant aftertaste.

**Arrowroot.** *Arrowroot* is a thickening agent derived from a tropical tuber. It is clear and tasteless and may be used in pie fillings, cobblers, and custards. Like cornstarch, arrowroot should be made into a slurry with water before being added to a mixture.

**Flour.** Flour is one of the most readily available thickening agents in the bakeshop. However, the protein content of flour inhibits its ability to thicken items as effectively as cornstarch or arrowroot. It requires twice as much flour to thicken the same amount of something compared to cornstarch or arrowroot. A low-protein flour works best for use as a thickening agent. Flour also turns a clear sauce opaque. However, it does not affect taste. Any product thickened with flour should be cooked thoroughly or the taste of raw flour will remain.

**Tapioca.** *Tapioca* is a thickening agent that is derived from the tropical cassava plant and is available in round granular form or ground flour form. Tapioca leaves filling looking clear and glossy and has twice the thickening power of flour. When using granular tapioca, it is important to have enough liquid in a filling to ensure that the granules dissolve. Tapioca is by far the best choice for thickening pies and cobblers.

**Gelatin.** *Gelatin* is a flavorless thickening agent made from animal protein. Gelatin is available in powdered and sheet forms. Powdered gelatin comes in 0.25-oz envelopes or in bulk form. Gelatin sheets are sold in a standard size that is equivalent to one teaspoon of powdered gelatin. Gelatin must be dissolved or soaked in cold water and then gently heated to fully cook before being added to a product. Gelatin is used in Bavarian cream desserts to help stabilize the whipped cream.

### Leavening Agents

A *leavening agent* is any ingredient that causes a bakery product to rise (leaven). *Leavening* is the process by which bakery products rise by the action of air, steam, chemicals, or yeast. The most basic way to leaven baked products is to incorporate air and steam into a dough or batter. Air pockets are created in products by whipping eggs or creaming fat and sugars. These pockets of air help a dough or batter to rise because the air is lighter than the surrounding dough or batter and forces the product up as it tries to escape. Steam is created when liquids in the dough or batter evaporate. Steam rising out of a dough or batter can fill air pockets and force a bread to expand.

Some of the leavening agents used in the bakeshop are natural, such as air and steam. Others are chemical leavening agents, such as baking soda, baking powder, cream of tartar, or ammonium bicarbonate. Yeast, a key leavening agent, is a living organism. ***See Figure 22-7.***

**Nutrition Note**

Edible gelatin is a regulated food product that is manufactured from collagen, a natural protein in animal bones. Gelatin is over 85% protein and contains all of the essential amino acids. It does not contain additives or preservatives, is low in calories, and does not contain cholesterol.

*Leavening Agents*

Baking Powder          Baking Soda          Cream of Tartar          Yeast

**22-7** *Leavening agents include chemical leaveners and yeast.*

**Baking Powder.** *Baking powder* is a chemical leavener that is a combination of baking soda and cream of tartar or sodium aluminum sulfate. The alkalinity of baking soda reacts with the acidity of the other ingredients to produce carbon dioxide. There are two types of baking powder: single acting and double acting. Single-acting baking powder reacts as soon as it encounters the liquid of the batter and must be baked immediately upon mixing. Double-acting baking powder does just as its name suggests—it acts twice. This type of baking powder reacts once when it is mixed with liquids, releasing some of its carbon dioxide, and then reacts again when it is heated in the oven. Double-acting baking powder is much more popular than single-acting and allows a batter to be prepared in advance without the risk that it will not rise when placed in the oven at a later time.

**Baking Soda.** *Baking soda (sodium bicarbonate)* is a powerful alkaline chemical leavener that reacts to an acidic dough or batter without the addition of heat by releasing carbon dioxide. Baking soda is used in batters in conjunction with acidic liquids such as buttermilk, lemon juice, or molasses. Too much baking soda in a product can produce a yellow color, cause brown spots, or lead to a bitter aftertaste. Baking soda should not be kept longer than a year, as it absorbs moisture from the air, which reduces its strength over time.

**Cream of Tartar.** *Cream of tartar* is a chemical leavener derived from tartaric acid, a by-product of wine production. When used in recipes, it is often combined with baking soda, which is an alkaline. When the two are combined, they neutralize each other, releasing carbon dioxide. In many egg foam recipes, such as angel food cake, cream of tartar is added because the acid helps to stabilize the egg whites.

**Ammonium Bicarbonate.** *Ammonium bicarbonate* is a chemical leavener used mainly in older cookie and cracker recipes. It reacts as soon as it is heated in the oven and does not require an acid or a base in order to react. It is important to bake any product containing ammonium bicarbonate thoroughly so that all of the gases are released. Otherwise, an ammonia smell can remain in the finished product.

**Yeast.** *Yeast* is a microscopic, living, single-celled fungus that releases carbon dioxide and alcohol through fermentation when provided with food (sugar) in a warm, moist environment. Yeast increases volume, improves flavor, and adds texture to dough. The fermentation process that yeast goes through makes grains and grain products easier to digest. Yeast is very sensitive to temperature and, depending on the type of yeast, only activates in environments ranging from 70°F to 130°F. *See Figure 22-8.* If the temperature is too cold, the yeast remains dormant. If the temperature is too hot, the yeast dies. The three common varieties of commercial yeast are compressed yeast, active dry yeast, and instant

*Chef's Tip:*
If active-dry yeast is purchased in ¼-oz packages (equivalent to ⅔ oz of compressed yeast), three packages must be used for every 2 oz of compressed yeast called for in the recipe.

active dry yeast.

- *Compressed (fresh) yeast* is a yeast that has approximately 70% moisture content and is available in 1-lb cakes or blocks. Compressed yeast produces the most carbon dioxide gas per cell of any yeast, lending itself to excellent dough-raising capabilities. Compressed yeast will always have a "use by" date on the package, and once opened, should be wrapped tightly in plastic wrap and stored under refrigeration. Once opened, it has a shelf life of two weeks. It can also be frozen for up to four months. When fresh, it should have a yeasty aroma, which will turn ammonia-like and sour when old or spoiled. The best method for using compressed yeast is to first dissolve it into double its weight of lukewarm water (90°F to 100°F) prior to adding it to a bread recipe.

- *Active dry yeast* is a form of yeast that has been dehydrated and looks like small granules. It is sold in ¼-oz packages or in 1-lb vacuum-sealed packages. Active dry yeast is convenient because yeast has a long shelf life and can be held for up to 1 year. Active dry yeast produces the least gas per cell among the different types of yeast. Because active dry yeast is lighter in weight due to drying, only half as much dry yeast (by weight) as fresh compressed yeast should be used in a recipe. The dough also needs to be proofed prior to baking. Active dry yeast produces the least gas per cell among the different types of yeast. Active dry yeast is best activated in warm water between 105°F and 115°F.

- *Instant dry yeast* is yeast that does not need to be hydrated. It may be directly added to a warm flour mixture. It also may be labeled as "quick-rise" or "rapid-rise" yeast. It is available in both dried and vacuum-packaged forms. Instant dry yeast is a much more active and aggressive yeast than either fresh compressed or active dry yeast so it works best for breads with short fermentation times. *Note:* Because it is so active, only one-fourth the amount of instant yeast is needed as compared to fresh compressed yeast and one-half the amount as compared to active dry yeast. Once opened, instant dry yeast can be stored under refrigeration for 3 to 4 months or in the freezer for up to 6 months.

- A *sourdough starter* is flour combined with nonchlorinated water. This mixture is left to sit at room temperature. The mixture creates its own yeast spores and grows its own bacteria from the grain (flour). It takes 2 to 3 weeks in order to grow enough yeast to leaven a loaf of bread. During this time, more flour must be added to the mixture as food for the bacteria and yeast spores. A portion of this water mixture is used to make sourdough bread and is the reason for the bread's distinct tangy flavor.

## Yeast Fermentation

### Yeast Action

| | |
|---|---|
| Storage stage | 30° to 40° |
| Slow action | 60° to 75° |
| Normal action | 80° to 85° |
| Fast action | 90° to 100° |
| Death | 140° |

| | |
|---|---|
| 140°F | Death |
| 130°F | |
| 120°F | |
| 110°F | |
| 100°F | |
| 90°F | Fast action |
| 80°F | Normal action |
| 70°F | Slow action |
| 60°F | |
| 50°F | |
| 40°F | |
| 30°F | Storage stage |

22-8  *Yeast activates when sugar is present at temperatures between 70°F and 130°F, causing the release of carbon dioxide and alcohol.*

## Flavorings

Flavorings are essential components in the bakeshop because of the zest they give to finished products. Flavorings are generally divided into two categories: artificial and natural. Natural flavorings impart a better flavor than artificial flavorings. Many different types of flavorings are used, including salt, extracts, emulsions, vanilla, and chocolate.

**Salt.** Salt is a flavoring ingredient that is used to season yeast dough products. When used in dough, salt improves the taste and controls yeast growth. Additionally, salt helps to balance sweetness in baked goods while adding additional flavor.

**Extracts.** An *extract* is a flavorful oil that has been mixed and dissolved with alcohol. Virtually any flavor is available in extract form. Vanilla, lemon, almond, coffee, mint, and rum are just a few flavors of extracts used in the bakeshop. Only small amounts of extract are needed in products, as the flavor of an extract is heavily concentrated.

**Emulsions.** An *emulsion* is an oil that has been mixed with water and an emulsifier. Lemon and orange emulsions are the most common emulsions used in the bakeshop and are much stronger than extracts. Emulsions should be used sparingly because they are quite strong.

**Vanilla.** *Vanilla* is a flavor derived from a dark brown pod that grows on a tropical orchid and is one of the most commonly used flavorings in the bakeshop. Vanilla is added to items such as cakes, cookies, and icings. Vanilla beans are expensive, but the flavor they lend is difficult to substitute. ***See Figure 22-9.*** When adding vanilla beans to a sauce, the bean is split open and the inside of the bean is scraped into the mixture. Pure vanilla extract is preferred over imitation vanilla extract, which may leave a chemical or bitter taste.

**Chocolate.** *Chocolate* is a flavoring ingredient made by roasting, skinning, crushing, and grinding cacao beans into a paste called chocolate liquor. This paste is then combined with various ingredients to create the desired flavor. The flavor of chocolate can range from bitter to sweet, depending on the type of chocolate being used. Types of chocolate commonly used in the bakeshop include cocoa powder, unsweetened chocolate, bittersweet or semisweet (dark) chocolate, milk chocolate, and white chocolate. ***See Figure 22-10.***

- *Cocoa powder* is a brown unsweetened powder extracted from ground cacao beans. It is reddish brown in color and has an intense flavor. Alkalized or *Dutch-processed cocoa powder* is a dark, unsweetened cocoa powder processed with an alkali to neutralize the natural acidity. This raises the pH of the cocoa from 5.5 to between 7 and 8. Dutch-processed powder adds a smooth taste to baked products. The two cocoas should not be substituted for each other as the difference in pH levels affects leavening agents differently.

**Chef's Tip:** Salt can inhibit yeast activation. Too much salt can destroy yeast, and too little salt results in bland flavor and products rising too quickly.

## Vanilla

Vanilla extract

Vanilla bean

**22-9** *Vanilla is a flavoring ingredient that comes from the seed pod of a tropical orchid.*

- *Unsweetened chocolate* is pure chocolate liquor. There are no sugar or milk products added to it, and it must contain 50% to 58% cocoa butter. Unsweetened chocolate is melted and used in many cake and frosting recipes with the addition of sweeteners such as sugar.

- Bittersweet and semisweet (dark) chocolate contain 35% chocolate liquor in addition to sugar, vanilla, and cocoa butter. Bittersweet chocolate is usually not as sweet as semisweet, but it is often difficult to tell the difference between the two. Both bittersweet and semisweet chocolate are used in a variety of baked goods including cookies and brownies.

- *Milk chocolate* is a chocolate product that contains at least 12% milk solids and 10% chocolate liquor. Milk chocolate cannot be substituted for dark chocolate in recipes as the milk content alters the flavor and may scald if melted.

- Premium *white chocolate* is a confectionary product made from cocoa butter, sugar, milk solids, and flavorings. White chocolate does not contain any cocoa solids. Less expensive white chocolate contains vegetable oils in place of cocoa butter.

## Chocolate

| Cocoa powder | White | Milk | Dark |

**22-10** *Chocolate is a flavoring ingredient that comes from the beans of a cacao plant.*

## BAKESHOP FORMULAS

A chef in a bakery or pastry shop relies on formulas in the same way that a scientist does in a laboratory. A *formula* is a format for a recipe where all ingredients are listed as a percentage relative to the main ingredient. **See Figure 22-11.** Formulas differ from recipes in that a formula lists the exact amount of each ingredient, by weight. Bakers and pastry chefs use formulas instead of recipes as precise measurement is crucial in baking because the complex chemical reactions needed to produce the desired result.

A formula ensures that consistent baked goods will be produced. A slight variance can result in a dramatically different final product, or even a completely unacceptable one. This is not true in a traditional recipe. For example, an extra half-pound of mirepoix can be added to a chicken stock without having too great an affect on the finished product. However, if one less egg than necessary is added to a cake batter, the cake will not rise properly and will be too dense.

Because slight changes in the proportions in a formula or in the method of preparation can have a big effect on the final product, it is imperative that each ingredient be measured accurately and the proper method of preparation be followed. When using a formula, all ingredients are weighed on a baker's scale before being added. Each ingredient in the formula plays an important role in the chemical reactions that occur both before a dough or batter enters the oven and again while it bakes. Mixing ingredients in the order listed on the formula is also crucial to the outcome of the final product. A formula must always be read thoroughly before beginning to mix the ingredients.

| Bakeshop Formulas—Hard Roll Dough | | | |
|---|---|---|---|
| Weight | Approximate Measure | Baker's Percentage | Ingredients |
| 7 lb 8 oz | 24 cup | 100% | breaded flour |
| 3 oz | 4 tbsp | 3% | salt |
| 3 ½ oz | 4 ⅔ tbsp | 3% | sugar |
| 3 oz | ¼ cup | 3% | shortening |
| 3 oz | ⅜ cup | 3% | egg whites |
| 4 lb 8 oz | 3 ⅜ gal. | 60% | water |
| 4 ½ oz | | 4% | compressed yeast |

**22-11** *In a formula, ingredients are listed as percentages of the main ingredient.*

## BAKESHOP MEASUREMENTS

In a bakeshop, ingredients are measured by weight or by volume. Although weight is more accurate than volume, it is important to understand both. Weight is measured in units such as grams (g), pounds (lb) and ounces (oz). Volume is measured in units such as liters (l), pints (pt), cups (c), tablespoons (tbsp), and teaspoons (tsp).

Because different ingredients have different densities, two items of the same volume can have different weight. For example, 1 c of brown sugar weighs more than 1 c of confectioner's sugar. Also, weight is measured in ounces (oz) and volume is measured in fluid ounces (fl oz). However, 1 oz does not equal 1 fl oz.

A baker's scale is used to measure most ingredients for formulas. Digital scales are also used in the bakeshop when weighing small amounts of an ingredient. Most digital scales can accurately measure weights as little as ⅛ oz and are much faster than using a baker's scale.

## BAKER'S PERCENTAGES

A *baker's percentage* is the expression of the amount of a particular ingredient as a percentage of the amount of the main ingredient of a formula. Flour is most commonly the main ingredient in a baked product and is therefore represented as 100% in a formula. If two different types of flour are used their total weight will be 100%. All other ingredients in the formula are expressed as a percentage of the amount of flour. For example, if a formula calls for 5% baking soda, 5% is the baker's percentage of baking soda needed.

It is important to note that a baker's percentage formula is quite different than a standard percentage formula that is learned in math class. In a standard mathematical percentage formula, the entire formula adds to equal 100%. Each ingredient is a percentage of that 100%. In a baker's percentage formula, the main ingredient is always equal to 100% and each other ingredient is expressed as a percentage relative to the main ingredient. Because a baker's percentage is a way of expressing ingredient proportions, and flour is already always expressed as 100%, the total of the percentages of the other ingredients in a particular formula is always greater than 100%.

To calculate the baker's percentage of an ingredient, apply the following formula:

$$BP = \left(\frac{WI}{WF}\right) \times 100$$

where

$BP$ = Baker's Percentage

$WI$ = weight of ingredient

$WF$ = weight of flour

$100$ = constant

For example, calculate the baker's percentage of 5 lb of sugar when a formula calls for 10 lb of flour.

$$BP = \left(\frac{WI}{WF}\right) \times 100$$

$$BP = \left(\frac{5}{10}\right) \times 100$$

$$BP = \textbf{50\%}$$

All ingredients must first be converted to the same units, such as pounds or grams, in order to figure out the exact baker's percentage of each item. For example, calculate the baker's percentage of 8 oz of baking soda when a formula calls for 10 lb of flour.

1. First convert all units to ounces. There are 16 oz in 1 lb. Ten pounds of flour equals 160 oz (10 × 16 = 160).

2. Apply the formula to determine the baker's percentage.

$$BP = \left(\frac{WI}{WF}\right) \times 100$$

$$BP = \left(\frac{8}{160}\right) \times 100$$

$$BP = 0.05 \times 100$$

$$BP = 5\%$$

Using this method, a recipe can easily be converted to a formula with baker's percentages. *See Figure 22-12.*

### Converting Recipes to Formulas

| Ingredients | Weight | Weight Converted to Ounces[*] | Baker's Percentage |
|---|---|---|---|
| Flour | 10 lb | 10 x 16 = **160 oz** | $\frac{160}{160 \times 100} = $ **100%** |
| Sugar | 5 lb | 5 x 16 = **80 oz** | $\frac{80}{80 \times 100} = $ **50 %** |
| Baking Soda | 8 oz | **8 oz** | $\frac{8}{160 \times 100} = $ **5%** |
| Salt | 4 oz | **4 oz** | $\frac{4}{160 \times 100} = $ **2.5%** |
| Shortening | 2 lb 8 oz | (2 x 16) + 8 = 32 + 8 = **40 oz** | $\frac{40}{160 \times 100} = $ **25%** |
| Milk | 3 lb | 3 x 16 = **40 oz** | $\frac{48}{160 \times 100} = $ **30%** |
| Eggs | 2 lb 4 oz | (2 x 16) + 4 = 32 + 4 = **36 oz** | $\frac{36}{160 \times 100} = $ **22.5%** |

[*] 1 lb = 16 oz

**22-12** *Converting a recipe to a formula with baker's percentages ensures a successful final product.*

When a formula does not contain flour, each ingredient in the formula will be expressed as a percentage of the main ingredient in the formula. The main ingredient will always be equal to 100%. For example, take a look at the following formula for apricot compote:

| Ingredients | Weight | Percentage |
|---|---|---|
| Dried Apricots | 10 lb | 100% |
| Granulated Sugar | 1 lb | 100% |
| Water | 2 lb | 20% |
| Almonds | 0.5 lb | 5% |
| TOTAL | 13.5 lb | 135% |

Note that the apricots are the main ingredient in this formula, so they are expressed as 100%. Every other ingredient is expressed as a percentage relative to that main ingredient.

Using baker's percentage allows bakers and pastry chefs to change the yield of a product while also controlling changes in flavor without risking a failed end product. For example, in the following two formulas for onion rye bread, one will have a greater onion flavor.

## Formula #1—Onion Rye Bread

| Ingredients | Weight | Percentage |
|---|---|---|
| Flour, white | 6 lb | 75% |
| Flour, rye | 2 lb | 25% |
| Water | 4 lb 8 oz | 56% |
| Onions, diced | 1 lb | 12.5% |
| Yeast, instant | 2 lb | 1.5% |
| Salt | 2 lb | 1.5% |
| Total | 13 lb 12 oz | 171.5% |

## Formula #2—Onion Rye Bread

| Ingredients | Weight | Percentage |
|---|---|---|
| Flour, white | 24 lb | 75% |
| Flour, rye | 8 lb | 25% |
| Water | 18 lb | 56% |
| Onions, diced | 2 lb 4 oz | 7% |
| Yeast, instant | 8 oz | 1.5% |
| Salt | 8 oz | 1.5% |
| Total | 53 lb 4 oz | 166% |

The second formula has more onions (2 lb 4 oz compared to 1 lb in the first formula). However, a look at the second formula percentages reveals that, while the second formula produces a greater amount of dough, the amount of onions is only 7%. The first formula contains 12.5% onions, which is a higher percentage of the main ingredient. Therefore, the first formula will produce a dough with a stronger onion flavor.

## YEAST BREAD PREPARATION

Bread is a versatile staple that can be served as part of any meal. ***See Figure 22-13.*** Most yeast bread doughs contain flour, liquid, shortening, sugar, eggs, salt, and yeast. Flour gives strength to the dough and acts as an absorbing agent. Liquid, usually milk or water, supplies moisture and helps form the gluten. Shortening supplies tenderness and improves the shelf life of the dough. Eggs supply structure to the dough and add color. Sugar acts as a stimulant to the yeast. Salt brings out the flavor in the dough. Fillings and toppings provide additional taste and eye appeal. Fillings are added prior to baking, while toppings are applied after baking.

### Yeast Breads

*Alpha Baking Co., Inc.*

*Alpha Baking Co., Inc.*

22-13 *Many types of rolls and breads are made from yeast doughs.*

Rolls, breads, and sweet doughs are yeast dough products commonly prepared in the bakeshop. There are three categories of yeast doughs: lean doughs, rich doughs, and rolled-in doughs.

- *Lean doughs* are doughs that are low in fat and sugar. Examples of lean dough breads include hard rolls, baguettes, and rye bread.

- *Rich doughs* are doughs that incorporate a lot of fat, sugar, and eggs into a heavy, soft structure. The finished product may be faintly yellow in color due to the large number of eggs that are used. Examples of rich doughs include those for cinnamon rolls and doughnuts.

- *Rolled-in doughs* are yeast doughs with a flaky texture that results from the incorporation of fat through a rolling and folding procedure. By alternating the layers of yeast dough, a very light and flaky texture is achieved in the finished product. Rolled-in doughs may be sweet, as for Danish rolls, or may not, as for croissants.

Yeast dough production involves twelve steps: scaling, mixing, kneading, fermentation, punching, scaling, rounding, panning, pan proofing, baking, cooling, and storing. These steps generally apply to all yeast doughs, although there may be slight variations depending on the product.

### Scaling

All ingredients must be scaled. A baker's scale is used to scale all ingredients. *See Figure 22-14.* The baker's scale must be properly balanced before use and again after adding measured ingredients.

*Scaling Ingredients*

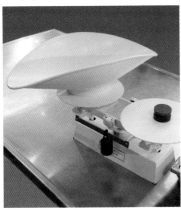

1–2. Set the large scoop on the left side of scale; balance the scale.

3–4. Adjust the scale to desired weight; add the ingredient to the scoop until scale is balanced.

**22-14** *A baker's scale is most often used to accurately measure bakeshop ingredients.*

## To use a baker's scale, apply the following procedure:

1. Set the scale scoop on the left side of the scale.
2. Balance the scale by placing weights on the right side until the scale is even.
3. Adjust the scale to equal the desired weight of the ingredient to be measured. This is done by adding the appropriate amount of weight to the right side of the scale and adjusting the ounce weight on the bar. For example, if a recipe calls for 1 lb 10 oz of flour, a 1-lb weight is added to the right side of the scale and the ounce bar is adjusted to 10 oz.
4. Add the ingredient to the scoop until the scale is balanced.

CULINARY PROCEDURES

## Mixing Yeast Doughs

A bench or floor mixer is typically used to mix yeast dough ingredients. The mixing step is very important for gluten development and uniform distribution of yeast throughout the dough. The three methods commonly used to mix yeast dough are the straight dough method, the modified straight dough method, and the sponge method.

- The *straight dough method* is the simplest yeast dough mixing method and involves combining all ingredients at once and mixing them together. However, the yeast may not be evenly distributed when the straight dough method is used. Combining the yeast with warm water before adding it to the rest of the ingredients ensures a more even distribution.

- The *modified straight dough method* is a yeast dough mixing method in which the ingredients are added in sequential steps.

- The *sponge method* is a two-step yeast dough mixing method that results in a lighter-textured product. In the first step, 50% of the flour is added to all of the liquid and yeast to form a batter that looks like a sponge. ***See Figure 22-15.*** The batter grows because of the buildup of gas. As the gas in the sponge forms and increases, the mixture of the flour, yeast, and liquid begins to ferment and grow or fill with gas, and air bubbles develop on the surface and throughout the batter. In the second step, the rest of the ingredients are added, including the rest of the flour and all of the fat, salt, and sugar. Baked products made using the sponge method usually have an intense flavor and a lighter texture. Rye bread is made using the sponge method.

*Sponge Method*

**22-15** *The sponge method of mixing yeast dough begins by mixing half of the flour with all of the wet ingredients to produce a spongy batter.*

## Kneading

Kneading is the process of pushing and folding dough until it is smooth and elastic to further develop gluten. Sweet dough may need to be kneaded longer than other doughs because high amounts of fat and sugar inhibit gluten development. If the air is humid, a little more flour may need to be added during the kneading stage.

### To knead yeast dough, apply the following procedure:

1. Place the dough on a lightly floured surface. ***See Figure 22-16.***
2. Push the dough down to the table and away from the body with the heel of the hands.
3. Fold the dough in half, bringing the farthest edge up and on top of the nearest edge.
4. Repeat steps 2 and 3 until the dough is smooth.

CULINARY PROCEDURES

*Kneading Yeast Dough*

*1–2. Push dough down and away with heel of hand.*

*3. Fold dough in half.*

*4. Repeat until dough is smooth.*

**22-16** *Dough is kneaded to develop gluten and produce a smooth texture.*

*Yeast Fermentation*

**22-17** *During fermentation, yeast dough doubles in size.*

### Fermentation

Fermentation is the process by which yeast converts sugar into carbon dioxide and alcohol. During the fermentation stage, the yeast dough should be placed in a lightly greased container, covered, and allowed to rest until it expands to double its original size or until it no longer springs back when pushed with the fingertips. ***See Figure 22-17.*** The gluten becomes much smoother and more elastic in the fermentation stage. Note: Fermentation takes place throughout the dough-making process until the dough reaches 140°F.

## Punching

After the dough has doubled in size, it must be gently deflated to relax the gluten, dispel some of the carbon dioxide, and redistribute the yeast. *Punching* refers to the process of folding over and pressing risen yeast dough to allow the built-up carbon dioxide to escape. ***See Figure 22-18.*** To punch a dough, the edges of the dough are pulled into the center of the dough, and the dough is then pressed to release gas. Punching also helps to redistribute the yeast.

## Scaling

*Scaling* is the process of weighing pieces of risen dough on a baker's scale to ensure products of consistent size. A dough cutter is used to cut the dough into pieces to the desired size before they are weighed on the scale. It is important to use a dough cutter to remove pieces of dough from the large dough mass, as tearing the dough destroys some of the gluten that has developed. Dough can be added or taken away until the desired weight is achieved for each individual piece of dough.

## Rounding

*Rounding* is the process of shaping scaled dough into smooth balls. Rounding helps the dough to proof evenly and have a smooth outer surface. To round dough, shape the dough into a round ball using the fingers to form the shape while rotating the dough on a table or flat surface and applying pressure. ***See Figure 22-19.***

*Punching*

22-18   *Punching allows built up carbon dioxide to be released from a risen yeast dough.*

*Rounding*

22-19   *Rounding dough into smooth balls helps the dough proof evenly and have a smooth outer surface.*

When shaping a rolled-out dough into a loaf or roll, it needs to be rolled tightly. Once rolled, the end of the roll, referred to as a seam, should be pinched tightly to seal the rolled shape and prevent it from unrolling while proofing or baking. The seam is placed facing the bottom of the pan.

## Panning

*Panning* is the process of placing rounded pieces of dough into the appropriate pans. Some yeast breads are baked in loaf pans, while rolls may be baked on sheet pans or, in the case of cloverleaf rolls, in muffin pans. It is important to place yeast breads seam-side down so the bread will not split during proofing or baking. Enough space must be left in the pans for the yeast bread to proof. Yeast breads typically double in size during the proofing stage. ***See Figure 22-20.***

## Proofing

*Proofing* is the process of letting yeast dough rise in a warm (85°F) and moist (80% humidity) environment until the dough doubles in size. A proofing cabinet helps reduce the time required for proofing and ensures a consistent product because the temperature and humidity are controlled. ***See Figure 22-21.*** Proofing is an extension of the fermentation process, but it is important not to confuse proofing with fermentation. Fermentation is done at a lower temperature than proofing. During proofing, the yeast bread doubles in size again and should spring back from a touch of the fingertips. It is important not to overproof a product as this results in a coarse texture. Underproofing results in a dense texture and poor volume.

*Panning*

22-20   *Yeast doughs are often portioned into or onto pans that allow space for the dough to proof.*

*Chef's Tip:*
*Proofing can take place at temperatures up to 115°F, but proofing at higher temperatures result in quicker fermentation and less development of flavor in the yeast bread.*

*Proofing*

22-21   *Proofing is typically done in a proofing cabinet that allows control of temperature and humidity.*

## Washing

**22-22** *Yeast bread doughs are washed to improve the outer color and sheen.*

After proofing, the bread is ready for baking. Some breads may require scoring, docking, or the application of a wash.

* A *wash* is a liquid brushed on the surface of a yeast dough product prior to baking. Washes are used to create color, make an outer surface shiny or dull, or make other ingredients such as seeds stick to the dough. The most common type of wash is an egg wash, which consists of whole eggs whisked with a little water. An egg wash is brushed onto the surface of dough to enhance the color of the bread and add a sheen to the baked product. ***See Figure 22-22.***

* Scoring is done on hard-crusted breads before they are baked. *Scoring* is the process of making shallow, angled cuts across the top of unbaked bread. This allows carbon dioxide to escape during baking and creates more volume in the finished product. If the dough is not scored, the bread may bulge unattractively at the sides. A sharp, unserrated knife works best for scoring. ***See Figure 22-23.***

* *Docking* is the process of making small holes in the dough before it is baked. This is primarily done in sweet, rich doughs for pastries. Docking allows steam to escape and promotes even baking. Docking can be done with the tines of a fork.

## Scoring

**22-23** *Slashing helps create more even volume in hard crusted breads.*

### Baking

Yeast bread expands quickly when first placed in a hot oven. *Oven spring* is the rapid expansion of yeast dough in the oven, resulting from the expansion of gases within the dough. When the temperature of the dough reaches 140°F, the yeast dies and there is no further expansion.

Some breads, such as French bread, benefit from the incorporation of steam in the oven. Steam results in the bread having a light, crispy crust, while the inner crumb remains moist and chewy. *Steam injection* is the process of adding water directly into the hot cavity of an oven so that steam is created. As the water sprays onto the interior surface of the oven, the water vaporizes and turns into steam. Bakeries typically have a steam injector connected to their ovens.

## Cooling

After baking is completed, the bread must be removed from the pans to allow for even cooling. If breads are left in the pans to cool, the hot bread sweats inside the pan and becomes soggy. Typically breads are placed on cooling racks so that air flow is maintained beneath the loaf. Similarly, hot bread should never be bagged until almost room temperature. If bread is bagged before it is sufficiently cooled, steam is trapped in the bag and the bread becomes soggy.

## Storing

Bread should be completely cooled before storing because excess moisture from trapped steam could cause the bread to stale. Cooled bread should be stored in an airtight container or plastic wrap. Hard-crusted bread should not be wrapped as doing so would cause the crust to become soft. Storing bread at room temperature or freezing bread is best. Refrigeration causes bread to stale, so it is best to avoid storing bread in the refrigerator.

### Historical Note

In 1928, a bakery in Chillicothe, Missouri was the first to use Otto Frederick Rohwedder's machine that both sliced and wrapped bread. His invention brought him recognition as the father of "sliced bread." Prior to this time, people believed that once bread was sliced it would become stale too quickly to be stored.

## QUICK BREAD PREPARATION

A *quick bread* is a baked product that is made with a quick-acting leavening agent such as baking powder, baking soda, or steam and that does not require any kneading or fermentation. Common quick bread preparations include biscuits, muffins, quick loaf breads such as banana nut bread, corn bread, and corn sticks. *See Figure 22-24.* A wide variety of ingredients such as vegetables, fruits, and nuts make quick breads one of the most versatile preparations in the professional kitchen.

Many terms are used to describe different aspects of mixing. Knowledge of these terms is essential to properly execute recipes for quick breads, cakes, pastries, and desserts.

- *Beating* is the process of vigorously agitating ingredients to incorporate air. Using a paddle attachment in a mixer works best.

- *Blending* is the process of mixing two or more ingredients together until they have been evenly distributed throughout the mixture. This is a much gentler method compared to beating and can be accomplished by using a rubber spatula, spoon, whisk, or a mixer with a paddle attachment.

- *Creaming* is the process of vigorously combining fat and sugar and vigorously to add air. Using a paddle attachment in a mixer that is placed on medium to high speed works best.

- *Cutting in* is the process of incorporating a cold solid fat into dry ingredients, such as flour, until pea-size lumps are formed.

- *Folding* is the process of gently incorporating light ingredients with heavier ones. Using a rubber spatula and a smooth fold-over motion is the best way to fold in ingredients.

- *Sifting* is the process of passing dry ingredients through a wire sieve to remove all of the lumps and add air to ingredients.

- *Stirring* is the process of gently mixing all ingredients until they are evenly combined.

- *Whipping* is the process of beating ingredients vigorously to add a lot of air to them. This is done best with the whisk attachment on a mixer.

## Types of Quick Breads

| Biscuits | Muffins | Quick Loaf Breads |

22-24  *Types of quick breads include biscuits, muffins, and quick loaf breads.*

### Mixing Methods

Quick breads are easier to prepare than yeast breads as the quick-acting leavening agents allow a quick bread to be mixed and baked without waiting for the dough to rise. Quick breads are made using one of three different mixing methods: the biscuit method, the muffin method, and the creaming method.

The *biscuit method* is a quick-bread mixing method used for recipes that include a cold solid fat, such as butter or lard. Solid fat helps produce a light and flaky product. The *muffin method* is a quick-bread mixing method that uses liquid fats, such as vegetable oil or melted butter, to produce a rich and tender product. The *creaming method* is a cake-mixing method that is sometimes applied to muffins and quick loaf breads. The creaming method is used for products with high amounts of granulated sugar and solid fat. These two ingredients are creamed together until smooth, which produces a uniformly blended mixture.

**Biscuit Method.** Biscuits can be made using many different recipes. A *biscuit* is a quick bread made with baking powder or baking soda and shortening is used instead of oil. However, all biscuits contain the same basic ingredients, with the amounts and procedures varying. Basic ingredients in biscuits include flour, baking powder or baking soda, shortening, milk, salt, and sometimes sugar. Butter may be added to improve the tenderness and richness of biscuits. Flavorings may be added to create specialty products, such as orange biscuits or cheese biscuits.

Flour supplies form and texture to biscuits. Baking powder and baking soda are quick-acting leavening agents that cause the dough to rise when liquid is added and heat is applied. Shortening is considered the most important ingredient in preparing biscuits because it creates tenderness. Milk provides the moisture, regulates the consistency of the dough, develops the flour, and causes the baking powder to generate its gas. Salt brings out the flavor and taste of the other ingredients. Sugar supplies sweetness, helps retain moisture, and produces a golden brown color.

## To prepare dough using the biscuit method, apply the following procedure:

1. Scale all ingredients.
2. Sift all dry ingredients together. *See Figure 22-25.*
3. Cut the fat into the dry ingredients, using a pastry blender, or a mixer with a paddle attachment.
4. Combine liquid ingredients in a separate bowl.
5. Add liquid ingredients to the dry ingredients and mix just until the mixture holds together. Do not overmix.
6. Place dough on bench and knead lightly for 30 sec to 45 sec. Dough should be soft, but should not stick to hands. Overkneading will toughen biscuits.
7. Roll dough out onto a floured surface.
8. Using a biscuit cutter, cut biscuits as close together as possible.
9. Place on greased or parchment-lined sheet pan for baking.
10. Brush tops of biscuits with melted butter or egg wash for added color.

CULINARY PROCEDURES

## Biscuit Mixing Method

1–2. Scale ingredients; sift the dry ingredients.

3. Cut in fat.

4–5. Combine liquid ingredients in separate bowl; add to dry ingredients.

6. Knead lightly 30 sec to 45 sec.

7–8. Roll out dough; cut with biscuit cutter.

9–10. Place on lined sheet pan; brush with melted butter or egg wash.

**22-25** Biscuit dough is made using the biscuit method of mixing.

**Muffin Method.** Muffins are a popular item that can be served on breakfast, lunch, and dinner menus. A *muffin* is a quick bread that is made with eggs, flour, and either oil, butter, or margarine. Muffins are shaped like cupcakes and have a pebbled top and a coarse inner texture. Muffins should have a cake-like texture and an even shape.

Following the proper mixing procedure is the key to successful muffins. A common fault in mixing muffin batter is mixing the batter to a point where the gluten within the flour becomes tough. To prevent the gluten from becoming tough, the dry and liquid ingredients should be mixed together in a mixer at slow speed using the paddle attachment. The batter should be mixed just until it is moistened, leaving it with a slightly rough appearance. The batter should not be smooth. Overmixing the ingredients creates air tunnels inside the muffins.

To prepare dough using the muffin method, apply the following procedure:

1. Scale all ingredients.
2. Sift dry ingredients together. **See Figure 22-26.**
3. Combine liquid ingredients, including any melted butter or oil.
4. Add the liquids to the dry ingredients and mix just until moistened. The batter should appear lumpy. Do not overmix.
5. Grease muffin pans or use paper liners.
6. Portion scoop batter into pans for uniform size.

## Muffin Mixing Method

1–2. Scale ingredients; sift the dry ingredients.

3–4. Combine liquid ingredients in a seperate bowl; add to dry ingredients.

5–6. Grease or line muffin pans; scoop batter for uniform size.

**22-26** *Muffins and some quick loaf breads are made using the muffin mixing method.*

**Creaming Method.** The creaming method starts by mixing room temperature fat and sugar in the mixer using a paddle attachment until the mixture is light and fluffy in texture. The eggs are then added one at a time and incorporated into the creamed fat and sugar. Next, liquids and dry ingredients are alternately added, a little at a time, and creamed. The creaming method produces a lighter product with a much finer crumb than the muffin method.

To prepare dough using the creaming method, apply the following procedure:

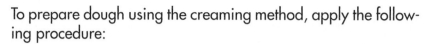

1. Scale all ingredients.
2. Combine the fat and sugar in a mixing bowl and mix with a paddle attachment on medium speed until light and fluffy. ***See Figure 22-27.***
3. Add eggs, one at a time, mixing well after each addition.
4. Sift dry ingredients.
5. Combine all liquid ingredients in a separate bowl.
6. Alternately add the dry and liquid ingredients to the creamed ingredients. Avoid overmixing.

## Creaming Method

*1–2. Scale ingredients; combine fat and sugar in mixer with paddle attachment.*

*3. Add eggs one at a time.*

*4. Sift dry ingredients.*

*5. Combine liquid ingredients in a seperate bowl.*

*6. Alternately add dry and wet ingredients to creamed ingredients.*

*7. Portion into greased or lined pans and bake.*

**22-27** *Some quick breads are mixed using the creaming method.*

## Quick Loaf Breads

A *quick loaf bread* is a quick bread that uses a batter similar to muffin batter. The shape is changed by baking the batter in a loaf pan. Quick loaf breads are so named because they contain a quick-acting leavening agent such as baking powder. They can be prepared using different fruits, vegetables, and nuts to produce a variety of flavors. Banana nut bread, pumpkin bread, and zucchini bread are examples of quick loaf breads. They are similar to muffins in quality and texture, but tend to stay fresh longer because of their size. Quick loaf breads can be served warm or cold. Some quick breads are topped with icing.

## Corn Bread and Corn Sticks

Corn bread and corn sticks are quick breads made from a batter containing cornmeal, eggs, milk, and oil. *Cornmeal* is dried, ground corn kernels. Available types include yellow corn or white corn, and coarse grain or fine grain. *Corn flour* is cornmeal that has been finely ground. Corn flour can be used in a variety of recipes including cakes, crêpes, and breads in combination with wheat flour.

Idaho Potato Commission

Most corn sticks are made by pouring corn bread batter into molds prior to baking. In some recipes, the consistency of the batter may be a little heavier when preparing corn sticks. The batter is made heavier by eliminating a small amount of the liquid. The recipes for preparing corn bread and corn sticks also vary in the mixing method. A recipe may call for a mixing method similar to that for muffins, or for a three-stage method of blending an egg-milk mixture into the dry ingredients and stirring in melted shortening.

Whichever method is used, the mixing should be done with the mixer on slow using the paddle attachment. The liquid must be added slowly because cornmeal does not absorb liquid quickly. If liquid is added too rapidly, lumps will form. Avoid overmixing by stirring the batter only until all the ingredients are mixed in. Do not beat a corn bread or corn stick mixture.

## Baking Quick Breads

Quick breads can be baked in a variety of loaf pans to create muffins, loaves, or sticks of various sizes. When baking quick breads or cakes, the pans should be filled only pan three-quarters full. If pans are overfilled, the batter will spill over the top as it rises during the baking process, ruining the final product. If the loaves are not served immediately, they should be refrigerated after baking. This allows them to be sliced more easily without crumbling. To extend the freshness of a quick breads, they should be wrapped tightly as soon as they cool. Quick breads can also be frozen with excellent results.

## SUMMARY

Bakers and pastry chefs need to be familiar with the variety of equipment, tools, bakeware, and ingredients used in a bakeshop. They must also understand how the combination of specific ingredients such as flours, fats, thickeners, flavorings, and sugars affect the flavor, texture, color, and nutritional value of baked products. The correct use of bakeshop formulas ensures that consistent baked goods will be produced, as each ingredient in the formula plays a critical role in the chemical reactions that occur both before a dough or batter enters the oven and again while it bakes.

Yeast-bread preparation consists of 12 specific steps: scaling, mixing, kneading, fermentation, punching, scaling, rounding, panning, pan proofing, baking, cooling, and storing. Similarly, quick-bread preparation requires the knowledge and application of fundamental mixing techniques such as creaming, kneading, and whipping. Quick breads are easier to prepare than yeast breads and are made using one of three mixing methods: the biscuit method, the muffin method, or the creaming method.

Refer to the CD-ROM for Quick Quiz® questions related to chapter content.

## Review Questions

1. Describe the use of a proofing cabinet.

2. Contrast six common types of flour.

3. Explain how sugar affects yeast dough.

4. What is the difference between confectioners' sugar and superfine sugar?

5. Contrast hydrogenated shortening and emulsified shortening.

6. Explain how milk and other dairy products affect bakery items.

7. List five common thickeners used in a bakeshop.

8. Describe the most basic way to leaven baked products.

9. Define the three common varieties of commercial yeast.

10. What is the difference between an extract and an emulsion?

11. How is a formula different from a recipe?

12. Describe how a baker's percentage formula is different than a standard percentage formula that is learned in a math class.

13. Give examples of products made from lean dough, rich dough, and rolled-in dough.

14. Identify the 12 steps to follow for producing yeast dough.

15. Describe the four-step procedure for using a baker's scale.

16. Describe the characteristics of breads made with the sponge method.

17. Identify the main factors to consider during the panning stage of yeast dough production.

18. Explain the cause of oven spring and identify the temperature at which it stops.

19. Explain how to store cooled bread.

20. Contrast the mixing methods of cutting and folding.

21. Describe the six-step procedure for preparing dough using the muffin method.

22. When making corn sticks, explain why the liquid must be added slowly to the mixer.

23. List guidelines to follow when baking and storing quick breads.

# Recipes

## 258

## Soft Dinner Rolls

*yield: 32 (2-oz rolls)*

| | |
|---|---|
| 2.75 oz | fresh yeast |
| 16 oz | lukewarm milk |
| 2 oz | sugar |
| 4 | eggs |
| 4 oz | butter, softened |
| ½ oz | salt |
| 2 lb 10 oz | bread flour |

*Egg wash*

| | |
|---|---|
| 1 | egg, beaten slightly |
| 1 tsp | water |

1. Place yeast and milk together in mixing bowl.
2. Slowly add sugar, then eggs, butter, and salt. Mix until the sugar dissolves.
3. Slowly add flour to the mixture until a smooth dough forms (approximately 10 min).
4. Proof dough in a greased bowl for approximately 1 hr.
5. Punch down dough, knead, and proof for 20 min.
6. Scale dough into desired roll shapes.
7. Mix egg and water and apply to surface of rolls.
8. Bake rolls at 425°F for 25 min.

### Nutrition Facts
**Per roll:** 162.4 calories; 25% calories from fat; 4.7 g total fat; 42.2 mg cholesterol; 190.9 mg sodium; 110.2 mg potassium; 24 g carbohydrates; 1.2 g fiber; 2.7 g sugar; 22.8 g net carbs; 5.8 g protein

## 259

## Hard Dinner Rolls

*yield: 48 (1.5-oz rolls)*

| | |
|---|---|
| 2.5 oz | fresh yeast |
| 1 lb | lukewarm milk |
| 4 oz | butter |
| 2 oz | sugar |
| ½ oz | salt |
| 2 | eggs |
| 2 | egg whites |
| 2 lb 8 oz | bread flour |

*Egg wash*

| | |
|---|---|
| 1 | egg |
| 1 tsp | water |

1. Dissolve yeast in half of the milk.
2. Cream sugar and butter in a mixing bowl using the paddle attachment.
3. Add salt, eggs, egg whites, bread flour, and the yeast mixture to the creamed mixture and mix until smooth (about 10 min).
4. Place dough in greased bowl and let proof until doubled in size (about 1 hr).
5. Punch down, and proof again for 20 min.
6. Scale into 1½-oz rolls.
7. Mix last egg with 1 tsp water and brush rolls with egg wash.
8. Proof for 15 min to 20 min.
9. Place a small pan of water in oven to create steam during baking.
10. Bake at 400°F for 25 min to 30 min or until golden brown.

### Nutrition Facts
**Per roll:** 105.1 calories; 24% calories from fat; 2.9 g total fat; 19.2 mg cholesterol; 126.3 mg sodium; 69 mg potassium; 15.9 g carbohydrates; 0.8 g fiber; 1.8 g sugar; 15.2 g net carbs; 3.7 g protein

## German Rye Bread

*yield: 4 (1-lb loaves)*

| | |
|---|---|
| 0.5 oz | active dry yeast |
| 4 oz | warm water |
| 13 oz | lukewarm milk |
| 5 oz | molasses |
| 11.5 oz | bread flour |
| 12.2 oz | rye flour |
| 1 oz | butter |
| 2 tbsp | sugar |
| 1 tsp | kosher salt |
| 1 | egg, beaten slightly |

1. Combine bread flour and rye flour in a bowl.
2. Combine molasses, water, and yeast and add 8 oz of the flour mixture. Mix for 2 min, cover, and set aside.
3. Let rise for approximately 1 hr until the mixture has doubled in size.
4. Add warm milk, butter, sugar, and salt to the mixture and mix.
5. Stir in the remaining flour mixture until the dough is stiff enough to knead.
6. Knead dough for 5 min in mixer using dough hook.
7. Transfer dough to a greased bowl, cover, and let rest until doubled in size (approximately 1 hr).
8. Punch down dough and divide to form round loaves.
9. Place on lightly greased sheet pan.
10. Make egg wash and brush loaves with egg wash.
11. Let dough rise until doubled in size.
12. Bake at 375°F for 30 min to 35 min until golden brown and crispy.

**Nutrition Facts**

**Per loaf:** 2503.7 calories; 4% calories from fat; 13.9 g total fat; 78 mg cholesterol; 756.4 mg sodium; 9390.2 mg potassium; 586.7 g carbohydrates; 15.3 g fiber; 345.7 g sugar; 571.4 g net carbs; 24.1 g protein

## White Bread

*yield: 12 (1-lb loaves)*

| | |
|---|---|
| 2.75 oz | fresh yeast |
| 70 oz | lukewarm milk |
| 9 oz | sugar |
| 2.25 oz | salt |
| 9 oz | shortening |
| 6 lb 8 oz | bread flour |

1. Dissolve yeast in 3 c milk. *Note:* Milk should be heated to 85°F to ensure proper activation.
2. Place remaining milk, sugar, salt, and shortening and mix until all ingredients are blended.
3. Alternately add the flour and yeast mixture until a smooth dough forms. Continue to mix for an additional 5–6 min.
4. Place dough in a large greased bowl and let rise for approximately 1 hr.
5. Punch down dough.
6. Scale and place in well-greased loaf pans. Let rise for approximately 30 min.
7. Bake at 450°F for 35 min or until golden brown.

**Nutrition Facts**

**Per loaf:** 1563.3 calories; 17% calories from fat; 30.7 g total fat; 28.5 mg cholesterol; 2141 mg sodium; 712 mg potassium; 269.4 g carbohydrates; 9.2 g fiber; 31.6 g sugar; 260.1 g net carbs; 47.4 g

# *Recipes*

## 262

## Whole-Wheat Bread

*yield: 5 (17-oz loaves)*

| | |
|---|---|
| 2 lb | whole-wheat flour |
| 1 lb | bread flour |
| 20 oz | lukewarm milk |
| 4 oz | shortening |
| 1 oz | salt |
| 2 oz | sugar |
| ½ oz | malt syrup |
| 2 oz | fresh yeast |
| 4 | eggs |

*Alpha Baking Co., Inc.*

1. Dissolve yeast in 5 oz of the lukewarm milk.
2. Place remaining ingredients in mixing bowl and mix at medium speed.
3. Slowly add the yeast mixture and mix for approximately 10 minutes until a smooth dough forms.
4. Place dough into a large, greased bowl or on a sheet pan and place in proofer. Let rise for approximately 1½ hr.
5. Punch down dough and scale out into 1-lb loaves.
6. Place in greased loaf pans, and proof for 30 min.
7. Bake at 400°F for 30 min to 35 min.

**Nutrition Facts**

**Per loaf:** 1345 calories; 22% calories from fat; 33.8 g total fat; 192.8 mg cholesterol; 2319.8 mg sodium; 1289.6 mg potassium; 221 g carbohydrates; 26.7 g fiber; 19.1 g sugar; 194.3 g net carbs; 49.2 g protein

## Vienna Bread

*yield: 7 lb (seven 1-lb loaves)*

*method: sponge*

| | |
|---|---|
| 9 oz | tepid water |
| 1 lb | bread flour |
| 2 oz | fresh yeast |
| ½ tbsp | malt |
| 1 lb 8 oz | warm water |
| 2 oz | softened margarine |
| 2 lb 10 oz | bread flour |
| 1.25 oz | salt |
| 1.25 oz | sugar |
| 3 tbsp | cornmeal |

1. Mix first four ingredients to form a fermented sponge and let rest for 30 min.
2. Add remainder of ingredients, except cornmeal, to mixer and mix for approximately 10 min until a smooth, stiff dough forms.
3. Place dough in a greased bowl or proofer, cover, and proof until doubled in size (approximately 1 hr).
4. Punch down, scale, and form dough into 1-lb long loaves and place on cornmeal-dusted sheet pans.
5. Score tops of loaves diagonally.
6. Proof for approximately 30 min.
7. Bake in 400°F oven for 30 min to 35 min or until golden brown and hollow sounding when tapped.

**Nutrition Facts**

**Per loaf:** 733.1 calories; 11% calories from fat; 9.8 g total fat; 0 mg cholesterol; 2050.4 mg sodium; 350.9 mg potassium; 135.3 g carbohydrates; 6 g fiber; 5.6 g sugar; 129.3 g net carbs; 23.9 g protein

## Baguettes

*yield: 3 loaves*

| 2 lb | bread flour |
| 1 lb 5 oz | water |
| .5 oz | fresh yeast |
| .4 oz | kosher salt |

1. Place flour, water, and yeast in mixing bowl and mix slowly until combined.
2. Add salt and increase speed to medium.
3. Knead with dough hook for approximately 10 min or until a smooth dough forms.
4. Transfer dough to lightly greased bowl, cover, and allow to proof until doubled in size, about 1½ hr.
5. Punch down the dough and let rest for a few minutes.
6. Scale the dough into three 18-oz portions.
7. Round into balls and proof again for 30 min.
8. Shape into three long loaves, approximately 20″ long, and proof again on a sheet pan dusted with cornmeal until doubled in size (approximately 15 min).
9. Brush with egg wash and score before baking.
10. Bake at 450°F for approximately 25 min to 30 min.

### Nutrition Facts

**Per loaf:** 1105.6 calories; 3% calories from fat; 5.2 g total fat; 0 mg cholesterol; 1444.8 mg sodium; 396.9 mg potassium; 221.1 g carbohydrates; 8.2 g fiber; 0.9 g sugar; 212.9 g net carbs; 38 g protein

## Croissants

*yield: 20 croissants*

| 1 oz | fresh yeast |
| 15 oz | unbleached flour |
| 1.75 oz | sugar |
| ½ oz | salt |
| 8.5 oz | milk |
| 1 lb | unsalted butter, softened |
| 1 | egg, beaten slightly |
| 1 tbsp | milk |

1. In a mixer with a dough hook, place the flour, sugar, and salt. Mix for 2 min on low speed.
2. Gently warm the 8.5 oz milk to approximately 85°F and stir in yeast.
3. Add yeast mixture to flour mixture and stir until combined. Knead in mixer for approximately 10 min.
4. Place dough in a greased bowl and let rise until doubled.
5. While dough is rising, place butter between two sheets of plastic wrap and, with a rolling pin, roll out the butter until it has formed a rectangle approximately 6″ × 8 ½″. Place in refrigerator to chill.
6. Punch down dough and roll it into an approximately 10″ × 15″ rectangle. Brush off excess flour. Unwrap butter and place in the middle of the dough. Fold the dough around the butter, enclosing it completely.
7. Brush off excess flour and lightly press with a rolling pin to push all the layers together. Roll the dough to form a new 10″ × 15″ rectangle.
8. Fold the dough into thirds. Wrap the dough in plastic and chill for 25 min to 30 min.
9. Repeat rolling out the dough and folding it into thirds two more times, chilling the dough between each turn.
10. After turning the third time, chill the dough overnight or for a minimum of 1 hr.
11. Remove dough from the refrigerator and slice into three pieces. Roll each piece into a rectangle about ¼″ thick.
12. Cut the rolled-out dough into uniform triangles. Roll each triangle up starting with the large end and place on a parchment-lined sheet pan with the tip tucked under and the ends slightly curved to form a crescent.
13. Mix egg and 1 tbsp milk to form an egg wash and brush the dough lightly with the wash.
14. Proof until the croissants double in size.
15. Bake at 375°F for 12 min to 15 min until golden.

### Nutrition Facts

**Per croissant:** 265.7 calories; 64% calories from fat; 19.4 g total fat; 60.7 mg cholesterol; 287.3 mg sodium; 79.6 mg potassium; 19.9 g carbohydrates; 0.9 g fiber; 3.3 g sugar; 19 g net carbs; 3.7 g protein

# *Recipes*

## 266

## Sweet Yeast Danish Dough

*yield: 30 (4-oz Danish)*

| | |
|---|---|
| 2 lb | lukewarm milk |
| 8 oz | sugar |
| 1 oz | salt |
| 4 lb 12 oz | bread flour |
| 2.5 oz | yeast |
| 8 oz | butter |
| 4 | eggs |
| ½ oz | lemon zest, grated |

*Egg wash*

| | |
|---|---|
| 1 | egg, beaten slightly |
| 1 tsp | water |

1. Dissolve the yeast in the milk and mix with 1 oz sugar and ⅓ of the flour. Let sit 20 min to form sponge.
2. Add the rest of the ingredients and mix to form a smooth but stiff dough.
3. Proof dough, covered, in a greased bowl for approximately 1 hr.
4. Punch dough down, roll out, and cut to form individual 4 oz sweet rolls.
5. Mix egg and water and whisk slightly.
6. Brush rolls with egg wash, and proof for 30 min.
7. Bake at 375°F for 35 min to 40 min.
8. After cooling, frost with buttercream icing.

**Nutrition Facts**

**Per Danish:** 380.4 calories; 21% calories from fat; 9.3 g total fat; 54.5 mg cholesterol; 393.5 mg sodium; 176.3 mg potassium; 62.1 g carbohydrates; 2.3 g fiber; 9.5 g sugar; 59.8 g net carbs; 11.6 g protein

## 267

## Cinnamon Rolls

*yield: 40 cinnamon rolls*

| | |
|---|---|
| ½ | recipe for Sweet Yeast Danish Dough |
| 8 oz | butter, softened |
| 8 oz | brown sugar |
| ½ oz | cinnamon |
| 16 oz | pecans (optional) |

1. Prepare Sweet Yeast Danish Dough recipe. After first proofing, roll out dough ½" thick into an even rectangular shape no more than 12" wide.
2. Brush softened butter evenly on surface of rolled-out dough.
3. Mix brown sugar, cinnamon, and pecan together and sprinkle evenly onto the buttered dough.
4. Slowly roll up the cinnamon roll dough, encasing cinnamon, sugar, nuts, and butter inside the rolled-up dough, occasionally pinching the ends to prevent the filling from falling out.
5. Cut roll into ½"-thick pieces and place in greased baking pan, allowing 1" between rolls.
6. Proof for 30 min or until almost doubled in size.
7. Bake at 375°F for 35 min to 40 min.
8. Remove from oven and let stand for 15 min before removing from the pan.

**Nutrition Facts**

**Per roll:** 146.1 calories; 75% calories from fat; 12.9 g total fat; 12.9 mg cholesterol; 7.8 mg sodium; 71.4 mg potassium; 8.2 g carbohydrates; 1.3 g fiber; 6 g sugar; 6.8 g net carbs; 1.2 g protein

## Baking Powder Biscuits

*yield: 41 (4-oz biscuits)*

| | |
|---|---|
| 5 lb | cake flour |
| 3 oz | baking powder |
| 2.5 oz | salt |
| 10 oz | sugar |
| 5 oz | milk powder |
| 2 lb | shortening |
| 2 lb 8 oz | water |

*Egg wash*

| | |
|---|---|
| 1 | egg, beaten slightly |
| 1 tsp | water |

1. Sift dry ingredients together. Cut in shortening until pea-size clumps form.
2. Add water and mix until smooth dough forms.
3. Roll dough out on lightly floured surface to ½″ thick.
4. Cut ½″ thick dough into biscuit shapes.
5. Place on greased pan and egg wash the tops of the biscuits. *Note:* If desired, biscuits can be brushed with melted butter instead of egg wash.
6. Bake at 400°F for 10 min to 15 min.

---

**Nutrition Facts**

**Per biscuit:** 446.2 calories; 47% calories from fat; 23.6 g total fat; 20.9 mg cholesterol; 907.8 mg sodium; 106.4 mg potassium; 52 g carbohydrates; 0.9 g fiber; 8.4 g sugar; 51.1 g net carbs; 5.6 g protein

---

## Buttermilk Biscuits

*yield: 24 (3-oz biscuits)*

| | |
|---|---|
| 2 lb 8 oz | all-purpose flour |
| 0.75 oz | salt |
| 2 oz | granulated sugar |
| 2 oz | baking powder |
| 1 lb | unsalted butter |
| 24 oz | milk |

1. Sift flour, salt, sugar, and baking powder.
2. Cut cold butter into cubes and add to flour mixture.
3. Cut the butter into the flour. *Note:* The butter should still be slightly visible, and the mixture should resemble coarse cornmeal.
4. Add the milk and mix until just combined.
5. Place the dough on a lightly floured surface and roll out to ½″ thick.
6. Cut into desired shapes.
7. Bake at 350°F for 15 to 20 min.

---

**Nutrition Facts**

**Per biscuit:** 336.1 calories; 43% calories from fat; 16.8 g total fat; 43.7 mg cholesterol; 609 mg sodium; 99.2 mg potassium; 40.5 g carbohydrates; 1.3 g fiber; 4.1 g sugar; 39.2 g net carbs; 6 g protein

---

## Blueberry Muffins

*yield: 30 (3-oz muffins)*

| | |
|---|---|
| 12 oz | sugar |
| 8 oz | eggs |
| ¼ oz | vanilla extract |
| 16 oz | milk |
| 9 oz | shortening |
| 1 lb 12 oz | cake flour |
| ¼ oz | salt |
| 1 oz | baking powder |
| 16 oz | fresh or frozen blueberries |

1. Cream sugar, eggs, vanilla extract extract, milk, and shortening together until smooth.
2. Sift cake flour, salt, and baking powder together. Slowly add to creamed ingredients and mix until moistened.
3. Fold in blueberries carefully until just mixed.
4. Pour batter into greased and floured muffin tins.
5. Bake at 375°F for 25 min to 30 min.

**Nutrition Facts**

**Per muffin:** 463.5 calories; 47% calories from fat; 24.4 g total fat; 682.7 mg cholesterol; 422.9 mg sodium; 277.8 mg potassium; 36.4 g carbohydrates; 0.8 g fiber; 15 g sugar; 35.6 g net carbs; 22.9 g protein

## Zucchini Bread

*yield: 10 (1-lb loaves)*

| | |
|---|---|
| 3 lb 2 oz | butter |
| 5 lb 15 oz | sugar |
| 1 lb 14 oz | whole eggs |
| 2 tbsp | vanilla extract |
| 6 lb 9 oz | zucchini, grated |
| 5 lb 15 oz | bread flour |
| 10 tsp | salt |
| 10 tsp | baking soda |
| 2 lb 10 oz | walnut pieces |

1. Grease and flour ten loaf pans. Set aside.
2. Melt butter and then remove saucepan from heat.
3. Add sugar to butter and stir in eggs, vanilla extract, and zucchini. Stir the mixture.
4. Sift flour, salt, and baking soda together.
5. Fold into zucchini mixture.
6. Fold walnut pieces into mixture.
7. Pour into greased pans to ¾ full.
8. Bake at 350°F for 50 min to 60 min.

**Nutrition Facts**

**Per loaf:** 5744.5 calories; 30% calories from fat; 206.1 g total fat; 660.4 mg cholesterol; 3755 mg sodium; 1739.7 mg potassium; 945.5 g carbohydrates; 17.7 g fiber; 732.4 g sugar; 927.8 g net carbs; 65.8 g protein

## Apricot and Cherry Scones

*yield: 25 (3-oz scones)*

| | |
|---|---|
| 6 oz | dried cherries, diced |
| 6 oz | dried apricots, diced |
| 1 lb 12 oz | bread flour |
| 3 tbsp | baking powder |
| 1 tsp | salt |
| 4 oz | honey |
| 3½ c | cream |
| 2 tbsp | coarse sugar |

1. Sift flour, baking powder, and salt together.
2. Mix fruit with flour mixture.
3. Mix in honey and all but 2 oz of cream. Mix just to incorporate, using caution to not overmix.
4. Rest mixture for 5 min.
5. Pour out onto floured surface, roll to 1½" thick.
6. Cut into triangle shapes.
7. Bake at 375°F for 10 min to 12 min or until golden brown.
8. Remove from oven, brush with remaining cream, and press lightly into coarse sugar.

### Nutrition Facts

**Per scone:** 229.4 calories; 26% calories from fat; 6.8 g total fat; 22.9 mg cholesterol; 277.3 mg sodium; 126.1 mg potassium; 38.5 g carbohydrates; 1.6 g fiber; 8.5 g sugar; 36.8 g net carbs; 4.6 g protein

## Pumpkin Cream Cheese Bread

*yield: 4 (1-lb loaves)*

| | |
|---|---|
| 1 lb 2 oz | sugar |
| 6 oz | oil |
| 3 | eggs |
| 3 oz | bread flour |
| 6 oz | cake flour |
| 14½ oz | canned pumpkin purée |
| ½ tsp | baking powder |
| ¼ oz | baking soda |
| 1 oz | pumpkin spice |
| 10 oz | nuts |

*Cream cheese filling*

| | |
|---|---|
| 12 oz | cream cheese |
| 1 oz | sugar |
| 1 | egg |
| 1 oz | cake flour |
| to taste | lemon extract |

1. To create bread batter, cream sugar, and pumpkin together.
2. Add oil and mix well.
3. Add eggs one at a time while mixing. Mix for 10 min.
4. Combine dry ingredients and mix on low speed 3 min to 5 min or until all ingredients are incorporated. Set batter aside.
5. To create cream cheese filling, cream the cream cheese with sugar in a mixer using the paddle attachment.
6. Alternately add eggs and flour until incorporated.
7. Add lemon extract and mix well.
8. Pour bread batter into greased pans to about half full.
9. Use a pastry bag to pipe cream cheese filling into the center of each pumpkin loaf from end to end.
10. Bake at 350°F for 1 hr 15 min or until the center is firm to the touch.

### Nutrition Facts

**Per loaf:** 2042.5 calories; 52% calories from fat; 124.8 g total fat; 305.1 mg cholesterol; 880.4 mg sodium; 817.9 mg potassium; 214.6 g carbohydrates; 10.1 g fiber; 141.1 g sugar; 204.5 g net carbs; 31.7 g protein

# Recipes

## 274

## Sour Cream Coffee Cake

*yield: 2 (1-lb coffee cakes; 24 slices per cake)*

| | |
|---|---|
| 4 oz | butter |
| 6 oz | sugar |
| 2 | eggs |
| 8 oz | sour cream |
| 6¾ oz | cake flour |
| ¼ oz | baking powder |

*Streusel topping*

| | |
|---|---|
| 2 oz | almond paste |
| 3 oz | butter |
| 5 oz | sugar |
| dash | nutmeg |
| 5 oz | bread flour |

1. Cream butter and sugar together until light and fluffy.
2. Add the eggs one at a time until incorporated.
3. Add the sour cream and mix well.
4. Sift together the flour and baking powder and add slowly to the mixture until evenly distributed. Set batter aside.
5. Mix all streusel ingredients together to form a coarse and crumbly mixture.
6. Pour batter into a greased 10″ cake pan.
7. Top cake batter with streusel topping.
8. Bake at 350°F for 30 min or until center is firm to the touch.

**Nutrition Facts**

**Per slice:** 197.2 calories; 44% calories from fat; 9.9 g total fat; 39.6 mg cholesterol; 43.5 mg sodium; 43.2 mg potassium; 25.2 g carbohydrates; 0.4 g fiber; 13.9 g sugar; 24.8 g net carbs; 2.5 g protein

## Banana Nut Bread

*yield: 13 (1-lb loaves)*

| | |
|---|---|
| 1 lb | shortening |
| 2 lb | sugar |
| 1 lb | butter |
| 1 oz | salt |
| 3 lb 6 oz | ripened bananas, mashed |
| 9 oz | eggs |
| ¾ oz | baking soda |
| 4 oz | water |
| 9 oz | buttermilk |
| 3 lb 6 oz | cake flour |
| 1 lb | walnuts, chopped |

1. Cream shortening, sugar, butter, and salt together.
2. Gradually add eggs, baking soda, water, bananas, and buttermilk. Mix just until smooth.
3. Slowly add cake flour and mix only until moistened.
4. Stir in chopped walnuts.
5. Scale into 1-lb loaves.

**Nutrition Facts**

**Per loaf:** 1630 calories; 48% calories from fat; 89.4 g total fat; 176.8 mg cholesterol; 1350.1 mg sodium; 766.8 mg potassium; 194.4 g carbohydrates; 7.4 g fiber; 86.5 g sugar; 187 g net carbs; 19.7 g protein

## Corn Bread

*yield: one 2" half pan (20 servings)*

| | |
|---|---|
| 8 oz | sugar |
| 8 oz | shortening |
| ½ oz | salt |
| 4 | eggs |
| 1 lb | cake flour |
| 1 lb 8 oz | cornmeal |
| 1½ oz | baking powder |
| 1 qt | milk |

1. Lightly cream sugar, shortening, and salt together.
2. Add eggs one at a time and mix thoroughly.
3. Sift flour, baking powder, and cornmeal together.
4. Add to creamed mixture and stir.
5. Place cornbread batter in greased 2" half pan.
6. Bake at 375°F for 25 min to 30 min or until toothpick inserted comes out clean.

### Nutrition Facts

**Per serving:** 345.4 calories; 35% calories from fat; 13.8 g total fat; 52.9mg cholesterol; 542 mg sodium; 172.8 mg potassium; 49.3 g carbohydrates; 2 g fiber; 14.2 g sugar; 47.3 g net carbs; 6.5 g protein

# CulinaryArts
## PRINCIPLES AND APPLICATIONS

# Cakes, Pastries, and Desserts

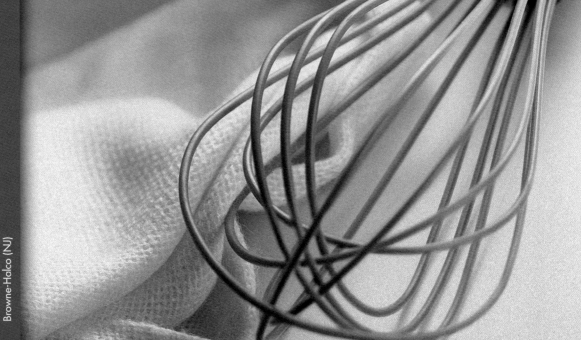

Browne-Halco (NJ)

## Chapter 23

*Cakes, pastries, and desserts* come in many different varieties. These sweet treats can be some of the most difficult and elaborate preparations to execute in the professional kitchen. Dessert comes from the French word *desservir*, which means "to clear the table." Dessert is the last course of a meal, and customers often relish the array of delights that await them at the end of a good meal. It is important for a chef to be knowledgeable about the preparation techniques involved in making cakes, pastries, and desserts. A pastry chef specializes in these skills and turns food into art.

## Chapter Objectives:

1. Identify common cake ingredients.
2. Demonstrate three methods for mixing cake batter.
3. Describe factors to consider when baking cakes.
4. Describe common types of icings.
5. Demonstrate the procedure for forming a paper pastry bag.
6. Describe how to use pastry tips, fill pastry bags, and pipe icing.
7. Describe six methods for preparing cookies.
8. Contrast the two major types of pie dough.
9. Use the formulas for determining baker's percentages.
10. Demonstrate how to mix and handle pie dough prior to baking.
11. Demonstrate the procedure for blind baking a pie shell.
12. Demonstrate how to prepare a fruit filling, a cream filling, and a chiffon filling.
13. Describe three common types of meringue.
14. Contrast common types of pastry.
15. Demonstrate how to make éclair paste.
16. Describe popular specialty desserts.
17. Demonstrate how to prepare individual soufflés, pastry cream, and sabayon.

## Key Terms

- cake flour
- genoise
- icing
- fondant
- soft cookies
- crisp cookies
- mealy pie dough
- flaky pie dough
- blind baking
- meringue
- phyllo dough
- overrun
- mouth feel
- granité
- parfait
- bombe
- coupe

## CAKES

Cakes of many different varieties can be baked using basic cake recipes and many different variations. ***See Figure 23-1.*** Although cake mixes are available, the best cakes are still made from scratch. A good cake recipe has a proper balance of ingredients and has been previously tested. Most bakers have a file of tested recipes obtained from trade publications, other bakers, and from baking-product companies. Most cakes are easy to prepare in quantity and can be stored successfully for fairly long periods of time.

### Cakes

Carlisle FoodService Products

 Cakes can be created in many flavors, shapes, and sizes to suit various occasions.

### Cake Ingredients

Common ingredients used in cakes include fat, cake flour, eggs, sugar, baking powder, liquid (milk or water), salt, and flavorings. Quality ingredients should be used to produce the proper taste, texture, and volume expected of the finished product. Ingredients may have only one function or may act in conjunction with other ingredients in the recipe to produce the desired results.

**Fat.** Types of fat used in cake making include butter, margarine, and shortening. When using shortening, a hydrogenated or emulsified shortening produces the most tender baked products. *Hydrogenated shortening* is a solid fat made from vegetable oils that have been hydrogenated. Hydrogenated shortening traps and holds a greater amount of air than other types of fat, which improves creaming qualities and tenderizes the baked *sified shortening* is a commercial shortening that is made from hydrogenated vegetable oils and is used in high-ratio cakes (cakes that contain more sugar than flour). Emulsified shortening can be used to produce products with a long shelf life.

**Cake flour.** Cake flour is a type of flour milled from a low-protein (6% to 8%), fine-textured wheat that is high in starch and contains very little gluten. Cake flour produces a less dense texture than other flours and a more tender, fine crumb. A good-quality cake flour is pure white with uniform granules and a high absorption rate. Cake flour is bleached with chlorine gas during processing, which results in flour that is more acidic. Acidic environments discourage the formation of gluten, which is important when making delicate baked goods. The acidity of the cake flour causes cakes to set faster because the starches begin to gelatinize in the oven. This in turn reduces baking time and results in a moister cake.

**Eggs.** Eggs are one of the most important ingredients used in making a cake, as they add moisture and nutritive value. Both fresh and frozen eggs can be used with good results. Frozen eggs are common in most bakeshops because of the convenience.

**Sugar.** Sugar is best known for its use as a sweetener. It also adds tenderness and color to baked goods through caramelization. Types of sugar commonly used in cake production include granulated, confectioners', and brown sugar. Superfine sugar is a type of granulated sugar with a very fine grain, which allows for quick dissolving and creaming.

**Baking Powder.** Baking powder is a common leavening agent used in baked goods. Double-acting baking powder is the most common type used in the professional kitchen. It releases leavening gases on contact with moisture and again during the baking process.

**Milk.** Milk used in cakes may be in dry or liquid form. Dry milk should blend quickly when mixed with water. Both types add moisture and tenderness to the cake.

**Flavorings.** Flavorings are very important to the taste of the cake. Quality flavorings are more expensive than imitations, but a pure flavoring yields a more pleasing flavor than an imitation flavoring. For example, pure vanilla extract produces a much better-tasting product than imitation vanilla.

## Mixing Methods

Before mixing a cake, all of the ingredients should be at room temperature. Each ingredient should always be weighed on a baker's scale for maximum accuracy. ***See Figure 23-2.*** It is important to weigh each ingredient separately before adding it to the batter. Each step of the mixing must carefully follow the recipe directions. The three methods of mixing cake batters are the two-stage (blending) method, the creaming method, and the sponge (whipping) method.

Weighing Ingredients

Edlund Co.

23-2    *Using a baker's scale helps to weigh ingredients accurately.*

**Two-Stage Method.** The two-stage (blending) method is the most basic mixing method and is the simplest to execute. After being carefully weighed, the dry ingredients are sifted together and blended with the fat and part of the liquid. In a separate bowl, the rest of the liquid and eggs are combined. The two mixtures are then mixed together until evenly distributed. *Note:* At intervals throughout the mixing process, the bowl must be scraped down so all ingredients are blended and a smooth batter is obtained.

To mix cake batter using the two-stage method, apply the following procedure:

1. Weigh all ingredients carefully.
2. Place all dry ingredients, fat, and part of the milk in the mixing bowl. Blend at slow speed for the required period. ***See Figure 23-3.***
3. In a separate bowl, blend eggs and the remaining milk.
4. Add egg mixture to batter in thirds, mixing well after each addition to ensure a smooth, uniform batter.

## Two-Stage Mixing Method

1. Weigh all ingredients carefully.

2. Blend all dry ingredients, fat, and part of the milk at slow speed.

3–4. In a separate bowl, blend eggs and remaining milk, add egg mixture to batter in thirds; mix well to ensure a smooth, uniform batter.

**23-3** In the two-stage mixing method, the dry mixture and the liquid mixture are made separately and then mixed together.

**Creaming Method.** The creaming method involves mixing the fat, sugar, salt, and spices together to a creamy consistency before any other ingredients are added. Then the eggs are added and blended well, before the wet ingredients are added. The flavorings are added last. There are some variations to this method of mixing, and it is important to follow the mixing instructions of each recipe carefully. *Note:* At intervals throughout the mixing process, the bowl must be scraped down so all ingredients are blended and a smooth batter is obtained.

When folding in whipped products such as egg whites, meringues, or whipped cream, it is important to first "temper" the mixture with a small amount of the whipped ingredient. This is done by completely folding in ⅓ of the whipped component into the base to start. After it is completely incorporated, fold in another third of the whipped component, taking care to not overwork or deflate the air that has been incorporated. After the second portion is incorporated, the final third can be added. Using a circular folding motion, gently incorporate the third portion.

**To mix cake batter using the creaming method, apply the following procedure:**

1. In a stainless-steel mixing bowl, cream the sugar, fat, salt, and spices. ***See Figure 23-4.*** During the mixing, small air cells are formed and incorporated into the mix. The volume increases and the mix becomes softer in consistency.

2. Add the eggs one at a time while continuing to mix at slow speed. At this stage, the eggs coat the air cells formed during the creaming stage and allow the air cells to expand and hold the liquid (when it is added) without curdling.

3. Add the liquid alternately with the sifted dry ingredients. Mix until a smooth batter is formed. During this stage, the liquid and dry ingredients are added alternately so the batter does not curdle. If all the liquid is added at one time, the cells coated by the eggs cannot hold all the moisture and curdling will occur.

4. Add the flavoring and blend thoroughly.

*Creaming Method*

1. Cream the sugar, fat, salt, and spices.

2. Add eggs one at a time.

3–4. Add liquid and dry ingredients alternately; add flavoring and blend thoroughly.

**23-4** The creaming method involves mixing fat and sugar together before any other ingredients are added.

**Sponge Method.** The sponge (whipping) method of cake mixing produces a light, fluffy batter. Although there are many variations of the sponge method, the most common is referred to as a *genoise*. In the genoise sponge method, the eggs and sugar are warmed and whipped to create volume and incorporate air before any other ingredients are added. This airy whipped yolk and sugar mixture is referred to as a foam. There are variations to this method of mixing, so it is important to follow the recipe instructions carefully.

To mix a cake batter using the sponge method, apply the following procedure:

1. Warm the eggs (whites, yolks, or whole eggs, as specified) and sugar to between 100°F and 105°F over a hot water bath while whisking continually. ***See Figure 23-5.*** This softens the eggs and dissolves the sugar, allowing for quicker whipping while adding volume.

2. Remove the mixture from the water bath and place on the mixer.

3. Whip on medium to high speed until the mixture appears to have peaked in volume (is no longer increasing in volume) and has developed a thick foam. Check to ensure that the mixture is thick enough to form a ribbon as it runs off the whip.

4. Slowly add any liquid and flavoring required by the recipe.

5. Gently fold in the sifted dry ingredients to ensure a smooth and uniform batter. Care should be taken to not overmix while folding, as doing so will deflate the batter and cause the resulting cake to be dense. *Note:* If fat is to be added, it should be done after the dry ingredients have been properly folded in. This will maximize the volume of the final product.

CULINARY PROCEDURES

## Sponge Method

1–3. Warm eggs and sugar to 100°F, whip until required volume or stiffness is obtained.

4. Slowly add any liquid or flavoring.

5. Gently fold in sifted flour.

**23-5** In the sponge (whipping) method, eggs and sugar are whipped to form a light, fluffy batter.

## Baking Cakes

The amount of cake batter required for a cake varies depending on the type and size of the cake. Most cake batters can be used to produce cakes in a variety of shapes and sizes. Baking time and temperature can vary depending on the size of the pan. ***See Figure 23-6.***

| Cake Baking Temperature | | | |
| --- | --- | --- | --- |
| Cake Type | Scaling Weight | Pan Size | Baking Temperature* |
| **Layer** | 13 to14 oz | 8″ dia | 375 to 385 |
| **Bar** | 5 to 6 oz | 2¾″ × 10″ | 390 to 400 |
| **Ring** | 10 to 14 oz | 6½″ dia | 390 to 400 |
| **Loaf** | 11 to 24 oz | 3¼″ to 7⅛″ | 385 to 400 |
| **Oval Loaf** | 8 oz | 6¼″ long | 385 to 400 |
| **Sheet** | 6 to 7 lb | 17″ × 25″ | 375 to 390 |

* in °F

**23-6** *Baking time and temperature depend on the size and shape of the cake pan.*

**Preparing Cake Pans.** Many cake pans are prepared by covering the bottom of the pan with parchment paper and greasing the sides lightly with shortening or pan grease (a mixture of one part flour to two parts shortening). A mixture of equal parts shortening and butter can be used in the preparation of pans to improve the flavor of the cake crust. Butter alone is not recommended for greasing cake pans, as the butter is prone to burning and might produce a bitter, burnt-tasting crust.

Whenever possible, cakes should be placed in the center of the oven where the heat is evenly distributed. If a number of cakes are being baked at one time, the pans should be placed so they do not touch one another or any part of the oven wall. Leaving a space of approximately 2″ to 3″ around each of the pans and between the pans and oven walls allows the heat to circulate freely around each pan. The oven must be preheated to the required temperature and checked periodically with an oven thermometer.

Generally, the larger the cake being baked and the richer the cake batter, the slower it should be heated. However, if the oven heat is set too low, the cake will rise and then fall, producing in a dense, heavy texture. If the oven heat is set too high, the outside of the cake will bake too rapidly and form a crust, which can burst when heat reaches the center of the cake and causes the cake to expand.

*Wisconsin Milk Marketing Board*

**Baking Time.** The baking time of a cake is divided into the following four stages of development:

- Stage 1—The cake is placed in the oven and starts to rise. At this stage, the lowest oven temperature called for in the baking instructions should be used to prevent overly quick browning and to keep a crust from forming.

- Stage 2—The cake continues to rise and the top surface begins to brown. The oven door should not be opened at this stage. The heat can be increased at this time if the recipe suggests it.

- Stage 3—The rising stops and the surface of the cake continues to brown. The oven door can be opened if necessary, and the heat can be reduced if the cake is browning too quickly.

- Stage 4—The cake starts to shrink, leaving the sides of the pan slightly. At this stage, the oven door can be opened and the cake can be tested for doneness.

**Baking at High Altitudes.** Most cake recipes are developed in low-altitude locations. The ingredients are balanced to produce good results when the cake is baked at or near sea level. If these recipes are used in high-altitude areas, adjustments to the recipe are required, such as reducing the baking powder and sugar and increasing the liquid. The exact amount of ingredients required can be determined by experimenting with the recipe or by consulting with baking-product manufacturers.

**Testing for Doneness.** Cakes are tested for doneness by inserting a wire tester or a toothpick into the center of the cake. The cake is done when the tester is dry when removed, with no batter adhering to it. Another method for testing doneness, the touch method, is used for heavier cakes such as fruitcakes. Using the touch method, the top surface of the cake is pressed with a finger. If the surface feels firm and the impression of the finger does not remain, the cake is done.

**Cooling Cakes.** When a cake is removed from the oven, it should be placed on a wire rack or shelf so that air circulates around the pan. The cake should be allowed to cool for a minimum of 5 min before it is removed from the pan. The pan can then be flipped over and the cake removed from the pan. If wax or parchment paper was used on the bottom of the pan, it is also removed at this time. The cake is then placed back on the wire rack or shelf and allowed to cool thoroughly.

**Potential Defects.** Defects that occur with baked cakes often involve the size, color, texture, and flavor of the cake. *See Figure 23-7.* Common defects include a cake turning out uneven, undersized, too dark or too light in color, tough, soggy, sticky, and too heavy or too light in texture. Many of these defects can be easily avoided by taking steps during the preparation of the recipe. Cakes prepared using either the two-stage or creaming methods may require different solutions than cakes prepared using the sponge method.

## Batter Cakes (Prepared with Two-Stage or Creaming Method)

| Defect | Possible Causes | Possible Remedies |
|---|---|---|
| **Dark crust color** | Oven too hot | Use correct baking temperature |
| | Too much top heat in oven | Check oven drafts |
| | Too much sugar, too much milk solids | Balance recipe |
| **Light crust color** | Oven too cool | Raise oven temperature |
| | Unbalanced recipe | Balance recipe |
| **Uneven baking** | Oven heat not uniform | Check oven drafts, flues, insulation |
| | Variations in baking pans | Use same type tins for entire batch |
| **Tough cakes** | Insufficient tenderizing | Increase sugar, shortening, or both |
| | Flour content too high | Balance recipe |
| | Wrong type of flour | Use soft wheat flour |
| **Thick, hard crust** | Oven too hot | Reduce oven temperature |
| | Cakes baked too long | Reduce baking time |
| | Slab-type cake tins not insulated | Use insulation around cake molds |
| **Sticky crust** | Sugar content too high | Balance recipe |
| | Improper mixing | Use care in mixing |
| **Soggy crust** | Cakes steam during cooling | Remove cakes from tins and allow to cool on rack; cool cakes before wrapping |
| **Crust cracks** | Oven too hot | Reduce oven temperature |
| | Stiff batter | Adjust flour and liquid contents |
| **Poor flavor** | Inferior materials used | Care in selecting materials |
| | Poor flavoring material or wrong combination | Use quality pure flavors; check flavor combinations |
| | Materials improperly stored | Material storage space should be free from foreign odors |
| **Lack of flavor** | Lack of salt | Use correct amount of salt |
| | Lack of flavoring materials or weak flavoring materials | Use sufficient flavoring and correct types |
| **Heavy cakes** | Too much sugar | Balance recipe |
| | Too much shortening | Balance recipe |
| | Liquid content high | Balance recipe |
| | Insufficient leavening | Balance recipe |
| | Too much leavening | Balance recipe |
| | Cakes underbaked | Bake out correctly |
| **Cakes too light and crumbly** | Batter overcreamed | Mix properly |
| | Leavening content high | Balance recipe |
| | Shortening content too high | Balance recipe |
| **Coarse grain** | Leavening content high | Balance recipe |
| | Separation of liquids and fats (curdled characteristic in batter) | Add liquids at proper temperatures and only as fast as it will emulsify well |
| **Tough-eating cakes** | Formula low in tenderizing materials, sugar, and shortening | Balance recipe |
| | Oven too hot | Regulate oven temperature |

**23-7**  *Problems with baked cakes often involve the size, color, texture, or flavor of the cake.* (continued on next page)

## Sponge Cakes

| Defect | Possible Causes | Possible Remedies |
|---|---|---|
| Dark crust color | Oven too hot | Regulate oven temperature |
| | Cakes overbaked | Give proper bake |
| | Excessive sugar content causing cake to have sugar crust | Balance recipe |
| Tough crust | Oven too hot | Regulate oven temperature |
| | Sugar content too high | Balance recipe |
| | Improper mixing | Exercise care in assembling batter |
| Thick and hard crust | Overbaking | Decrease baking time |
| | Cold oven | Regulate oven temperature |
| Strong flavor | Off-flavored materials | Check storage space of materials for foreign odors |
| | Poor flavoring materials | Use only top-quality flavors |
| | Cakes burned or overbaked | Exercise care in baking |
| Lack of flavor | Insufficient salt in recipe | Increase salt content |
| | Poor flavor combination | Use proper flavor blends |
| | Poor-quality flavoring materials used | Use only top-quality materials |
| Heavy cakes | Overbeaten or underbeaten eggs | Beat eggs to wet peak |
| | Overmixing after flour has been added | Fold flour in just enough to incorporate |
| | Too much sugar | Balance recipe |
| | Too high a baking temperature | Regulate oven temperature |
| Coarse grain | Cold oven | Regulate oven temperature |
| | Overbeaten whites | Whip to wet peak |
| | Insufficiently mixed batter | Fold until smooth |
| Tough cakes | Overmixing ingredients | Mix properly |
| | Excessive sugar content | Balance recipe |
| | Bakes too hot | Regulate oven temperature |
| | Flour content high or wrong type flour used | Balance recipe; use soft wheat flour |
| Dry cakes | Low sugar content | Balance recipe |
| | Overbaking | Lessen baking time |
| | Eggs overbeaten | Whip to wet peak |
| | Flour content too high | Balance recipe |

**23-7** *(continued from previous page)*

## ICINGS

Like cake recipes, variations of icing recipes are used to obtain different flavors and textures. An *icing* is a sugar-based coating often spread on the outside or between layers of a baked good. Icings, also known as frostings, have three main functions when applied to a baked product such as a cake. They form a protective coating to seal in moisture and flavor, they improve the taste, and they add eye appeal. ***See Figure 23-8.*** There are seven basic types of icing: buttercream, flat, foam, fudge, fondant, royal, and glaze.

## Buttercream

Buttercream is one of the most popular types of icing. It is simple to prepare, easy to store, and adds flavor and eye appeal to a dish. Buttercream icing is made by creaming together shortening or butter, powdered sugar, and, in some cases, eggs. It is light and airy because lots of air cells are trapped in the icing when mixing using the creaming method. Buttercream icing should be stored in a cool place and covered with plastic wrap or wax paper to avoid crusting. For best results, buttercream icing should not be stored in the refrigerator, as refrigeration causes it to harden. If the shortening hardens, considerable mixing is required to return the icing to the appropriate consistency for spreading.

## Flat Icing

Flat icing is the simplest icing to prepare. It is applied to sweet rolls, doughnuts, Danish pastry, and other baked goods. Flat icing is usually prepared by combining water, powdered sugar, corn syrup, and flavoring and then heating the mixture to approximately 100°F. It should always be heated in a double boiler because direct heat or overheating causes it to lose its gloss when it cools. Flat icing should always be covered with a damp cloth when not in use. To store, flat icing should be covered with a thin coating of water, plastic wrap, or wax paper. Previously stored flat icing must be heated to approximately 100°F over a water bath before use.

## Foam Icing

Foam (boiled) icing is prepared by combining sugar, glucose (sweetener), and water, boiling the mixture to approximately 240°F, and adding the resulting syrup to an egg white meringue while still hot. If heavy syrup is added to the meringue, a heavy icing is produced. If thin syrup is added, the result is a thin icing. Foam icing may be colored slightly, and must be applied the same day it is prepared. Only the amount needed should be prepared, as it breaks down if held overnight. This icing should be applied in generous amounts and worked into peaks.

## Fudge Icing

Fudge icing is a rich, heavy-bodied icing that is usually prepared by adding a hot

## Icings

Carlisle FoodService Products

**23-8** *Icings add eye appeal and flavor to baked goods such as cakes and petit fours.*

*Chef's Tip:*
Mix buttercream icing at medium speed. Increasing the mixing time will aerate the icing and increase the volume.

Carlisle FoodService Products

liquid or syrup to the other ingredients called for in the recipe while whipping to obtain smoothness. Fudge icing should be used warm; if cool, it should be reheated in a double boiler before applying. Fudge icing is generally used to ice layer cakes, loaf cakes, sheet cakes, and cupcakes. It dries rapidly when stored. To store, fudge icing should be covered with plastic wrap, stored in the refrigerator, and reheated over a water bath before use.

## Fondant

*Fondant* is a rich, white, cooked icing that hardens when exposed to the air. It is used mainly on small cakes (petit fours) that are eaten with the fingers. It is prepared by cooking sugar, glucose, and water to a temperature of 240°F, letting it cool to 150°F, and then mixing it until it is creamy and smooth. Fondant is the most difficult and time-consuming icing to prepare; most bakers purchase a ready-made fondant or a powdered product known as drifond® from a baker's supply house. The ready-made product is usually purchased in a 40-lb tin and stores well in a cool place covered with a damp cloth or a small amount of water to keep it from drying out. Drifond® needs only water and a small amount of glucose added to produce an excellent fondant. When using drifond®, the amount needed can be prepared in a very short time.

Chef's Tip:
Exercise caution when heating fondant. If it is heated over 100°F, the product loses its gloss, resulting in a dull finish when it hardens.

When needed for use, fondant is heated to about 100°F in a double boiler while stirring constantly. This thins the icing so that it flows freely over the item to be covered. The secret to successfully covering an item with fondant is the consistency of the icing. If the fondant is too heavy after it is heated, it can be thinned down by using a glaze consisting of one part glucose to two parts water or a regular simple syrup may be used. The fondant can be colored and flavored as desired for use as a base for other icings. Fondant icing, like flat icing, must be kept covered with a very thin coating of water, plastic wrap, or wax paper in a cool place. It can be refrigerated, but may lose some gloss when reheated in a double boiler or over a hot water bath.

## Royal Icing

Royal icing is similar to flat icing, but the addition of egg whites produces a thicker icing that hardens to a brittle texture. Royal icing hardens when exposed to air, so it must be kept covered with a damp towel when not in use. It is used for making decorative flowers for cakes used in window displays and for show work such as sugar sculptures.

## Glaze

A glaze is a thin, transparent coating of icing that is usually poured over a baked product. Most glazes are made by thinning down and heating fruit purées, fruit juices, chocolate, or coffee. Apricot glaze is the most common fruit glaze used in the professional bakeshop. It has a mild flavor and a nice sheen. Glazes give shine to bakery products such as coffee cakes and extend shelf life by sealing in moisture.

## Icing Preparation

The best ingredients should be used when preparing icings. It is also important to obtain the proper consistency before applying any icing. The consistency needed depends on the intended use of the icing. In most cases, the consistency can be controlled by adding or eliminating an amount of powdered sugar.

Icing is often colored to attract the eye. Standard principles regarding color used in icings include:

- Colors in paste and powder forms produce better results than liquid food colorings.

- Certain colors can be blended to create other colors. For example, red and blue create violet and blue and yellow create green.

- Two or more colors can be used if desired by placing a layer of each color of icing side by side in a pastry bag.

- Pastel colors are pleasing to the eye.

- Red and yellow colors create a hungry feeling.

**Pastry Bags.** A pastry bag is made of parchment paper, silicon, canvas, or plastic. Most decorators prefer to make their own pastry bags from parchment paper. Paper cones are simple to make, easier to handle, and a separate cone can be made for each color of icing used. When finished decorating, paper cones can be discarded, whereas other types of pastry bags must be washed.

To form a paper cone pastry bag, apply the following procedure:

1. Cut a square of parchment paper into a large triangle. *See Figure 23-9.*

2. With the long edge on top, start to roll the paper by turning one short edge towards the centerline of the triangle.

3. Continue rolling the paper into a cone shape across to the far short edge.

4. Tuck the top corner of the far short edge into an overlapping edge of the cone to secure. Cut off excess paper at the top of the cone.

5. Cut off the bottom tip of the cone so a pastry tip can be inserted. *Note:* If too much paper is cut away, the resulting hole will be too large to hold the pastry tip inside the cone.

## Forming Paper Cone Pastry Bags

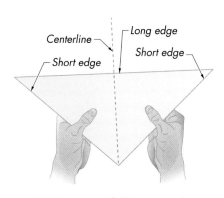

Centerline
Long edge
Short edge
Short edge

1. Cut a square of silicon or parchment paper into a large triangle.

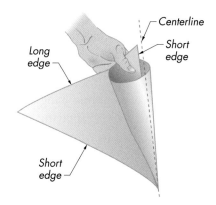

Centerline
Short edge
Long edge
Short edge

2. Roll paper by turning one short edge toward centerline.

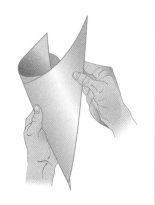

3. Continue rolling paper into a cone shape.

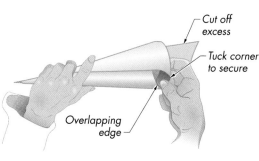

Cut off excess
Tuck corner to secure
Overlapping edge

4. Tuck top corner of far short edge into overlapping edge of cone to secure; cut off excess paper.

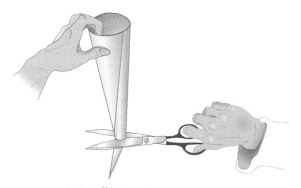

5. Cut off bottom tip.

**23-9** A pastry bag can easily be made from a piece of silicon or parchment paper.

## Filling a Pastry Bag

Deposit icing away from sides

Tip

**23-11** When filling a pastry bag, care should be taken to eliminate air pockets.

**Pastry Tips.** In order to create decorative patterns and shapes with icing, a plastic or metal pastry tip is inserted into the pastry bag before it is filled. Many different pastry tips are used to make various designs. **See Figure 23-10.** For example, a round pastry tip is used for writing.

**Filling Pastry Bags.** To fill a pastry bag with icing, hold the bag in the left hand (if right-handed) at the middle using a very light grip. Fold the top half of the bag is back (or cuff) over the left hand (if right-handed). Insert icing into the center of the bag using a spatula. As the spatula is withdrawn, the hand holding the bag can gently squeeze the icing off the spatula and into the bag. **See Figure 23-11.** After the pastry bag is filled approximately halfway up, it is tightly twisted just above the filling. The twisted portion of the bag is held in the hand as a handle while piping out the icing. As the bag empties, the bag can again be twisted to push the icing to the tip, similar to wringing out a wet rag to expel the water.

## Pastry Tips

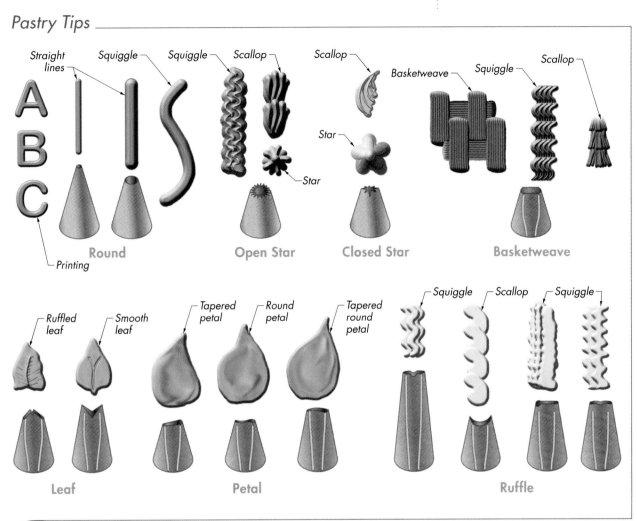

Air pockets within the icing or along the sides of the pastry bag can result in bursts of air that can easily ruin a decorating job. To reduce the risk of air pockets, the icing should be deposited down to the tip of the bag and away from the top sides. Once the bag is filled, the icing should be squeezed and compacted down to the tip to push out any remaining air pockets. The pastry bag should always be tested for air pockets by piping out a sample of icing before applying icing onto a baked good.

**Piping Icing.** When decorating, the bag is held with the right hand (if right-handed) at the top of the bag and the left hand lightly gripping the lower half. The hand at the top of the bag applies the pressure to cause the icing to flow. The hand on the lower half is used as a guide. In all decorating tasks, the two most important factors are holding the bag at the correct angle and applying the correct pressure to obtain a smooth, even flow of icing. ***See Figure 23-12.***

### Using a Pastry Bag

23-12 Using the correct angle and amount of pressure is key when using a pastry bag.

## COOKIES

Cookies are a popular dessert that can be served alone or with ice creams, sherbets, puddings, or fruit. The ingredients used in cookies are similar to the ingredients used in cakes, although cookie doughs and batters generally have higher fat content and lower moisture content than cake batters. In addition, cookies are usually baked at higher temperatures for less time than cakes.

Cookies can be classified by texture as either soft or crisp. *Soft cookies* are cookies prepared from dough that contains a great deal of moisture. *Crisp cookies* are cookies prepared from dough that contains a high percentage of sugar. The following tips are helpful when preparing cookies:

- Follow mixing instructions carefully to avoid overmixing or undermixing.

- Use recipes that specify measurements by weight rather than by volume. Weights are more exact than volumes.

- Weigh all ingredients carefully. Too much or too little of an ingredient can result in a finished product of poor quality.

- Form cookies in uniform size so they bake evenly.

- Bake cookies on pans covered with parchment paper.

- Bake according to recipe instructions. Check the baking progress periodically for necessary adjustments. Baking time for cookies varies depending on size, thickness of dough, and quality of ingredients.

- Place an extra sheet pan underneath the pan that is holding the cookies if they are getting too much bottom heat. This can be detected if the edges of the cookies brown rapidly.

- Store cookies properly. Crisp cookies should be stored in a tin container with a loose-fitting top and placed in a dry location. Crisp cookies should be warmed in a 225°F oven for 5 min just prior to serving. Soft cookies should be placed in an airtight tin container. Adding a slice of fresh bread to the tin will help keep soft cookies fresh.

Chef's Tip:
Overmixing cookie doughs or batters result in a coarse, hard-to-handle product. Undermixing cookie batter or dough causes excessive spreading when baking.

### Cookie Preparation Methods

Cookies are categorized by the method used in preparation. Six common methods for preparing cookies include icebox, rolled, pressed, sheet, bar, and drop methods. *See Figure 23-13.* The method to use is determined by the consistency of the cookie dough or batter.

## Cookie Preparation Methods

| Method | | Procedure | Dough or Batter Consistency |
|--------|--|-----------|------------------------------|
| **Sheet** | | Batter is spread into a sheet pan, baked, and cut into squares | Moist, soft batter |
| **Rolled** | | Dough is refrigerated, rolled out, and cut into desired shapes | Stiff, dry dough |
| **Icebox** | | Dough is rolled into 1 lb to 1½ lb lengths, refrigerated, and sliced to desired size | Stiff, fairly dry dough |
| **Bar** | | Dough is refrigerated and rolled into a sheet pan, flattened, baked, and cut to desired size | Stiff, fairly dry dough |
| **Pressed** | | Dough is squeezed through a pastry tube or cookie press | Moist, soft dough |
| **Drop** | | Batter is dropped into spoonfuls onto sheet pan at room temperature | Moist, soft batter |

*Wisconsin Milk Marketing Board*

**23-13** *Cookies can be categorized by preparation method.*

**Sheet.** The sheet method of cookie preparation involves a moist, soft batter that is poured into a parchment-lined sheet pan, baked, and cut to size. The baked cookies are typically cut into square or rectangular units.

**Rolled.** The rolled method of cookie preparation is most often used when a specific shape of cookie is desired. The cookies are prepared from stiff, dry dough. The dough is refrigerated until chilled thoroughly and then rolled out on a floured piece of canvas to a thickness of approximately ⅛″.

## Rolled Method

Cut to desired shapes

**23-14** *Cookies are cut out with a cookie cutter when prepared with the rolled method.*

## Icebox Method

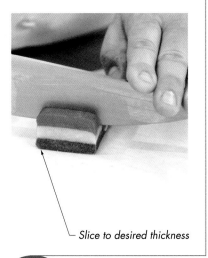

Slice to desired thickness

**23-15** *Icebox cookies are formed into a round length, chilled overnight, and then cut to desired thickness.*

Cookies are cut into desired shapes and sizes with a cookie cutter or knife, placed on sheet pans covered with parchment paper, and then baked. **See Figure 23-14.**

**Icebox.** The icebox method of cookie preparation involves refrigerating the dough overnight and then cutting the cookies to desired size. The cookies are prepared from stiff, fairly dry dough. The dough is scaled into units of 1 lb to 1½ lb, rolled into lengths of approximately 16″, wrapped in parchment paper, and refrigerated overnight. The next day, the dough is sliced into units typically ¼″ thick (or as desired), placed on sheet pans covered with parchment paper, and then baked. Thinner slices produce crispier cookies. **See Figure 23-15.**

**Bar.** In the bar method of cookie preparation, dough is formed into flat lengths, baked, and then cut to size. The cookies are prepared from stiff, fairly dry dough. The dough is scaled into 1-lb units, refrigerated until thoroughly chilled, and rolled into round strips the length of a sheet pan. Three strips are placed on each parchment-lined pan, leaving a space between each strip. The strips are flattened by hand, brushed with egg wash, baked, and cut into bars. **See Figure 23-16.**

## Bar Method

**23-16** *Using the bar preparation method, dough is formed into flat lengths, baked, and then cut to size.*

**Pressed.** In the pressed method of cookie preparation, dough is pressed through a die or pastry tube tip to produce a cookie with a three-dimensional shape. Traditional cookie press dies form cookies in the shape of a leaf, snowflake, or tree depending on the die. The cookies are prepared from moist, soft dough. The dough is placed in a cookie press or a pastry bag, squeezed or piped onto sheet pans covered with parchment paper, and then baked. **See Figure 23-17.**

## Pressed Method

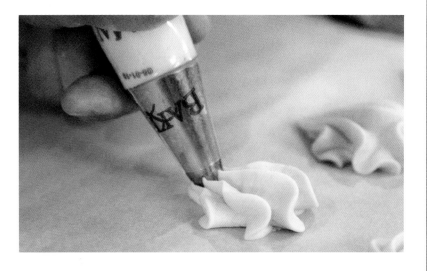

23-17 *Dough can be pressed through a die or pastry tip to produce a cookie with a three-dimensional shape.*

**Drop.** In the drop method of cookie preparation, dough is formed into consistent-size balls with a spoon. The cookies are prepared from a moist, soft batter. The batter should be at room temperature, and is dropped by uniform spoonfuls onto parchment-covered sheet pans and then baked. *See Figure 23-18.*

## PIES

Pie is one of the most popular types of dessert. The two basic types of pies are single-crust and double-crust. A single-crust pie consists of one crust on the bottom and a filling. A double-crust pie consists of two crusts, one on the bottom and one on the top, and a filling. *See Figure 23-19.*

In both types of pies, the tenderness of the crust often determines how the pie is received. The filling may be of excellent quality, but a tough crust may cause a customer to reject the whole preparation. Different types of pie dough vary in preparation techniques but have very similar ingredients. The mixing processes for pie dough are critical procedures that are learned with practice.

## Types of Pie Doughs

Most pie doughs are made primarily of flour and fat, but can differ in how the flour and fat are mixed together and in the amount of liquid added to the dough. The best method of learning how pie dough is properly mixed

## Drop Method

23-18 *Drop cookies are formed by dropping consistent-size balls of dough from a portion scoop.*

is by using the hands to rub the dough during the mixing process. The ingredients of pie dough are rubbed (cut in) with the palm of the hands until the proper consistency is obtained. Knowing the amount of rubbing required for proper consistency is a skill acquired through experience. A mixing machine is often used for large quantities of pie dough but can easily overmix the dough. A mixing machine also creates more heat in the dough as it mixes, which can cause the fat to break down.

## Pie Types

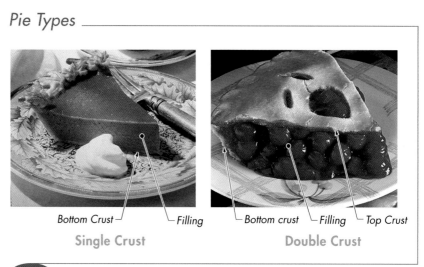

Bottom Crust — Filling     Bottom crust — Filling — Top Crust

**Single Crust**      **Double Crust**

**23-19** *Pies are either single crust or double crust.*

Pie doughs are classified into two basic types: mealy pie dough and flaky pie dough. Flaky pie dough can be made to produce a short- or long-flake dough. **See Figure 23-20.** The basic difference between these two types of dough is how the fat is combined with the flour. The type of filling to be added determines which type of dough is the best choice.

**Mealy Pie Dough.** A *mealy pie dough* is a pie dough that absorbs the least amount of liquid because the flour and fat are rubbed together until the flour is completely covered with shortening and the mixture resembles fine cornmeal. The flour is then unable to absorb a large amount of liquid or moisture from the filling because the flour granules are coated in fat. Mealy pie dough is used mostly for bottom-crust pies, especially fruit or custard pies, because it has high resistance to moisture and is less likely to become soggy.

**Flaky Pie Dough.** A *flaky pie dough* (commonly referred to as *pâte brisée*) is a pie dough prepared by cutting fat into flour until pea- or hazelnut-size particles of fat are formed. The flour in flaky pie dough is not completely mixed in or blended with the fat as in a mealy dough. A short-flake crust is the most common type of flaky pie dough. There are two types of flaky pie crust: short-flake crust and long-flake crust.

| Pie Crust Types | | | |
|---|---|---|---|
| Type | Characteristics | Mixing | Refrigeration |
| **Mealy** | • High resistance to absorbing moisture<br>• Used for fruit or custard bottom-crust pies | Rub flour and fat together until flour is completely covered | 45 min to 1 hr |
| **Short Flake** | • Most common pie crust | Rub flour and fat until fat is pea size and there are no flour spots | 45 min to 1 hr |
| **Long Flake** | • Absorbs the most moisture of any pie crust | Rub flour and fat together until fat is walnut size | 3 hr to 12 hr |

**23-20** *Two basic types of pie dough are mealy pie dough and flaky pie dough.*

A *short-flake crust* is a pie crust that absorbs a slightly larger amount of liquid than mealy pie dough because the flour and fat are rubbed only until no flour spots are evident and the fat is in pea-size particles. The flour of a short-flake crust is not coated with fat to the same degree as in a mealy pie dough, so the flour is able to absorb slightly more liquid. Short-flake crust is commonly used for lattice tops of pies and for the bottom crust of a prebaked pie shell (commonly referred to as a *blind-baked pie shell*) that will be filled with a cooked, cooled filling.

A *long-flake crust* is a pie crust that absorbs the greatest amount of liquid because the flour and fat are rubbed together less than in a mealy or short-flake crust. In a long-flake crust, the flour and fat are rubbed together very lightly, leaving the fat in chunks about the size of a walnut. Long-flake pie crust is best used for pie top crusts as well as prebaked pie shells. Short-flake and long-flake pie dough can be used interchangeably depending on the preference of the pastry chef because they are fairly similar.

## Pie Dough Ingredients

The ingredients used in most pie dough recipes are flour, fat, liquid salt, and sugar. Each ingredient plays an important part in creating the finished product.

**Flour.** Pastry flour contains the ideal gluten content for pie dough and produces the best results. Flour that contains too much or too little gluten results in tough or sticky dough. If pastry flour is not available, a blend of 60% cake flour and 40% bread flour can be used. Pastry flour has a tendency to pack and form lumps, so it should be sifted before use. Lumps of flour do not absorb liquid as readily as sifted flour and can lead to overmixing, resulting in a tough crust. Pastry flour is also used for dusting the bench when the pie dough is rolled out. Some bakers and pastry chefs chill the pastry flour in the refrigerator in order to keep the pie dough below 70°F during the mixing period. This method is especially useful if mixing is done in a hot kitchen environment.

**Fat.** The fat used in pie dough is usually hydrogenated shortening or butter. Hydrogenated shortening is used often because it has no taste and has a plastic consistency that is ideal when cutting flour into fat. If butter is used, it should be blended with hydrogenated shortening using a ⅓ : ⅔ ratio of butter to shortening. This blend must be chilled in the refrigerator and allowed to harden slightly before it is cut into the flour, as butter tends to soften when mixed with shortening. The use of butter may not pay off as the flavor of the butter may be overpowered by the pie filling. If lard is used for pie dough, a high-quality lard is required. Most types of lard add a flavor that can be objectionable if the filling does not overpower it, so lard is seldom used.

Wisconsin Milk Marketing Board

**Liquid.** The liquid used in preparation of pie dough may be water or milk, depending on the recipe being used. It is important that the liquid used is at a very cold temperature (at least 40°F or below). This helps to keep the fat particles hard, preventing the dough from becoming too soft. Milk produces a richer dough and a better-colored crust. If dry milk is used, it must be dissolved in water before being added to the flour mixture. The amount of liquid required depends on the type of pie dough being prepared.

**Salt.** Salt is used to enhance the flavor of the dough. The salt should be dissolved in the liquid ingredients to ensure even distribution and prevent burnt spots.

**Sugar.** Sugar adds sweetness and color to a baked crust. Several types of sugar can be used, depending on the recipe. Like salt, sugar should be dissolved in the liquid to ensure a complete and even distribution.

**Thickening Agents.** Starches and flours are used to thicken pie fillings. Starches are used more often than flours because they produce a better sheen and do not discolor or become heavy. Common types of starch include cornstarch, tapioca starch, and rice starch. Products such as waxy maize starch (ClearJel®) or modified starch are blends of starch and vegetable gum that produce a finished product with a high sheen. Blended starches that include vegetable gum gelatinize quickly when cooked and offset the action of fruit acids, whereas other starches, such as cornstarch, may give the product a cloudy or milky appearance. These blended products maintain fruit flavor and color, produce a smooth consistency, and do not cloud when refrigerated.

Starch or flour is typically added to a cooked filling while boiling. Starch or flour begins to swell at approximately 160°F to 170°F. The swelling is complete when the temperature reaches 200°F to 205°F. The amount of starch used in a recipe depends on both the gelling properties of the starch and the acidity of the fruit and juice being thickened. Usually 3 oz to 5 oz of starch is required for each quart of liquid. More thickener is required for fruits and juices with high acidity, such as citrus fruits. Flour or starch should be diluted in cold water or juice before being added to a boiling filling. When the starch is added, the filling should be whipped vigorously to ensure that the filling remains smooth and creamy.

A *pregelatinized starch* is a starch that can thicken a liquid without additional cooking. Pregelatinized starch is precooked and does not require heat to absorb liquid and gelatinize. It thickens quickly when blended with sugar and added to a liquid. When using pregelatinized starch, always follow manufacturer recommendations.

**Baker's Percentages.** The flakiness of pie dough depends on how much or how little the fat is mixed into the flour. If the fat is left in larger-size chunks (pea to hazelnut size), the dough will be much flakier than if the fat were mixed with the flour to produce a cornmeal-size particle. Because the fat is combined more completely in mealy dough, less water is absorbed. Because the crust is less absorbent, occasionally a pastry chef will add a small percentage less water and fat (up to 5% less than in a flaky dough) to a mealy dough. This is optional and is a matter of preference. The higher the percentage of fat to flour by weight, the more tender the resulting crust.

Recommended baker's percentages of ingredients for an average pie dough recipe are the following:

| Mealy Pie Dough (Yield 1 lb) | | | Flaky Pie Dough (Yield 1 lb) | | |
| --- | --- | --- | --- | --- | --- |
| Ingredients | Weight | Percentage | Ingredients | Weight | Percentage |
| Pastry flour | 16 oz | 100% | Pastry flour | 16 oz | 100% |
| Shortening | 10.4 oz | 65% | Shortening | 11.2 oz | 70% |
| Water | 4 oz | 25% | Water | 4.8 oz | 30% |
| Salt | 0.3 oz | 2% | Salt | 0.3 oz | 2% |
| Sugar | 0.8 oz | 5%* | Sugar | 0.8 oz | 5%* |
| Dry milk | 0.24 oz | 1%–2%* | Dry milk | 0.24 oz | 1%–2%* |

* optional

*Note:* Although the percentages are slightly different in the two examples, many pastry chefs use the ratio for a flaky dough for both versions of dough and simply change the way the fat is incorporated into the flour to yield the different pie doughs.

Pastry flour is the main ingredient in pie dough and equals 100% in a baker's percentage. All other ingredient percentages are relative to the amount of flour. To determine baker's percentages of ingredients used in a pie dough recipe, first, all weights must be reduced to the smallest common units, such as ounces. Then the weight of each ingredient is divided by the weight of flour. The baker's percentages of a recipe are determined by applying the following formulas:

1. Reduce all weights to the smallest common units. Weight in pounds is reduced to ounces by applying the following formula:

$TW = (WL \times 16) + WO$

where

$TW$ = total weight (in oz)

$WL$ = weight of ingredient (in whole lb)

16 = constant

$WO$ = weight of ingredient (in oz)

2. Divide the weight of each ingredient by the weight of flour. Baker's percentages are determined by applying the following formula

$$BP = \frac{WI}{WF}$$

where

$BP$ = baker's percentage

$WI$ = weight of ingredient

$WF$ = weight of flour

For example, what are the baker's percentages for a pie dough recipe that uses 10 lb flour, 7 lb 4 oz shortening, 3 lb 8 oz water, 4.8 oz salt, 2.4 oz dry milk, and 2.4 oz sugar?

1. Reduce the weight of each ingredient to ounces.

**Flour**

$TW = (WL \times 16) + WO$

$TW = (10 \times 16) + 0$

$TW = 160$ oz of flour

**Shortening**

$TW = (WL \times 16) + WO$

$TW = (7 \times 16) + 4$

$TW = 112 + 4$

$TW = 116$ oz of shortening

**Water**

$TW = (WL \times 16) + WO$

$TW = (3 \times 16) + 8$

$TW = 48 + 8$

$TW = 56$ oz of water

2. Find the baker's percentage of each ingredient.

**Shortening**

$$BP = \frac{WI}{WF} \times 100$$

$$BP = \frac{116}{160} \times 100$$

$$BP = 0.725 \times 100$$

$$BP = 72.5\% \text{ shortening}$$

**Water**

$$BP = \frac{WI}{WF} \times 100$$

$$BP = \frac{56}{160} \times 100$$

$$BP = 0.35 \times 100$$

$$BP = 35\% \text{ water}$$

**Salt**

$$BP = \frac{WI}{WF} \times 100$$

$$BP = \frac{4.8}{160} \times 100$$

$$BP = 0.03 \times 100$$

$$BP = 3\% \text{ salt}$$

**Dry Milk**

$$BP = \frac{WI}{WF} \times 100$$

$$BP = \frac{2.4}{160} \times 100$$

$$BP = 0.015 \times 100$$

$$BP = 1.5\% \text{ dry milk}$$

**Sugar**

$$BP = \frac{WI}{WF} \times 100$$

$$BP = \frac{2.4}{160} \times 100$$

$$BP = 0.015 \times 100$$

$$BP = 1.5\% \text{ sugar}$$

*Note:* The ingredients in the golden pie dough recipe are consistent with the suggested baker's percentages and should produce excellent pie dough.

Chef's Tip:
When making large quantities of pie dough in a bench or floor mixer, it is best to use the paddle attachment and keep the speed at its lowest setting.

## Pie Dough Preparation

Pie dough must be mixed properly to achieve good results. The correct amount and mixing of ingredients produces a good pie crust. The salt, sugar, and cold liquid should be blended together until the salt and sugar are thoroughly dissolved. The liquid mixture is then poured into the flour mixture and mixed only until the flour absorbs the liquid.

Most faults develop as the pie dough is being mixed. First, the flour should be sifted before the fat is rubbed or cut in. It is difficult to make pie dough in a mixer because the flour usually becomes coated with fat too quickly. Too much gluten can develop when mixing pie dough with a mixer, resulting in dough that is not as tender as desired. Another mistake is adding extra flour or water to pie dough, as doing so causes overmixing and results in toughness. **See Figure 23-21.**

| Causes and Remedies for Faulty Pies | | |
|---|---|---|
| Problem | Possible Causes | Possible Remedies |
| **Excessive shrinkage of crusts** | Not enough shortening | Increase the shortening |
| | Too much water | Cut quantity of water |
| | Dough worked too much | Do not overmix |
| | Flour too strong | Use a weaker flour or increase shortening content |
| **Crust not flaky** | Dough mixed too warm | Have water cold |
| | Shortening too soft | Have shortening at right temperature |
| **Bottom crust soaks too much juice** | Rubbing flour and fat too much | Do not rub too much |
| | Insufficient baking | Bake longer |
| | Crust too rich | Reduce amount of shortening |
| | Too cool an oven | More bottom heat |
| **Tough crust** | Flour too strong | Increase the shortening |
| | Dough overmixed | Just incorporate the ingredients |
| | Overworking the dough | Work dough as little as possible |
| | Too much water | Reduce amount of water |
| **Soggy crust** | Not enough bottom heat | Regulate oven correctly |
| | Oven too hot | Regulate oven correctly |
| | Having filling hot | Use only cold filling |
| **Fruit boils out** | Oven too cold | Regulate oven temperature |
| | Fruit slightly sour | Use more sugar |
| | No holes in top crust | Have a few openings in top crust |
| | Crust not properly sealed | Seal bottom and top crust on edges |
| **Custard pies curdle** | Overbaked | Take out of oven as soon as set |
| **Blisters on pumpkin pies** | Oven too hot | Regulate oven temperature |
| | Too long baking | Take out of oven as soon as set |
| **Bleeding of meringue** | Moisture in egg whites | Use a stabilizer in the meringue |
| | Poor egg whites | Check egg whites for body |
| | Grease in egg whites | Be sure equipment is free from grease |

**23-21** *Weighing ingredients accurately, following proper mixing procedures, and having the correct oven temperature help to produce a successful pie crust.*

**Mealy and Flaky Crusts.** A meaty crust and short-flake crust are handled the same way after mixing. Once the dough is mixed, it is wrapped with plastic wrap and refrigerated until it is firm enough to be rolled out (about 45 min to 1 hr). Long-flake crust must be refrigerated for a longer period, usually several hours or overnight. If the pie dough is not refrigerated long enough, it will be too soft and difficult to roll out. When the pie dough is firm, it is removed it from the refrigerator and divided into 8-oz units. Any unneeded 8-oz units should be returned to the refrigerator to be kept firm until ready to be rolled out.

Each 8-oz unit provides enough dough for one bottom or one top crust for an 8″ or 9″ pie. Only one unit of dough should be rolled out at a time. The rolling should be done on a bench dusted with pastry flour. In some cases, bakers roll pie dough on a floured piece of canvas to keep the dough from sticking to the bench. After the dough is rolled and the bottom or top crust is formed, any remaining scraps can be pressed together and reused. *See Figure 23-22.*

*Rolling Pie Dough* _____

**23-22** *Pie dough should be rolled out on a floured surface.*

**Crumb Crusts.** A *crumb crust* is a pie crust made with crumbled cookie or graham cracker pieces that are pressed together with melted butter to form a solid shell. Crumb crusts can be made from various ingredients including graham cracker crumbs, wafer crumbs, ginger snaps, or vanilla and chocolate crumbs. Ground nuts are sometimes added along with sugar. Crumb crusts are often baked for a short amount of time to make the crust more stable. Crumb crusts are most often used for unbaked pies such as cream or chiffon pies and are also popular for cheesecakes.

**Specialty Pie Crusts.** Specialty pie crust may be prepared by adding cheese, spices, ground nuts, or other products to the pie dough. The added ingredient usually replaces up to 20% of the flour, except in the case of spices.

### Baking Pie Crusts

Sometimes pie dough is baked for a short period before it is filled. *Blind baking* is a term for baking pie shells 10 min to 15 min before a filling is added. Blind-baked crusts are often used for pies with unbaked fillings, such as chiffon or cream.

**To blind bake a pie shell, apply the following procedure:**

1. Roll dough out to desired size and thickness and place in the bottom of pie pan. ***See Figure 23-23.***
2. Dock the dough with a fork or dough docker by pricking small holes into the pie dough to allow steam to escape. Docking prevents the dough from rising in the oven.
3. Cover the dough with parchment paper and press the paper down against the edges and sides of the dough.
4. Fill the pan with pie weights or dried beans to prevent the crust from rising in the pan.
5. Bake the crust at 350°F for 10 min to 15 min, or until light golden brown.
6. Cool and fill with desired filling.

*CULINARY PROCEDURES*

---

*Blind Baking Pie Shells*

*1. Roll out dough to desired thickness and place in pie pan.*

*2–5. Dock dough with fork or dough docker; cover dough with parchment paper and fill pie shell with weights or dried beans; bake crust at 350°F.*

*6. Let the crust cool and then fill with desired filling.*

**23-23** *Blind baking is a term for baking pie shells before a filling is added.*

## PIE FILLINGS

For best results, pie fillings must meet the same high quality standards as pie crusts. A pie filling must be thickened to the proper consistency and flavored and seasoned appropriately. In addition to thickness and flavor, the appearance of the filling is also important. Fillings for dessert pies are commonly divided into four types: fruit, cream, chiffon, and soft fillings. Other fillings, such as ice cream and nesselrode (custard and fruit filling flavored with rum and chestnuts), are considered specialty fillings.

Filling the pie shell with a uniform amount of filling is an essential step to produce a consistent product. To accomplish this, the proper amount of filling required for each pie is determined by weight. The prepared pie shell is placed on a baker's scale and the scale is balanced. The baker's scale is then set for the required amount of filling. Filling is then added to the pie shell until the scale balances a second time. This procedure is repeated for each pie.

### Fruit Fillings

The most popular type of pie filling is fruit filling. ***See Figure 23-24.*** Fruit used in fruit fillings can be frozen, canned, fresh, or dried. Each form of fruit requires different preparation for use in a pie filling. A fruit filling recipe used should state whether the fruit is frozen, canned, dried, or fresh.

*Fruit Pies*

Wisconsin Milk Marketing Board

**23-24** *Fruit fillings can be made from fresh, dried, frozen, or canned fruits.*

**Frozen-Fruit Fillings.** Frozen fruit is the most common type of fruit used in commercial pie fillings. Frozen fruit will have the same flavor characteristics and appearance as fresh fruit, with the major difference being that it is available year-round. Frozen fruit is produced when fruit is at peak freshness and ripeness. The fruit is cleaned, cut into the appropriate size, blanched, and frozen immediately. Fruit is frozen either in raw form or parboiled (partly cooked) as soon as possible after picking. The fruit is mixed with natural juices and sugar, and in some cases additional color, and then packed in plastic tubs, buckets, or vacuum-sealed plastic and flash frozen. Frozen fruit is commonly available in 30-lb buckets, although smaller amounts (6½-lb and 10-lb containers) are sometimes available.

Frozen fruit must be completely defrosted before it is used in a pie filling. The best method for defrosting frozen fruit is to place the unopened container in the refrigerator. Defrosting usually takes about one day. To defrost faster, an opened container of fruit can be set in a hot water bath and stirred. When defrosting frozen fruit using a hot water bath, the fruit must be constantly stirred to ensure the fruit is completely defrosted before use. After defrosting, the fruit is strained to remove the juice. The strained juice can be reserved and then thickened, with sugar added if desired, and added back to the fruit. The fruit must be completely defrosted or it bleeds (continues to release juice) and causes the filling to separate. Thickening the juices and adding them back to the fruit prevents the fruit from bleeding.

**Canned-Fruit Fillings.** Canned fruit is commonly used in pie fillings because it is available year-round and the cans (No. 10) are easy to store. Canned fruit can be purchased packed in water or syrup or in a solid pack. A *solid pack* is a canned product, such as fruit, with little or no water added. A water or syrup pack contains less fruit and a higher percentage of juice and sugar than a solid pack. A solid pack is preferred for pie fillings because it has more fruit and less sugar. This permits more sugar to be added after the juice is thickened, which produces better results.

**Dried-Fruit Fillings.** Dried fruit, such as apricots, apples, and raisins, is occasionally used for pie fillings. Dried fruit must be soaked in water to restore natural moisture that was removed in the drying process. In some cases, the liquid and fruit may be brought to a boil. Boiling restores moisture to the fruit, making it soft and plump. The dried fruit is then soaked as it cools. After soaking, the liquid is drained from the fruit, thickened, flavored, and poured back over the fruit.

**Fresh-Fruit Fillings.** Fresh fruit is the best choice for achieving the best flavor. Nevertheless, fresh fruit is used less often for pie fillings than frozen or canned fruit because frozen or canned fruit is more convenient and often costs less. Fresh fruit requires more preparation time than frozen or canned fruit. However, when a fruit is in season, it may be just as economical to purchase fresh as it is when frozen or canned.

The amount of water and sugar to use in a fresh fruit filling is based on the amount of fresh fruit used, the type of fruit, and the natural sweetness of the fruit. Usually, 65% to 70% water, based on the weight of the fresh fruit being used, is sufficient. For example, if 10 lb of fruit is used, 6½ lb to 7 lb of water is needed.

Fresh-fruit fillings can be prepared using the cooked-fruit method or the cooked-juice method. The cooked-fruit method is recommended when firmer fruits require some cooking before being used as a filling.

Chef's Tip:

Although fresh apples are firm, they are seldom cooked using the cooked fruit method. Fresh apples are commonly sautéed briefly with butter. When the apples have softened slightly, they are tossed with sugar and a bit of lemon juice.

Most fresh-fruit fillings (except those made with fresh berries) are prepared using the cooked-fruit method. Dried fruits, such as raisins, currents, or figs used as a filling, are also prepared using the cooked-fruit method. These fruits need to be softened by cooking before being baked in a pie shell.

To prepare a fruit filling using the cooked-fruit method, apply the following procedure:

1. Place fruit, sugar, and a small amount of juice in saucepan with the desired or required spices. Bring to a boil. *See Figure 23-25.*
2. Dissolve starch in cold water and pour slowly into the boiling fruit and juice mixture while stirring constantly.
3. Bring the mixture back to a boil and cook until clear.
4. Add salt, and color (if desired), and stir until thoroughly blended.
5. Cool slightly and pour the filling into unbaked pie shells.

*Cooked Fruit Method*

1. Bring fruit, sugar, and a small amount of juice to a boil.

2–3. Add dissolved starch and boil until clear.

4–5. Add salt and color and stir; cool slightly and pour into unbaked pie shells.

**23-25** *The cooked fruit method is recommended for rhubarb, apple, and cranberry-apple pies.*

The cooked-juice method is recommended when using fruit that is slightly soft, such as frozen fruit and fresh ripe berries, as only the juice from the fruit is cooked. Frozen fruit is slightly cooked prior to freezing, which results in a slightly softer flesh. If it were cooked again, it would become too soft and would smash or break apart easily. Fresh berry fillings prepared using the cooked-juice method will usually have a small amount of the berries cooked to soften, puréed, and then thickened with a starch. The resulting gel is then added to the remainder of the fresh berries.

CULINARY PROCEDURES

To prepare a berry fruit filling using the cooked-juice method for fresh berries, apply the following procedure:

1. Clean and remove leaves and stems from berries.
2. Separate about 20% of the berries and purée until smooth.
3. Place the puréed berries in a saucepan and bring to a simmer. **See Figure 23-26.**
4. Mix the starch, sugar, and salt with the appropriate amount of water to result in the desired volume of filling. Slowly add the starch mixture to the simmering fruit purée.
5. Continue to simmer and stir until the mixture thickens.
6. Carefully pour the thickened mixture over the remainder of the fresh berries and stir gently to avoid crushing or breaking the fruit.

## Cooked Juice Method

*1–4. Bring puréed berries to simmer; add starch mixture to fruit purée.*

*5. Simmer and stir until mixture thickens.*

*6. Pour syrup over drained fruit and stir gently; cool and pour into unbaked pie shells.*

**23-26** *The cooked juice method is recommended for cherry, blueberry, peach, apricot, and blackberry pies.*

### Cream Fillings

Cream fillings are simple to prepare, but care must be taken to achieve a smooth, flavorful filling. The most popular cream pies are chocolate, vanilla, coconut, butterscotch, and banana. After the filling is prepared, it is placed in a prebaked pie shell and topped with meringue or a cream topping.

To prepare a cream filling, apply the following procedure:

1. Place milk in the top of a double boiler and heat. **_See Figure 23-27._**

2. In a separate container, beat eggs and add sugar, salt, and starch or flour. Add cold milk while stirring constantly until a thin paste forms.

3. Add the scalding milk to the thin paste, whipping constantly until the mixture thickens and becomes smooth. _Note:_ It is important to beat the filling vigorously once the starch or flour starts to thicken. Not doing so will cause the filling to become lumpy.

4. Cook until starch or flour is completely incorporated. Remove from heat. _Note:_ Undercooking will cause a cream filling to have a floury or starchy taste.

5. Add flavoring and butter or shortening.

6. Pour into prebaked pie shells and let cool.

7. Top with meringue or whipped topping and serve.

*CULINARY PROCEDURES*

## Preparing a Cream Filling

1–2. Beat eggs, sugar, salt, and starch or flour; add cold milk to form thin paste.

3–4. Add scalding milk to thin paste; cook until starch is incorporated; add flavoring and fat.

6–7. Pour into prebaked pie shells and cool; top with merigue or whipped cream.

**23-27** Cream fillings are poured into baked pie shells and topped with meringue.

## Soft Fillings

A _soft filling_ is a pie filling that is uncooked and is baked in an unbaked pie crust. Soft-filling pies are the most difficult pies to make. The difficulty lies in baking the filling and crust to the proper temperature without overbaking or underbaking either part. Common soft-filling pies include pumpkin pie, custard pie, and pecan pie. Common mistakes made when making soft-filling pies include the following:

- The crust is soggy. Roll out the pie dough on graham cracker crumbs instead of flour or sprinkle some crushed cake crumbs across the inside bottom of the pie shell prior to filling.

- The filling is runny or separates. Use precooked or pregelatinized starch instead of cornstarch to bind the filling.

- The filling is watery. Use an egg-white stabilizer (¼ oz per quart of filling) or tapioca flour (1 oz per quart of milk) to bind the filling and improve the appearance of the finished product.

Generally, soft pies are baked at 400°F for the first 10 min to 15 min of the baking time. The temperature can be reduced to 325°F or 350°F for the remainder of the time. The pie is removed from the oven as soon as the filling sets. This helps to set the bottom crust and cook the bottom of the soft filling before it can make the bottom crust soggy. To check the filling for doneness, use either of the following methods:

- Insert a knife 1″ from the center. If the knife comes out clean, the filling is done.

- Gently shake the pie. If the filling is no longer liquid, the pie is done. The center may move slightly, but carryover heat continues to cook the filling after the pie is removed from the oven.

## Chiffon Fillings

A *chiffon filling* is a light, fluffy filling prepared by folding (blending one mixture over another) a meringue into a fruit or cream pie filling. In most cases, a small amount of plain gelatin is added to the fruit- or custard-based filling to help the chiffon filling set when cooled.

### To prepare a chiffon filling, apply the following procedure:

1. Prepare either a fruit or cream filling. ***See Figure 23-28.***
2. Soak plain gelatin for 5 min in cold water. Add gelatin to the hot fruit or cream filling, stirring until the gelatin is thoroughly dissolved.
3. Place the filling in a fairly shallow pan and let cool. Refrigerate until the filling begins to set. *Note:* If the filling starts to set completely it will be difficult to fold in the egg whites uniformly. Stir the filling occasionally while cooling so it cools uniformly.

4. Prepare a meringue by whipping egg whites and sugar together until the mixture forms stiff peaks.

5. Fold the meringue into the jellied fruit or cream mixture gently, preserving as many of the air cells as possible. This step should be done quickly so that the gelatin does not set before the folding is finished.

6. Pour the chiffon filling into a baked pie shell and refrigerate until set.

7. Top with whipped cream or whipped topping.

**Chef's Tip:**
It is important to not use too much gelatin in a dessert, as the resulting dessert will have a rubbery, unpleasant texture. A dessert with the right amount of gelatin will be just firm enough to retain its shape while still being tender and smooth.

## Preparing Chiffon Filling

1–3. Prepare a cream or fruit filling; soak plain gelatin in cold water; add to filling; cool, stirring constantly.

4–5. Prepare a meringue; fold meringue into filling.

6–7. Pour into prebaked shells; refrigerate until set; top with whipped cream or whipped topping.

**23-28** *Chiffon filling is a gently folded mixture of fruit or cream filling and meringue.*

## MERINGUES

Meringues are most often used to top pies. A *meringue* is a mixture of egg whites and sugar. Meringues can be soft or hard, depending on the ratio of sugar to egg whites. The greater the amount of sugar in proportion to egg whites, the harder (more stable) the resulting meringue. A soft meringue typically has a 1:1 ratio, one part sugar to one part egg whites, while a hard meringue has a 2:1 ratio, two parts sugar to one part egg whites. Soft meringues are often used as a dessert topping, such as lemon meringue pie.

If a meringue-topped pie is to be held for some time prior to eating, a hard meringue with a ratio of 1½:1 (1½ parts sugar to 1 part egg white) would work best as it will not weep. Hard meringues are used to make

crisp shells or disks, such as dacquoise disks, that hold soft desserts. The following tips are helpful for making meringues:

- Egg whites must be beaten in grease-free bowls. Fat of any kind, especially egg yolk, prohibits egg whites from whipping and increasing in volume. This occurs because fats coat the protein molecules in the egg white and prevent them from taking on adequate air and forming stable bubbles (foam). Instead of whipping properly, fat-coated protein molecules expand slightly with air and then collapse immediately. Because of this, it is important that all items (e.g., bowls, whisks) used in making a meringue are clean and free of residue from fats, including oils or egg yolks.

- Allow egg whites to stand at room temperature for at least 30 min before whipping. Egg whites at room temperature whip up faster.

- Add cream of tartar at the beginning of the beating stage. The acid in cream of tartar helps stabilize and add volume to egg whites.

Three common types of meringue are common (French) meringue, Swiss meringue, and Italian meringue. *See Figure 23-29.* These three types of meringue are used in a variety of desserts. A common (French) meringue is made simply by whipping egg whites with a bit of acid (lemon juice or cream of tartar) and gradually adding sugar until desired stiffness is achieved.

| Types of Meringue | | |
|---|---|---|
| Type | Preparation | Stability |
| Common | Egg whites are whipped with an acid (lemon juice or cream of tartar) and sugar is gradually added | Least stable |
| Swiss | Egg whites and sugar are warmed over a hot water bath and then beaten to desired stiffness | Moderately stable |
| Italian | Sugar and water are heated to 240°F, cooled to 220°F, poured into egg whites, and whipped to desired stiffness | Most stable |

**23-29** *Three common types of meringue include common (French), Swiss, and Italian.*

**Swiss Meringue.** A Swiss meringue is made by first gently warming the egg whites and sugar over a hot water bath and then beating them until desired stiffness is achieved. Swiss meringue is a more difficult meringue to make than common meringue. After the whites and sugar have been heated in a double boiler or bain marie to about 110°F, the sugar dissolves thoroughly, making this meringue more stable and less prone to weeping. However, if the egg white mixture becomes too hot, the egg whites will not whip properly, resulting in a runny meringue.

**Italian Meringue.** An Italian meringue is made by heating a mixture of sugar, water, and a touch of cream of tartar to a temperature of 242°F. This mixture is then cooled to 220°F and poured in a slow, steady stream into egg whites while beating continuously. Italian meringues are most commonly used in making buttercream frosting. The cooking of the egg whites with the very hot sugar mixture make this meringue the most stable. However, cooking the whites also results in a taffy-like texture that is not as light and airy as the texture of a common or Swiss meringue, making it less desirable on its own.

## PASTRIES

Pastries are commonly thought of as something sweet made with a light, flaky bread or crust. However, pastries can also be savory dishes depending on the ingredients used. Once a chef understands the procedure for making pastry items, many different sweets, hors d'oeuvres, and savory dishes can be prepared. The word pastry comes from the word "paste," which means a mixture of flour, liquid, and fat, and typically refers to different types of products made from pastes. The most common types of pastries include basic puff pastry (pâte feuilletée), éclair paste (pâte à choux), strudel, and phyllo. ***See Figure 23-30.***

*Pastries* _____

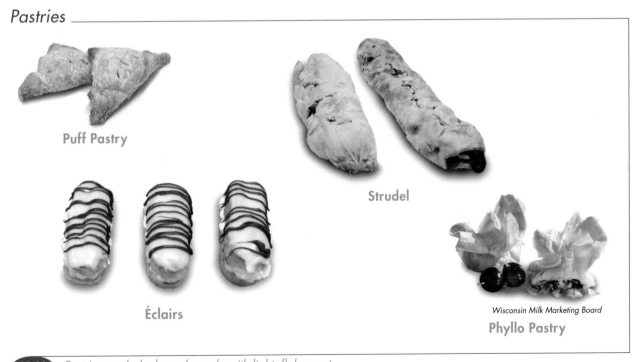

Puff Pastry

Éclairs

Strudel

Wisconsin Milk Marketing Board
Phyllo Pastry

(23-30) *Pastries are baked goods made with light, flaky crust.*

## Puff Pastry (Pâte Feuilletée)

A *puff pastry* is a type of dough that does not contain sugar or a leavening agent but increases greatly in size when heat is applied. The French term for puff pastry is *pâte feuilletée* (pronounced paht foo-yee-TAY). Puff pastry dough is made by rolling and folding alternate layers of fat and dough, in the same manner as used to make Danish dough or croissant dough. While Danish and croissant doughs use yeast to make the dough rise, puff pastry does not contain a leavening agent. Instead, puff pastry rises as a result of the steam created when moisture in the layers of dough is released during baking.

## Éclair Paste (Pâte à Choux)

*Éclair paste* is a dough made from beating eggs into a paste of boiled water, butter, and flour. It is used to make cream puffs, éclairs, beignets, and profiteroles. Éclair paste is also known as *pâte à choux* (pronounced paht ah shoo). The high egg content in éclair paste makes the dough rise and gives it a distinct character that makes it easily distinguishable from any other type of pastry. Cream puffs are round pastries filled with sweetened whipped cream. Éclairs are oblong pastries that are filled with pastry cream and topped with a chocolate glaze. Beignets are strips of éclair paste that are sweetened with sugar, deep-fried, and dusted with powdered sugar. Profiteroles are round pastries filled with ice cream or other savory or sweet fillings.

**Action Lab**

Make one recipe of pâte à choux using whole milk and another recipe using water. Evaluate the results. The pâte à choux with the milk will have a softer and more delicate crust than the batch made with water, which will be crisper and crumblier.

### To make éclair paste, apply the following procedure:

1. Combine the liquid, fat, salt, and sugar (if desired) and bring to a boil. *Note:* Ensure the liquid reaches a rapid boil in order for the fat to disperse properly. ***See Figure 23-31.***
2. Add flour and stir until the paste pulls away from the sides of the pan and forms a ball.
3. Allow the paste to rest and cool for a few minutes. This ensures the eggs do not immediately cook when added to the paste.
4. Add the eggs one at a time, vigorously beating after each addition. If the eggs are added too quickly, the batter may not be as smooth as desired. *Note:* The eggs might not all be used, depending upon the moisture content of the flour and the size of each egg. Stop adding eggs when the dough begins to fall away from the beaters.
5. Fill a pasty bag with paste and pipe the paste into desired shapes onto parchment-lined pans.
6. Bake the paste until completely dry. *Note:* If it is not completely dry when removed from the oven, the pastry may collapse.
7. Fill with desired filling and serve as soon as possible. *Note:* Filled éclairs should be eaten within a few hours or they become soggy.

CULINARY PROCEDURES

1. Combine liquid, fat, salt, and sugar; bring to a boil.

2–3. Add flour and stir until paste forms a ball; allow the paste to rest and cool for a few minutes.

4. Add eggs one at a time, beating vigorously after each addition.

5. Pipe in desired shapes on parchment-lined pans.

6–7. Bake until completely dry; fill with desired filling and serve.

**23-31** *Éclair paste is used to make cream puffs, éclairs, beignets, and profiteroles.*

## Strudel and Phyllo Dough

*Strudel dough* is a dough that is made from high-gluten flour and rolled paper thin. The high gluten content of the flour results in an elastic dough with many flaky layers. Strudel dough is rolled out, filled with a sweet or savory filling, and then rolled up. *Phyllo dough* is a dough consisting of very thin sheets of unleavened dough. Each sheet is buttered and layered to produce a pastry consisting of many thin, flaky, buttery layers. Phyllo is a staple in Greek and Turkish cuisine. Many strudels are even made with phyllo dough.

It takes an extremely skilled chef to make phyllo dough from scratch. It is more common to purchase commercially made phyllo dough for use in recipes calling for strudel or phyllo dough. Commercially made phyllo dough is available frozen and must be thawed before use. Phyllo dough must be kept moist either by brushing each layer with butter or covering the dough with a damp cloth. Working with one sheet at a time is best as the dough dries out in minutes.

## SPECIALTY DESSERTS

Custards, soufflés, and frozen desserts are all types of specialty desserts that are not classified as cakes, pies, or cookies. Ice creams and sherbets are very popular and can be made in a variety of flavors. They can be served alone or with cake, cookies, or other dessert items. Specialty desserts using ice cream combined with fruits, fruit sauces, and liqueurs create eye-appealing parfaits and coupes. Flambés (flaming desserts) are ignited tableside for the customer's enjoyment.

### Custards

A *custard* is a liquid that is thickened by cooking and coagulating egg proteins that are part of the mixture. There are three varieties of cooked custards: baked, souffléed, and stirred.

**Baked Custards.** Baked custards (also called baked puddings) are most often baked in a water bath at a temperature of 275°F–300°F until the custard has set, but not completely cooked. Custard continues to cook after it has been removed from the oven. A water bath is used to prevent custard from drying out during cooking. Baked custard is generally a solid form that retains the shape of its container and is firm enough to slice. Another method used to prepare baked custard is to cook it in a steamer. Baked custards include desserts such as cheesecake, bread pudding, rice pudding, crème brûlée, and crème caramel.

- *Cheesecake* is a variety of baked custard that most commonly contains cream cheese. Cheesecake is typically made with a graham cracker or cookie crust in a springform pan. There are two common varieties of cheesecake available. A creamy version known as Italian cheesecake is light, fluffy, and creamy, while a drier, denser, and richer filling is common for a New York cheesecake. Other varieties of cheesecake include unbaked versions that are not custard-based desserts. Unbaked cheesecakes rely on gelatin to set the ingredients and make the cheesecake firm.

- *Bread pudding* is a baked custard that is made by pouring a custard mixture over bread chunks and baking it in the oven. Bread pudding can be sweet, where pieces of day-old Danish or sweet rolls are used and the custard is flavored with cinnamon, nutmeg, or other sweet spices and baked. It can also be savory, where sugar is omitted from the custard and savory ingredients such as herbs and meats are added to the bread chunks and baked.

*Planet Hollywood International, Inc.*

- *Rice pudding* is a baked custard made from cooked rice combined with a sweet custard and often dried fruits such as raisins or currants.

- *Crème brûlée* is a baked custard that still possesses the texture of a spoonable stirred custard beneath the surface. Here a sweet custard is poured into an oven-proof, individual-size casserole dish and baked in a water bath. After it has set, it is removed and cooled completely. When needed for service, crème brûlée is topped with granulated sugar, which is caramelized with a pastry torch before serving. The caramelized sugar produces a thin, glasslike candy covering over the surface of a creamy, spoonable custard. A crème brûlée can be made savory with the addition on spices, herbs, or other savory ingredients. Crème brûlée is served in the dish (often a ramekin) in which it was prepared.

- *Crème caramel,* also called flan, is a baked custard in which a small amount of hot caramelized sugar is poured into a small serving dish. The sweet, uncooked custard is poured into the dish on top of the caramelized sugar. It is then baked in a water bath and removed when set. To serve, the custard is unmolded and served upside down on a plate so that the caramel sauce runs down over the baked custard. ***See Figure 23-32.***

## Crème Caramel

**23-32** *Crème caramel (flan) is served upside down so the caramel sauce can run down over the baked custard.*

**Steamed Puddings.** Steamed puddings are similar to baked custards except that they are cooked using the steaming method. This moist-heat cooking method keeps steamed puddings from drying out as happens occasionally with a baked custard. Steamed puddings are typically made with fruit, suet (animal fat), flour, eggs, bread crumbs, baking soda for leavening, and brown sugar or molasses. The use of a dark-colored sweetener gives these puddings a characteristic dark appearance. Steamed puddings are highly flavored with such spices as ginger, mace, nutmeg, and allspice. Rum, brandy, or both are used to provide aroma and flavor.

Steamed puddings can be prepared in large or individual metal containers that are covered in foil and placed in a steamer or by baking in a water bath in a 350°F oven covered with a damp cloth for approximately 2 hr to 3 hr. Another method of preparing steamed puddings is to place the pudding in a damp muslin cloth dusted with flour, tie the ends of the cloth loosely to allow for expansion, and suspend it just above hot water to cook the pudding with steam. Steamed puddings have a heavy texture and are commonly served hot with a warm sauce that complements the pudding's flavor and color.

**Soufflés.** Soufflé custards are commonly referred to as soufflés. Although soufflés are baked in the oven, they are quite different from other baked custards. Sweet soufflés are prepared by using a cooked custard base (pastry cream) that is combined with different flavors such as chocolate, vanilla, or fruit. This base is then lightened with egg whites that have been whipped with a portion of sugar and carefully folded in. It is important that the flavored custard base and the whipped egg whites both be at room temperature before mixing. Egg whites whip better when not chilled, and having them at the same temperature as the base allows the base and the whites to mix more easily.

## Soufflés

Daniel NYC

**23-33** *Soufflés are delicate desserts that must be served before they collapse.*

This light and fluffy soufflé base is then put into straight-side ramekins that have been buttered and coated with granulated sugar all the way to the rim. As they bake, soufflés grow tall from the air expansion in the egg foam when heated, similar to cakes. However, soufflés are not stable like cakes and easily collapse once removed from the oven. For this reason, soufflés can only be made to order and served immediately. *See Figure 23-33.*

When preparing a soufflé, the beaten egg whites must be folded into the custard base gently and baked carefully to prevent the soufflé from becoming heavy and soggy. A properly baked soufflé rises high above the rim of the ramekin and should have a golden brown surface. If any part of the interior or rim of the ramekin is not buttered and sugared, the soufflé sticks and does not rise properly. To serve, sweet soufflés are commonly split open at the table and served with a slightly beaten cream that is flavored with a liquor or fruit purée.

To prepare individual soufflés, apply the following procedure:

1. Prepare ramekins in advance by buttering the interior and rim of each dish completely and then thoroughly coating the buttered surface in granulated sugar. *Note:* If there is any spot that is not completely buttered and sugared, it will prevent the soufflé from rising.

2. Mix desired flavoring base into prepared pastry cream.

3. Begin whipping egg whites. When soft peaks have formed, begin adding sugar slowly while continually whipping until all the sugar has been added and stiff peaks have formed.

4. Carefully fold approximately ⅓ of the whipped egg whites into the soufflé base and mix well to incorporate and lighten the base. Continue to fold in the remaining whipped whites and mix gently to prevent a loss of volume.

5. Pour the soufflé mixture into the prepared ramekins, filling to about ¼″ from the top and bake immediately.

6. Remove soufflés when they have risen well above the surface of the ramekin and turned a golden brown. Serve immediately.

**Stirred Custards.** Stirred custards can be creamy and thick but pourable, or they can be thicker and heavier but still spoonable. Pourable stirred custards are often served as a sauce for a dessert. Stirred custard that is thick and spoonable may be referred to as pudding.

Stirred custards need to be cooked carefully to avoid unintentionally scrambling the eggs from heating to too high a temperature. When cooked alone, egg whites begin to coagulate (become solid) at 145°F, and yolks coagulate at 155°F. When additional ingredients such as sugar or starch are added to eggs, the temperature at which the eggs coagulate increases. This means that an egg and sugar or starch mixture can be heated to a temperature above 155°F before starting to coagulate. Two stirred custards every chef should know how to prepare are crème anglaise and pastry cream. Sabayon is another popular stirred custard made using a slightly different procedure than the other two.

**Crème Anglaise.** *Crème anglaise* is vanilla custard sauce that is made by whisking in sugar in two stages, half into whisked egg yolks and the other half into scalded milk or cream. Crème anglaise should be made in a non-reactive saucepan or stainless steel double boiler to prevent the eggs from scrambling or curdling. The scalded milk and sugar mixture is then whisked rapidly in three intervals into the sugar and yolk mixture in order to temper the yolks and prevent curdling. The mixture is then put back on the stove and heated until thick and smooth. Crème anglaise is fully cooked when it reaches 175°F and should be nappe, or able to smoothly coat the back of a spoon. Crème anglaise is well known for its use as a beautiful vanilla custard sauce served to accompany many dessert dishes. Because this stirred custard sauce is rich and silky smooth, it freezes well.

## To prepare crème anglaise, apply the following procedure:

1. Place milk (and cream or half-and-half, if specified in recipe), half of the required amount of granulated sugar, and salt in a nonreactive saucepan and bring just to a scald to completely dissolve the sugar. *Note:* If a whole vanilla bean is to be used, it can be split and scraped, and the seeds and pod can be added to the milk in this step. If an extract is to be added, it can be added just prior to straining.

2. In a stainless steel mixing bowl, whisk egg yolks and remainder of granulated sugar until light and airy.

3. Temper the yolks by adding about ⅓ of the hot milk mixture to the yolk and sugar mixture while whisking rapidly.

4. When milk and yolks are mixed well, add another ⅓ of the hot milk mixture to the yolks while whisking, and repeat with the remainder of the milk.

5. Pour the complete mixture back into the sauce pan and continue to cook while whisking until the mixture reaches 175°F or coats the back of a spoon.

6. Add vanilla extract (or another type of extract for additional flavoring).

7. Strain the sauce into a stainless steel bain marie to remove any lumps and immediately place in an ice bath until cooled completely. Keep cold until needed.

CULINARY PROCEDURES

**Pastry Cream.** *Pastry cream* is a custard sauce made with egg yolks, sugar, milk, and a starch (usually cornstarch or flour). Unlike crème anglaise, pastry cream can be brought to a boil and simmered because the starch protects the egg yolks from curdling. Boiling allows the starch to fully gelatinize and removes any raw starch taste from the finished product. A common use of pastry cream is to fill a tart shell and then top it with sliced fresh fruit. It is also commonly used as a filling for napoleons, cream puffs, and éclairs, as well as many other sweet pastries. When properly prepared, it should have the consistency and appearance of a smooth pudding.

To prepare pastry cream, apply the following procedure:

1. Place ¾ of the required milk, half of the required sugar, and the salt in a nonreactive saucepan and bring just to a scald in order to completely dissolve the sugar. *Note:* If a whole vanilla bean is to be used, it can be split in half and scraped, and the seeds and pod can be added to the milk in this step. If an extract is to be added, it can be added just prior to straining.

2. Place yolks and half of the required amount of sugar in mixer. Using the whip attachment, whip on high speed until mixture turns pale yellow and thick.

3. Whisk flour or cornstarch into yolk mixture and mix well. Add remaining ¼ of the required milk and mix well until smooth.

4. Pour about ⅓ of the hot milk mixture into the yolk mixture while whisking continuously to temper the yolks.

5. Add the tempered yolk mixture back to the saucepan of milk while stirring rapidly. Return to the heat while stirring constantly.

6. Bring mixture to a boil while stirring vigorously. *Note:* As the mixture approaches boiling, it will begin to appear lumpy. Don't be alarmed. As soon as it reaches the boiling point, the continuous whisking will break apart the lumps and the pastry cream will appear smooth. Continue to whisk and boil for 1 min.

7. Remove the cooked pastry cream from the heat and place in a stainless steel bain marie or mixing bowl. Carefully fold in the softened butter, using caution to not overmix the pastry cream as doing so would cause thinning.

8. Cover pastry cream immediately with plastic wrap and place in an ice bath to cool completely. Keep cold until needed. *Note:* When covering pastry cream with plastic wrap, the plastic should be pressed down to make contact with the surface to prevent the pastry cream from forming a skin.

**Sabayon.** Sabayon is actually a foam, but is often grouped under stirred custards as it is made in a similar manner. Sabayon is prepared by whisking the sugar and yolks over a double boiler until they are whipped into a thin, silky foam. Champagne or a high-quality sweet wine, such as Marsala, is commonly added to increase the flavor of the sauce. Heating the eggs while whisking cooks them and makes the resulting foam stable enough to retain a shape. Sabayon is commonly served as a sauce and can be served warm or cold. A classic use for sabayon is over a dish of fresh berries, as the rich, sweet sauce is balanced by the acidic quality of the berries.

## To prepare sabayon, apply the following procedure:

1. Place sugar, yolks, salt, and champagne or Marsala wine in a mixer and whisk until light colored and thick.

2. Remove bowl from mixer and place over a pot of simmering water. Whisk continuously until mixture is thick and much lighter colored and forms a ribbon. At this stage, the mixture should have a temperature of approximately 175°F.

3. Serve immediately while warm, or return the bowl to the mixer and whip until completely cooled.

4. When cool, remove the sabayon and place in a nonreactive bowl covered with plastic wrap pressed against the surface to prevent a skin from forming.

CULINARY PROCEDURES

## Creams

Creams are light and airy desserts where whipped cream or whipped egg whites have been folded in to increase volume. Some of these have gelatin added to aid in retaining their shape. Cream desserts include Bavarian creams, chiffons, and mousses.

**Bavarian.** A Bavarian (or Bavarian cream) is a flavored custard sauce that has been stabilized with gelatin while still warm. After it has almost completely cooled, whipped cream is folded in. The mixture is then placed in a mold to set before being served. Too much gelatin in a Bavarian could make it overly firm and rubbery, while too little gelatin could make it too soft and it could lose its shape after being removed from the mold.

Once a Bavarian has set completely and is firm, the solid portion of the Bavarian mold is dipped in warm water for a few seconds to loosen the Bavarian from the mold. The mold can then be turned over onto a plate. A Bavarian is commonly sliced to order. Caution should be taken with highly acidic fruits such as pineapple because a high acid content weakens the strength of gelatin. When a high-acid fruit is used, the recipe may require more gelatin to set properly.

**Chiffon.** A chiffon is similar to a Bavarian, but instead of folding in whipped cream, whipped egg whites are folded into the cooled custard sauce. Chiffon can be placed into a mold but is most often used as a pie filling, as in a lemon chiffon pie, or folded into cakes such as an Italian cream cake.

**Mousse.** A mousse is similar to a chiffon and a Bavarian in the way that it is produced. A mousse is lightened with whipped cream, whipped egg whites, or both. The main difference is that a mousse contains only a small amount of gelatin, if any at all. A mousse can be served as a dessert on its own or can be used as a cake filling.

## Frozen Desserts

Frozen desserts are a popular dessert choice. They may be served alongside or on top of other dessert dishes or eaten as an individual dish. Common frozen dessert items used in food service include ice creams, gelati, sorbets and sherbets, parfaits, bombes, sundaes, and flambés.

**Ice Cream.** Ice cream can be the main attraction for a dessert course or can be a wonderful dessert accompaniment, as with pie *à la mode* (served with a scoop of ice cream). *See Figure 23-34.* There are many different flavors of ice cream available commercially, but the range of flavors that a chef can produce is limited only by his or her imagination. Many chefs are combining flavors such as pistachio and white chocolate, ginger and lemongrass, or banana and coconut. Many ice creams also contain sweet or savory additions such as chocolate chips, peanut butter, nuts, pretzels, or caramel. Whatever the flavor, it is important that the chef has an understanding of how the ingredients of an ice cream base interact with each other.

*Pie à la Mode*

**23-34** *Pie slices are often serve à la mode (with a scoop of ice cream).*

There are two general types of ice cream base. The first incorporates eggs to make a crème anglaise ice cream base and produces a rich and creamy dessert. The second variety is a simple mixture of milk, cream, sugar, and flavorings that are heated to dissolve the sugar and then chilled prior to churning. Whichever ice cream base is used, it is important to refrigerate it for at least a few hours to allow excess moisture to be absorbed and bound to the sugar. Doing so results in fewer ice crystals being formed during the freezing process.

Ice cream increases in volume and becomes lighter in texture when air is churned into it. *Overrun* is the increase in volume of ice cream or other frozen products as a result of the incorporation of air during churning and freezing. Overrun of 100% would mean that an ice cream doubled in size during churning and freezing and contains 50% air by volume. The FDA allows a maximum of 100% overrun, although better-quality ice cream generally has less. Too much overrun makes ice cream feel frothy and thin in the mouth, rather than creamy and smooth. Some premium ice creams sold commercially have overrun of around 20%, meaning that they contain 10% air. A lower overrun gives ice cream a richer mouth feel.

Ice cream should be smooth, creamy, and rich, with enough of the main flavoring ingredient to be flavorful but not overpowering. The proper balance of ingredients works together to form the overall texture and flavor that makes ice cream so popular. For example, eggs add richness to ice cream but also act as emulsifiers in the base. Excess moisture is bound by the lecithin in egg yolk, which helps to prevent ice crystals from forming on the surface and throughout the ice cream. Fewer ice crystals allow for a smoother and creamier mouth feel. *Mouth feel* is the sensation a food or beverage creates in the mouth, other than that created by the item's flavor.

Texture, consistency, and temperature all affect mouth feel. Both the cream and eggs allow for air to be whipped into the product during the churning process, giving the end product a smoother, creamier mouth feel. If air were not whipped into the mixture, it would freeze into a dense, hard ice cube rather than smooth, creamy ice cream.

In addition to adding sweetness, sugar also helps to lower the freezing point of an ice cream base, keeping it from freezing rock hard. However, too much sugar can prevent an ice cream from freezing, and too little sugar results in an icy, unpleasant mouth feel and a product that is difficult to scoop.

**Gelato.** *Gelato* is an Italian version of ice cream, with a creamier, denser texture than that of typical American ice cream. Gelato is churned for a shorter period, incorporating less air and leading to a denser product.

**Sorbet.** *Sorbet* is a refreshing frozen dessert made of fresh fruit and simple syrup. Sorbet is traditionally served before the main course at formal dinners to cleanse the palate of flavors from prior courses. It is churned in an ice cream maker in the same way as ice cream. Sorbet is naturally

lower in calories and fat than ice cream because of the absence of dairy products, especially heavy cream and egg yolks. Many different types of fruit may be used in the making of sorbets. Pears, watermelon, blackberries, and raspberries are a few popular varieties. The fruit in a sorbet is always puréed before being frozen in an ice cream maker.

**Sherbet.** *Sherbet* is similar to sorbet—in fact the word "sherbet" is actually derived from a poor translation of the French word "sorbet." Along with the slight change in the name came a slight change to the recipe for sorbet. Sherbet is a creamier version of traditional icy sorbet, with some milk added to it during the churning process. It is lower in fat than ice cream but, because of the presence of milk, it still has a slightly creamy texture compared to the icy texture of sorbet. Some chefs add a small amount of egg whites to sherbet as another way of giving sherbet a smoother texture. If egg whites are to be used, only pasteurized whites are acceptable, as they will not be cooked.

**Granité.** *Granité* is a frozen dessert made by frequently stirring a mixture of water, sugar, and flavorings, such as fruit juice or wine, as it is freezing. It has a grainy texture, with ice crystals larger than those of a sorbet. Granité is a popular dish that is often used to cleanse the palate between courses or served as a refreshing dessert on a hot summer day.

**Parfaits.** A *parfait* is a dessert prepared by alternating layers of ice cream and toppings, such as a sauce or fruit, and is topped with whipped cream. ***See Figure 23-35.*** Parfaits are often served in a fairly narrow glass. Parfaits can be served immediately or frozen and held for service at a later date. Preparing parfaits in advance is helpful for large group service. The following are preparation instructions for common types of parfaits:

- Parfait Crème de Menthe—Alternate layers of crème de menthe and vanilla ice cream. Garnish with whipped cream, chopped nuts, and a maraschino cherry.

- Rainbow Parfait—Alternate layers of strawberry sauce with vanilla, strawberry, and chocolate ice cream. Garnish with fresh strawberries, whipped cream, and chopped nuts.

- Parfait Melba—Alternate layers of melba sauce with vanilla ice cream. Garnish with fresh raspberries and whipped cream.

- Pineapple Parfait—Alternate layers of crushed pineapple with vanilla ice cream or lemon sherbet. Garnish with whipped cream, chopped nuts, and a maraschino cherry.

- Chocolate Parfait—Alternate layers of chocolate syrup with chocolate or vanilla ice cream. Garnish with hot fudge, whipped cream, and a maraschino cherry.

- Strawberry Parfait—Alternate layers of strawberry sauce with vanilla ice cream. Garnish with fresh strawberries, whipped cream, and chopped nuts.

*Parfaits*

**23-35** *Parfaits are served in parfait glasses to highlight the layers of ice cream and toppings.*

- Butterscotch Parfait—Alternate layers of butterscotch sauce with vanilla ice cream. Garnish with whipped cream, chopped nuts, and a maraschino cherry.

**Bombes.** A *bombe*, or bombe glacée, is a French ice cream dessert where at least two varieties of ice cream are layered inside a spherical mold, frozen, and then unmolded and decorated for service. Some pastry chefs dip the frozen unmolded bombe in tempered chocolate to create a shell over its surface. When three ice cream varieties are layered in a rectangular mold, the item is referred to as a Neapolitan. A Neapolitan is commonly unmolded and sliced for service. The most common combination of ice creams in a Neapolitan is strawberry, chocolate, and vanilla, although any varieties may be used.

**Sundaes.** *Coupe* is the French word for the dessert commonly known as a sundae. A *sundae* is a dessert made with one or more scoops of ice cream or sherbet, along with one or more liqueurs, sauces, or pieces of fruit. ***See Figure 23-36.*** They are often topped with whipped cream, chopped nuts, and a maraschino cherry. Sundaes can be topped with hot fudge, caramel, butterscotch, or any sort of fruit sauce, and quickly made to order. Classic versions of the sundae are as follows:

- Coupe Melba—Vanilla ice cream covered with a peach half, topped with melba sauce, and garnished with whipped cream and a sliced peach.

- Coupe Helene—Vanilla ice cream covered with a Bartlett pear half, topped with chocolate sauce, and garnished with whipped cream and a maraschino cherry.

- Coupe Savory—Mocha ice cream covered with assorted diced fresh fruits flavored with anisette liqueur, and garnished with whipped cream and chopped nuts.

- Strawberry Coupe—Vanilla ice cream covered with fresh strawberries that have been tossed in curaçao liqueur, and garnished with whipped cream and a fresh strawberry.

- Pineapple Coupe—Vanilla ice cream covered with diced pineapple flavored with Kirschwasser, and garnished with whipped cream and a maraschino cherry

**Flambés.** Flambé desserts are flaming desserts that are served tableside. Cherries jubilee and crêpes Suzette are both showcase flambé desserts. These special desserts are made in front of the customer by pouring a liqueur over a fruit sauce that is layered over ice cream or crêpes and setting the liqueur on fire. The flame is extinguished when all the alcohol has been burned off. The flavor of the liqueur remains and blends with the ice cream, which is only slightly melted on the surface. *Note:* A liqueur burns best if it is approximately 100 proof and slightly warm. Liqueur that is labeled "100 proof" contains 50% alcohol by volume. Brandy, rum, and cognac are the liqueurs used most often for flambé because of their high alcohol content and flavor.

*Sundae*

Wisconsin Milk Marketing Board

23-36　Sundaes can be made with a number of fruits, syrups, and nuts to suit individual tastes.

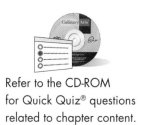

## SUMMARY

Desserts are the last course of a meal and therefore leave the last impression. A well-executed dessert can be the highlight of a meal. Whether preparing cakes, pastries, or desserts, a pastry chef must have a thorough understanding of the basic procedures involved. Mastering the professional techniques involved in the preparation of cakes, pastries, pies, cookies, and specialty desserts enables a chef to produce sweet finales for wonderful meals and dazzle customers with his or her artistic talents and skills.

## Review Questions

1. Identify the most common ingredients used in cakes.

2. Describe the two-stage method of mixing cake batters.

3. Describe the creaming method of mixing cake batters.

4. Describe the sponge method of mixing cake batters.

5. Describe how to test cakes for doneness.

6. Identify seven types of icing.

7. List common materials that a pastry bag can be made from.

8. Describe the icebox method of cookie preparation.

9. What are the disadvantages of using a mixing machine to mix pie doughs?

10. Contrast the characteristics of mealy pie dough and flaky pie dough.

11. Identify the main ingredients of pie dough.

12. Name the main ingredient in pie dough that equals 100% in a baker's percentage.

13. Explain the purpose of blind baking.

14. List four categories of fruit fillings for pies.

15. Describe the main characteristics of a chiffon filling.

16. List three common problems to avoid when making soft-filling pies.

17. How does the ratio of sugar to egg whites affect meringues?

18. List four pastries made with éclair paste.

19. Contrast strudel dough and phyllo dough.

20. Describe the qualities of baked custard.

21. Describe the six-step procedure for making individual soufflés.

22. Describe the main steps for making crème anglaise.

23. Describe two types of ice cream base.

24. How does gelato differ from ice cream?

25. Describe how parfaits are made and served.

# Recipes

## 277

## Carrot Cake

*yield: 3 (8-in round cakes; 12 slices per cake)*

| | |
|---|---|
| 1 lb | sugar |
| 4 | eggs |
| 8 oz | vegetable oil |
| 1 lb 5 oz | all-purpose flour |
| ½ oz | baking powder |
| ¼ oz | salt |
| 1 tsp | cinnamon |
| 1 tsp | nutmeg |
| 1 lb 8 oz | shredded carrots |
| 8 oz | chopped walnuts or pecans; optional |

1. Cream the sugar and eggs in mixer until creamy and smooth.
2. Add vegetable oil and continue mixing until well combined.
3. Add flour, baking powder, salt, cinnamon, and nutmeg. Mix well.
4. Stir in shredded carrots and nuts.
5. Pour into a greased and floured baking pan. Bake in a preheated 350°F oven for 35 min to 40 min, or until cake springs back to the touch.

### Nutrition Facts

**Per slice:** 222.4 calories; 43% calories from fat; 11.2 g total fat; 23.5 mg cholesterol; 139.3 mg sodium; 114.3 mg potassium; 28.1 g carbohydrates; 1.4 g fiber; 13.7 g sugar; 26.7 g net carbs; 3.5 g protein

## 278

## Pound Cake

*yield: 5 (1-lb loaves; 8 slices per loaf)*

| | |
|---|---|
| 1 lb | butter |
| 1 lb | shortening |
| 1 lb | granulated sugar |
| 1 pt | whole eggs |
| ¼ oz | salt |
| ¼ oz | vanilla extract |
| 1 lb 2 oz | cake flour |
| ½ oz | baking powder |
| 1 | lemon, squeezed |

1. Cream the butter and shortening until creamy and smooth.
2. Add sugar, and continue to cream until light and fluffy.
3. Add eggs gradually and continue creaming.
4. Add the salt, vanilla, and lemon juice. Mix well.
5. Gradually stir in sifted flour and baking powder.
6. Scale 1 lb 2 oz of dough into greased loaf pans and bake at 350°F for 45 min to 50 min.

### Nutrition Facts

**Per slice:** 287.2 calories; 67% calories from fat; 21.8 g total fat; 82.1 mg cholesterol; 124.8 mg sodium; 33.3 mg potassium; 20.5 g carbohydrates; 0.2 g fiber; 11.5 g sugar; 20.3 g net carbs; 2.6 g protein

## White Cake

*yield: 1 (9-in double-layer cake; 12 slices per cake)*

| | |
|---|---|
| 9 oz | cake flour |
| 6.3 oz | emulsified shortening |
| 11.25 oz | sugar |
| 0.34 oz | salt |
| 0.56 oz | baking powder |
| 3.15 oz | water |
| 0.56 oz | nonfat dry milk |
| 2.25 oz | eggs |
| 3.6 oz | egg whites |
| 3.6 oz | water |
| to taste | vanilla extract |

1. Mix flour, shortening, sugar, salt, baking powder, water, and dry milk with a paddle attachment in a mixer on slow speed for 5 min. Scrape bowl.
2. Mix eggs, water, and vanilla in a separate bowl. Add half of egg mixture to dry ingredients. Mix at slow speed until smooth. Scrape bowl.
3. Add the rest of the egg mixture and mix at slow speed for 3 min. Scrape bowl again.
4. Pour into greased and floured baking pans. Bake in a preheated 350°F oven for 35 min to 40 min, or until cake springs back to the touch.

**Nutrition Facts**

**Per slice:** 312.2 calories; 40% calories from fat; 14.1 g total fat; 27.5 mg cholesterol; 470.8 mg sodium; 40.8 mg potassium; 43.7 g carbohydrates; 0.4 g fiber; 26.8 g sugar; 43.3 g net carbs; 3 g protein

**Recipe Variations:**

**280 Yellow Cake:** Omit egg whites and adjust the amount of ingredients as follows: 4.95 oz shortening, 0.23 oz salt, 0.39 oz baking powder, 4.5 oz water, 0.9 oz nonfat dry milk, 5.85 oz eggs, 2.7 oz water.

## Angel Food Cake

*yield: 3 cakes (12 slices per cake)*

| | |
|---|---|
| 1 qt | egg whites |
| 1 lb | granulated sugar |
| 12 oz | cake flour |
| 1 lb | powdered sugar |
| ½ oz | cream of tartar |
| ¼ oz | salt |
| ¼ oz | vanilla extract |
| ¼ oz | almond extract |
| 2 | lemons, grated zest |

1. Beat egg whites until foamy. Add cream of tartar and mix at high speed with a paddle attachment.
2. Begin gradually adding granulated sugar, vanilla, and almond extract.
3. Beat until eggs are glossy and stiff peaks form.
4. Sift powdered sugar, cake flour, and salt together.
5. When stiff peaks have formed, fold the remaining ingredients into the egg mixture.
6. Place in 9″ tube pans and bake at 375°F for 30 min to 35 min.
7. Cool upside down for 30 min to 40 min.

**Nutrition Facts**

**Per slice:** 147.6 calories; 0% calories from fat; 0.1 g total fat; 0 mg cholesterol; 121.6 mg sodium; 119.6 mg potassium; 33.1 g carbohydrates; 0.2 g fiber; 25.4 g sugar; 32.9 g net carbs; 3.7 g protein

# Recipes

## Buttercream Icing 282

*yield: 1.25 qt*

| | |
|---|---|
| 5 oz | shortening |
| 5 oz | lightly salted butter, softened |
| 0.25 oz | salt |
| 2.5 oz | nonfat dry milk |
| 2 oz | water |
| ½ tsp | vanilla extract |
| 1 lb 8 oz | confectioners' sugar |

*Carlisle FoodService Products*

1. Place all ingredients in mixing bowl and mix at slow speed using paddle attachment for 5 min.
2. Whip at medium speed for 10 min to 15 min to acquire desired lightness.

### Nutrition Facts
**Per ounce:** 4295.6 calories; 1% calories from fat; 7.2 g total fat; 5.3 mg cholesterol; 90.3 mg sodium; 55 mg potassium; 1085.2 g carbohydrates; 0 g fiber; 1066.8 g sugar; 1085.2 g net carbs; 0.7 g protein

**Recipe Variations:**

283 **Nut Icing:** Add 4 oz chopped nuts.

284 **Raisin Icing:** Add 4 oz ground raisins.

285 **Cherry Icing:** Add 4 oz chopped cherries.

286 **Candied Fruit Icing:** Add 4 oz chopped fruit.

287 **Marmalade Icing:** Add 4 oz marmalade.

288 **Coconut Icing:** Add 6 oz coconut.

289 **Fondant Icing:** Add 1 lb 4 oz fondant.

290 **Peppermint Candy Icing:** Add 4 oz crushed peppermint candy.

291 **Chocolate Icing:** Add 2.5 oz cocoa and 2.5 oz water.

292 **Lady Baltimore Filling:** Add 8 oz chopped cherries, nuts, and raisins.

## Flat Icing 293

*yield: 1 qt*

| | |
|---|---|
| 2 lb 8 oz | confectioners' sugar |
| 4 oz | corn syrup |
| 2 oz | pasteurized egg whites |
| 8 oz | hot water |

1. Place confectioners' sugar, corn syrup, and egg whites in mixing bowl. Mix until smooth at slow speed using paddle attachment while adding hot water.
2. When ready to use, heat the amount of icing needed in a double boiler and use to glaze cakes or petit fours.

### Nutrition Facts
**Per ounce:** 149.3 calories; 0% calories from fat; 0 g total fat; 0 mg cholesterol; 9.3 mg sodium; 4.3 mg potassium; 38.1 g carbohydrates; 0 g fiber; 37.5 g sugar; 38.1 g net carbs; 0.2 g protein

## Foam (Boiled) Icing 294

*yield: 1 qt*

| | |
|---|---|
| 0.5 oz | sugar |
| 6 oz | glucose |
| 8 oz | water |
| 16 oz | egg whites |
| 1 pinch | salt |
| ⅛ tsp | cream of tartar |

1. Place sugar, glucose, and water in a saucepan. Heat to 242°F.
2. Remove from heat and let cool to 220°F.
3. In the meantime, beat egg whites and salt in the mixing bowl of the electric mixer until soft peaks form.
4. When syrup reaches 220°F, begin pouring hot syrup into egg whites slowly in a fine stream while whisking rapidly.
5. Continue whipping until mixture has cooled to room temperature.

### Nutrition Facts
**Entire ounce:** 19.4 calories; 1% calories from fat; 0 g total fat; 0 mg cholesterol; 33.8 mg sodium; 25.1 mg potassium; 3.2 g carbohydrates; 0 g fiber; 3.2 g sugar; 3.2 g net carbs; 1.5 g protein

## Fudge Icing

*yield: 5 lb*

| | |
|---|---|
| 7 | egg yolks |
| 6 oz | sugar |
| 16 oz | heavy cream |
| 1 lb 8 oz | chocolate |
| 10 oz | butter |
| 7 oz | corn syrup |
| 3 oz | sour cream |
| 1½ tsp | vanilla extract |

1. Place chocolate and butter in a large stainless-steel mixing bowl.
2. Combine yolks, sugar, and cream in a saucepan. Heat, stirring constantly until mixture is hot and coats the spoon. *Note:* Do not allow the mixture to boil.
3. Pour mixture over chocolate and butter, and stir until melted.
4. Whisk in corn syrup, sour cream, and vanilla.
5. Cool to spreading consistency.

**Nutrition Facts**

**Per ounce:** 109.5 calories; 64% calories from fat; 8.3 g total fat; 34.6 mg cholesterol; 7.9 mg sodium; 8.7 mg potassium; 9.8 g carbohydrates; 0.5 g fiber; 4.1 g sugar; 9.3 g net carbs; 0.8 g protein

## Chocolate Chip Cookies

*yield: 6 dozen*

*Wisconsin Milk Marketing Board*

| | |
|---|---|
| 1 lb 8 oz | sugar |
| 1 lb | shortening |
| 8 oz | eggs |
| ½ oz | salt |
| ¼ oz | baking soda |
| 1 lb 8 oz | all-purpose flour |
| 8 oz | pecans (optional) |
| ¼ oz | vanilla extract |
| 1 lb 12 oz | chocolate chips |

1. In a mixer with a paddle attachment, cream the shortening and sugar until light and fluffy.
2. Add eggs and vanilla and scrape down mixer. Cream until fluffy.
3. In a separate bowl, mix flour, salt, and baking soda.
4. Add to creamed mixture until fully incorporated.
5. Stir in chocolate chips and (if desired) pecans.
6. Drop cookies onto a parchment-lined sheet pan. Bake at 350°F for 12 min to 15 min.
7. Remove from oven and let cool.

**Nutrition Facts**

**Per cookie:** 207.2 calories; 51% calories from fat; 12.3 g total fat; 16.9 mg cholesterol; 109.1 mg sodium; 27.6 mg potassium; 24.1 g carbohydrates; 1.2 g fiber; 9.6 g sugar; 22.9 g net carbs; 2.1 g protein

## Cream Cheese Icing

*yield: 2 lb*

| | |
|---|---|
| 13 oz | cream cheese |
| 1 lb 2 oz | powdered sugar |
| 1 tsp | vanilla extract |
| 1 oz | milk |

1. Combine all ingredients in mixer. Use a paddle attachment to mix until thoroughly combined and a smooth frosting forms.
2. Add additional milk if frosting needs to be thinned out.

**Nutrition Facts**

**Per ounce:** 95.9 calories; 37% calories from fat; 4 g total fat; 12.8 mg cholesterol; 34.5 mg sodium; 15.5 mg potassium; 14.5 g carbohydrates; 0 g fiber; 14.2 g sugar; 14.5 g net carbs; 0.9 g protein

## 298

## Sugar Cookies

*yield: 6 dozen*

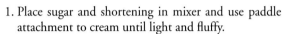

| | |
|---|---|
| 2 lb | sugar |
| 1 lb 8 oz | vegetable shortening |
| ¾ oz | salt |
| 2 lb 12 oz | cake flour |
| 1½ oz | baking powder |
| 8 oz | eggs |
| 8 oz | skim milk |

1. Place sugar and shortening in mixer and use paddle attachment to cream until light and fluffy.
2. Add eggs, scraping down the sides of the bowl, and mix until light and fluffy.
3. Add in flour, baking powder, salt, and milk.
4. Mix until dough forms, scraping down the sides of the bowl as needed.
5. Place cookie dough in refrigerator overnight or for at least 1 hr.
6. Roll out cookie dough onto floured surface and cut out desired shapes.
7. Bake on sheet pan at 375°F for 7 min to 8 min.

### Nutrition Facts

**Per cookie:** 194.3 calories; 41% calories from fat; 9 g total fat; 18.2 mg cholesterol; 183.5 mg sodium; 28 mg potassium; 26.4 g carbohydrates; 0.3 g fiber; 12.8 g sugar; 26.1 g net carbs; 1.9 g protein

## 299

## Icebox Cookies

*yield: 5 dozen*

| | |
|---|---|
| 3 lb 6 oz | confectioners' sugar |
| 2 lb 8 oz | shortening |
| 2 lb | butter |
| 10 | eggs |
| 7 lb 8 oz | cake flour |
| ½ oz | salt |
| 1 | grated lemon zest |

1. Cream sugar, shortening, and butter in a mixer at medium speed using a paddle attachment.
2. Add eggs one at a time, scraping down the sides of mixer as needed, and mix thoroughly.
3. Add flour, salt, and lemon zest. Mix until dough forms.
4. Shape cookie dough into desired shape. Let stand in refrigerator overnight.
5. Remove from the refrigerator, slice, and bake at 375°F for 7 min to 8 min.

### Nutrition Facts

**Per cookie:** 578.8 calories; 47% calories from fat; 30.7 g total fat; 77.3 mg cholesterol; 106.3 mg sodium; 75 mg potassium; 69.7 g carbohydrates; 1 g fiber; 25.2 g sugar; 68.8 g net carbs; 5.8 g protein

## Chocolate Brownies

300

*yield: 1 full-size sheet pan (120 small brownies)*

| | |
|---|---|
| 4 lb 12 oz | granulated sugar |
| 1 lb 12 oz | shortening |
| 1½ oz | salt |
| 1 oz | vanilla extract |
| 2 lb 8 oz | eggs |
| 24 oz | glucose |
| 2 lb 10 oz | cake flour |
| 1 lb | cocoa |
| 2 lb 4 oz | pecans, chopped (optional) |

1. Cream the sugar, shortening, and vanilla.
2. Add eggs and glucose alternately, mixing well.
3. Sift together cake flour, cocoa, and salt. Add and mix until smooth.
4. Stir in pecans (if desired).
5. Spread mixture on greased and floured sheet pan.
6. Bake at 350°F for 45 min to 50 min.

**Nutrition Facts**

**Per brownie:** 259.3 calories; 45% calories from fat; 13.6 g total fat; 43.3 mg cholesterol; 155.6 mg sodium; 119.9 mg potassium; 33.8 g carbohydrates; 2.2 g fiber; 18.5 g sugar; 31.5 g net carbs; 3.5 g protein

## French Butter Cookies

301

*yield: 12 dozen*

| | |
|---|---|
| 3 lb | almond paste |
| 3 lb | confectioners' sugar |
| 2 lb | butter |
| 4 lb | shortening |
| ½ oz | salt |
| 3 lb | egg whites |
| 1 oz | vanilla extract |
| 6 lb | bread flour |
| 1 lb 8 oz | pastry flour |

1. Mix the almond paste, confectioners' sugar, butter, shortening, and salt until a smooth paste forms.
2. Gradually add the egg whites and vanilla, and cream until light and fluffy.
3. Sift together both flours and stir into mixture.
4. Place cookie dough into pastry bag with desired tip and press out onto a parchment-lined sheet pan.
5. Bake at 375°F for 7 min to 8 min.

**Nutrition Facts**

**Per cookie:** 329.4 calories; 55% calories from fat; 20.7 g total fat; 20.6 mg cholesterol; 56 mg sodium; 70.9 mg potassium; 31.4 g carbohydrates; 1 g fiber; 12.8 g sugar; 30.4 g net carbs; 4.6 g protein

# Recipes

## 302

## Mealy Pie Dough

*yield: 2 lb (4 crusts)*

| | |
|---|---|
| 16 oz | pastry flour |
| 10.4 oz | shortening |
| 4 oz | water |
| 0.3 oz | salt |
| 0.8 oz | sugar |
| 0.24 oz | dry milk |

1. Place flour and shortening in a large bowl. Cut shortening into flour with pastry blender until the flour is completely covered with shortening and the mixture is mealy in appearance.
2. Place the water, salt, sugar, and dry milk in a stainless steel bowl and mix well.
3. Pour the liquid mixture over the flour/shortening mixture and mix gently until the liquid is absorbed by the flour. Do not overmix.
4. Place the dough on a sheet pan. Cover and refrigerate until firm enough to roll out.
5. Remove from refrigerator, scale into 8-oz units, and refrigerate again until ready to be rolled out.

**Nutrition Facts**

**Per crust:** 1039.9 calories; 58% calories from fat; 68 g total fat; 38.9 mg cholesterol; 837.4 mg sodium; 142 mg potassium; 94.8 g carbohydrates; 1.9 g fiber; 6.7 g sugar; 92.9 g net carbs; 9.7 g protein

## Flaky Pie Dough

*yield: 2 lb (4 crusts)*

| | |
|---|---|
| 16 oz | pastry flour |
| 11.2 oz | shortening |
| 4.8 oz | water |
| 0.3 oz | salt |
| 0.8 oz | sugar |
| 0.24 oz | dry milk |

1. Place flour and shortening in a large bowl. Cut shortening into flour until the flour is completely covered with shortening and the mixture is mealy in appearance.
2. Place the water, salt, sugar, and dry milk in a stainless steel bowl and mix well.
3. Pour the liquid mixture over the flour/shortening mixture and mix gently until the liquid is absorbed by the flour. Do not overmix.
4. Place the dough on a sheet pan. Cover and refrigerate until firm enough to roll out.
5. Remove from refrigerator, scale into 8-oz units, and refrigerate again until ready to be rolled out.

**Nutrition Facts**

**Per crust:** 1086 calories; 60% calories from fat; 73.1 g total fat; 41.8 mg cholesterol; 838.4 mg sodium; 142 mg potassium; 94.8 g carbohydrates; 1.9 g fiber; 6.7 g sugar; 92.9 g net carbs; 9.7 g protein

## 304

# Basic Crumb Crust

*yield: 1 (9-in pie crust)*

| | |
|---|---|
| 8 oz | cookie crumbs (such as graham crackers or vanilla wafers) |
| 3 oz | granulated sugar |
| 2 oz | unsalted butter, melted |

1. Combine all ingredients in a bowl.
2. Press mixture firmly onto bottom and up sides of pie pan.
3. Bake at 350°F for 10 min to 15 min or until golden brown and toasted.

**Nutrition Facts**

**Entire recipe:** 1695.4 calories; 35% calories from fat; 68.9 g total fat; 122 mg cholesterol; 1378.4 mg sodium; 321.5 mg potassium; 259.3 g carbohydrates; 6.3 g fiber; 155.5 g sugar; 252.9 g net carbs; 16.1 g protein

# Cheesecake

*yield: 1 (10-in cake ; 16 slices)*

American Egg Board

| | |
|---|---|
| 1 lb 8 oz | cream cheese |
| 6½ oz | sugar |
| 5 | eggs |
| 12 oz | sour cream |
| to taste | vanilla extract |

1. Cream the cream cheese at room temperature in a mixer using a paddle attachment and scrape bowl.
2. Add sugar and cream together with cream cheese, mixing thoroughly. Scrape down bowl.
3. Slowly add eggs at room temperature one at a time and mix until smooth, scraping bowl as needed.
4. Add sour cream and mix until smooth.
5. Add vanilla extract.
6. Place in a 10″ spring form pan lined with foil to prevent leaking.
7. Bake in a water bath at 325°F for 1 hr 15 min, or until

**Nutrition Facts**

**Per slice:** 261.8 calories; 70% calories from fat; 20.8 g total fat; 122.2 mg cholesterol; 159 mg sodium; 102.6 mg potassium; 13.7 g carbohydrates; 0 g fiber; 11.8 g sugar; 13.7 g net carbs; 5.8 g protein

# Recipes

## Cherry Pie

*yield: 2 (8-in pies; 8 slices per pie)*

| | |
|---|---|
| 1 | Mealy Pie Dough recipe |
| 2 lb | frozen cherries, thawed and drained |
| 14 oz | cherry juice |
| 2 oz | cornstarch |
| 6 oz | granulated sugar |
| 3 oz | corn syrup |

1. Prepare Mealy Pie Dough recipe and blind bake two crusts. Set aside.
2. Thaw the frozen cherries.
3. Drain the juice from the cherries by placing them in a colander. Save all of the juice and add water to the juice if needed to equal 6 oz
4. Place all of the juice, reserving 1 c cherry juice in a saucepan and bring to a boil.
5. In a stainless steel bowl, dissolve the cornstarch in remaining cherry juice. Add the cornstarch mixture to the heated cherry juice and whisk vigorously.
6. Return mixture to a boil and cook until thickened and clear.
7. Add the corn syrup and stir until thoroughly blended. Remove from the heat.
8. Add the cherries, folding them into the thickened juice gently to avoid breaking or crushing the fruit. Let cool about 30 min.
9. Fill unbaked pie shells with filling.
10. Place a top crust or lattice crust over the filing; seal and flute the edges. If a top crust is used, slits should be cut into it to allow for the release of steam. Lattice crusts can be brushed with an egg wash and sprinkled with sugar.
11. Bake at 400°F for 50 min to 60 min until crust is golden brown.

### Nutrition Facts
**Per slice:** 394.8 calories; 38% calories from fat; 17.1 g total fat; 9.7 mg cholesterol; 218 mg sodium; 148.9 mg potassium; 58 g carbohydrates; 1.7 g fiber; 30.7 g sugar; 56.3 g net carbs; 3.1 g protein

## 307

## Apple Pie

*yield: 2 (8-in pies; 8 slices per pie)*

| | |
|---|---|
| 1 | Mealy Pie Dough recipe |
| 2 lb | fresh apples (preferably tart apples such as Granny Smiths), peeled, cored, and either sliced thin or cut into 1" cubes |
| 7 oz | water or apple juice |
| 6 oz | granulated sugar |
| 2 tsp | lemon juice |
| ⅛ tsp | salt |
| ⅛ tsp | cinnamon |
| pinch | nutmeg |
| 1 oz | cornstarch |

1. Prepare Mealy Pie Dough recipe and blind bake two crusts. Set aside.
2. Place the apples, sugar, cinnamon, salt, and nutmeg in a nonreactive saucepan.
3. Dissolve the cornstarch in the lemon juice and add it to the apples.
4. Cover and simmer until apples have softened. Cool slightly until lukewarm.
5. Place the apple mixture into the baked pie shells and cover with a full top crust or lattice crust. Bake at 400°F until the filling is bubbly and topping is light brown, approximately 25 min to 30 min.

### Nutrition Facts
**Per slice:** 341.5 calories; 44% calories from fat; 17.1 g total fat; 9.7 mg cholesterol; 228.6 mg sodium; 104.2 mg potassium; 44.8 g carbohydrates; 1.3 g fiber; 18 g sugar; 43.5 g net carbs; 2.6 g protein

## Pumpkin Pie

*yield: 6 (9-in pies; 8 slices per pie)*

| | |
|---|---|
| 2 | Mealy Pie Dough recipe |
| 6 lb | canned pumpkin |
| 1 lb | brown sugar |
| 2 lb | granulated sugar |
| 3 qt | whole milk |
| 1 lb 6 oz | eggs |
| 4 oz | cake flour |
| ½ oz | cinnamon |
| ¼ oz | nutmeg |
| ¼ oz | ginger |
| ¼ oz | salt |

1. Prepare a double recipe of Mealy Pie Dough and blind bake six pie crusts.
2. Mix pumpkin, sugar, salt, flour, and spices in mixer at medium speed until thoroughly mixed.
3. Add the eggs, one at a time, alternately with the milk at low speed until all ingredients have been thoroughly blended.
4. Pour mixture into prepared pie shells and bake at 375°F for 45 min to 50 min until filling is set.

### Nutrition Facts

**Per slice:** 370.3 calories; 36% calories from fat; 15.1 g total fat; 76.5 mg cholesterol; 382.9 mg sodium; 286.1 mg potassium; 53.5 g carbohydrates; 2.2 g fiber; 32.4 g sugar; 51.3 g net carbs; 6.3 g protein

## Cream Pie

*yield: 3 (8-in pies; 8 slices per pie)*

| | |
|---|---|
| 3 | Basic Crumb Crust recipe |
| 2 qt | whole milk |
| 18 oz | sugar |
| 10 | egg yolks |
| 3 | whole eggs |
| 4 oz | cornstarch |
| 2 oz | cake flour |
| 5 oz | unsalted butter |
| ¼ tsp | salt |
| 1 tbsp | vanilla extract |

1. Prepare three Basic Crumb Crusts and pan each pie shell in an 8″ pie pan.
2. In a heavy saucepan, dissolve the sugar in the milk, bringing the mixture just to a boil, being careful not to scorch the milk.
3. Whisk together eggs and sugar in a separate bowl and then sift in the flour and cornstarch.
4. Temper the egg mixture by pouring in half of the hot milk mixture while whisking rapidly.
5. Stir the warmed egg mixture back into the remaining milk and bring to a boil while stirring constantly.
6. Allow to simmer for only 30 sec, until mixture is smooth and thick.
7. Remove from heat and stir in butter, salt, and vanilla extract.
8. Pour cream filling into pie shells. Refrigerate until cold.

### Nutrition Facts

**Per slice:** 445.5 calories; 36% calories from fat; 18.6 g total fat; 149.9 mg cholesterol; 242.4 mg sodium; 177.9 mg potassium; 63.9 g carbohydrates; 0.9 g fiber; 45.1 g sugar; 63 g net carbs; 6.8 g protein

### Recipe Variations:

**310 Banana Cream Pie:** Cover the bottom of a baked pie shell or crumb crust with cream pie filling. Place sliced bananas on the surface of the pie filling. Cover the banana slices with enough pie filling to fill pie shell. Let cool completely and either top with heavy cream that has been whipped with sugar and cream of tartar or top with meringue. Bake in a 450°F oven for 2 min to 3 min until topping browns.

**311 Coconut Cream Pie:** Stir 2 lb of coconut into the cream pie filling and pour into baked pie shells. Let the filling cool and top with meringue. Bake at 375°F for 2 min to 3 min until topping browns.

**312 Chocolate Cream Pie:** Stir 12 oz to 14 oz dark or bittersweet chocolate into the cream mixture after all other ingredients have been incorporated.

# Recipes

## 313

## Lemon Chiffon Pie

*yield: 3 (8-in pies; 8 slices per pie)*

| | |
|---|---|
| 1 | Flaky Pie Dough recipe |
| 1.5 lb | water |
| 1 lb | sugar |
| 2 tsp | salt |
| 1 oz | grated lemon zest |
| ½ lb | egg yolks |
| ½ lb | lemon juice |
| 5 oz | cornstarch |
| ¾ oz | gelatin |
| ½ lb | hot water |
| 1 lb | egg whites |
| 12 oz | granulated sugar |

1. Prepare Flaky Pie Dough recipe and blind bake three pie shells. Set aside to cool.
2. Place 1.5 lb water, 1 lb sugar, salt, and lemon zest in a heavy saucepan and bring to a boil.
3. Whisk together the egg yolks, lemon juice, and cornstarch in a stainless steel mixing bowl.
4. Temper the egg mixture gradually by adding half of the hot mixture to the eggs while whisking rapidly. Pour back into the hot mixture while continually whipping mixture until thickened. Remove from heat.
5. Dissolve gelatin in hot water and gently stir into the lemon mixture.
6. Begin whipping egg whites in another bowl and slowly add 12 oz of sugar until stiff peaks form a meringue.
7. Gently fold meringue into the hot lemon filling.
8. Pour the filling into baked pie shells and cool in refrigerator until set.

**Nutrition Facts**

**Per slice:** 378 calories; 35% calories from fat; 14.7 g total fat; 123.6 mg cholesterol; 374.2 mg sodium; 79.3 mg potassium; 56.5 g carbohydrates; 0.5 g fiber; 35.4 g sugar; 56 g net carbs; 5.3 g protein

## Pecan Pie

*yield: 8 (9-in pies; 8 slices per pie)*

| | |
|---|---|
| 3 | Mealy Pie Dough recipe |
| 9 lb | light corn syrup |
| 6 oz | pastry flour |
| 6 oz | sugar |
| 3 lb 4 oz | eggs |
| ½ oz | salt |
| 8 oz | melted butter |
| 1 oz | vanilla extract |
| 1 lb 12 oz | chopped pecans |

1. Prepare a triple recipe of Mealy Pie Dough and line eight pie pans.
2. Place the flour, sugar, syrup in mixer with paddle attachment. Blend together at low speed.
3. Set mixer at low speed and stir in eggs, one at a time, followed by salt, melted butter, and vanilla.
4. Stir in pecans.
5. Fill pie shells with pecan mixture and bake at 325°F for 45 min to 50 min.

**Nutrition Facts**

**Per slice:** 549.6 calories; 43% calories from fat; 27 g total fat; 116.9 mg cholesterol; 354.3 mg sodium; 116.7 mg potassium; 75.3 g carbohydrates; 1.6 g fiber; 55.4 g sugar; 73.7 g net carbs; 6.2 g protein

## Puff Pastry (Pâte Feuilletée)

*yield: 5 lb 10 oz (24 puff pastries)*

| | |
|---|---|
| 1 lb 8 oz | bread flour |
| 8 oz | pastry flour |
| 4 oz | softened butter |
| ½ oz | salt |
| 1 lb 2 oz | cold water |
| 2 lb | butter |
| 3 oz | bread flour (as needed) |

1. Sift together bread and cake flour and place in bowl of mixer.
2. Add 4 oz softened butter and mix using paddle attachment just until mixture resembles coarse marbles of fat.
3. Mix salt and water together and pour into flour and butter mixture.
4. Mix at slow speed with a dough hook for about 3 min, until dough is smooth and ingredients appear evenly distributed.
5. Remove dough from mixer and form into a rectangular shape on a parchment-lined sheet pan. Cover dough with plastic wrap and let rest under refrigeration for 1 hr.
6. To prepare the roll-in butter, mix the 2 lb butter and (as needed) 3 oz bread flour in a mixer with a paddle attachment on low speed until smooth and well mixed.
7. Remove the roll-in butter mixture from the mixer and place on a sheet of parchment paper. Place another sheet on top and roll the butter mixture to a rectangle approximately 8″ × 12″ and between ¼″ and ½″ thick. Square off the edges of the butter, cover, and refrigerate for 10 min to firm slightly.
8. While the roll-in butter mixture is cooling, begin to roll the dough mixture out on a floured work surface to approximately 16″ × 24″ while keeping the sides straight.
9. Place the roll-in butter mixture on one half of the dough, leaving about a ½″ margin around the edge of the dough.
10. Fold the other half of the dough over the roll-in butter mixture and seal the edges of the dough together tightly. Rotate the dough 90° and roll it out once again to approximately 16″ × 24″, keeping the sides straight. *Note:* Check often to make sure that the dough is not sticking to the work surface.
11. Fold the dough in thirds (known as a three-fold) by folding one-third of the dough toward the center and then the remaining one-third over the first third. This process of making the three-fold is known as a "turn." *Note:* Each time the dough is turned, make an impression with the fingertip to keep track of how many turns have been done. The second time will get two finger impressions, the third three, and so on.
12. Brush the dough free of excess flour, gently press one finger in the center of the dough, just enough to leave a slight impression, and cover the dough with plastic wrap.
13. Let rest for 30 min in the refrigerator.
14. Remove the dough from the refrigerator and give it a half turn (90° turn) so the former length becomes the width.
15. Roll out dough into a rectangle equal to the original (prefolded) size, or 16″ × 24″.
16. Fold the dough using a second three-fold, brush off excess flour, wrap, and store for 30 min.
17. Repeat the three-fold procedure three more times, or five times in all.
18. After rolling the dough for the fifth time, the dough is ready to be made into desired units.
19. Allow the units to rest for 30 min before baking.
20. Bake puff pastry units in a preheated oven at 400°F for approximately 20 min.

**Nutrition Facts**

**Per pastry:** 454.3 calories; 67% calories from fat; 35.1 g total fat; 91.4 mg cholesterol; 235.3 mg sodium; 52.1 mg potassium; 30.5 g carbohydrates; 0.9 g fiber; 0.2 g sugar; 29.6 g net carbs; 5 g protein

# Recipes

## Éclair Paste (Pâte à Choux)

*yield: 4.5 lb (60 shells)*

| | |
|---|---|
| 16 oz | water |
| 16 oz | milk |
| ¼ oz | salt |
| 16 oz | butter |
| 16 oz | bread flour |
| 32 oz | eggs |

1. In a saucepan, bring milk, water, salt, and butter to a rolling boil.
2. Remove from heat and vigorously mix in flour with a high-heat spatula.
3. Return the saucepan to the burner and heat while mixing vigorously until mixture forms a ball and pulls away from the sides of the pan.
4. Transfer the dough to a mixing bowl with a paddle attachment and allow to cool until about 140°F. Begin adding the eggs, one at a time, while mixing at medium speed. The dough should begin to pull away from the sides of the bowl and the eggs should be completely absorbed into the paste. *Note:* It may not be necessary to use all the eggs, depending on the moisture of the flour mixture.
5. Fill a pastry bag with prepared paste. Pipe onto a parchment-lined sheet pan in desired shape.
6. Bake at 425°F for 10 min. *Note:* Do not open the oven door during baking, as this may cause pâte à choux to fall.
7. Reduce oven temperature to 375°F and bake for an additional 10 min to 12 min or until pâte à choux is lightly browned and sounds hollow when tapped lightly.
8. Remove from oven and allow to cool to room temperature.
9. Fill with desired filling.

### Nutrition Facts
**Per shell:** 108.6 calories; 65% calories from fat; 8 g total fat; 81 mg cholesterol; 72.4 mg sodium; 41.2 mg potassium; 6 g carbohydrates; 0.2 g fiber; 0.6 g sugar; 5.8 g net carbs; 3.1 g protein

---

## 317

## Meringues

*yield: 3 lb*

| | |
|---|---|
| 16 oz | egg whites |
| 32 oz | sugar |

1. Using a whisk attachment on a mixer, beat egg whites at medium speed until frothy, then increase speed to high until the egg whites form soft peaks.
2. Gradually add sugar and whip until the mixture forms stiff peaks. Use caution to not overwhip as the whites can break.
3. Place finished whipped meringue in a pastry bag and pipe onto parchment-lined paper into desired shapes.
4. Bake at 200°F until crisp but not browned. This will take approximately 1 hr to 3 hr, depending on size. Use caution when removing from the oven as finished meringues are fragile.

### Nutrition Facts
**Per ounce:** 78.1 calories; 0% calories from fat; 0 g total fat; 0 mg cholesterol; 15.7 mg sodium; 15.8 mg potassium; 19 g carbohydrates; 0 g fiber; 19 g sugar; 19 g net carbs; 1 g protein

---

## Italian Meringue

*yield: 3 lb 8 oz*

| | |
|---|---|
| 2 lb | granulated sugar |
| 8 oz | water |
| 6 oz | egg whites |

1. Place water and sugar in a saucepan along with a candy thermometer and bring to a boil.
2. Heat until all the sugar is dissolved, continually washing down sides of saucepan with a pastry brush dipped in cold water.
3. Continue to cook until the mixture reaches 240°F.
4. When the sugar mixture reaches 240°F, remove from the heat and allow to cool to 220°F.
5. Meanwhile, place the egg whites in the bowl of a mixer with a whip attachment and whip until soft peaks are formed.
6. When sugar mixture cools to 220°F, turn mixer to high speed and slowly pour the sugar mixture in a fine stream into the egg whites while continuing to mix.
7. Once all of the sugar has been added, continue to whip until the mixture has cooled to room temperature.

### Nutrition Facts
**Per ounce:** 64.1 calories; 0% calories from fat; 0 g total fat; 0 mg cholesterol; 5.4 mg sodium; 5 mg potassium; 16.2 g carbohydrates; 0 g fiber; 16.2 g sugar; 16.2 g net carbs; 0.3 g protein

## Crème Anglaise

*yield: 2½ qt (4-oz servings)*

| | |
|---|---|
| 1 qt | cream |
| 1 qt | whole milk |
| 2 | vanilla beans, split in half lengthwise and scraped |
| 8 oz | sugar |
| 8 oz | egg yolks |
| ¼ tsp | salt |

1. In a heavy saucepan, bring the cream, milk, and vanilla bean just to a boil.
2. Whisk the egg yolks, sugar, and salt together in a nonreactive bowl.
3. Slowly temper the egg mixture with approximately one-third of the hot cream mixture, whisking constantly.
4. Return the mixture to the saucepan, whisking the entire mixture together.
5. Return to heat and cook on medium until the mixture reaches the ribbon stage and is thick enough to coat the back of a spoon, approximately 175°F.
6. Transfer to a bowl and place over an ice bath to cool. *Note:* Cover the surface with plastic wrap before refrigerating.

---

**Nutrition Facts**

**Per serving:** 192.1 calories; 61% calories from fat; 13.4 g total fat; 177.6 mg cholesterol; 63.1 mg sodium; 100.3 mg potassium; 14.6 g carbohydrates; 0 g fiber; 14 g sugar; 14.6 g net carbs; 3.9 g protein

---

## 320

## Pastry Cream

*yield: 25 (4-oz servings)*

| | |
|---|---|
| 14 oz | sugar |
| 5.4 oz | cornstarch |
| ½ tsp | salt |
| 13 oz | egg yolks |
| 2 qt | milk |
| 1 | vanilla bean, split in half lengthwise |
| 4 oz | butter |

1. Whisk the sugar, cornstarch, and salt together in a nonreactive bowl.
2. Whisk the egg yolks into the sugar mixture and mix well.
3. In a medium saucepan, bring the milk and vanilla bean to a boil. As soon as it comes to a boil, slowly pour about ⅓ of the hot mixture into the egg mixture to temper it while whisking constantly.
4. Pour the tempered mixture back into the pot with the remainder of the hot milk while whisking continuously.
5. Return the mixture to medium heat, whisking constantly, until the mixture just reaches a boil.
6. Remove from the heat immediately.
7. Strain the mixture into a bowl over an ice bath, and whisk in the butter.
8. Cover with plastic wrap to prevent a skin from forming and place in refrigerator until needed.
9. Remove vanilla bean seeds from pod and discard pod. Stir seeds into pastry cream.

---

**Nutrition Facts**

**Per serving:** 211.6 calories; 42% calories from fat; 10.1 g total fat; 199.5 mg cholesterol; 85.9 mg sodium; 129.3 mg potassium; 25.5 g carbohydrates; 0.1 g fiber; 20.1 g sugar; 25.5 g net carbs; 4.9 g protein

---

# Recipes

## White Chocolate Banana Cream Tart `321`

*yield: 1 (9-in tart; 8 servings)*

*Crust*

| | |
|---|---|
| 4 oz | butter |
| 3 oz | sugar |
| 1 | egg |
| 5 oz | flour |

*Filling*

| | |
|---|---|
| 3 | egg yolks |
| 1 oz | sugar |
| 2 tbsp | cornstarch |
| 8 oz | milk |
| ½ | vanilla bean, split |
| 3 oz | white chocolate |
| ½ oz | butter |
| 4 oz | whipping cream |
| 3 | bananas |
| ½ oz | banana liqueur |
| ½ oz | lemon juice |
| 3 oz | white chocolate shavings |

1. To begin the crust, cream butter and sugar until smooth.
2. Add egg, and beat until blended. Add flour and mix until everything is incorporated.
3. Gather mixture into a ball and flatten into a disk.
4. Wrap with plastic and refrigerate for at least 30 min.
5. Roll out thin and place in a 10 tart pan. Prick sides and bottom with a fork.
6. Place foil and pie weights in bottom of pan.
7. Blind bake for 10 min at 375°F.
8. To begin the filling, whisk yolks, sugar, and cornstarch in bowl until combined.
9. Pour milk into small saucepan and scrape in seeds from vanilla bean. Bring to a boil.
10. Slowly add milk mixture into beaten egg mixture while whisking.
11. Return to stove and bring to a boil, stirring constantly.
12. Remove from heat and strain mixture into large bowl.
13. Add white chocolate and butter. Mix until melted.
14. Chill mixture until completely cooled.
15. Whip cream until stiff peaks have formed. Fold in the chilled white chocolate pastry cream.
16. Cut bananas into thick slices and toss with liqueur and lemon juice. Fold into pastry cream.
17. Spoon filling into cooled tart shell. Garnish with white chocolate shavings.

### Nutrition Facts

**Per serving:** 511.6 calories; 49% calories from fat; 28.6 g total fat; 167.9 mg cholesterol; 50 mg sodium; 360.5 mg potassium; 57.5 g carbohydrates; 2.6 g fiber; 34.5 g sugar; 54.9 g net carbs; 7.3 g protein

## Vanilla Pudding `322`

*yield: 32 (4-oz servings)*

| | |
|---|---|
| 3 qt | whole milk |
| 13 oz | sugar |
| 6 oz | cornstarch |
| 9 oz | egg yolks |
| ½ tsp | salt |
| 3 oz | butter |
| ½ oz | vanilla extract |

1. Place the milk and half of the sugar in a double boiler and heat to a scald.
2. Place the remaining sugar, cornstarch, egg yolks, and salt in a nonreactive bowl.
3. Gradually add half of the scalded milk until a thin paste forms.
4. Pour the paste mixture back into the remaining scalded milk and whisk until smooth.
5. Turn heat back to medium and whisk until the mixture begins to thicken.
6. Remove from the heat and stir in butter and vanilla.
7. Pour into desired serving dishes and chill until cold.

### Nutrition Facts

**Per serving:** 165.7 calories; 38% calories from fat; 7.3 g total fat; 113.3 mg cholesterol; 77.6 mg sodium; 141.2 mg potassium; 20.8 g carbohydrates; 0 g fiber; 16.4 g sugar; 20.8 g net carbs; 4.2 g protein

## 323

## Chocolate Mousse

*yield: 10 (3-oz servings)*

| | |
|---|---|
| 8 oz | bittersweet chocolate |
| 2 oz | unsalted butter |
| 3 oz | pasteurized egg yolks |
| 1 oz | rum (or ¼ tsp rum extract and 1 oz milk or cream) |
| 5 oz | pasteurized egg whites |
| 1.5 oz | sugar |
| 8.5 oz | heavy cream |

1. Melt chocolate and butter over double boiler.
2. When chocolate has melted, remove from heat and whisk in egg yolks rapidly, one at a time, until all are incorporated. Add rum and stir.
3. In a separate mixing bowl, whisk egg whites to medium peaks form, then gradually begin adding sugar and whip until stiff peaks form.
4. Add about one-quarter of the egg whites into the chocolate to lighten it and then carefully fold in the remaining whites. Allow to cool to room temperature.
5. In a separate bowl, whip cream until firm peaks form.
6. Gently fold into the cooled chocolate mixture.
7. Spoon into serving bowls, or chill and pipe into baked tartlet shells.

### Nutrition Facts

**Per serving:** 294.2 calories; 67% calories from fat; 23 g total fat; 151.8 mg cholesterol; 41.7 mg sodium; 136.9 mg potassium; 19.8 g carbohydrates; 0 g fiber; 4.4 g sugar; 19.8 g net carbs; 4.5 g protein

## 324

## Sabayon

*yield: 20 oz (ten 2-oz servings)*

| | |
|---|---|
| 6 oz | champagne |
| 4 | egg yolks |
| pinch | salt |
| 4 oz | granulated sugar |
| 4.25 oz | cream |

1. Whisk together champagne, egg yolks, salt, and sugar in a nonreactive bowl.
2. Place bowl over simmering water and whisk vigorously until the mixture is thick and pale yellow in color, approximately 160°F.
3. Remove from heat and serve immediately if serving warm.
4. For chilled sabayon, place over an ice bath and whisk until completely cooled. In separate bowl, whisk cream until it reaches soft peaks.
5. Gently fold whipped cream into the egg mixture and chill until ready to serve.

### Nutrition Facts

**Per serving:** 121.4 calories; 47% calories from fat; 6.5 g total fat; 101.2 mg cholesterol; 38 mg sodium; 31.3 mg potassium; 12.1 g carbohydrates; 0 g fiber; 11.4 g sugar; 12.1 g net carbs; 1.4 g protein

### Recipe Variations:

**325** **Grapefruit Sabayon:** Substitute pink grapefruit juice for the champagne.

# Recipes

## 326

## Chocolate Soufflé

*yield: 5 (4-oz servings)*

| | |
|---|---|
| 4 oz | unsalted butter |
| 7 oz | bittersweet chocolate |
| 0.75 oz | cocoa powder |
| 1 oz | egg yolks |
| 1 tsp | vanilla |
| 5.75 oz | egg whites |
| ¼ tsp | salt |
| 3.5 oz | sugar |

*Daniel NYC*

1. Prepare soufflé cups by thoroughly coating them with butter and then dusting them with granulated sugar, making sure to pour out any loose sugar.
2. Melt the unsalted butter and chocolate in a double boiler over low heat.
3. Whisk the cocoa, egg yolks, and vanilla into the butter and chocolate mixture.
4. In a separate bowl, beat the egg whites until foamy and add the salt. Continue beating until soft peaks form.
5. Add sugar, one tablespoon at a time, and beat until the egg-white mixture is shiny and stiff peaks have formed.
6. Fold egg whites into the chocolate mixture.
7. Pour the chocolate mixture into soufflé cups, not filling more than three-quarters full.
8. Bake at 400°F for 5 min to 6 min or until the soufflés rise at least 1″ over the rim of the soufflé molds.
9. Serve immediately with crème anglaise, chocolate sauce, or a Chantilly crème.

### Nutrition Facts
**Per serving:** 474.9 calories; 57% calories from fat; 32.3 g total fat; 118.8 mg cholesterol; 176.5 mg sodium; 271.8 mg potassium; 47.9 g carbohydrates; 1.4 g fiber; 20.3 g sugar; 46.4 g net carbs; 6.9 g protein

## 327

## Mocha Bavarian

*yield: 10 (3-oz servings)*

| | |
|---|---|
| 16 oz | prepared crème anglaise |
| 1 oz | espresso |
| 0.25 oz | vanilla extract |
| 0.5 oz | gelatin |
| 12 oz | cream |

1. Gently warm the crème anglaise.
2. Mix espresso and vanilla together over a double boiler and sprinkle gelatin over the top.
3. Melt gelatin until completely dissolved in espresso.
4. Whisk the gelatin mixture into the crème anglaise. Place the mixture over an ice bath and continue whisking.
5. Chill in the refrigerator until thick but not completely set.
6. Beat cream to soft peaks and gently fold into chilled custard mixture.
7. Pour into molds and refrigerate until completely set.

### Nutrition Facts
**Per serving:** 257.5 calories; 75% calories from fat; 22.1 g total fat; 166.1 mg cholesterol; 62.3 mg sodium; 97.4 mg potassium; 12 g carbohydrates; 0 g fiber; 10.6 g sugar; 12 g net carbs; 3.4 g protein

## Chocolate and Peppermint Soufflé

*yield: 10 (4-oz servings)*

| | |
|---|---|
| 8 oz | milk |
| 10 | eggs separated |
| 6 oz | granulated sugar |
| 1 oz | all-purpose flour |
| 5 oz | bittersweet chocolate, finely chopped |
| 2 oz | peppermint candy, crushed |

1. Prepare ramekins in advance by buttering the interior of the dish and the entire rim completely and then thoroughly coating the buttered surface in granulated sugar. *Note:* If there is any spot that is not completely buttered and sugared, it will prevent the soufflé from rising.
2. Bring 6 oz milk, salt, and 3 oz sugar to a scald.
3. In separate stainless steel mixing bowl, whisk together yolks and 3 oz of sugar until light in color and very thick.
4. Add flour and remaining 2 oz milk and whisk.
5. Temper yolks mixture by adding hot milk in one-third increments to yolks while whisking rapidly until all milk has been added.
6. Return the mix to the saucepan and cook over low heat until mixture is thick and coats the back of a spoon. Simmer for 1 min more to cook out flour taste.
7. Remove from heat, incorporate chocolate, and stir well.
8. Cover tightly with plastic wrap pressed to the surface until mixture cools to room temperature.
9. When cooled, add peppermint candy to the mixture and stir well. *Note:* This base can be held for up to a day.
10. Begin whipping egg whites. When soft peaks have formed, add sugar slowly while continually whipping until all the sugar has been added and stiff peaks have formed.
11. Carefully fold approximately one-third of the whipped egg whites into the soufflé base and mix well to incorporate and lighten the base.
12. Continue to fold in the remaining whipped whites and fold gently to prevent any loss of volume.
13. Using a pastry bag or a spoon, pour the mixture into the prepared ramekins, filling to about ¼″ from the top and bake immediately in a 400°F oven.
14. Remove when soufflés have risen well above the surface of the ramekin and the surface has turned a golden brown. Serve immediately.

### Nutrition Facts

**Per serving:** 254.1 calories; 33% calories from fat; 10 g total fat; 213.9 mg cholesterol; 83.6 mg sodium; 156.9 mg potassium; 35.2 g carbohydrates; 0.1 g fiber; 22.2 g sugar; 35.1 g net carbs; 8 g protein

## Mango Sorbet

*yield: 1½ quarts (twelve 4-oz servings)*

| | |
|---|---|
| 3 c | water |
| 1 c | sugar |
| 5–6 | ripe mangoes |
| ¼ c | fresh lemon juice |
| ¼ c | Cointreau or orange-flavored liqueur |

1. Combine water and sugar in medium-size saucepan. Cook until sugar dissolves and liquid becomes clear.
2. Transfer simple syrup to a mixing bowl and cool immediately in an ice bath.
3. Pare mangoes scraping the fruit from inside; discard the pits.
4. Purée fruit along with ½ c of the simple syrup in a blender until smooth.
5. Combine the rest of the simple syrup, mango purée, lemon juice, and liquor in a large bowl.
6. Transfer and freeze in an ice cream maker according to manufacturer's instructions.

### Nutrition Facts

**Per serving:** 139.5 calories; 1% calories from fat; 0.2 g total fat; 0 mg cholesterol; 3 mg sodium; 141.2 mg potassium; 33.4 g carbohydrates; 1.6 g fiber; 29.5 g sugar; 31.8 g net carbs; 0.5 g protein

# Recipes

## 330

## Ice Cream Base

*yield: 1 gal (4-oz servings)*

| | |
|---|---|
| 10 c | heavy cream |
| 5 c | whole milk |
| ½ oz | salt |
| 3 c | sugar |
| 3 | vanilla beans, split and scraped |
| 24 | egg yolks |

*California Fresh Apricot Council*

1. Combine heavy cream, milk, and vanilla bean in a saucepan and bring to a rolling boil.
2. In a separate mixing bowl, whisk egg yolks along with granulated sugar.
3. Add approximately 8 oz of the hot mixture to the whisked egg yolks and sugar to temper the yolks. Add the tempered egg mixture back to the saucepan and return to medium heat.
4. Cook the cream and egg mixture over medium heat until it reaches 180°F to 185°F or coats the back of a spoon easily. *Note:* This is referred to as the ribbon stage.
5. Immediately pour through a fine mesh sieve into a clean bowl. Place the bowl immediately into an ice bath and cool completely.
6. Place mixture in bowl of electric ice cream machine and mix according to manufacturer's instructions.

### Nutrition Facts
**Per serving:** 265.3 calories; 61% calories from fat; 18.4 g total fat; 212.3 mg cholesterol; 207.2 mg sodium; 96.8 mg potassium; 22 g carbohydrates; 0 g fiber; 20.9 g sugar; 22 g net carbs; 4 g protein

## 331

## Tangerine Granité

*yield: 1 pt (four 4-oz servings)*

| | |
|---|---|
| 1 tbsp | honey |
| 2 c | tangerine juice |
| 1½ oz | Cointreau or orange-flavored liqueur |

1. Dissolve honey in tangerine juice and combine with Cointreau or orange-flavored liqueur.
2. Freeze the mixture in a stainless hotel pan until solid (approximately 3 hr).
3. Serve by scraping the surface with a metal spatula, breaking up large ice crystals.
4. Garnish with tangerine segments and a sprig of mint.

### Nutrition Facts
**Per serving:** 108.7 calories; 1% calories from fat; 0.2 g total fat; 0 mg cholesterol; 1.4 mg sodium; 222.6 mg potassium; 20.4 g carbohydrates; 0.3 g fiber; 16.5 g sugar; 20.1 g net carbs; 0.6 g protein

## Cherries Jubilee

*yield: 4 (4-oz servings)*

| | |
|---|---|
| 2 tbsp | butter |
| 1 pt | pitted Bing cherries and juice |
| ¼ cup | sugar |
| ¼ tsp | arrowroot |
| ¼ tsp | cold water |
| 2 oz | Kirsch (or orange juice) |

1. Melt butter in sauté pan and add cherries.
2. When cherries are tender, add sugar and heat to dissolve and begin to caramelize.
3. Deglaze pan with Kirsch (or orange juice) and flambé and reduce by half. *Note:* If orange juice is used instead of Kirsch, the cherries will not be able to be flambéed.
4. Mix arrowroot and cold water to make a slurry, then add to fruit and bring back to a boil.
5. Simmer for 30 sec until mixture thickens and becomes clear.

**Nutrition Facts**

**Per serving:** 219.8 calories; 21% calories from fat; 5.8 g total fat; 15.3 mg cholesterol; 4.6 mg sodium; 166.4 mg potassium; 34.6 g carbohydrates; 1.9 g fiber; 27.9 g sugar; 32.7 g net carbs; 1.2 g protein

## Pâte Dough

*yield: 12 oz (twelve 1-oz servings)*

| | |
|---|---|
| ½ c | unsalted butter |
| ½ c | pastry flour |
| 1 tbsp | sugar |
| ¼ tsp | salt |
| 8 tbsp | ice water |

1. Cut the butter into small cubes.
2. Cover and freeze for 1 hr.
3. Sift together flour, sugar, and salt.
4. Combine frozen butter in a pastry blender until pea size.
5. Slowly add half the ice water and mix well.
6. Continue to add water until a dough consistency is achieved.
7. Lightly knead dough and form a small ball. Wrap in plastic wrap and chill for 1 hr.

**Nutrition Facts**

**Per serving:** 92.5 calories; 73% calories from fat; 7.7 g total fat; 20.3 mg cholesterol; 49.8 mg sodium; 8.3 mg potassium; 5.5 g carbohydrates; 0.1 g fiber; 1.1 g sugar; 5.4 g net carbs; 0.5 g protein

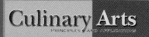

## USDA Institutional Meat Purchase Specifications (IMPS) . . .

| Series | Item No. | Product | Item No. | Product |
|--------|----------|---------|----------|---------|
| **100** (fresh beef) | 100 | Carcass | 132 | Triangle |
| | 101 | Side | 134 | Beef bones |
| | 102 | Forequarter | 135 | Diced beef |
| | 103–112 | Rib | 136 | Ground beef |
| | 113–114 | Chuck | 138–139 | Beef trimmings |
| | 117 | Foreshank | 155 | Hindquarter |
| | 118–120 | Brisket | 157 | Hindshank |
| | 121 | Plate, short plate | 158–170 | Round |
| | 122 | Plate, full | 172 | Loin, full loin, trimmed |
| | 123 | Short ribs | 173 | Loin, short loin |
| | 124 | Rib, back ribs | 175 | Loin, strip loin |
| | 125 | Chuck, armbone | 181–185 | Loin, sirloin |
| | 127 | Chuck, cross-cut | 189–192 | Loin, tenderloin |
| | 130 | Chuck, short ribs | 193 | Flank, flank steak |
| **200** (lamb) | 200 | Carcass | 231–232 | Loins |
| | 202 | Foresaddle | 233–234 | Legs |
| | 203 | Bracelet | 235–236 | Back |
| | 204 | Rack | 238–239 | Trimmings |
| | 206 | Shoulders | 242–243 | Loins, full |
| | 207–208 | Shoulders, square-cut | 244 | Loin, boneless, 3-way |
| | 209 | Breast | 245 | Sirloin |
| | 210 | Foreshank | 246 | Tenderloin |
| | 229 | Hindsaddle, long-cut | 295 | Lamb for stewing |
| | 230 | Hindsaddle | 296 | Ground lamb |
| **300** (veal and calf) | 300 | Carcass | 337 | Hindshank |
| | 303 | Side | 338–339 | Trimmings |
| | 304 | Foresaddle, 11 ribs | 341 | Back, 9 ribs, trimmed |
| | 304A | Forequarter, 11 ribs | 342 | Back, strip, boneless |
| | 306 | Hotel rack, 7 ribs | 344 | Loin, strip loin, boneless |
| | 307 | Rack, ribeye, 7 ribs | 346 | Leg, butt tenderloin, defatted |
| | 308–311 | Chuck | 347 | Loin, short tenderloin |
| | 312 | Foreshank | 349–363 | Leg, top round, cap on |
| | 313–314 | Breast | 389 | Mixed bones |
| | 323 | Veal short ribs | 390 | Marrow bones |
| | 330 | Hindsaddle, 2 ribs | 391 | Marrow |
| | 331–332 | Loins | 395 | Veal for stewing |
| | 334–336 | Legs | 396–397 | Ground veal |
| **400** (fresh pork) | 400 | Carcass | 419 | Jowl |
| | 401–402 | Leg (fresh ham) | 420 | Pig's feet, front |
| | 403–407 | Shoulder | 421 | Neck bones |
| | 408–409 | Belly | 422 | Loin, back ribs |
| | 410–414 | Loin | 423 | Loin, country style ribs |
| | 415 | Tenderloin | 424 | Loin, riblet |
| | 416 | Spareribs | 435 | Diced pork |
| | 417 | Shoulder hocks | 496 | Ground pork |
| | 418 | Trimmings | | |

## . . . USDA Institutional Meat Purchase Specifications (IMPS) . . .

| Series | Item No. | Product | Item No. | Product |
|---|---|---|---|---|
| 500 (cured, smoked, and fully-cooked pork) | 500–512 | Ham | 548 | Pork center-cut loin, 8 ribs (cured and smoked) |
| | 514 | Pork, diced (cured) | 550 | Canadian style bacon (cured and smoked), unsliced |
| | 515–530 | Pork shoulder | | |
| | 531 | Pork Boston butt | 555 | Jowl butts, cellar trim (cured) |
| | 535, 537 | Belly | 556 | Jowl squares (cured and smoked) |
| | 536, 538 | Bacon, slab | 558–559 | Spareribs |
| | 539–541 | Bacon, sliced | 560 | Hocks, ham (cured and smoked) |
| | 545–546 | Pork loin (cured and smoked) | 561 | Hocks, shoulder (cured and smoked) |
| | 547 | Pork center loin, boneless, (cured and smoked) | 562 | Clear fatback (cured) |
| | | | 563 | feet, front (cured) |
| 600 (cured, dried, and smoked beef) | 600–601 | Beef brisket | 619–620 | Sliced dried beef |
| | 602–603 | Beef knuckle | 621 | Beef, cooked, cured, chunked, and formed |
| | 604–605 | Beef top (inside) round | 622 | Beef, sliced, cooked, cured, chunked, and formed |
| | 606–607 | Beef bottom, (gooseneck) round | 623 | Beef top (inside) round, cooked |
| | 608 | Beef outside round, corned | 624 | Beef outside round, corned, cooked |
| | 609 | Beef rump butt, corned | 625–626 | Brisket, boneless, deckle off, corned, cooked |
| | 611 | Beef pastrami | | |
| | 612 | Beef fajita strips | 627 | Beef knuckle, peeled, cooked |
| | 613–614 | Beef tongue, cured, trimmed | 628 | Beef loin, top sirloin butt, center cut, boneless, cooked |
| | 617 | Processed dried beef | 629 | Bottom (gooseneck) round, heel out, cooked |
| | 618 | Sliced processed dried beef | 630 | Beef ribeye roll, boneless, cooked |
| | | | 631 | Charbroiled beef patties |
| 700 (variety meats) | 701–703 | Beef liver | 722 | Beef, lamb, or pork kidney |
| | 704–705 | Calf liver | 723 | Cheek meat |
| | 707–708 | Veal liver | 724 | Head meat |
| | 710 | Pork liver | 725 | Beef, lamb, or pork brains |
| | 713 | Lamb liver | 726 | Beef tripe, scalded, bleached (denuded) |
| | 715 | Veal sweetbreads | 727 | Beef tripe, honeycomb, bleached |
| | 716 | Beef tongue, short cut | 728 | Pork chitterlings |
| | 717 | Tongue, Swiss cut | 729 | Pork stomach (maws), scalded |
| | 720 | Beef heart, trimmed | 731 | Edible tallow |
| | 721 | Beef Oxtail, trimmed | 732 | Lard (edible) |
| 800 (sausage) | 800 | Frankfurters | 813 | Polish sausage |
| | 801 | Bologna | 814 | Meat loaves |
| | 802 | Pork sausage | 815 | Meat food product loaves |
| | 803 | Liver sausage | 816 | Knockwurst |
| | 804 | Cooked salami | 817 | Breakfast sausage, cooked |
| | 805 | Minced luncheon meat | 818 | Italian sausage |
| | 806 | Lebanon bologna | 819 | Ham links |
| | 807 | Thuringer | 820 | Head cheese |
| | 808 | Dry salami | 821 | Pepperoni |
| | 809 | Cervelat | 822 | Bratwurst |
| | 810 | Breakfast sausage | 824 | Pork rib shape patty |
| | 811 | Smoked sausage | 825 | Canned luncheon meat |
| | 812 | New England brand sausage | 826 | Scrapple |

| Series | Item No. | Product | Item No. | Product |
|---|---|---|---|---|
| 900 | | Goat | | |
| 1100 (fresh beef portion cuts) | 1100–1101 | Cubed steak | 1150 | Top side steak, boneless |
| | 1102 | Braising steak, Swiss | 1167 | Round, knuckle steak |
| | 1103 | Rib, rib steak | 1169 | Round, top (inside) round steak |
| | 1112 | Rib, ribeye roll steak | 1170 | Round, bottom (gooseneck) round steak |
| | 1114 | Chuck, shoulder-clod, top blade steak | 1173 | Loin, porterhouse steak |
| | 1116 | Chuck, chuck eye roll steak | 1174 | Loin, T-bone steak |
| | 1121 | Plate, skirt steak | 1179–1180 | Loin, strip loin steak |
| | 1123 | Short ribs, flanken style | 1184 | Loin, top sirloin butt steak, boneless |
| | 1136–1137 | Ground beef patties | 1185 | Loin, bottom sirloin butt steak |
| | 1138 | Beef steaks, flaked and formed, frozen | 1189–1190 | Loin, tenderloin steak |
| 1200 (lamb portion cuts) | 1200–1201 | Cubed steak | 1232 | Loin chops |
| | 1202 | Braising steak, Swiss | 1234 | Leg chops, boneless |
| | 1204 | Rib chops | 1296 | Ground lamb patties |
| | 1207 | Shoulder chops | 1297 | Lamb steaks, flaked and formed, frozen |
| 1300 (veal and calf portion cuts) | 1300–1301 | Cubed steak | 1336 | Cutlets |
| | 1302 | Veal slices | 1337 | Osso buco, hindshank |
| | 1306 | Rack, rib chops | 1338 | Veal steak, flaked and formed, frozen |
| | 1309 | Chuck, shoulder arm chops | 1349 | Leg, top round, cap off, cutlets |
| | 1312 | Osso buco, foreshank | 1396–1397 | Ground veal patties |
| | 1332 | Loin chops | | |
| 1400 (fresh pork portion cuts) | 1400–1401 | Steak cubed | 1410–1413 | Loin chops |
| | 1402 | Cutlets | 1438 | Steaks, flanked and formed, frozen |
| | 1406 | Boston butt steaks | 1495 | Coarse chopped pork |
| | 1407 | Shoulder butt steaks, boneless | 1496 | Ground pork patties |
| 1500 (cured, smoked, and fully-cooked pork portion cuts) | 1513 | Ham patties (cured), fully-cooked | | |
| | 1531 | Ham steaks (cured and smoked), boneless | | |
| | 1545 | Pork loin chops (cured and smoked) | | |
| | 1548 | Pork loin chops, boneless, center cut (cured and smoked) | | |
| | 1596 | Pork patty, pre-cooked | | |

# Glossary

## A

**abalone:** A univalve contained in a bowl-shaped shell with a brown exterior and an iridescent multicolored interior.

**Achiote (annatto) seeds:** Red, corn-kernel-shaped seeds of the annatto tree, native to South America.

**active dry yeast:** A form of yeast that has been dehydrated and looks like small granules.

**active listener:** A person who provides feedback to the speaker via slight gestures such as direct eye contact, nodding the head to show understanding, and not interrupting until the person is finished.

**adductor muscle:** A solid section of cream-colored flesh that is lean, juicy, possesses a sweet, delicate flavor, and is the portion of the scallop most commonly eaten.

**adequate intake (AI):** The dietary intake value that is used when no DRI has been established, and there is no estimated average requirement calculated for a nutrient.

**aitch bone (pronounced like the letter "h"):** The buttock or rump bone, located at the top of the leg.

**à la carte menu:** A menu that prices all food and beverage items separately.

**albumen:** The clear portion of the raw egg; makes up two-thirds of the egg and consists mostly of ovalbumin protein.

**allemande (pronounced ah-leh-MAHND) sauce:** Sauce made by adding fresh lemon juice and a yolk-and-cream liaison to velouté.

**all-purpose flour:** A type of flour that is a mixture of both hard and soft wheat flours.

**all-purpose shortening:** A solid white product made from hydrogenated vegetable oils.

**allspice (Jamaican pepper):** The dried, unripened fruit of a small pimiento tree that flourishes in Jamaica.

**almond:** The edible kernel of a small, lozenge-shaped fruit that grows on small trees native to the Mediterranean region.

**American cheese:** Processed cheese produced by mixing shredded varieties of various cheeses with some dairy or nondairy products, melting it together, and pouring it into molds to shape it.

**ammonium bicarbonate:** A chemical leavener used mainly in older cookie and cracker recipes.

**amuse bouche:** A very petite hors d'oeuvre served as a complimentary first course by many chefs at upscale restaurants.

**anadromous fish:** A saltwater fish that migrates into freshwater to spawn.

**anise:** A small annual plant from the parsley family that produces a comma-shaped seed known as the anise seed.

**antipastos:** The Italian term for serving small portions of foods as either a starter course or a complete meal.

**appetizer:** Food typically served as the first course of a seated meal.

**appetizer salad:** A salad that is served as a starter to a meal.

**apple:** A hard round pome that can range in flavor from sweet to tart and in color from pale yellow to dark red.

**apprentice:** Someone in a formal training program who shadows a professional to learn directly under the professional's supervision and guidance.

**apricot:** A fruit that has pale orange-yellow skin with a fine, downy texture and a sweet and aromatic flesh.

**aquafarming:** The raising of fish or other seafood in a controlled environment.

**aromatic:** An ingredient such as an herb, spice, or vegetable added to a food to enhance its natural flavors and aromas.

**arrowroot:** A thickening agent derived from a tropical tuber.

**artichoke:** The unopened flower bud of a thistle plant.

**asiago cheese:** A grating cheese with a nutty, toastlike flavor.

**asparagus:** A vegetable consisting of an edible stalk that is most often referred to as a spear.

**aspic:** A savory jelly made from clarified meat, fish, or vegetable stock and gelatin that is used to glaze foods.

**as-purchased (AP) cost:** The entire original cost of a bulk item.

**assistant manager:** A person who helps the general manager carry out all the affairs of the operation and assists in many of the functions of the manager.

**Atlantic (Eastern or American) oyster:** An oyster with a distinctive salty flavor and plump, tender, meaty texture; contained in a flatter shell than other oyster varieties.

**automatic coffee brewer:** A coffee brewer that dispenses coffee into a glass pot held on a warming plate.

**automatic coffee urn:** A large, fully automatic coffee-brewing unit that is connected to a water supply line.

**avocado (alligator pear):** A pear-shaped fruit that grows on trees native to Central and South America.

## B

**baby beef:** A term applied to beef from cattle less than 18 months of age.

**baby T-bone steak:** A 6 oz to 8 oz steak cut from the loin of veal.

**back:** A rack and loin still joined together and well trimmed.

**back strap:** A tough tendon that runs parallel to the vertebrae.

**bacon:** The cured and smoked belly of a hog.

**bacteria:** Single-celled microorganisms that live in soil, water, organic matter, or the bodies of plants and animals and receive their nourishment by supplying their own food, absorbing dissolved organic matter, or obtaining food from a host.

**bain-marie:** A round stainless steel food storage container with high walls used for holding sauces or soups in a hot or cold water bath or steam table.

**bake pan:** A rectangular aluminum pan with 2″ tall sides and loop handles.

**baker:** A person responsible for the operation of the bakery, usually under the supervision of a pastry chef.

**baker's cheese:** A fresh skim-milk cheese that is much like cottage cheese but is softer and finer grained.

**baker's percentage:** The expression of the amount of a particular ingredient as a percentage of the amount of the main ingredient of a formula.

**bakery and pastry section:** An area containing assorted pieces of equipment for producing all of the baked goods and pastries for the operation.

**baker's scale (balance scale):** A scale with two platforms that uses a counterbalance system to measure weight.

**baking:** Cooking by surrounding the item with dry heat in an oven.

**baking powder:** A chemical leavener that is a combination of baking soda and cream of tartar or sodium aluminum sulfate; commonly used to make a batter light and porous.

**baking soda (sodium bicarbonate):** A powerful alkaline chemical leavener that reacts to an acidic dough or batter without the addition of heat by releasing carbon dioxide.

**ballotine:** A poultry leg that is boned out, stuffed with forcemeat, poached or braised until tender, sliced, and served hot.

**balsamic vinegar:** A vinegar made by aging red wine vinegar in wooden casks for many years.

**bamboo shoot:** An immature shoot of the bamboo plant, harvested shortly after it emerges from the soil and reaches approximately 6″ in length.

**banana:** An elongated yellow fruit that grows in a hanging bunch on a banana plant.

**banquet chef:** A person in charge of all parties and banquet functions.

**banquet manager:** A person responsible for the management of all banquets and food-related functions in the operation.

**banquet section:** An area where preparation and production for private dining functions for large or small groups takes place.

**banquette service:** A term that refers to serving customers seated in a banquette.

**barding:** The process of laying a piece of pork fat back across the surface of a lean cut of meat, such as a roast, to add moisture and flavor.

**barrow:** A castrated male hog.

**basil:** The pointy green leaf from a plant of the same name.

**bass:** A common variety of spiny-finned fish with white-colored lean flesh that produces a sweet-tasting, delicate fillet.

**basted eggs:** Fried eggs with unbroken yolks that are cooked the same as sunny-side up eggs, but with the tops slightly cooked by tilting the pan and basting the eggs with hot butter from the pan.

**basting:** The process of continually brushing or ladling juices and fat over an item during the cooking process to aid in moisture retention.

**batonnet cut:** A cut that produces a stick-shaped item with dimensions of ¼″ × ¼″ thick × 2″ long.

**battering:** The process of dipping an item in a wet mixture of flour, liquid, and fat for deep-frying.

**bay leaf:** The thick, aromatic leaf of the evergreen bay laurel tree grown in the Mediterranean.

**beating:** The process of vigorously agitating ingredients to incorporate air.

**béchamel (cream sauce or white sauce):** A mother sauce that is made by thickening milk with a white roux and seasonings.

**beet:** A round root vegetable with a deep reddish purple color.

**bell pepper (sweet pepper):** A fruit-vegetable that contains hundreds of seeds in its inner cavity.

**bel paese:** A semisoft cheese with a buttery flavor and light color that melts easily.

**bench brush:** A brush with long bristles set in vulcanized rubber attached to a wood handle; used to brush excess flour from the bench (baker's table) when working with pastry or bread dough.

**berry:** A type of fruit that is small and has many tiny seeds.

**beurre:** The French word for butter.

**beurre manié:** A mixture made with equal amounts by weight of pastry or cake flour and softened butter that is whisked into a sauce just before service.

**beverage section:** An area containing items needed for the service staff to provide beverages to the guests.

**biological hazard:** Various forms of microorganisms such as bacteria, parasites, viruses, and fungi that can cause a foodborne illness.

**biotin:** B vitamin that aids in cell growth, the production of fatty acids, metabolism, the transfer of carbon dioxide, and hair and nail growth.

**biscuit:** A quick bread made with baking powder or baking soda and shortening; used instead of oil.

**biscuit method:** A quick bread mixing method used for recipes that include a cold solid fat, such as butter or lard.

**bisque:** Form of cream soup that is typically made from shellfish.

**bivalves:** Mollusks that have a top shell and a bottom shell connected by a central hinge, which the animal can close for protection.

**blackberry:** A sweet, dark purple to black, aggregate fruit that grows on a bramble bush.

**blade:** The sharp part of the knife; used for cutting.

**blanching:** A quick moist-heat cooking method used to partially cook an item.

**blending:** The process of mixing two or more ingredients together until they have been evenly distributed throughout the mixture.

**bleu cheese:** A semisoft blue-veined cheese made from cow's milk that is characterized by the presence of green-blue mold.

**blind baking:** A term for baking pie shells 10 min to 15 min before a filling is added.

**blueberry:** A small, dark blue berry that grows on a shrub.

**blue-veined cheese:** A cheese produced by inserting harmless live mold spores into the center of ripening cheese.

**boiling:** A moist-heat cooking method that uses liquid heated to the boiling point as a convection medium to heat food.

**bok choy (pak choi):** A member of the cabbage family that has tender white ribs and bright green leaves.

**bolster:** A thicker band of metal located where the blade of a knife joins the handle.

**bombe (bombe glacée):** A French ice cream dessert where at least two varieties of ice cream are layered inside a spherical mold, frozen, and then unmolded and decorated for service.

**boneless fish:** Fish shaped like round fish; however, they have thick cartilage skeletons instead of bones.

**boning knife:** A short, thin knife with a 6″ to 8″ pointed blade.

**booth service:** A term that refers to serving customers who are seated in a booth.

**Boston butt:** A square, compact cut of hog shoulder located just above the lower half of the shoulder (picnic ham).

**bouillon:** The liquid that remains after simmering meats.

**bound salad:** A combination of a main item, flavoring ingredients, and seasonings held together with a binding agent.

**bouquet garni:** A mixture of fresh vegetables or herbs that are tied into a small bundle with butcher's twine.

**bowl scraper:** Curved plastic scraping tool used to scrape food items from curved surfaces.

**box grater:** A stainless steel box with grids of various sizes used to cut food into small pieces.

**bracelet:** A hotel rack (double rack) of meat with the breast still attached.

**braising:** A combination cooking method for larger (roast size) pieces of meat.

**bran:** The tough outer layer of grain that covers the endosperm.

**branded beef:** Beef with a trademark or trade name that is used by some packers to indicate their own grades.

**Brazil nuts:** Contained in the seeds of very large fruit grown on trees native to the rainforests of South America.

**bread flour (hard wheat flour):** A type of flour that has a high protein content and that produces the most gluten when water is added.

**breading:** A three-step process used to coat and seal an item in preparation for deep frying.

**bread pudding:** A baked custard that is made by pouring a custard mixture over bread chunks and baking it in the oven.

**breast: 1.** The top front portion of the meat above the rib cage, consisting of white meat in flightless birds, or dark meat in birds that fly. **2.** A thin, flat cut of meat located under the shoulder and ribs in veal and lamb.

**breast quarter:** A half of a breast, a wing, and a portion of the back.

**brick cheese:** A washed-rind semisoft cheese made from cow's milk with a mild, sweet flavor and a texture that is firm yet elastic, with many small holes.

**brie:** A soft cheese with a strong odor, a sharp taste, and a creamy white color.

**brisket (breast):** A thin section of meat that contains the ribs, the breastbone, and layers of lean muscle, fat, and connective tissue.

**broccoli:** A member of the cabbage family with tight clusters of dark green florets on top of a pale green stalk with dark green leaves.

**brochette:** An hors d'oeuvre consisting of food that is speared onto wooden, metal, or natural skewers and then broiled or grilled.

**broiler:** A large piece of cooking equipment in which the heat source is located above the food instead of below it.

**broiler cook:** A person responsible for preparing all broiled foods, such as steaks, fish, and chicken.

**broiling:** Cooking with a direct heat source above the item.

**broth:** A flavorful liquid made by simmering stock along with meat or vegetables and seasonings.

**brunoise cut:** A cut that produces a cube-shaped item with six equal sides measuring ⅛″.

**brussels sprout:** A member of the cabbage family that consists of very small round heads of tightly packed leaves that grow along an upright stalk.

**buckwheat:** An herb with edible seeds.

**buffalo chopper (food chopper):** An appliance used to process larger amounts of a product into roughly equal-size pieces.

**buffet service:** A style of service where all the food is arranged on a table and customers walk up to the table to serve themselves.

**bulbs:** Strongly flavored vegetables that grow underground.

**bull:** A sexually mature, uncastrated male cattle.

**bus person:** The person responsible for removing dirty dishes from tables, setting tables, and taking dirty dishes to the dishwashing area.

**butcher's knife:** A heavy knife with a curved, pointed blade that is from 7″ to 14″ in length.

**butler service:** A style of service where servers carry beautifully arranged hors d'œuvres on a silver tray or small elegant platter.

**butter:** A fat product made from cream that has a butterfat content of at least 80%.

**butterflied fillet:** Two single fillets from a dressed fish that are held together by the uncut back or belly of the fish.

**butterfly cut:** A cut used for a thicker meat, fish, or poultry item to separate the item almost completely in half horizontally.

## C

**cabbages:** A variety of vegetables with edible flowers, leaves, or heads.

**cajun spice:** A spice blend that may consist of red and black pepper, oregano, salt, thyme, garlic, ground fennel seeds, and paprika.

**cake flour (soft wheat flour):** A type of flour characterized by low amounts of protein and high amounts of starch.

**cake pan:** A baking pan used for making cakes and similar baked goods.

**calcium:** A macromineral that is used in muscle contraction, blood vessel contraction and expansion, secretion of hormones and enzymes, and sending messages to the nervous system.

**California menu:** A menu that offers all of the food and beverage selections for any meal served throughout the day.

**calorie:** A measure of the amount of energy a food item provides to the body.

**camembert:** A soft cheese made from cow's milk with a yellow color and a waxy, creamy consistency.

**Canadian bacon:** A sliced boneless loin of pork that is trimmed, pressed, and smoked.

**canapé:** An hors d'oeuvre that looks like a miniature open-faced sandwich.

**candy/deep fry thermometers:** Thermometers used to measure the temperatures of hot substances as they are cooking.

**canned fuel:** A gelled, flammable fuel form that is placed beneath the chafing dish; when lit it will provide hours of heat.

**can rack shelving:** Shelving with rails in which cans of product can be loaded from the top.

**cantaloupe:** An orange-fleshed melon with rough, deeply grooved skin.

**capers:** The unopened flower buds of a shrub that grows wild in the Mediterranean.

**capon:** A surgically castrated male chicken.

**caramelization:** A process that adds a rich color to the crust of bakery items containing milk or dairy products.

**caramelize:** To heat sugar (or sugars naturally present in foods) until it liquefies and turns golden brown in color.

**caraway seeds:** The small crescent-shaped brown seeds of the caraway plant.

**carbohydrate: 1.** The complex chemical substance found in food that is the primary energy source for the human body. **2.** A nutrient that provides the body with energy in the form of sugars and starches.

**cardamom:** The dried, immature fruit of a tropical bush in the ginger family.

**carotenoid:** An organic pigment found in orange or yellow vegetables such as carrots, yellow squash, tomatoes, or red peppers.

**carpaccio:** A term used to describe meats or seafood that are sliced thin and served raw.

**career objective:** Single sentence stating what position a candidate is interested in.

**carrot:** An orange-colored root vegetable that is rich in vitamin A.

**carryover cooking:** The rise in internal temperature of an item after it is removed from the oven due to residual heat on the surface of the item.

**cashew nut:** The butter-flavored kidney-shaped kernel of the fruit of the cashew tree, grown in India and Africa.

**cashier:** The person responsible for handling payment in situations where the server does not collect on the bill.

**casserole:** A baked dish of mixed foods containing a starch (such as potatoes, pasta, or rice), other ingredients (such as meat or vegetables), and a sauce.

**cassia (Chinese cinnamon):** The bark of a small evergreen tree that is thicker and darker in color than cinnamon.

**cast iron skillet:** A shallow-walled pan made of thick, heavy iron to withstand extreme high heat.

**catering director:** A person responsible for all functions taking place in an operation.

**catering service:** A food service operation that brings prepared meals and service staff to a client's location.

**catfish:** Freshwater fish named for the whiskerlike barbels that protrude off the sides of the fish's face.

**caul fat:** A meshlike fatty membrane that surrounds sheep or pig intestines.

**cauliflower:** A member of the cabbage family with a head of tightly packed white florets on a short, white-green stalk with large pale green leaves.

**caviar:** The harvested roe (eggs) of sturgeon fish.

**cayenne pepper:** A pungent powder ground from the small pods of certain varieties of hot peppers in the capsicum family.

**celery:** A stalk vegetable that is 12″ to 20″ in length.

**celery root (celeriac):** The knobby root of a type of celery grown for its root rather than its stalk.

**celery seed:** The tiny brown seeds from the lovage herb.

**cephalopod:** A mollusk with tentacles that extend from the base of a distinct head.

**ceviche:** A dish originating in Peru, made from raw fish or shellfish, lemons or limes, chopped onion, and minced chilies.

**chafing dish:** A hotel pan warmer with a reservoir for heated water and a heat source below.

**chain muscle:** A thin strip of tender meat surrounded with fat; located next to the tenderloin.

**chalazae:** The two small stringy parts of the egg white that anchor the yolk to the white.

**channel knife:** A cutting tool with a thin metal blade with one raised channel used to remove a large string from the external surface of a food item.

**charcuterie (pronounced shar-coo-ta-REE):** The production of pâtés, terrines, galantines, sausages, and other products traditionally made from pork.

**chaud (pronounced show):** French term meaning "hot."

**chaud froid platter:** A platter of hot food items presented cold.

**chayote:** A gourd native to Mexico and Central America.

**cheddar:** A hard, mild to sharp-tasting aged cheese.

**cheese:** A dairy product consisting of the coagulated (thickened), compressed, and usually ripened curd of milk that has been separated from the whey.

**cheesecake:** A variety of baked custard that most commonly contains cream cheese.

**cheesecloth:** Loosely woven cotton gauze used to strain stocks and fine sauces.

**chef:** A general term that may refer to many different specialty and culinary management positions in a food service operation.

**chef de partie:** Any of the station chefs that are responsible for a particular area.

**chef instructor:** A person who teaches culinary arts after having worked in the industry as a chef for many years.

**chemical hazard:** Any hazardous substance in chemical form.

**cherry:** A round, smooth-skinned fruit that grows in a cluster on a cherry tree.

**cherry pitter (olive pitter):** A specialty tool used to remove the pits from small fruits such as cherries and olives.

**chervil:** An herb with dark green, curly leaves that is native to Russia.

**cheshire cheese:** The oldest of the named English cheeses; classified as a hard cheese.

**chestnuts:** The fruit of the chestnut tree.

**chèvre:** French term for "goat," that is used for any variety of goat's milk cheese.

**chèvre frais:** Fresh goat's milk cheese.

**chicory:** A hearty, flavorful, and slightly bitter variety of leafy vegetable.

**chiffonade cut:** A cut that produces thinly sliced or shredded leafy vegetables or herbs.

**chiffon filling:** A light, fluffy filling prepared by folding (blending one mixture over another) a meringue into a fruit or cream pie filling.

**chile (hot pepper):** A pepper with a very distinct mild to hot flavor. Varieties include the jalapeño, habañero, poblano, and serrano.

**chili powder:** A spice blend consisting of Mexican peppers, oregano, cumin, garlic, and other spices.

**chill drawer:** A refrigerated pull-out drawer that is located beneath a work surface to hold items to be cooked to order.

**china cap:** A perforated cone-shaped metal strainer used to strain gravies, soups, stocks, sauces, and other liquids.

**chinese five-spice powder:** A combination of equal proportions of ground Szechwan pepper, cloves, star anise, cinnamon, and fennel seeds.

**chinois:** A china cap that strains liquids through a fine-mesh screen.

**chirashi sushi (scattered sushi):** Sushi consisting of assorted toppings that are either scattered over or mixed into vinegar-seasoned rice.

**chitterlings:** The large and small intestines of the hog.

**chive:** A very delicate herb that has a mild onion flavor.

**chlorophyll:** An organic pigment found in green vegetables, such as spinach, broccoli, and asparagus.

**chocolate:** A flavoring ingredient made by roasting, skinning, crushing, and grinding cacao beans into a paste called chocolate liquor.

**cholesterol:** An odorless white waxy substance present in all cells of the body.

**chopping:** Rough-cutting an item so that there are relatively same-size small pieces throughout, although there is no uniformity in shape required.

**chowder:** A very hearty soup with large chunks of potatoes and other main ingredients.

**chow mein:** Noodle dish created with mian.

**chuck (shoulder clod chuck):** The shoulder of an animal.

**cider vinegar:** Vinegar made by fermenting unpasteurized apple juice or cider until the sugars are converted into alcohol.

**cilantro (Chinese parsley):** The stem and leaves of the cilantro plant.

**cinnamon:** The dried, thin, inner bark of a small evergreen tree native to India and Sri Lanka.

**citrus:** A type of fruit that grows on thorny trees or shrubs in tropical regions; has a thick rind and pulpy meat.

**clam knife:** A specialty tool with a short, flat, round-tipped blade and a sharp edge.

**clams:** Bivalves found in both freshwater and saltwater.

**clarify:** To remove impurities, sediment, cloudiness, and particles to leave a very clear and pure liquid.

**classical cuisine:** A cooking style where the quality of ingredients and the use of refined preparation techniques are emphasized.

**Classical French Service:** A style of service where some of the food items are either finished or fully prepared tableside in front of the guest.

**clearmeat:** Very cold, lean ground meat, fish, or poultry that is combined with an acid such as wine, lemon juice, or tomato product and cold ground mirepoix, egg whites, and an oignon brûlé (pronounced onion brew-lay).

**clear (unthickened) soup:** A stock-based soup with a thin, watery consistency.

**cleaver:** A heavy, rectangular-bladed knife.

**clove:** The dried, unopened bud of a tropical evergreen tree primarily found in the East Indies and on islands off the coast of Africa.

**coagulation:** The process of changing from a high-moisture (liquid) state to a low-moisture (semi-liquid or solid) state.

**cockle:** A bivalve mollusk with a 1″-wide shell with deep, straight ridges that surround a small, fleshy meat.

**cocoa powder:** A brown unsweetened powder extracted from ground cacao beans.

**coconut:** The large fruit of the coconut palm tree.

**colander:** A bowl-shaped perforated metal strainer usually made from stainless steel or aluminum.

**cold-pack cheese:** A processed or blended cheese made from pasteurized milk without the aid of heat.

**cold wrap:** A variety of cold sandwich in which a flat bread or tortilla is coated with a spread, topped with one or more fillings, and rolled tightly.

**collagen:** A soft, white, connective tissue that breaks down into gelatin when heated.

**collard greens:** Greens in the kale family that have large, green leaves with a thick, white vein.

**combination cooking:** Any cooking method that uses both moist and dry cooking methods.

**combi oven:** An oven that cooks food by surrounding it with moist, hot air and combines the convection cooking method at the same time.

**commis (apprentice cook):** A person who is currently enrolled in a culinary school and who desires to excel in the food service and hospitality industry.

**complete proteins:** Proteins that contain all nine essential amino acids and are found in animal-derived foods such as beef, chicken, fish, eggs, cheese, and other dairy products.

**complex carbohydrate (starch):** A carbohydrate that provides energy to the body.

**composed salad:** A salad consisting of a base, body, garnish, and dressing carefully composed and arranged attractively on a plate.

**compound butter:** A flavorful butter made by working flavoring ingredients such as fresh herbs, vegetable purées, dried fruits, preserves, or wine reductions into whole butter.

**compressed (fresh) yeast:** A yeast that has approximately 70% moisture content and is available in 1-lb cakes or blocks.

**conch:** A univalve that has a pinkish-orange shell and resembles a large snail.

**Concord grape:** A seeded grape with a deep black color.

**condiments:** Savory, spicy, or salty accompaniments to food, such as a relish or a sauce.

**conduction:** A method of heat transfer where heat is passed from one object to another through physical contact.

**confectioners' (powdered) sugar:** Granulated sugar that has been ground into a very fine powder.

**confit (pronounced cone-FEE):** A French term for meat that has been cooked and preserved in its own fat.

**consommé (pronounced con-so-may):** A very rich and flavorful broth that has been further clarified to remove any impurities or particles that could cloud the finished product.

**contaminant:** Any microorganism or substance that can contaminate food or preparation equipment.

**contamination:** The state of food or equipment being potentially hazardous, resulting from unsafe organisms or other items coming in contact with food or preparation equipment.

**convection oven:** An oven with an interior fan that circulates the dry, hot air throughout the cabinet.

**conventional oven:** An enclosed heating cabinet typically located beneath a range or within a wall unit.

**conversion factor:** The number of units of one measurement that equals one unit of another measurement.

**converted rice:** Specially processed long-grain rice that is parboiled to remove the surface starch and hull, dried, and further milled to produce either brown or white rice.

**cooking:** The process of subjecting foods to heat in order to accomplish three main goals: to make them taste better, to make them easier to digest, and to kill harmful microorganisms that may be present in the food.

**cooking-loss yield test:** A procedure used to determine the total weight as served of a food product.

**cooling wand:** A heavy-gauge, hollow plastic paddle with a screw-on cap at the top of the handle that is filled with water and frozen prior to use.

**copper:** A trace mineral that is used with iron in the formation of hemoglobin to make red blood cells.

**coriander:** The ridged seeds of the coriander plant.

**corn flour:** Cornmeal that has been finely ground.

**cornmeal:** Dried, ground corn kernels.

**corn syrup:** A sweet syrup made by adding an enzyme to starch extracted from corn kernels.

**costing:** The process of determining the total cost of preparing a recipe based on the individual costs of all ingredients.

**cottage cheese:** A fresh cheese that is the simplest of all cheeses.

**cottage ham:** Smoked, boneless meat extracted from the blade section of the Boston butt.

**coulis:** A sauce typically made from either raw or cooked puréed fruits or vegetables.

**count:** The measurement of whole items or the actual number of items being used.

**coupe:** The French word for the dessert commonly known as a sundae.

**court bouillon (French for "short broth"):** A highly flavored and aromatic vegetable broth made from simmering vegetables with herbs and a small amount of an acidic liquid (usually vinegar or wine).

**cow:** A mature female cattle that has borne one or more calves.

**cranberry:** A small round fruit that is red in color and sour tasting.

**crayfish (crawfish, crawdad):** A freshwater crustacean that resembles a small lobster.

**cream cheese:** A soft, fresh cheese with a rich, mild flavor.

**creaming:** The process of vigorously combining fat and sugar to incorporate air.

**creaming method:** A cake-mixing method that is sometimes applied to muffins and quick loaf breads.

**cream of tartar:** A chemical leavener derived from tartaric acid, a by-product of wine production.

**crème anglaise:** Vanilla custard sauce made by whisking sugar in two stages; half into whisked egg yolks and the other half into scalded milk or cream.

**crème brulée:** A baked custard that still possesses the texture of a spoonable stirred custard beneath the surface.

**crème caramel (flan):** A baked custard in which a small amount of hot caramelized sugar is poured into a small serving dish.

**crêpes:** French pancakes that are lighter and thinner than basic pancakes.

**crêpe pan:** A small skillet with very short sloped sides.

**crisp cookies:** Cookies prepared from dough that contains a high percentage of sugar.

**cross-contamination:** Contamination that occurs when a biological, chemical, or physical contaminant is transferred from one item to another through an intermediate carrier.

**cross-hatch markings:** Crisscrossed charred lines created when food comes in contact with a hot grill.

**crown roast:** An unsplit rack of lamb with the rib ends Frenched and the ribs formed into a circle to resemble a crown.

**crudités (pronounced crew-dee-tay):** A raw vegetable platter, French word for "raw."

**crumb crust:** A pie crust made with crumbled cookie or graham cracker pieces that are pressed together with melted butter to form a solid shell.

**crushed red pepper (crushed chilies or chili flakes):** A blend of crushed hot chili peppers.

**crustacean:** A shellfish that has a segmented, hard external shell.

**cucumbers:** Fruit-vegetables that are often eaten raw or pickled.

**cumin:** The dried, aromatic seeds of a plant that is a member of the parsley family.

**curd:** The thick, casein-rich part of coagulated milk.

**curing:** The processing of an item with salt or chemicals to retard the action of bacteria and to preserve the meat

**currant:** A small red or black berry that grows on bushes native to western Europe.

**curry leaf (neem leaf):** A shiny green leaf from a small tree native to southern India.

**curry powder:** A blend of a number of spices, typically including combinations of cloves, cumin, coriander, black pepper, red pepper, mustard, cinnamon, nutmeg, ginger, cardamom, fenugreek, and turmeric.

**custard:** A liquid that is thickened by cooking and coagulating egg proteins that are part of the mixture.

**cutlet:** A thin, boneless slice of veal.

**cutting board:** A cutting surface designed to protect work surfaces from cuts and scratches.

**cutting in:** The process of incorporating a solid cold fat into dry ingredients, such as flour, until pea-sized lumps are formed.

**cuttlefish:** A cephalopod with well-developed eyes, a cuttlebone, and ten tentacles.

**cycle menu:** Common in institutional food service settings such as hospitals and school cafeterias.

## D

**dandelion green:** A green from the dandelion plant, usually considered a weed, with yellow flowers and long, dark green, serrated leaves.

**dark brown sugar:** A moist sugar product that contains approximately 7% molasses.

**dark corn syrup:** A corn syrup with added caramel color and flavor to give it an aroma similar to molasses.

**deck oven:** A drawerlike oven that is commonly stacked one on top of another, providing multiple-temperature baking shelves.

**decline stage:** The phase in which the bacteria die and leave behind high levels of toxins.

**deep fat fryer:** A cooking unit used to cook foods by submersion in hot fat.

**deep frying:** A cooking method that involves submerging foods completely in very hot fat, usually between 350°F and 375°F.

**degreasing:** The process of removing surface fat from a food.

**demi-glace:** Sauce made by adding equal parts of espagnole and brown stock together and reducing the mixture by half.

**dessert salad:** A sweet salad usually consisting of nuts, fruits, and sweeter vegetables such as carrots, and may be bound by gelatin.

**diagonal cut:** A cut that produces oval slices.

**dietary reference intake (DRI):** A quantity of a particular nutrient that is sufficient to meet the nutrient requirements of 97% to 98% of all healthy individuals in a group.

**digital scale:** A scale with a spring-loaded steel platform and a digital display.

**dill:** A member of the parsley family with feathery, blue-green-colored leaves.

**dim sum:** The Chinese term for serving small portions of foods as either a starter course or a complete meal.

**dip:** A creamy sauce or condiment that is served with hard food such as bread, vegetables, or chips.

**direct contamination:** Contamination that occurs when uncooked foods, or the plants or animals that the foods are made with, are contaminated in their natural environment.

**discrimination:** Unfair treatment of people based on characteristics such as gender, race, ethnicity, age, religion, appearance, or disability.

**dishwasher:** A person who operates the dishwashing machine, which washes all china, glassware, and silverware (called flatware), while keeping breakage to a minimum.

**docking:** The process of making small holes in the dough before it is baked.

**double boiler:** A round stainless steel pot that sits on top of another pot containing simmering water.

**double lamb chop:** A rib chop cut to a thickness equal to that of two standard rib chops.

**dough cutter (bench scraper):** A flat stainless steel blade attached to a sturdy handle.

**dough docker:** An aluminum, heavy plastic, or stainless steel roller with stainless steel pins used to perforate dough.

**drawn fish:** Fish that have only the viscera removed.

**dredging:** The process of lightly dusting an item in seasoned flour or fine bread crumbs for deep frying.

**dressed:** A term used to describe an animal that has been slaughtered and cleaned for consumption by removing all blood and inedible parts.

**dressed fish:** Scaled fish with the viscera, gills, and fins removed.

**drumstick:** The lower section of a leg of poultry; located below the knee joint.

**drupe:** A fruit that contains only one seed or pit.

**dry aging:** The process of aging larger cuts of meat that are hung in a very well-controlled environment where temperature, humidity, and constant air flow are monitored around the clock for up to 6 weeks.

**dry-heat cooking:** Any cooking method that uses hot air, hot metal, a flame, or hot fat to conduct heat to the food without any moisture.

**dry measuring cup:** A metal cup used to measure dry ingredients.

**dry-rind cheese:** A cheese that is allowed to ripen with its exterior exposed to air.

**dry storage:** A clean and secure area where dry goods are kept before use.

**dungeness crab:** A Pacific crab that has desirable, sweet-tasting meat.

**dunnage rack shelving:** Shelving consisting of reinforced stainless steel platforms that serve to store items at least 6″ above the floor.

**dutch-processed cocoa powder:** A dark, unsweetened cocoa powder processed with an alkali to neutralize the natural acidity.


## E

**éclair paste:** A dough made from beating eggs into a paste of boiled water, butter, and flour.

**edam:** A waxed-rind semisoft cheese made from cow's milk with a firm, crumbly texture; usually shaped like a ball with a slightly flattened top and bottom and a red wax coating.

**edible portion (EP):** The amount of a food item that can be used in a recipe after unusable parts are trimmed away.

**edible portion cost:** The cost of the usable part of a bulk item.

**education:** Résumé category that lists the education a job candidate has obtained.

**eel:** A long, slender fish with a body that resembles that of a snake.

**eggplant:** A fruit-vegetable with deep purple, edible skin and yellow to white, spongy flesh.

**egg slicer:** A slicing tool with a series of tightly pulled wires that, when pressed against a peeled hard-cooked egg, will slice the egg into consistently thick slices.

**elastin:** A tough, rubbery, yellowish connective tissue that does not break down with heat.

**electric slicer:** A tool used to slice foods such as meats and cheese into uniform slices.

**electronic probe thermometer (thermocouple thermometer):** A thin stainless steel stem attached by wires to a battery-operated readout device.

**employee handbook:** A written document containing all of the official policies and procedures of an establishment.

**employment application:** A standard form that requests basic information from a job applicant, such as name, address, relevant work experience, educational background, position desired, and availability.

**emulsified shortening:** A commercial shortening made from hydrogenated vegetable oils that is used in high-ratio cakes (cakes that contain more sugar than flour).

**emulsifier:** A substance that enables two substances that would usually not mix well, such as oil and water, to blend together into a smooth substance.

**emulsion: 1.** A mixture of two typically unmixable liquids (such as oil and water) that are forced to bond with each other to result in a creamy, smooth product with a uniform appearance. **2.** An oil that has been mixed with water and an emulsifier.

**endosperm:** The largest component of a grain kernel; milled to produce flours and other products.

**English lamb chop:** A 2″-thick cut taken along the entire length of the unsplit loin.

**escargot:** The French term for "snail"; refers to any variety of land snail fit for human consumption.

**espagnole (pronounced ess-spah-nyol) sauce:** A mother sauce made from a full-bodied brown stock, brown roux, tomato purée, and a hearty caramelized mirepoix.

**essence:** A fish stock similar to a fumet that uses a greater amount of aromatic ingredients such as celery, morels, a bouquet garni, a sachet, and even fennel root.

**ethylene gas:** An odorless gas that a fruit emits as it ripens and that encourages other surrounding fruits to ripen.

**European-style butter:** A type of butter that contains 82% to 86% butterfat.

**eviscerate:** To remove all entrails and organs from a carcass.

**executive chef (head chef):** The person in charge of the kitchen in a large operation.

**expediter (food checker):** A person responsible for all food that leaves the kitchen and the look of each plate.

**extract:** A flavorful oil that has been mixed and dissolved with alcohol.

**extrusion:** The process of shaping pasta by pushing the dough through dies that create various shapes.

## F

**fabricated (portion-controlled) cut:** A small, ready-to-cook cut that is packaged to certain specifications for quality, size, and weight.

**facultative bacteria:** Bacteria that can survive either with or without oxygen.

**family-style service:** A style of service where food is served on platters and placed directly on the table.

**farro (spelt):** An ancient Italian grain that is high in protein and similar in taste to barley.

**fat:** A nutrient that provides energy, promotes healthy skin, and carries fat-soluble vitamins such as A, D, E, and K throughout the body.

**fats:** Organic compounds found in plants and animals that later become energy when consumed as food.

**fat cap:** A thick layer of fat that surrounds a muscle.

**fat-soluble vitamins:** Vitamins that dissolve in fat.

**fell:** The thin, paperlike covering on the outside of a lamb carcass.

**fennel:** A celery-like stalk with overlapping leaves that grow out of a large bulb at its base.

**fennel seed:** The small seedlike fruit of the fennel plant.

**fenugreek:** A spice that is the pepple-shaped seed of a plant in the pea family.

**feta:** A fresh cheese of Greek origin made from sheep's or goat's milk.

**fiber:** A form of a complex carbohydrate that is nondigestible and nonnutritive, but essential in a healthy diet.

**fig:** The small, pear-shaped fruit of the fig tree.

**file powder:** Made from the ground leaves of the sassafras plant.

**fillet:** The fleshy side of a fish, cut lengthwise away from the backbone.

**filling:** The main ingredient in a sandwich.

**fines herbes:** The combination of parsley, chives, tarragon, and chervil used in classical French cuisine.

**finger cot:** A protective sleeve placed over the finger to prevent contamination of a cut.

**fingerling:** A small, tapered, waxy potato.

**fire-suppression system:** An automatic fire extinguishing system that is activated by the intense heat generated by a fire.

**first in, first out (FIFO):** The process of dating items as they are received, and rotating older items to the front while placing the new items behind the old.

**first-line manager:** An employee responsible for the day-to-day supervision of hourly employees.

**fish:** A classification for aquatic animals that have fins for moving through the water, gills for breathing, and an internal bone structure with a backbone (or vertebrae).

**fish poacher:** A long, thin style of pot with loop handles specifically designed to poach fish.

**fish steak:** A cross-sectional slice of a larger-size dressed fish.

**fixed cost:** A cost that does not change as sales increase or decrease.

**fixed menu:** A menu that is developed and then rarely changes.

**flaky pie dough (pâte brisée):** A dough prepared by cutting the fat into the flour until pea- or hazelnut-size particles of fat are formed.

**flambé (pronounced flahm-bay):** A French term meaning "to flame" which refers to the procedure of heating alcohol and igniting it to deglaze a pan.

**flank:** A thin, flat section of the beef hindquarters located beneath the loin.

**flashbake oven:** An oven that uses both infrared and visible light waves to cook foods quickly and evenly.

**flatfish:** Thin, wide fish with both eyes located on one side of the head.

**flat wing tip (paddle):** The second section of wing located between the two wing joints.

**flavonoid:** An organic pigment found in purple, dark red, and white vegetables such as red cabbage, beets, parsnips, and cauliflower.

**flavored vinegars:** Vinegars made from any traditional vinegar in which other items such as herbs, spices, fruits, vegetables, or flowers are added.

**flavorings:** Items added to food that alter the natural flavor of food so much that the food will be described with a flavoring in its name, such as vanilla ice cream or raspberry vinaigrette.

**flounder:** A lean-fleshed saltwater flatfish.

**flour:** The fine powder that is left after grinding grain.

**foie gras (pronounced fwah grah):** The fattened liver of a duck or goose.

**folate:** A B vitamin that is used in cell production and maintenance, is particularly important during infancy and pregnancy, and is found in leafy green vegetables.

**folded omelet:** An omelet that is cooked until nearly done and then folded before serving.

**folding:** The process of gently incorporating light ingredients with heavier ones.

**fondant:** A rich, white, cooked icing that hardens when exposed to the air.

**fontina:** A waxed-rind semisoft cheese made from cow's milk.

**foodborne illness:** An illness that is carried or transmitted to two or more people through contact with or consumption of contaminated food.

**Food Code:** Establishes standards that assist food control jurisdictions at the national, state, and local levels in regulating the food service industry.

**food mill:** A hand-cranked kitchen tool used to purée soft or cooked foods.

**food processor:** An appliance used to purée, chop, grate, slice, and shred food.

**food safety:** The practice of handling food in ways to prevent contamination or spoilage.

**food service buyer:** A person at a food service establishment who handles the selection and acquisition of food service items.

**food service director:** A person responsible for all budgets and top management decisions, as well as supervision of other managers and supervisors employed in the establishment.

**food service distributor:** An organization that sells food, supplies, or equipment to food service clients.

**food service sales:** A career area that allows someone interested in the food service industry to work for a distributor.

**food waste disposer:** A food grinder mounted beneath warewashing sinks to eliminate solid food material.

**foresaddle:** The front half of the carcass consisting of the shoulder, rack, breast, and shank.

**forequarter:** The front quarter of beef consisting of the rib, chuck, shank, brisket, and short plate.

**formal evaluation:** A scheduled evaluation conducted at a predetermined time, such as every six months, where an employee meets with the manager for an evaluation that is documented and becomes part of an employee's permanent record.

**formula:** A format for a recipe where all ingredients are listed as a percentage relative to the main ingredient.

**French grill:** A cast iron plate with raised ridges on one side and a completely flat cooking surface on the other side.

**frenching:** A method of removing the meat and fat from the end of a rib bone, and is generally applied to chops.

**French knife (chef's knife):** A large, multipurpose knife with a tapering blade used for slicing, chopping, mincing, and dicing.

**frittata:** A traditional folded omelet, but is served open-faced after being browned under a broiler or in a hot oven.

**froid (pronounced fwa):** French term meaning "cold."

**fruit corer:** A cylindrical tool that is used to remove the center core from fruits such as apples or pineapples.

**fruits:** The edible, ripened ovaries of flowering plants that contain one or more seeds.

**fruit-vegetable:** A fruit that is served as a vegetable.

**fry cook (short order cook):** A person responsible for work performed around the range and deep fat fryer.

**fumet:** A concentrated stock made from fish bones or shellfish shells and vegetables that is used to make soups and sauces.

**fungi (singular fungus): 1.** An extremely large group of plants that range in size from tiny microorganisms to large mushrooms. **2.** Plantlike organisms such as mushrooms and truffles.

**funnel:** A tapered bowl and tube used to pour a liquid from a larger container into a smaller container.

**fusion cuisine:** A cooking style that blends characteristics of two or more ethnic cuisines.

## G

**galantine:** A boned-out cut of meat, poultry, or seafood that is stuffed with forcemeat, poached in stock, cooled, sliced, and glazed with aspic.

**game bird:** A wild bird, such as quail, partridge, pheasant, or grouse, that is hunted for human consumption.

**garde manger (pantry and cold kitchen chef):** A person responsible for the cold food department of a professional kitchen.

**garde manger section:** An area of the professional kitchen where salads, salad dressings, deli and cheese trays, cold appetizers, cold sandwiches, buffet showpieces or centerpieces, cold platters, and charcuterie are prepared.

**garlic: 1.** A member of the lily or onion family and has a strongly flavored and aromatic bulb, commonly called a head. **2.** A bulb vegetable made up of several small cloves that are enclosed in a thin, pale, husklike skin.

**gastrique:** A sugar syrup made by caramelizing a small amount of granulated sugar in a saucepan and deglazing the pan with a small amount of vinegar.

**gelatin:** A flavorless thickening agent made from animal protein.

**gelatinization:** The process of a starch absorbing moisture.

**gelatin salad:** A salad made from flavored gelatin formed in a mold.

**gelato:** An Italian version of ice cream, with a creamier, denser texture than that of typical American ice cream.

**general manager:** A person who conducts and directs all affairs of the operation and oversees food production, beverage sales, and customer service.

**genoise:** The most common sponge method where the eggs and sugar are warmed and whipped to create volume and incorporate air before any other ingredients are added.

**genuine spring lamb:** A lamb marketed between the age of 3 and 5 months that has been fattened primarily on its mother's milk.

**germ:** The smallest part of a grain kernel and contains a small amount of natural oils as well as vitamins and minerals.

**giblets:** The collective name for a bird's neck, heart, liver, and gizzard (the bird's second stomach).

**gilt:** An immature female hog.

**ginger:** A spice that comes from the bumpy root of a tropical plant grown in China, India, and Jamaica.

**glace (pronounced glahss):** A highly reduced stock that results in an intense flavor.

**glace de poisson:** Fish glaze.

**glace de veau:** Veal glaze.

**glace de viande:** Beef glaze.

**glace de volaille:** Chicken glaze.

**glazing:** The process of covering an item with water to form a protective coating of ice before the item is frozen.

**glucose syrup:** A type of corn syrup that contains a high amount of glucose (sugar).

**gluten:** A combination of proteins in flour that provides the strength to hold the shape, form, and texture of bakery products when cooked. The protein in flour that, when combined with water, gives structure, strength, and elasticity to a baked product.

**gorgonzola:** A blue-veined cheese that is mottled with characteristic blue-green veins produced by a mold known as penicillium glaucum.

**gouda:** A semisoft waxed-rind cheese that is similar to Edam cheese, but contains more fat.

**grain:** The edible fruit (kernel or seed) of a grass.

**grain-fed beef:** Meat obtained from cattle that were grain-fed for a period of 90 days to a year.

**grana (meaning "grain"):** Group of Italian cheeses, Parmesan being one example.

**Grande cuisine:** A style of food preparation involving intricate, elaborate cuisine and strict adherence to elaborate preparation methods and culinary principles.

**granité:** A frozen mixture made by frequently stirring a mixture of water, sugar, and flavorings such as fruit juice or wine as it is freezing.

**granulated sugar:** An all-purpose, white sugar composed of small, uniform-size crystals.

**grape:** A type of fruit that has a smooth skin and grows on woody vines in large clusters.

**grapefruit:** A round fruit with a thick, yellow outer rind and rather tart flesh.

**grass-fed beef:** Meat obtained from cattle that were raised on grass with little or no special feed.

**gratin:** Any dish prepared using the gratinée method.

**gratinée (pronounced grah-teen-YAY):** The process of topping a dish with a thick sauce, cheese, or bread crumbs and then browning it in a broiler or high-temperature oven.

**grating cheese:** A hard, crumbly, dry cheese used grated or shaved onto food prior to serving.

**griddle:** A cooking surface made of metal on which foods are cooked.

**griddling:** The process of cooking foods on a solid metal cooking surface called a griddle.

**grill:** A cooking unit consisting of a large metal grate, also referred to as a grill, placed over a heat source.

**grilling:** The process of cooking food over a heat source on open metal grates.

**grits:** A type of meal made from ground corn or hominy.

**grooved griddle:** A griddle with raised ridges that create grill marks on foods.

**grouper:** A lean, white-fleshed fish that has excellent eating qualities, similar to those of a red snapper.

**gruyère:** A hard cheese made in Switzerland that is similar in many ways to Emmenthaler, but has smaller holes and a sharper taste.

**guava:** A small oval-shaped fruit, usually 2″ to 3″ in diameter, with thin edible skin that can be yellow, red, or green.

**gueridon:** A cart that can be wheeled to a table and is typically equipped with a carving station or a sautéing and flambéing unit called a réchaud (pronounced ray-show).

**guild:** An organization of craftsmen having exclusive control of the production of a particular craft and the distribution of its products.

**H**

**HACCP:** A systemic approach to the identification, evaluation, and control of food safety hazards.

**halibut:** A very large flatfish that resembles a giant flounder; only swordfish, tuna, and some sharks reach a larger size.

**ham:** The thigh and buttock of the hog.

**ham hock:** The knee joint of a hog.

**hand tools:** Handheld implements used in cutting, preparing, and serving food.

**harassment:** Behavior that is found to be threatening or disturbing and that is not considered acceptable by the general public.

**hard cheese:** A firm, somewhat pliable and supple cheese with a slightly dry texture and buttery flavor.

**hard-shell clam (quahog):** A clam with a blue-grey shell that contains a chewy flesh.

**havarti:** A Danish dry-rind semisoft cheese made from cow's milk with a buttery, somewhat sharp flavor.

**hazelnut (filbert):** A marble-sized nut from the hazel tree.

**head cabbage:** A member of the cabbage family that consists of many layers of thick leaves that form a head.

**head cheese:** The jellied, spiced, pressed meat from the head of a hog.

**heifer:** A young female cattle that has not borne a calf.

**hen (stewing chicken):** A female chicken that has laid eggs for one or more seasons and is usually more than 10 months old.

**herbs:** A group of aromatic plants whose leaves, stems, or flowers are used to add flavor and aroma to food.

**Herbes de Provence:** A classical mixture of herbs used in the cuisine of southern France.

**herring:** A long, thin fish that is somewhat fatty with shiny silvery-blue skin.

**highly susceptible populations:** People more likely to experience foodborne illness than others, such as infants and young children, pregnant women, the elderly, and people who are seriously ill, who take certain medications, or who have life-threatening allergies.

**hindsaddle:** The tail or rear half of the carcass consisting of the loin and leg.

**hindquarter:** The back quarter of meat consisting of the short loin, flank, sirloin, rump, and round.

**hollandaise:** A mother sauce that is thickened with egg yolks.

**hominy:** A corn product treated with lye, which causes the kernels to swell to twice their normal size before the hull is stripped.

**honey:** A sweet, thick fluid made by honey bees from flower nectar and is 1½ times as sweet as granulated sugar.

**honeycomb tripe:** The lining of the second stomach in cattle.

**honeydew:** A melon with a smooth outer skin that changes from a pale green color to a creamy yellow color as it ripens and has a mild, sweet flavor.

**honing (truing):** The process of aligning a blade's edge and removing any burrs or rough spots on the blade.

**hors d'oeuvre:** An elegant, bite-size portion of food, creatively presented and meant to be served apart from a seated meal.

**horseradish:** A spice obtained by peeling and grating horseradish root.

**host or hostess:** The person responsible for welcoming and seating customers in the food service establishment.

**hotel pan:** A stainless steel pan used to cook, serve, or hold food.

**hotel rack:** An unsplit rib section of a lamb carcass.

**hot foods section:** An area containing equipment such as a broiler, deep fryer, and open burner range as well as the refrigeration and storage needed to operate the section.

**husk (hull):** The inedible, protective outer covering of grain.

**hydrogenated shortening:** A commercial shortening made from vegetable oils that have been hydrogenated to produce a solid fat with a flexible melting point.

**hydrogenation:** The change that occurs when additional hydrogen atoms are forced to bond with molecules found in plant oils such as corn oil; changes the molecular structure of liquid fat into a creamy or solid state.

**hygroscopic:** The ability to attract water.

**icing:** A sugar-based coating often spread on the outside or between layers of a baked good.

**Inari sushi:** Vinegar-seasoned rice and toppings stuffed into a small purse of fried tofu.

**incomplete proteins:** Proteins that do not contain all nine essential amino acids, but if combined correctly with other proteins, a complete protein can be obtained.

**individually quick-frozen (IQF):** A designation for products preserved using a method whereby each item is glazed with a thin layer of water and frozen individually.

**induction range:** An electric range that uses a magnetic coil below the surface of the burner to heat food rapidly.

**informal evaluation:** An evaluation done on a daily basis by a manager, providing constructive feedback on daily skills, duties, and overall performance.

**infrared thermometer:** A thermometer that measures the surface temperature of an item through the use of infrared laser technology.

**insoluble fiber:** Fiber that will not dissolve in water.

**instant dry yeast:** Yeast that does not need to be hydrated first.

**instant read thermometer (stem thermometer):** A thermometer with a long stainless steel stem attached to either a digital or mechanical display.

**institutional cook:** A person who prepares large quantities of prepackaged and prepared foods in an institutional setting such as a school cafeteria, hospital, or the military.

**institutional cooking:** Cooking at an operation where all food is prepared in large quantities.

**institutional food service:** A type of food service that features a cafeteria-style operation or a limited menu and is usually found in a university, hospital, or school.

**Institutional Meat Purchase Specifications (IMPS):** Specifications published by the United States Department of Agriculture (USDA) for commonly purchased meats and meat products.

**insulated carrier:** An insulated container made of heavy polyurethane or other plastic material designed to hold hotel pans of hot or cold foods for transport.

**interpersonal skill:** An ability to work with and treat people with respect and as individuals.

**iodine:** A trace mineral that helps control energy use in the body and helps regulate the thyroid gland, which is what controls metabolism.

**iron:** A trace mineral that is necessary to hemoglobin production, the substance in blood that carries oxygen throughout the body.

**irradiation:** A process where food is subjected to ionizing radiation to kill insects, bacteria, and parasites.

## J

**Jerusalem artichoke (sunchoke):** A tuber with a thin, brown, knobby-looking skin, similar to that of ginger root. Jerusalem artichokes are related to sunflowers.

**jícama:** A round tuber between 6″ and 8″ in length that is native to Mexico and Central America.

**job description:** A summary of an open position that accurately describes skills, qualities, education, and previous work experience desired in an applicant, as well as a summary of the duties and responsibilities of the job.

**job interview:** A face-to-face meeting between a manager and a job candidate that allows the manager to find out more about the candidate, provide information about the job, and answer any questions that the candidate may have.

**jowl:** The cured and smoked cheek meat of a large hog. It is sometimes referred to as jowl bacon.

**juicer:** An electric or manual device used to extract juice from fruits and vegetables.

**julienne cut:** A cut that produces a stick-shaped item with dimensions of ⅛″ × ⅛″ thick × 2″ long.

**juniper berries:** Small, purple berries with an aromatic flavor similar to that of rosemary.

**jus lié (also called a fond lié):** Made from equal parts of espagnole and brown stock.

## K

**kale:** A member of the cabbage family that has large, curly leaves that vary in color.

**kamut:** An ancient variety of wheat that has a natural sweetness and a buttery quality.

**kasha:** Toasted buckwheat.

**ketchup (catsup):** A thick, tomato-based sauce.

**king crabs:** The largest-size variety of crabs, typically weighing between 6 lb and 20 lb and measuring as much as 6′ from the tip of one leg to the tip of the opposite leg.

**kitchen brigade system:** A structured chain of command with specific job titles and duties that streamline food production in the professional kitchen.

**kitchen fork (meat fork or chef's fork):** A large fork with two long prongs.

**kitchen sheers:** Heavy-gauge scissors used in the preparation of foods.

**kitchen spoon:** A large spoon made of stainless steel.

**kitchen timer:** A time measuring tool that indicates the amount of time that has passed, or sounds an alarm when a specified time period has ended.

**kiwifruit:** A small, barrel-shaped fruit, approximately 3″ long and weighing between 2 oz and 4 oz.

**kneading:** The process of developing the gluten in a dough.

**kohlrabi:** A hybrid vegetable created by crossbreeding cabbages and turnips; has a pale green or purple bulbous stem and dark green leaves.

**kosher salt:** A type of salt used for curing, seasoning, and preparing kosher foods.

## L

**ladle:** A stainless steel cuplike bowl attached to a long handle and used to stir or serve soups, stocks, dressings, and sauces.

**lag phase:** The adjustment phase where bacteria reproduce slowly.

**lamb:** The meat of immature sheep slaughtered before one year of age.

**langoustine:** A lobster that is similar in appearance to a shrimp, except it has very long front arms with long, thin claws.

**lard:** A 100% fat product made from rendered pork fat.

**larding:** The process of inserting thin strips of pork fat into lean meat with a larding needle.

**large volume measure:** A large, graduated aluminum container used to measure volume.

**lavender:** A member of the evergreen family with bluish-green, spiky stems and purple flowering tops.

**leadership skill:** An ability to influence others in a way that results in changes reflecting the values, goals, and vision of an individual or organization.

**lean doughs:** Doughs that are low in fat and sugar.

**leavening:** The process by which bakery products rise by the action of air, steam, chemicals, or yeast.

**leavening agent:** Any ingredient that causes a bakery product to rise (leaven).

**lecithin:** A natural emulsifier found in egg yolks.

**leek:** A long, white vegetable with a cylindrical bulb, similar in appearance to a scallion but much larger.

**leg:** A poultry cut that consists of a drumstick and a thigh.

**leg quarter:** A thigh, a drumstick, and a portion of the back.

**legume:** A plant that produces pods that contain a row of edible seeds.

**lemon:** A tart yellow citrus fruit with high acidity.

**lemongrass (citronella grass):** The fibrous stalk of a tropical grass.

**lemon zest:** The grated peel and pith of a lemon that is often used as a flavor enhancer.

**lettuce:** A member of the category of greens known as "salad greens."

**liaison:** A mixture of egg yolks and heavy cream that is used to thicken sauces.

**light brown sugar:** A moist sugar product that contains approximately 3.5% molasses.

**limburger:** A washed-rind semisoft cheese with a characteristic strong aroma and flavor.

**lime:** A small citrus fruit that can range in color from dark green to yellow-green.

**liquid measuring cup:** A plastic or glass cup used to measure various volumes of liquid ingredients.

**loaf pan:** A short, deep, rectangular pan typically used for baking loaves of bread.

**log phase:** The rapid growth period for bacteria.

**loin backs:** The rib bones removed from the loin.

**loin (saddle cut):** A primal cut located between the primal rack and leg, and includes the 12th and 13th rib in beef and veal (the 13th rib in lamb), the loin eye muscle, the center section of the tenderloin, the strip loin, and some flank meat.

**long-flake crust:** A pie crust that absorbs the greatest amount of liquid because the flour and fat are rubbed together less than in a mealy or short-flake crust.

**lotte:** Monkfish in French.

**lowboy unit:** A type of reach-in unit located beneath a work surface.

**lychee:** A fruit with a thin, red, inedible shell and a light pink to whitish flesh that is refreshing, juicy, and sweet.

## M

**macadamia nuts:** Marble-sized nuts from an evergreen tree.

**mace:** The lacy red-orange covering of the nutmeg kernel. Mace turns dark yellow when dried.

**mackerel:** A long, streamlined fish that tapers toward the rear with firm, grayish pink flesh that is rich in flavor.

**magnesium:** A macromineral that is essential for muscle contraction, normal bowel functions, bone and tooth structure, and nerve transmission.

**mahi-mahi:** A high-quality lean fish that has colorful skin and firm, pink flesh with a sweet flavor.

**main course salad:** A fairly complete and balanced dish with protein (such as meat, seafood, eggs, or legumes) and an assortment of vegetables.

**Maine lobster:** A cold-water lobster that has a dark bluish-green shell, two large heavy claws, medium-size antennae, and four slender legs on each side of the body.

**maître d':** The head of the dining room service.

**Maki sushi:** Vinegar-seasoned cooked rice layered with ingredients, rolled in a dried seaweed paper called nori.

**malt vinegar:** Vinegar made from malted barley.

**manchego:** A hard, sheep's milk cheese produced in the La Mancha region of Spain.

**mandarin:** A small, dark-orange-colored citrus fruit that is closely related to the orange but is more fragrant, whose varieties include tangerines, clementines, satsumas, tangors, and tangelos.

**mandoline:** A manual slicing tool made of a very durable plastic or stainless steel.

**mango:** An oval or kidney-shaped fruit with orange to orange-yellow flesh.

**mangosteen:** A round sweet, juicy fruit, about the size of an orange, with a hard, thick, dark purple rind that is inedible.

**maple syrup:** A sweet syrup derived from the sap of maple trees.

**marbling:** Trace amounts of fat speckled throughout the muscle.

**margarine:** A fat product made from hydrogenated vegetable oils.

**marinade:** A liquid commonly used to soak meat to add moisture and flavor to the meat.

**marjoram (sweet marjoram):** An herb with short, oval leaves from a plant in the mint family.

**mark:** To cook an item on a broiler or grill just long enough to produce crosshatch marks, and then finish the item in an oven later, just prior to service.

**market menu:** A menu that changes often to coincide with changes in available product.

**mascarpone cheese:** A cream cheese of Italian origin.

**matignon:** A uniformly cut mixture of onions, carrots, and celery, and may also contain smoked bacon or ham.

**mayonnaise:** The thick, uncooked emulsion formed by combining salad oil with egg yolks, vinegar, and seasonings.

**meal-specific menu:** A menu written to be inclusive of a specific meal only, such as breakfast, lunch, or dinner.

**mealy pie dough:** A dough that absorbs the least amount of liquid because the flour and fat are rubbed together until the flour is completely covered with shortening and the mixture resembles fine cornmeal.

**measurement equivalent:** The amount of one form of measure that is equal to another form of measure.

**measuring spoon:** A spoon used to measure a small volume of ingredients.

**meat mallet (meat tenderizer):** A hammer-like tool used to pound and break the connective tissues and muscle fibers in tough cuts of meat.

**melon:** A type of fruit that has a hard-skin and a soft inner flesh that contains many seeds.

**mentor:** A trusted and experienced person who can coach and guide a new employee.

**meringue:** A mixture of egg whites and sugar.

**mesclun (spring mix):** A blend of young salad greens.

**mezzes:** The Greek term for serving small portions of foods as either a starter course or a complete meal.

**microgreens:** Very small immature greens that are tender and flavorful.

**microorganism:** A single-celled plant or animal that can only be seen through a microscope.

**microplane:** A razor-sharp hand-held grater that shaves food into fine or very fine pieces.

**microwave oven:** A cooking unit using electronically generated microwaves to heat water molecules within foods to cook them.

**middle manager:** A person who directly supervises first-line managers.

**mignonette:** A spicy cocktail sauce or a vinaigrette-style dressing served with fresh seafood such as oysters on the half shell, shrimp, crab, and lobster.

**milk chocolate:** A chocolate product that contains at least 12% milk solids and 10% chocolate liquor.

**millet:** A small, round, slightly nutty grain that is very high in protein.

**milling:** The successive grinding of grain into a fine meal and finally into a fine powder.

**mincing:** Finely chopping an item to yield a very finely cut product.

**minerals:** Inorganic substances that are used in various processes throughout the body.

**mint:** A versatile herb with soft, bright green leaves that can be used both in sweet and savory dishes as well as in teas and condiments.

**mirepoix: 1.** A rough cut of carrots, onions, and celery measuring approximately ¾″ × ¾″ × ¾″, used to make soups and stocks. **2.** A rough-cut vegetable mixture consisting of 25% carrots, 50% onions, and 25% celery cut into approximately 1″ pieces and used to flavor stocks and sauces.

**mise en place:** A French term meaning "put in place," and refers to having the tools needed to perform a task properly, having the ingredients for a recipe properly prepared ahead of time, and even having a station set up properly so that it is prepared to accommodate orders coming into the kitchen.

**mixing bowl:** A large, smooth bowl used for mixing small or large amounts of ingredients.

**mixing paddle:** A long-handled paddle used to stir foods in deep pots or steam kettles.

**modern American plated service:** A style of service where all food is prepared, portioned, plated, and garnished by the kitchen staff.

**modified straight dough method:** A yeast dough mixing method in which the ingredients are added in sequential steps.

**moist-heat cooking:** Any cooking method that uses liquid (including stocks and sauces) or steam as the cooking medium.

**molasses:** The dark residual syrup that is left after sugar cane has been processed.

**mold:** A cotton-like fungus that are visible to the eye in large clusters.

**molds:** Pans that have distinctive shapes.

**mollusk:** A shellfish with a soft unsegmented body.

**molting:** The process of a crab shedding its hard shell to grow a larger shell.

**monkfish:** A lean saltwater fish with a sweet, firm-textured flesh that when cooked has qualities similar to those of lobster meat.

**monounsaturated fat:** A fat found in plant-derived foods and oils such as olives, olive oil, peanuts, peanut oil, and canola oil.

**Monterey Jack:** A dry-rind semisoft cheese that displays a smooth texture, a creamy white color, and a mild taste.

**mother sauce:** One of five sauces from which all classical sauces described by Escoffier are produced.

**mouth feel:** The sensation a food or beverage creates in the mouth, other than that created by the item's flavor.

**mozzarella:** A very tender cheese with a soft, plastic-like curd.

**mozzarella di bufala (buffalo mozzarella):** An unripened fresh cheese made from water buffalo's milk, or a combination of cow's milk and water buffalo's milk.

**muenster cheese:** A washed-rind semisoft cheese with a flavor between that of brick cheese and Limburger.

**muffin:** A quick bread that is made with eggs, flour, and either oil, butter, or margarine.

**muffin method:** A quick bread mixing method that uses liquid fats, such as vegetable oil or melted butter, to produce a rich and tender product.

**muffin pan:** A large rectangular pan with numerous round inserts.

**muskmelon:** A round, orange-fleshed melon with beige or brown, netted skin.

**mussel:** A freshwater or saltwater bivalve mollusk with whiskerlike threads (byssal threads) that extend outside of the shell that allow the animal to attach to items underwater for protection.

**mustard:** A pungent powder or paste made from the seeds of the mustard plant for use as a seasoning or condiment.

**mustard greens:** The large, dark green leaves of the mustard plant.

**mustard seed:** The extremely tiny seeds of the mustard plant, available in three different forms: yellow, brown, and black.

**mutton:** Meat of sheep slaughtered over one year of age.

## N

**nage:** A very aromatic, concentrated, and slightly reduced broth, often court bouillon, that is used as a finished sauce.

**napa cabbage (celery cabbage):** A member of the cabbage family that has an elongated head of tightly packed, crinkly, yellow-green leaves with a thick, white center vein.

**nappe:** The consistency of a liquid that is thick enough to coat the back of a spoon.

**nectarine:** A sweet, slightly tart, orange to yellowish fruit.

**networking:** A means of using personal connections through friends, teachers, or people in the industry to locate possible employment opportunities.

**Neufchâtel:** A soft, fresh cheese that originated in the Normandy region of France.

**New American cuisine:** A food preparation style that emphasizes the use of locally grown foods native to America.

**new potato (early crop potato):** Any variety of potato that is harvested before its starches fully develop.

**niacin:** A B vitamin that aids in metabolism and is found in chicken, fish, milk, eggs, yeast, nuts, fruits, and vegetables.

**Nigiri sushi:** Small hand-formed mounds of vinegar-seasoned cooked rice garnished with a topping.

**noisette (pronounced nwa-ZET):** A small, round, boneless medallion of meat cut from the rib.

**nonperishable items:** Food items that have a long shelf life and typically can be kept for six months to a year.

**nori:** A dried seaweed paper.

**Nouvelle cuisine (new cuisine):** A cooking style where foods are cooked quickly, seasoned lightly, and presented simply.

**NSF International:** An organization involved with standards development, product certification, education, and risk management for public health and safety.

**nutmeg:** The kernel from a large, yellow, nectarine-shaped fruit of a large tropical evergreen.

**nut oils:** Relatively expensive forms of vegetable oil that are named for the nuts from which the oil is extracted.

**nutrient:** A substance that provides nourishment.

**nutrition:** The study of food and nourishment that focuses on how food is taken in and utilized.

**nuts:** Hard-shelled, dry fruits or seeds that contain an inner kernel.

## O

**oblique cut:** A cut that produces small, rounded pieces with two angled sides.

**obsolescence:** The removal of an item from the menu while ingredients for that particular item are still in stock.

**octopus:** A cephalopod with a defined head, well-developed eyes, eight tentacles, and a beak that is used to crack open the shells of its prey.

**offset spatula (offset turner):** A tool with a wide metal blade that is bent upward and back toward a handle.

**oignon brûlé:** Half an onion that is charred on the cut side in a heavy-bottomed pan.

**oil:** A fat that remains in a liquid state at room temperature.

**okra:** A fibrous pod vegetable containing round, white seeds.

**omelet:** An egg dish made with beaten eggs and cooked into a solid form.

**omelet pan:** The smallest of the sauteuse skillets; has a nonstick finish.

**onion:** A bulb vegetable made up of many concentric layers of fleshy, juicy leaves.

**on the half shell:** A phrase referring to bivalves such as oysters and clams that are opened and served on one half of the shell.

**orange:** A sweet, round, orange-colored citrus fruit.

**orange roughy:** A saltwater fish with orange to light gray skin that has a rough appearance.

**oregano:** An herb with dark green leaves and a pungent, peppery flavor.

**orientation:** A period during which new employees are introduced to the policies and procedures of an establishment, the roles and responsibilities of a job, the people they will be working with, and any other information needed to be successful in the new job.

**oven spring:** The rapid expansion of yeast dough in the oven resulting from the expansion of gases within the dough.

**over-easy eggs:** Fried eggs with unbroken yolks that are flipped over to the other side and cooked easy (lightly), medium, or hard (well).

**overhead shelf:** A shelf mounted on the wall or above a work surface.

**overhead warmer:** A heat source located above a hot food service area to keep plates of prepared food hot for service to a customer.

**overrun:** The increase in volume of ice cream or other frozen products as a result of the incorporation of air during churning and freezing.

**oxtail:** The tail from a beef carcass.

**oyster:** A saltwater bivalve mollusk with a very rough shell that is coated with calcium deposits.

**oyster knife:** A specialty tool with a short, slightly thin, dull-edged blade with a tapered point used to open oysters.

## P

**Pacific (Japanese) oyster:** A large oyster with a thick, curvy shell that yields a briny, sweet, and mild-tasting meat.

**palette knife (cake spatula):** A long, flat, narrow blade with a rounded edge that is most often used for icing cakes.

**panbroiling:** The process of cooking food in a small amount of fat in a skillet or sauté pan.

**pan-frying:** A cooking method that involves cooking food in a smaller amount of fat than deep-frying.

**panini grill:** An Italian clamshell-style grill made specifically to cook grilled sandwiches.

**panning:** The process of placing rounded pieces of dough in the appropriate pans.

**pantothenic acid:** A B vitamin that aids in metabolism and is found in whole-grain cereals, legumes, eggs, and meat.

**papaya:** A pear- or cylinder-shaped tropical fruit weighing between 1 lb and 2 lb, with inside flesh that ranges in color from orange to reddish yellow.

**papillae:** Small bumps that cover the tongue.

**paprika:** A spice made by grinding paprika peppers after the seeds and stems have been removed.

**parasite:** A small living organism that needs a living host to survive.

**parcooked:** A term for slightly or partially cooked.

**parfait:** A dessert prepared by alternating layers of ice cream and toppings such as a sauce or fruit and topped with whipped cream.

**paring knife (vegetable knife):** A short knife with a 2″ to 4″ very stiff blade.

**parisienne scoop (melon baller):** A scoop with a stainless steel blade typically formed into a round half-ball cup attached to a handle.

**parmesan:** A grating cheese that originated in Parma, Italy.

**parsley:** An herb with dark green leaves that is used as both a flavoring and a garnish.

**parsnip:** An off-white root vegetable, similar in shape to a carrot.

**par stock:** The amount of inventory needed on hand to provide adequate supply from one delivery to the next.

**partial carcass cut:** A cut of meat that is a primary division of a whole carcass.

**passion fruit:** A small oval-shaped fruit, typically weighing 2 oz to 3 oz, with firm, inedible skin that can be either yellow or purple.

**pasta:** A term for rolled or extruded products made from a dough of flour, water, salt, oil, and sometimes eggs.

**pasta extruder:** A machine that forces dough through a shaped die to create pasta of various shapes.

**pasta machine:** A motor-driven or hand-cranked machine with a set of rollers through which pasta dough is passed to flatten the dough to a desired thickness.

**pastry bag:** A cone-shaped paper, canvas, or plastic bag with two open ends, the smaller of which is fitted with a pastry tip.

**pastry brush:** A small, narrow brush used to apply liquids such as egg wash or butter onto baked products.

**pastry chef:** A person who supervises the pastry department, prepares dessert menus and baked goods, schedules work performed in the pastry department, and often decorates cakes and special pastries.

**pastry cloth:** A large piece of canvas that is used as a surface for rolling out dough, since dough does not stick to it easily.

**pastry cream:** A custard sauce made with egg yolks, sugar, milk, and a starch (usually cornstarch or flour).

**pastry tip:** A cone-shaped tip that is fitted into the narrow end of a pastry bag and used to create decorative patterns and shapes with icing.

**pastry wheel:** A dough-cutting tool with a stainless steel rotating disk attached to a handle.

**pâté (pronounced pah-tay):** A preparation made from a finely ground meat layered with additional garnishes or ingredients and baked in a loaf-shaped pan.

**pâte à choux (pronounced paht ah shoo):** Éclair paste.

**pâté en croûte (pronounced pah-tay ahn crewt):** Made with pork forcemeat (finely ground meat) that is wrapped in pastry dough and baked in a loaf-shaped pan.

**pâte feuilletée (pronounced paht foo-yee-TAY):** The French term for puff pastry.

**pathogens:** Infectious and toxigenic microorganisms that cause disease.

**paysanne cut:** A cut that produces a thin, square, tile-shaped cut with dimensions of ½″ × ½″ × ⅛″ thick.

**peach:** A sweet, orange to yellowish fruit with downy or fuzzy skin.

**peanut:** A nut that is not a true nut but is actually a legume that grows underground.

**pear:** A bell-shaped pome with a thin peel and sweet flesh.

**pecan:** The nut from the pecan tree; native to the Mississippi River valley.

**pectin:** A chemical found in plants that, when combined with an acid, forms a clear gel used as an edible thickening agent.

**peel: 1.** A long, flat, narrow piece of wood or metal shaped like a wide, thin paddle. **2.** A brightly colored, thick outer rind of a citrus fruit.

**peeled and deveined:** A term used to describe shrimp that has been peeled and has had the sand vein (intestinal tract) removed prior to freezing.

**penicillium glaucum:** The mold produced by Gorgonzola cheese mottled with characteristic blue-green veins.

**perch:** Freshwater fish native to the Great Lakes and northern Canada.

**perishable items:** Food items that have a short shelf life and should be purchased frequently in relatively small amounts.

**permanent emulsion:** A mixture of two typically unmixable liquids forced to bond permanently with the aid of an emulsifier, such as egg yolk.

**persimmon:** A bright-orange fruit that is similar in shape to a tomato and very sweet when ripe.

**petit-gris:** The most popular variety of snails, originating in France.

**phosphorus:** A macromineral that is used in the structure of DNA and cellular membranes.

**phyllo dough:** A dough consisting of very thin sheets of unleavened dough.

**physical activity:** Any bodily movement produced by skeletal muscles resulting in energy expenditure.

**physical hazard:** Any hazard that is not biological or chemical and can lead to a physical injury.

**pickling salt (canning salt):** A very pure form of salt that has no residual dust, iodine, or other additives.

**pickling spice:** A blend of several whole spices such as cloves, cinnamon, nutmeg, peppercorns, allspice, bay leaves, ginger, mace, crushed red pepper, dill seed, fennel, and mustard seed, and is primarily used in pickling.

**picnic (callie) ham:** The lower half of the shoulder of the hog carcass.

**pie marker (cake marker):** A round, heavy wire tool with guide bars used for marking pies or round cakes for cutting.

**pie pan:** A round, somewhat shallow pan with sloped sides used for baking pies or pie crusts.

**pike:** A lean fish with a long, lean body, a long, broad, flat snout, and bands of sharp teeth.

**pineapple:** A sweet, acidic tropical fruit with a prickly, pinecone-like exterior and juicy, yellow flesh.

**pine nut:** The nut of various types of pine trees.

**pistachio:** The pale green nut from the pistachio tree.

**pith:** The white layer just beneath the peel of a citrus fruit.

**plantain:** A close relative of the banana that is larger and has a dark brown skin.

**plum:** An oval-shaped fruit that grows on trees in warm temperate climates with a flesh that can vary from reddish orange to yellow or greenish yellow.

**poach:** To cook foods in a shallow amount of liquid held between 160°F and 180°F.

**pod and seed vegetables:** All varieties of legumes (such as peas and beans), corn, and okra.

**polyunsaturated fat:** A fat found in plant-derived foods such as corn oil, soybean oil, sunflower oil, and sesame oil.

**pome:** A type of fruit that contains a core of seeds and an edible peel.

**pomegranate:** A round, bright red fruit with a hard, thick outer skin; measures about 3″ in diameter.

**poppy seed:** A very small, light blue-colored seed of a specially cultivated poppy plant grown chiefly in Holland, India, and the Middle East.

**pork:** The meat of hogs usually slaughtered less than a year old.

**portfolio:** Collection of items that depict skills, accomplishments, and overall abilities.

**portion control scoop (disher):** A stainless-steel scoop of a specific size with a thumb-operated release lever used to serve food in accurate amounts.

**portion scale (spring scale):** A scale with a spring-loaded platform and a mechanical dial display.

**portion size:** The actual size of an individual serving.

**portion size conversion:** A method of recipe conversion where both the total yield and portion size of the recipe are converted.

**Port Salut:** A washed-rind semisoft cheese that has a soft, smooth, orange-colored rind and a glossy ivory-colored interior.

**positive reinforcement:** A technique of giving positive feedback to an employee when he or she accomplishes a task to expected standards.

**posole:** Soup that uses hominy.

**pot and pan rack shelving:** Overhead storage that suspends pots and pans from a rack.

**potassium:** A mineral that works with sodium to maintain blood pressure and fluid levels in the body.

**potato:** A round, oval-shaped, or elongated tuber that grows underground.

**potentially hazardous food:** Food that requires temperature control because it is in a form that is capable of supporting the rapid and progressive growth of infectious or toxigenic microorganisms.

**poultry:** The collective term for the various kinds of birds raised for human consumption.

**poultry half:** Consists of a full half-length of bird split down the breast and spine.

**poultry seasoning:** A ground mixture often containing sage, thyme, marjoram, savory, pepper, onion powder, and celery salt.

**pregelatinzed starch:** A starch that can thicken a liquid without additional cooking.

**prep cook:** A person who prepares or "preps" a majority of the foods used on the line.

**preparation and processing areas:** Areas in a professional kitchen are where food items are cut, trimmed, prepared, sliced, and cooked.

**prep section:** An area containing tools and equipment needed to prepare basic items that may be needed by multiple stations.

**prickly pear:** A fruit with protruding prickly fibers that is a member of the cactus family.

**primal leg:** A hind leg of veal or lamb.

**primal (wholesale) cut:** A division of a partial carcass cut.

**prix fixe menu:** A menu that offers a complete meal with all courses included at a set price.

**processed cheese:** A mixture of any variety of finely ground cheese scraps that are mixed with nondairy ingredients and melted together.

**processed cheese food:** A cheese-based product that may contain as little as 51% cheese.

**proofing:** The process of letting yeast dough rise in a warm (85°F), moist environment until the dough doubles in size.

**proofing cabinet (holding cabinet):** A tall, narrow, stainless steel box in which temperature and humidity can be controlled.

**prosciutto ham:** A type of dry, hard-cured Italian ham.

**protein: 1.** A complex organic compound found in living plant and animal cells. **2.** A complex chain of amino acids that is used by the body to help build and repair body tissues.

**provolone:** A hard cheese with a mild to sharp taste (depending on age) and an elastic texture.

**puff pastry:** A type of dough that does not contain sugar or a leavening agent but increases greatly in size when heat is applied.

**Pullman loaf:** A long, rectangular loaf of bread that has a flat top, uniform slices, and a consistently fine texture from one end to the other.

**pulpo (Spanish for "octopus"):** Cold cooked octopus that is commonly used to make a salad.

**pummelo:** A pale green to yellow citrus fruit that is larger than a grapefruit, with a thick, spongy rind.

**pumpernickel flour:** A meal form of rye made by grinding the whole grain.

**pumpkin seed:** A seed that comes from many varieties of pumpkins.

**punching:** The process of folding over and pressing risen yeast dough to allow the built-up carbon dioxide to escape.

**purchase specification:** An accurate description of an item noting the grade, quality, packaging, and unit size required.

**purchasing:** The selection and procurement of goods, including all steps from selecting goods, to verifying the correct goods are received, and finally paying for the goods.

**purple potato (blue potato):** A potato with smooth, thin, purple skin and deep purple flesh.

**purveyor:** A company or person who makes food, ingredients, or supplies available to a business for purchase.

## Q

**quarter:** One side of a carcass divided into two parts between the 12th and 13th rib bones.

**quick bread:** A baked product that is made with a quick-acting leavening agent such as baking powder, baking soda, or steam that does not require any kneading or fermentation.

**quick loaf bread:** A quick bread that uses a batter similar to muffin batter.

**quince:** A hard yellow fruit that grows in warm climates.

**quinoa:** A gluten-free grain that is high in vitamins and minerals and forms a complete protein.

## R

**rack:** A complete rib from a whole veal carcass containing two unsplit primal ribs.

**radiation:** A method of heat transfer from a heat source to an item through heat waves.

**radish:** A small, round root vegetable that has a peppery taste.

**rambutan:** Sweet, fragrant fruit covered on the outside with soft spikes.

**range (stove top):** A large appliance with surface burners used to cook food.

**raspberry:** A subtly tart, red fruit that grows on vines.

**ratite:** Any of a large variety of flightless birds that have small wings in relation to body size and flat breastbones.

**raw bar:** A section of countertop or buffet line devoted to raw, or cooked and chilled, fish and seafood.

**reach-in unit:** A temperature-controlled cabinet for storing food items.

**receiving area: 1.** The entry point where anything ordered by the food service operation, such as food, supplies, or equipment, enters the building. **2.** The area of the professional kitchen where all delivered items are checked for freshness, appropriate amounts ordered, temperature, and price.

**recipe conversion:** The process of increasing or decreasing a recipe to produce a larger or smaller yield than that of the original recipe.

**red flame grape:** A seedless hybrid grape that ranges from a light purple-red color to a darker purple color.

**red potato:** A round, red-skinned, waxy potato excellent for roasting, broiling, or grilling.

**red snapper:** A lean fish with a juicy, fine-flavored pink flesh that becomes pearly white when cooked and flakes easily.

**reduction:** The process of gently simmering water out of a liquid through evaporation to concentrate the flavor and thicken the texture of the food.

**reel oven:** An oven that has shelves that rotate in a manner similar to a Ferris wheel.

**refined grain:** A grain that has undergone processing to remove components such as the germ or bran.

**remouillage (French for "rewetting"):** A stock made from using bones that have already been used once to make a stock.

**requisition:** An internally generated invoice that is used to aid in tracking inventory as it moves from storage to production.

**restaurateur:** The term used to describe the manager or owner of a restaurant.

**restaurer:** French term meaning "to restore."

**résumé:** A document listing an applicant's education, experience, and interests.

**rib eye:** A large, eye-shaped muscle within the rib that is a continuation of the sirloin muscle.

**riblet:** A rectangular strip of meat containing part of a rib bone, cut from the breast of lamb.

**riboflavin:** A B vitamin that aids in metabolism, red blood cell formation, respiration, antibody production, and growth regulation and is found in dairy products, leafy green vegetables, liver, yeast, almonds, and soybeans.

**rice pudding:** A baked custard made from cooked rice combined with a sweet custard and often dried fruits such as raisins or currants.

**ricer:** A sieve with a plunger used to purée food by pushing the food through a perforated metal plate.

**rice vinegar:** Vinegar made from rice wine.

**rich doughs:** Doughs that incorporate a lot of fat, sugar, and eggs into a heavy, soft structure.

**ricotta:** A white and creamy cheese that is somewhat similar to cottage cheese, yet it is made from the whey of other cheeses instead of milk.

**rings:** Pans without bottoms that are used to give products a round shape.

**rivets:** Metal fasteners used to attach the tang of a knife to the handle.

**roasting:** Cooking by surrounding food with dry, indirect heat (heated air).

**roasting pan:** A rectangular pan with deep (4″ to 5″) sides.

**rolled (French) omelet:** An egg mixture that is cooked and folded, then cooked filling-ingredients are added through a slit cut in the omelet prior to serving.

**rolled-in doughs:** Special yeast doughs with a flaky texture that results from the incorporation of fat through a rolling and folding procedure.

**rolling pins:** Cylinders used to roll out many types of bread, pie, cookie, and pastry dough.

**roll-in unit:** An individual refrigeration unit that allows speed racks to be rolled in and out of the unit through a door opening that is just above floor height.

**romano:** A grating cheese that is similar to Parmesan but softer in texture.

**rondeau:** A wide, shallow-walled, round pot used for braising, stewing, and searing meats.

**rondelle cut:** A cut that produces disk-shaped slices.

**roots:** Vegetables that extend deep into the soil to reach water and nutrients.

**Roquefort:** A blue-veined cheese characterized by a sharp, tangy flavor and by blue-green veins that flow through the white curd.

**rosemary:** An herb that is a member of the evergreen family with needlelike leaves that have an aroma of fresh pine and a hint of mint.

**rotisserie:** A sideways broiler.

**round:** A very large grouping of muscles that represent the hind leg (hip and thigh) of the beef carcass.

**round fish:** Fish with a long, cylindrical body and an eye located on each side of the head.

**rounding:** The process of shaping scaled dough into smooth balls.

**roux (pronounced roo):** A cooked mixture of equal amounts by weight of flour and fat that is used to thicken sauces and soups.

**rubber spatula:** A scraping tool consisting of a wide rubber blade attached to a long handle.

**rump:** The tail section of the beef carcass, located behind the sirloin and above the round.

**russet potato:** A potato with a reddish brown, thin skin, a long shape, and shallow eyes.

**Russian service:** A style of service where platters of food are prepared, portioned, plated, and garnished by the kitchen staff and served using a special technique of holding a spoon and fork in one hand to serve food directly to the customers' empty plates.

**rutabaga:** A round root vegetable derived from a cross between a savoy cabbage and a turnip.

**rye flour:** A dark-colored flour milled from rye seeds.

**S**

**sachet d'épices (French for a "sack of spices"):** A piece of cheesecloth filled with aromatic ingredients such as herbs, spices, mushroom stems, leeks, and other aromatic vegetables.

**saddle:** A complete loin from a whole veal carcass containing two unsplit loins.

**saffron:** A spice made from the dried, bright red stigmas of the purple flower of the saffron plant.

**sage:** A fragrant herb with narrow, velvety, greenish-grey leaves.

**salad dressing:** A cooked, mayonnaise-like product usually made from distilled vinegar, vegetable oil, water, sugar, mustard, salt, modified cornflour, xanthan gum and guar gum (as stabilizers), and riboflavin (for coloring).

**salad greens:** A category of vegetable consisting of edible leaves, so called because they are used almost exclusively in raw salads or as a garnish.

**salamander:** A small overhead broiler that is usually attached to an open burner range.

**salmon:** An anadromous fish found in both the northern Atlantic Ocean and the Pacific Ocean.

**Salmonella:** A bacteria found in the intestinal tracts of all chickens that can cause food poisoning.

**salt pork:** Fat back or clear plate that contains lean meat and has been cured in salt.

**sanitation and safety areas:** Locations where sanitation or safety equipment is kept.

**sanitation section:** Contains cleaning supplies and the large commercial dish machine and dish racks for washing all serviceware items and kitchen smallwares.

**sanitizer:** A chemical agent used to sterilize food-contact surfaces.

**sanitizing:** The process of reducing the number of microorganisms on a clean surface to safe levels.

**santuko knife:** An Asian-style knife with a razor-sharp edge and a heel that is perpendicular to the spine.

**sashimi:** Raw fish eaten without rice.

**saturated fat:** A fat found in animal-derived foods such as beef, whole milk, cream, cheese, butter, eggs, poultry skin, and fats such as shortening and lard used in baked goods.

**saucepan:** A small, slightly shallow pan with straight or slightly sloped sides.

**saucepot:** A small form of a stockpot.

**sautéing:** Cooking food quickly in a small amount of fat.

**sauté pan:** A general term used to describe round, shallow-walled pans used for sauté applications.

**sauteuse:** A round, sloped, shallow-walled pan with a long handle.

**sautoir:** A round, shallow pan with straight sides.

**savory:** A member of the mint family with smooth, slightly narrow leaves.

**scaling:** The process of weighing pieces of risen dough on a baker's scale to ensure even-size products.

**scallion:** A bulb vegetable with a slightly swollen base and long, bright green leaves that are slender and hollow.

**scallop:** A bivalve with a highly regular fan-shaped shell and a well-developed adductor muscle.

**scallopine:** A small, ¼″-thick slice of veal (generally leg meat) about 2″ to 3″ in diameter.

**scoring:** The process of making shallow, angled cuts across the top of unbaked bread.

**scrambled eggs:** Eggs that are whisked while raw to combine the yolk and white and then sautéed to produce fluffy curds.

**searing:** The process of browning the surface of a food item quickly and with high heat to seal in the juices.

**sea salt:** Salt produced through the evaporation of seawater.

**sea scallop:** A large scallop found in deep saltwater with a coarse texture.

**seasoned salt:** A generic term for a blend of salt and other savory ingredients.

**seasoning:** The process of preparing a cooking surface for use to prevent food from sticking.

**seasonings:** Items added to enhance the natural flavor of foods.

**security cage:** Lockable wire cage storage unit used to hold expensive or dangerous items such as fine china or hazardous chemicals.

**selenium:** A trace mineral that is needed in small amounts to make selenoproteins, which are antioxidants.

**self-rising flour:** A type of flour that that contains measured amounts of baking powder and salt so additional baking powder and salt do not need to be added.

**semi-à la carte menu:** A menu that has the entrée accompanied by and priced to include the starch and vegetable and usually a choice of soup or salad, while the appetizer and dessert courses are presented and charged separately.

**semisoft cheese:** Cheese that is firmer than soft cheese but is not as hard as hard cheese.

**sensory perception:** The ability of the eyes, ears, nose, mouth, and skin to gather information and evaluate the environment.

**separate course salad:** A salad served as a course of its own.

**server:** The person responsible for serving all food and beverages to the guests.

**sesame seeds:** Small, flat, honey-colored seeds with a nutty flavor that come from pods on the sesame plant.

**sexual harassment:** Unwelcome verbal or physical conduct of a sexual nature.

**shallot:** A small bulb vegetable that is similar in shape to a bulb of garlic with two or three cloves inside.

**shank:** A cut of meat from the lower front or rear leg.

**shark:** A boneless fish with a skeleton composed of cartilage rather than bone.

**sharpening steel:** A round steel rod approximately 18″ long with a handle that is used to maintain an edge on a knife.

**sheeter:** A piece of equipment used to roll out large pieces of dough.

**sheet pan:** An aluminum pan with very low sides that is used to bake large amount of products at once.

**shell:** The thin hard covering of an egg; composed of calcium carbonate.

**shellfish:** A classification for aquatic invertebrates that may have a hard external skeleton or shell.

**shell membrane:** A thin, skin-like material located directly under the egg shell.

**sherbet:** A creamier version of traditional icy sorbet, with some milk added to it during the turning process.

**shirred (baked) eggs:** Eggs that are prepared using an oven.

**shocking (refreshing):** The technique of quickly stopping the cooking process in foods by plunging them into ice water.

**short-flake crust:** A pie crust that absorbs a slightly larger amount of liquid than mealy pie dough because the flour and fat are rubbed only until no flour spots are evident and the fat is in pea-size particles.

**short loin:** The first primal cut in the hindquarter; located just to the rear of the primal rib.

**short order section:** An area that makes casual foods that are cooked quickly.

**short plate:** A thin portion of the beef forequarter that lies just beneath the rib cut.

**shrinkage (S):** Weight or volume that is lost from a food item during the cooking process.

**side:** A half of a complete carcass split along the backbone.

**side salad:** A small salad served to accompany a main course.

**sieve:** A fine-mesh or perforated strainer used to purée or sift, or to remove liquid from a food.

**sifter:** A cylindrical metal container with a hand crank and a woven metal screen stretched across the bottom that is used to aerate and remove lumps from dry ingredients.

**sifting:** The process of passing dry ingredients through a wire sieve to remove all of the lumps and add air to ingredients.

**silicone baking mat:** A woven silicone nonstick mat.

**silverskin:** Very tough, silver-colored elastin that does not break down while cooking.

**simmering:** A moist-heat cooking method that uses a liquid heated below the boiling point as a convection medium to heat food.

**sirloin:** A primal cut of beef situated just behind the short loin and accounts for approximately 7% of the carcass weight.

**skate (ray):** A lean, boneless fish with two winglike sides.

**skimmer:** A flat, stainless steel perforated disk connected to a long handle, used to skim impurities or food from soups, stocks, and sauces.

**sleeper:** A lobster that is dying.

**slicer:** A knife with a narrow, flexible blade usually between 10″ and 14″ long.

**slipper lobster:** A lobster with large disk-shaped antennae that protrude from the front of the head.

**slurry:** A mixture of equal parts of cool water and cornstarch.

**smallwares:** Items that are used to store or hold foods at some stage of the preparation process.

**smelt:** A small, lean member of the salmon family having a slender body, a long, pointed head, a large mouth, and a deeply forked tail.

**smoke point:** The temperature at which fats begin to break down under heat and begin to smoke or burn.

**snow (spider) crab:** Similar to the king crab, but smaller, lower in cost, and available in greater supply. It is sold as cooked, frozen leg clusters.

**sodium:** A mineral that the body uses to help maintain blood pressure levels and fluid balance.

**soft cheese (rind-ripened cheese):** A cheese that has been sprayed with a harmless mold to produce a thin skin.

**soft cookies:** Cookies prepared from dough that contains a great deal of moisture.

**soft filling:** A pie filling that is uncooked and baked in an unbaked pie crust.

**soft-shell (long-neck, steamer) clam:** A clam with a thin, brittle shell that breaks easily.

**solanine:** A natural toxin that is not harmful when consumed in small amounts but can be harmful if eaten in large quantities.

**sole:** Lean flatfish regarded as the most flavorful and smoothest textured of all flatfish.

**solid pack:** A canned product, such as fruit, with little or no water added.

**soluble fiber:** Fiber that dissolves in water.

**sorbet:** A refreshing frozen dessert made of fresh fruit and simple syrup.

**sorrel:** A green with large leaves that are acidic in flavor.

**soy sauce:** A dark brown sauce that is made from mashed soybeans, roasted barley, salt, and water.

**sourdough starter:** Flour combined with nonchlorinated water.

**sous chef:** A person responsible for carrying out the chef's orders each day, instructing personnel in the preparation of some foods, and assisting the chef in directing all kitchen production and service.

**spaghetti squash:** A winter squash with flesh that can be separated into spaghetti-like strands after it is cooked.

**spareribs:** The whole rib section removed from the belly (side) of the hog carcass.

**specification:** A particular characteristic of a product that is required to meet a specific production or service need.

**speed rack (tallboy):** A tall cart with rails intended to hold entire sheet pans of food.

**spices:** Derivatives of bark, seeds, roots, buds, or berries of very aromatic plants.

**spinach:** A dark green, leafy vegetable that is rich in vitamin A, folate, potassium, and magnesium.

**spiny (rock) lobster:** A lobster with many prominent spines on the body and legs, long, slender antennae, very small claws, and five very slender legs on each side of the body.

**sponge method:** A two-step yeast dough mixing method that results in a lighter-textured product.

**springform pan:** A pan with a metal clamp on the side of the pan that allow the sides of the pan to be removed.

**spring lamb:** A lamb marketed between the age of 5 and 10 months.

**squab:** A young pigeon bred and raised for human consumption.

**squid:** A cephalopod having an internal cuttlebone, a distinct head, a beak, well-developed eyes, fins, and ten tentacles.

**stag:** A male cattle that is castrated after it has become sexually mature.

**standardized recipe:** A set of directions for preparing a particular menu item to ensure consistent quality, portion size, and cost.

**star fruit (carambola):** A tart tropical fruit that is shaped like a star when cut crosswise.

**stationary phase:** The period of time following the log phase when bacteria reproduce to such an extent that they actually begin to crowd themselves and fight for food, moisture, and space.

**steaming:** A moist-heat cooking method that uses steam as a convection medium to heat food.

**steam injection:** The process of adding water directly into the hot cavity of an oven so that steam is created.

**steam-jacketed kettle:** A large cooking kettle that has a hollow lining into which steam is pumped.

**steam table:** An open-top table with heated wells that are filled with water.

**steer:** A male calf that is castrated prior to reaching sexual maturity.

**stewing:** A combination cooking method for bite-sized or slightly larger pieces of meat that typically incorporates a thick and flavorful sauce.

**stilton:** A blue-veined cheese made from cow's milk with a flavor that is milder than Roquefort or Gorgonzola.

**stirring:** The process of gently mixing all ingredients until they are evenly combined.

**stock:** The strained liquid that results from cooking seasonings and vegetables or the bones of meat, poultry, or fish.

**stockpot:** A large, round, high-walled pot that is taller than it is wide.

**stone crab:** An Atlantic crab with a brownish-red shell and large claws of unequal size.

**storage areas:** Designated areas within the professional kitchen where food items are stored.

**straight dough method:** The simplest yeast dough mixing method and involves combining all ingredients at once and mixing them together.

**strainer:** A bowl-shaped woven mesh screen tool with a handle.

**strawberry:** A bright red, heart-shaped fruit covered with tiny black seeds.

**strudel dough:** A dough made from high-gluten flour that is rolled paper thin.

**sturgeon:** A saltwater fish with a wedge-shaped snout and a body style similar to that of a shark.

**suckling pig:** A pig between 4 weeks and 6 weeks old.

**sundae:** A dessert made with one or more scoops of ice cream or sherbet, along with one or more liqueurs, sauces, or pieces of fruit.

**sunflower seed:** A seed from the sunflower plant.

**sunny-side up eggs:** Lightly cooked fried eggs with unbroken yolks that are never flipped over to cook the other side.

**superfine sugar:** Granulated sugar with very small crystals.

**supervisor:** A person who oversees the overall operation of a food service establishment and confers with and directs managers of each operation.

**suprême sauce:** A chicken velouté with the addition of cream.

**surimi:** A fish product that looks, cooks, and tastes like crabmeat.

**sushi:** A vinegar-seasoned rice dish garnished with items such as raw fish, cooked seafood, egg, or vegetables.

**sweetbreads:** The thymus glands of an animal, located in the throat of the animal.

**sweet potato: 1.** A tuber that grows on a vine. **2.** A low-starch/high-moisture tuber that has sweet, orange-colored flesh and a thin skin.

**swiss chard:** A green with large, dark green leaves that grow on white or reddish stalks.

**swiss cheese:** A term for various varieties of hard cheese with large holes (eyes).

**swordfish:** A very large, fatty-fleshed fish that has a long, swordlike bill extending from its head.

**szechwan pepper:** A spice made from the dried, ground berries of an ash tree native to China.

# T

**Tabasco® sauce:** A very hot, vinegar-based sauce made from red peppers, vinegar, and salt.

**table d'hôte menu:** A menu that identifies the item that will be served for each course as determined by the chef.

**tamis (drum sieve):** A flat, round sieve with a wood or aluminum round frame and a mesh screen bottom.

**tamarind:** A spice that comes from long pods that grow on the tamarind tree.

**tang:** The unsharpened tail of the blade that extends into the handle of a knife.

**tapas:** The Spanish term for serving small portions of foods as either a starter course or a complete meal.

**tapioca:** A thickening agent that is derived from the tropical cassava plant and is available in round granular form or ground flour form.

**tartare:** A preparation of freshly ground or chopped raw meat that is seasoned and then served as a small mound.

**tart pan: 1.** A round, shallow baking pan, often equipped with a removable bottom. **2.** Shallow pans with short sides that are used to bake tarts.

**tarragon:** An herb native to Siberia.

**taste buds:** The hundreds of minute cells each of the papillae is covered with.

**tea sandwich:** A petite and delicate sandwich that originated in England as a snack served with afternoon tea.

**technical skill:** An ability to successfully perform a task of a trade.

**temperature danger zone:** A range of temperature between 41°F and 135°F.

**temporary emulsion:** A temporary mixture of two typically unmixable liquids that eventually separate into their original state when allowed to rest.

**tender:** A small strip of breast meat.

**tenderloin: 1.** The inner pectoral muscle that runs alongside the breastbone. **2.** An eye-shaped muscle running from the primal rib cut into the primal leg.

**terre (pronounced tair):** French word meaning "earth."

**terrine:** A preparation made from either coarsely or finely ground meat, seafood, poultry, or vegetables and baked in a loaf-shaped pan.

**thermal server system:** A coffee-brewing system that brews and dispenses coffee into insulated servers that may be stored at convenient locations throughout a dining establishment.

**thiamin:** A B vitamin that aids in metabolism and is found in legumes, pork, pecans, spinach, oranges, cantaloupe, milk, and eggs.

**thick soup:** A soup having a thick texture and consistency.

**thigh:** The upper section of leg located below the hip and above the knee joint.

**Thompson grape:** A seedless white grape that is pale to light green in color.

**thyme:** An herb whose leaves and tender stems are picked just before its blossoms start to bloom.

**tiered steamer:** A round, stainless steel pot with a perforated steamer insert.

**tilapia:** A lean, firm, white-meat fish that is primarily aqua-farmed, as wild tilapia tends to have a muddy taste.

**tip:** The outermost section of a wing.

**tomatillo (Mexican husk tomato):** A bright green fruit-vegetable the size of a small tomato.

**tomato:** A juicy fruit-vegetable that contains many edible seeds.

**tomato sauce:** A mother sauce made by sautéing mirepoix and tomatoes, adding white stock, and thickening with a roux.

**tongs:** A spring-type metal tool consisting of two grippers.

**top manager:** An administrator or owner of a food service establishment.

**toque:** The traditional tall, pleated hat worn by chefs.

**tossed salad:** A mixture of leafy greens such as lettuce, chicory, spinach, or fresh herbs mixed together with other items such as meats, vegetables, fruits, nuts, cheese, and/or croutons, and served with a dressing.

**total weight as served (AS):** The weight of a food product after fabrication and/or loss from cooking.

**total yield conversion:** A method of recipe conversion where the total yield of the recipe is converted regardless of portion size.

**tourné knife:** A small knife that is similar to a paring knife but has a curved blade that looks like a bird's beak.

**training program:** A comprehensive plan of study containing all materials and information necessary for an employee to learn the requirements of a job.

**trans fat:** A solid fat made from liquid vegetable oil through a process called hydrogenation.

**trim loss:** Unusable volume, weight, or count that is removed from an as-purchased product prior to preparation.

**tripe:** The muscular inner lining of the stomach of an animal.

**tropical fruit:** A type of fruit that comes from a hot humid location.

**trout:** Round, fat fish with tender flesh that is rich yet delicate tasting.

**truffle:** An edible fungus with a distinctive taste.

**trunnion kettle:** A small steam-jacketed kettle. Instead of a drain on the bottom of the unit, a trunnion kettle can be tilted to empty it by pulling a lever or turning a wheel.

**tube pan:** A round baking pan with a second round tube in the center, resulting in a ring-shaped area to hold batter.

**tubers:** Short, thick, oblong vegetables that are part of a plant stem.

**tuna:** Very large fish that are members of the mackerel family and have a low to moderate fat content and bright red flesh.

**turbinado sugar:** A pale brown sugar that is purified and cleaned through a steaming process before being dried into coarse-grain crystals.

**turmeric:** A spice made from the root of a lily-like plant in the ginger family.

**turnip:** A round, fleshy root vegetable that is purple and white in color.

**turnip greens:** Greens of the turnip plant.

# U

**unit cost:** The cost of a bulk item per unit of measure, such as pound, ounce, gallon, cup, tablespoon, or count.

**univalves:** Mollusks that have a single solid shell and a single foot.

**unsweetened chocolate:** Pure chocolate liquor.

**utensil rack shelving:** Overhead storage with suspended hooks to allow kitchen tools to be easily accessible.

**utility knife:** A multipurpose knife used for anything from cleaning and cutting fruits and vegetables to carving roasted poultry.

# V

**vanilla:** A flavor derived from a dark brown pod that grows on a tropical orchid and is one of the most commonly used flavorings in the bakeshop.

**variable cost:** A cost that increases or decreases in proportion to the volume of production.

**variety meat (offal):** An edible part of an animal that is not part of a wholesale or primal cut.

**veal:** The meat of young, usually male, milk-fed calves.

**vegetable oils:** A group of oils that are extracted from vegetables, cottonseed, corn, grape seeds, peanuts, sesame seeds, or soybeans.

**vegetable peeler:** A cutting tool with a metal blade attached to a metal handle used to remove the skin or thin peel from fruits and vegetables.

**vegetables:** The edible leaves, stalks, roots, bulbs, seeds, and flowers of non-woody plants.

**velouté (pronounced veh-loo-TAY):** A mother sauce made from a flavorful light stock (either veal, chicken, or mild fish stock) and a blonde roux.

**ventilation system:** A large exhaust system that sucks heat, smoke, and fumes out of the kitchen and into the outside air.

**vertical cutter/mixer (VCM):** An appliance used to cut and mix foods simultaneously for fast volume production.

**viande:** The French term for meat.

**vinegars:** Very sour, clear liquids used in cooking, marinades, and salad dressings.

**Virginia ham:** A ham that is cured in salt for a period of about 7 weeks.

**vitamin A:** A fat-soluble vitamin that is crucial for good eye health and cell division and reproduction.

**vitamin B$_6$:** An essential nutrient for protein and red blood cell metabolism; helps in hemoglobin production, immune system maintenance, and blood sugar level maintenance; and is found in beans, fish, meat, poultry, and in some fruits and vegetables.

**vitamin B$_{12}$:** Nutrient that helps maintain healthy nerve cells and red blood cells and is needed in the construction of DNA; it is found in fish, meat, poultry, eggs, and dairy products.

**vitamin C (ascorbic acid):** A vitamin that aids in the formation of collagen, ligaments, joints, bones, and teeth and is found in fruits and vegetables.

**vitamin D:** A fat-soluble vitamin responsible for the formation of healthy bones and teeth and helps the body use calcium and phosphorus properly.

**vitamin E:** A fat-soluble vitamin that protects cells from oxygen damage.

**vitamin K:** A fat-soluble vitamin that has an essential role in the formation of blood clots.

**vitamins:** Nutrients needed in very small quantities to nourish the body.

**volume:** The amount of space that an item occupies.

## W

**walk-in unit:** A room-size insulated storage unit used to store whole cases of product as they are delivered.

**walleye pike (jack salmon):** A lean fish that resembles a pike but is a member of the perch family.

**walnut:** The fruit of the walnut tree.

**wasabi:** The light-green-colored root of an Asian plant with a hot and tangy flavor similar to horseradish.

**wash:** A liquid brushed on the surface of a yeast dough product prior to baking.

**washed-rind cheese:** A semisoft cheese that has an exterior rind that is washed or rinsed with a fluid such as brine, wine, olive or nut oil, or fruit juice.

**watermelon:** A sweet, extremely juicy melon, round or oblong in shape, with pink or red flesh and green skin.

**water-soluble vitamins:** Vitamins that dissolve in water.

**waxed-rind cheese:** A cheese produced by dipping a wheel of freshly made cheese into a liquid wax and allowing the wax to harden.

**weight:** A measure of the heaviness or mass of an object.

**wet aging:** The process of aging meat in vacuum-sealed plastic.

**whetstone:** A stone used to grind the edge of the blade to the proper angle for sharpness.

**whey:** The watery part of milk.

**whipping:** The process of beating ingredients vigorously to add a lot of air to them.

**white:** The white portion of the egg; makes up two-thirds of the egg and consists mostly of albumin protein.

**white chocolate:** A confectionary product made from cocoa butter, sugar, milk solids, and flavorings.

**white mirepoix:** A mirepoix with onions and celery, but the carrots are replaced with parsnips and leeks.

**white pepper:** A spice made from mature berries of the pepper plant.

**white potato:** An oblong starchy potato with a thin, white or light brown skin and tender white flesh.

**white stock:** Stock produced by gently simmering raw poultry, beef, veal, pork, or fish bones in water with vegetables and herbs.

**whole carcass:** A whole animal with the hide, head, entrails, and feet removed.

**whole fish:** Fish marketed the way they are taken from the water. Nothing has been done to process them.

**whole-wheat flour:** Unbleached flour that contains the bran, endosperm, and germ of the wheat kernel.

**wine vinegar:** A vinegar produced from red and white wines, champagne, and sherry.

**wing drumette:** The innermost section of a wing; located between the first wing joint and the shoulder.

**wire shelving unit:** A shelving unit made of stainless steel wire.

**wire whisk:** A mixing tool made of many stainless steel wires bent into loops and sealed into a stainless steel handle.

**wood-burning oven:** An oven with a curved hearth made of masonry that is heated with wood.

**work experience:** A résumé category that describes professional experience related to the job being applied for.

**working chef:** A person who assists in production by working and cooking where needed in addition to having the regular responsibilities of a chef.

**work section:** An area where kitchen professionals are all working for the same purpose and at the same time.

**work station:** An area in a work section for a specific task to be performed by a specific person.

**worcestershire sauce:** A pungent, dark-colored sauce.

## Y

**yam:** A low-starch/high-moisture tuber that is not nearly as sweet as a sweet potato.

**yearling lamb:** A lamb marketed between the age of 10 and 12 months.

**yeast: 1.** A form of fungi that can be found in foods. **2.** A microscopic, living single-celled fungus that releases carbon dioxide and alcohol through fermentation when provided with food (sugar) in a warm, moist environment.

**yellow potato:** An oval waxy potato with yellowish skin and flesh and pink eyes.

**yield:** The amount of a food or beverage product that results from a specific recipe after fabrication, preparation, and cooking.

**yield percentage:** The ratio of the edible portion of a food to the as-purchased amount.

**yield tests:** Commonly used to determine raw yields, cooking-loss yields, and shrinkage yields.

**yolk:** The yellow portion of the egg where all of the fat as well as much of the protein, vitamins, and minerals are found.

## Z

**zest:** The colored outermost layer of the peel of a citrus fruit such as a lemon or lime.

**zester:** A small hand tool that has a stainless blade with five or six sharpened holes in its end.

**zinc:** A trace mineral that helps in the growth and maintenance of all tissues and helps the body manufacture fats, carbohydrates, and proteins.

**zushi:** The Japanese term for cold, vinegar-seasoned, cooked rice.

fondant, 888
fond lié, 552
fontina, 380
food
    contamination, 73–76, 75. *See also*
        foodborne illness
    cooking methods
        combination, *255,* 255–257, *257*
        dry-heat, 239–249, *240, 243–244*
        moist-heat, *250–252,* 250–254
    flavor, 284–286
    heating and cooling, *92,* 92–93
    issuing, 229
    presentation, 284, *285*
    receiving, 228
    storage, 109, 228–229
        cold, 111–113, *112*
        dry, 109–111, *110*
        hot, *114,* 114–119, *115, 117, 118*
    texture, 286, *287*
    transport, 93–94
    waste, 229
foodborne illness, 76–82, *77–78*
    bacteria, 78–80
    highly susceptible populations, 81–82
    potentially hazardous food, 81
food checker, 16
food chopper, 125
food cost, 215
    edible portion cost, *218,* 218–219
    managing, 224
    portion control, 226
    recipe cost, 223–224, *224*
    unit cost, 215–217, *216*
    yield percentage, 219–224
Food and Drug Administration (FDA), 72
food processor, *124,* 125
food safety, 72, *73*
    guidelines, 82–83
    hazard analysis and critical control
        (HACCP), 83–86, *84, 86*
    potentially hazardous food (PHF), 81
    sanitation practices, 87–94, *89, 91*
food service buyer, 18
food service director, 16–17
food service distributor, *18*
food service industry
    career opportunities, 12–18
        customer service, *13,* 13–14, 19
        food production, *7,* 14–16
        management, *12,* 16–18
        other opportunities, *18*
    current industry, 8–9, *9, 10,* 10–12, *11*
    history, *4,* 4–8
        classic trends, 5–8, *6, 7*
        food and culture evolution, *4*
        influential chefs, 5–7

restaurant origin, 4–5, *5*
food service sales, 18
food waste disposer, 134, *135*
food mill, *161,* 162
forequarter, *658*
foresaddle, *658*
formulas, 216–223, 845–849, *846, 848,*
    900–901
French grill, *156*
Frenching, 673
French knife, *174, 178, 187*
French omelet, 330–331, *331*
French rolling pin, *168*
French toast, *338*
fresh cheese, *375,* 375–377
frittata, *332,* 332–333
frog legs, *742*
fruit, 276, 341, 362, *464*
    classifications, *464*
        berry, *468,* 468–469
        citrus, *464,* 464–467, *465, 466*
        drupe, *475,* 475–476
        fruit-vegetable, *484*
        grape, *469,* 469–470
        melon, *477,* 477–478
        pome, *470,* 470–474, *471–472, 473*
        tropical fruit, 478–483, *479*
    cooking methods, 486
        bake, 487
        broil, 487
        deep-fry, 486
        grill, *487*
        poach, 487
        roast, 487
        sauté, 486
        simmer, 487
    platter, *388*
    purchasing, 484
        canned, 485
        dried, *486*
        frozen, 485
        grading, 484
        ripening, 484–485
fruit corer, 167
fruit juice, 342
fruit salad, 372
fruit-vegetable, *484,* 510, 510–514
fry cook, 15
fryer, 127
fumet, 539–541, *541*
fungi, 75, *502,* 502–503
funnel, *170*
fusion cuisine, 8, *9*

G

galantine, 391
game bird, *610,* 610–613

garde manger, 349. *See also* cold foods
garde manger chef, 14
garde manger section, 121–122
garlic, 292, 294, 497, *498*
garnish, 394, 415, *416*
garnishing tools, 179–180
gastrique, 553
gelatin, *840,* 841
gelatinization, *542*
gelatin salad, *373,* 373–374
gelato, 923
general manager, 16
genoise, 882
genuine spring lamb, 654
geoduck, 737
germ (grain), *804,* 805
giblets, 605–606, *606*
gilt, 653
ginger, *299,* 300
glace, 537, *538*
glace de poisson, 537
glace de veau, 537
glace de viande, 537
glace de volaille, 537
glaze, 888–889
glazing, 723
glucose syrup, 837
gluten, 794, 835
goose, 607
Gorgonzola, 381
Gouda, 380
gourd, 506
government guidelines, 272
    dietary guidelines, 274
    MyPyramid Food Guidance System,
        274, *275*
    nutrition facts label, 272–274, *273*
grading stamp, *675, 679, 747*
grain, 276, *804*
    composition, *804,* 804–805
    cooking methods, 811–812
        pilaf method, 812–813, *814*
        risotto method, 812–813, *813*
        simmer, 812
    processing, *805,* 805–806
    storage, 811
    types, 806
        corn, 808–809
        other, 810–812
        rice, 806–808, *807*
        wheat, 809–810, *810*
grain-fed beef, 651
grana, 384
grande cuisine, 5–6, *6*
granité, 924

rice vinegar, 312
rich doughs, 850
ricotta, 376
ring, *832, 834*
risotto, 812, *813*
rivets, *175,* 176
roasting, 246–248
roasting pan, 154
rolled-in doughs, 850–851
rolled omelet, 330–331, *331*
rolling pin, *168,* 169, 834
    French, *832*
roll-in unit, 113
romaine lettuce, 352
Romano, 384
rondeau, 157
rondelle cut, *188, 189*
root, *503,* 503–504
Roquefort, 381
rosemary, *293, 295,* 356
rotisserie, 131
round fish, 714, *715*
    cutting steaks, *751*
    filleting, *752*
    skinning, *753*
rounding, *854,* 854–855
round roast, 700, *701*
round (beef), 667, *669*
roux, 543–546, *544, 545*
rubber spatula, *165, 174*
rump, *667*
    roast, *667*
russet potato, *782*
Russian service, 64
rutabaga, *503,* 505
rye flour, 835


S

sabayon, 921
sachet d'épices, *531*
saddle, 672
saffron, *299,* 302
sage, *293, 295,* 356
salad, *350,* 350–351
    additional ingredients, 361
      fruits and nuts, 362
      proteins, 362
      starches, 362
      vegetables, 361
    nutrition, 351
    dressings, *319,* 363
      mayonnaise, 367, *368*
      vinaigrette, *364,* 364–367, *365, 366*
      vinegar and oil selection, 363
    greens, 351–356
      chicory, *353,* 353–354

      herbs and specialties, *356*
      lettuces, 351–353, *352*
      other greens, 354–356, *355*
      preparation, 357–361, *358, 359, 360*
      storage, 360–361
    varieties, 367–374
      bound, *371*
      composed, *370,* 370–371
      fruit, *372,* 372–373
      gelatin, *373,* 373–374
      tossed, *368,* 368–369, *369*
      vegetable, *372*
salad dressing, *313,* 314. *See also* salad:
    dressings
salad greens, 351
    chicory, *353,* 353–354
    herbs and specialties, *356*
    lettuce, 351–353, *352*
    other, 354–355, *355*
    preparation, 357, *359,* 359–360, *360*
    sanitation practice, 357–358, *358*
    storage, 360–361
salamander, *131*
salesperson, 18
salmon, 720, *722,* 722–723
salmonella, *77,* 323. *See also* foodborne
    illness
salsa, 560
salt, *289,* 289–290, 844, 898
salt pork, 672
sandwich, *410,* 410–411
    components, *410*
      bread, *411,* 411–412, *412*
      filling, *414,* 414–415
      garnish, 415, *416*
      spread, 413–414
    large quantity preparation, 429–430, *430*
    nutrition, 415–416
    styles, 416–425
sandwich station, 426, *427*
    equipment, 428
    ingredients, 426–427, *427*
sandwich styles, 416–417
    cold, *417*
      multidecker, 418, *419*
      open-faced, 418, *420*
      simple cold closed, 417
      tea sandwich, 418, *420*
      wrap, 418–419, *421*
    hot, *422*
      deep-fried, 425
      grilled and griddled, *424*
      open-faced, *423*
      simple hot closed, *422*
      wrap, 425
sanitation practice, 87
    food transport, 93–94
    handwashing, *89,* 89–90

heating and cooling, 92–93
    internal cooking temperatures, 92
    personal hygiene, 87–88
    warewashing, 90–92, *91*
sanitation and safety areas, 103
sanitation section, 123
sanitizing, 87
santoku knife, *178,* 179
sardine, *725*
sashimi, 441
saturated fat, 264
sauce, 550
    butter, 556–560, *557, 558*
    contemporary, 560–562
    mother, 550, *555*
      béchamel, 550–551
      espagnole, 551–552
      hollandaise, 553–556, *555*
      tomato sauce, 552–553, *553*
      velouté, 551
    tomato, 552–553, *553*
sauce chef, 14
saucepan, 155, *156*
saucepot, *157,* 157–158
sausage, 338–339, *339*
sauté chef, 14
sautéed eggs, 327–328
sautéing, 244
sauté pan, 155, *156*
sauteuse, 155, *156*
sautoir, 155, *156*
savory, *293,* 295
scale, *107, 108,* 144
scaling (dough), *854*
scallion, 498
scallop, 739–740
scallopine, 672
scoop, 145, *146*
scoring, *856*
scrambled eggs, 327–328
seafood, 714
    escargot, 741
    fish, 714–729, *715. See also* fish
    frog legs, *742*
    shellfish, 729–742. *See also* shellfish
searing, 245
sea salt, 290
seasoned salt, *303,* 304
seasonings, 246, 287–290. *See also*
    flavorings
    cooking techniques, 304–305
    defined, 287
    lemon zest, *290*
    pepper, 287–288, *288*
    salt, *289,* 289–290
security cage, *110,* 111
seeds, *308*
selenium, 270

## USING THE *CULINARY ARTS: PRINCIPLES AND APPLICATIONS* CD-ROM

*Before removing the CD-ROM from the protective sleeve, please note that the book cannot be returned for refund or credit if the CD-ROM sleeve seal is broken.*

### System Requirements

The *Culinary Arts: Principles and Applications* CD-ROM is designed to work best on a computer meeting the following hardware/software requirements:
- Intel® Pentium® (or equivalent) processor
- Microsoft® Windows® XP, Windows® 2000, Windows ® NT, Windows® Me, Windows 98® SE, Windows® 98, or Windows® 95 operating system
- 128 MB of available system RAM (256 MB or more recommended)
- 90 MB of available disk space
- 1024 × 768 16-bit (thousands of colors) color display or better
- Sound output capability and speakers
- CD-ROM drive

### Opening Files

Insert the CD-ROM into the computer CD-ROM drive. Within a few seconds, the home screen will be displayed allowing access to all features of the CD-ROM. Information about the usage of the CD-ROM can be accessed by clicking on USING THIS CD-ROM. The Quick Quizzes®, Illustrated Glossary, Media Clips, Culinary Flash Cards, Culinary Math, and ATPeResources can be accessed by clicking on the appropriate button on the home screen. Clicking on the American Tech web site button (www.go2atp.com) accesses information on related educational products. Unauthorized reproduction of the material on this CD-ROM is strictly prohibited.

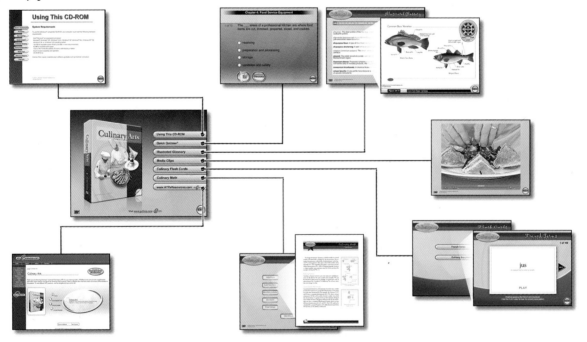